Linear Algebra

An Introduction Using Mathematica®

Linear Algebra

An Introduction Using Mathematica®

Fred Szabo

Concordia University
Montreal, Quebec, Canada

A Harcourt Science and Technology Company

San Diego San Francisco New York Boston
London Toronto Sydney Tokyo

Cover image of Riemann surface $w(z) = \sqrt[6]{z^2 - 1}$ over the complex z-plane created by Michael Trott using Mathematica. Copyright © 1999. Used with permission.

Academic Press
A Harcourt Science and Technology Company
525 B Street, Suite 1900, San Diego, CA 92101-4495 USA
http://www.academicpress.com

Academic Press
Harcourt Place, 32 Jamestown Road, London NW1 7BY, UK
http://www.hbuk.co.uk/ap/

Harcourt/Academic Press
200 Wheeler Road, Burlington, MA 01803 USA
http://www.harcourt-ap.com

Library of Congress Catalog Card Number: 99-68559
ISBN 13: 978-0-123-81401-2

Transferred to Digital Printing 2009

CONTENTS

APPENDICES

PREFACE

This book is a text for a first undergraduate course in linear algebra. It is written for college and university students who have completed two semesters of college-level mathematics. All students majoring in mathematics and client disciplines such as computer science, engineering, biology, chemistry, physics, economics, statistics, actuarial mathematics, and so on should be able to take this course. The presentation is matrix-based and the covered material includes the standard topics for a first course recommended by the Linear Algebra Curriculum Study Group. The aim of the book is to make linear algebra accessible to all college majors through a focused presentation of the material that is enriched by the opportunities for interactive teaching and learning afforded by the *MATHEMATICA* computing system.

The bookends for the course are Gaussian elimination and singular value decomposition. This text begins with the problem of finding the exact solutions of a linear system $A\mathbf{x} = \mathbf{b}$, using elementary row operations. It ends with the solution $\widehat{\mathbf{x}} = A^+\mathbf{b}$, where A^+ is the pseudoinverse of A, derived from the singular value decomposition of the matrix A. If A is invertible, then $\widehat{\mathbf{x}} = A^+\mathbf{b}$ is the exact solution $A^{-1}\mathbf{b}$. If A is not invertible, then $\widehat{\mathbf{x}}$ is a best approximation. Most of the material in between leads up to and is unified in the singular value decomposition. At every step along the way, *MATHEMATICA* serves to remove arithmetical obstacles, to build confidence, to illustrate concepts and constructions, and to facilitate teaching and learning.

MATHEMATICA is an integrated electronic environment in which mathematics can be studied, discovered, and presented. It provides an wonderful framework for computer-assisted symbolic and numerical computation. While this textbook takes advantage of this technology, most of the material is not dependent on it. *MATHEMATICA* routines can usually be converted into manual steps. Moreover, the details of the presented calculations are sufficient to allow students to rework *MATHEMATICA* examples by hand, if such practice is needed. In most cases, it is left to the instructor to decide where exercises should be done with pencil and paper, or with a computer. At times, both methods are appropriate.

TO THE TEACHER

A prominent mathematical and pedagogical feature of this book is the systematic integration of *MATHEMATICA* into the course. The benefits are many. For example, rather than being a distraction, the *syntax* of *MATHEMATICA* provides opportunities for cultivating mathematical rigor. Since matrices are central to the course, the **Array** and **Tables** routines are frequently used. However, these routines serve not only to create matrices but also to encourage students to think about functions and functional expressions. The dual computational power of *MATHEMATICA*, through its distinction between symbolic and numerical calculations, helps to lay the foundation for an appreciation of the difference between exact and approximate solutions of mathematical problems. As such, it gives concrete meaning to the leading problem around which this course is built: the problem of solving linear systems.

One of the great benefits of using *MATHEMATICA* for the teaching of linear algebra is that it eliminates the arithmetical tedium associated with certain aspects of the subject. Students are able to tackle realistic problems and can be sure of the accuracy of their solutions. Furthermore, *MATHEMATICA* provides the opportunity to engage in modest forms of research, so that individuals as well as groups can work on meaningful projects. However, this book is not about programming or computational complexity. All exercises can be solved with the functions built into *MATHEMATICA*, supplemented with the additional routines supplied in the packages specified in Appendix E.

TO THE STUDENT

One of the goals of this book is to stimulate you intellectually and to encourage you to appreciate the benefits of mathematical rigor. The systematic interplay between theory and practice throughout the text is designed for this purpose. In this connection, technology acts both as a catalyst and a facilitator. You will find it easier to discuss mathematical ideas with your classmates since you will develop confidence in the computational correctness of your work.

You will find that learning and retention can be reinforced through easy electronic experimentation. Theoretical material therefore becomes more accessible. By using *MATHEMATICA* to replicate the worked examples and illustrations in this book, you can intersperse your studies with periods of playful exploration. As you study difficult concepts, theorems, and definitions, you can use *MATHEMATICA* to generate examples and counterexamples to test and deepen your understanding of challenging material, even when some of the subject matter may not yet be within your grasp.

You will also find that this book presents linear algebra as a unified subject, leading to a complete solution of a well-defined and easily understood problem: the problem of solving linear systems. You will discover that Gaussian elimination can be used to find exact solutions whenever they exist. The singular value decomposition, on the other hand, yields best approximations when exact solutions are impossible. The text begins with Gaussian elimination and ends with singular value decomposition. Most of the material in between serves to build a bridge between these two methods for solving linear systems.

ORGANIZATION

Chapter 1 introduces systems of linear equations and their matrix forms. We cover Gaussian elimination applied to augmented matrices and define the matrix form of linear systems. We discuss basic and free variables, pivots, the row echelon and reduced row echelon forms, and the solution of linear systems by back substitution. We also cover the definition of matrices and their *MATHEMATICA* representations. We then discuss homogeneous linear systems and uses them to study dependence relations among the columns of a matrix. Applications at the end of the chapter include the balancing of chemical equations, the analysis of resistor circuits, the rewriting of higher-order differential equations as systems of first-order equations, heat transfer problems, and population dynamics.

Chapter 2 explores the algebra of matrices. We cover addition, scalar multiplication, matrix products, matrix transposition, matrix inversion, and the trace. In addition, we present a lexicon of the special matrices encountered in the course and establish some of their distinguishing properties. We also discuss the *LU* and *PLU* decompositions of matrices and use them to solve linear systems. Furthermore, we present several tests for matrix invertibility. One of the key results proved in this chapter is the fact that invertible matrices are products of elementary matrices. The chapter ends with applications from geometry, computer animation, coding theory, economics, and graph theory.

Chapter 3 deals with determinants. We use the Laplace expansion to define them. However, we also show that this definition is equivalent to the definition based on permutations. We use this second approach to prove elementary properties of determinants. This part of the course also encourages students to reflect on the usually unstated importance of order in linear algebra. The evaluation of determinants is based on ordered lists of scalars. The row echelon reduction of matrices involves the order of rows. Matrix representations of linear transformations are built on ordered sets of vectors, and so on. The chapter ends with a variety of applications, including the use of determinants in geometry and calculus.

Chapter 4 introduces real vector spaces. It deals with linear combinations and the linear dependence and independence of sets of vectors. To provide a tangible conceptual framework, we classify vector spaces into coordinate spaces, polynomial spaces, matrix spaces, and others. We define bases, standard bases, and dimension. We then discuss in detail the idea of coordinate vectors and coordinate conversion. Moreover, we spend considerable time on the four fundamental subspaces of a rectangular matrix and illustrate their importance in linear algebra. The chapter ends with a discussion of vector spaces as direct sums of subspaces.

Chapter 5 deals with linear transformations and their basic properties. We discuss their images and kernels, as well as the singularity and invertibility of linear transformations. We prove that every finite-dimensional real vector space is isomorphic to a coordinate space. We use this fact to link matrices and linear transformations, and to explain why linear transformations can be regarded as matrix transformations. We then show that matrix multiplication corresponds to functional composition. We conclude the chapter with a discussion of similar matrices and show that similar matrices are connected by coordinate conversion matrices.

Chapter 6 deals with eigenvalues and eigenvectors. We discuss the mapping properties of diagonal matrices and study diagonalizable linear transformations. In addition, we analyze the connection between the eigenvalues of a linear transformation and invertibility. We also discuss the use of Hessenberg matrices and Householder transformations for finding characteristic polynomials. Two key results, proved in this chapter, are that eigenvectors belonging to distinct eigenvalues are linearly independent and that a linear transformation is diagonalizable if and only if it has enough linearly independent eigenvectors. Applications at the end of the chapter deal with the eigenvalues of D_x, third-order differential equations, biological rates of change, discrete dynamical systems, and Markov chains.

Chapter 7 introduces the basic geometric concepts of length, distance, and angle. The chapter begins with a discussion of vector and matrix norms, and their relation to length. We discuss the one-norm, two-norm, infinity-norm, and Frobenius norm. This work lays the foundation both for the geometry of real vector spaces and for the applications of the singular value decomposition in Chapter 9. We then show how inner products provide the unifying structure for studying length, distance, and angles. Applications integrated throughout the chapter include angles in statistics, the principal axis theorem for quadratic forms applied to conic sections, and Sylvester's theorem for quadratic forms applied to quadric surfaces.

Chapter 8 is devoted to orthogonality. The basic construction around which this chapter is organized is that of an orthogonal projection. We introduce the concept of an orthogonal basis and show how the Gram-Schmidt process can be used to construct orthogonal bases. Applications throughout the chapter include proof that real inner products are dot products relative to orthonormal bases, the QR decomposition of matrices, proof that the fundamental subspaces determined by a matrix are orthogonal in pairs, and the method of least squares. A key result established in this chapter is that inner-product-preserving linear transformations are self-adjoint linear transformations, represented by symmetric matrices. We use this fact to prove the spectral theorem for symmetric matrices, the basis for the singular value decomposition.

Chapter 9 consolidates the ideas and techniques studied in this course. We bring them together in the definition and characterization of the singular values and singular vectors of real rectangular matrices. We show that this topic uses most of the properties of finite-dimensional linear algebra covered in the preceding chapters: linear systems, eigenvalues and eigenvectors, diagonalization, inner products, norms, orthogonal vectors and orthonormal bases, and the orthogonal diagonalization of real symmetric matrices. We discuss the computational properties of invertible matrices, as well as the two-norm and Frobenius norm of a square matrix in terms of singular values. In the Applications section, we use singular value decomposition to construct orthonormal bases for fundamental subspaces. We also use the singular value decomposition to define the pseudoinverse of a matrix and use it to show that all linear systems have solutions, either exact or best possible approximations. Other applications deal with the effective rank of a matrix, the role of orthogonal matrices in computer graphics, and the condition number of a matrix.

NOTATION

We use capital letters $A, B, C, \ldots$ for matrices and bold lowercase letters $\mathbf{u}, \mathbf{v}, \mathbf{w}, \mathbf{x}, \mathbf{y}, \mathbf{z}, \ldots$ for vectors. Identity matrices are always denoted by I.

Examples and theorems are numbered within chapters. To distinguish them more easily, we end examples with the symbol ◄► and proofs of theorems with the symbol ■. If the proof of a theorem has been omitted, we terminate the statement of the theorem with the symbol ■.

APPENDICES

This book ends with short appendices on several topics.

Appendix A consists of a brief discussion of complex numbers and a statement of the fundamental theorem of algebra. This theorem is required in Chapter 8 to prove that real symmetric matrices have real eigenvalues.

Appendix B discusses numerical calculations and illustrates round-off errors.

Appendix C introduces the principle of mathematical induction. Mathematical induction is used to prove the uniqueness of the reduced row echelon form, to prove that square row echelon matrices are upper triangular, to define the determinant of a square matrix, to prove that eigenvectors belonging to distinct eigenvalues are linearly independent, to define the Gram-Schmidt orthogonalization algorithm, and to prove that a self-adjoint linear transformation on a finite-dimensional real inner product space can be represented by an orthogonal matrix.

Appendix D explains the use of the sigma notation and ellipses.

Appendix E itemizes the *MATHEMATICA* packages used in this book. Also included is a complete list of the defined functions used to solve many of the exercises.

Appendix F provides answers to selected exercises.

ACKNOWLEDGMENTS

Teaching has never been as rewarding or as much fun as it is now. In response to the enthusiasm of my students and colleagues, I have written this book as the background material for an interactive first course in linear algebra. The encouragement and positive feedback that I have received during the design and development of the book have given me the energy required to complete the project. I am grateful to several of my colleagues, especially Jean Turgeon, for reading earlier versions of the manuscript and for providing encouraging comments. I would also like to thank all of my students who have joyfully suffered through the evolution of this project and have contributed to its success. Many of them are now using linear algebra successfully in their daily work.

I also owe a profound debt of gratitude to the reviewers for their incisive comments and constructive criticism. The book is much the better due to their input.

I am particularly grateful to David Coyle of Dawson College for working out the answers to the exercises. I am also grateful to Andreas Soupliotis of Softimage for many discussions on computer applications of linear algebra. I owe a further debt of gratitude to my son Stuart for working through most of the examples and many of the exercises. He has critiqued them with the keen eye of a Harvard freshman and has helped to make this book more accessible.

I must also acknowledge the numerous fruitful discussions on the teaching of linear algebra with Michael Keeffe and Maggie Lattuca of the Open and Distance Learning Office of Concordia University. I would also like to thank my student Peter Wulfraat for his contribution to interactive linear algebra and for helpful comments and graphics in *MATHEMATICA*.

I would like to express a special thanks to Robert Ross of Harcourt/Academic Press for launching the project, and to my editor, Mike Sugarman of Harcourt/Academic Press, for sharing my vision and seeing the project through to its completion. Additional thanks go to Victor Curran, Diane Grossman, and Angela Dooley of Harcourt/Academic Press as well as to Amy Mayfield for their technical and production assistance.

Thanks are also due to Amy Hendrickson for assistance with the TeXnological aspects of the production of the book. The manuscript was written and typeset in Scientific WorkPlace, the TeX-based document preparation system, produced by MacKichan Software Inc. I would like to thank Roger Hunter of MSI for having been instrumental in creating this wonderful writing tool for mathematics. The final version of the book was produced in LaTeX by Integre Technical Publishing.

Montreal, Quebec, Canada   *Fred Szabo*

USING *MATHEMATICA*

In this introduction, we present some of the basics of *MATHEMATICA* required for this text. We assume that you are familiar enough with computers to be able to launch the program, use the keyboard and mouse, manipulate menus, and save and reload your work. The text was written for *MATHEMATICA* versions 3.0 and higher.

GETTING STARTED

When we first launch *MATHEMATICA*, we are presented with an empty work space. *MATHEMATICA* has created a file, called a **notebook**. When we try to save this file, we are prompted for a file name. We might choose to call our first file **linearalgebra**. *MATHEMATICA* would then save the file with the name **linearalgebra.nb**. The extension "nb" is added automatically and links the file to *MATHEMATICA*. In many computer systems we can double-click on the file name, and by doing so launch *MATHEMATICA* and load the file into memory.

As Figure 1 shows, *MATHEMATICA* notebooks are divided into sequences of nested **cells**. Each cell is bounded by a square bracket and has a certain property. It may be a text cell or an input-output cell. By default, notebook cells are input-output cells unless this is changed in the program preferences. This means that any input is interpreted as an instruction to carry out a mathematical operation.

The first cell in Figure 1 is a text cell. It contains the title *Gaussian Elimination*. The next four cells are input-output cells. The sixth cell is another text cell. It contains the title *Singular Value Decomposition*. The last four cells are again input-output cells. As we can see, the input-output cells are numbered consecutively.

In the following table, we show different input-output possibilities.

Label	Input	Label	Output
In[1]:=	**3+4**	Out[1]=	7
In[2]:=	**3*4**	Out[2]=	12
In[3]:=	**3/4**	Out[3]=	$\dfrac{3}{4}$
In[4]:=	**N[1/3]**	Out[4]=	0.3333
In[5]:=	**x^3**	Out[5]=	x^3

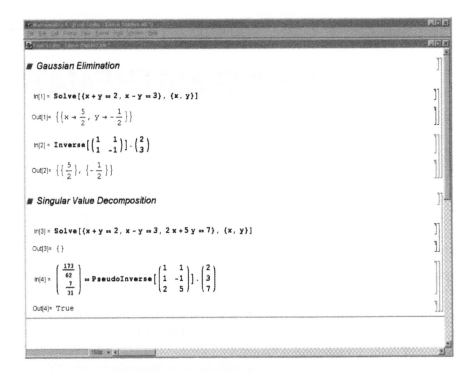

FIGURE 1   A *MATHEMATICA* Notebook Screen.

If the inputs **3+4**, **3 * 4**, **3 / 4**, **N[3/4]**, and $\mathbf{x}^3$ are executed in sequence, *MATHEMATICA* labels them consecutively from In[1] to In[5] and labels the corresponding outputs Out[1] to Out[5]. The first and second outputs are the sums and products of the numbers in the input expressions. The third output, the fraction 3/4, is an exact, or **symbolic**, value associated with the division of 3 by 4. The fourth output is an approximation of 1/3, and is also called a **numeric** output.

The next table provides an example of how *MATHEMATICA* interprets a space between variables.

Label	Input	Label	Output
In[1]:=	**x=3;y=7;**		
In[2]:=	**xy**	Out[2]=	xy
In[3]:=	**x y**	Out[3]=	21

As we see in the first row, a semicolon placed at the end of an input suppresses the output. We can see in the second row that although the variables *x* and *y* were assigned the values 3 and 7 (row 1), the expression **[xy]** is not interpreted as multiplication, but is output unchanged. The third row shows that a space between the variables signals multiplication: the input **[x y]** returns the value 21.

In the next table, we show how *MATHEMATICA* outputs exact and approximate values of an expression involving a variable. We also show the effect of clearing the assigned value of that variable.

Label	Input	Label	Output
In[1]:=	x=7	Out[1]=	7
In[2]:=	x^3	Out[2]=	343
In[3]:=	N[x^3]	Out[3]=	343.
In[4]:=	Clear[x]		
In[5]:=	x^3	Out[5]=	x^3

We can see that if x has the value 7, *MATHEMATICA* replaces every subsequent occurrence of x by 7 until the current value of x has been cleared.

Assumption

To simplify the presentation of the examples in this text, we assume that all values of the variables used have been cleared at the beginning of a session. The first input label in any example is therefore always In[1]. In the answers to the exercises at the end of the book, we show how variables with more than one value can coexist in the same session provided that the values are localized by means of the *Box* construction.

Execution

MATHEMATICA does not execute an input command until it has been told to do so. This is done by placing the cursor anywhere on the line containing the command and pressing the keys *Shift + KeypadEnter*. The same can be achieved by invoking the menu commands *Kernel > Evaluation > Evaluate Next Input.*

CREATING MATRICES

The most frequently encountered mathematical objects in this text are matrices. A matrix is a collection of cells, arranged in rows and columns. The cells usually contain real or complex numbers, both exact and approximate, and sometimes polynomials and other functions. *MATHEMATICA* provides us with several options for creating such matrices.

The most obvious and natural way of building a matrix is to use the menus *Input* and *Create Table/Matrix/Palette* to create a matrix template. Figure 2 shows a matrix palette for constructing a 3×4 matrix.

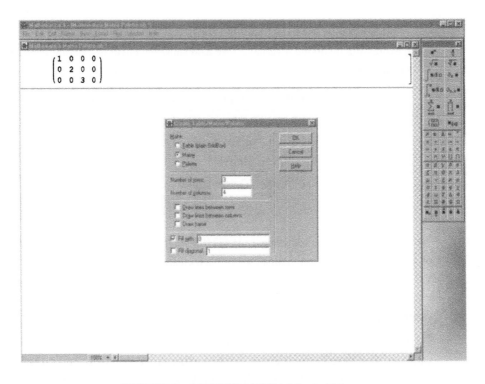

FIGURE 2   *MATHEMATICA* Matrix Palette.

MATHEMATICA then places a matrix template on the screen. We can use the cursor to select the cells and modify their entries. Here is an example of the result of such a construction.

$$\text{In}[1]:= A = \begin{pmatrix} -85 & -55 & -37 & -35 \\ 97 & 50 & 79 & 56 \\ 49 & 63 & 57 & -59 \end{pmatrix}$$

Out[1]= {{-85,-55,-37,-35},{97,50,79,56},{49,63,57,-59}}

The output of this operation is the list

$$\{\{-85, -55, -37, -35\}, \{97, 50, 79, 56\}, \{49, 63, 57, -59\}\}$$

made up of the three lists

$$\{-85, -55, -37, -35\}, \quad \{97, 50, 79, 56\}, \quad \text{and} \quad \{49, 63, 57, -59\}$$

This output explains how *MATHEMATICA* stores matrices. *MATHEMATICA* creates a sequence, called a ***list***, whose terms are the rows of the given matrix. Since each row is also a list, the matrix A is stored as a list of three lists.

EXAMPLE 1   ■ Two Lists of Lists

As a *MATHEMATICA* structure, the list {{1, 2, 3}} is a matrix consisting of one row and three columns, whereas the list {{1}, {2}, {3}} is a matrix consisting of three rows and one column. ◀▶

As is to be expected, we can also input a matrix by typing out its list-of-lists form.

The MatrixForm Function

We can always instruct *MATHEMATICA* to display matrices in ***rectangular form*** using the **MatrixForm** function.

EXAMPLE 2   ■ A Matrix Output in Rectangular Form

```
In[1]:= MatrixForm[A={{-85,-55,-37,-35},{97,50,79,56},
{49,63,57,-59}}]
```

Out[1]//MatrixForm=

$$\begin{pmatrix} -85 & -55 & -37 & -35 \\ 97 & 50 & 79 & 56 \\ 49 & 63 & 57 & -59 \end{pmatrix}$$

MATHEMATICA has produced a matrix, displayed as a rectangular array. ◀▶

As shown in the next example, we need to be careful when using the **MatrixForm** function for purposes of computation.

EXAMPLE 3   ■ A Matrix Output in Graphic Form

In this example, we show that the order in which the **MatrixForm** function is applied has an impact on the nature of the output.

```
In[1]:= MatrixForm[A={{1,2},{3,4}}]
```

Out[1]//MatrixForm=

$$\begin{pmatrix} 1 & 2 \\ 3 & 4 \end{pmatrix}$$

```
In[2]:= B=MatrixForm[{{a,b},{c,d}}]
```

Out[2]//MatrixForm=

$$\begin{pmatrix} a & b \\ c & d \end{pmatrix}$$

We seem to have obtained two matrices *A* and *B* of the same size. We should therefore be able to add them. However, the input command

```
In[3]:= A+B
```

produces the perhaps unexpected result

Out[3]//MatrixForm=

$$\left\{ \left\{ 1+\begin{pmatrix} a & b \\ c & d \end{pmatrix}, 2+\begin{pmatrix} a & b \\ c & d \end{pmatrix} \right\}, \left\{ 3+\begin{pmatrix} a & b \\ c & d \end{pmatrix}, 4+\begin{pmatrix} a & b \\ c & d \end{pmatrix} \right\} \right\}$$

which shows that *A* is a mathematical object, whereas *B* is only a graphical object. ◄►

Another way to construct a matrix and display its rectangular form at the same time is to concatenate two commands, as illustrated in the next example.

EXAMPLE 4   ■ Matrix Output in Rectangular Form

```
In[1]:= A={{-85,-55,-37,-35},{97,50,79,56},{49,63,57,-59}};

MatrixForm[A]
```

Out[1]//MatrixForm=

$$\begin{pmatrix} -85 & -55 & -37 & -35 \\ 97 & 50 & 79 & 56 \\ 49 & 63 & 57 & -59 \end{pmatrix}$$

The command

```
A={{-85,-55,-37,-35},{97,50,79,56},{49,63,57,-59}};
```

creates the matrix. The semicolon at the end of the command tells *MATHEMATICA* to suppress the output. The command **MatrixForm[A]** that follows instructs *MATHEMATICA* to output a printable version of the matrix. ◄►

Automating the Rectangular Form Outputs

If we need to see all matrices during a particular *MATHEMATICA* session in rectangular form, we can instruct *MATHEMATICA* to output all matrices automatically in that form. The command

$$\texttt{\$Post:=If[MatrixQ[\#],MatrixForm[\#],\#]\&}$$

has this effect.

EXAMPLE 5 ■ **Automatic Rectangular Form**

```
In[1]:= $Post:=If[MatrixQ[#],MatrixForm[#],#]&
```

```
In[2]:= A={{12,4},{4,5}}
Out[2]//MatrixForm=
```

$$\begin{pmatrix} 12 & 4 \\ 4 & 5 \end{pmatrix}$$

As we can see, the displayed matrix is in rectangular form. ◄►

PACKAGES

MATHEMATICA is a rich and varied environment for technical computing. However, not all features of the program are needed by all users at all times. The program is therefore divided into a core, called the ***kernel***, and more specialized parts, called ***packages***. Whenever a particular package is required, it must be loaded separately. The specific packages used in this book are **AffineMaps**, **Arrow**, **Combinatorica**, **GaussianElimination**, **ImplicitPlot**, **Legend**, **MatrixManipulation**, **MultiDescriptiveStatistics**, **Orthogonalization**, and **Rotations**. They are stored in specific folders in the *MATHEMATICA* directory and are loaded by referencing them by their location and file name. Here are some examples.

Package	Loading Command
Arrow	`<<Graphics`Arrow``
GaussianElimination	`<<LinearAlgebra`GaussianElimination``
ImplicitPlot	`<<Graphics`ImplicitPlot``
Orthogonalization	`<<LinearAlgebra`Orthogonalization``

We can see from this table that in the usual configuration, the **Arrow** and **ImplicitPlot** packages are located in the Graphics folder, and the **GaussianElimination** and **Orthogonalization** packages are located in the LinearAlgebra folder. The packages are loaded with the command `<<Folder`Package`. The package name must be surrounded with the backquote character `.

HELP

Whenever *MATHEMATICA* is launched, the kernel is loaded and all features implemented in the kernel can be used. In addition, the entire interactive *MATHEMATICA* manual is available online through the *Help* menu. Also online is a *Getting Started* section. Beginning users should familiarize themselves with these features of *MATHEMATICA*.

FUNCTIONS AND DEFINED FUNCTIONS

The Help menu contains a complete list of all mathematical functions built into the *MATHEMATICA* kernel. The next table shows how the cosine, exponential, and logarithmic functions are represented.

Function	Input	Output
Cosine	`Cos[`π`]`	-1
Exponential	`Exp[3]`	e^3
Logarithm	`N[Log[45]]`	3.8066

Note that the *MATHEMATICA* names for the three functions shows in this table are all capitalized. This is no accident—all function names in *MATHEMATICA* begin with capital letters.

In addition to having an extensive library of built-in functions, *MATHEMATICA* allows us to create new functions as needed in specific contexts. Several syntactic devices are available.

The input statement `f[x_]:=3x+5` defines a function f whose value at the input x is $3x + 5$. The variable in the definition `f[x_]` must be followed by an underscore and $f[x\_]$ itself must be followed by a colon. Until cleared, the function f has the defined meaning. Thus $f[7]$ is 26, and so on.

The same syntax can be used to define functions in more than one variable. The input statement `g[x_,y_]:=3x+7y+5`, for example, defines a function g whose value at the input (x, y) is $3x + 7y + 5$. Therefore $g[1, 2]$ is 22, and so on.

Here are two examples of defined functions required by users of *MATHEMATICA* version 3.0. Both functions compute the trace of a square matrix. Although such a function is provided in version 4, users of earlier versions of *MATHEMATICA* have to define their own trace function.

EXAMPLE 6 ■ **First Trace Function**

The function

```
MatrixTrace[A_]:=Sum[A[[i,i]],{i,Length[A]}]
```

computes the trace of a square matrix. ◀▶

EXAMPLE 7 ■ **Second Trace Function**

The function

```
mtrace[A_]:=If[SquareMatrixQ[A],

   Block[{n=Length[A]},Sum[A[[i,i]],{i,n}],

   ''Error:Not a square matrix.'']
```

also computes the trace of a square matrix, and it does more. It checks whether the matrix is square. If yes, the trace is computed. If no, a message is printed. ◀▶

Pure Functions

The construction **Function[**{x}*,body*] allows us to define a function in the variable x without naming the function. For example, the definition

```
Function[{x},3x+5]
```

yields a function that produces the output 23 if x is assigned the value 6. The command

```
Function[{x},3x+5][6]
```

evaluates the function at $x = 6$. The expression **Function[**{x}*,body*] is known as a ***pure function***. The same function can be defined more simply by writing **body&**. The input

```
(3x+5)&[6]
```

also produces the output 23.

MATHEMATICA RULES

In its simplest form, linear algebra deals with linear equations and systems of linear equations. The statement $x = 5$ is a linear equation. So is the statement $x + y = 5$. *MATHEMATICA* provides various interpretations of such statements. In mathematical logic and computer science, these interpretations are known as ***rewrite rules***.

1. The statement $x = 5$ can be regarded as a ***global rule***. It says that wherever the variable x is encountered in an expression, *MATHEMATICA* is to rewrite the expression by replacing all occurrences of x by 5. Thus the input statements $x = 5; x + 7$ produce the output 12. This global rule applies until the value assigned to x has been cleared by a **Clear** command.

2. The statement $x = 5$ can also be regarded as a statement that the variable x has the same value as the constant 5. If we were to rewrite an expression containing x by replacing all occurrences of x by 5, we would get the same result as if we were to rewrite the expression by replacing all occurrences of 5 by x. The expression **x==5** represents this interpretation. Thus *MATHEMATICA* responds to the input commands **x=2+5;x==7** with the output True. The input statement **x** produces the output 12. In *MATHEMATICA*, all mathematical equations must therefore be written with a double equal sign.

3. The statement $x = 5$ can also be interpreted as a ***local rule***. In that case, it is written as **{{x→5}}**. This time, *MATHEMATICA* does not assign 5 as a value to x. It merely tells us that if we assign the value 5 to x, then a certain equation involving x will become true. For example,

$$\textbf{x+2==7/.x} \rightarrow \textbf{5}$$

yields the output True. The required arrow can either be constructed from the hyphen and the greater-than sign, or it can be inserted from the general input palette. The operator "/." is the substitution operator. The command

$$\textbf{Solve[x+2==7,x]}$$

produces the local rule $\{\{x \rightarrow 5\}\}$, and the substitution statement

$$\textbf{x+2==7/.x} \rightarrow \textbf{5}$$

yields the value True. The brackets around $x \rightarrow 5$ indicate that the output is a local rule.

4. Delayed assignments are another way of using the equation $x = 5$. The command **x:=5** is useful for creating procedures for carrying out a variety of computations later. It tells *MATHEMATICA* to assign the value 5 to the variable x whenever it encounters x next. Therefore, the input **x:=5** produces no output. However, any occurrence of x in subsequent commands is replaced by 5.

BLOCKS

The **Block** function allows us to use the same variable names with different meanings in different contexts. The function **Block[{x, y, ...}, expr]** specifies that occurrences of the variables $x, y, ...$ in *expr* are ***local***. New values can be assigned to the variables in a **Block** calculation. However, they are replaced by the original values of the variables once the calculation has been completed. Here is an example.

EXAMPLE 8   ■ A Block Calculation

```
In[1]:= x:=5;Block[{x},x=3]
Out[1]= 3
```

```
In[2]:= x
Out[2]= 5
```

```
In[3]:= x:=5;Clear[y];Block[{x,y},x+y]
Out[3]= 5+y
```

```
In[4]:= y
Out[4]= y
```

This shows that the use of x and y inside the **Block** structure had no effect on the values of x and y outside of the block. ◄►

The **Module** function has a similar purpose. The function **Module[**$\{x, y, \ldots\}$**,** *expr***]** also specifies that occurrences of the variables x, y, and so on in *expr* are local. However, in the **Module** construction, the variables are first replaced by new ones and a **Block** routine is then carried out. Once a calculation has been completed, the original variables and their original values are restored. However, multiple uses of the **Module** function lead to the proliferation of auxiliary variables. The **Module** function therefore should be used sparingly in interactive work.

REPETITIONS

The **Map** function is a helpful device for solving multiple problems of the same kind. Suppose, for example, that we need to calculate the determinants of three matrices. Instead of writing three separate statements involving the **Det** function, we can write one statement using **Det** and **Map**. Here is an example.

EXAMPLE 9   ■ Using the Det and Map Functions

We use the **Map** function to find the determinants of three matrices.

$$\text{In[1]:= } \textbf{Map}\left[\textbf{Det,} \left\{\begin{pmatrix} 1 & 2 \\ 3 & 4 \end{pmatrix}, \begin{pmatrix} -1 & 7 \\ 7 & 2 \end{pmatrix}, \begin{pmatrix} 2 & 0 \\ 0 & 3 \end{pmatrix}\right\}\right]$$

$$\text{Out[1]= } \{-2,-51,6\}$$

As we can see, a single use of the function **Det,** coupled with **Map,** produced the required determinants. ◄►

CONDITIONAL STATEMENTS

The command **If[φ,t,f]** outputs t if the condition φ evaluates to True and to f if φ evaluates to False. Moreover, the input statements

$$\textbf{i:=3;j:=3;If[i==j,7,12]}$$

produce the output 7. On the other hand, the input statements

$$\textbf{j:=4;If[i==j,7,12]}$$

yield the output 12.

NEGATION

Occasionally, we need to negate a *MATHEMATICA* expression. The exclamation mark serves this purpose. We negate an expression φ by writing !φ. The following examples illustrate the use of negation. The input statements

$$\textbf{i:=3;If[!i==4,7,12]}$$

yield the output 7, whereas the statements

$$\textbf{i:=3;If[!i==3,7,12]}$$

yield the output 12.

SPECIAL SYMBOLS

Traditionally, the constants e, i, and π have been represented in *MATHEMATICA* by **E, I,** and **Pi**. However, *MATHEMATICA* now has special built-in symbols for these constants that resemble the usual notation. We therefore use the standard symbols e, i, and π for these constants in most places in this text.

INTEGRATED TUTORIAL

In this text, *MATHEMATICA* is integrated into both the elementary and the more advanced aspects of the subject. Students can therefore develop simultaneously the manipulative and conceptual experience required to use *MATHEMATICA* with confidence for more difficult problems later in the text. This approach eliminates the need for a separate *MATHEMATICA* tutorial. All step-by-step solutions are complete and can easily be replicated by students new to *MATHEMATICA*. Whenever manual practice is also required, the given examples and exercises can usually be solved independently with pencil and paper. The details in the worked examples suggest the necessary steps.

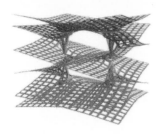

1

LINEAR SYSTEMS

In this chapter, we will use the letters from the end of the alphabet as **variables**, and letters from the beginning of the alphabet as **constants**. Unless we specify the contrary, we assume constants to be real numbers and variables to range over the set $\mathbb{R}$ of real numbers. We will use subscripts to enlarge our supply of variables and constants. Thus

$$x, y, z, x_1, y_1, z_1, x_2, y_2, z_2, x_3, y_3, z_3, x_4, y_4, z_4, \ldots$$

denote variables and

$$a, b, c, d, a_1, a_2, a_3, b_1, b_2, b_3, c_1, c_2, c_3, d_1, d_2, d_3, \ldots$$

denote constants. Occasionally, we also use the letter t as a variable and the letters m and r as constants. The context always makes it clear whether a letter is used as a variable or a constant. Symbols such as 3, 5, and π of course also denote constants. The difference is that $a, b, c, \ldots$ stand for fixed but arbitrary numbers, whereas $3, 5, \pi, \sqrt{2}$, and others stand for specific numbers. We will use variables and constants to construct linear equations. Depending on the context, we will sometimes refer to variables as **unknowns**.

LINEAR EQUATIONS

Linear algebra begins with the study of linear equations. Consider, for example, the equation

$$ax + by = c$$

in the variables x and y and the real numbers a, b, c. This equation is linear in an algebraic sense since the variables x and y are synonymous with their first powers x^1 and y^1. It is also linear in a geometric sense. If $b \neq 0$, we can solve the equation for y and obtain

$$y = -\frac{a}{b}x + \frac{c}{b}$$

This equation determines a straight line in the plane, with slope $-a/b$ and y-intercept c/b.

Two real numbers r_1 and r_2 are considered to be a solution of $ax + by = c$ if $ar_1 + br_2 = c$. In plane geometry, the solutions of this equation correspond to the points of the line determined by the equation.

Linear equations can be formed with any number of variables.

DEFINITION 1.1   A *linear equation* in the variables $x_1, \ldots, x_n$ is an equation of the form

$$a_1 x_1 + \cdots + a_n x_n = b$$

where $a_1, \ldots, a_n$ and b are real or complex numbers. The numbers $a_1, \ldots, a_n$ are the **coefficients** of the equation and b is the **constant term**.

Occasionally, it will be convenient to write the equation $a_1 x_1 + \cdots + a_n x_n = b$ in the equivalent form

$$a_1 x_1 + \cdots + a_n x_n - b = 0$$

Here is an example of a linear equation. The equation

$$3x_1 + 5x_2 - \pi x_3 + 45x_4 - 27x_5 = 18$$

is a linear equation in the variables x_1, x_2, x_3, x_4, and x_5. Its coefficients a_1, a_2, a_3, a_4, and a_5 are $3, 5, -\pi, 45$, and -27, and its constant term b is 18.

We refer to the process of finding real numbers r_1, r_2, r_3, r_4, r_5 for the variables x_1, x_2, x_3, x_4, x_5 such that

$$3r_1 + 5r_2 - \pi r_3 + 45r_4 - 27r_5 = 18$$

as *solving the equation*, and we call the list $(r_1, r_2, r_3, r_4, r_5)$ a solution of the equation. For example, the lists $(1, 3, 0, 0, 0)$ and $(0, 0, 0, 1, 1)$ are solutions of the last equation.

DEFINITION 1.2   A list $(r_1, \ldots, r_n)$ of real numbers is a **solution** of the linear equation $a_1 x_1 + \cdots + a_n x_n = b$ if $a_1 r_1 + \cdots + a_n r_n = b$.

Two steps are required to express and solve linear equations in *MATHEMATICA*. First we must write the equation in *MATHEMATICA* form. We will see that *MATHEMATICA* allows us to enter equations in both regular and what we will call *matrix form*. Then we can use the **Solve** function to find the solutions. If an equation contains more than one variable, we need to specify the variables for which the equation should be solved. The following examples illustrate the syntax.

EXAMPLE 1.1   ■ **Writing a Linear Equation in *MATHEMATICA***

Write the linear equation $a_1 x_1 + a_2 x_2 = b$ in *MATHEMATICA* notation.

Solution. The variables x_1 and x_2 are written as **x[1]** and **x[2]**, the constants a_1 and a_2 as **a[1]** and **a[2]**, and the equation as **a[1]x[1]+a[2]x[2]==b**. The two equal signs indicate to *MATHEMATICA* that the expression is an equation. ◄►

EXAMPLE 1.2 ■ **Using *MATHEMATICA* to Solve Linear Equations in Regular Form**

Write the following linear equations in *MATHEMATICA* notation and find their solutions.

$$\text{a. } 5x_1 = 16 \qquad\qquad \text{b. } 5x_1 - x_2 = 16$$
$$\text{c. } x_1 + 45x_2 + 7x_3 = 0 \qquad \text{d. } x_1 + 5x_2 + 5x_3 - \pi x_4 = 1$$

Solution. Equation (a) is written as `5x[1]==16`. We then use the **Solve** function to solve for x_1.

```
In[1]:= Solve[5x[1]==16]

Out[1]= {{x[1]→ 16/5 }}
```

The output statement $\{\{\texttt{x[1]} \rightarrow \frac{16}{5}\}\}$ provides us with a rule for replacing `x[1]` in the equation `5x[1]==16` that produces the true statement that $5 \times 16/5 = 16$.

Equation (b) contains two variables. We use the **Solve** function to solve for x_1 in terms of x_2.

```
In[2]:= Solve[5x[1]-x[2]==16,{x[1]}]

Out[2]= {{x[1] → 1/5 (16+x[2])}}
```

Equation (c) contains three variables. We use the **Solve** function to solve for x_1 in terms of x_2 and x_3.

```
In[3]:= Solve[x[1]+45x[2]+7x[3]==0,{x[1]}]

Out[3]= {{x[1]→ -45x[2]-7x[3]}}
```

Equation (d) contains four variables, and we use the **Solve** function again to solve for x_1 in terms of x_2, x_3, and x_4.

```
In[4]:= Solve[x[1]+5x[2]+5x[3]-Pi x[4]==1,{x[1]}]

Out[4]= {{x[1]→1-5x[2]-5x[3]+πx[4]}}
```

Most of the time, we will write linear equations in ***matrix form*** (the reason will become clear in the next chapter). To do this, we use the coefficients of an equation to form a row of constants, and the variables of the equation to form a corresponding column. For example, the equation $a_1x_1 + a_2x_2 = b$ in matrix form is written as

$$\begin{pmatrix} a_1 & a_2 \end{pmatrix} \begin{pmatrix} x_1 \\ x_2 \end{pmatrix} = b$$

We use the name *matrix form* since the expressions

$$\begin{pmatrix} a_1 & a_2 \end{pmatrix} \quad \text{and} \quad \begin{pmatrix} x_1 \\ x_2 \end{pmatrix}$$

are special cases of what we will later call **matrices** and since **MatrixForm** is the name of the *MATHEMATICA* function used to produce a matrix format.

The *MATHEMATICA* notation for the matrix form of the equation $a_1x_1 + a_2x_2 = b$ is either

$$\{a[1],a[2]\}.\{x[1],x[2]\}==b$$

or

$$Dot[\{a[1],a[2]\},\{x[1],x[2]\}]==b$$

MATHEMATICA provides both the infix notation $\{a[1],a[2]\}.\{x[1],x[2]\}$ and the outfix notation $Dot[\{a[1],a[2]\},\{x[1],x[2]\}]$ for the sum $a_1x_1 + a_2x_2$. Most of the time, we will use the infix notation. However, in Chapter 7 we sometimes use both notations to stress the fact that **Dot** is a special case of a more general function and to improve the readability of complicated expressions.

EXAMPLE 1.3 ■ **Using *MATHEMATICA* to Solve a Linear Equation in Matrix Form**

Express the linear equation $5x_1 - x_2 = 16$ in matrix form and find its solutions.

Solution. The matrix form of the equation $5x_1 - x_2 = 16$ is

$$\begin{pmatrix} 5 & -1 \end{pmatrix} \begin{pmatrix} x_1 \\ x_2 \end{pmatrix} = 16$$

This equation contains the variables x_1 and x_2. We use *MATHEMATICA* to solve for x_1 in terms of x_2.

```
In[2]:= Solve[{5,-1}.{x[1],x[2]}==16,{x[1]}]

Out[2]= {{x[1] → 1/5 (16+x[2])}}
```

As we can see, the solutions of the matrix equation agree with the solutions found for the corresponding linear equation in Example 1.2. ◄►

EXAMPLE 1.4 ■ **Converting a Matrix Equation to a Linear Equation**

Convert the matrix equation

$$\begin{pmatrix} 3 & 9 \end{pmatrix} \begin{pmatrix} x \\ y \end{pmatrix} = 6$$

to a linear equation in regular form.

Solution. In this equation, the coefficient of x is 3 and the coefficient of y is 9. Therefore, $3x + 9y = 6$ is the associated linear equation. ◀▶

EXAMPLE 1.5   ■ Converting a Linear Equation to a Matrix Equation

Convert the equation $x_1 + x_2 - 7x_3 = 15$ to a matrix equation.

Solution. The list of coefficients of the variables in $x_1 + x_2 - 7x_3 = 15$ is $(1, 1, 7)$. Therefore,

$$\begin{pmatrix} 1 & 1 & -7 \end{pmatrix} \begin{pmatrix} x_1 \\ x_2 \\ x_3 \end{pmatrix} = 15$$

is the associated matrix equation. ◀▶

EXERCISES 1.1

1. Form all possible linear equations $a_1 x_1 + a_2 x_2 = 5$ using the constants 0, 1, and 2 as coefficients.

2. Rewrite the equations found in Exercise 1 in matrix form.

3. Set up the *MATHEMATICA* command for solving the equation $3x_1 + 7x_2 - 9x_3 = 12$ in terms of x_2 and x_3, but do not solve.

4. Rewrite the command in Exercise 3 in matrix form.

5. Use *MATHEMATICA* to solve the equation in Exercise 3 for x_1 in terms of x_2 and x_3.

6. Use *MATHEMATICA* to solve the equation in Exercise 3 for x_2 in terms of x_1 and x_3.

7. Use the variables x and y to construct a linear equation for which $x = 9$ and $y = -7$ is a solution.

8. Find coefficients a and b for which the equation $ax + by = 12$ has the solution $x = 1$ and $y = 1$.

9. Find coefficients a and b for which the equation $ax + by = 12$ has the solution $x = 1$ and $y = 0$.

10. Why is it impossible to find coefficients a and b for which the equation $ax + by = 12$ has the solution $x = 0$ and $y = 0$?

Points, Vectors, and Linear Equations

In geometry, linear equations and their solutions are pictured as sets of points. Relative to a fixed coordinate system, a *point* in the plane is labeled with an ordered pair of real numbers (x, y) and a point in space with an ordered triple of real numbers (x, y, z). Points are considered to be geometric objects. Ordered pairs and ordered triples of real numbers, on the other hand, are algebraic objects.

Linear algebra provides a second perspective. Relative to a fixed set of coordinate axes, every point in the plane and every point in space can be pictured as the tip of an *arrow* that starts at the origin of the coordinate system. Thus the set of all points also determines a set of arrows. These arrows are examples of what we will later call *vectors*.

Vector geometry is based on the addition, subtraction, expansion, and contraction of vectors. When thinking of points as vectors, we usually write them in column form. In this text, the set

$$\left\{ \begin{pmatrix} x \\ y \end{pmatrix} : x, y \in \mathbb{R} \right\}$$

of all vectors with two real coordinates will be denoted by $\mathbb{R}^2$. Similarly, the set

$$\left\{ \begin{pmatrix} x \\ y \\ z \end{pmatrix} : x, y, z \in \mathbb{R} \right\}$$

of all vectors with three real coordinates will be denoted by $\mathbb{R}^3$. We refer to the real numbers x, y, and z as the *coordinates* or *components* of the vectors. When a vector is written in column form, we refer to it as a *column vector*.

Algebraically, there is no reason why column vectors should only have two or three coordinates. We therefore define $\mathbb{R}^n$ as the set of all column vectors $\mathbf{x}$ with n real coordinates, where $n = 1, 2, 3, 4, \ldots$ is any positive integer. In this notation $\mathbb{R} = \mathbb{R}^1$. The usefulness of this construction will become clear as we proceed.

Point Form of Vectors

To save space and to draw attention to their geometric interpretation, we often write the vectors $\mathbf{x} \in \mathbb{R}^n$ in *point form* as $(x_1, \ldots, x_n)$. We use boldface letters $\mathbf{u}, \mathbf{v}, \mathbf{w}, \mathbf{x}, \mathbf{y}, \mathbf{z}, \ldots$ to name vectors and italic letters $u, v, w, x, y, z, \ldots$ to denote vector components.

The Algebra of Vectors

Vector algebra begins with the observation that vectors can be added and multiplied by constants in a geometrically meaningful way. In Figure 1 we picture the sum

$$\mathbf{u} + \mathbf{v} = \begin{pmatrix} x \\ y \end{pmatrix} + \begin{pmatrix} x' \\ y' \end{pmatrix} = \begin{pmatrix} x + x' \\ y + y' \end{pmatrix}$$

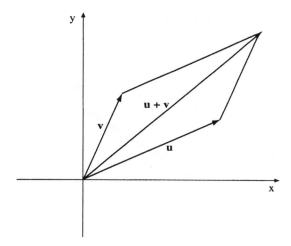

FIGURE 1   Vector Addition.

of the vectors **u** and **v** as the arrow formed by one of the diagonals of the parallelogram determined by **u** and **v**.

In Figure 2 we show how the result

$$a\mathbf{u} = a \begin{pmatrix} x \\ y \end{pmatrix} = \begin{pmatrix} ax \\ ay \end{pmatrix}$$

of multiplying the vector **u** by a real number a can be interpreted geometrically. The vector $a\mathbf{u}$ can be depicted as an arrow lying on the same line as the vector **u**. If $a > 0$, then $a\mathbf{u}$ points in the same direction as **u**. If $a < 0$, then $a\mathbf{u}$ points in the opposite direction.

The two illustrated operations satisfy certain algebraic laws that will be discussed in Chapter 4. They are special cases of *vector addition* and *scalar multiplication*. Many other types of mathematical objects and operations satisfy the same laws. Later we will therefore use the collective terms vector, vector addition, and scalar multiplication for all of these different kinds of objects and corresponding operations.

Vectors in *MATHEMATICA*

In *MATHEMATICA*, vectors are represented as lists. If x, y, and z are arbitrary real numbers, the lists $\{\{x\}, \{y\}\}$ and $\{\{x\}, \{y\}, \{z\}\}$, for example, represent the vectors

$$\begin{pmatrix} x \\ y \end{pmatrix} \quad \text{and} \quad \begin{pmatrix} x \\ y \\ z \end{pmatrix}$$

respectively. The lists $\{x, y\}$ and $\{x, y, z\}$ also represent vectors. In Chapter 2 we will discuss the different interpretations of such lists.

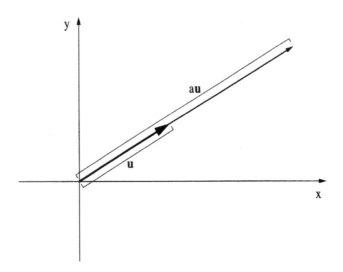

FIGURE 2   Scalar Multiplication.

Geometry and Linear Equations

The graphic capability of *MATHEMATICA* allows us to illustrate the geometric properties of the solutions of linear equations in $\mathbb{R}^2$ and $\mathbb{R}^3$. Here are some examples.

EXAMPLE 1.6   ■ Graphing a Linear Equation in $\mathbb{R}^2$

Use *MATHEMATICA* to graph the solutions of the equation $x + 2y - 7 = 0$.

Solution. We use the **ImplicitPlot** function to plot the graph. First we load the **ImplicitPlot** package.

```
In[1]:= <<Graphics`ImplicitPlot`
```

We now instruct *MATHEMATICA* to plot the equation. We assume that the variable x ranges over the interval $[-10, 10]$. The associated values for y are implicitly determined.

```
In[2]:= ImplicitPlot[{x+2y-7==0},{x,-10,10}]
```

The result is the following graph.

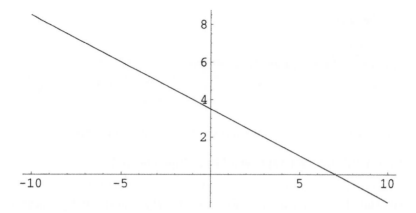

```
Out[2]=  - Graphics -
```

As we can see, the solution $x = 1$ and $y = 3$ corresponds to a point on the graph. ◀▶

EXAMPLE 1.7 ■ Finding Points on the Graph of a Linear Equation in $\mathbb{R}^2$

Use *MATHEMATICA* to find two points on the graph of the equation $x + 2y - 7 = 0$.

Solution. We decide, arbitrarily, to find the points $(7, y)$ and $(-15, y)$.

```
In[1]:= x=7;Solve[{x+2y-7==0},{y}]

Out[1]= {{y→0}}
```

This shows that the point $(7, 0)$ lies on the graph of the given equation. Next we find the point $(-15, y)$.

```
In[2]:= x=-15;Solve[{x+2y-7==0},{y}]

Out[2]= 11
```

This shows that the point $(-15, 11)$ also lies on the graph of the given equation. ◀▶

EXAMPLE 1.8 ■ Finding a Line in $\mathbb{R}^2$ Determined by Two Points

Use *MATHEMATICA* to find the line in $\mathbb{R}^2$ determined by the points $(1, 1)$ and $(3, 5)$.

Solution. Let $y = mx + b$ be the equation of the line determined by $(1, 1)$ and $(3, 5)$. We need to find the slope m and the y-intercept b of the line.

The calculation

```
In[1]:=  Solve[{1==m+b,5==3m+b}]

Out[1]= {{b→ -1,m→ 2}}
```

tells us that $y = 2x - 1$ is the equation of the required line. ◀▶

EXAMPLE 1.9 ■ **Graphing a Linear Equation in $\mathbb{R}^3$**

Consider the equation $x + 2y - 3z - 7 = 0$ in the variables x, y, and z and the real constants 1, 2, −3, and −7. This equation is algebraically and geometrically linear. Three real numbers r_1, r_2, and r_3 are a solution of $x + 2y - 3z - 7 = 0$ if $ar_1 + br_2 + cr_3 + d = 0$.

In geometry, the solution $x = r_1$, $y = r_2$, and $z = r_3$ corresponds to the point (r_1, r_2, r_3) in the plane $x + 2y - 3z - 7 = 0$. The following graph displays the solutions of the equation.

```
In[1]:=  Plot3D[1/3(x+2y-7),{x,-10,10},{y,-10,10}]
```

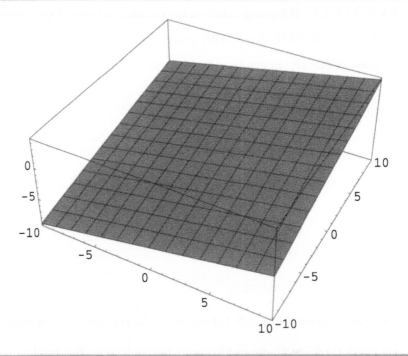

```
Out[1]=  - SurfaceGraphics -
```

As we can see, the solutions of $x + 2y - 3z - 7 = 0$ form a plane in space. ◀▶

EXAMPLE 1.10   ■ Verifying That a Point Lies on a Plane

Use *MATHEMATICA* to determine which of the points $(1, 2, 3)$ and $(2, 4, 1)$ lie in the plane $x + 2y - 3z - 7 = 0$.

Solution. We assign the values $1, 2$, and 3 to x, y, and z and evaluate $x + 2y - 3z - 7$.

```
In[1]:= x=1;y=2;z=3;x+2y-3z-7

Out[1]= -11
```

This tells us that the point $(1, 2, 3)$ does not lie in the plane $x + 2y - 3z - 7 = 0$ since the calculated value of $x + 2y - 3z - 7$ is not 0. We therefore investigate the second point.

```
In[2]:= x=2;y=4;z=1;x+2y-3z-7

Out[2]= 0
```

Hence the point $(2, 4, 1)$ lies in the given plane. ◄►

Parametric Equations

We can also use the vectors in $\mathbb{R}^2$ or $\mathbb{R}^3$ to define new kinds of equations known as *parametric equations*. For example, we can form the equation

$$\begin{pmatrix} x \\ y \end{pmatrix} = \begin{pmatrix} 1 \\ 2 \end{pmatrix} + t \begin{pmatrix} 5 \\ 8 \end{pmatrix}$$

in which x, y, and t are variables. The variable t is called a *parameter*. The solutions of this equation are determined by letting t take on all real numbers as values. It is easy to see that the set of solutions of this equation is a straight line in the plane.

Parametric equations are usually written in point form. The vector multiplied by the parameter t is called a *directional arrow*. For any arrow **a** and directional arrow **d** starting at the origin, the equation $\mathbf{x} = \mathbf{a} + t\mathbf{d}$ is called the *parametric representation* of the line determined by **a** and **d**.

EXAMPLE 1.11   ■ Parametric Representation of a Straight Line in $\mathbb{R}^2$

Show that the parametric equation $(x, y) = (2, 3) + t(5, 7)$ represents the straight line through the point $(2, 3)$ and in the direction of the arrow determined by the points $(0, 0)$ and $(5, 7)$.

Solution. We first use the point-slope formula

$$y - y_1 = m(x - x_1)$$

to find the equation of the straight line determined by the given point and arrow. Since the arrow starts at $(0, 0)$ and ends at $(5, 7)$, we know that it determines a line whose slope is $7/5$. Therefore

$$y - 3 = \frac{7}{5}(x - 2)$$

is the required equation. We can rewrite this equation as $7x - 5y + 1 = 0$.

On the other hand, the parametric equation

$$(x, y) = (2, 3) + t(5, 7)$$

corresponds to the two equations $x = 2 + 5t$ and $y = 3 + 7t$. If we solve both equations for t and equate the results, we get

$$\frac{x - 2}{5} = t = \frac{y - 3}{7}$$

This equation converts to $7x - 5y + 1 = 0$. ◄►

In the case of $\mathbb{R}^2$, we have a choice of defining a line in parametric or nonparametric form. In the case of $\mathbb{R}^3$, however, we have no such choice. A linear equation of the form $ax + by + cz + d = 0$ always determines a plane rather than a straight line. We must therefore use the parametric form to define a line in $\mathbb{R}^3$.

EXAMPLE 1.12   ■ Parametric Equation of a Straight Line in $\mathbb{R}^3$

Find the parametric equation of the straight line determined by the arrow $\mathbf{a} = (1, 2, 3)$ and the directional arrow $\mathbf{d} = (5, -2, 6)$.

Solution. In the parameter t, the required vector equation is

$$(x, y, z) = (1, 2, 3) + t(5, -2, 6)$$

We can rewrite this equation by separating out its components:

$$\begin{cases} x = 1 + 5t \\ y = 2 - 2t \\ z = 3 + 6t \end{cases}$$

In its component form, the straight line is represented by three linear equations. ◄►

EXERCISES 1.2

1. Graph the vectors $(1, 2)$, $(4, 5)$, $(-6, 3)$, $(1, 2) + (4, 5)$, and $(1, 2) + (-6, 3)$.

2. Write the vectors in Exercise 1 as column vectors and graph them as arrows.

3. Write the vectors in Exercise 1 in *MATHEMATICA* notation.

4. Write the following column vectors in point form.

$$\text{a. } \begin{pmatrix} 8 \\ -2 \end{pmatrix} \quad \text{b. } \begin{pmatrix} 7 \\ 3 \end{pmatrix} \quad \text{c. } \begin{pmatrix} -15 \\ -1 \end{pmatrix}$$

5. Write the column vectors in Exercise 4 in *MATHEMATICA* notation.

6. Compute the vector $3(1, 2) + 0(4, 5) - 2(-6, 3)$.

7. Compute the column vector

$$4 \begin{pmatrix} 8 \\ -2 \end{pmatrix} + 4 \begin{pmatrix} 7 \\ 3 \end{pmatrix} + 4 \begin{pmatrix} -15 \\ -1 \end{pmatrix}$$

8. Use the **ImplicitPlot** function to graph the following lines in $\mathbb{R}^2$, using $-5 \leq x, y \leq 5$.

   a. $x + y = 0$ b. $x - y = 0$

   c. $x + y = 1$ d. $3x - 4y = 2$

9. Use the **Plot3D** function to graph the following planes in $\mathbb{R}^3$, using $-5 \leq x \leq 5$ and $-10 \leq y \leq 10$.

   a. $x + y + z = 0$ b. $x - y + z = 0$

   c. $x + y - z = 1$ d. $3x - 4y + 5z = 2$

10. Use *MATHEMATICA* to find the lines in $\mathbb{R}^2$ determined by the following pairs of points.

$$\text{a. } \begin{pmatrix} 2 \\ 2 \end{pmatrix} \begin{pmatrix} 1 \\ -1 \end{pmatrix} \quad \text{b. } \begin{pmatrix} 2 \\ 2 \end{pmatrix} \begin{pmatrix} -1 \\ 1 \end{pmatrix}$$

$$\text{c. } \begin{pmatrix} 3 \\ 7 \end{pmatrix} \begin{pmatrix} 2 \\ 2 \end{pmatrix} \quad \text{d. } \begin{pmatrix} 0 \\ 5 \end{pmatrix} \begin{pmatrix} 8 \\ 1 \end{pmatrix}$$

11. Use *MATHEMATICA* to graph the planes in $\mathbb{R}^3$ determined by the following triples of points.

$$\text{a. } \begin{pmatrix} 2 \\ 2 \\ 3 \end{pmatrix} \begin{pmatrix} 7 \\ 6 \\ 5 \end{pmatrix} \begin{pmatrix} 1 \\ -1 \\ 1 \end{pmatrix} \quad \text{b. } \begin{pmatrix} 2 \\ 2 \\ -2 \end{pmatrix} \begin{pmatrix} 7 \\ 6 \\ 5 \end{pmatrix} \begin{pmatrix} 1 \\ -1 \\ 1 \end{pmatrix}$$

$$\text{c. } \begin{pmatrix} 2 \\ 2 \\ 6 \end{pmatrix} \begin{pmatrix} 7 \\ 6 \\ 5 \end{pmatrix} \begin{pmatrix} -1 \\ 1 \\ 0 \end{pmatrix} \quad \text{d. } \begin{pmatrix} 0 \\ 5 \\ 0 \end{pmatrix} \begin{pmatrix} 8 \\ 1 \\ 3 \end{pmatrix} \begin{pmatrix} 7 \\ 6 \\ 5 \end{pmatrix}$$

12. Use *MATHEMATICA* to determine which of the points $(0, 5)$, $(8, 3)$, and $(6, 5)$ lie on the line $-28x + 32y + 128 = 0$.

13. Use *MATHEMATICA* to determine which of the three points $(8, 3, 1)$, $(0, 5, -72)$, and $(6, 5, -30)$ lie in the plane $-28x + 32y + 4z + 128 = 0$.

14. Find the equations of the straight lines determined by the following pairs of points. Write each equation in the form $ax + by = c$.

$$\text{a. } (3, 4) \text{ and } (6, 2) \qquad \text{b. } (3, 4) \text{ and } (-6, 2)$$
$$\text{c. } (3, 4) \text{ and } (6, -2) \qquad \text{d. } (3, 4) \text{ and } (0, 0)$$

15. Determine which of the following sets of points satisfy the same linear equation. Give a geometric explanation of your answer.

$$\text{a. } (1, 10), (-2, 1), (3, 16) \qquad\qquad \text{b. } (1, -2), (-5, 4), (1, 2)$$
$$\text{c. } (0, -\tfrac{100}{3}, -16), (-2, 0, 2), (-1, -7, -5) \qquad \text{d. } (1, 1, -9), (3, -1, -7), (0, -4, -16)$$

16. Find the parametric equation of the straight line determined by the point $a = (4, -2)$ and the directional arrow $(7, 12)$. Graph the straight line.

17. Find the parametric equation of the straight line determined by the point $a = (4, -2, 6)$ and the directional arrow $(7, 12, -5)$.

18. Use *MATHEMATICA* to graph the following linear equations and describe how they differ.

$$\text{a. } 4x + 3y = 7 \quad \text{b. } -4x + 3y = 7 \quad \text{c. } 4x - 3y = 7$$

19. Use *MATHEMATICA* to graph the following linear equations by treating z as a function of x and y. Describe the graphs.

$$\text{a. } x + y + z = 7 \qquad \text{b. } 3x + y + z = 7$$
$$\text{c. } x + 3y + z = 7 \qquad \text{d. } x + y + 3z = 7$$

LINEAR SYSTEMS

Linear algebra provides a unifying language, a simplified notation, and comprehensive techniques for solving systems of linear equations.

DEFINITION 1.3 *A **linear system** in the variables $x_1, \ldots, x_n$ is a sequence*

$$\begin{cases} a_{11}x_1 + \cdots + a_{1n}x_n = b_1 \\ \qquad\qquad\vdots \\ a_{m1}x_1 + \cdots + a_{mn}x_n = b_m \end{cases}$$

*of equations. The real numbers $a_{11}, \ldots, a_{mn}$ are the **coefficients** of the system and the real numbers $b_1, \ldots, b_m$ are its **constant terms**.*

We refer to a linear system in m equations and n variables as an $m \times n$ linear system. If $m = n$, we call the system **determined**. If $m < n$, we call it **underdetermined**. If $m > n$, we call it **overdetermined**. The coefficients and constant terms of a linear system will be assumed to be real numbers, unless otherwise specified.

One of our goals is to develop systematic ways for finding the solutions of linear systems. A list $(r_1, \ldots, r_n)$ of real numbers is a **solution** of an $m \times n$ linear system if it is a solution of each equation in the linear system. We will see later that all linear systems have either 0, 1, or infinitely many solutions. Table 1 lists the three possibilities.

TABLE 1   Solutions of Linear Systems.

Linear Systems	Equations/Variables	Possible Solutions
Determined	$m = n$	$0, 1, \infty$
Overdetermined	$m > n$	$0, 1, \infty$
Underdetermined	$m < n$	$0, \infty$

If a linear system has at least one solution, we call the system **consistent**. If it has no solution, we call it **inconsistent**.

Determined Linear Systems

The linear system

$$\begin{cases} x + y = 1 \\ x + y = 2 \end{cases}$$

is a 2×2 determined linear system. It has no solution and is therefore inconsistent. The 2×2 determined linear system

$$\begin{cases} x + 2y = 5 \\ 2x + y = 7 \end{cases}$$

on the other hand, has the unique solution $(3, 1)$ since $3 + 2 \times 1 = 5$ and $2 \times 3 + 1 = 7$. It is therefore consistent. The 2×2 determined linear system

$$\begin{cases} x + y = 1 \\ -x - y = -1 \end{cases}$$

is consistent since it has infinitely many solutions. For every real number s, the point $(s, 1-s)$ is a solution.

We can use *MATHEMATICA* to illustrate these three linear systems graphically using the **ImplicitPlot** package, which allows us to plot linear systems in two variables.

```
In[1]:= <<Graphics`ImplicitPlot`
```

```
In[2]:= ImplicitPlot[{x+y==1,x+y==2},{x,-5,10},{y,-4,4}]
```

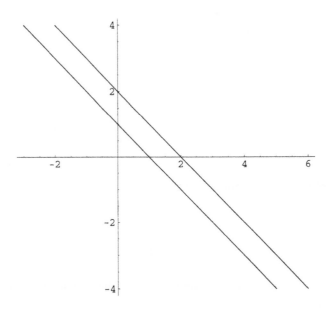

```
Out[2]= - Graphics -
```

```
In[3]:= ImplicitPlot[{x+y==7,x-y==3},{x,-5,10},{y,-4,10}]
```

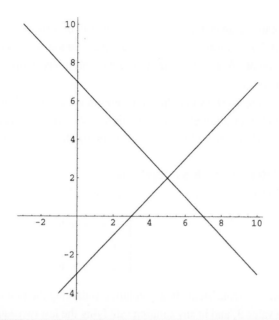

Out[3]= - Graphics -

In[4]:= **ImplicitPlot[{x+y==1,-x-y==-1},{x,-5,10},{y,-4,10}]**

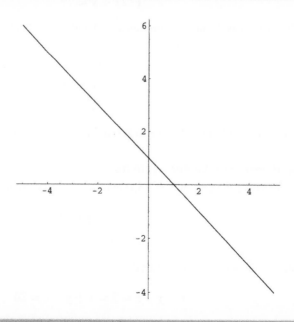

Out[4]= - Graphics -

As we can see, an inconsistent 2×2 linear system corresponds to two parallel nonintersecting lines. A 2×2 linear system with a unique solution corresponds to two lines intersecting at a single point. A 2×2 linear system with infinitely many solutions corresponds to two overlapping lines.

These examples show that there are determined linear systems with no solution, with one solution, and with infinitely many solutions. We will see later on that there are no other possibilities. There are no linear systems with exactly two solutions, for example.

Overdetermined Linear Systems

Let us consider the case of 3×2 linear systems. The system

$$\begin{cases} x + 2y = 5 \\ x - 2y = 1 \\ 2x + y = 6 \end{cases}$$

for example, is inconsistent. In any solution satisfying the first two equations, x would have to take the value 3, and in any solution satisfying the last two equations, x would have to take the value $13/5$. Hence there is no common value for x that satisfies all three equations.

The consistent system

$$\begin{cases} x + 2y = 5 \\ x - 2y = 1 \\ 2x + y = 7 \end{cases}$$

on the other hand, has the unique solution $x = 3$ and $y = 1$. On the other hand, the consistent system

$$\begin{cases} x + 2y = 5 \\ 3x + 6y = 15 \\ -x - 2y = -5 \end{cases}$$

has infinitely many solutions. For any real number s, the point $(-2s + 5, s)$ is a solution.

Underdetermined Linear Systems

Now let us consider the case of 2×3 linear systems. The system

$$\begin{cases} x + y + z = 25 \\ x + y + z = 2 \end{cases}$$

for example, is inconsistent. It implies that

$$x + y + z = 5 = x + y + z = 20$$

This is obviously false. Graphically, this inconsistency is illustrated by two nonintersecting planes, as shown in Figure 3.

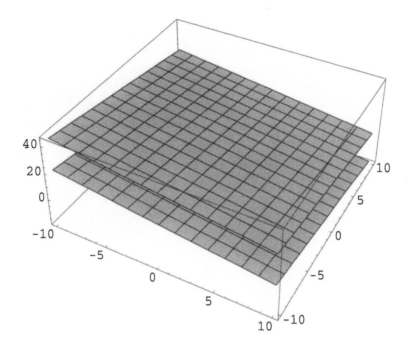

FIGURE 3   Two Parallel Planes.

The system

$$\begin{cases} x + y + z = 25 \\ x + 9y + z = 2 \end{cases}$$

on the other hand, has infinitely many solutions. For every real number s, the point $(s, -s, 1)$ is a solution. Figure 4 shows that the set of solutions forms a straight line, determined by the intersection of the planes $x + y + z = 5$ and $x + y + 2z = 20$.

EXERCISES 1.3

1. Construct linear systems with the following properties and explain why the systems have these properties.

    a. Consistent and 2×2 b. Inconsistent and 2×2

    c. Consistent and 2×3 d. Inconsistent and 2×3

    e. Unique solution and 3×3 f. Infinitely many solutions and 3×3

    g. Inconsistent and 4×3 h. Infinitely many solutions and 4×4

2. Find a linear system determined by four numbers whose sum is 100, with the first three numbers adding up to 50, and the last three numbers adding up to 75. Explain why the system is an underdetermined linear system with infinitely many solutions.

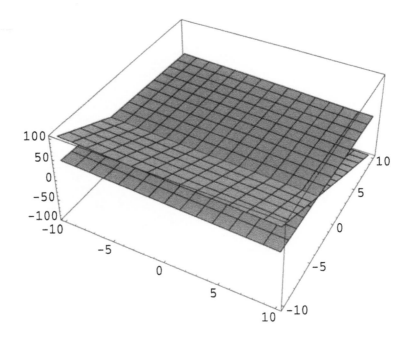

FIGURE 4   Two Intersecting Planes.

3. Show that all $m \times n$ linear systems whose equations are of the form $a_1x_1 + \cdots + a_nx_n = 0$ are consistent.

4. Determine by inspection which of the following linear systems are consistent.

a. $\begin{cases} x + 2y = 2 \\ x - y = 4 \end{cases}$
b. $\begin{cases} x + 2y - z = 2 \\ x - y + z = 4 \\ 2x + y - z = 5 \end{cases}$
c. $\begin{cases} x + 2y - z = 2 \\ x - y + z = 4 \end{cases}$

5. Classify the following linear systems as determined, overdetermined, or underdetermined. Explain why the systems are inconsistent.

a. $\begin{cases} 2x + y - z = 2 \\ 2x + y - z = 6 \end{cases}$
b. $\begin{cases} 2x + y - z = 2 \\ x - 3y + z = 1 \\ 2x + y - z = 6 \end{cases}$
c. $\begin{cases} 2x + 3y = 4 \\ x - y = 1 \\ x + y = 3 \end{cases}$

6. Classify the following linear systems as determined, overdetermined, or underdetermined. Determine by inspection which of the systems are consistent and have one solution, and

which are consistent and have infinitely many solutions.

a. $\begin{cases} 3x + y = 6 \\ x - 2y = 3 \end{cases}$

b. $\begin{cases} 3x + y = 6 \\ 6x + 2y = 12 \end{cases}$

c. $\begin{cases} 3x + y = 6 \\ 6x + 2y = 1 \\ x + y = 7 \end{cases}$

d. $\begin{cases} 3x + y - z = 6 \\ x - 2y + z = 3 \\ 4x - 8y + 2z = 12 \end{cases}$

e. $\begin{cases} 3x + y - z + w = 6 \\ x - 2y + z - w = 3 \\ 4x - 8y + 2z = 12 \end{cases}$

f. $\begin{cases} x - 2y + z = 3 \\ 4x - 8y + 2z = 12 \end{cases}$

7. Use the **ImplicitPlot** function of *MATHEMATICA* to describe the geometric properties of the following inconsistent overdetermined 3×2 linear systems.

a. $\begin{cases} x + 2y = 5 \\ x - 2y = 1 \\ 2x + y = 6 \end{cases}$

b. $\begin{cases} x + 2y = 5 \\ x - 2y = 1 \\ 2x + y = 7 \end{cases}$

c. $\begin{cases} x + 2y = 5 \\ 3x + 6y = 15 \\ -x - 2y = -5 \end{cases}$

Geometry and Linear Systems

We conclude this section with some additional examples from three-dimensional geometry.

EXAMPLE 1.13   ■ **Three Intersecting Planes**

Use *MATHEMATICA* to verify that $(11, 8, -18)$ is the point of intersection of the three planes given by the equations $a + b + c = 1$, $3a + 5b + 4c = 1$, and $-a + 6b + 2c = 1$.

Solution. We first assign the values $11, 8$, and -18 to a, b, and c, respectively.

```
In[1]:= a=11;b=8;c=-18;
```

```
In[2]:= a+b+c==1
Out[2]= True
```

```
In[3]:= 3a+5b+4c==1
Out[3]= True
```

```
In[4]:= -a+6b+2c==1
Out[4]= True
```

Our calculations show that $a = 11$, $b = 8$, and $c = -18$ is a solution of the linear system

$$\begin{cases} a + b + c = 1 \\ 3a + 5b + 4c = 1 \\ -a + 6b + 2c = 1 \end{cases}$$

and that $(11, 8, -18)$ is therefore the point of intersection of the three given planes. ◄►

EXAMPLE 1.14 ■ Four Points and a Plane

Show that there is no plane containing the points $(1, 1, 1)$, $(3, 5, 4)$, $(-1, 6, 2)$, and $(1, 2, 3)$.

Solution. Suppose that the given points lie on the plane $ax + by + cz + d = 0$. Then the linear system

$$\begin{cases} a + b + c + d = 0 \\ 3a + 5b + 4c + d = 0 \\ -a + 6b + 2c + d = 0 \\ a + 2b + 3c + d = 0 \end{cases}$$

must have a common solution. If we examine these equations more closely, we discover that the only common solution is $a = b = c = d = 0$. Thus the plane containing the four given points would have to be $0x + 0y + 0z + d = 0$. This is impossible. ◄►

EXAMPLE 1.15 ■ Two Parallel Planes

Show that the two planes $x + 2y + 3z = 1$ and $x + 2y + 3z = 2$ have no point of intersection.

Solution. Suppose that the point (a, b, c) lies on the given planes. Algebraically this means that the linear system

$$\begin{cases} a + 2b + 3c = 1 \\ a + 2b + 3c = 2 \end{cases}$$

has a solution. This implies that $a + 2b + 3c = 1$ and that $a + 2b + 3c = 2$. In other words, it implies that $1 = 2$. This is a contradiction. ◄►

EXAMPLE 1.16 ■ Four Points and a Line

Show that there is no line in the $\mathbb{R}^2$ containing the points $(1, 1)$, $(3, 5)$, $(-1, 6)$, and $(7, 2)$.

Solution. Suppose that there is a line $y = mx + b$ containing the four points. Since each point must satisfy this equation, the given points determine a linear system

$$\begin{cases} 1 = m + b \\ 5 = 3m + b \\ 6 = -m + b \\ 2 = 7m + b \end{cases}$$

consisting of four equations in two unknowns. If we add the first equation to the third, we get $7 = 2b$. If we subtract the first equation from the second, we get $5 = -2m$. This means that b must be $7/2$ and m must be $-5/2$. If we substitute these values in the fourth equation, we get $2 = 7(-5/2) + 7/2 = -14$. This is a contradiction. ◄►

EXERCISES 1.4

1. Use the techniques discussed in this section to find the plane determined by the three points $(1, 1, 1)$, $(3, 5, 2)$, and $(7, 2, 5)$.

2. Find a plane that is parallel to the plane $x - y + 2z = 5$ and contains the point $(3, 4, 5)$. Explain why the linear system determined by the two planes is inconsistent.

3. Use *MATHEMATICA* to graph the solutions of the following linear systems.

   a. $\begin{cases} x + 2y = 2 \\ x - y = 4 \end{cases}$ b. $\begin{cases} x + y = 2 \\ -2x - 2y = -4 \end{cases}$ c. $\begin{cases} x + 2y = 2 \\ 6x + 12y = 18 \end{cases}$

4. Rewrite the linear system

   $$\begin{cases} x = 4 + 3t \\ y = 8 + 2t \\ z = 3 - 7t \end{cases}$$

   as a parametric equation of a straight line in $\mathbb{R}^3$.

5. Convert the parametric equation of the straight line in $\mathbb{R}^5$, defined by the arrow $\mathbf{a} = (1, 2, 3, 4, 5)$ and the directional arrow $\mathbf{d} = (3, -1, 2, -6, 9)$, to a linear system.

MATHEMATICA AND LINEAR SYSTEMS

We now develop a language and general techniques for finding and describing the solutions of consistent systems. As we have already seen, we can use *MATHEMATICA* to solve these systems. We can also use *MATHEMATICA* to verify that solutions found by other means are correct. For example, the calculation

```
In[1]:= Solve[{x+y==5,x-y==3},{x,y}]

Out[1]= {{x→4,y→1}}
```

produces the solution $(x, y) = (4, 1)$ for the linear system

$$\begin{cases} x + y = 5 \\ x - y = 3 \end{cases}$$

If a manual attempt to solve the system had produced the assignments $x = 5$ and $y = 1$, we could use *MATHEMATICA* to show that this is incorrect.

```
In[2]:= x=5;y=1;x+y==5

Out[2]= False
```

MATHEMATICA is telling us that the claim that $x + y$ equals 5 is false.

EXAMPLE 1.17   ■ A Consistent System with One Solution

Use *MATHEMATICA* to show that the linear system

$$\begin{cases} x + y - z - 4 = 0 \\ 2x - y - 3z + 6 = 0 \\ -x + y + 6z = 0 \end{cases}$$

has one solution.

Solution. We write the given system as a list of equations and apply the **Solve** function.

```
In[1]:= Solve[{x+y-z-4==0,2x-y-3z+6==0,-x+y+6z==0},{x,y,z}]

Out[1]= {{x → -30/13, y → 66/13, z → -16/13}}
```

The *MATHEMATICA* output shows that the given system has one solution. ◄►

EXAMPLE 1.18   ■ A Consistent System with Infinitely Many Solutions

Use *MATHEMATICA* to show that the linear system

$$\begin{cases} x + y - z = 2 \\ x - y + z = 4 \\ 2x + y - z = 5 \end{cases}$$

has infinitely many solutions.

Solution. Again we use the **Solve** function.

```
In[1]:= Solve[{x+y-z==2,x-y+z==4,2x+y-z==5}]

Solve::svars:  Equations may not give solutions for all

''solve'' variables.

Out[1]= {{x → 3, y → -1+z}}
```

MATHEMATICA is telling us that although the system consists of three equations in three unknowns, two of the variables depend on the third. If we replace z by 10, for example, the replacements of x by 3 and of y by $-1 + 10$ produce a solution of the system. Since we have infinitely many choices for z, the system has infinitely many solutions. The message

```
Equations may not give solutions for all ''solve'' variables.
```

tells us that the system is underdetermined. If we were to instruct *MATHEMATICA* to solve the system for two of the three variables, say for x and y, the message would disappear:

```
In[1]:= Solve[{x+y-z==2,x-y+z==4,2x+y-z==5},{x,y}]
Out[1]= {{x → 3,y → -1+z}}
```

As expected, the solution $x = 3$ and $y = -1 + z$ is a function of z. ◄►

EXAMPLE 1.19   ■ An Underdetermined Linear System

Use *MATHEMATICA* to solve the system

$$\begin{cases} x + y - z = 2 \\ x - y + z = 4 \end{cases}$$

obtained from the previous example by deleting the third equation.

Solution. Again we use the **Solve** function. Since the system has two equations, we solve for two of the three variables.

```
In[1]:= Solve[{x+y-z==2,x-y+z==4},{x,y}]
Out[1]= {{x → 3,y → -1+z}}
```

As we can see, the system has the same solutions as that in the previous example. ◄►

The next example explains the behavior of *MATHEMATICA* when it encounters an inconsistent system.

EXAMPLE 1.20   ■ An Inconsistent Linear System

Use *MATHEMATICA* to show that the system

$$\begin{cases} x + y - z = 2 \\ x - y + z = 4 \\ 2x + y - z = 6 \end{cases}$$

is inconsistent.

Solution. We try to use **Solve** function.

```
In[1]:= Solve[{x+y-z==2,x-y+z==4,2x+y-z==6}]
Out[1]= {}
```

MATHEMATICA returns the empty set since the system has no solution. ◄►

We conclude this section by considering an inconsistent overdetermined system.

EXAMPLE 1.21   ■ An Overdetermined Linear System

Use *MATHEMATICA* to show that the system

$$\begin{cases} x + y = 2 \\ x - y = 4 \\ 2x + y = 6 \end{cases}$$

has no solution.

Solution. *MATHEMATICA* treats an overdetermined system like any other system. Since the system has at least as many equations as variables, we can use the **Solve** function without having to specify any variables to solve for.

```
In[1]:= Solve[{x+y==2,x-y==4,2x+y==6}]
Out[1]= {}
```

The empty output tells us that the system is inconsistent. ◄►

Gaussian Elimination

The basic steps used to find the solutions of a linear system are known as *Gaussian elimination*. The method is based on three principles of equality.

1. The multiplication of both sides of an equation by the same nonzero constant preserves the solutions of the equation.

2. The addition of the same quantity to both sides of an equation preserves the solutions of the equation.

3. The replacement of a quantity in an equation by an equal quantity preserves the solutions of the equation.

Gauss discovered that these principles could be used to devise a systematic procedure for solving linear systems. Suppose that

$$\begin{cases} a_{11}x_1 + \cdots + a_{1n}x_n = b_1 \\ \vdots \\ a_{m1}x_1 + \cdots + a_{mn}x_n = b_m \end{cases}$$

is a linear system in the variables $x_1, \ldots, x_n$. The solutions of this system can be found by using the following three operations based on the principles of equality.

1. *Multiplying all terms of an equation of the system by a nonzero constant.* This operation is justified by the fact that the equations $x + 2y = 7$ and $sx + 2sy = 7s$ have the same solutions for all nonzero constants s.

2. *Adding a multiple of one equation of the system to another equation of the system.* The logical basis for this operation is the fact that if $x + 2y = 7$ and $3x - y = 5$, then $(x + 2y) + s(3x - y) = 7 + 5s$ for all constants s.

3. *Changing the order in which the equations of a system are listed.*

The third operation systematizes the solution process, but has no effect on the equations themselves.

In general, the Gaussian elimination operations can be applied to a linear system in many different ways. One of the objectives of this text is to develop a systematic approach to the choice and order of application of these rules.

EXAMPLE 1.22 ■ A Gaussian Elimination

Use elementary operations to solve the system

$$\begin{cases} x + 2y - z = 2 \\ x - y + z = 4 \\ 2x + y - z = 5 \end{cases}$$

Solution. Here is one possible sequence of operations that converts the given system to a new system from which a solution for the given system can easily be calculated.

1. We begin by subtracting the first equation from the second:

$$\begin{cases} x + 2y - z = 2 \\ x - y + z = 4 \\ 2x + y - z = 5 \end{cases} \rightarrow \begin{cases} x + 2y - z = 2 \\ -3y + 2z = 2 \\ 2x + y - z = 5 \end{cases}$$

2. We then subtract twice the first equation from the third:

$$
\begin{cases}
x + 2y - z = 2 \\
-3y + 2z = 2 \\
2x + y - z = 5
\end{cases}
\rightarrow
\begin{cases}
x + 2y - z = 2 \\
-3y + 2z = 2 \\
-3y + z = 1
\end{cases}
$$

3. Next we subtract the third equation from the second:

$$
\begin{cases}
x + 2y - z = 2 \\
-3y + 2z = 2 \\
-3y + z = 1
\end{cases}
\rightarrow
\begin{cases}
x + 2y - z = 2 \\
z = 1 \\
-3y + z = 1
\end{cases}
$$

4. At this stage, we can see that in any solution, z must be 1.

5. We then subtract the second equation from the third:

$$
\begin{cases}
x + 2y - z = 2 \\
z = 1 \\
-3y + z = 1
\end{cases}
\rightarrow
\begin{cases}
x + 2y - z = 2 \\
z = 1 \\
-3y = 0
\end{cases}
$$

6. Next we divide the third equation by -3:

$$
\begin{cases}
x + 2y - z = 2 \\
z = 1 \\
-3y = 0
\end{cases}
\rightarrow
\begin{cases}
x + 2y - z = 2 \\
z = 1 \\
y = 0
\end{cases}
$$

7. Hence y must be 0.

8. We then replace y by 0 and z by 1 in the first equation:

$$
\begin{cases}
x + 2y - z = 2 \\
z = 1 \\
y = 0
\end{cases}
\rightarrow
\begin{cases}
x - 1 = 2 \\
z = 1 \\
y = 0
\end{cases}
$$

9. By adding 1 to both sides of the first equation, we get the solution of the system: $x = 3$, $y = 0$, and $z = 1$. ◄►

Example 1.22 shows that the values of the variables x, y, and z can either be read off directly from an intermediary system, or they can be obtained from one of the systems by substituting the values for some variables back into the system, as in steps 8 and 9. The latter method is known as ***back substitution***.

EXERCISES 1.5

1. Use Gaussian elimination and back substitution to solve the following linear systems.

a. $\begin{cases} x + 2y = 2 \\ x - y = 4 \end{cases}$
   b. $\begin{cases} x + 2y - z = 2 \\ x - y + z = 4 \\ 2x + y - z = 5 \end{cases}$
   c. $\begin{cases} x + 2y - z = 2 \\ x - y + z = 4 \end{cases}$

2. Use the **Solve** function of *MATHEMATICA* to solve the following linear systems.

a. $\begin{cases} 2x + y - z = 2 \\ x - 3y + z = 1 \\ 2x + y - 2z = 6 \end{cases}$
   b. $\begin{cases} 2x + y - z = 2 \\ x - 3y + z = 1 \\ 2x + y - z = 6 \end{cases}$
   c. $\begin{cases} 2x + y - z = 2 \\ x - 3y + z = 1 \end{cases}$

3. Use *MATHEMATICA* to find the points of intersection of the following pairs of lines.

   a. The lines $x + y = 5$ and $x - y = 3$
   b. The lines $2x - 7y = 1$ and $x + 3y = 4$
   c. The lines $x = 5$ and $x + y = 7$
   d. The lines $x - y = 0$ and $y = 9$
   e. The lines $x = 5$ and $y = 21$

4. Use Gaussian elimination to find the points of intersection of the following sets of planes.

   a. The planes $x + y + z = 5$, $x - y + 2z = 3$, and $x + 3y - z = 4$
   b. The planes $2x - 7y = 1$, $x - z = 0$, and $x + 3y - z = 4$
   c. The planes $14x + 7y + z = 5$, $y = 3$, and $x + y + z = 7$
   d. The planes $x - y + 4z = 12$, $y = 7$, and $x + y = 9$
   e. The planes $x - y - z = 5$, $x - 2y + 7z = 21$, and $y - z = 1$

MATRICES AND LINEAR SYSTEMS

We can associate matrices with linear systems in two basic ways. We can form the *augmented matrix* of a system or we can rewrite the system as a *matrix equation*.

In this section, we first discuss the idea of converting a linear system to an augmented matrix. We define the row echelon and reduced row echelon forms of a matrix, and show how to adapt the Gaussian elimination process to a procedure for converting augmented matrices to row echelon and reduced row echelon form. We show that the solutions of the linear system can be obtained by *back substitution* using echelon matrices.

We then discuss the conversion of a linear system to a matrix equation and show how matrix algebra can be used to solve the system.

Let us begin with a general introduction to matrices and their *MATHEMATICA* form.

DEFINITION 1.4 *A **matrix** is a rectangular array of mathematical objects.*

Most matrices studied in this text are rectangular arrays of real numbers. The array

$$A = \begin{pmatrix} 1 & 2 & -1 & \pi \\ 1 & -1 & 1.5 & 4 \\ 2 & 1 & -1 & 5 \end{pmatrix}$$

for example, is a matrix of real numbers. The objects in the cells of the array are called the **entries** of the matrix.

If the entries of a matrix are all zero, we call the matrix a **zero matrix** and denote it by **0**. We use the same symbol no matter how many rows and columns a zero matrix may have. The meaning of the symbol **0** will always be clear from the context.

MATHEMATICA provides several options for defining matrices. We can write them as a sequence of rows of entries; we can use matrix palettes as entry devices; or we can use functions and expressions, in combination with the built-in **Array** and **Table** functions, to create them. The first two methods are described in Appendix E.

Arrays and Functions

Suppose that $f[i, j]$ is a function in two variables, either built into *MATHEMATICA* or user-defined. Then we can use the **Array** function to generate a matrix whose ijth entry is $f[i, j]$.

EXAMPLE 1.23 ■ **Using Array to Generate a Matrix**

1. The **Max** function is a built-in *MATHEMATICA* function whose value at $[i, j]$ is the maximum of i and j. We can combine it with the **Array** function to generate a matrix.

```
In[1]:= Array[Max, {3,4}]
Out[1]= {{1,2,3,4},{2,2,3,4},{3,3,3,4}}
```

The output is a matrix consisting of the three rows {1, 2, 3, 4}, {2, 2, 3, 4}, and {3, 3, 3, 4}.

2. The **Random** function is a built-in *MATHEMATICA* function whose values are pseudorandom real numbers of a specified type (integer, real, or complex) lying in a specified range $\{min, max\}$. The default outputs are real numbers between 0 and 1. We can combine the **Array** function with the **Random** function to generate a **random matrix**. Here is an example of a random matrix with integer entries between -8 and 8.

```
In[2]:= Array[Random[Integer, {-8,8}]&, {2,3}]
Out[2]= {{-7,-6,0},{4,8,-8}}
```

3. The next example is a random matrix with real entries between 0 and 1.

```
In[3]:= Array[Random[]&, {3,4}]
Out[3]= {{0.324862,0.945044,0.0499139,0.0230458},
{0.635336,0.240203,0.0979025,0.72275},
{0.119975,0.22964,0.194659,0.267442}}
```

4. We can also generate a matrix by combining the **Array** function with a user-defined function.

```
In[4]:= f[i_,j_]:=3i+5j;
```

```
In[5]:= Array[f, {3,4}]
Out[5]= {{8,13,18,23},{11,16,21,26},{14,19,24,29}}
```

The output is a matrix consisting of the three rows $\{8, 13, 18, 23\}$, $\{11, 16, 21, 26\}$, and $\{14, 19, 24, 29\}$.

5. We can also define a function by specifying its values, one by one, and then using its name to generate a matrix.

```
In[6]:= g[1,1]:=5;g[1,2]:=8;g[2,1]:=9;g[2,2]:=11;
```

```
In[7]:= Array[g, {2,2}]
Out[7]= {{5,8},{9,11}}
```

The output is a matrix whose two rows are $\{5, 8\}$ and $\{9, 11\}$.

6. The entries of a matrix can also be nonnumerical mathematical objects. The following matrix, for example, has polynomials as entries.

```
In[8]:= g[1,1]:=1-t;g[1,2]:=t^2;g[2,1]:=9;g[2,2]:=11+t-t^3;
```

```
In[9]:= Array[g,{2,2}]
Out[9]= {{1-t,t^2},{9,11+t-t^3}}
```

The most appropriate choice of one of these method for generating a particular matrix depends on the context. ◄►

EXERCISES 1.6

1. Use the **Array** function, together with a built-in *MATHEMATICA* function in two variables, to generate a matrix with four rows and five columns.

2. Use the **Array** function, together with a user-defined function in two variables, to generate a matrix with four rows and five columns.

3. Use the **Array** function, together with a function in two variables defined by listing its values, to generate a matrix with four rows and five columns.

4. Use the **Array** function, together with the **Random** function, to generate a matrix with four rows and five columns.

5. Use the **Array** function, together with a user-defined function in two variables, to generate a matrix with four rows and five columns whose entries are polynomials.

Tables and Expressions

Suppose that *expr* is an expression in two variables such as $3i + 5j$. Then we can use the **Table** function to generate a matrix whose ijth entry is $3i + 5j$.

EXAMPLE 1.24 ■ **Using Table to Generate an Exact Matrix**

Use the **Table** function to generate a matrix with three rows and four columns whose ijth entry is $3i + 5j$.

Solution. We first define the expression $3i + 5j$ and then use it as an argument in the **Table** function.

```
In[1]:= expr:=3i+5j
```

```
In[2]:= Table[expr,{i,3},{j,4}]
Out[2]= {{8,13,18,23},{11,16,21,26},{14,19,24,29}}
```

The output is a matrix consisting of the three rows $\{8, 13, 18, 23\}$, $\{11, 16, 21, 26\}$, and $\{14, 19, 24, 29\}$. ◄►

Suppose that *expr* is an expression in two variables such as `N[Exp[i]+Log[j]]`. Then we can use the **Table** function again to generate a matrix whose *ij*th entry is `N[Exp[i]+Log[j]]`.

EXAMPLE 1.25 ■ **Using Table to Generate an Approximate Matrix**

```
In[1]:= Table[N[Exp[i]+Log[j]],{i,3},{j,4}]
Out[1]= {{2.71828,3.41143,3.81689,4.10458},
{7.38906,8.0822,8.48767,8.77535},
{20.0855, 20.7787,21.1841,21.4718}}
```

The output is a 3×4 matrix whose entries are approximations of sums of values of the exponential and logarithmic functions. ◄►

We can also use the **Table** function in combination with random expressions.

EXAMPLE 1.26 ■ **Using Table to Generate a Random Matrix**

```
In[1]:= Table[Random[],{i,3},{j,4}]
Out[1]= {{0.270095,0.540834,0.601887,0.438495},
{0.782561,0.94476,0.347765,0.615154},
{0.940499,0.982093,0.911094,0.560458}}
```

The output is a matrix whose three rows consist of random numbers between 0 and 1. ◄►

EXERCISES 1.7

1. Write three expressions in i and j, and use them in conjunction with the **Table** function to generate three matrices with four rows and five columns.

2. Use three expressions based on built-in *MATHEMATICA* functions, together with the **Table** function, to generate three matrices with four rows and five columns.

3. Use the **Table** function, together with expressions built with the **Random** function, to generate three matrices with four rows and five columns.

Matrices as Functions

If we take a closer look at a system such as

$$\begin{cases} x + 2y - z = 2 \\ x - y + z = 4 \\ 2x + y - z = 5 \end{cases}$$

we realize that it is uniquely determined by the coefficients of its variables and the constants on the right-hand side of the equations. We can record this information in a matrix

$$A = \begin{pmatrix} 1 & 2 & -1 & 2 \\ 1 & -1 & 1 & 4 \\ 2 & 1 & -1 & 5 \end{pmatrix}$$

The matrix consists of three rows and four columns. We know that the coefficients of x are listed in the first column, the coefficients of y in the second column, and the coefficients of z in the third column. The last column contains the constants occurring on the right-hand side of the equations.

If we write $A[i, j]$ for the information in the cell located in the ith row and jth column of the matrix A, then

$$A = \begin{pmatrix} A[1, 1] & A[1, 2] & A[1, 3] & A[1, 4] \\ A[2, 1] & A[2, 2] & A[2, 3] & A[2, 4] \\ A[3, 1] & A[3, 2] & A[3, 3] & A[3, 4] \end{pmatrix}$$

with $A[1, 1] = 1$, $A[1, 2] = 2$, $A[1, 3] = -1$, and so forth. Since $A[i, j]$ occurs in the ith row and jth column of A, we call it the ijth **entry** of A. In *MATHEMATICA* it is denoted by **A[[i,j]]**. If we use the list $\{1, 2, 3\}$ to label the rows of A and the list $\{1, 2, 3, 4\}$ to label the columns, we have an address system for A where each cell is labeled by a pair of integers taken from the set

$$\{1, 2, 3\} \times \{1, 2, 3, 4\}$$

consisting of the ordered pairs $(1, 1)$, $(1, 2)$, $(1, 3)$, $(1, 4)$, $(2, 1)$, $(2, 2)$, $(2, 3)$, $(2, 4)$, $(3, 1)$, $(3, 2)$, $(3, 3)$, and $(3, 4)$. We refer to lists such as $\{1\}$, $\{1, 2\}$, $\{1, 2, 3\}$, $\{1, 2, 3, 4\}$ as **standard lists**. We write **n** for the standard list $\{1, 2, \ldots, n\}$ consisting of the integers from 1 and n. (We use a boldface **n** to distinguish the list $\{1, 2, \ldots, n\}$ from its largest element n.) As we can see, the location of a cell in a matrix with m rows and n columns can be specified by a pair of integers taken from the set

$$\mathbf{m} \times \mathbf{n} = \{1, 2, \ldots, m\} \times \{1, 2, \ldots, n\}$$

We call $m \times n$ the **dimension** of A.

Since the entry $A[i, j]$ of a matrix A is uniquely identified by the pair (i, j), we can think of a matrix with m rows and n columns as a function from $\mathbf{m} \times \mathbf{n}$ to the mathematical system

$\mathbb{X}$ from which the entries $A[i, j]$ are taken. In this text, typical systems in which matrices take their values are the set $\mathbb{R}$ of real numbers and the set $\mathbb{R}[t]$ of polynomials in t with coefficients in $\mathbb{R}$. In some of the calculus applications, the system $\mathbb{X}$ is more general.

DEFINITION 1.5   *An $\mathbb{X}$-**valued matrix** of dimension $m \times n$ is a function $A : \mathbf{m} \times \mathbf{n} \to \mathbb{X}$.*

We call this the ***functional form*** of A. We assume that $\mathbb{X} = \mathbb{R}$, unless otherwise specified, and refer to such matrices as ***real matrices***. We often simplify the terminology by speaking simply of *matrices* if it is clear from the context that their entries are intended to be real numbers. If $m = n = 1$, we usually identify the matrix $[A[1, 1]]$ with its entry $A[1, 1]$.

EXAMPLE 1.27   ■ **An $\mathbb{R}$-Valued 2 × 3 Matrix**

Suppose that $A : \mathbf{2} \times \mathbf{3} \to \mathbb{R}$ is the function defined by $A[i, j] = \text{Max}[i, j]$. Then

$$A = \begin{pmatrix} A[1, 1] & A[1, 2] & A[1, 3] \\ A[2, 1] & A[2, 2] & A[2, 3] \end{pmatrix} = \begin{pmatrix} 1 & 2 & 3 \\ 2 & 2 & 3 \end{pmatrix}$$

We see that the matrix A is a rectangular array with real entries, arranged in two rows and three columns. ◄►

EXAMPLE 1.28   ■ **An $\mathbb{R}[t]$-Valued 2 × 2 Matrix**

Suppose that $A : \mathbf{2} \times \mathbf{2} \to \mathbb{R}[t]$ is the function defined by t if $i + j$ is even and t^3 otherwise. Then

$$A = \begin{pmatrix} A[1, 1] & A[1, 2] \\ A[2, 1] & A[2, 2] \end{pmatrix} = \begin{pmatrix} t & t^3 \\ t^3 & t \end{pmatrix}$$

Here the matrix A is a rectangular array with polynomial entries, arranged in two rows and two columns. ◄►

MATHEMATICA has a built-in routine for creating the standard lists used in the definition of matrices.

EXAMPLE 1.29   ■ **Standard Lists**

Use *MATHEMATICA* to create the standard lists **3** and **4**.

Solution. The **Range** function creates standard lists.

```
In[1]:= Range[3]

Out[1]= {1,2,3}
```

```
In[2]:= Range[4]

Out[2]= {1,2,3,4}
```

As we can see, the outputs are the standard lists **3** and **4**. ◄►

In the next example, we show that *MATHEMATICA* has built-in functions for computing the dimension of a matrix.

EXAMPLE 1.30 ■ The Dimension of a Matrix

Use the **Dimensions** and **Length** functions of *MATHEMATICA* to calculate the dimension of the matrix

$$A = \begin{pmatrix} 1 & 2 & 3 & 4 & 5 \\ 6 & 7 & 8 & 9 & 10 \\ 11 & 12 & 13 & 14 & 15 \end{pmatrix}$$

in two different ways.

Solution. First we use the **Create Matrix** menu to define A. Then we apply the **Dimensions** function to A.

```
In[1]:= A= ( 1   2   3   4   5
            6   7   8   9   10  ) ;
           11  12  13  14  15
```

```
In[2]:= Dimensions[A]

Out[2]= {3,5}
```

It is understood that the first number in the list $\{3, 5\}$ counts the number of rows of A and the second number counts the number of columns.

Next we use the **Length** function, which counts the number of elements of a list. Since each element $\{1, 2, 3, 4, 5\}$, $\{6, 7, 8, 9, 10\}$, $\{11, 12, 13, 14, 15\}$ of A is a row of A, the **Length** function computes the number of rows of A.

```
In[3]:= Length[A]

Out[3]= 3
```

To find the number of columns of A, we form the transpose of A.

```
In[4]:= Transpose[A]

Out[4]= {{1,6,11},{2,7,12},{3,8,13},{4,9,14},{5,10,15}}
```

Each element of this list denotes a column of A and we can use the **Length** function again to compute the number of columns of A.

```
In[5]:= Length[Transpose[A]]
Out[5]= 5
```

Thus *MATHEMATICA* provides us with several ways of computing the dimension of a matrix. ◄►

EXERCISES 1.8

1. Use the **Create Matrix** menu and the **BasicInput** palette to create the following matrices.

$$
a. \begin{pmatrix} 4 \\ 5 \\ 6 \end{pmatrix} \quad
b. \begin{pmatrix} 1 & 2 & 3 & 4 & 5 \\ -9 & 3 & 7 & 7 & 8 \\ 0 & 12 & 45 & \pi & 27 \end{pmatrix} \quad
c. \begin{pmatrix} 1 & 0 & 0 \\ 6 & 0 & 0 \\ 2 & 0 & 0 \end{pmatrix}
$$

2. Use the list of lists method in *MATHEMATICA* to create the following matrices.

$$
a. \begin{pmatrix} 1 & 0 \\ 0 & 1 \end{pmatrix} \quad
b. \begin{pmatrix} -85 \\ 97 \\ 49 \end{pmatrix} \quad
c. \begin{pmatrix} -85 & 97 & 49 \end{pmatrix} \quad
d. \begin{pmatrix} 0 & 1 & 0 \\ 1 & 0 & 0 \\ 0 & 0 & 1 \end{pmatrix}
$$

3. Write the following matrices in rectangular form.

   a. The matrix $A : 2 \times 3 \to \mathbb{R}$ defined by

   $A[1, 1] = 3, A[1, 2] = -7, A[1, 3] = \pi, A[2, 1] = 2.8, A[2, 2] = 21, A[2, 3] = -100$

   b. The matrix $A : 3 \times 2 \to \mathbb{R}$ defined by

   $A[1, 1] = 3, A[1, 2] = 2.8, A[2, 1] = -7, A[2, 2] = 21, A[3, 1] = \pi, A[3, 2] = -100$

   c. The matrix $A : 1 \times 2 \to \mathbb{R}$ defined by $A[1, 1] = 3, A[1, 2] = 2.8$

   d. The matrix $A : 2 \times 1 \to \mathbb{R}$ defined by $A[1, 1] = 3, A[2, 1] = 2.8$

   e. The matrix $A : 4 \times 4 \to \mathbb{R}$ defined by $A[i, j] = 1$ if $i = j$ and 0 otherwise

   f. The matrix $A : 4 \times 4 \to \mathbb{R}$ defined by $A[i, j] = i$ if $i = j$ and 0 otherwise

   g. The matrix $A : 4 \times 4 \to \mathbb{R}$ defined by $A[i, j] = \sqrt{j}$ if $i = j$ and 0 otherwise

   h. The matrix $A : 3 \times 3 \to \mathbb{R}$ defined by $A[i, j] = i + j$ for $1 \leq i, j \leq 3$

   i. The matrix $A : 1 \times 1 \to \mathbb{R}$ defined by $A[1, 1] = 99$

   j. The matrix $A : 3 \times 3 \to \mathbb{R}$ defined by $A[i, j] = i^2$ if $i + j$ is even and 0 otherwise

4. Write the following matrices in functional form.

$$
\text{a.} \begin{pmatrix} -85 & -55 & -37 & -35 \\ 97 & 50 & 79 & 56 \\ 49 & 63 & 57 & -59 \end{pmatrix} \quad \text{b.} \begin{pmatrix} 1 & 0 & 0 \\ 0 & 1 & 0 \\ 0 & 0 & 1 \end{pmatrix} \quad \text{c.} \begin{pmatrix} 3 & 4 \\ 2 & 3 \\ 1 & 2 \end{pmatrix}
$$

5. Explain why the set $\{A[1, 1] = 7, A[2, 2] = 5, A[2, 3] = 0\}$ is not a matrix.

6. Use the **Range** function of *MATHEMATICA* to generate the following standard lists.

a. $\mathbf{1} = \{1\}$ b. $\mathbf{2} = \{1, 2\}$ c. $\mathbf{5} = \{1, 2, 3, 4, 5\}$ d. $\mathbf{6} = \{1, 2, 3, 4, 5, 6\}$

7. Convert the following matrices to lists of lists.

$$
\text{a.} \begin{pmatrix} -85 & 0 & -37 \\ 97 & 50 & 4 \end{pmatrix} \quad \text{b.} \begin{pmatrix} t & 0 \\ 0 & t^2 - 1 \\ 3 & 0 \end{pmatrix} \quad \text{c.} \begin{pmatrix} 2 & -3 & 4 \\ 3 & 4 & 7 \\ 4 & 5 & 6 \end{pmatrix}
$$

8. Use the **Create Matrix** menu and the **BasicInput** palette to create the matrices defined in Exercise 3.

Vectors as Matrices

Now that we have the language of matrices available to us, we can redefine vectors as special matrices. We will use this opportunity to define two types of vectors, column vectors and row vectors.

DEFINITION 1.6   *A column vector is a matrix of the form $A : \mathbf{m} \times \mathbf{1} \to \mathbb{X}$.*

We call m the **height** of A. Usually we drop any reference to the set $\mathbf{1}$ and write the matrix A as

$$
\begin{pmatrix} a_1 \\ \vdots \\ a_m \end{pmatrix}
$$

with $a_1 = A[1, 1], a_2 = A[2, 1], \ldots, a_m = A[m, 1]$. We refer to the entries a_i as the **components** of the vector A. In *MATHEMATICA*, this vector corresponds to the list

$$
\{\{A[[1, 1]]\}, \ldots, \{A[[m, 1]]\}\}
$$

We write vectors as $\mathbf{x}, \mathbf{y}, \mathbf{z}, \mathbf{u}, \mathbf{v}, \mathbf{w}, \ldots$ to distinguish them notationally from more general matrices.

EXAMPLE 1.31   ■ A Column Vector

Write the column vector $A : \{1, 2, 3\} \times \{1\} \to \mathbb{R}$, defined by $A[1, 1] = 1$, $A[2, 1] = 5$, $A[3, 1] = 9$, as a *MATHEMATICA* expression and output the vector in matrix form.

Solution. We define A as a **list of lists** and request the output to be in matrix form.

```
In[1]:= A={{1},{5},{9}}//MatrixForm

Out[1]//MatrixForm=
                     ⎛ 1 ⎞
                     ⎜ 5 ⎟
                     ⎝ 9 ⎠
```

As we can see, *MATHEMATICA* displays A in column format. ◄►

DEFINITION 1.7   *A **row vector** is a matrix $A : \mathbf{1} \times \mathbf{n} \to \mathbb{X}$.*

We call n the **length** of A. Usually we drop the reference to the set $\mathbf{1}$ and write A as

$$\left(\begin{array}{ccc} a_1 & \cdots & a_n \end{array} \right)$$

with $a_1 = A[1, 1], a_2 = A[1, 2], \ldots, a_n = A[1, n]$. As in the case of column vectors, we refer to the entries a_i of A as the **components** of A. In *MATHEMATICA*, this vector corresponds to the list

$$\{\{A[[1, 1]], \ldots, A[[1, n]]\}\}$$

EXAMPLE 1.32   ■ A Row Vector

Write the row vector $A : \{1\} \times \{1, 2, 3\} \to \mathbb{R}$, defined by $A[1, 1] = 1$, $A[1, 2] = 5$, $A[1, 3] = 9$, as a *MATHEMATICA* expression and output the vector in matrix form.

Solution. As in the previous example, we write A as a list of lists and request the output to be in matrix form.

```
In[1]:= A= {{1,5,9}}//MatrixForm

Out[1]//MatrixForm=
                     ( 1  5  9 )
```

This time *MATHEMATICA* produces a row vector. ◄►

We use the term *vector* ambiguously for both row and column vectors. However, in most cases, vectors will be column vectors. It will always be clear from the context whether the vectors involved are row or column vectors. We denote vectors by $\mathbf{x}, \mathbf{y}, \mathbf{z}, \mathbf{u}, \mathbf{v}, \mathbf{w}, \ldots$ to distinguish them notationally from other matrices.

EXERCISES 1.9

1. Find *MATHEMATICA* expressions for the following row vectors.

$$\text{a. } (1, 2, -1) \quad \text{b. } (-8, 2, 99) \quad \text{c. } (x_1, x_2, x_3)$$

2. Use *MATHEMATICA* to compute the row vector

$$3(1, 2, -1) - 7(-8, 2, 99) + (1, 0, -100)$$

3. Find *MATHEMATICA* expressions for the following column vectors.

$$\text{a. } \begin{pmatrix} 1 \\ 2 \\ -1 \end{pmatrix} \quad \text{b. } \begin{pmatrix} -8 \\ 2 \\ 99 \end{pmatrix} \quad \text{c. } \begin{pmatrix} x_1 \\ x_2 \\ x_3 \end{pmatrix} \quad \text{d. } \begin{pmatrix} 0 \\ 2 \\ 0 \end{pmatrix} \quad \text{e. } \begin{pmatrix} 0 \\ 0 \\ 0 \end{pmatrix}$$

4. Use *MATHEMATICA* to compute the column vector

$$3 \begin{pmatrix} 1 \\ 2 \\ -1 \end{pmatrix} - 7 \begin{pmatrix} -8 \\ 2 \\ 99 \end{pmatrix} + \begin{pmatrix} 1 \\ 0 \\ -100 \end{pmatrix}$$

5. Rewrite the row vectors in Exercise 1 as matrices in functional form.

6. Rewrite the column vectors in Exercise 3 as matrices in functional form.

AUGMENTED MATRICES

As we have seen, a linear system is determined by the coefficients of its variables and its equational constants. We can arrange these data in matrix form and develop a procedure for finding the solutions of a linear system by manipulating the associated matrices.

DEFINITION 1.8   *The **augmented matrix** of the linear system*

$$\begin{cases} a_{11}x_1 + \cdots + a_{1n}x_n &= b_1 \\ &\vdots \\ a_{m1}x_1 + \cdots + a_{mn}x_n &= b_m \end{cases}$$

is the $m \times (n + 1)$ matrix

$$(A \mid \mathbf{b}) = \begin{pmatrix} a_{11} & \cdots & a_{1n} & b_1 \\ \vdots & \vdots & \vdots & \vdots \\ a_{m1} & \cdots & a_{mn} & b_m \end{pmatrix}$$

EXAMPLE 1.33   ■ An Augmented Matrix

Find the augmented matrix $(A \mid \mathbf{b})$ of the linear system

$$\begin{cases} x + 2y - z = 2 \\ x - y + z = 4 \\ 2x + y - z = 5 \end{cases}$$

Solution. Since the given system consists of three equations in three unknowns, the augmented matrix $(A \mid \mathbf{b})$ will be a 3×4 matrix. Its first column will contain the coefficients of x, its second column the coefficients of y, and its third column the coefficients of z. The last column will contain the constant terms. The matrix

$$\begin{pmatrix} 1 & 2 & -1 & 2 \\ 1 & -1 & 1 & 4 \\ 2 & 1 & -1 & 5 \end{pmatrix}$$

clearly satisfies these conditions. ◀▶

The rectangular array made up of the first three columns of A is the **_coefficient matrix_** of the given system and the fourth column is the **_vector of constants_** of the system. The **MatrixManipulation** package makes it easy to create augmented matrices.

```
In[1]:= <<LinearAlgebra`MatrixManipulation`
```

With this package, we can now build the matrix A and the vector $\mathbf{b}$ in Example 1.33.

```
In[2]:= X=LinearEquationsToMatrices[{x+2y-z==2,x-y+z==4,
2x+y-z==5},{x,y,z}]

Out[2]= {{{1,2,-1},{1,-1,1},{2,1,-1}},{2,4,5}}
```

```
In[3]:= A=X[[1]]

Out[3]= {{1,2,-1},{1,-1,1},{2,1,-1}}
```

```
In[4]:= b=Partition[X[[2]],1]

Out[4]= {{2},{4},{5}}
```

```
In[5]:= MatrixForm[AppendRows[A,b]]

Out[5]//MatrixForm=
```

$$\begin{pmatrix} 1 & 2 & -1 & 2 \\ 1 & -1 & 1 & 4 \\ 2 & 1 & -1 & 5 \end{pmatrix}$$

Gaussian elimination steps can be translated into operations on augmented matrices:

1. The multiplication of an equation of a system by a nonzero constant translates into the multiplication of every entry in a row of a matrix by a constant.

2. The addition of a multiple of one equation of a system to another equation of the system translates into multiplying each entry in a row by the same constant and adding the new entries to the corresponding entries of another row.

3. The permutation of the order of two equations of a system translates into the permutation of two rows of an augmented matrix.

In the next section, we will introduce appropriate notation for these operations, known as *elementary row operations*.

We now illustrate how these operations can be used to solve a linear system.

EXAMPLE 1.34   ■ Gaussian Elimination and Augmented Matrices

Use elementary row operations on augmented matrices to convert the system

$$\begin{cases} x + 2y - z = 2 \\ x - y + z = 4 \\ 2x + y - z = 5 \end{cases}$$

to the system

$$\begin{cases} x = 3 \\ y = 0 \\ z = 1 \end{cases}$$

Solution. We saw earlier that the augmented matrix $(A \mid \mathbf{b})$ of the given system is

$$\begin{pmatrix} 1 & 2 & -1 & 2 \\ 1 & -1 & 1 & 4 \\ 2 & 1 & -1 & 5 \end{pmatrix}$$

We use elementary row operations to convert $(A \mid \mathbf{b})$ to a matrix $(B \mid \mathbf{c})$ and then rewrite this matrix in the form of a linear system.

1. First we subtract the first row from the second:

$$
\begin{pmatrix}
1 & 2 & -1 & 2 \\
1 & -1 & 1 & 4 \\
2 & 1 & -1 & 5
\end{pmatrix}
\rightarrow
\begin{pmatrix}
1 & 2 & -1 & 2 \\
0 & -3 & 2 & 2 \\
2 & 1 & -1 & 5
\end{pmatrix}
$$

2. Then we subtract twice the first row from the third:

$$
\begin{pmatrix}
1 & 2 & -1 & 2 \\
0 & -3 & 2 & 2 \\
2 & 1 & -1 & 5
\end{pmatrix}
\rightarrow
\begin{pmatrix}
1 & 2 & -1 & 2 \\
0 & -3 & 2 & 2 \\
0 & -3 & 1 & 1
\end{pmatrix}
$$

3. Next we subtract the third row from the second:

$$
\begin{pmatrix}
1 & 2 & -1 & 2 \\
0 & -3 & 2 & 2 \\
0 & -3 & 1 & 1
\end{pmatrix}
\rightarrow
\begin{pmatrix}
1 & 2 & -1 & 2 \\
0 & 0 & 1 & 1 \\
0 & -3 & 1 & 1
\end{pmatrix}
$$

4. Then we subtract the second row from the third:

$$
\begin{pmatrix}
1 & 2 & -1 & 2 \\
0 & 0 & 1 & 1 \\
0 & -3 & 1 & 1
\end{pmatrix}
\rightarrow
\begin{pmatrix}
1 & 2 & -1 & 2 \\
0 & 0 & 1 & 1 \\
0 & -3 & 0 & 0
\end{pmatrix}
$$

5. Next we multiply the third row by $1/3$:

$$
\begin{pmatrix}
1 & 2 & -1 & 2 \\
0 & 0 & 1 & 1 \\
0 & -3 & 0 & 0
\end{pmatrix}
\rightarrow
\begin{pmatrix}
1 & 2 & -1 & 2 \\
0 & 0 & 1 & 1 \\
0 & 1 & 0 & 0
\end{pmatrix}
$$

6. Then we add the second row to the first:

$$
\begin{pmatrix}
1 & 2 & -1 & 2 \\
0 & 0 & 1 & 1 \\
0 & 1 & 0 & 0
\end{pmatrix}
\rightarrow
\begin{pmatrix}
1 & 0 & 0 & 3 \\
0 & 0 & 1 & 1 \\
0 & 1 & 0 & 0
\end{pmatrix}
$$

7. We conclude by exchanging the second row and the third:

$$
\begin{pmatrix}
1 & 0 & 0 & 3 \\
0 & 0 & 1 & 1 \\
0 & 1 & 0 & 0
\end{pmatrix}
\rightarrow
\begin{pmatrix}
1 & 0 & 0 & 3 \\
0 & 1 & 0 & 0 \\
0 & 0 & 1 & 1
\end{pmatrix}
$$

The last matrix is the matrix $(B \mid \mathbf{c})$. It is the augmented matrix of

$$\begin{cases} x = 3 \\ y = 0 \\ z = 1 \end{cases}$$

This completes the conversion. ◄►

In the next example, we show how a linear system can be solved by simplifying its augmented matrix to the point where we can read off the values of some of the variables. By back substituting these values into the remaining equations, we can find a solution for the entire system.

EXAMPLE 1.35 ■ Augmented Matrices and Back Substitution

Use back substitution to solve the linear system determined by the augmented matrix

$$(A \mid \mathbf{b}) = \begin{pmatrix} -5 & 1 & 7 & 3 \\ 9 & 0 & 9 & 2 \\ 4 & 3 & 5 & 1 \end{pmatrix}$$

Solution. We first use the described row operations to convert $(A \mid \mathbf{b})$ to

$$B = \begin{pmatrix} -5 & 1 & 7 & 3 \\ 0 & \frac{9}{5} & \frac{108}{5} & \frac{37}{5} \\ 0 & 0 & -35 & -\frac{110}{9} \end{pmatrix}$$

From the third row of B we know that $-35z = -\frac{110}{9}$. Hence $z = 22/63$. By substituting this value of z back into the equation

$$\frac{9}{5}y + \frac{108}{5}z = \frac{37}{5}$$

we get

$$\frac{9}{5}y + \left(\frac{108}{5}\right)\left(\frac{22}{63}\right) = \frac{37}{5}$$

Hence $y = -5/63$. We then substitute the values of z and y back into the equation

$$-5x + y + 7z = 3$$

and obtain $x = -8/63$. ◄►

EXAMPLE 1.36 ■ Gaussian Elimination in *MATHEMATICA*

Use *MATHEMATICA* to illustrate a set of individual Gaussian elimination steps that convert the augmented matrix

$$(A \mid \mathbf{b}) = \begin{pmatrix} 3 & 2 & 2 & 3 \\ 1 & 0 & 3 & 2 \\ 3 & 5 & 0 & 4 \end{pmatrix}$$

to the augmented matrix of the system

$$\begin{cases} 3x + 2y + 2z = 3 \\ -2y + 7z = 3 \\ 17z = 11 \end{cases}$$

Solution. The required matrix is

$$(B \mid \mathbf{c}) = \begin{pmatrix} 3 & 2 & 2 & 3 \\ 0 & -2 & 7 & 3 \\ 0 & 0 & 17 & 11 \end{pmatrix}$$

We convert $(A \mid \mathbf{b})$ to $(B \mid \mathbf{c})$ in six steps. First we create the augmented matrix $(A \mid \mathbf{b})$.

```
In[1]:= (A|b)= ⎛ 3 2 2 3 ⎞
               ⎜ 1 0 3 2 ⎟
               ⎝ 3 5 0 4 ⎠

Out[1]= {{3,2,2,3},{1,0,3,2},{3,5,0,4}}
```

Then we apply appropriate row operations to $(A \mid \mathbf{b})$.

```
In[2]:= {{3,2,2,3},{1,0,3,2},{3,5,0,4} - {3,2,2,3}}

Out[2]= {{3,2,2,3},{1,0,3,2},{0,3,-2,1}}
```

```
In[3]:= {{3,2,2,3},{1,0,3,2}-⅓{3,2,2,3},{3,5,0,4}}

Out[3]= {{3,2,2,3},{0,-⅔,7/3,1},{0,3,-2,1}}
```

```
In[4]:= {{3,2,2,3},3{0,-⅔,7/3,1},{0,3,-2,1}}

Out[4]= {{3,2,2,3},{0,-2,7,3},{0,3,-2,1}}
```

```
In[5]:= {{3,2,2,3},{0,-2,7,3},{0,3,-2,1}+3/2{0,-2,7,3}}

Out[5]= {{3,2,2,3},{0,-2,7,3},{0,0,17/2,11/2}}
```

In[6]:= {{3,2,2,3},{0,-2,7,3},2{0,0,$\frac{17}{2}$,$\frac{11}{2}$}}

Out[6]= {{3,2,2,3},{0,-2,7,3},{0,0,17,11}}

We conclude by converting this output to matrix form.

In[7]:= Out[6]//MatrixForm

Out[7]//MatrixForm=

$$\begin{pmatrix} 3 & 2 & 2 & 3 \\ 0 & -2 & 7 & 3 \\ 0 & 0 & 17 & 11 \end{pmatrix}$$

As we can see, the final output is the matrix $(B \mid \mathbf{c})$. ◄►

In the next example, we show that augmented matrices and Gaussian elimination can be used to calculate the coefficients of interpolating polynomials. A polynomial $p(t) = a_0 + a_1 t + \cdots + a_{n-1} t^{n-1}$ is an *interpolating polynomial* for the points (x_1, y_1), (x_2, y_2), ..., (x_n, y_n) if $p(x_i) = y_i$ for all $1 \le i \le n$, since the graph of $p(t)$ passes through all of the given points. We assume that $x_i \ne x_j$ for $i \ne j$. If the coordinates x_i are $1, 2, 3, \ldots, n$, for example, and we know the points $(2, y_2)$ and $(3, y_3)$, we can use $p(2.5)$ as an interpolating value between y_2 and y_3 relative to $p(t)$ since the point $(2.5, p(2.5))$ lies on the graph of the polynomial. It is an easy exercise to show that any n points (x, y) with distinct x-coordinates determine a unique interpolating polynomial of degree $n - 1$.

EXAMPLE 1.37 ■ Augmented Matrices and Interpolating Polynomials

Use augmented matrices and Gaussian elimination to find the interpolating polynomial passing through the points $(-1, 5)$, $(2, 7)$, and $(3, 4)$.

Solution. Let $p(t) = a_0 + a_1 t + a_2 t^2$ be the required polynomial. If the given points lie on the graph of $p(t)$, then $p(-1) = 5$, $p(2) = 7$, and $p(3) = 4$. This information determines the linear system

$$\begin{cases} a_0 + a_1(-1) + a_2(-1)^2 = 5 \\ a_0 + a_1(2) + a_2(2)^2 = 7 \\ a_0 + a_1(3) + a_2(3)^2 = 4 \end{cases}$$

whose augmented matrix $(A \mid \mathbf{b})$ is

$$\begin{pmatrix} 1 & -1 & 1 & 5 \\ 1 & 2 & 4 & 7 \\ 1 & 3 & 9 & 4 \end{pmatrix}$$

We can use elementary row operations to convert $(A \mid \mathbf{b})$ to the augmented matrix

$$(B \mid \mathbf{c}) = \begin{pmatrix} 1 & -1 & 1 & 5 \\ 0 & 3 & 3 & 2 \\ 0 & 0 & 12 & -11 \end{pmatrix}$$

of the linear system

$$\begin{cases} a_0 - a_1 + a_2 = 5 \\ 3a_1 + 3a_2 = 2 \\ 12a_2 = -11 \end{cases}$$

By back substitution we can determine that the coefficients of $p(t)$ are $a_0 = \frac{15}{2}$, $a_1 = \frac{19}{12}$, and $a_2 = -\frac{11}{12}$. Hence $p(t) = \frac{15}{2} + \frac{19}{12}t - \frac{11}{12}t^2$ is the required polynomial. ◀▶

EXERCISES 1.10

1. Find the augmented matrices of the following linear systems.

a. $\begin{cases} 2x + y - z = 2 \\ x - 3y + z = 1 \\ 2x + y - 2z = 6 \end{cases}$ b. $\begin{cases} 2x + y - z = 2 \\ x - 3y + z = 1 \\ 2x + y - z = 6 \end{cases}$ c. $\begin{cases} 2x + y - z = 2 \\ x - 3y + z = 1 \end{cases}$

d. $\begin{cases} x - z + w = 2 \\ -y + z - 7w = 4 \\ 2x + y = 5 \end{cases}$ e. $\begin{cases} x + 2y - z = 2 \\ -y + z = 4 \\ y + 2x = 5 \end{cases}$ f. $\begin{cases} x = 2 \\ y = 4 \\ y + 2x = 5 \end{cases}$

2. Construct linear systems with the following augmented matrices.

a. $\begin{pmatrix} 3 & 2 & 3 \\ 5 & 4 & 2 \\ 5 & 2 & 1 \\ 1 & 1 & 3 \end{pmatrix}$ b. $\begin{pmatrix} 3 & 5 & 5 & 1 \\ 2 & 4 & 2 & 1 \\ 3 & 2 & 1 & 3 \end{pmatrix}$ c. $\begin{pmatrix} 3 & 5 & 5 & 1 \\ 2 & 4 & 2 & 1 \end{pmatrix}$

3. Use *MATHEMATICA* to create the augmented matrices of the following linear systems.

a. $\begin{cases} x + y = 3 \\ y = \frac{37}{5} \end{cases}$ b. $\begin{cases} x + y + z = 3 \\ y + z = \frac{37}{5} \end{cases}$ c. $\begin{cases} x + 2y - z = 2 \\ x - y + z = 4 \\ 2x + y - z = 5 \end{cases}$

4. Use *MATHEMATICA* and augmented matrices to solve the following linear systems.

a. $\begin{cases} x + 2y - z = 2 \\ x - y + z = 4 \\ 2x + y - z = 5 \end{cases}$ b. $\begin{cases} x - y + z = 2 \\ x - y + z = 4 \\ 2x + y - z = 5 \end{cases}$ c. $\begin{cases} x + 2y - z + w = 2 \\ x - y + z = 4 \\ 2x + y - z = 5 \end{cases}$

5. Use augmented matrices and Gaussian elimination to find the following interpolating polynomials.

  a. The polynomial $p(t) = a_0 + a_1 t + a_2 t^2$ passing through the points $(-1, 3)$, $(0, 5)$, $(2, 1)$

  b. The polynomial $p(t) = a_0 + a_1 t + a_2 t^2 + a_3 t^3$ passing through the points $(-1, 3)$, $(0, 5)$, $(2, 1)$, $(7, 4)$

  c. The polynomial $p(t) = a_0 + a_1 t + a_2 t^2 + a_3 t^3 + a_4 t^4$ passing through the points $(-1, 3)$, $(0, 5)$, $(2, 1)$, $(7, 4)$, $(8, 1)$

6. Show that if $(x_1, y_1), \ldots, (x_n, y_n)$ are n points in $\mathbb{R}^2$ with distinct x-coordinates, then there exists a unique polynomial $p(t)$ of degree $n - 1$ for which $p(x_i) = y_i$.

ROW ECHELON MATRICES

Gaussian elimination can be used to convert augmented matrices to *row echelon form*. The basic strategy in the reduction is to convert the augmented matrices to new matrices containing as many zeros as possible, where these zeros occur in a prescribed order.

The definition of a row echelon matrix uses the ideas of a zero row and the leading entry of a row. Consider any $m \times n$ matrix A of the form

$$
\begin{pmatrix}
a_{11} & a_{12} & \cdots & a_{1n} \\
 & & \vdots & \\
a_{i1} & a_{i2} & \cdots & a_{in} \\
 & & \vdots & \\
a_{m1} & a_{m2} & \cdots & a_{mn}
\end{pmatrix}
=
\begin{pmatrix}
\mathbf{r}_1(A) \\
\vdots \\
\mathbf{r}_i(A) \\
\vdots \\
\mathbf{r}_m(A)
\end{pmatrix}
$$

with $\mathbf{r}_i(A) = (a_{i1} \cdots a_{in})$ denoting the ith row of A. We call $\mathbf{r}_i(A)$ a *zero row* of A if $a_{i1} = \cdots = a_{in} = 0$. All rows that are not zero rows are called *nonzero rows*. A *leading entry* of a nonzero row is the leftmost nonzero entry of the row.

DEFINITION 1.9 *An $m \times n$ matrix is in **row echelon form** if it has the following three properties: 1. All zero rows occur below all nonzero rows. 2. All entries below a leading entry are zero. 3. The leading entry of a nonzero row occurs in a column to the right of the column containing the leading entry of the row above it.*

Elementary Row Operations

The operations required to reduce a matrix to row echelon form are the *elementary row operations*. They correspond to the operations on linear systems and augmented matrices described earlier. Suppose that A is an $m \times n$ matrix with rows $\mathbf{r}_1(A), \ldots, \mathbf{r}_m(A)$.

1. The operation $(R_i \rightarrow sR_i)$, for $s \neq 0$, replaces the row $\mathbf{r}_i(A)$ with the row $(b_{i1} \cdots b_{in})$ in which $b_{ik} = sa_{ik}$ for $1 \leq k \leq n$. We write

$$A \ (R_i \rightarrow sR_i) \ B$$

to express the fact that the matrix B has been obtained from the matrix A by the operation $(R_i \rightarrow sR_i)$.

2. The operation $(R_i \rightleftharpoons R_j)$ interchanges the rows $\mathbf{r}_i(A)$ and $\mathbf{r}_j(A)$ of A. We write the expression

$$A \ (R_i \rightleftharpoons R_j) \ B$$

to indicate that the matrix B has been obtained from the matrix A by the operation $(R_i \rightleftharpoons R_j)$.

3. The operation $(R_i \rightarrow R_i + sR_j)$ replaces the row $\mathbf{r}_i(A)$ with the row $(b_{i1} \cdots b_{in})$ in which $b_{ik} = a_{ik} + sa_{jk}$ for $1 \leq k \leq n$. We write

$$A \ (R_i \rightarrow R_i + sR_j) \ B$$

to express the fact that the matrix B has been obtained from the matrix A by the operation $(R_i \rightarrow R_i + sR_j)$.

EXAMPLE 1.38   ■ Reducing a Matrix to Row Echelon Form

Use elementary row operations to reduce the matrix

$$A = \begin{pmatrix} 0 & 2 & 3 & 4 \\ 0 & 0 & 0 & 0 \\ 6 & 7 & 0 & 8 \\ 0 & 4 & 1 & 8 \end{pmatrix}$$

to row echelon form.

Solution. The following steps convert A to row echelon form.

1. First we move all zero rows to the bottom of the matrix.

2. Then we modify the nonzero rows using interchanges and row additions until we have the desired form. The following are three possible steps achieving the conversion.

   a. We interchange the second and fourth rows:

   $$\begin{pmatrix} 0 & 2 & 3 & 4 \\ 0 & 0 & 0 & 0 \\ 6 & 7 & 0 & 8 \\ 0 & 4 & 1 & 8 \end{pmatrix} \ (R_2 \rightleftharpoons R_4) \ \begin{pmatrix} 0 & 2 & 3 & 4 \\ 0 & 4 & 1 & 8 \\ 6 & 7 & 0 & 8 \\ 0 & 0 & 0 & 0 \end{pmatrix}$$

b. We interchange the first and third rows:

$$\begin{pmatrix} 0 & 2 & 3 & 4 \\ 0 & 4 & 1 & 8 \\ 6 & 7 & 0 & 8 \\ 0 & 0 & 0 & 0 \end{pmatrix} (R_3 \rightleftharpoons R_1) \begin{pmatrix} 6 & 7 & 0 & 8 \\ 0 & 4 & 1 & 8 \\ 0 & 2 & 3 & 4 \\ 0 & 0 & 0 & 0 \end{pmatrix}$$

c. We subtract a multiple of the second row from the third row:

$$\begin{pmatrix} 6 & 7 & 0 & 8 \\ 0 & 4 & 1 & 8 \\ 0 & 2 & 3 & 4 \\ 0 & 0 & 0 & 0 \end{pmatrix} \left(R_3 \rightarrow R_3 - \frac{1}{2}R_2 \right) \begin{pmatrix} 6 & 7 & 0 & 8 \\ 0 & 4 & 1 & 8 \\ 0 & 0 & \frac{5}{2} & 0 \\ 0 & 0 & 0 & 0 \end{pmatrix}$$

As we can see, the last matrix is in row echelon form. ◄►

If we wanted a fraction-free row echelon form of the matrix A in Example 1.38, we would conclude the reduction with a fourth step:

$$\begin{pmatrix} 6 & 7 & 0 & 8 \\ 0 & 4 & 1 & 8 \\ 0 & 0 & \frac{5}{2} & 0 \\ 0 & 0 & 0 & 0 \end{pmatrix} (R_3 \rightarrow 2R_3) \begin{pmatrix} 6 & 7 & 0 & 8 \\ 0 & 4 & 1 & 8 \\ 0 & 0 & 5 & 0 \\ 0 & 0 & 0 & 0 \end{pmatrix}$$

Although elementary row operations are not built-in *MATHEMATICA* functions, we can define them. For example, the functions **erop1**, **erop2**, and **erop3**, defined in Appendix E, are *MATHEMATICA* versions of $(R_i \rightarrow sR_i)$, $(R_i \rightleftharpoons R_j)$, and $(R_i \rightarrow R_i + sR_j)$. Using these functions, we can reproduce the previous example in *MATHEMATICA* as follows:

```
In[1]:= erop1[A_,i_,s_]:=If[s!=0,Block[{lst=Part[A,i]},

ReplacePart[A,s*lst,i]],''Error; scalar must be non-
zero''];
```

```
In[2]:= erop2[A_,i_,j_]:=Block[{lst1=Part[A,i],lst2=Part[A,j]},

ReplacePart[ReplacePart[A,lst1,j],lst2,i]];
```

```
In[3]:= erop3[A_,i_,j_,s_]:=Block[{lst=Part[A,i]+s*Part[A,j]},

ReplacePart[A,lst,i]];
```

In[4]:= A={{0,2,3,4},{0,0,0,0},{6,7,0,8},{0,4,1,8}};

In[5]:= A1=erop2[A,2,4];

In[6]:= A2=erop2[A1,3,1];

In[7]:= A3=erop3[A2,3,2,-1/2];

In[8]:= erop1[A3,3,2]

Out[8]= {{6,7,0,8},{0,4,1,8},{0,0,5,0},{0,0,0,0}}

EXERCISES 1.11

1. Extract the rows of the following matrices. Label each row as $\mathbf{r}_i(A)$ or $\mathbf{r}_i(B)$, as appropriate.

$$A = \begin{pmatrix} 1 & 0 & 0 & 0 & 3 \\ 1 & 2 & 0 & 5 & 3 \\ 0 & 0 & 0 & 4 & -1 \end{pmatrix} \qquad B = \begin{pmatrix} 1 & 0 & 0 & 0 & 3 \\ 0 & 2 & 0 & 5 & 3 \\ 0 & 0 & 0 & 0 & 0 \end{pmatrix}$$

2. Identify the zero rows of each of the matrices in Exercise 1.

3. For each matrix M in Exercise 1, carry out the following operations on the rows of the matrix.

   a. $\mathbf{r}_i(M) \rightarrow 4[\mathbf{r}_i(M)]$
   b. $\mathbf{r}_2(M) \rightarrow \mathbf{r}_2(M) + 7[\mathbf{r}_1(M)]$
   c. $\mathbf{r}_2(M) \rightleftharpoons \mathbf{r}_3(M)$

4. Use elementary row operations to convert the following matrices to row echelon matrices.

   a. $\begin{pmatrix} -1 & 1 & 4 & 1 \\ -3 & 3 & 5 & 5 \\ -2 & -3 & 3 & -3 \end{pmatrix}$ b. $\begin{pmatrix} 2 & -3 & 0 \\ 0 & 4 & 3 \\ 1 & 0 & 5 \\ 2 & 5 & 2 \end{pmatrix}$ c. $\begin{pmatrix} 0 & 20 & -40 \\ 0 & -405 & -499 \\ 0 & 5840 & 4522 \end{pmatrix}$

   Explain why the resulting matrices are in row echelon form.

5. Repeat Exercise 4 in *MATHEMATICA* using the functions **erop1**, **erop2**, and **erop3**, as needed.

6. Use elementary row operations to convert the matrix A to the matrix B.

$$A = \begin{pmatrix} 4 & -13 & 5 & 0 & 5 \\ 4 & 4 & 2 & 5 & -5 \\ 1 & 0 & -3 & -5 & -8 \\ -5 & 4 & 7 & -9 & 5 \end{pmatrix} \qquad B = \begin{pmatrix} 4 & -13 & 5 & 0 & 5 \\ 0 & 68 & -12 & 20 & -40 \\ 0 & 0 & -250 & -405 & -499 \\ 0 & 0 & 0 & 5840 & 4522 \end{pmatrix}$$

7. Repeat Exercise 6 in *MATHEMATICA* using the functions **erop1**, **erop2**, and **erop3**, as needed.

8. Determine which of the following are row echelon matrices. Explain your answers.

a. $\begin{pmatrix} 1 & 0 & 0 & 0 & 3 \\ 1 & 2 & 0 & 5 & 3 \\ 0 & 0 & 0 & 4 & -1 \end{pmatrix}$ b. $\begin{pmatrix} 1 & 0 & 0 & 0 & 3 \\ 0 & 2 & 0 & 5 & 3 \\ 0 & 0 & 0 & 4 & -1 \end{pmatrix}$ c. $\begin{pmatrix} 0 & 2 & 0 & 5 & 3 \\ 1 & 0 & 0 & 0 & 3 \\ 0 & 0 & 0 & 4 & -1 \end{pmatrix}$

Inverse of Elementary Row Operations

Some of our later work will depend on the fact that elementary row operations are invertible.

1. If $s \neq 0$ and B was obtained from A by $(R_i \rightarrow sR_i)$, then A can be recovered from B by applying $(R_i \rightarrow \frac{1}{s}R_i)$ to B.

2. If B was obtained from A by $(R_i \rightarrow R_i + sR_j)$, then A can be recovered from B by applying $(R_i \rightarrow R_i - sR_j)$ to B.

3. If B was obtained from A by $(R_i \rightleftharpoons R_j)$, then A can be recovered from B by applying the operation $(R_i \rightleftharpoons R_j)$ to B.

The invertibility of elementary row operations provide the rationale for introducing the following relationship among matrices.

DEFINITION 1.10   *A matrix A is **row equivalent** to a matrix B if we can convert A to B by a finite number of elementary row operations.*

In the exercises, we prove that row equivalence is an equivalence in the sense that it is a reflexive, symmetric, and transitive relation. What makes row equivalence useful is the fact that the linear systems determined by row equivalent augmented matrices have the same solutions. This follows immediately from the fact that Gaussian elimination preserves the solutions of linear systems.

EXAMPLE 1.39 ■ **Row Equivalent Matrices**

Show that the matrices

$$A = \begin{pmatrix} 1 & 2 & 3 \\ 4 & 5 & 6 \\ 7 & 8 & 9 \end{pmatrix} \quad \text{and} \quad B = \begin{pmatrix} 39 & 48 & 57 \\ -8 & -10 & -12 \\ 1 & 2 & 3 \end{pmatrix}$$

are row equivalent.

Solution. We use elementary row operations to convert A to B. The following are three possible steps converting A to B.

1. We interchange the first and third rows:

$$\begin{pmatrix} 1 & 2 & 3 \\ 4 & 5 & 6 \\ 7 & 8 & 9 \end{pmatrix} (R_1 \rightleftharpoons R_3) \begin{pmatrix} 7 & 8 & 9 \\ 4 & 5 & 6 \\ 1 & 2 & 3 \end{pmatrix}$$

2. We multiply the second row by a nonzero real number:

$$\begin{pmatrix} 7 & 8 & 9 \\ 4 & 5 & 6 \\ 1 & 2 & 3 \end{pmatrix} (R_2 \rightarrow -2R_2) \begin{pmatrix} 7 & 8 & 9 \\ -8 & -10 & -12 \\ 1 & 2 & 3 \end{pmatrix}$$

3. We subtract a multiple of the second row from the first row:

$$\begin{pmatrix} 7 & 8 & 9 \\ -8 & -10 & -12 \\ 1 & 2 & 3 \end{pmatrix} (R_1 \rightarrow R_1 - 4R_2) \begin{pmatrix} 39 & 48 & 57 \\ -8 & -10 & -12 \\ 1 & 2 & 3 \end{pmatrix}$$

Since B can be obtained from A by three elementary row operations, A and B are row equivalent. ◀▶

Every matrix can be reduced to row echelon form. The proof of the next theorem provides an algorithm for doing so.

THEOREM 1.1 (Gaussian elimination theorem) *Every matrix is row equivalent to a row echelon matrix.*

Proof. Let A be a given matrix. We first use the operations $\mathbf{r}_i(A) \rightleftharpoons \mathbf{r}_j(A)$ to move all zero rows of A below all nonzero rows. We then use the same operations to move $\mathbf{r}_i(A)$ below $\mathbf{r}_j(A)$ if the first nonzero element of $\mathbf{r}_i(A)$ occurs to the right of the first nonzero element of $\mathbf{r}_j(A)$. We continue this process until all such permutations have been completed. If the resulting matrix has a row $\mathbf{r}_r(A)$ above a row $\mathbf{r}_s(A)$ whose first nonzero entry a_{rp} is in the same position as the first nonzero entry a_{sp} of $\mathbf{r}_s(A)$, we use the operation $\mathbf{r}_s(A) \rightarrow \mathbf{r}_s(A) + c\,[\mathbf{r}_r(A)]$, with an appropriate scalar c, to replace $\mathbf{r}_s(A)$ by a row in which $a_{sp} = 0$. We continue until all such replacements have been completed. By construction, the resulting matrix is in echelon form. ■

EXERCISES 1.12

1. Use elementary row operations to show that the following matrices are row equivalent.

a.
$$\begin{pmatrix} 4 & -13 & 5 & 0 & 5 \\ 4 & 4 & 2 & 5 & -5 \\ 1 & 0 & -3 & -5 & -8 \\ -5 & 4 & 7 & -9 & 5 \end{pmatrix}$$
b.
$$\begin{pmatrix} 4 & -13 & 5 & 0 & 5 \\ 0 & 68 & -12 & 20 & -40 \\ 0 & 0 & -250 & -405 & -499 \\ 0 & 0 & 0 & 5840 & 4522 \end{pmatrix}$$

c.
$$\begin{pmatrix} 8 & -9 & 7 & 5 & 0 \\ 4 & 4 & 2 & 5 & -5 \\ 9 & -9 & 4 & 0 & -8 \\ -5 & 4 & 7 & -9 & 5 \end{pmatrix}$$
d.
$$\begin{pmatrix} 8 & -9 & 7 & 5 & 0 \\ 0 & 9 & -31 & -45 & -64 \\ 0 & 0 & 52 & -126 & -59 \\ 0 & 0 & 0 & 5840 & 4522 \end{pmatrix}$$

2. Repeat Exercise 1 in *MATHEMATICA* using the functions **erop1**, **erop2**, and **erop3**, as needed.

3. Use the **RowReduce** function to reduce the following matrices to row echelon form. Describe what you think is special about the resulting matrices.

a.
$$\begin{pmatrix} 2 & -3 & 0 \\ 0 & 4 & 3 \\ 1 & 0 & 5 \\ 2 & 5 & 2 \end{pmatrix}$$
b.
$$\begin{pmatrix} 0 & 0 & 0 & 0 & 3 \\ 0 & 2 & 0 & 5 & 3 \\ 1 & 0 & 0 & 4 & -1 \end{pmatrix}$$
c.
$$\begin{pmatrix} 0 & 2 & 0 & 5 & 3 \\ 0 & 0 & 0 & 0 & 0 \\ 0 & 0 & 0 & 4 & -1 \end{pmatrix}$$

d.
$$\begin{pmatrix} 2 & 2 & 2 & 2 & 2 \\ 2 & 2 & 2 & 2 & 2 \\ 2 & 2 & 2 & 2 & 2 \end{pmatrix}$$
e.
$$\begin{pmatrix} 1 & 0 & 0 & 0 & 3 \\ 1 & 2 & 0 & 5 & 3 \end{pmatrix}$$
f.
$$\begin{pmatrix} 0 & 0 & 3 & 0 & 2 \\ 0 & 2 & 2 & 2 & 4 \\ 2 & 0 & 0 & 8 & 0 \end{pmatrix}$$

4. Show that row equivalence is a reflexive, symmetric, and transitive relation on matrices. It is *reflexive* in the sense that every matrix is row equivalent to itself. It is *symmetric* in the sense that if A is row equivalent to B, then B is row equivalent to A. It is also *transitive* in the sense that if A is row equivalent to B and B is row equivalent to C, then A is row equivalent to C.

5. Show that if A and B are row equivalent augmented matrices, then the linear systems determined by A and B have the same solutions.

REDUCED ROW ECHELON MATRICES

It is often helpful to use elementary row operations to reduce a row echelon matrix further to a unique, still simpler form in which all diagonal entries are 1. By the *diagonal entries* of a matrix $A = (a_{ij})$, we mean all entries a_{ij} for which $i = j$. The next example illustrates this.

EXAMPLE 1.40 ■ **Reducing a Matrix to a Matrix with 1s on the Diagonal**

Use elementary row operations to reduce the matrix

$$A = \begin{pmatrix} 1 & -3 & 0 & 2 \\ 2 & 6 & 0 & 7 \\ 0 & 3 & 0 & 5 \end{pmatrix}$$

to the row echelon form

$$B = \begin{pmatrix} 1 & -3 & 0 & 2 \\ 0 & 1 & 0 & \frac{5}{3} \\ 0 & 0 & 0 & -17 \end{pmatrix}$$

Solution. The following steps achieve the desired reduction.

1. We subtract a multiple of the first row from the second row:

$$\begin{pmatrix} 1 & -3 & 0 & 2 \\ 2 & 6 & 0 & 7 \\ 0 & 3 & 0 & 5 \end{pmatrix} (R_2 \rightarrow R_2 - 2R_1) \begin{pmatrix} 1 & -3 & 0 & 2 \\ 0 & 12 & 0 & 3 \\ 0 & 3 & 0 & 5 \end{pmatrix}$$

2. We add a multiple of the third row to the second row:

$$\begin{pmatrix} 1 & -3 & 0 & 2 \\ 0 & 12 & 0 & 3 \\ 0 & 3 & 0 & 5 \end{pmatrix} (R_2 \rightarrow R_2 - 4R_3) \begin{pmatrix} 1 & -3 & 0 & 2 \\ 0 & 0 & 0 & -17 \\ 0 & 3 & 0 & 5 \end{pmatrix}$$

3. We interchange the second and third rows:

$$\begin{pmatrix} 1 & -3 & 0 & 2 \\ 0 & 0 & 0 & -17 \\ 0 & 3 & 0 & 5 \end{pmatrix} (R_2 \rightleftharpoons R_3) \begin{pmatrix} 1 & -3 & 0 & 2 \\ 0 & 3 & 0 & 5 \\ 0 & 0 & 0 & -17 \end{pmatrix}$$

4. We multiply the third row by a nonzero real number:

$$
\begin{pmatrix} 1 & -3 & 0 & 2 \\ 0 & 3 & 0 & 5 \\ 0 & 0 & 0 & -17 \end{pmatrix} (R_2 \rightarrow \frac{1}{3}R_2) \begin{pmatrix} 1 & -3 & 0 & 2 \\ 0 & 1 & 0 & \frac{5}{3} \\ 0 & 0 & 0 & -17 \end{pmatrix}
$$

As we can see, the diagonal entries of B are 1. ◀▶

Pivots

If we take a closer look at the matrix B in Example 1.40, we see that B can be reduced further. We can eliminate the -3 in the first row by adding 3 times the second row to the first:

$$
\begin{pmatrix} 1 & -3 & 0 & 2 \\ 0 & 1 & 0 & \frac{5}{3} \\ 0 & 0 & 0 & -17 \end{pmatrix} (R_1 \rightarrow R_1 + 3R_2) \begin{pmatrix} 1 & 0 & 0 & 7 \\ 0 & 1 & 0 & \frac{5}{3} \\ 0 & 0 & 0 & -17 \end{pmatrix} = C
$$

The matrix C that results has certain special features. We describe them in terms of **pivots**.

DEFINITION 1.11   *The first nonzero entry (from the left) of a row in a row echelon matrix is a **pivot** of the matrix.*

DEFINITION 1.12   *A row echelon matrix is in **reduced row echelon form** if it is either a zero matrix or if all of its pivots are 1 and all entries above its pivots are 0.*

Although zero matrices have no pivots, it is convenient to consider them to be in reduced row echelon form. The reduced row echelon form of a row echelon matrix is obtained by applying elementary row operations to the matrix. Since these operations preserve the column positions of the pivots of a matrix, we refer to these positions as **pivot positions**.

DEFINITION 1.13   *A **pivot column** of a matrix is a column with a **pivot position**.*

EXAMPLE 1.41   ■ **The Pivots of a Matrix**

Use *MATHEMATICA* to generate a random 4×5 matrix A and find the pivot positions and pivot columns of A.

Solution. First we generate a random matrix A.

```
In[1]:= A=Table[Random[Integer,{-100,100}],{3},{4}];
MatrixForm[A]

Out[1]//MatrixForm=
```
$$
A = \begin{pmatrix} -50 & -12 & -18 & 31 & -26 \\ -62 & 1 & -47 & -91 & -47 \\ -61 & 41 & -58 & -90 & 53 \\ -1 & 94 & 83 & -86 & 23 \end{pmatrix}
$$

The five columns c_1, c_2, c_3, c_4, and c_5 of A are

$$\begin{pmatrix} -50 \\ -62 \\ -61 \\ -1 \end{pmatrix}, \quad \begin{pmatrix} -12 \\ 1 \\ 41 \\ 94 \end{pmatrix}, \quad \begin{pmatrix} -18 \\ -47 \\ -58 \\ 83 \end{pmatrix}, \quad \begin{pmatrix} 31 \\ -91 \\ -90 \\ -86 \end{pmatrix}, \quad \text{and} \quad \begin{pmatrix} -26 \\ -47 \\ 53 \\ 23 \end{pmatrix},$$

respectively. We find the reduced row echelon form of A as follows:

```
In[2]:= MatrixForm[RowReduce[A]]

Out[2]//MatrixForm=
```

$$\begin{pmatrix} 1 & 0 & 0 & 0 & \frac{29617237}{32143690} \\ 0 & 1 & 0 & 0 & \frac{66706161}{32143690} \\ 0 & 0 & 1 & 0 & -\frac{45096177}{32143690} \\ 0 & 0 & 0 & 1 & \frac{10223688}{16071845} \end{pmatrix}$$

This shows that the pivot positions of A are a_{11}, a_{22}, a_{33}, and a_{44} and that the columns c_1, c_2, c_3, and c_4 are therefore pivot columns. ◀▶

Pivot columns are not always contiguous. Here is an example.

EXAMPLE 1.42 ■ Noncontiguous Pivot Columns

Construct a 4×5 matrix A whose pivot columns are c_1, c_2, and c_4.

Solution. First we create two columns c_1 and c_2 with the property that $ac_1 + bc_2 = 0$ if and only if $a = b = 0$. For example, if

$$c_1 = \begin{pmatrix} 0 \\ 1 \\ 2 \\ 3 \end{pmatrix} \quad \text{and} \quad c_2 = \begin{pmatrix} -5 \\ -2 \\ 0 \\ 4 \end{pmatrix}$$

then

$$a \begin{pmatrix} 0 \\ 1 \\ 2 \\ 3 \end{pmatrix} + b \begin{pmatrix} -5 \\ -2 \\ 0 \\ 4 \end{pmatrix} = \begin{pmatrix} -5b \\ a - 2b \\ 2a \\ 3a + 4b \end{pmatrix} = \begin{pmatrix} 0 \\ 0 \\ 0 \\ 0 \end{pmatrix}$$

if and only if $a = b = 0$. We therefore choose two nonzero values for a and b, the values $a = 2$ and $b = 3$, for example, and let $c_3 = 2c_1 + 3c_2$. We continue to experiment with the columns c_1 and c_2, as well as a possible candidate for c_4, until we have found a column c_4 with the property that $ac_1 + bc_2 + cc_4 = \mathbf{0}$ if and only if $a = b = c = 0$. Column c_5 can be arbitrary. Here is the result of one such choice.

Let A be the matrix

$$\begin{pmatrix} 0 & -5 & -15 & 4 & 7 \\ 1 & -2 & -4 & 3 & 6 \\ 2 & 0 & 4 & 2 & 1 \\ 3 & 4 & 18 & 1 & 4 \end{pmatrix}$$

Then its reduced row echelon form is

$$\begin{pmatrix} 1 & 0 & 2 & 0 & -\frac{25}{4} \\ 0 & 1 & 3 & 0 & 4 \\ 0 & 0 & 0 & 1 & \frac{27}{4} \\ 0 & 0 & 0 & 0 & 0 \end{pmatrix}$$

The pivot positions of A are a_{11}, a_{22}, and a_{34} and its pivot columns are c_1, c_2, and c_4. ◄►

These examples suggest a method for converting matrices to reduced row echelon matrices. The conversion of a matrix to its reduced row echelon form is known as **Gauss-Jordan elimination**.

THEOREM 1.2 (Row echelon form theorem) *Every matrix is row equivalent to a matrix in reduced row echelon form.*

Proof. Suppose that A is a given matrix. We first use Gaussian elimination to convert A to a matrix B in echelon form. We then convert B to a matrix C whose pivots are 1 by applying the operations $\mathbf{r}_i(A) \rightarrow s\,[\mathbf{r}_i(A)]$ where required. If C has a row $\mathbf{r}_j(A)$ containing a pivot, and if a row $\mathbf{r}_i(A)$ above $\mathbf{r}_j(A)$ has a nonzero entry above the pivot, we use the operation $\mathbf{r}_i(A) \rightarrow \mathbf{r}_i(A) + s\left[\mathbf{r}_j(A)\right]$ to replace $\mathbf{r}_i(A)$ by a row that has a 0 above the pivot of $\mathbf{r}_j(A)$. We continue until all entries above a pivot are 0. By construction, the resulting matrix is in reduced row echelon form. ■

We can use the existence of the reduced row echelon form of matrices to characterize consistent linear systems.

THEOREM 1.3 *An $m \times n$ linear system in the variables $x_1, \ldots, x_n$ with augmented $m \times (n + 1)$ matrix $(A \mid \mathbf{b})$ is consistent for all $\mathbf{b} \in \mathbb{R}^m$ if and only if the matrix A has m pivots.*

Proof. By Theorem 1.2 we may assume that the augmented matrix $(A \mid \mathbf{b}) = \left(a_{ij}\right)$ is in reduced row echelon form. Suppose that $(A \mid \mathbf{b})$ has m pivots. Since $(A \mid \mathbf{b})$ is in reduced row echelon form, $a_{ii} = 1$ and $a_{ij} = 0$ for all $1 \leq i \leq m$ for which $i \neq j$. It follows, therefore, from the definition of augmented matrices, that all assignments $x_i = a_{i(n+1)} \in \mathbb{R}$ yield solutions of the given system.

To prove the converse, we show the joint assumptions that the system is consistent and that $(A \mid \mathbf{b})$ does not have m pivots lead to a contradiction.

Suppose that the system is consistent for all $\mathbf{b} \in \mathbb{R}^m$ and that the ith row of $(A \mid \mathbf{b})$ does not contain a pivot position. Then we can choose a nonzero entry $a_{i(n+1)}$ for which $\begin{pmatrix} 0 & \cdots & 0 & a_{i(n+1)} \end{pmatrix}$ is the ith row. By the definition of augmented matrices, $(A \mid \mathbf{b})$ is then the augmented matrix of an inconsistent system. This proves the theorem. ∎

Basic and Free Variables

We can use the pivot columns of the augmented matrix of a linear system to classify the variables of the system.

DEFINITION 1.14 *The variables corresponding to the pivot columns of the augmented matrix of a linear system are called **basic variables**. All other variables are called **free variables** of the system.*

Here is an example. The augmented matrix

$$A = \begin{pmatrix} \mathbf{c}_1(A) & \mathbf{c}_2(A) & \mathbf{c}_3(A) & \mathbf{c}_4(A) & \mathbf{c}_5(A) \end{pmatrix}$$

of the linear system

$$\begin{cases} -5x_2 - 15x_3 + 4x_4 = 7 \\ x_1 - 2x_2 - 4x_3 + 3x_4 = 6 \\ 2x_1 + 4x_3 + 2x_4 = 1 \\ 3x_1 + 4x_2 + 18x_3 + x_4 = 4 \end{cases}$$

in Example 1.42 has three basic variables and one free variable. As we saw, the reduced row echelon form of A is

$$B = \begin{pmatrix} 1 & 0 & 2 & 0 & -\frac{25}{4} \\ 0 & 1 & 3 & 0 & 4 \\ 0 & 0 & 0 & 1 & \frac{27}{4} \\ 0 & 0 & 0 & 0 & 0 \end{pmatrix}$$

The pivot columns of A are therefore $\mathbf{c}_1(A)$, $\mathbf{c}_2(A)$ and $\mathbf{c}_4(A)$. This means that the basic variables of the system are x_1, x_2, and x_4, and the free variable is x_3. In other words, we are free to choose any value for x_3. Once this choice is made, the values of x_1, x_2, and x_4 that produce a solution of the system are uniquely determined. The linear system

$$\begin{cases} x_1 = -2x_3 - \frac{25}{4} \\ x_2 = -3x_3 + 4 \\ x_3 = \frac{27}{4} \end{cases}$$

shows why this is the case. It is called the **general solution** of the original system since it provides explicit formulas for finding all solutions.

EXERCISES 1.13

1. Determine which of the following matrices are in reduced row echelon form. Explain your answers.

a.
$$\begin{pmatrix} 1 & 0 & 0 & 0 & 3 \\ 1 & 2 & 0 & 5 & 3 \\ 0 & 0 & 0 & 4 & -1 \end{pmatrix}$$
b.
$$\begin{pmatrix} 1 & 0 & 0 & 0 & 3 \\ 0 & 1 & 0 & 5 & 3 \\ 0 & 0 & 0 & 1 & -1 \end{pmatrix}$$
c.
$$\begin{pmatrix} 0 & 1 & 0 & 5 & 3 \\ 1 & 0 & 0 & 0 & 3 \\ 0 & 0 & 0 & 1 & -1 \end{pmatrix}$$

d.
$$\begin{pmatrix} 1 & 2 & 2 & 2 & 2 \\ 0 & 1 & 2 & 2 & 2 \\ 0 & 0 & 1 & 2 & 2 \end{pmatrix}$$
e.
$$\begin{pmatrix} 1 & 0 & 0 & 0 & 3 \\ 0 & 1 & 0 & 5 & 3 \end{pmatrix}$$
f.
$$\begin{pmatrix} 1 & 0 & 0 & 0 & 3 \end{pmatrix}$$

2. Use elementary row operations to reduce the matrix

$$\begin{pmatrix} -5 & -1 & 9 & 2 & 1 \\ 5 & 1 & 5 & 9 & 2 \\ 7 & -2 & 4 & -8 & 3 \end{pmatrix}$$

to reduced row echelon form. Use *MATHEMATICA* to verify each step of the reduction.

3. Apply the **RowReduce** function of *MATHEMATICA* to the following matrices and explain why the resulting matrices are in reduced row echelon form.

a.
$$\begin{pmatrix} -1 & 1 & 4 & 1 \\ -3 & 3 & 5 & 5 \\ -2 & -3 & 3 & -3 \end{pmatrix}$$
b.
$$\begin{pmatrix} 2 & -3 & 0 \\ 0 & 4 & 3 \\ 1 & 0 & 5 \\ 2 & 5 & 2 \end{pmatrix}$$
c.
$$\begin{pmatrix} 5 & -1 & 3 \\ -1 & 0 & -4 \\ 3 & -4 & -5 \end{pmatrix}$$

4. Find the pivot columns of the following matrices. Explain your answers.

a.
$$\begin{pmatrix} 1 & 0 & 0 & 0 & 3 \\ 1 & 2 & 0 & 5 & 3 \\ 0 & 0 & 0 & 4 & -1 \end{pmatrix}$$
b.
$$\begin{pmatrix} 0 & 0 & 0 & 0 & 3 \\ 0 & 2 & 0 & 5 & 3 \\ 0 & 0 & 0 & 4 & -1 \end{pmatrix}$$
c.
$$\begin{pmatrix} 0 & 2 & 0 & 5 & 3 \\ 1 & 0 & 0 & 0 & 3 \\ 0 & 0 & 0 & 4 & -1 \end{pmatrix}$$

d.
$$\begin{pmatrix} 2 & 2 & 2 & 2 & 2 \\ 2 & 2 & 2 & 2 & 2 \\ 2 & 2 & 2 & 2 & 2 \end{pmatrix}$$
e.
$$\begin{pmatrix} 1 & 0 & 0 & 0 & 3 \\ 0 & 2 & 0 & 5 & 3 \end{pmatrix}$$
f.
$$\begin{pmatrix} 1 & 0 & 0 & 0 & 3 \end{pmatrix}$$

MATRIX EQUATIONS

We now introduce a second matrix form for linear systems in which we can calculate the solutions of the systems.

The Systems $A\mathbf{x} = \mathbf{b}$

We recall from our earlier discussion that in addition to being able to represent the system

$$\begin{cases} 3x + 4y - 7z = 1 \\ 2x + y = 2 \\ 8x - 6y - z = 15 \end{cases}$$

by the augmented matrix

$$(A \mid \mathbf{b}) = \begin{pmatrix} 3 & 4 & -7 & 1 \\ 2 & 1 & 0 & 2 \\ 8 & -6 & -1 & 15 \end{pmatrix}$$

we can also write it as

$$\begin{pmatrix} 3 & 4 & -7 \\ 2 & 1 & 0 \\ 8 & -6 & -1 \end{pmatrix} \begin{pmatrix} x \\ y \\ z \end{pmatrix} = \begin{pmatrix} 1 \\ 2 \\ 15 \end{pmatrix}$$

We call this equation the **matrix equation** of the system and write it as $A\mathbf{x} = \mathbf{b}$. The operation producing the column vector

$$A\mathbf{x} = \begin{pmatrix} 3x + 4y - 7z \\ 2x + y \\ 8x - 6y - z \end{pmatrix}$$

from

$$A = \begin{pmatrix} 3 & 4 & -7 \\ 2 & 1 & 0 \\ 8 & -6 & -1 \end{pmatrix} \quad \text{and} \quad \mathbf{x} = \begin{pmatrix} x \\ y \\ z \end{pmatrix}$$

will later be called **matrix multiplication**. The general form of this operation will be discussed in Chapter 2.

DEFINITION 1.15   *The **matrix equation** $A\mathbf{x} = \mathbf{b}$ determined by the $m \times n$ linear system*

$$\begin{cases} a_{11}x_1 + \cdots + a_{1n}x_n = b_1 \\ \vdots \\ a_{m1}x_1 + \cdots + a_{mn}x_n = b_m \end{cases}$$

is the equation

$$\begin{pmatrix} a_{11} & \cdots & a_{1n} \\ \vdots & \ddots & \vdots \\ a_{m1} & \cdots & a_{mn} \end{pmatrix} \begin{pmatrix} x_1 \\ \vdots \\ x_n \end{pmatrix} = \begin{pmatrix} b_1 \\ \vdots \\ b_m \end{pmatrix}$$

where $A\mathbf{x}$ *denotes the vector*

$$\begin{pmatrix} a_{11}x_1 + \cdots + a_{1n}x_n \\ \vdots \\ a_{m1}x_1 + \cdots + a_{mn}x_n \end{pmatrix}$$

If we assign the constants $c_1, \ldots, c_n$ to the variables $x_1, \ldots, x_n$ and

$$\widehat{\mathbf{x}} = \begin{pmatrix} c_1 \\ \vdots \\ c_n \end{pmatrix}$$

then $A\widehat{\mathbf{x}}$ denotes the constant vector

$$\begin{pmatrix} a_{11}c_1 + \cdots + a_{1n}c_n \\ \vdots \\ a_{m1}c_1 + \cdots + a_{mn}c_n \end{pmatrix}$$

DEFINITION 1.16   *A vector $\widehat{\mathbf{x}}$ is a solution of the matrix equation $A\mathbf{x} = \mathbf{b}$ if the vectors $A\widehat{\mathbf{x}}$ and $\mathbf{b}$ are equal.*

EXAMPLE 1.43   ■ **A Linear System in Matrix Form**

Use *MATHEMATICA* to convert the linear system

$$\begin{cases} x + 3y - 7z + w - 45 = 0 \\ 2x - z + 8w - 1 = 0 \\ x + y + z + 9 = 0 \\ x - 8y - 8z + 12w = 0 \end{cases}$$

to a system in matrix form.

Solution. The **LinearEquationsToMatrices** function of *MATHEMATICA* makes it easy to convert linear systems of equations to their matrix form.

```
In[1]:= <<LinearAlgebra`MatrixManipulation`
```

```
In[2]:= M=LinearEquationsToMatrices

[{x+3y-7z+w-45= =0,2x-z+8w-1= =0,x+y+z+9= =0,

x-8y-8z+12w= =0},{x,y,z,w}]

Out[2]= {{{1,3,-7,1},{2,0,-1,8},{1,1,1,0},

{1,-8,-8,12}},{45,1,-9,0}}
```

The output is a list $\{M[[1]], M[[2]]\}$ consisting of the coefficient matrix $M[[1]]$ and the vector of constants $M[[2]]$. We use $M[[1]]$ and $M[[2]]$ to construct the required matrix equation $Ax = \mathbf{b}$.

```
In[3]:= MatrixForm[M[[1]]].MatrixForm[{{x},{y},{z},{w}}]
==MatrixForm[M[[2]]]
```

$$\text{Out[3]=} \begin{pmatrix} 1 & 3 & -7 & 1 \\ 2 & 0 & -1 & 8 \\ 1 & 1 & 1 & 0 \\ 1 & -8 & -8 & 12 \end{pmatrix} \cdot \begin{pmatrix} x \\ y \\ z \\ w \end{pmatrix} == \begin{pmatrix} 45 \\ 1 \\ -9 \\ 0 \end{pmatrix}$$

As we can see, the result is a linear system in the matrix form. ◄►

We saw in Example 1.37 that in certain applications, the coefficient matrix A of a linear system $Ax = \mathbf{b}$ has the special form

$$\begin{pmatrix} 1 & x_1 & x_1^2 & \cdots & x_1^{n-1} \\ 1 & x_2 & x_2^2 & \cdots & x_2^{n-1} \\ & & \vdots & & \\ 1 & x_{n-1} & x_{n-1}^2 & \cdots & x_{n-1}^{n-1} \\ 1 & x_n & x_n^2 & \cdots & x_n^{n-1} \end{pmatrix}$$

This type of matrix is known as a ***Vandermonde matrix***. Among other things, it provides a template for generating interpolating polynomials. Suppose, for example, that two parallel scientific experiments involve taking five readings and that we would like to use the graphs of the associated interpolating polynomials to compare the results of the two experiments. The next example illustrates how Vandermonde matrices can be used for this purpose.

EXAMPLE 1.44   ■ A Vandermonde Matrix

Suppose that the observation times of two parallel experiments are 1, 2, 3, 4, and 5. The results of the experiments are shown in the following chart.

Time	1	2	3	4	5
Experiment 1	3	4	2	6	7
Experiment 2	2	3	2	6	8

Use a Vandermonde matrix to find the interpolating polynomials for the experiments and overlap the graphs of the polynomials.

Solution. First we define the matrix A. Then we calculate the vector $A\mathbf{x}$, since it is the same for both experiments.

```
In[1]:= x₁=1;x₂=2;x₃=3;x₄=4;x₅=5;
```

$$
In[2]:= A =
\begin{pmatrix}
1 & x_1 & x_1^2 & x_1^3 & x_1^4 \\
1 & x_2 & x_2^2 & x_2^3 & x_2^4 \\
1 & x_3 & x_3^2 & x_3^3 & x_3^4 \\
1 & x_4 & x_4^2 & x_4^3 & x_4^4 \\
1 & x_5 & x_5^2 & x_5^3 & x_5^4
\end{pmatrix};
$$

```
In[3]:= MatrixForm[A]
```

Out[3]//MatrixForm=

$$
\begin{pmatrix}
1 & 1 & 1 & 1 & 1 \\
1 & 2 & 4 & 8 & 16 \\
1 & 3 & 9 & 27 & 81 \\
1 & 4 & 16 & 64 & 256 \\
1 & 5 & 25 & 125 & 625
\end{pmatrix}
$$

We now find $A\mathbf{x}$ for $\mathbf{x} = (a_0, a_1, a_2, a_3, a_4)$.

```
In[4]:= Ax=A.{a₀,a₁,a₂,a₃,a₄}
Out[4]= {a₀+a₁+a₂+a₃+a₄,a₀+2a₁+4a₂+8a₃+16a₄,
a₀+3a₁+9a₂+27a₃+81a₄,a₀+4a₁+16a₂+64a₃+256a₄,
a₀+5a₁+25a₂+125a₃+625a₄}
```

This tells us that

$$
A\mathbf{x} =
\begin{pmatrix}
1 & 1 & 1 & 1 & 1 \\
1 & 2 & 4 & 8 & 16 \\
1 & 3 & 9 & 27 & 81 \\
1 & 4 & 16 & 64 & 256 \\
1 & 5 & 25 & 125 & 625
\end{pmatrix}
\begin{pmatrix}
a_0 \\ a_1 \\ a_2 \\ a_3 \\ a_4
\end{pmatrix}
=
\begin{pmatrix}
a_0 + a_1 + a_2 + a_3 + a_4 \\
a_0 + 2a_1 + 4a_2 + 8a_3 + 16a_4 \\
a_0 + 3a_1 + 9a_2 + 27a_3 + 81a_4 \\
a_0 + 4a_1 + 16a_2 + 64a_3 + 256a_4 \\
a_0 + 5a_1 + 25a_2 + 125a_3 + 625a_4
\end{pmatrix}
$$

Next we solve the equations $A\mathbf{x} = \mathbf{b}$ and $A\mathbf{x} = \mathbf{c}$, where

$$\mathbf{b} = \begin{pmatrix} 3 \\ 4 \\ 2 \\ 6 \\ 7 \end{pmatrix} \quad \text{and} \quad \mathbf{c} = \begin{pmatrix} 2 \\ 3 \\ 2 \\ 6 \\ 8 \end{pmatrix}$$

are the vectors containing the data of two experiments.

In[5]:= $\texttt{Solve[\{a_0+a_1+a_2+a_3+a_4==3,a_0+2a_1+4a_2+8a_3+16a_4==4,}$

$\texttt{a_0+3a_1+9a_2+27a_3+81a_4==2,a_0+4a_1+16a_2+64a_3+256a_4==6,}$

$\texttt{a_0+5a_1+25a_2+125a_3+625a_4==7\},\{a_0,a_1,a_2,a_3,a_4\}]}$

Out[5]= $\{\{a_0 \to -28, a_1 \to \frac{119}{2}, a_2 \to -\frac{147}{4}, a_3 \to 9, a_4 \to -\frac{3}{4}\}\}$

In[6]:= $\texttt{Solve[\{a_0+a_1+a_2+a_3+a_4==2,a_0+2a_1+4a_2+8a_3+16a_4==3,}$

$\texttt{a_0+3a_1+9a_2+27a_3+81a_4==2,a_0+4a_1+16a_2+64a_3+256a_4==6,}$

$\texttt{a_0+5a_1+25a_2+125a_3+625a_4==8\},\{a_0,a_1,a_2,a_3,a_4\}]}$

Out[6]= $\{\{a_0 \to -22, a_1 \to 46, a_2 \to -\frac{341}{12}, a_3 \to 7, a_4 \to -\frac{7}{12}\}\}$

Therefore, the interpolating polynomial for Experiment 1 is

$$p(x) = -28 + \frac{119}{2}x - \frac{147}{4}x^2 + 9x^3 - \frac{3}{4}x^4$$

and that for Experiment 2 is

$$q(x) = -22 + 46x - \frac{341}{12}x^2 + 7x^3 - \frac{7}{12}x^4$$

We now graph the two polynomials.

In[7]:= $\texttt{p=-28 + \frac{119}{2}x - \frac{147}{4}x^2+9x^3 - \frac{3}{4}x^4;}$

In[8]:= $\texttt{q=-22 + 46x - \frac{341}{12}x^2+7x^3 - \frac{7}{12}x^4;}$

```
In[9]:= Plot[{p,q},{x,0,5}]
```

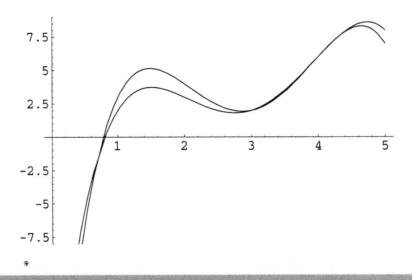

```
Out[9]= - Graphics -
```

The two overlapping graphs illustrate the comparative results of Experiments 1 and 2. ◄►

EXERCISES 1.14

1. Rewrite the following linear systems as matrix equations.

a. $\begin{cases} 2x + y - z = 2 \\ x - 3y + z = 1 \\ 2x + y - 2z = 9 \end{cases}$ b. $\begin{cases} 2x + y - z - 5w = 0 \\ x - 3y + z = 0 \\ 2x + y - z + w = 0 \end{cases}$ c. $\begin{cases} 2x + y - z = 1 \\ x - 3y + z = 2 \\ 6x - 3y + 7z = 3 \\ x - 3y + z = 4 \end{cases}$

d. $\begin{cases} 2x + y - z = 2 \\ z - y + x = 1 \\ 2x + z - 2y = 9 \end{cases}$ e. $\begin{cases} 2x + y - z = 1 \\ x - 3y + x = 2 \\ 6z - 3y + 7w = 3 \\ w - 3y + z = 4 \end{cases}$ f. $\begin{cases} 6z - 3y + 7w = 3 \\ w - 3y + z = 4 \end{cases}$

2. Use the Vandermonde matrix to find the interpolating polynomial of degree 5 for the following experimental data.

Time	1	2	3	4	5	6
Experiment	1	−1	2	3	−2	5

3. Rewrite the parametric equation $(x, y, z) = (4, 8, 3) + t(3, 2, 7)$ of a straight line in $\mathbb{R}^3$ as a matrix equation of the form $A\mathbf{x} = \mathbf{b}$, where $\mathbf{x} = (x, y, z)$.

4. Rewrite the parametric equation $(x, y, z) = (4, 8, 3) + t(3, 2, 7)$ of a straight line in $\mathbb{R}^3$ as a matrix equation of the form $A\mathbf{x} = \mathbf{b}$, where $\mathbf{x} = (x, y, z, t)$.

The Systems $A\mathbf{x} = \mathbf{0}$

We conclude this chapter with a discussion of the linear systems of the form $A\mathbf{x} = \mathbf{0}$. These systems are called **homogeneous** and will arise in many contexts. Here are two examples.

In Chapter 4, we will use homogeneous systems to show that a set $\{\mathbf{x}_1, \ldots, \mathbf{x}_n\}$ of column vectors is **linearly independent** by forming the matrix $(\mathbf{x}_1 \ \cdots \ \mathbf{x}_n)$ and showing that the equation

$$(\mathbf{x}_1 \ \cdots \ \mathbf{x}_n) \begin{pmatrix} a_1 \\ \vdots \\ a_n \end{pmatrix} = \mathbf{0}$$

has only the solution $a_1 = \cdots = a_n = 0$.

In Chapter 6, we will study equations of the form $A\mathbf{x} = \lambda\mathbf{x}$. To do so, we rewrite them as homogeneous systems of the form $(A - \lambda I)\mathbf{x} = 0$. We then find the scalars λ, called the **eigenvalues** of the matrix A, by analyzing the **determinant** of the matrix $A - \lambda I$.

We begin our analysis of homogeneous linear systems by showing that they are consistent since they always have the solution $\mathbf{x} = \mathbf{0}$. This solution is called the **trivial solution.**

THEOREM 1.4   *The zero vector is a solution of the homogeneous linear system $A\mathbf{x} = \mathbf{0}$.*

Proof. It follows from the definition of $A\mathbf{0}$ that

$$A\mathbf{0} = A \begin{pmatrix} 0 \\ \vdots \\ 0 \end{pmatrix} = \begin{pmatrix} a_{11}0 + \cdots + a_{1n}0 \\ \vdots \\ a_{m1}0 + \cdots + a_{mn}0 \end{pmatrix} = \begin{pmatrix} 0 \\ \vdots \\ 0 \end{pmatrix} = \mathbf{0}$$

Hence the zero vector is a solution of the system. ∎

All nonzero solutions of a homogeneous linear system are called **nontrivial**. Systems that are not homogeneous are called **nonhomogeneous.**

Homogeneous linear systems are consistent since they always have the trivial solution. For underdetermined homogeneous systems, more is true.

THEOREM 1.5   *Every underdetermined homogeneous linear system has a nontrivial solution.*

Proof. A linear system is underdetermined if it contains more variables than equations. Suppose that the system $A\mathbf{x} = \mathbf{0}$ has m equations in n variables and that $m < n$. This means that A is a nonzero $m \times n$ matrix. Let $(A \mid \mathbf{0})$ be the associated augmented matrix and let $(B \mid \mathbf{0})$ be the reduced row echelon form of $(A \mid \mathbf{0})$. Since A is a nonzero, the matrix B must have some nonzero pivots. Suppose, therefore, that B has p pivots associated with p basic variables in $\mathbf{x}$. Since the number of rows of B is less than the number of its columns, p must be less than n. Hence the system must have $n - p > 0$ free variables. Each nonzero assignment of values to these variables yields a nontrivial solution of the given homogeneous system. ∎

EXAMPLE 1.45   ■ An Underdetermined Homogeneous System

Use *MATHEMATICA* to show that the underdetermined homogeneous system

$$\begin{cases} 3x - y + z = 0 \\ x + 2y - 4z = 0 \end{cases}$$

has infinitely many solutions.

Solution. Since the given system consists of two equations in three variables, we elect to solve it for two of the three variables.

```
In[1]:= Solve[{3x-y+z==0,x+2y-4z==0},{x,y}]

Out[1]= {{x → 2z/7, y → 13z/7}}
```

This shows that for any assignment of a value to z we obtain a solution. Since we have infinitely many choices of values for z, the given system has infinitely many solutions. ◀▶

The next two examples show that the number of equations does not determine the number of solutions of a homogeneous linear system. A system with the same number of equations as variables may have one or infinitely many solutions.

EXAMPLE 1.46   ■ A Homogeneous System with One Solution

Use *MATHEMATICA* to show that the homogeneous system

$$\begin{cases} 2x + y - z = 0 \\ x - y + z = 0 \\ x + 2y - z = 0 \end{cases}$$

consisting of three equations in three variables has only the trivial solution.

Solution. We use the **Solve** function.

```
In[1]:= Solve[{2x+y-z==0,x-y+z==0,x+2y-z==0},{x,y,z}]

Out[1]= {x → 0, y → 0, z → 0}
```

This shows that the system $A\mathbf{x} = \mathbf{0}$ has only the trivial solution. ◀▶

EXAMPLE 1.47 ■ A Homogeneous System with Infinitely Many Solutions

Use *MATHEMATICA* to show that the homogeneous system

$$\begin{cases} x + y - 2z = 0 \\ x - 2y + z = 0 \\ 2x - y - z = 0 \end{cases}$$

of three equations in three unknowns has infinitely many solutions.

Solution. We begin by computing the number of basic variables of the system.

```
In[1]:= RowReduce[{{1,1,-2},{1,-2,1},{2,-1,-1}}]//MatrixForm
Out[1]//MatrixForm=
```

$$\begin{pmatrix} 1 & 0 & -1 \\ 0 & 1 & -1 \\ 0 & 0 & 0 \end{pmatrix}$$

This shows that the variables x and y of the system $A\mathbf{x} = \mathbf{0}$ are the basic variables. We therefore solve for x and y in terms of z.

```
In[2]:= Solve[{x+y-2z==0,x-2y+z==0,2x-y-z==0},{x,y}]
Out[2]= {{x→z,y→z}}
```

Since we are free to assign any value to z, the system has infinitely many solutions of the form $x = y = z$. ◄►

One of the characteristic features of homogeneous linear systems is that their sets of solutions contain the trivial solution and are closed under addition and multiplication by constants. In Chapter 4, we will interpret this property as saying that the set of solutions of a homogeneous linear system forms a ***vector space***.

EXAMPLE 1.48 ■ A Sum of Solutions

Use *MATHEMATICA* to show that if **u** and **v** are solutions of

$$\begin{cases} 3x - y + z - w = 0 \\ x + 2y - 4z + w = 0 \end{cases}$$

and a and b are two arbitrary real numbers, then $a\mathbf{u} + b\mathbf{v}$ is a solution of the given system.

Solution. First we find the solutions of the given system. Since we have two equations in four variables, we solve for two of them.

```
In[1]:= Solve[{3x-y+z-w==0,x+2y-4z+w==0},{x,y}]
Out[1]= {{x→ 1/7 (w+2z),y→ 1/7(-4w+13z)}}
```

Let

$$\mathbf{u} = \left(\tfrac{1}{7}, -\tfrac{4}{7}, 0, 1\right) \quad \text{and} \quad \mathbf{v} = \left(\tfrac{11}{7}, \tfrac{40}{7}, 4, 3\right)$$

be the solutions of the system determined by the assignment $w = 1$ and $z = 0$, and the assignment $w = 3$ and $z = 4$. We show that for any scalars a and b, the vector $a\mathbf{u} + b\mathbf{v}$ is a solution of the given system. We begin by inputting A, $a\mathbf{u}$ and $b\mathbf{v}$.

```
In[2]:= A= ( 3 -1  1 -1 ) ; u=a ( 1/7   ) ; v=b ( 11/7  ) ;
           ( 1  2 -4  1 )       ( -4/7  )       ( 40/7  )
                                 (  0   )       (  4   )
                                 (  1   )       (  3   )
```

```
In[3]:= Simplify[A.(u+v)]
Out[3]= {0,0}
```

This tells us that

$$\begin{pmatrix} 3 & -1 & 1 & -1 \\ 1 & 2 & -4 & 1 \end{pmatrix} \left[a \begin{pmatrix} \tfrac{1}{7} \\ -\tfrac{4}{7} \\ 0 \\ 1 \end{pmatrix} + b \begin{pmatrix} \tfrac{11}{7} \\ \tfrac{40}{7} \\ 4 \\ 3 \end{pmatrix} \right] = \begin{pmatrix} 0 \\ 0 \end{pmatrix}$$

Hence $a\mathbf{u} + b\mathbf{v}$ is a solution of the system. ◄►

Linear Dependence Relations and Homogeneous Systems

The left-hand side of $A\mathbf{x} = \mathbf{0}$ can be written as a sum of column vectors. Thus the system

$$\begin{cases} a_{11}x_1 + \cdots + a_{1n}x_n = 0 \\ a_{21}x_1 + \cdots + a_{2n}x_n = 0 \\ \quad \vdots \\ a_{m1}x_1 + \cdots + a_{mn}x_n = 0 \end{cases}$$

can be rewritten as $x_1\mathbf{c}_1 + x_2\mathbf{c}_2 + \cdots + x_n\mathbf{c}_n = \mathbf{0}$, where

$$\mathbf{c}_i = \begin{pmatrix} a_{1i} \\ \vdots \\ a_{mi} \end{pmatrix}$$

and $1 \le i \le n$. We call this result a **vector equation**.

The trivial solution and the nontrivial solutions of a vector equation provide examples of two of the most fundamental relationships among vectors. The general importance of these relationships in linear algebra will become obvious in Chapter 4.

DEFINITION 1.17   *The column vectors* $\mathbf{c}_1, \ldots, \mathbf{c}_n$ *are **linearly independent** if the vector equation* $x_1\mathbf{c}_1 + x_2\mathbf{c}_2 + \cdots + x_n\mathbf{c}_n = \mathbf{0}$ *has only the trivial solution. They are **linearly dependent** if they are not linearly independent.*

DEFINITION 1.18   *A **linear dependence relation** among the columns vectors* $\mathbf{c}_1, \ldots, \mathbf{c}_n$ *is any nontrivial solution of the equation* $x_1\mathbf{c}_1 + x_2\mathbf{c}_2 + \cdots + x_n\mathbf{c}_n = \mathbf{0}$.

EXAMPLE 1.49   ■ **Linear Dependence Relations**

Use *MATHEMATICA* to show that the linear dependence relations of the system

$$\begin{pmatrix} -5 & 5 & 1 & 8 \\ 6 & -1 & 2 & 3 \\ 3 & 0 & 1 & 2 \end{pmatrix} \begin{pmatrix} a \\ b \\ c \\ d \end{pmatrix} = \begin{pmatrix} 0 \\ 0 \\ 0 \end{pmatrix}$$

are the same as those of the reduced row echelon form of the system.

Solution. Since we have three equations in four unknowns, we select three of its variables as basic variables and solve the system for them. We choose a, b, and d as basic variables.

```
In[1]:= Solve[{{-5,5,1,8},{6,-1,2,3},
{3,0,1,2}}.{a,b,c,d}=={0,0,0},{a,b,d}]
Out[1]= {{a→ -c/19, b→ 8c/19, d→ -8c/19}}
```

Next we reduce A to row echelon form.

```
In[2]:= A= ( -5  5  1  8
              6 -1  2  3
              3  0  1  2 );
```

```
In[3]:= MatrixForm[B=RowReduce[A]]

Out[3]//MatrixForm=
```

$$\begin{pmatrix} 1 & 0 & 0 & -\frac{1}{8} \\ 0 & 1 & 0 & 1 \\ 0 & 0 & 1 & \frac{19}{8} \end{pmatrix}$$

We now find the solutions the homogeneous linear system determined by this matrix for the same variables a, b, and d.

```
In[3]:= Solve[{{1,0,0,-1/8},{0,1,0,1},
       {0,0,1,19/8}}.{a,b,c,d}=={0,0,0},{a,b,d}]

Out[3]= {{a → -c/19, b → 8c/19, d → -8c/19}}
```

This shows that the two system have the same solutions. They therefore have the same linear dependence relations. ◀▶

Since any homogeneous system $A\mathbf{x} = \mathbf{0}$ can be written in the form

$$x_1\mathbf{c}_1 + x_2\mathbf{c}_2 + \cdots + x_n\mathbf{c}_n = \mathbf{0}$$

it is clear that the linear dependence relations among the columns of A correspond precisely to the nontrivial solutions of the equation $A\mathbf{x} = \mathbf{0}$. We simplify this statement by saying that a nontrivial solution of a homogeneous system is a linear dependence relation of the system.

THEOREM 1.6   *If a homogeneous linear system results from another system by elementary row operations, then the two systems have the same linear dependence relations.*

Proof. Since elementary row operations preserve the solutions of the linear systems involved, a system $B\mathbf{x} = \mathbf{0}$ that is row equivalent to a system $A\mathbf{x} = \mathbf{0}$ has the same linear dependence relations as $A\mathbf{x} = \mathbf{0}$. ∎

Theorem 1.6 tells us that if

$$A = (\mathbf{c}_1(A) \ \cdots \ \mathbf{c}_n(A)) \quad \text{and} \quad B = (\mathbf{c}_1(B) \ \cdots \ \mathbf{c}_n(B))$$

are two row equivalent matrices, then

$$x_1\mathbf{c}_1(B) + \cdots + x_n\mathbf{c}_n(B) = \mathbf{0}$$

if and only if

$$x_1\mathbf{c}_1(A) + \cdots + x_n\mathbf{c}_n(A) = \mathbf{0}$$

We can also say that the columns of A are linearly independent if and only if those of B are. A basic property of real matrices is that their pivot columns are linearly independent and that their nonpivot columns are linearly dependent on their pivot columns.

EXAMPLE 1.50   ■ Linear Independence of Pivot Columns

Show that the pivot columns of the reduced row echelon matrix

$$A = \begin{pmatrix} 1 & 0 & 4 & -2 \\ 0 & 1 & 5 & 0 \end{pmatrix}$$

are linearly independent.

Solution. It is clear from the definition that the pivot columns of A are

$$\mathbf{c}_1(A) = \begin{pmatrix} 1 \\ 0 \end{pmatrix} \quad \text{and} \quad \mathbf{c}_2(A) = \begin{pmatrix} 0 \\ 1 \end{pmatrix}$$

They are linearly independent since

$$a \begin{pmatrix} 1 \\ 0 \end{pmatrix} + b \begin{pmatrix} 0 \\ 1 \end{pmatrix} = \begin{pmatrix} a \\ b \end{pmatrix} = \begin{pmatrix} 0 \\ 0 \end{pmatrix}$$

if and only if $a = b = 0$. The nonpivot columns of A are

$$\mathbf{c}_3(A) = \begin{pmatrix} 4 \\ 5 \end{pmatrix} \quad \text{and} \quad \mathbf{c}_4(A) = \begin{pmatrix} -2 \\ 0 \end{pmatrix}$$

As we can see, $\mathbf{c}_3(A)$ and $\mathbf{c}_4(A)$ are the unique sums $4\mathbf{c}_1(A) + 5\mathbf{c}_2(A)$ and $-2\mathbf{c}_1(A) + 0\mathbf{c}_2(A)$ of the pivot columns $\mathbf{c}_1(A)$ and $\mathbf{c}_2(A)$. ◄►

Uniqueness of the Reduced Row Echelon Form

Several results about linear systems are established using the fact that there is only one matrix in reduced row echelon form that is row equivalent to a given matrix.

THEOREM 1.7 (Reduced row echelon form theorem)   *The reduced row echelon form of a matrix is unique.*

Proof. Let A be an $m \times n$ matrix. We proceed by induction on the number of columns of A.

If $n = 1$, then A consists of a single column. There are only two reduced row echelon matrices possible, namely, the column vectors

$$(a_{i1}) = \begin{pmatrix} 0 \\ 0 \\ \vdots \\ 0 \end{pmatrix} \quad \text{and} \quad (b_{i1}) = \begin{pmatrix} 1 \\ 0 \\ \vdots \\ 0 \end{pmatrix}$$

with $a_{i1} = 0$ for all $1 \leq i \leq n$, and $b_{i1} = 1$ if $i = 1$ and $b_{i1} = 0$ otherwise. Clearly these matrices are not row equivalent and cannot both come from A by elementary row operations. This establishes the result for $n = 1$.

Now let $n = k + 1$, and assume that the theorem is true for matrices of dimension $m \times k$. Let B and C be two reduced row echelon forms of A. We want to show that $B = C$. First we note that B and C are row equivalent since both are obtained from A by elementary row operations. Let $\mathbf{c}_{k+1}(A)$, $\mathbf{c}_{k+1}(B)$ and $\mathbf{c}_{k+1}(C)$ be the last columns of A, B, and C, respectively. If we partition the matrices A, B, and C into

$$A = (A_k \ \mathbf{c}_{k+1}(A)) \quad B = (B_k \ \mathbf{c}_{k+1}(B)) \quad C = (A_k \ \mathbf{c}_{k+1}(C))$$

by separating out their last columns, then A_k, B_k, and C_k are in reduced row echelon form. Moreover, the elementary row operations that change A into B change A_k into B_k, while the elementary row operations that change A into C change A_k into C_k. By the induction hypothesis, we have $B_k = C_k$. It therefore follows from Theorem 1.6 that the columns of B satisfy the same linear dependence relations as the columns of C. Thus B and C have the same pivot columns. Therefore, $\mathbf{c}_{k+1}(B)$ is a pivot column if and only if $\mathbf{c}_{k+1}(C)$ is a pivot column.

First consider the case where $\mathbf{c}_{k+1}(B)$ is a pivot column. Its leading 1 must occur in the row corresponding to the top zero row of B_k. The same is true of the leading 1 of $\mathbf{c}_{k+1}(C)$. Thus the leading 1s of $\mathbf{c}_{k+1}(B)$ and $\mathbf{c}_{k+1}(C)$ occur in the same row. Hence $\mathbf{c}_{k+1}(B)$ must equal $\mathbf{c}_{k+1}(C)$. Therefore, $B = C$.

Next consider the case where $\mathbf{c}_{k+1}(B)$ is not a pivot column. Then $\mathbf{c}_{k+1}(C)$ cannot be a pivot column either. Therefore, $\mathbf{c}_{k+1}(B)$ is a unique sum of multiples of the pivot columns of B_k and $\mathbf{c}_{k+1}(B)$ is a similar unique sum of multiples of the pivot columns of C_k. Since $B_k = C_k$ and since B and C are row equivalent and are both in reduced row echelon form, these linear combinations must be identical. Hence $\mathbf{c}_{k+1}(B) = \mathbf{c}_{k+1}(C)$. Therefore, $B = C$.

This completes the induction. ∎

EXERCISES 1.15

1. Find the pivot columns of the following matrices. If a column is not a pivot column, explain why.

$$\text{a.} \begin{pmatrix} 1 & 3 & 3 & 1 \\ 5 & 0 & 0 & 3 \\ 3 & 1 & 3 & 0 \\ 3 & 4 & 2 & 3 \end{pmatrix} \quad \text{b.} \begin{pmatrix} 2 & 3 & 3 & 2 \\ 5 & 5 & 4 & 5 \\ 0 & 2 & 1 & 1 \end{pmatrix} \quad \text{c.} \begin{pmatrix} 4 & 0 & 4 & 1 & 4 \\ 5 & 4 & 3 & 3 & 5 \\ 3 & 5 & 1 & 2 & 1 \end{pmatrix}$$

2. Find linear systems whose augmented matrices are the matrices in Exercise 1. Explain the different possible choices of basic and free variables in each case.

3. Use the following matrices to create homogeneous linear systems $A\mathbf{x} = \mathbf{0}$. Determine in each case whether the set of solutions is a point, line, plane, or all of $\mathbb{R}^3$.

$$\text{a. } A = \begin{pmatrix} -8 & -5 & 7 \\ 9 & 0 & 9 \\ 1 & 2 & 3 \end{pmatrix} \quad \text{b. } A = \begin{pmatrix} 0 & 0 & 0 \\ 0 & 0 & 0 \\ 0 & 0 & 0 \end{pmatrix} \quad \text{c. } A = \begin{pmatrix} 1 & 0 & 0 \\ 0 & 1 & 0 \\ 0 & 0 & 1 \end{pmatrix}$$

$$\text{d. } A = \begin{pmatrix} 1 & 7 & -3 \\ 9 & 9 & 6 \\ 8 & 2 & 9 \end{pmatrix} \quad \text{e. } A = \begin{pmatrix} 3 & 2 & 1 \\ 4 & 3 & 2 \end{pmatrix} \quad \text{f. } A = \begin{pmatrix} 1 & -5 & 7 \\ 2 & -10 & 14 \\ 3 & -15 & 21 \end{pmatrix}$$

4. Use Theorem 1.5 to show that for the vectors $\mathbf{x} = (1, 2)$, $\mathbf{y} = (3, 4)$, and $\mathbf{z} = (-2, 7)$ there exist real numbers a, b, c, not all zero, for which $a\mathbf{x} + b\mathbf{y} + c\mathbf{z} = \mathbf{0}$.

5. Find a real number λ for which the homogeneous system $(A - \lambda I)\mathbf{x} = \mathbf{0}$ determined by

$$A = \begin{pmatrix} 3 & 2 \\ 4 & 3 \end{pmatrix} \quad \text{and} \quad \mathbf{x} = \begin{pmatrix} x \\ y \end{pmatrix}$$

has a nontrivial solution.

6. Use *MATHEMATICA* to show that the homogeneous systems

$$\begin{cases} 5x + 4y + 2z = 0 \\ 5x + 2y + z = 0 \end{cases} \quad \text{and} \quad \begin{cases} 5x + 4y + 2z + s\,(5x + 2y + z) = 0 \\ 5x + 2y + z = 0 \end{cases}$$

have the same solutions for all scalars s.

7. Use *MATHEMATICA* to show that the homogeneous system

$$\begin{cases} 3x - y + z = 0 \\ x + 2y - 4z = 0 \end{cases}$$

and the system

$$\begin{cases} x - \frac{2}{7}z = 0 \\ y - \frac{13}{7}z = 0 \end{cases}$$

obtained from the first system by Gauss-Jordan elimination have the same solution.

8. Use Theorem 1.6 to show that the homogeneous systems

$$\begin{pmatrix} 2 & -1 & 1 & 4 \\ 1 & -3 & 3 & 5 \\ 5 & -2 & -3 & 3 \end{pmatrix} \begin{pmatrix} w \\ x \\ y \\ z \end{pmatrix} = \begin{pmatrix} 0 \\ 0 \\ 0 \end{pmatrix}$$

and

$$\begin{pmatrix} 2 & -1 & 1 & 4 \\ 0 & -5 & 5 & 6 \\ 0 & 0 & 25 & 32 \end{pmatrix} \begin{pmatrix} w \\ x \\ y \\ z \end{pmatrix} = \begin{pmatrix} 0 \\ 0 \\ 0 \end{pmatrix}$$

have the same dependence relations.

9. Show that there exist no linear dependence relations among the columns of the matrix

$$\begin{pmatrix} 1 & 0 & 0 \\ 0 & 1 & 0 \\ 0 & 0 & 1 \end{pmatrix}$$

10. Show that the column vectors determined by the following matrices are linearly independent.

a. $\begin{pmatrix} 1 & 3 & 3 & 1 \\ 5 & 0 & 0 & 3 \\ 3 & 1 & 3 & 0 \\ 3 & 4 & 2 & 3 \end{pmatrix}$
b. $\begin{pmatrix} 2 & 3 & 3 \\ 5 & 5 & 4 \\ 0 & 2 & 1 \end{pmatrix}$
c. $\begin{pmatrix} 1 & 3 & 3 & 1 \\ 5 & 0 & 0 & 3 \\ 3 & 1 & 3 & 0 \\ 3 & 4 & 2 & 3 \\ 2 & 1 & 3 & 5 \end{pmatrix}$

11. Show that the column vectors determined by the following matrices are linearly dependent.

a. $\begin{pmatrix} 1 & 3 & 4 & 1 \\ 5 & 0 & 5 & 3 \\ 3 & 1 & 4 & 0 \\ 3 & 4 & 7 & 3 \end{pmatrix}$
b. $\begin{pmatrix} 2 & 3 & 3 & 1 & 0 \\ 5 & 5 & 4 & 3 & 2 \\ 0 & 2 & 1 & 6 & 5 \end{pmatrix}$
c. $\begin{pmatrix} 4 & 0 & 8 \\ 5 & 4 & 2 \\ 3 & 5 & -4 \end{pmatrix}$

APPLICATIONS

Chemical Equations

Stoichiometry is the study of quantitative relationships in chemical reactions. One of the steps in the standard procedure for solving problems in stoichiometry is to write balanced chemical equations. Suppose that a chemical reaction is described by the equation

$$H_2S + O_2 \rightarrow H_2O + SO_2$$

expressing the fact that if oxygen (O) is added to a compound made up of hydrogen (H) and sulfur (S), then water (H_2O) and sulfur dioxide (SO_2) result. We would like to solve the mass-mass problem of finding positive integers $a, b, c,$ and d that balance the equation. The chemicals H_2S and O_2 are called the **reactants** and H_2O and SO_2 are called the **products**. The balancing integers a, b, c, and d are called **stoichiometric coefficients**.

In the language of linear algebra, this problem consists of finding constants a, b, c, and d for which the following two vector sums are equal:

$$a \begin{pmatrix} 2H \\ 1S \\ 0O \end{pmatrix} + b \begin{pmatrix} 0H \\ 0S \\ 2O \end{pmatrix} = c \begin{pmatrix} 2H \\ 0S \\ 1O \end{pmatrix} + d \begin{pmatrix} 0H \\ 1S \\ 2O \end{pmatrix}$$

This is equivalent to solving the linear system

$$\begin{cases} 2a - 2c = 0 \\ a - d = 0 \\ 2b - c - 2d = 0 \end{cases}$$

A solution is easily read off the equations. If we let $a = c = d = 2$ and $b = 3$, then we get a balanced chemical equation. Hence

$$2H_2S + 3O_2 \rightarrow 2H_2O + 2SO_2$$

is a feasible mass-mass relationship between the chemical substances involved in the process.

Electric Circuits

Systems of linear equations can be used to determine the properties of simple steady-state circuits involving elements such as resistors, capacitors, and batteries. Such circuits can be analyzed through the use of two sets of principles:

1. **Ohm's law** $V = IR$, which describes a voltage V, measured in **volts**, as the amount of a current I, measured in **amps**, passing through a resistor R, whose capacity is measured in **ohms**.

2. **Kirchhoff's laws** for parallel and series combinations of the circuit elements. Kirchhoff's laws simplify the analysis of complicated circuits and make it possible to calculate the current flowing through the circuit by representing the circuit with a linear system according to the following two principles.

   a. Currents entering any junction of a circuit must equal the sum of the currents leaving the junction. This principle expresses the law of the conservation of charge. Consider the junction in Figure 5. By Kirchhoff's current law, $i_1 = i_2 + i_3$.

   b. The sum of the changes in potential voltage across the elements around any closed circuit loop must be zero. This is a statement of the law of conservation of energy.

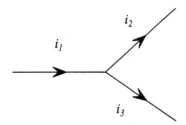

FIGURE 5   Electric Circuit Junction.

In a circuit containing resistors and electromotive forces, certain conventions determine the signs of the potential differences.

1. If a resistor is traversed in the direction of the current, as shown in Figure 6(a), then the potential difference across the resistor is $-IR$.

2. If a resistor is traversed in the direction opposite to the current, as shown in Figure 6(b), then the potential difference across the resistor is $-IR$.

3. If a source of an electromotive force is traversed in the direction of the force, as shown in Figure 6(c), then the potential difference is $+E$.

4. If a source of an electromotive force is traversed in the opposite direction of the force, as shown in Figure 6(d), then the potential difference is $-E$.

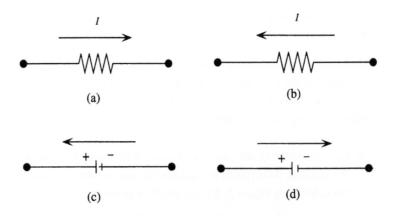

FIGURE 6   Electric Current Flow.

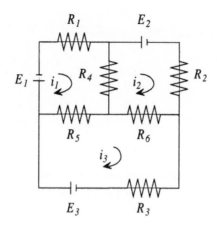

FIGURE 7   Electrical Resistor Circuit.

EXAMPLE 1.51 ■ A Resistor Circuit

Use *MATHEMATICA* to calculate the currents in the circuit shown in Figure 7 for $R_1 = R_1 = R_5 = 1$, $R_2 = R_4 = R_6 = 2$, and $E_1 = 2$, $E_2 = 3$, and $E_3 = 6$.

Solution. By Kirchhoff's voltage law, the currents i_1, i_2, and i_3 are determined by the linear system

$$
\begin{cases}
(R_1 + R_4 + R_5)i_1 - R_4 i_2 - R_5 i_3 = E_1 \\
-R_4 i_1 + (R_2 + R_4 + R_6)i_2 - R_6 i_3 = E_2 \\
-R_5 i_1 - R_6 i_2 + (R_3 + R_5 + R_6)i_3 = E_3
\end{cases}
$$

We solve this system for i_1, i_2, and i_3 by applying Gaussian elimination to the augmented matrix of the system. For the given values for R_i and E_j, the augmented matrix of the system is

$$
\begin{pmatrix}
R_1 + R_4 + R_5 & -R_4 & -R_5 & E_1 \\
-R_4 & R_2 + R_4 + R_6 & -R_6 & E_2 \\
-R_5 & -R_6 & R_3 + R_5 + R_6 & E_3
\end{pmatrix}
=
\begin{pmatrix}
4 & -2 & -1 & 2 \\
-2 & 6 & -2 & 3 \\
-1 & -2 & 4 & 6
\end{pmatrix}
$$

We begin by defining the matrix A.

```
In[1]:= A = ( 4  -2  -1  2
             -2   6  -2  3
             -1  -2   4  6 );
```

Next we row-reduce A.

```
In[2]:= B=MatrixForm[RowReduce[A]]
```

Out[2]//MatrixForm=

$$\begin{pmatrix} 1 & 0 & 0 & \frac{13}{5} \\ 0 & 1 & 0 & \frac{5}{2} \\ 0 & 0 & 1 & \frac{17}{5} \end{pmatrix}$$

It follows from the definition of augmented matrices that $i_1 = 13/5 \approx 2.6$ amps, $i_2 = 5/2 \approx 2.5$ amps, and $i_3 = 17/5 = 3.4$ amps. ◄►

Higher-Order Differential Equations

Suppose that $y^{(1)}(t)$, $y^{(2)}(t)$, ..., $y^{(n)}(t)$ are the first n derivatives of a function $y(t)$. Our goal is to rewrite any nth order differential equation involving these functions as a linear system of first-order differential equations. Here is the method, described for $n = 3$. Consider the differential equation

$$y^{(3)}(t) + a_3 y(t) + a_2 y^{(1)}(t) + a_1 y^{(2)}(t) = 0$$

We rewrite this equation as

$$y^{(3)}(t) = -a_3 y(t) - a_2 y^{(1)}(t) - a_1 y^{(2)}(t)$$
$$= -a_3 y_1(t) - a_2 y_2(t) - a_1 y_3(t)$$
$$= \begin{pmatrix} -a_3 & -a_2 & -a_1 \end{pmatrix} \begin{pmatrix} y_1(t) \\ y_2(t) \\ y_3(t) \end{pmatrix}$$

where $y_1(t) = y(t)$, $y_2(t) = y^{(1)}(t)$, and $y_3(t) = y^{(2)}(t)$. In this notation, $y_3^{(1)}(t) = y^{(3)}(t)$.

Using these equations, we form the matrix equation

$$\begin{pmatrix} y_1^{(1)}(t) \\ y_2^{(1)}(t) \\ y_3^{(1)}(t) \end{pmatrix} = \begin{pmatrix} 0 & 1 & 0 \\ 0 & 0 & 1 \\ -a_3 & -a_2 & -a_1 \end{pmatrix} \begin{pmatrix} y_1(t) \\ y_2(t) \\ y_3(t) \end{pmatrix}$$

The result is a linear system of first-order equations.

This method works in general. Let $y_1(t) = y(t)$, and let $y_i(t) = y_{i-1}^{(1)}(t)$ be the ith derivative of a function $y(t)$, for $i = 2, \ldots, n$, and that suppose that

$$y^{(n)}(t) + a_1 y^{(n-1)}(t) + \cdots + a_{n-1} y^{(1)}(t) + a_n y(t) = 0$$

is a homogeneous nth order linear differential equation. We use the functions $y_i(t)$ to form the matrix equation

$$
\begin{pmatrix}
y_1'(t) \\
y_2'(t) \\
\vdots \\
y_{n-1}'(t) \\
y_n'(t)
\end{pmatrix}
=
\begin{pmatrix}
0 & 1 & 0 & \cdots & 0 \\
0 & 0 & 1 & \cdots & 0 \\
\vdots & \vdots & \vdots & \vdots & \vdots \\
0 & 0 & 0 & \cdots & 1 \\
-a_n & -a_{n-1} & -a_{n-2} & \cdots & -a_1
\end{pmatrix}
\begin{pmatrix}
y_1(t) \\
y_2(t) \\
\vdots \\
y_{n-1}(t) \\
y_n(t)
\end{pmatrix}
$$

The result is a linear system of first-order differential equations corresponding to the given nth order equation. In Section 6 we will show that this reduction can be used to solve certain higher-order differential equations. The associated coefficient matrix is discussed briefly in Exercises 6.8.

EXAMPLE 1.52 ■ Reducing a Differential Equation to a First-Order System

Find a first-order linear system for the third-order differential equation

$$
y^{(3)}(t) + 5y^{(2)}(t) - 2y(t) = 0
$$

and write the result in matrix form.

Solution. We let $y_1(t) = y(t)$, $y_2(t) = y_1^{(1)}(t)$, and $y_3(t) = y_2^{(1)}(t)$. Using these definitions, we rearrange the given equation as

$$
\begin{aligned}
y^{(3)}(t) &= 2y(t) - 5y^{(2)}(t) \\
&= 2y(t) + 0y^{(1)}(t) - 5y^{(2)}(t) \\
&= 2y_1(t) + 0y_2(t) - 5y_3(t)
\end{aligned}
$$

We then form linear system

$$
\begin{cases}
y_1^{(1)}(t) = y_2(t) \\
y_2^{(1)}(t) = y_3(t) \\
y_3^{(1)}(t) = 2y_1(t) + 0y_2(t) - 5y_3(t)
\end{cases}
$$

and write it as the matrix equation

$$
\begin{pmatrix}
y_1^{(1)}(t) \\
y_2^{(1)}(t) \\
y_3^{(1)}(t)
\end{pmatrix}
=
\begin{pmatrix}
0 & 1 & 0 \\
0 & 0 & 1 \\
2 & 0 & -5
\end{pmatrix}
\begin{pmatrix}
y_1(t) \\
y_2(t) \\
y_3(t)
\end{pmatrix}
$$

The result is the required first-order linear system in matrix form. ◄►

Heat Transfer

In physics and engineering, linear systems are used to solve heat transfer problems. The method is based on the assumption that the temperature at a point in a thin metal plate, for example, is approximately equal to the average of the temperatures of surrounding points. The approximation improves with the number of surrounding points chosen.

EXAMPLE 1.53 ■ **Heat Transfer**

Consider a thin rectangular metal plate as shown in Figure 8.

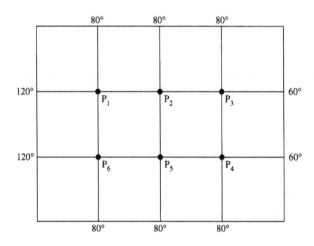

FIGURE 8   Cross Section of a Metal Plate.

Calculate the temperature of the plate at points $P_1, \ldots, P_6$.

Solution. It is clear from the discussion preceding this example that if t_i is the temperature of the plate at point P_i, then

$$\begin{cases} t_1 = \frac{1}{4}(80 + t_2 + t_6 + 120) \\ t_2 = \frac{1}{4}(80 + t_3 + t_5 + t_1) \\ t_3 = \frac{1}{4}(80 + 60 + t_4 + t_2) \\ t_4 = \frac{1}{4}(t_3 + 60 + 80 + t_5) \\ t_5 = \frac{1}{4}(t_2 + t_4 + 80 + t_6) \\ t_6 = \frac{1}{4}(t_1 + t_5 + 80 + 120) \end{cases}$$

We use *MATHEMATICA* to solve this system.

```
In[1]:= NSolve[{4t1==80+t2+t6+120,4t2==80+t3+t5+t1,
4t3==80+60+t4+t2,4t4==t3+60+80+t5,4t5==t2+t4+80+t6,
4t6==t1+t5+80+120},{t1,t2,t3,t4,t5,t6}]
Out[1]={{t1→94.2857,t2→82.8571,t3→74.2857,
t4→74.2857,t5→82.8571,t6→94.2857}}
```

Thus the temperature of the plate at points P_1 and P_6 is approximately 94 degrees, at P_2 and P_5 approximately 83 degrees, and at P_3 and P_4 approximately 74 degrees. ◄►

Population Dynamics

Suppose that the females in family of animals are divided into n age groups, with y_i the number of females in the ith group. The length of the equal time intervals will depend on the nature of the animals. Suppose also that the probability that a female in the ith age group will be still be alive in the $(i + 1)$th age group is p_i, and let a_i be the average number of females born to a female in the ith group. Then

$$
\mathbf{y} = \begin{pmatrix} y_1 \\ y_2 \\ \vdots \\ y_{n-1} \\ y_n \end{pmatrix} = \begin{pmatrix} a_1 & a_2 & \cdots & a_{n-1} & a_n \\ p_1 & 0 & \cdots & & 0 \\ 0 & p_2 & \cdots & & 0 \\ \vdots & \vdots & & & \vdots \\ 0 & 0 & & p_{n-1} & 0 \end{pmatrix} \begin{pmatrix} x_1 \\ x_2 \\ \vdots \\ x_{n-1} \\ x_n \end{pmatrix} = A\mathbf{x}
$$

where the vector $\mathbf{x} = \mathbf{x}(0)$ represents the population distribution of the family at time $t = 0$, and $\mathbf{y} = \mathbf{y}(1)$ represents the distribution at time $t = 1$. After k time intervals, the population distribution is given by the vector $\mathbf{y}(k) = A^k\mathbf{x}(0)$. The matrix A is called the **Leslie matrix** for the given family.

EXAMPLE 1.54 ■ Population Dynamics

Use a Leslie matrix to calculate the female population after five time intervals of a family divided into three age groups, with the first family consisting of 300 females, the second family of 700 females, and the third family of 500 females. The probability that a female in the first group will be alive in the second group is $1/2$, and the probability that a female in the second group will be alive in the third group is $3/4$. Suppose that no female in the first age group gives birth to an offspring, that the average number of female offspring of a female in the second group is 7, and that in the third group is 3.

Solution. We know from the discussion above that the population vector

$$\mathbf{y}(5) = \begin{pmatrix} 0 & 7 & 3 \\ 0.5 & 0 & 0 \\ 0 & 0.75 & 0 \end{pmatrix}^5 \begin{pmatrix} 300 \\ 700 \\ 500 \end{pmatrix}$$

yields the required data. We use *MATHEMATICA* to calculate $\mathbf{y}(5)$.

```
In[1]:= MatrixPower[{{0,7,3},{0.5,0,0},{0,0.75,0}},5].
{300,700,500}

Out[1]:= {82534.4,8193.75,8526.56}
```

After five time intervals, the family therefore consists of approximately 82,534 females in the first age group, 8,194 females in the second group, and 8,527 females in the third group.
◀▶

EXERCISES 1.16

1. Balance the following chemical equations.

   a. $N_2 + H_2 \rightarrow NH_3$

   b. $C_3H_8 + O_2 \rightarrow CO_2 + H_2O$

   c. $CH_4 + NH_3 + O_2 \rightarrow HCN + H_2O$

2. Consult a textbook on chemistry and use linear algebra to solve additional stoichiometric problems found there.

3. Use Ohm's and Kirchhoff's laws to verify the equation for calculating the current shown in Figure 9.

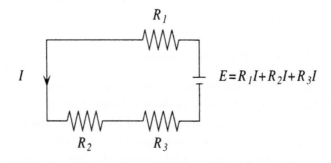

FIGURE 9   A Three-Resistor Circuit.

4. Use Ohm's and Kirchhoff's laws, in conjunction with Gaussian elimination, to calculate the currents in the loops in Figure 10.

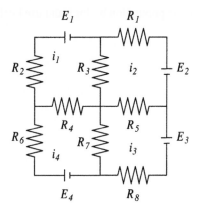

FIGURE 10   An Eight-Resistor Circuit.

5. Consult a textbook on physics and electrical engineering and use linear algebra to solve additional electrical circuit problems found there.

6. Reduce the following higher-order differential equations to systems of first-order equations.

a. $y^{(4)}(t) + y^{(3)}(t) - 2y(t) = 0$ b. $y^{(2)}(t) - 8y^{(1)}(t) + y(t) = 0$
c. $y^{(3)}(t) + y^{(2)}(t) + y(t) = 0$ d. $y^{(5)}(t) + 5y^{(3)}(t) - 6y(t) = 0$

7. Consult a textbook on differential equations and use linear systems to reduce higher-order homogeneous linear differential equations found there to systems of first-order differential equations.

8. Use a linear system to calculate the average temperature at sixteen equidistant interior points of a thin square metal plate whose boundary temperatures are 100°C on the top and bottom of the plate, 60°C on the left side of the plate, and 120°C on the right side of the plate.

9. Suppose that the females of a certain family of animals are divided into five age groups and that the Leslie matrix for this family is

$$A = \begin{pmatrix} 0.0 & 3.0 & 2.0 & 1.5 & 0.1 \\ 0.9 & 0.0 & 0.0 & 0.0 & 0.0 \\ 0.0 & 0.6 & 0.0 & 0.0 & 0.0 \\ 0.0 & 0.0 & 0.5 & 0.0 & 0.0 \\ 0.0 & 0.0 & 0.0 & 0.5 & 0.0 \end{pmatrix}$$

Calculate the number of females of the different age groups after eight years if at time $t = 0$ each age group contains 1,000 females.

10. Consult a textbook on population biology and use Leslie matrices to solve additional population problems.

REVIEW

KEY CONCEPTS ▶ Define and discuss each of the following.

Equivalence
Row equivalence.

Geometry
Arrow, line, plane, point.

Linear Systems
Basic variable, free variable. Consistent system, determined system, inconsistent system, overdetermined system, underdetermined system. Back substitution, Gaussian elimination, Gauss-Jordan elimination. Homogeneous linear system, linear dependence relation, nontrivial solution, trivial solution.

Lists
List of lists, standard lists.

Matrices
Augmented matrix, coefficient matrix, row echelon matrix, reduced row echelon matrix, Vandermonde matrix. Dimension. Entry. Pivot, pivot column, pivot position.

Matrix Operations
Elementary row operation.

Vectors
Column vector, row vector, zero vector.

KEY FACTS ▶ Explain and illustrate each of the following.

1. Every matrix is row equivalent to a row echelon matrix.

2. Every matrix is row equivalent to a matrix in reduced row echelon matrix.

3. The zero vector is a solution of the homogenous linear system $A\mathbf{x} = \mathbf{0}$.

4. An $m \times n$ linear system in the variables $x_1, \ldots, x_n$ with augmented $m \times (n + 1)$ matrix $(A \mid \mathbf{b})$ is consistent for all $\mathbf{b} \in \mathbb{R}^m$ if and only if the matrix A has m pivots.

5. Every underdetermined homogeneous linear system has a nontrivial solution.

6. If a homogeneous linear system results from another system by elementary row operations, then the two systems have the same linear dependence relations.

7. The reduced row echelon form of a matrix is unique.

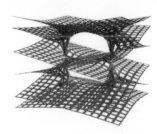

2

MATRIX ALGEBRA

INTRODUCTION

As we saw in Chapter 1, it is natural to represent mathematical data for linear systems in matrix form. The following examples illustrate additional instances where matrix algebra is natural and useful.

EXAMPLE 2.1 ■ A Cost Analysis

A furniture manufacturer has received orders to furnish several houses in a new housing development, 5 in an Alpine style, 7 in a Colonial style, and 14 in a Spanish style. To produce this furniture, the manufacturer needs different quantities of wood, metal, plastic, glass, and leather. The unit costs of the raw materials are $120, $150, $40, $80, and $200, respectively. Table 1 lists the quantities of raw materials required for the three types of house.

TABLE 1   Required Raw Materials.

	Wood	Metal	Plastic	Glass	Leather
Alpine	20	8	2	18	4
Colonial	18	7	4	15	2
Spanish	15	9	3	21	18

Use matrix algebra to find the total cost of the raw material required for this project.

Solution. First we convert the given table into the matrix

$$A = \begin{pmatrix} 20 & 8 & 2 & 18 & 4 \\ 18 & 7 & 4 & 15 & 2 \\ 15 & 9 & 3 & 21 & 18 \end{pmatrix}$$

If each type of house requires 10 pieces of comparable furniture, then the manufacturer must produce 50 pieces of Alpine furniture, 70 pieces of Colonial furniture, and 140 pieces of Spanish furniture. The vector $\mathbf{x} = (50, 70, 140)$ lists the number of different pieces of furniture of each style to be produced. The matrix

$$\mathbf{x}A = \begin{pmatrix} 50 & 70 & 140 \end{pmatrix} \begin{pmatrix} 20 & 8 & 2 & 18 & 4 \\ 18 & 7 & 4 & 15 & 2 \\ 15 & 9 & 3 & 21 & 18 \end{pmatrix}$$

$$= \begin{pmatrix} 4360 & 2150 & 800 & 4890 & 2860 \end{pmatrix}$$

tabulates the total number of units of wood, metal, plastic, glass, and leather required for the entire project. Moreover, the vector $\mathbf{y} = (120 \quad 150 \quad 40 \quad 80 \quad 200)$ lists the unit costs of these raw materials. The product

$$\mathbf{x}A\mathbf{y} = \begin{pmatrix} 4360 & 2150 & 800 & 4890 & 2860 \end{pmatrix} \begin{pmatrix} 120 \\ 150 \\ 40 \\ 80 \\ 200 \end{pmatrix} = 1{,}840{,}900$$

shows that the total cost of the raw materials for the entire project is $1,840,900. ◄►

EXAMPLE 2.2 ■ Mortgage Payments

Use matrix algebra to compute the monthly payments required for a $100,000 mortgage if the annual interest rate is 9% per year and the mortgage is to be paid off in 30 years.

Solution. Suppose that M is the required monthly payment, that $\mathbf{x}$ is the vector

$$\begin{pmatrix} 100{,}000 & 100{,}000 & -M \end{pmatrix}$$

whose first two components are the amount of the mortgage and whose third component is $-M$, and that A is the 3×3 matrix

$$\begin{pmatrix} 1 & 1 & 0 \\ \left(\frac{1}{100}\right)\left(\frac{9}{12}\right) & \left(\frac{1}{100}\right)\left(\frac{9}{12}\right) & 0 \\ 1 & 1 & 1 \end{pmatrix}$$

containing the monthly rate interest. The product

$$\mathbf{x}A = \begin{pmatrix} 100{,}000 & 100{,}000 & -M \end{pmatrix} \begin{pmatrix} 1 & 1 & 0 \\ \left(\frac{1}{100}\right)\left(\frac{9}{12}\right) & \left(\frac{1}{100}\right)\left(\frac{9}{12}\right) & 0 \\ 1 & 1 & 1 \end{pmatrix}$$

of $\mathbf{x}$ and A produces the vector

$$\mathbf{y} = \begin{pmatrix} 100{,}750 - M & 100{,}750 - M & -M \end{pmatrix}$$

whose first component

$$100{,}750 - M = 100{,}000 + \left(100{,}000 \times \left(\frac{1}{100}\right)\left(\frac{9}{12}\right)\right) - 12M$$

is the amount of the mortgage still owed after one month. If we now use the vector **y** in place of **x** and calculate the vector-matrix product

$$\begin{pmatrix} 100{,}750 - M & 100{,}750 - M & -M \end{pmatrix} \begin{pmatrix} 1 & 1 & 0 \\ \left(\frac{1}{100}\right)\left(\frac{9}{12}\right) & \left(\frac{1}{100}\right)\left(\frac{9}{12}\right) & 0 \\ 1 & 1 & 1 \end{pmatrix}$$

we obtain the vector

$$\mathbf{z} = \begin{pmatrix} 1.0151 \times 10^5 - 2.0075M & 1.0151 \times 10^5 - 2.0075M & -1.0M \end{pmatrix}$$

whose first component $1.0151 \times 10^5 - 2.0075M$ represents the amount of the mortgage still owed after two months. The vector **z** can also be calculated directly from **x** and A by using **x** and the *square* of A.

The first component of the vector-matrix product

$$\begin{pmatrix} 100{,}000 & 100{,}000 & -M \end{pmatrix} \begin{pmatrix} 1 & 1 & 0 \\ \left(\frac{1}{100}\right)\left(\frac{9}{12}\right) & \left(\frac{1}{100}\right)\left(\frac{9}{12}\right) & 0 \\ 1 & 1 & 1 \end{pmatrix}^2$$

is also **z**.

Using these ideas, we can express the amount of the mortgage owed after $12 \times 30 = 360$ months as the first component

$$\begin{pmatrix} 1.4731 \times 10^6 - 1830.7M & 1.4731 \times 10^6 - 1830.7M & -1.0M \end{pmatrix}$$

of the vector-matrix product

$$\begin{pmatrix} 100{,}000 & 100{,}000 & -M \end{pmatrix} \begin{pmatrix} 1 & 1 & 0 \\ \left(\frac{1}{100}\right)\left(\frac{9}{12}\right) & \left(\frac{1}{100}\right)\left(\frac{9}{12}\right) & 0 \\ 1 & 1 & 1 \end{pmatrix}^{360}$$

Since the mortgage is paid off after that time,

$$1.4731 \times 10^6 - 1830.7M = 0$$

If we solve this equation for M, we get $M = 804.66$. Our monthly payments are therefore approximately \$805.00. We will see later that the particular forms of the vectors **x**, **y**, **z** and of the matrix A are required because of the nature of the rules for matrix multiplication. ◄►

Real Matrices

We usually simplify the exposition by writing the entries of matrices as integers. Occasionally, however, *MATHEMATICA* needs to be told explicitly that the entries of a matrix are *real numbers* since the algorithms used for computing with integers and real numbers often are not the same.

Telling *MATHEMATICA* that the entries of a matrix are real numbers is very simple: We place a decimal point behind one of the entries. As far as *MATHEMATICA* is concerned, the matrix

$$A = \begin{pmatrix} 20 & 8 & 2 & 18 & 4 \\ 18. & 7 & 4 & 15 & 2 \\ 15 & 9 & 3 & 21 & 18 \end{pmatrix}$$

is a matrix with real number entries since one of its entries is written as "18." We could have written "18.0" in place of "18." and achieved the same result. We also could have used the built-in real number function N instead. The expression **N[18]** also denotes the real number 18.

Certain *MATHEMATICA* functions automatically produce real numbers. The **Random** function, for example, is programmed to generate real numbers by default. The command

$$\textbf{Table[Random[],\{m\},\{n\}]}$$

generates an $m \times n$ matrix with random real numbers as entries. If we want to generate a matrix with integer entries, we need to specify this explicitly. The command

$$\textbf{Random[Integer,\{-10,10\}]}$$

generates a random integer between -10 and 10.

From a computational point of view, *MATHEMATICA* distinguishes between computations with integers and computations with real numbers. Computations with integers are exact, or symbolic, whereas computations with numbers explicitly specified as real are approximate. This does not mean, however, that *MATHEMATICA* treats all real numbers as approximate numbers. If we use the constant **Pi**, *MATHEMATICA* computes with π symbolically as if it had infinite precision. If we use the expression **N[Pi]** instead, *MATHEMATICA* usually computes with its five-place approximation, 3.14159. The distinction between exact and approximate computation will be particularly important in Chapters 6 and 9.

EXAMPLE 2.3   ■ A Random Matrix with Real Entries

Use *MATHEMATICA* to generate a random 3×4 matrix with real entries.

Solution. We use the **Table** and **Random** functions to generate the desired matrix.

```
In[1]:= Table[Random[],{3},{4}]

Out[1]= {{0.0652009,0.248906,0.188679,0.125429},

{0.539365,0.336125,0.717192,

0.288962},{0.245198,0.204198,

0.617031,0.166143}}
```

The output is a random 3×4 matrix in list form. By adding the **MatrixForm** option, we get a rectangular array.

```
In[2]:= Table[Random[],{3},{4}]//MatrixForm

Out[2]//MatrixForm=
```

$$\begin{pmatrix} 0.584581 & 0.999948 & 0.345947 & 0.430273 \\ 0.132221 & 0.338007 & 0.924586 & 0.927451 \\ 0.0458219 & 0.339871 & 0.276449 & 0.902633 \end{pmatrix}$$

The result is a real 3×4 matrix, written as a rectangular array. ◄►

EXERCISES 2.1

1. Set up a total cost matrix for the basic ingredients of three meals at a major resort hotel. One night the hotel serves filet mignon with vegetables. The meal requires 80 lb of meat, 100 lb of vegetables, 2 lb of spices, and 3 lb of other ingredients. The next night the hotel serves Irish stew. The meal requires 100 lb of meat, 300 lb of vegetables, 4 lb of spices, and 5 lb of other ingredients. The third night the hotel serves Hungarian goulash. The meal requires 150 lb of meat, 350 lb of vegetables, 5 lb of spices, and 7 lb of other ingredients. The average cost per pound of the meat is $8.00, of the vegetables $3.00, of the spices $9.00, and of the other ingredients $4.00. Compute the total cost of the three meals.

2. A businessman returned from a European business trip with the following currencies: 100 German marks, 7000 French francs, 100,000 Italian lire, and 120 British pounds. The next week he had to visit Canada and decided to convert these currencies to Canadian dollars. At the time, one Canadian dollar was worth 1.0776 German marks, 3.61496 French francs, 1066.51257 Italian lira, and 0.38288 British pounds.

   a. Write a row vector $(q_1 \ q_2 \ q_3 \ q_4)$ for the quantity of each currency the businessman brought back from Europe.

b. Express the values $c_1, c_2, c_3,$ and c_4 of one unit of each currency in Canadian dollars as a column vector.

c. Show that the amount of Canadian dollars the businessman obtained for the European currencies is given by the product

$$\begin{pmatrix} q_1 & q_2 & q_3 & q_4 \end{pmatrix} \begin{pmatrix} c_1 \\ c_2 \\ c_3 \\ c_4 \end{pmatrix} = q_1c_1 + q_2c_2 + q_3c_3 + q_4c_4$$

3. Use the **Table** and **Random** functions of *MATHEMATICA* to generate a 3×4, a 2×5, and a 4×4 random matrix with integer entries.

4. Use the **MatrixForm** function to reformat the matrices obtained in Exercise 2 in rectangular form.

5. Use the **Table** and **Random** functions to generate a 3×4, a 2×5, and a 4×4 random matrix with real entries.

6. Use the **MatrixForm** function to reformat the matrices obtained in Exercise 4 in rectangular form.

7. Use the ideas discussed in Example 2.2 to calculate the monthly payments required for a $95,000 mortgage to be paid off in 25 years if the annual interest rate is 6%. Use the *MATHEMATICA* help facility to find out how to set up the necessary vector-matrix equation.

BASIC MATRIX OPERATIONS

In this section, we discuss the *sum $A + B$* of two matrices, the *scalar multiples sA* of a scalar and a matrix, the *product AB* of two matrices, and the *transpose A^T* of a matrix.

Matrix Addition

One of most basic operation on matrices of the same dimension is their addition.

DEFINITION 2.1   *If $A = (a_{ij})$ and $B = (b_{ij})$ are two $m \times n$ matrices, the **sum** $A + B$ of A and B is the matrix $(c_{ij}) = (a_{ij} + b_{ij})$.*

The operation of forming the sum of two matrices is called *matrix addition*. This definition also covers the *subtraction* of matrices since we can think of $A - B$ as $A + (-B)$, where $-B = (-b_{ij})$. The matrix $A - B$ is called the *difference* of A and B.

EXAMPLE 2.4 ■ **A Sum of Matrices**

Consider the two 3×4 matrices

$$A = \begin{pmatrix} 1 & 2 & 0 & 3 \\ 5 & 3 & 0 & 1 \\ 9 & 2 & 1 & 6 \end{pmatrix} \quad \text{and} \quad B = \begin{pmatrix} 1 & 2 & 3 & 4 \\ 2 & 6 & 7 & 8 \\ 9 & 10 & 11 & 12 \end{pmatrix}$$

The matrix

$$A + B = \begin{pmatrix} 2 & 4 & 3 & 7 \\ 7 & 9 & 7 & 9 \\ 18 & 12 & 12 & 18 \end{pmatrix}$$

is the sum of the matrices A and B. ◄►

Scalar Multiplication

Another basic operation on matrices is the multiplication of each entry of a matrix by the same scalar.

DEFINITION 2.2   *If $A = (a_{ij})$ is an $m \times n$ matrix and s is a scalar, then the **scalar multiple** sA of A is the matrix (sa_{ij}).*

The operation of forming a scalar multiple of a matrix is called *scalar multiplication*.

EXAMPLE 2.5 ■ **A Scalar Multiple of a Matrix**

Let $A = \begin{pmatrix} 1 & 2 & 0 & 3 \\ 5 & 3 & 0 & 1 \\ 9 & 2 & 1 & 6 \end{pmatrix}$ and $s = 7$. Then

$$sA = 7A = \begin{pmatrix} 7\times1 & 7\times2 & 7\times0 & 7\times3 \\ 7\times5 & 7\times3 & 7\times0 & 7\times1 \\ 7\times9 & 7\times2 & 7\times1 & 7\times6 \end{pmatrix} = \begin{pmatrix} 7 & 14 & 0 & 21 \\ 35 & 21 & 0 & 7 \\ 63 & 14 & 7 & 42 \end{pmatrix}$$

The matrix $7A$ is the scalar multiple of 7 and A. ◄►

Matrix Multiplication

We will see later that, from a functional point of view, the multiplication of matrices of compatible size is the most important operation. If $A = (a_{ij})$ is an $m \times n$ matrix and $B = (b_{jk})$ is an $n \times p$ matrix, we will define the **product** AB of A and B. This operation is based on the idea of a row-column product.

If $\mathbf{r}_i(A)$ and $\mathbf{c}_j(B)$ are the ith row A and the jth columns of B, then the **row-column product** of $\mathbf{r}_i(A)$ and $\mathbf{c}_j(B)$ is defined to be

$$\mathbf{r}_i(A)\,\mathbf{c}_j(B) = \begin{pmatrix} a_{i1} & \cdots & a_{in} \end{pmatrix} \begin{pmatrix} b_{1j} \\ \vdots \\ b_{nj} \end{pmatrix} = a_{i1}b_{1j} + \cdots + a_{in}b_{nj}$$

EXAMPLE 2.6 ■ A Row-Column Product

Suppose that $\mathbf{x} = (a\ b\ c)$ is a row vector of length 3, and that

$$\mathbf{y} = \begin{pmatrix} x \\ y \\ z \end{pmatrix}$$

is a column vector of height 3. Then the product of $\mathbf{x}$ and $\mathbf{y}$ is

$$\mathbf{xy} = \begin{pmatrix} a & b & c \end{pmatrix} \begin{pmatrix} x \\ y \\ z \end{pmatrix} = ax + by + cz$$

If a, b, c, x, y, z are constants, then $\mathbf{xy}$ is also a constant. ◂▸

DEFINITION 2.3   *The **product** AB of an $m \times n$ matrix A and $n \times p$ matrix B is the $m \times p$ matrix*

$$\begin{pmatrix} \mathbf{r}_1(A)\,\mathbf{c}_1(B) & \cdots & \mathbf{r}_1(A)\,\mathbf{c}_p(B) \\ & \vdots & \\ \mathbf{r}_m(A)\,\mathbf{c}_1(B) & \cdots & \mathbf{r}_m(A)\,\mathbf{c}_p(B) \end{pmatrix}$$

whose ijth entry is row-column product $\mathbf{r}_i(A)\,\mathbf{c}_j(B)$.

The operation of forming the product of two matrices is called **matrix multiplication**.

EXAMPLE 2.7 ■ A Product of Matrices

Calculate the matrix products AB and BA for

$$A = \begin{pmatrix} 1 & 2 \\ 3 & 4 \end{pmatrix} \quad \text{and} \quad B = \begin{pmatrix} a & b \\ c & d \end{pmatrix}$$

Solution. We begin by calculating AB. By definition,

$$AB = \begin{pmatrix} \mathbf{r}_1(A) \\ \mathbf{r}_2(A) \end{pmatrix} \begin{pmatrix} \mathbf{c}_1(B) & \mathbf{c}_2(B) \end{pmatrix} = \begin{pmatrix} \mathbf{r}_1(A)\mathbf{c}_1(B) & \mathbf{r}_1(A)\mathbf{c}_2(B) \\ \mathbf{r}_2(A)\mathbf{c}_1(B) & \mathbf{r}_2(A)\mathbf{c}_2(B) \end{pmatrix}$$

Therefore

$$AB = \begin{pmatrix} \begin{pmatrix} 1 & 2 \end{pmatrix} \begin{pmatrix} a \\ c \end{pmatrix} & \begin{pmatrix} 1 & 2 \end{pmatrix} \begin{pmatrix} b \\ d \end{pmatrix} \\ \begin{pmatrix} 3 & 4 \end{pmatrix} \begin{pmatrix} a \\ c \end{pmatrix} & \begin{pmatrix} 3 & 4 \end{pmatrix} \begin{pmatrix} b \\ d \end{pmatrix} \end{pmatrix}$$

$$= \begin{pmatrix} a + 2c & b + 2d \\ 3a + 4c & 3b + 4d \end{pmatrix}$$

Next we calculate BA. By definition,

$$BA = \begin{pmatrix} \mathbf{r}_1(B) \\ \mathbf{r}_2(B) \end{pmatrix} \begin{pmatrix} \mathbf{c}_1(A) & \mathbf{c}_2(A) \end{pmatrix}$$

Therefore

$$BA = \begin{pmatrix} \begin{pmatrix} a & b \end{pmatrix} \begin{pmatrix} 1 \\ 3 \end{pmatrix} & \begin{pmatrix} a & b \end{pmatrix} \begin{pmatrix} 2 \\ 4 \end{pmatrix} \\ \begin{pmatrix} c & d \end{pmatrix} \begin{pmatrix} 1 \\ 3 \end{pmatrix} & \begin{pmatrix} c & d \end{pmatrix} \begin{pmatrix} 2 \\ 4 \end{pmatrix} \end{pmatrix}$$

$$= \begin{pmatrix} a + 3b & 2a + 4b \\ c + 3d & 2c + 4d \end{pmatrix}$$

The matrices AB and BA are equal if and only if

$$\begin{cases} a + 2c = a + 3b \\ b + 2d = 2a + 4b \\ 3a + 4c = c + 3d \\ 3b + 4d = 2c + 4d \end{cases}$$

It is easy to assign values to a, b, c, d for which $AB \neq BA$. ◄►

EXAMPLE 2.8 ■ Two Matrices for Which $AB = 0$ and $BA \neq 0$

Show that if

$$A = \begin{pmatrix} 0 & 1 \\ 0 & 2 \end{pmatrix} \quad \text{and} \quad B = \begin{pmatrix} 3 & 4 \\ 0 & 0 \end{pmatrix}$$

then $AB = 0$ and $BA \neq 0$.

Solution. By the definition of matrix products,

$$AB = \begin{pmatrix} 0 & 1 \\ 0 & 2 \end{pmatrix} \begin{pmatrix} 3 & 4 \\ 0 & 0 \end{pmatrix} = \begin{pmatrix} 0 & 0 \\ 0 & 0 \end{pmatrix}$$

and

$$BA = \begin{pmatrix} 3 & 4 \\ 0 & 0 \end{pmatrix} \begin{pmatrix} 0 & 1 \\ 0 & 2 \end{pmatrix} = \begin{pmatrix} 0 & 11 \\ 0 & 0 \end{pmatrix}$$

As we can see, $AB = 0$, but $BA \neq 0$. ◄►

EXAMPLE 2.9   ■ The Product of a 2 × 4 and a 4 × 2 Matrix

Use *MATHEMATICA* to find the products AB and BA, where

$$A = \begin{pmatrix} 1 & 2 & 0 & 3 \\ 5 & 3 & 0 & 1 \end{pmatrix} \quad \text{and} \quad B = \begin{pmatrix} 1 & 2 \\ 2 & 6 \\ 9 & 10 \\ 3 & 8 \end{pmatrix}$$

Solution. We use the **Create Matrix** palette to input A and B.

In[1]:= A= $\begin{pmatrix} 1 & 2 & 0 & 3 \\ 5 & 3 & 0 & 1 \end{pmatrix}$; B= $\begin{pmatrix} 1 & 2 \\ 2 & 6 \\ 9 & 10 \\ 3 & 8 \end{pmatrix}$;

In[2]:= **MatrixForm[A.B]**

Out[2]//MatrixForm=

$$\begin{pmatrix} 14 & 38 \\ 14 & 36 \end{pmatrix}$$

In[3]:= **MatrixForm[B.A]**

Out[3]//MatrixForm=

$$\begin{pmatrix} 11 & 8 & 0 & 5 \\ 32 & 22 & 0 & 12 \\ 59 & 48 & 0 & 37 \\ 43 & 30 & 0 & 17 \end{pmatrix}$$

As expected, the matrices AB and BA are not the same. ◄►

It is often helpful to express a product AB in terms of A and the columns of B.

THEOREM 2.1 (Matrix-column product theorem) *If A is an $m \times n$ matrix and if B is an $n \times p$ matrix whose columns are $c_1(B), \ldots, c_p(B)$, then $AB = (Ac_1(B) \quad \cdots \quad Ac_p(B))$.*

Proof. This result follows from the fact that

$$A\, c_i(B) = \begin{pmatrix} r_1(A) \\ \vdots \\ r_m(A) \end{pmatrix} c_i(B) = \begin{pmatrix} r_1(A)\, c_i(B) \\ \vdots \\ r_m(A)\, c_i(B) \end{pmatrix}$$

for all $1 \le i \le p$. ■

COROLLARY 2.2 *If A is an $m \times n$ matrix and*

$$b = \begin{pmatrix} b_1 \\ \vdots \\ b_n \end{pmatrix}$$

an $n \times 1$ column vector, then $Ab = b_1 c_1(A) + \cdots + b_n c_n(A)$.

Proof. Let $A = (a_{ij})$. Then it follows from Theorem 2.1 that

$$Ab = \begin{pmatrix} r_1(A) \\ \vdots \\ r_m(A) \end{pmatrix} \begin{pmatrix} b_1 \\ \vdots \\ b_n \end{pmatrix} = \begin{pmatrix} a_{11}b_1 + \cdots + a_{1n}b_n \\ \vdots \\ a_{m1}b_1 + \cdots + a_{mn}b_n \end{pmatrix}$$

and

$$\begin{pmatrix} a_{11}b_1 + \cdots + a_{1n}b_n \\ \vdots \\ a_{m1}b_1 + \cdots + a_{mn}b_n \end{pmatrix} = \begin{pmatrix} a_{11}b_1 \\ \vdots \\ a_{m1}b_1 \end{pmatrix} + \cdots + \begin{pmatrix} a_{1n}b_n \\ \vdots \\ a_{mn}b_n \end{pmatrix}$$

$$= b_1 c_1(A) + \cdots + b_n c_n(A)$$

This proves the corollary. ■

COROLLARY 2.3 *If $a = (a_1 \quad \cdots \quad a_n)$ is a $1 \times n$ row vector and $B = (b_{ij})$ is an $n \times p$ matrix, then $aB = a_1 r_1(B) + \cdots + a_n r_n(B)$.*

Proof. It follows immediately from the definition of matrix products that

$$aB = (a_1 \quad \cdots \quad a_n) \begin{pmatrix} r_1(B) \\ \vdots \\ r_n(B) \end{pmatrix} = a_1 r_1(B) + \cdots + a_n r_n(B)$$

This proves the corollary. ■

COROLLARY 2.4   *For all matrices A and compatible sums $a\mathbf{x} + b\mathbf{y}$ of vectors,*

$$A\,(a\mathbf{x} + b\mathbf{y}) = a\,A\mathbf{x} + b\,A\mathbf{y}$$

Proof. Suppose that $A = (\mathbf{a}_1 \;\cdots\; \mathbf{a}_n)$ is an $m \times n$ matrix with columns $\mathbf{a}_i$, and that

$$a\mathbf{x} + b\mathbf{y} = a\begin{pmatrix} x_1 \\ \vdots \\ x_n \end{pmatrix} + b\begin{pmatrix} y_1 \\ \vdots \\ y_n \end{pmatrix} = \begin{pmatrix} ax_1 + by_1 \\ \vdots \\ ax_n + by_n \end{pmatrix}$$

Theorem 2.1 and Corollary 2.2 imply that

$$A\,(a\mathbf{x} + b\mathbf{y}) = A\begin{pmatrix} ax_1 + by_1 \\ \vdots \\ ax_n + by_n \end{pmatrix}$$

$$= \begin{pmatrix} \mathbf{a}_1 & \cdots & \mathbf{a}_n \end{pmatrix}\begin{pmatrix} ax_1 + by_1 \\ \vdots \\ ax_n + by_n \end{pmatrix}$$

$$= (ax_1 + by_1)\,\mathbf{a}_1 + \cdots + (ax_n + by_n)\,\mathbf{a}_n$$

$$= (ax_1\mathbf{a}_1 + \cdots + ax_n\mathbf{a}_n) + (by_1\mathbf{a}_1 + \cdots + by_n\mathbf{a}_n)$$

$$= a\,(x_1\mathbf{a}_1 + \cdots + x_n\mathbf{a}_n) + b\,(y_1\mathbf{a}_1 + \cdots + y_n\mathbf{a}_n)$$

$$= a\,A\mathbf{x} + b\,A\mathbf{y}$$

This proves the corollary. ■

The following theorems describe basic properties of matrix products.

THEOREM 2.5 (Composition theorem)   *For all matrices A and B and column vectors $\mathbf{x}$ of compatible dimension, $A(B\mathbf{x}) = (AB)\mathbf{x}$.*

Proof. Suppose that A is an $m \times n$ matrix, that $B = \begin{pmatrix} \mathbf{b}_1 & \cdots & \mathbf{b}_p \end{pmatrix}$ is an $n \times p$ matrix, and that

$$\mathbf{x} = \begin{pmatrix} x_1 \\ \vdots \\ x_p \end{pmatrix}$$

is a $p \times 1$ column vector. Then it follows from Theorem 2.1 and Corollary 2.2 that

$$B\mathbf{x} = x_1\mathbf{b}_1 + \cdots + x_p\mathbf{b}_p$$

and

$$A\,(B\mathbf{x})\ = A\,\left(x_1\mathbf{b}_1 + \cdots + x_p\mathbf{b}_p\right)$$
$$= x_1 A\mathbf{b}_1 + \cdots + x_p A\mathbf{b}_p$$
$$= \left(\begin{array}{ccc} A\mathbf{b}_1 & \cdots & A\mathbf{b}_p \end{array}\right)\left(\begin{array}{c} x_1 \\ \vdots \\ x_p \end{array}\right)$$
$$= (AB)\mathbf{x}$$

This proves the theorem. ■

THEOREM 2.6   *For all matrices* A, B, C *of compatible dimension and all scalars* s,

1. $(AB)C = A\,(BC)$

2. $A(B + C) = AB + AC$

3. $(A + B)\,C = AC + BC$

4. $s(AB) = (sA)B = A\,(sB)$

5. $0A = 0$

6. $A0 = 0$

Proof. We show that $(AB)C = A(BC)$. Suppose that $C = \left(\begin{array}{ccc} \mathbf{c}_1 & \cdots & \mathbf{c}_p \end{array}\right)$. Then it follows from Theorem 2.1 and Theorem 2.5 that

$$A\,(BC) = A\left(B\left(\begin{array}{ccc} \mathbf{c}_1 & \cdots & \mathbf{c}_p \end{array}\right)\right)$$
$$= A\left(\begin{array}{ccc} B\mathbf{c}_1 & \cdots & B\mathbf{c}_p \end{array}\right)$$
$$= \left(\begin{array}{ccc} A\,(B\mathbf{c}_1) & \cdots & A\left(B\mathbf{c}_p\right) \end{array}\right)$$
$$= \left(\begin{array}{ccc} (AB)\mathbf{c}_1 & \cdots & (AB)\mathbf{c}_p \end{array}\right)$$
$$= (AB)C$$

We leave the proofs of statements 2 through 4 as exercises. ■

Although the intended meaning of the equation $A0 = 0$ in Theorem 2.6 is clear, the notation is ambiguous. The two zero matrices in the equation may in fact be of different dimension. Here is an example:

$$\left(\begin{array}{cccc} 1 & 0 & 7 & 3 \\ 6 & 8 & 5 & 8 \\ 1 & 9 & 5 & 3 \end{array}\right)\left(\begin{array}{cc} 0 & 0 \\ 0 & 0 \\ 0 & 0 \\ 0 & 0 \end{array}\right) = \left(\begin{array}{cc} 0 & 0 \\ 0 & 0 \\ 0 & 0 \end{array}\right)$$

We use Theorem 2.6 to show that the set of solutions of a ***homogeneous linear system*** has the following basic closure property.

THEOREM 2.7 (Homogeneous systems theorem)   *If* $\mathbf{y}$ *and* $\mathbf{z}$ *are solutions of a homogeneous system* $A\mathbf{x} = \mathbf{0}$ *and* a *and* b *are scalars, then* $a\mathbf{y} + b\mathbf{z}$ *is also a solution of* $A\mathbf{x} = \mathbf{0}$.

Proof. Suppose that $A\mathbf{y} = A\mathbf{z} = \mathbf{0}$, and let a and b be two arbitrary scalars. Then it follows from Theorem 2.6 that

$$A(a\mathbf{y} + b\mathbf{z}) = A(a\mathbf{y}) + A(b\mathbf{z}) = a(A\mathbf{y}) + b(A\mathbf{z}) = a\mathbf{0} + b\mathbf{0} = \mathbf{0}$$

Hence $a\mathbf{y} + b\mathbf{z}$ is a solution of $A\mathbf{x} = \mathbf{0}$. ∎

Matrix Transpose

With every matrix we can associate another matrix obtained by interchanging the rows and columns of the given matrix.

DEFINITION 2.4   *The **transpose** of an* $m \times n$ *matrix* $A = (a_{ij})$ *is the* $n \times m$ *matrix* $A^T = (b_{ji})$ *whose entries are linked to those of* A *by the equations* $b_{ji} = a_{ij}$ *of* A.

EXAMPLE 2.10   ■ The Transpose of a Matrix

Use *MATHEMATICA* to find the transpose of the matrix

$$A = \begin{pmatrix} 1 & 2 & 9 & 3 \\ 2 & 6 & 7 & 8 \end{pmatrix}$$

Solution. We use the **Transpose** function.

```
In[1]:= Transpose[A= ( 1 2 9 3
                        2 6 7 8 )]//MatrixForm

Out[1]//MatrixForm=
                        ( 1  2
                          2  6
                          9  7
                          3  8 )
```

The result is the transpose A^T of the matrix A. ◄►

From Example 2.10, we can see the motivation behind linking the entries of A^T to those of A in terms of the location indices i and j. If

$$A = \begin{pmatrix} 1 & 2 & 9 & 3 \\ 2 & 6 & 7 & 8 \end{pmatrix} = \begin{pmatrix} a_{11} & a_{12} & a_{13} & a_{14} \\ a_{21} & a_{22} & a_{23} & a_{24} \end{pmatrix}$$

then

$$A^T = \begin{pmatrix} 1 & 2 \\ 2 & 6 \\ 9 & 7 \\ 3 & 8 \end{pmatrix} = \begin{pmatrix} b_{11} & b_{12} \\ b_{21} & b_{22} \\ b_{31} & b_{32} \\ b_{41} & b_{42} \end{pmatrix} = \begin{pmatrix} a_{11} & a_{21} \\ a_{12} & a_{22} \\ a_{13} & a_{23} \\ a_{14} & a_{24} \end{pmatrix}$$

The claim that $b_{ij} = a_{ji}$ for all $1 \le i \le 4$ and $1 \le j \le 2$ establishes the necessary connection between A and A^T.

Matrix Multiplication and the Transpose

We frequently need to combine various matrix operations using the transpose. The following relationship between row-column products links matrix products and matrix transposes.

LEMMA 2.8   *For any $m \times n$ matrix $A = (a_{ij})$ and any $n \times p$ matrix $B = (b_{jk})$, the row-column products $\mathbf{r}_i(A)\mathbf{c}_j(B)$ and $\mathbf{r}_j(B^T)\mathbf{c}_i(A^T)$ are equal.*

Proof. By the definition of row-column products,

$$\mathbf{r}_i(A)\mathbf{c}_j(B) = (a_{i1} \cdots a_{in}) \begin{pmatrix} b_{1j} \\ \vdots \\ b_{nj} \end{pmatrix} = a_{i1}b_{1j} + \cdots + a_{in}b_{nj}$$

and

$$\mathbf{r}_j(B^T)\mathbf{c}_i(A^T) = \begin{pmatrix} b_{1j} & \cdots & b_{nj} \end{pmatrix} \begin{pmatrix} a_{i1} \\ \vdots \\ a_{in} \end{pmatrix} = b_{1j}a_{i1} + \cdots + b_{nj}a_{in}.$$

Therefore, $\mathbf{r}_i(A)\mathbf{c}_j(B) = \mathbf{r}_j(B^T)\mathbf{c}_i(A^T)$ since $a_{ik}b_{kj} = b_{kj}a_{ik}$ for all i, j, and k. ■

We illustrate Lemma 2.8 with an example.

EXAMPLE 2.11   ■ $\mathbf{r}_i(A)\mathbf{c}_j(B) = \mathbf{r}_j(B^T)\mathbf{c}_i(A^T)$

Show that $\mathbf{r}_1(A)\mathbf{c}_2(B) = \mathbf{r}_2(B^T)\mathbf{c}_1(A^T)$, where

$$A = \begin{pmatrix} a_{11} & a_{12} & a_{13} \\ a_{21} & a_{22} & a_{23} \end{pmatrix} \quad \text{and} \quad B = \begin{pmatrix} b_{11} & b_{12} \\ b_{21} & b_{22} \\ b_{31} & b_{32} \end{pmatrix}$$

Solution. By the definition of the transpose,

$$A^T = \begin{pmatrix} a_{11} & a_{21} \\ a_{12} & a_{22} \\ a_{13} & a_{23} \end{pmatrix} \quad \text{and} \quad B^T = \begin{pmatrix} b_{11} & b_{21} & b_{31} \\ b_{12} & b_{22} & b_{32} \end{pmatrix}$$

Therefore,

$$\mathbf{r}_1(A)\mathbf{c}_2(B) = \begin{pmatrix} a_{11} & a_{12} & a_{13} \end{pmatrix} \begin{pmatrix} b_{12} \\ b_{22} \\ b_{32} \end{pmatrix} = b_{12}a_{11} + b_{22}a_{12} + b_{32}a_{13}$$

and

$$\mathbf{r}_2(B^T)\mathbf{c}_1(A^T) = \begin{pmatrix} b_{12} & b_{22} & b_{32} \end{pmatrix} \begin{pmatrix} a_{11} \\ a_{12} \\ a_{13} \end{pmatrix} = b_{12}a_{11} + b_{22}a_{12} + b_{32}a_{13}$$

This shows that $\mathbf{r}_1(A)\mathbf{c}_2(B) = \mathbf{r}_2(B^T)\mathbf{c}_1(A^T)$. ◄►

The following theorem summarizes the interactions between sums, scalar multiples, products, and transposes.

THEOREM 2.9 (Transpose theorem)   *For any two square matrices A and B of compatible dimension,*

$$
\begin{array}{llcl}
1. & (A^T)^T & = & A \\
2. & (sA)^T & = & sA^T \\
3. & (A + B)^T & = & A^T + B^T \\
4. & (AB)^T & = & B^T A^T
\end{array}
$$

Proof. We prove the theorem in four steps.

1. Suppose $A = (a_{ij})$ is an $m \times n$ matrix. By definition, A^T is the $n \times m$ matrix (a'_{ji}) in which $a'_{ji} = a_{ij}$. Therefore, A^{TT} is an $m \times n$ matrix (a''_{ij}) in which $a''_{ji} = a_{ij}$. Hence $a''_{ij} = a_{ij}$ for all i and j. This means that $A^{TT} = A$.

2. For any scalar s, $(sA)^T = (sa_{ij})^T = (sb_{ji}) = s(b_{ji}) = s(A^T)$. Hence $(sA)^T = s(A^T)$.

3. Now suppose that B is another $m \times n$ matrix (b_{ij}), and that $B^T = (b'_{ji})$. Then

$$(A + B)^T = (a_{ij} + b_{ij})^T = (b'_{ji} + a'_{ji}) = (a'_{ji} + b'_{ji}) = A^T + B^T$$

4. It remains to prove that $(AB)^T = B^T A^T$. We simplify the notation by assuming that A has two rows and B has three columns.

$$A = \begin{pmatrix} \mathbf{r}_1(A) \\ \mathbf{r}_2(A) \end{pmatrix} \quad \text{and} \quad B = \begin{pmatrix} \mathbf{c}_1(B) & \mathbf{c}_2(B) & \mathbf{c}_3(B) \end{pmatrix}$$

By the definition of matrix products and transposes,

$$AB = \begin{pmatrix} \mathbf{r}_1(A)\mathbf{c}_1(B) & \mathbf{r}_1(A)\mathbf{c}_2(B) & \mathbf{r}_1(A)\mathbf{c}_3(B) \\ \mathbf{r}_2(A)\mathbf{c}_1(B) & \mathbf{r}_2(A)\mathbf{c}_2(B) & \mathbf{r}_2(A)\mathbf{c}_3(B) \end{pmatrix}$$

and

$$(AB)^T = \begin{pmatrix} \mathbf{r}_1(A)\mathbf{c}_1(B) & \mathbf{r}_2(A)\mathbf{c}_1(B) \\ \mathbf{r}_1(A)\mathbf{c}_2(B) & \mathbf{r}_2(A)\mathbf{c}_2(B) \\ \mathbf{r}_1(A)\mathbf{c}_3(B) & \mathbf{r}_2(A)\mathbf{c}_3(B) \end{pmatrix}$$

It therefore follows from Lemma 2.8 that

$$\begin{aligned}(AB)^T &= \begin{pmatrix} \mathbf{r}_1(B^T)\mathbf{c}_1(A^T) & \mathbf{r}_1(B^T)\mathbf{c}_2(A^T) \\ \mathbf{r}_2(B^T)\mathbf{c}_1(A^T) & \mathbf{r}_2(B^T)\mathbf{c}_2(A^T) \\ \mathbf{r}_3(B^T)\mathbf{c}_1(A^T) & \mathbf{r}_3(B^T)\mathbf{c}_2(A^T) \end{pmatrix} \\ &= \begin{pmatrix} \mathbf{r}_1(B^T) \\ \mathbf{r}_2(B^T) \\ \mathbf{r}_3(B^T) \end{pmatrix} \begin{pmatrix} \mathbf{c}_1(A^T) & \mathbf{c}_2(A^T) & \mathbf{c}_3(A^T) \end{pmatrix} \\ &= B^T A^T \end{aligned}$$

This proves the theorem. ∎

EXAMPLE 2.12 ■ Matrix Operations and the Transpose

Use *MATHEMATICA* to generate two random 3×3 matrices and use them to illustrate the previous theorem.

Solution. We use the **Array** and **Random** functions to generate two 3×3 matrices A and B.

```
In[1]:= A=Array[Random[Integer,{-5,5}]&,{3,3}]

Out[1]= {{-3,-1,-2},{-4,-4,3},{-2,0,1}}
```

```
In[2]:= B=Array[Random[Integer,{-5,5}]&,{3,3}]

Out[2]= {{4,-5,-1},{5,2,4},{0,0,-1}}
```

We begin by showing that $(A^T)^T = A$.

```
In[3]:= Transpose[Transpose[A]]

Out[3]= {{-3,-1,-2},{-4,-4,3},{-2,0,1}}
```

Next we show that $(sA)^T = s(A^T)$.

```
In[4]:= s Transpose[A]

Out[4]= {{-3 s,-4 s,-2 s},{-s,-4 s,0},{-2 s,3 s,s}}
```

```
In[5]:= Transpose[s A]

Out[5]= {{-3 s,-4 s,-2 s},{-s,-4 s,0},{-2 s,3 s,s}}
```

We now show that $(A + B)^T = A^T + B^T$.

```
In[6]:= Transpose[A+B]

Out[6]= {{1,1,-2},{-6,-2,0},{-3,7,0}}
```

```
In[7]:= Transpose[A]+Transpose[B]

Out[7]= {{1,1,-2},{-6,-2,0},{-3,7,0}}
```

We conclude by showing that $(AB)^T$ and $B^T A^T$.

```
In[8]:= Transpose[A.B]//MatrixForm

Out[8]//MatrixForm=
```

$$\begin{pmatrix} -17 & -36 & -8 \\ 13 & 12 & 10 \\ 1 & -15 & 1 \end{pmatrix}$$

```
In[9]:= Transpose[B].Transpose[A]//MatrixForm

Out[9]//MatrixForm=
```

$$\begin{pmatrix} -17 & -36 & -8 \\ 13 & 12 & 10 \\ 1 & -15 & 1 \end{pmatrix}$$

As we can see, all identities hold. ◄►

The Trace

The sum of the diagonal entries of a square matrix often gives us useful information about the properties the matrix. For any $n \times n$ matrix $A = (a_{ij})$ with entries in a mathematical system $\mathbb{X}$ equipped with addition, the function

$$\text{trace } A = a_{11} + \cdots + a_{nn}$$

produces an element in the system $\mathbb{X}$.

DEFINITION 2.5   *The **trace** of an $n \times n$ matrix $A = (a_{ij})$ is the sum $a_{11} + \cdots + a_{nn}$ of the diagonal entries of A.*

EXAMPLE 2.13   ■ The Trace of a Matrix

Suppose that

$$A = \begin{pmatrix} a_{11} & a_{12} & a_{13} \\ a_{21} & a_{22} & a_{23} \\ a_{31} & a_{32} & a_{33} \end{pmatrix} = \begin{pmatrix} 1 & 2 & 3 \\ 5 & 6 & 7 \\ 9 & 10 & 11 \end{pmatrix}$$

Then trace $A = a_{11} + a_{22} + a_{33} = 1 + 6 + 11 = 18.$ ◄►

Next we describe the interaction between trace and basic matrix operations.

THEOREM 2.10 (Trace theorem)   *If A and B are two $n \times n$ matrices, then*

1. $\text{trace } A^T = \text{trace } A$
2. $\text{trace } sA = s \text{ trace } A$
3. $\text{trace}(A + B) = \text{trace } A + \text{trace } B$
4. $\text{trace } AB = \text{trace } BA$

Proof. Suppose that $A = (a_{ij})$ and $B = (b_{ij})$ are two $n \times n$ matrices. Since A and A^T have identical diagonals, it is clear that they have the same trace. Moreover, the trace of sA is s times the trace of A since

$$\text{trace } sA = sa_{11} + \cdots + sa_{nn}$$
$$= s(a_{11} + \cdots + a_{nn})$$
$$= s \text{ trace } A$$

The fact that $\text{trace}(A + B) = \text{trace } A + \text{trace } B$ is also immediate since

$$\text{trace } A + \text{trace } B = (a_{11} + \cdots + a_{nn}) + (b_{11} + \cdots + b_{nn})$$
$$= (a_{11} + b_{11}) + \cdots + (b_{nn} + a_{nn})$$
$$= \text{trace}(A + B)$$

It remains for us to consider the case of matrix products. By the definition of the trace and Lemma 2.8,

$$\text{trace } AB = \mathbf{r}_1(A)\mathbf{c}_1(B) + \cdots + \mathbf{r}_n(A)\mathbf{c}_n(B)$$
$$= \mathbf{r}_1(B)\mathbf{c}_1(A) + \cdots + \mathbf{r}_n(B)\mathbf{c}_n(A)$$
$$= \text{trace } BA$$

This proves the theorem. ∎

In Version 4.0 of *MATHEMATICA*, the function computing the trace of a matrix is **Tr**. In earlier versions, a corresponding function must be defined by the user. It should be noted that the *MATHEMATICA* function **Trace** serves a different purpose and does not compute the trace of a matrix.

EXAMPLE 2.14 ■ Computing the Trace of a Matrix

Use *MATHEMATICA* to compute the trace of a general 4×4 matrix $A = (a_{ij})$ and show that it is the trace of A.

Solution. Let **A** be the matrix defined as follows.

```
In[1]:= A=Table[a[i,j],{i,1,4},{j,1,4}];MatrixForm[A]

Out[1]//MatrixForm=

        ( a[1,1]   a[1,2]   a[1,3]   a[1,4] )
        ( a[2,1]   a[2,2]   a[2,3]   a[2,4] )
        ( a[3,1]   a[3,2]   a[3,3]   a[3,4] )
        ( a[4,1]   a[4,2]   a[4,3]   a[4,4] )
```

Then trace $A = a[1, 1] + a[2, 2] + a[3, 3] + a[4, 4]$. The next calculation shows that **Tr** produces this value.

```
In[2]:= Tr[A]

Out[2]= a[1,1]+a[2,2]+a[3,3]+a[4,4]
```

As expected, *MATHEMATICA* outputs the trace of A. ◄►

EXERCISES 2.2

1. Compute the sums $A + B$ and $B + A$ of the matrices

$$A = \begin{pmatrix} 1 & 2 & 3 & 4 & 5 \\ 2 & -1 & 0 & 5 & 6 \\ -6 & 4 & 3 & 1 & 2 \end{pmatrix} \quad B = \begin{pmatrix} 2 & 5 & 6 & -2 & 0 \\ 2 & 2 & 7 & 4 & 3 \\ 2 & 1 & 8 & -8 & 6 \end{pmatrix}$$

and explain why $A + B = B + A$.

2. Compute the matrices $3A - 5B$ and $-(5B - 3A)$ determined by the matrices A and B in Exercise 1 and explain why the two matrices are equal.

3. Write out the rows $\mathbf{r}_i(A)$ and the columns $\mathbf{c}_j(B)$ of the matrices

$$A = \begin{pmatrix} 1 & 2 & 3 & 4 & 5 \\ 2 & 1 & 0 & 5 & 6 \\ 6 & 4 & 3 & 1 & 2 \end{pmatrix} \quad B = \begin{pmatrix} 2 & 5 & 6 \\ 2 & 2 & 7 \\ 2 & 1 & 8 \\ 3 & 8 & 2 \\ 3 & 9 & 4 \end{pmatrix}$$

Use the row-column products $\mathbf{r}_i\mathbf{c}_j = \mathbf{r}_i(A)\mathbf{c}_j(B)$ to compute AB in the form of

$$\begin{pmatrix} \mathbf{r}_1\mathbf{c}_1 & \mathbf{r}_1\mathbf{c}_2 & \mathbf{r}_1\mathbf{c}_3 \\ \mathbf{r}_2\mathbf{c}_1 & \mathbf{r}_2\mathbf{c}_2 & \mathbf{r}_2\mathbf{c}_3 \\ \mathbf{r}_3\mathbf{c}_1 & \mathbf{r}_3\mathbf{c}_2 & \mathbf{r}_3\mathbf{c}_3 \end{pmatrix}$$

4. Compute the product BA of the matrices A and B in Exercise 3. Explain why $AB \neq BA$.

5. Try to use *MATHEMATICA* to find the matrix product AB of a 3×5 matrix A and a 4×3 matrix B. Explain the resulting error message "Tensors have incompatible shapes."

6. Try to form the matrix product BA using the matrices in Example 5. Explain the result.

7. Find two nonzero 3×3 matrices A and B for which $AB = BA$, and two other nonzero matrices A and B for which $AB \neq BA$.

8. Use the **Array** function to generate a 3×5 matrix A and a 5×3 matrix B. Then use *MATHEMATICA* to find the products AB and BA.

9. Suppose that

$$A = \begin{pmatrix} 1 & 2 & 3 \\ 2 & 1 & 0 \\ 6 & 4 & 3 \end{pmatrix} \quad B = \begin{pmatrix} 3 & 4 & 5 \\ 0 & 5 & 6 \\ 3 & 1 & 2 \end{pmatrix} \quad \mathbf{x} = \begin{pmatrix} x \\ y \\ y \end{pmatrix}$$

Compute $B\mathbf{x}$, $A(B\mathbf{x})$, AB, and $(AB)\mathbf{x}$. Explain why $A(B\mathbf{x}) = (AB)\mathbf{x}$.

10. Use the matrices A and B in Exercise 9 and compute $(AB)^T$, A^T, and B^T. Explain why $B^T A^T = (AB)^T$.

11. Suppose that

$$A = \begin{pmatrix} 1 & 2 & 3 \\ 2 & 1 & 0 \end{pmatrix} \quad \mathbf{x} = \begin{pmatrix} 3 \\ 0 \\ 3 \end{pmatrix} \quad \mathbf{y} = \begin{pmatrix} 5 \\ 6 \\ 2 \end{pmatrix}$$

Use *MATHEMATICA* to show that $A(a\mathbf{x} + b\mathbf{y}) = aA\mathbf{x} + bA\mathbf{y}$ for all a, b.

12. Suppose that

$$A = \begin{pmatrix} 1 & 2 & 3 \\ 2 & 1 & 0 \\ 6 & 4 & 3 \end{pmatrix} \quad \mathbf{x} = \begin{pmatrix} x \\ y \\ z \end{pmatrix}$$

Show that

$$A\mathbf{x} = x \begin{pmatrix} 1 \\ 2 \\ 6 \end{pmatrix} + y \begin{pmatrix} 2 \\ 1 \\ 4 \end{pmatrix} + z \begin{pmatrix} 3 \\ 0 \\ 3 \end{pmatrix}$$

13. Show that if $\mathbf{0}$ is the $m \times n$ zero matrix, then $A + \mathbf{0} = A$ for all $m \times n$ matrices A.

14. Show that for all matrices A and B of compatible dimension, $A + B = B + A$.

15. Show that for all matrices A and B of compatible dimension and all scalars s, $s(A + B) = sA + sB$.

16. Show that for all matrices A and all scalars r and s, $(r + s)A = rA + sA$.

17. Show that for all matrices A and all scalars r and s, $(rs)A = r(sA)$.

18. Show that for all matrices A, B, C of compatible dimension, $(A + B) + C = A + (B + C)$.

19. Show that if A, B, C are matrices of compatible dimension, then $A(B + C) = AB + AC$.

20. Show that if A, B, C are matrices of compatible dimension, then $(A + B)C = AC + BC$.

21. Show that for all matrices A and B of compatible dimension and all scalars s, $s(AB) = (sA)B = A(sB)$.

22. Show that for any $m \times n$ matrix A, the matrix $0A$ is the $m \times n$ zero matrix.

23. Show that for any $m \times n$ matrix A and any $n \times p$ zero matrix $\mathbf{0}$, the matrix $A\mathbf{0}$ is the $m \times p$ zero matrix.

24. Use *MATHEMATICA* to compute the trace of a random real square matrix of dimensions 12×12.

25. Show that the function **MatrixTrace** defined by

```
MatrixTrace[A_]:=Sum[A[[i,i]],{i,Length[A]}]
```

computes the trace of the square matrix A.

26. Use the trace function **Tr** of *MATHEMATICA* to compute trace A, trace B, trace$(A + B)$, and trace AB, where

$$A = \begin{pmatrix} -91 & -47 & -61 \\ 41 & -58 & -90 \\ 53 & -1 & 94 \end{pmatrix} \quad B = \begin{pmatrix} 83 & -86 & 23 \\ -84 & 19 & -50 \\ 88 & -53 & 85 \end{pmatrix}$$

27. Show that, in general, trace $AB \neq$ trace A trace B.

A LEXICON OF MATRICES

In this section, we define and illustrate the basic types of matrices encountered in this text.

Square Matrices

A matrix $A : \mathbf{m} \times \mathbf{n} \to \mathbb{X}$ is *square* if $\mathbf{m} = \mathbf{n}$. All other matrices will be referred to as *rectangular* matrices.

EXAMPLE 2.15 ■ A Square Matrix with Entries in $\mathbb{R}$

Let $\mathbf{m} = \mathbf{n} = \{1, 2\}$ and $A[1, 1] = 1$, $A[1, 2] = 2$, $A[2, 1] = 5$, $A[2, 2] = \pi$. Then

$$A = \begin{pmatrix} a_{11} & a_{12} \\ a_{21} & a_{22} \end{pmatrix} = \begin{pmatrix} 1 & 2 \\ 5 & \pi \end{pmatrix}$$

is a square matrix with real entries. ◀▶

EXAMPLE 2.16 ■ A Square Matrix with Entries in $\mathbb{R}[t]$

Let $\mathbf{m} = \mathbf{n} = \{1, 2, 3\}$ and $A[1, 1] = 1$, $A[1, 2] = t^2$, $A[1, 3] = t + 3$, $A[2, 1] = 5t^2 - t + 1$, $A[2, 2] = -6$, $A[2, 3] = 7t$, $A[3, 1] = 9t^3$, $A[3, 2] = 10$, and $A[3, 3] = t - \pi$. Then

$$A = \begin{pmatrix} a_{11} & a_{12} & a_{13} \\ a_{21} & a_{22} & a_{23} \\ a_{31} & a_{32} & a_{33} \end{pmatrix} = \begin{pmatrix} 1 & t^2 & t + 3 \\ 5t^2 - t + 1 & -6 & 7t \\ 9t^3 & 10 & t - \pi \end{pmatrix}$$

is a square matrix with polynomial entries. ◀▶

Lower-Triangular Matrices

A square matrix $A : \mathbf{n} \times \mathbf{n} \to \mathbb{X}$ is *lower triangular* if $A[i, j] = 0$ for all $i < j$.

EXAMPLE 2.17 ■ **A Lower-Triangular Matrix with Entries in $\mathbb{R}$**

Let $\mathbf{n} = \{1, 2, 3\}$ and $A[1, 1] = 1$, $A[1, 2] = 0$, $A[1, 3] = 0$, $A[2, 1] = 5$, $A[2, 2] = 6$, $A[2, 3] = 0$, $A[3, 1] = 9$, $A[3, 2] = 10$, and $A[3, 3] = 11$. Then

$$A = \begin{pmatrix} a_{11} & a_{12} & a_{13} \\ a_{21} & a_{22} & a_{23} \\ a_{31} & a_{32} & a_{33} \end{pmatrix} = \begin{pmatrix} 1 & 0 & 0 \\ 5 & 6 & 0 \\ 9 & 10 & 11 \end{pmatrix}$$

is a lower-triangular matrix with real entries. ◄►

EXAMPLE 2.18 ■ **A Lower-Triangular Matrix with Entries in $\mathbb{R}[t]$**

Let $\mathbf{n} = \{1, 2\}$ and $A[1, 1] = t^2$, $A[1, 2] = 0$, $A[2, 1] = t$, $A[2, 2] = t^3 - 6$. Then

$$A = \begin{pmatrix} a_{11} & a_{12} \\ a_{21} & a_{22} \end{pmatrix} = \begin{pmatrix} t^2 & 0 \\ t & t^3 - 6 \end{pmatrix}$$

is a lower-triangular matrix with polynomial entries. ◄►

Upper-Triangular Matrices

A square matrix $A : \mathbf{n} \times \mathbf{n} \to \mathbb{X}$ is *upper triangular* if $A[i, j] = 0$ for all $i > j$.

EXAMPLE 2.19 ■ **An Upper-Triangular Matrix with Entries in $\mathbb{R}$**

Let $\mathbf{n} = \{1, 2\}$ and $A[1, 1] = 1$, $A[1, 2] = 5$, $A[2, 1] = 0$, $A[2, 2] = 6$. Then

$$A = \begin{pmatrix} a_{11} & a_{12} \\ a_{21} & a_{22} \end{pmatrix} = \begin{pmatrix} 1 & 5 \\ 0 & 6 \end{pmatrix}$$

is an upper-triangular matrix with real entries. ◄►

EXAMPLE 2.20 ■ **An Upper-Triangular Matrix with Entries in $\mathbb{R}[t]$**

Let $\mathbf{n} = \{1, 2, 3\}$ and $A[1, 1] = 3t^5 - 77$, $A[1, 2] = 5t^3 + t^2 - 1$, $A[1, 3] = 9t^{33}$, $A[2, 1] = 0$, $A[2, 2] = -t^3 + 6t + 44$, $A[2, 3] = 10$, $A[3, 1] = 0$, $A[3, 2] = 0$, and $A[3, 3] = t - 11$. Then

$$A = \begin{pmatrix} a_{11} & a_{12} & a_{13} \\ a_{21} & a_{22} & a_{23} \\ a_{31} & a_{32} & a_{33} \end{pmatrix} = \begin{pmatrix} 3t^5 - 77 & 5t^3 + t^2 - 1 & 9t^{33} \\ 0 & -t^3 + 6t + 44 & 10 \\ 0 & 0 & t - 11 \end{pmatrix}$$

is an upper-triangular matrix with polynomial entries. ◄►

Diagonal Matrices

The sequence $(a_{11}, a_{22}, \ldots, a_{nn})$ of entries of an $n \times n$ matrix $A = (a_{ij})$ is the **diagonal** of A. An entry on the diagonal is called a **diagonal entry** of A. An $n \times n$ matrix (a_{ij}) is **diagonal** if $a_{ij} = 0$ for all $i \neq j$. We sometimes write a diagonal matrix $A = (a_{ij})$ as $\operatorname{diag}(a_{11}, \ldots, a_{nn})$.

EXAMPLE 2.21 ■ **A Diagonal Matrix with Entries in** $\mathbb{R}$

The matrix

$$\begin{pmatrix} a_{11} & a_{12} & a_{13} & a_{14} \\ a_{21} & a_{22} & a_{23} & a_{24} \\ a_{31} & a_{32} & a_{33} & a_{34} \\ a_{41} & a_{42} & a_{43} & a_{44} \end{pmatrix} = \begin{pmatrix} 6 & 0 & 0 & 0 \\ 0 & 8 & 0 & 0 \\ 0 & 0 & 3 & 0 \\ 0 & 0 & 0 & -12 \end{pmatrix}$$

is a diagonal matrix with diagonal $(a_{11}, a_{22}, a_{33}, a_{44}) = (6, 8, 3, -12)$. ◄►

Symmetric Matrices

A matrix A is **symmetric** if $A = A^T$.

EXAMPLE 2.22 ■ **A Symmetric Matrix**

The matrix

$$\begin{pmatrix} a_{11} & a_{12} & a_{13} \\ a_{21} & a_{22} & a_{23} \\ a_{31} & a_{32} & a_{33} \end{pmatrix} = \begin{pmatrix} 6 & 5 & -1 \\ 5 & 8 & \pi \\ -1 & \pi & 3 \end{pmatrix}$$

is a symmetric matrix with entries in $\mathbb{R}$. ◄►

Identity Matrices

A diagonal matrix $\operatorname{diag}(1, \ldots, 1)$ whose diagonal entries are all 1 is called an **identity matrix**. We write I_n for the $n \times n$ identity matrix. If n is clear from the context, we use I in place of I_n.

EXAMPLE 2.23 ■ **The 4 × 4 Identity Matrix**

Use *MATHEMATICA* to generate the 4×4 identity matrix I_4.

Solution. We can use the built-in **IdentityMatrix** function.

$$\text{In[1]}:=\ \text{I4}=\begin{pmatrix} 1 & 0 & 0 & 0 \\ 0 & 1 & 0 & 0 \\ 0 & 0 & 1 & 0 \\ 0 & 0 & 0 & 1 \end{pmatrix}$$

Out[1]= {{1,0,0,0},{0,1,0,0},{0,0,1,0},{0,0,0,1}}

As we can see, $a_{11} = a_{22} = a_{33} = a_{44} = 1$, and all other entries are 0. ◀▶

It is clear from the definition that the $n \times n$ identity matrix I_n is in reduced row echelon form and that each of its columns is a pivot column. It therefore has n pivots since each of its diagonal entries is a pivot.

Elementary Matrices

An $n \times n$ **elementary matrix** is a matrix E obtained from the identity matrix I_n by one elementary row operation. It is clear that identity matrices are elementary since the operations $R_i \rightarrow 1R_i$ preserve identity matrices. Gaussian and Gauss-Jordan elimination can be described in terms of left-multiplication by elementary matrices.

EXAMPLE 2.24 ■ Elementary Matrices

Use *MATHEMATICA* to compute the elementary matrices corresponding to the elementary row operation $(R_1 \rightleftharpoons R_3)$, $(R_2 \rightarrow -6R_2)$, $(R_1 \rightarrow R_1 + -3R_2)$.

Solution. First we input the identity matrix I_4.

In[1]:= Id4=IdentityMatrix[4];

Next we create the matrix E_1 by applying the operation $(R_1 \rightleftharpoons R_3)$ to I_4.

In[2]:= MatrixForm[E1=Id4[[{3,2,1,4}]]]

Out[2]//MatrixForm=
$$\begin{pmatrix} 0 & 0 & 1 & 0 \\ 0 & 1 & 0 & 0 \\ 1 & 0 & 0 & 0 \\ 0 & 0 & 0 & 1 \end{pmatrix}$$

Now we reset the value of **Id4** and create the matrix E_2 corresponding to the operation $(R_2 \rightarrow -6R_2)$.

```
In[3]:= M=Id4=IdentityMatrix[4];
```

```
In[4]:= M[[2]]=-6M[[2]];M
Out[4]= {{1,0,0,0},{0,-6,0,0},{0,0,1,0},{0,0,0,1}}
```

```
In[5]:= MatrixForm[E2=M]
Out[5]//MatrixForm=
```
$$\begin{pmatrix} 1 & 0 & 0 & 0 \\ 0 & -6 & 0 & 0 \\ 0 & 0 & 1 & 0 \\ 0 & 0 & 0 & 1 \end{pmatrix}$$

We conclude by constructing the matrix E_3 obtained from I_4 by applying the operation $(R_1 \rightarrow R_1 + -3R_2)$.

```
In[6]:= P=IdentityMatrix[4]; P[[1]]=P[[1]]+(-3)P[[2]];P
Out[6]= {{1,-3,0,0},{0,1,0,0},{0,0,1,0},{0,0,0,1}}
```

```
In[7]:= MatrixForm[E3=P]
Out[7]//MatrixForm=
```
$$\begin{pmatrix} 1 & -3 & 0 & 0 \\ 0 & 1 & 0 & 0 \\ 0 & 0 & 1 & 0 \\ 0 & 0 & 0 & 1 \end{pmatrix}$$

It is easy to check that the matrices E_1, E_2, and E_3 have the specified properties. ◄►

Elementary Permutation Matrices

Among the elementary row operations are the row interchanges. An *elementary permutation matrix* is an elementary matrix obtained from an identity matrix by a single operation of the form $(R_i \rightleftharpoons R_j)$.

EXAMPLE 2.25 ■ **Elementary Permutation Matrices**

Use *MATHEMATICA* to create three 3×3 elementary permutation matrices.

Solution. We choose to create the matrices E_1, E_2, and E_3 corresponding to the elementary row operations $(R_1 \rightleftharpoons R_2)$, $(R_3 \rightleftharpoons R_1)$, and $(R_2 \rightleftharpoons R_3)$.

```
In[1]:= E1=IdentityMatrix[3];MatrixForm[E2=E1[[{2,1,3}]]]

Out[1]//MatrixForm=
```
$$\begin{pmatrix} 0 & 1 & 0 \\ 1 & 0 & 0 \\ 0 & 0 & 1 \end{pmatrix}$$

```
In[2]:= F1=IdentityMatrix[3];MatrixForm[F2=F1[[{3,2,1}]]]

Out[2]//MatrixForm=
```
$$\begin{pmatrix} 0 & 0 & 1 \\ 0 & 1 & 0 \\ 1 & 0 & 0 \end{pmatrix}$$

```
In[3]:= G1=IdentityMatrix[3];MatrixForm[G2=G1[[{1,3,2}]]]

Out[3]//MatrixForm=
```
$$\begin{pmatrix} 1 & 0 & 0 \\ 0 & 0 & 1 \\ 0 & 1 & 0 \end{pmatrix}$$

It is easy to see that the matrices $E_1 A$, $E_2 A$, and $E_3 A$ are the same as the matrices obtained from A by given permutations of rows. ◄►

The next example shows that *left-multiplication* by an elementary permutation matrix corresponds to the interchange of the corresponding rows of the matrix.

EXAMPLE 2.26 ■ **Left-Multiplication by an Elementary Permutation Matrix**

Use *MATHEMATICA* to illustrate the effect of a left-multiplication by an elementary permutation matrix.

Solution. The following calculation shows that left-multiplication by the elementary matrix

$$E_1 = \begin{pmatrix} 0 & 1 & 0 \\ 1 & 0 & 0 \\ 0 & 0 & 1 \end{pmatrix}$$

permutes the first two rows of the matrix A.

```
In[1]:= A= ( 1  2  3 )    E1= ( 0  1  0 )    MatrixForm[E1.A]
        ( 4  5  6 );       ( 1  0  0 );
        ( 7  8  9 )        ( 0  0  1 )

Out[1]//MatrixForm=
                          ( 4  5  6 )
                          ( 1  2  3 )
                          ( 7  8  9 )
```

As we can see, applying the elementary row operation $(R_1 \rightleftharpoons R_2)$ to the matrix A has the same effect as forming the $E_1 A$. ◄►

Right-multiplication by an elementary permutation matrix corresponds to an ***elementary column operation.***

EXAMPLE 2.27 ■ Right-Multiplication by an Elementary Permutation Matrix

Use *MATHEMATICA* to illustrate the effect of right-multiplication by an elementary matrix.

Solution. We let A and E_1 be the matrices in Example 2.26 and form the matrix product $A E_1$.

```
In[1]:= A= ( 1  2  3 )    E1= ( 0  1  0 )    MatrixForm[A.E1]
        ( 4  5  6 );       ( 1  0  0 );
        ( 7  8  9 )        ( 0  0  1 )

Out[1]//MatrixForm=
                          ( 2  1  3 )
                          ( 5  4  6 )
                          ( 8  7  9 )
```

As we can see, $A E_1$ results from A by the interchange of the first two columns. ◄►

We summarize the connection between elementary row operations and elementary matrices as a theorem.

THEOREM 2.11 *If the matrix B is the result of an elementary row operation applied to a matrix A, then there exists an elementary matrix E for which $B = EA$.*

Proof. We must consider three cases.

1. If A $(R_i \rightleftharpoons R_j)$ B, we let E be the matrix obtained by I $(R_i \rightleftharpoons R_j)$ E.
2. If A $(R_i \rightarrow sR_i)$ B, we let E be the matrix obtained by I $(R_i \rightarrow sR_i)$ E.
3. If A $(R_i \rightarrow R_i + sR_j)$ B, we let E be the matrix obtained by I $(R_i \rightarrow R_i + sR_j)$ E.

In each case, it follows from the definition of matrix multiplication that $EA = B$. ■

COROLLARY 2.12 *A matrix A is row equivalent to a matrix B if and only if there exist elementary matrices $E_1, \ldots, E_p$ such that $A = E_1 \cdots E_p B$.*

Proof. This proof is left as an exercise. ■

Permutation Matrices

A *permutation matrix P* is a product $E_1 \cdots E_n$ of elementary permutation matrices.

EXAMPLE 2.28   ■ A Permutation Matrix

Find two elementary permutation matrices E_1 and E_2 whose product $P = E_2 E_1$ converts the matrix

$$A = \begin{pmatrix} a & b & c & d \\ e & f & g & h \\ i & j & k & l \end{pmatrix}$$

to the matrix

$$PA = \begin{pmatrix} i & j & k & l \\ a & b & c & d \\ e & f & g & h \end{pmatrix}$$

Solution. We can convert A to PA by first applying the operation $(R_1 \rightleftharpoons R_3)$ and then the operation $(R_2 \rightleftharpoons R_3)$. We know from Example 2.25 that the elementary permutation matrices corresponding to these operations are

$$E_1 = \begin{pmatrix} 0 & 0 & 1 \\ 0 & 1 & 0 \\ 1 & 0 & 0 \end{pmatrix} \quad \text{and} \quad E_2 = \begin{pmatrix} 1 & 0 & 0 \\ 0 & 0 & 1 \\ 0 & 1 & 0 \end{pmatrix}$$

We therefore form the product

$$P = E_2 E_1 = \begin{pmatrix} 1 & 0 & 0 \\ 0 & 0 & 1 \\ 0 & 1 & 0 \end{pmatrix} \begin{pmatrix} 0 & 0 & 1 \\ 0 & 1 & 0 \\ 1 & 0 & 0 \end{pmatrix} = \begin{pmatrix} 0 & 0 & 1 \\ 1 & 0 & 0 \\ 0 & 1 & 0 \end{pmatrix}$$

and use *MATHEMATICA* to confirm that *PA* is the intended matrix.

In[1]:= MatrixForm$\left[\begin{pmatrix} 0 & 0 & 1 \\ 1 & 0 & 0 \\ 0 & 1 & 0 \end{pmatrix} \cdot \begin{pmatrix} a & b & c & d \\ e & f & g & h \\ i & j & k & l \end{pmatrix} \right]$

Out[1]//MatrixForm=
$$\begin{pmatrix} i & j & k & l \\ a & b & c & d \\ e & f & g & h \end{pmatrix}$$

As we can see, the matrix *P* represents the appropriate permutation. ◄►

Generalized Diagonal Matrices

In Chapter 9, we will study the decomposition of rectangular matrices into products of simpler matrices. To be able to describe this decomposition elegantly, we extend the idea of a diagonal matrix to the rectangular case. We say that an $m \times n$ matrix $A = (a_{ij})$ is **generalized diagonal** if $a_{ij} = 0$ for all $i \neq j$. If no confusion arises, we refer to a generalized diagonal matrix simply as a diagonal matrix.

EXAMPLE 2.29   ■ A 2 × 3 Generalized Diagonal Matrix

The matrix

$$A = \begin{pmatrix} a_{11} & a_{12} & a_{13} \\ a_{21} & a_{22} & a_{23} \end{pmatrix} = \begin{pmatrix} -1 & 0 & 0 \\ 0 & 4 & 0 \end{pmatrix}$$

is diagonal since $a_{12} = a_{13} = a_{21} = a_{23} = 0$. ◄►

EXAMPLE 2.30   ■ A 3 × 2 Generalized Diagonal Matrix

The matrix

$$A = \begin{pmatrix} a_{11} & a_{12} \\ a_{21} & a_{22} \\ a_{31} & a_{32} \end{pmatrix} = \begin{pmatrix} -1 & 0 \\ 0 & 4 \\ 0 & 0 \end{pmatrix}$$

is diagonal since $a_{12} = a_{21} = a_{31} = a_{32} = 0$. ◄►

A square diagonal matrix is obviously a generalized diagonal matrix and so is every rectangular zero matrix.

EXAMPLE 2.31 ■ A 2 × 4 Diagonal Zero Matrix

The matrix

$$A = \begin{pmatrix} a_{11} & a_{12} & a_{13} & a_{14} \\ a_{21} & a_{22} & a_{23} & a_{24} \end{pmatrix} = \begin{pmatrix} 0 & 0 & 0 & 0 \\ 0 & 0 & 0 & 0 \end{pmatrix}$$

is diagonal since $a_{12} = a_{13} = a_{14} = a_{21} = a_{23} = a_{24} = 0$. ◄►

Submatrices

A matrix obtained from a matrix A by deleting some of its rows and columns is called a *submatrix* of A. In this text, submatrices are required, for example, to define and compute determinants, to define and compute eigenvalues, and to define and compute pseudoinverses of rectangular matrices.

MATHEMATICA has two built-in operations for extracting submatrices. The command

TakeMatrix[A, {p,q}, {r,s}]

extracts from the matrix $A = (a_{ij})$, the matrix consisting of all entries determined by a_{pq} and a_{rs}. The command

SubMatrix[A, {p,q}, {r,s}]

on the other hand, extracts from A the $r \times s$ matrix whose top left-hand corner is a_{pq}.

EXAMPLE 2.32 ■ A Submatrix Determined by Two Entries

Use *MATHEMATICA* to extract from the matrix

$$A = \begin{pmatrix} 0 & 0 & 1 & 1 & 5 & 3 & 8 \\ 5 & 7 & 6 & 5 & 7 & 1 & 9 \\ 3 & 9 & 2 & 0 & 7 & 8 & 0 \\ 8 & 5 & 3 & 8 & 2 & 2 & 3 \end{pmatrix}$$

the submatrix B determined by a_{23} and a_{44}.

Solution. We first load the **MatrixManipulation** package and use the **TakeMatrix** function.

```
In[1]:= <<LinearAlgebra`MatrixManipulation`
```

Next we use the **Create Matrix** menu to input the matrix A.

$$\text{In[2]:=} \quad \mathbf{A} = \begin{pmatrix} 0 & 0 & 1 & 1 & 5 & 3 & 8 \\ 5 & 7 & 6 & 5 & 7 & 1 & 9 \\ 3 & 9 & 2 & 0 & 7 & 8 & 0 \\ 8 & 5 & 3 & 8 & 2 & 2 & 3 \end{pmatrix}$$

Out[2]= {{0,0,1,1,5,3,8},{5,7,6,5,7,1,9},

{3,9,2,0,7,8,0},{8,5,3,8,2,2,3}}

We now extract the matrix B. In the **TakeMatrix[A,{2,3},{4,4}]** command, we specify the addresses of the top left-hand entry $A[2, 3]$ and the bottom right-hand entry $A[4, 4]$ of B.

In[3]:= **MatrixForm[B=TakeMatrix[A,{2,3},{4,4}]]**

Out[3]//MatrixForm=

$$\begin{pmatrix} 6 & 5 \\ 2 & 0 \\ 3 & 8 \end{pmatrix}$$

The matrix B is the 2×3 matrix consisting of all entries between a_{23} and a_{44}. ◀▶

EXAMPLE 2.33 ■ A Submatrix Determined by an Entry and a Dimension

Illustrate the extraction of the submatrix B in Example 2.32 by replacing the **TakeMatrix** function by the **SubMatrix** function.

Solution. The top left-hand entry of the matrix B is a_{23} and the matrix has three rows and two columns. In the **SubMatrix[A,{2,3},{3,2}]** command, we specify the address of the top left-hand entry $A[2, 3]$ of A and indicate the number of rows and columns {3, 2} of B. The calculation

In[1]:= **B=SubMatrix[A,{2,3},{3,2}];MatrixForm[B]**

Out[1]//MatrixForm=

$$\begin{pmatrix} 6 & 5 \\ 2 & 0 \\ 3 & 8 \end{pmatrix}$$

yields the matrix B in Example 2.32. ◀▶

EXERCISES 2.3

1. Determine which of the following matrices are upper triangular, lower triangular, diagonal, or symmetric. Justify your answers by appealing to the definitions of the matrix properties involved. Keep in mind that a matrix may have several or none of these properties.

a.
$$\begin{pmatrix} 1 & 0 & -2 & 6 \\ 0 & 5 & 7 & 1 \\ 0 & 0 & 0 & 1 \\ 0 & 0 & 0 & 1 \end{pmatrix}$$
b.
$$\begin{pmatrix} 0 & 0 & 0 & 0 \\ 0 & 0 & 0 & 0 \\ 0 & 0 & 0 & 0 \\ 0 & 0 & 0 & 0 \end{pmatrix}$$
c.
$$\begin{pmatrix} 45 & -8 & -93 \\ 8 & -62 & 77 \\ 93 & -77 & 99 \end{pmatrix}$$

d.
$$\begin{pmatrix} 1 & 0 & 0 & 0 \\ 0 & 1 & 0 & 0 \\ 0 & 0 & 1 & 0 \\ 0 & 0 & 0 & 1 \end{pmatrix}$$
e.
$$\begin{pmatrix} 1 & 0 & 0 & 0 \\ 0 & 3 & 0 & 0 \\ 0 & 0 & 1 & 0 \\ 0 & 0 & 0 & -11 \end{pmatrix}$$
f.
$$\begin{pmatrix} 1 & 0 & 0 & 9 \\ 0 & 1 & 7 & 0 \\ 0 & 7 & 1 & 0 \\ 9 & 0 & 0 & 1 \end{pmatrix}$$

2. Use *MATHEMATICA* to create 4×4 square matrices with integer entries, with real entries, and with polynomial entries.

3. Use *MATHEMATICA* to create lower-triangular 5×5 matrices with integer entries, with real entries, and with polynomial entries.

4. Use *MATHEMATICA* to create upper-triangular 5×5 matrices with integer entries, with real entries, and with polynomial entries.

5. Use *MATHEMATICA* to create 6×6 diagonal matrices with integer entries, with real entries, and with polynomial entries.

6. Prove that if A and B are two $n \times n$ diagonal matrices, then the matrix AB is diagonal.

7. Use *MATHEMATICA* to create symmetric 5×5 matrices with integer entries, with real entries, and with polynomial entries.

8. Use *MATHEMATICA* to create two real generalized diagonal 7×4 and 5×6 matrices with integer entries.

9. For any matrix $A = (a_{ij})$, the **TakeMatrix[$A, \{p, q\}, \{r, s\}$]** function of *MATHE-MATICA* in the **GaussianElimination** package produces the submatrix of A of all entries between a_{pq} and a_{rs}. The **SubMatrix[$A, \{p, q\}, \{r, s\}$]** function, on the other hand, produces the $r \times s$ submatrix of A starting at entry a_{qp}. Use the **TakeMatrix** and **SubMatrix** functions of *MATHEMATICA* to find three submatrices of the matrix

$$A = \begin{pmatrix} 3 & 0 & 1 & 7 & 1 \\ -1 & 2 & 0 & -1 & 1 \\ 0 & -2 & 21 & 1 & 1 \end{pmatrix}$$

and explain why the resulting matrices are submatrices.

10. Use *MATHEMATICA* to find two 2×2 matrices A and B for which $AB \neq BA$.

11. Use *MATHEMATICA* to find two 2×2 matrices A and B for which $AB = BA$.

12. Use *MATHEMATICA* to find two 3×3 matrices A and B for which $AB \neq BA$.

13. Use *MATHEMATICA* to find two 3×3 matrices A and B for which $AB = BA$.

14. Determine which of the following matrices are elementary matrices and explain why.

a. $\begin{pmatrix} 1 & 0 & 0 & 0 \\ 0 & 1 & 0 & 0 \\ 0 & 0 & 1 & 0 \\ 0 & 0 & 0 & 1 \end{pmatrix}$
b. $\begin{pmatrix} 0 & 0 & 1 & 0 \\ 0 & 1 & 0 & 0 \\ 1 & 0 & 0 & 0 \\ 0 & 0 & 0 & 1 \end{pmatrix}$
c. $\begin{pmatrix} 0 & 0 & 0 & 1 \\ 0 & 0 & 1 & 0 \\ 0 & 1 & 0 & 0 \\ 1 & 0 & 0 & 0 \end{pmatrix}$

15. Describe the effect of multiplying the matrix

$$A = \begin{pmatrix} 45 & -8 & -93 & 92 \\ 43 & -62 & 77 & 66 \\ 54 & -5 & 99 & -61 \\ -50 & -12 & -18 & 31 \end{pmatrix}$$

on the right by the matrices in Exercise 14. Repeat this exercise by multiplying A on the left by these matrices.

16. Use *MATHEMATICA* to generate a random 7×7 matrix and find its transpose.

17. Use *MATHEMATICA* to generate a random 6×6 matrix A and show that $A^T A$ and $A A^T$ are symmetric.

18. Find two nonzero 10×10 matrices A and B for which $A + B = 0$ and $AB \neq 0$.

19. Find two 3×3 matrices A and B for which $AB = 0$ and $BA \neq 0$.

20. Prove that a matrix A is row equivalent to a matrix B if and only if there exist elementary matrices $E_1, \ldots, E_p$ such that $A = E_1 \cdots E_p B$.

21. A square matrix A is **skew-symmetric** if $A = -A^T$. Use *MATHEMATICA* to verify that the matrix A

$$\begin{pmatrix} a_{11} & a_{12} & a_{13} & a_{14} \\ a_{21} & a_{22} & a_{23} & a_{24} \\ a_{31} & a_{31} & a_{33} & a_{34} \\ a_{41} & a_{42} & a_{43} & a_{44} \end{pmatrix} = \begin{pmatrix} 0 & -2 & 9 & -3 \\ 2 & 0 & -7 & -8 \\ -9 & 7 & 0 & 5 \\ 3 & 8 & -5 & 0 \end{pmatrix}$$

is skew-symmetric.

22. Prove that every square matrix is the sum of a symmetric and a skew-symmetric matrix.

23. Use *MATHEMATICA* to show that $AA^T = (AA^T)^T$ for any 3×4 matrix

$$A = \begin{pmatrix} a_{11} & a_{12} & a_{13} & a_{14} \\ a_{21} & a_{22} & a_{23} & a_{24} \\ a_{31} & a_{32} & a_{33} & a_{34} \end{pmatrix}$$

24. Show that the left-multiplication of a matrix by an elementary matrix corresponds to carrying out the corresponding elementary row operation on the matrix.

25. Show that the right-multiplication of a matrix by an elementary matrix corresponds to carrying out the corresponding elementary column operation on the matrix.

26. Find all 3×3 submatrices of the matrix

$$A = \begin{pmatrix} 19 & 12 & 17 & 13 \\ 25 & 12 & 20 & 12 \\ 17 & 20 & 12 & 18 \\ 24 & 24 & 22 & 13 \end{pmatrix}$$

27. Prove that every square row echelon matrix is upper triangular.

28. Prove that if A and B are two diagonal $n \times n$ matrices, then $AB = BA$.

29. Use the **Table** and **KroneckerDelta** functions to generate the 5×5 identity matrix.

INVERTIBLE MATRICES

Often we can solve a linear system $Ax = b$ by finding a unique matrix, denoted by A^{-1}, that has the property that $x = A^{-1}b$. We now study this possibility.

DEFINITION 2.6 *An $n \times n$ matrix A is **invertible** if there exists a matrix B such that* $AB = BA = I_n$.

If A and B are two $n \times n$ matrices and AB is the $n \times n$ identity matrix I_n, then A is called a *left inverse* of B and B is called a *right inverse* of A. We will show that left and right inverses are unique and that a square matrix has a left inverse if and only if it has a right inverse. We will also show that the left inverse, if it exists, equals the right inverse. Hence we call the matrix B for which $AB = BA = I_n$ *the inverse* of A. For algebraic reasons, we denote the inverse of a matrix A by A^{-1}. The operation of forming the inverse of a matrix is called *matrix inversion*. A square matrix that is not invertible is called *noninvertible* or *singular*. In this terminology, an invertible matrix is also called *nonsingular*.

At various stages in our work, we will give criteria for invertibility. For example, we will show that an $n \times n$ matrix is invertible if and only if it has n pivots. This is equivalent to showing that A is row equivalent to I_n. In Chapter 7, we will also show that the invertibility of a matrix A is equivalent to A preserving lengths and angles. Several other characterizations are given when they are needed. In Chapter 4, for example, we show that A is invertible if and only if the columns of A are linearly independent. In this section, we limit our discussion to showing that matrix inverses are unique and that the product of two invertible matrices is again an invertible matrix. We begin by proving two uniqueness properties of inverses.

THEOREM 2.13   *If B is a right inverse of A and C is a left inverse of A, then $B = C$.*

Proof. Suppose that $AB = CA = I$. Then $B = IB = (CA)B = C(AB) = CI = C$. ∎

THEOREM 2.14   *If B and C are inverses of A, then $B = C$.*

Proof. Since B and C are inverses, B is certainly a right inverse and C is a left inverse. Therefore $B = C$ by Theorem 2.13. ∎

One of the important properties of invertible matrices is their closure under multiplication.

THEOREM 2.15   *If A and B are invertible matrices, then so is AB.*

Proof. Suppose that A^{-1} and B^{-1} are the inverses of A and B. We show that $B^{-1}A^{-1}$ is both a left inverse and a right inverse of AB. Since $B^{-1}A^{-1}AB = B^{-1}B = I$, the matrix $B^{-1}A^{-1}$ is a left inverse. Moreover, since $ABB^{-1}A^{-1} = A^{-1}A = I$, it is also a right inverse. Hence it is an inverse. By Theorem 2.14, $B^{-1}A^{-1}$ is the only matrix with this property. ∎

Invertibility and Gaussian Elimination

Elementary row operations can be used to find the inverse A^{-1} of an invertible $n \times n$ matrix A. The procedure is based on Theorem 2.17. It consists of writing the matrix A and the identity matrix I_n side by side, and applying the same sequence of elementary row operations to A and to I_n, until A has been reduced to the identity matrix. This process transforms I_n to A^{-1}. The following example illustrates the procedure.

EXAMPLE 2.34   ■ **Elementary Row Operations and Matrix Inversion**

Use elementary row operations to find the inverse of the matrix

$$A = \begin{pmatrix} 1 & 2 \\ 3 & 4 \end{pmatrix}$$

Solution. We apply elementary row operations to the following pair of matrices:

$$[A, I_2] = \left[\begin{pmatrix} 1 & 2 \\ 3 & 4 \end{pmatrix}, \begin{pmatrix} 1 & 0 \\ 0 & 1 \end{pmatrix} \right]$$

$$\rightarrow \left[\begin{pmatrix} 1 & 2 \\ 0 & -2 \end{pmatrix}, \begin{pmatrix} 1 & 0 \\ -3 & 1 \end{pmatrix} \right]$$

$$\rightarrow \left[\begin{pmatrix} 1 & 0 \\ 0 & -2 \end{pmatrix}, \begin{pmatrix} -2 & 1 \\ -3 & 1 \end{pmatrix} \right]$$

$$\rightarrow \left[\begin{pmatrix} 1 & 0 \\ 0 & 1 \end{pmatrix}, \begin{pmatrix} -2 & 1 \\ \frac{3}{2} & -\frac{1}{2} \end{pmatrix} \right] = [I_2, B]$$

Let us use *MATHEMATICA* to confirm that the matrix B is the inverse of A.

```
In[1]:= A= ( 1  2
            3  4 ); B=Inverse[A];
```

```
In[2]:= A.B==IdentityMatrix[2]

Out[2]= True
```

```
In[3]:= B.A==IdentityMatrix[2]

Out[3]= True
```

Since $AB = BA = I_2$, the matrix B is the inverse of the matrix A. ◄►

Invertibility and Elementary Matrices

In this section, we show that elementary matrices are the building blocks of invertible matrices: Every invertible matrix is a product of elementary matrices. We begin by showing that elementary matrices are invertible.

THEOREM 2.16   *Every elementary matrix is invertible.*

Proof. Let E be an elementary matrix. To prove that E is invertible we consider three cases, depending on which type of elementary row operation was used to create E.

1. If E is the result of the operation $(R_i \rightleftharpoons R_j)$, then $E^{-1} = E$.

2. If E is the result of the operation $(R_i \rightarrow sR_i)$, then E^{-1} is the matrix obtained from I by the operation $(R_i \rightarrow \frac{1}{s}R_i)$.

3. If E is the result of the operation $(R_i \rightarrow R_i + sR_j)$, then E^{-1} is the matrix obtained from the identity matrix I by the operation $(R_i \rightarrow R_i - sR_j)$.

It is easy to see that the matrices defined in this way have the required properties. ∎

EXAMPLE 2.35 ■ **Inverses of Elementary Matrices**

We illustrate the steps involved in the previous proof by examining the case of 3×3 elementary matrices.

Solution. The following three cases illustrate how elementary matrices are inverted.

```
In[1]:= E1= ( 0 1 0 )      ; E1.E1==IdentityMatrix[3]
         ( 1 0 0 )
         ( 0 0 1 )

Out[1]= True
```

```
In[2]:= E2= ( 1 0 0 )      ; E3= ( 1 0 0 )      ;
         ( 0 3 0 )          ( 0 ⅓ 0 )
         ( 0 0 1 )          ( 0 0 1 )
       E2.E3==IdentityMatrix[3]

Out[2]= True
```

```
In[3]:= E4= ( 1 0 0 )      ; E5= ( 1 0 0 )      ;
         ( 0 1 0 )          ( 0 1 0 )
         (-5 0 1 )          ( 5 0 1 )
       E4.E5==IdentityMatrix[3]

Out[3]= True
```

We can see that in each case, the second matrix is the inverse of the first. ◄►

Example 2.35 suggests the following useful test for matrix invertibility.

THEOREM 2.17   *An $n \times n$ matrix A is invertible if and only if it is row equivalent to I_n.*

Proof. Suppose that A is invertible. Theorem 1.2 tells us that A is row equivalent to a matrix B in reduced row echelon form. By Corollary 2.12, there exist elementary matrices $E_1, \ldots, E_p$ for which $E_1 \cdots E_p A = B$. Since all of the matrices $E_1, \ldots, E_p$, and A are invertible and since, by Theorem 2.15, the product of invertible matrices is invertible, the matrix B is invertible. If B is in reduced row echelon form and is not I_n, then B must have a zero row. This contradicts the fact that B is invertible. Hence $B = I_n$. Conversely, suppose that A is row equivalent to I_n. Then we know from Corollary 2.12 that A is a product $E_1 \cdots E_p$ of elementary matrices. Since elementary matrices are invertible, it follows that A is a product of invertible matrices and is therefore invertible. ∎

COROLLARY 2.18 *An $n \times n$ matrix A is invertible if and only if its reduced row echelon form is I_n.*

Proof. By Theorem 2.17, A is invertible if and only if it can be reduced to I_n by elementary row operations. On the other hand, I_n is a matrix in reduced row echelon form. By Theorem 1.7, the reduced row echelon form of A is unique. Hence A is invertible if and only if its reduced row echelon form is I_n. ∎

Theorem 2.17 also characterizes invertible matrices in terms of elementary matrices.

COROLLARY 2.19 *A matrix A is invertible if and only if there exist elementary matrices $E_1, \ldots, E_p$ for which $A = E_1 \cdots E_p$.*

Proof. This corollary follows immediately from the proof of Theorem 2.17. ∎

COROLLARY 2.20 *An $n \times n$ matrix A is invertible if and only if it has n pivots.*

Proof. This corollary follows from Theorem 2.17 since the identity matrix I_n has n pivots. ∎

The next theorem establishes a connection between invertibility and homogeneous linear systems.

THEOREM 2.21 *An $n \times n$ matrix A is row equivalent to I_n if and only if the system $A\mathbf{x} = \mathbf{0}$ has only the trivial solution.*

Proof. Suppose that $A\mathbf{x} = \mathbf{0}$ and that A is row equivalent to I_n. Then there exist elementary matrices $E_1, \ldots, E_p$ for which $E_1 \cdots E_p A = I_n$. Suppose further that $A\mathbf{x} = \mathbf{0}$. Then

$$E_1 \cdots E_p A\mathbf{x} = I_n\mathbf{x} = \mathbf{x} = E_1 \cdots E_p\mathbf{0} = \mathbf{0}$$

Therefore $\mathbf{x} = \mathbf{0}$. Conversely, if $A\mathbf{x} = \mathbf{0}$ has only the trivial solution, then A must be row equivalent to a reduced row echelon matrix B with n leading 1s. Hence $B = I_n$. ∎

COROLLARY 2.22 *An $n \times n$ matrix A is invertible if and only if the equation $A\mathbf{x} = \mathbf{0}$ has only the trivial solution.*

Proof. This corollary follows immediately from the last two theorems. ∎

Next we establish the useful fact that the requirement that a matrix is invertible if and only if it has both a left and a right inverse can be weakened. The existence of one of these inverses is sufficient for invertibility. We first prove a lemma.

LEMMA 2.23 *Suppose that A is an $n \times n$ real matrix. Then the linear system $Ax = b$ has a solution for every $\mathbf{b} \in \mathbb{R}^n$ if and only if A is row equivalent to I_n.*

Proof. Let $\mathbf{b}$ be any vector in $\mathbb{R}^n$ and convert the augmented matrix $(A \mid \mathbf{b})$ to its unique reduced row echelon form $(B \mid \mathbf{c})$. If B is the $n \times n$ identity matrix, then $\widehat{\mathbf{x}} = \mathbf{c}$ is a solution of $Ax = \mathbf{b}$.

Conversely, suppose that the matrix B is not the identity. Then the last row of B must be a zero row. Since the matrix $(B \mid \mathbf{c})$ was obtained from $(A \mid \mathbf{b})$ by row reduction, there exist elementary matrices $E_1, \ldots, E_p$ for which $E_1 \cdots E_p A = B$. Let $\mathbf{e}_n = (\delta_i) \in \mathbb{R}^n$ in which $\delta_i = 0$ if $i \neq n$ and $\delta_i = 1$ if $i = n$. Using the fact that elementary matrices are invertible, we form the vector $\mathbf{b}_0 = (E_1, \ldots, E_p)^{-1} \mathbf{e}_n \in \mathbb{R}^n$. Since $E_1 \cdots E_p A = B$, it follows from the definition of $\mathbf{e}_n$ that matrix $E_1 \cdots E_p (A \mid \mathbf{b}_0) = (B \mid \mathbf{e}_n)$. This system is inconsistent since the last row is of the form $(0\ 0\ \cdots\ 0\ 1)$. Hence B is not the identity. The system $Ax = \mathbf{b}_0$ therefore has no solution for $\mathbf{b}_0$. ∎

We use this lemma to show that any matrix with a right inverse is invertible.

THEOREM 2.24 *An $n \times n$ matrix A is invertible if and only if there exists a matrix B such that $AB = I_n$.*

Proof. Suppose that A is invertible. Then by the definition of invertibility, there exists a matrix B such that $BA = AB = I_n$.

Conversely, suppose that $AB = I_n$ and consider the linear system $Ax = \mathbf{c}$ for an arbitrary vector $\mathbf{c} \in \mathbb{R}^n$. Since $(AB)\mathbf{c} = A(B\mathbf{c}) = I_n \mathbf{c} = \mathbf{c}$, the vector $\widehat{\mathbf{x}} = B\mathbf{c}$ is a solution of the equation $Ax = \mathbf{c}$. By Lemma 2.23, the matrix A is therefore row equivalent of I_n. Hence Theorem 2.17 tells us that A is invertible. ∎

The Matrix Inversion Algorithm

Example 2.34 illustrates how Corollaries 2.18 and 2.20 can be used to develop a procedure for finding the inverse of a matrix. The following theorem is the basis for this method. Let $(A \mid I_n)$ be the matrix obtained from A by adding the columns of I_n to A, and let $(I_n \mid B)$ be the matrix obtained from I_n by adding to columns of A to I_n.

THEOREM 2.25 (Matrix inversion theorem) *A matrix B is the inverse of a matrix A if and only if the matrices $(A \mid I_n)$ and $(I_n \mid B)$ are row equivalent.*

Proof. Suppose that A is an $n \times n$ invertible matrix. By Corollary 2.18, the reduced row echelon form of A is I_n. Since each step in the row reduction of A to I_n corresponds to left-multiplication by an elementary matrix, the reduction of A to I_n is of the form

$$A \to E_1 A \to \cdots \to (E_p \cdots E_1) A = I_n$$

for some elementary matrices $E_1, \ldots, E_p$. Therefore, $E_p \cdots E_1 = A^{-1}$. This tell us that the reduction

$$I_n \to E_1 I_n \to \cdots \to (E_p \cdots E_1) I_n = A^{-1}$$

yields A^{-1}. Therefore, the elementary row operations corresponding to $E_1, \ldots, E_p$, applied to the matrix $(A \mid I_n)$, result in the matrix $(I_n \mid B)$. By retracing our steps, we get the converse. ■

EXAMPLE 2.36 ■ Inverting a Matrix

Use *MATHEMATICA* and elementary row reduction to compute the inverse of the matrix

$$A = \begin{pmatrix} 1 & 2 & 3 \\ 0 & 1 & 0 \\ 2 & 0 & 12 \end{pmatrix}$$

Solution. We form the matrix $(A \mid I_n)$ and reduce it to $(I_n \mid B)$. By Theorem 2.25, the matrix B will be A^{-1}.

```
In[1]:= <<LinearAlgebra`MatrixManipulation`
```

```
In[2]:= A1= ⎛ 1  2   3 ⎞
            ⎜ 0  1   0 ⎟ ;
            ⎝ 2  0  12 ⎠
```

```
In[3]:= A2=IdentityMatrix[3]
Out[3]= {{1,0,0},{0,1,0},{0,0,1}}
```

```
In[4]:= A3=AppendRows[A1,A2]
Out[4]= {{1,2,3,1,0,0},{0,1,0,0,1,0},{2,0,12,0,0,1}}
```

```
In[5]:= A4=RowReduce[A3]
Out[5]= {{1,0,0,2,-4,-1/2},{0,1,0,0,1,0},{0,0,1,-1/3,2/3,1/6}}
```

```
In[6]:= A5=TakeColumns[A4,{4,6}]
Out[6]= {{2,-4,-1/2},{0,1,0},{-1/3,2/3,1/6}}
```

```
In[7]:= A1.A5==IdentityMatrix[3]

Out[7]= True
```

This shows that *A4* is the *MATHEMATICA* form of the matrix $(I_n \mid B)$ and that *A5* is A^{-1}. ◄►

If the matrix A is invertible, we can use matrix inversion to solve a linear system $A\mathbf{x} = \mathbf{b}$. This approach is helpful if we need to solve the system for several constant vectors $\mathbf{b}_1, \ldots, \mathbf{b}_n$. We compute A^{-1} once, and then find the solutions of the systems by multiplying each of the vectors $\mathbf{b}_1, \ldots, \mathbf{b}_n$ by A^{-1}.

EXAMPLE 2.37   ■ Using Matrix Inversion to Solve $A\mathbf{x} = \mathbf{b}$

Use *MATHEMATICA* and matrix inversion to solve the matrix equations

$$\begin{pmatrix} 3 & 2 & 3 \\ 5 & 4 & 2 \\ 5 & 2 & 1 \end{pmatrix} \begin{pmatrix} x \\ y \\ z \end{pmatrix} = \begin{pmatrix} 4 \\ 5 \\ 6 \end{pmatrix} \quad \text{and} \quad \begin{pmatrix} 3 & 2 & 3 \\ 5 & 4 & 2 \\ 5 & 2 & 1 \end{pmatrix} \begin{pmatrix} x \\ y \\ z \end{pmatrix} = \begin{pmatrix} -2 \\ -8 \\ 7 \end{pmatrix}.$$

Solution. We begin by creating the matrix A.

```
In[1]:= A= ( 3  2  3
            5  4  2
            5  2  1 );
```

Next we compute the inverse of A.

```
In[2]:= B=Inverse[A];
```

We conclude by computing the products **B.{4,5,6}** and **B.{-2,-8,7}**.

```
In[3]:= {B.{4,5,6},B.{-2,-8,7}}

Out[3]= {{7/5, -7/10, 2/5}, {22/5, -149/20, -1/10}}
```

This solves the two matrix equations. ◄►

EXERCISES 2.4

1. Show that each of the following matrices is row equivalent to the identity matrix I_2.

a. $\begin{pmatrix} 6 & 0 \\ 0 & 6 \end{pmatrix}$ b. $\begin{pmatrix} 6 & 0 \\ 4 & 6 \end{pmatrix}$ c. $\begin{pmatrix} 6 & 3 \\ 4 & 6 \end{pmatrix}$ d. $\begin{pmatrix} 0 & -1 \\ 1 & 0 \end{pmatrix}$

2. Use the matrix inversion algorithm to find the inverses of the matrices in Exercise 1.

3. Show that each matrix in Exercise 1 is a product of elementary matrices.

4. Show that each matrix in Exercise 1 has two pivots.

5. Show that for each matrix A in Exercise 1, the homogeneous linear system $A\mathbf{x} = \mathbf{0}$ has only the trivial solution.

6. Find the inverse of the product of the matrices in Exercise 1.

7. Show that each of the following matrices is row equivalent to the identity matrix I_3.

a. $\begin{pmatrix} 4 & 0 & 0 \\ 0 & 5 & 0 \\ 0 & 0 & 4 \end{pmatrix}$ b. $\begin{pmatrix} 4 & 0 & 0 \\ 2 & 5 & 0 \\ 3 & 2 & 4 \end{pmatrix}$ c. $\begin{pmatrix} \frac{1}{\sqrt{2}} & \frac{1}{\sqrt{2}} & 0 \\ -\frac{1}{\sqrt{2}} & \frac{1}{\sqrt{2}} & 0 \\ 0 & 0 & 1 \end{pmatrix}$

8. Use the matrix inversion algorithm to find the inverses of the matrices in Exercise 7.

9. Show that each matrix in Exercise 7 is a product of elementary matrices.

10. Show that each matrix in Exercise 7 has three pivots.

11. Show that for each matrix A in Exercise 7, the homogeneous linear system $A\mathbf{x} = \mathbf{0}$ has only the trivial solution.

12. Find the inverse of the product of the matrices in Exercise 7.

13. Use *MATHEMATICA* to create a random 4×4 matrix A with four pivots. Verify that A is invertible.

14. Use *MATHEMATICA* and matrix inversion to solve the matrix equation

$$\begin{pmatrix} 0 & 3 & -4 & -5 \\ -2 & -1 & -9 & 5 \\ 8 & 4 & -3 & 8 \\ -9 & 7 & 5 & 0 \end{pmatrix} \begin{pmatrix} w \\ x \\ y \\ z \end{pmatrix} = \begin{pmatrix} 1 \\ 2 \\ 3 \\ 4 \end{pmatrix}$$

15. Construct a 3×2 matrix A and a 2×3 matrix B for which $BA = I_2$ and $AB \neq I_3$. Explain why the existence of A and B does not contradict Theorem 2.24.

Matrix Inverses, Transposes, and Products

We conclude this section by establishing a link between matrix inversion, matrix transposes, and matrix products.

THEOREM 2.26 *If A is an invertible matrix, then A^T is invertible.*

Proof. Suppose that $B = (A^{-1})^T$. Then Theorem 2.9 implies that

$$A^T B = A^T (A^{-1})^T = (A^{-1}A)^T = I^T = I$$

Hence A^T has a right inverse. It follows from Theorem 2.24 that A^T is invertible. By Theorem 2.14, the matrix B is the inverse of A^T. ∎

We can use Theorems 2.24 and 2.26 to prove that a matrix A is invertible if and only if it has a left inverse.

COROLLARY 2.27 *An $n \times n$ matrix A is invertible if and only if there exists a matrix B such that $BA = I$.*

Proof. Suppose that A is invertible and $B = A^{-1}$. Then $BA = A^{-1}A = I$. Conversely, suppose that $BA = I$. Then $(BA)^T = A^T B^T = I^T = I$. Therefore, B^T is a right inverse of A^T, and Theorem 2.24 guarantees that A^T is invertible. It therefore follows from Theorem 2.26 that $A^{TT} = A$ is also invertible. ∎

The next theorem tells us that the product of two invertible matrices is invertible.

THEOREM 2.28 *If A and B are invertible, then AB is invertible.*

Proof. Suppose that A and B are invertible and that A^{-1} and B^{-1} are their inverses. Then

$$(AB)(B^{-1}A^{-1}) = AA^{-1} = I$$

Hence the matrix $B^{-1}A^{-1}$ is a right inverse of the matrix AB. By Theorem 2.24, AB is therefore invertible. The uniqueness of matrix inverses implies that $(AB)^{-1} = B^{-1}A^{-1}$. ∎

The proof of Theorem 2.28 is an example of a *constructive* proof. It not only tells us that the product of two invertible matrices is invertible, but also tells us how to construct $(AB)^{-1}$ from A^{-1} and B^{-1}. It is quite remarkable that the converse of Theorem 2.28 holds since its proof is highly *nonconstructive*. It requires the use of the law of contradiction in both directions.

THEOREM 2.29 (Invertible product theorem) *If A and B are real $n \times n$ matrices and AB is invertible, then so are A and B.*

Proof. Suppose that AB is invertible and that B is not invertible. Then Theorem 2.21 guarantees that there exists a nonzero vector $\mathbf{x}$ for which $B\mathbf{x} = \mathbf{0}$. Since $(AB)\mathbf{x} = A(B\mathbf{x}) = A(\mathbf{0}) = \mathbf{0}$, the same vector $\mathbf{x}$ is a nonzero solution of the equation $(AB)\mathbf{x} = \mathbf{0}$. Hence AB is not invertible. This is a contradiction.

Suppose therefore that B is invertible and that A is not. Then by the same theorem there also exists a nonzero vector $\mathbf{y}$ for which $A\mathbf{y} = \mathbf{0}$. Consider the vector $B^{-1}\mathbf{y}$. If $B^{-1}\mathbf{y} = \mathbf{0}$, then $\mathbf{y} = BB^{-1}\mathbf{y} = B\mathbf{0} = \mathbf{0}$. This contradicts the fact that $\mathbf{y}$ is a nonzero solution of $A\mathbf{y} = \mathbf{0}$. Hence $B^{-1}\mathbf{y}$ is nonzero. But since $(AB)(B^{-1}\mathbf{y}) = A\mathbf{y} = \mathbf{0}$, the equation $(AB)\mathbf{z} = \mathbf{0}$ has the nonzero solution $\mathbf{z} = B^{-1}\mathbf{y}$. This contradicts Theorem 2.21 since AB is assumed to be invertible. Hence both A and B are invertible. ■

EXERCISES 2.5

1. Show that in each of the following examples, the matrices A and B are inverses.

a. $A = \begin{pmatrix} 0 & 1 \\ -1 & 0 \end{pmatrix}$ $B = \begin{pmatrix} 0 & -1 \\ 1 & 0 \end{pmatrix}$

b. $A = \begin{pmatrix} 2 & -3 \\ 0 & -3 \end{pmatrix}$ $B = \begin{pmatrix} \frac{1}{2} & -\frac{1}{2} \\ 0 & -\frac{1}{3} \end{pmatrix}$

2. Show that if a and b are any two real numbers and $b \neq 0$, then the matrix

$$A = \begin{pmatrix} a & \dfrac{1-a^2}{b} \\ b & -a \end{pmatrix}$$

is its own inverse.

3. Suppose that A, B, and C are real $n \times n$ matrices with the property that $AB = CA$. Does it follow that $A^{-1} = B = C$? Explain your answer.

4. Use *MATHEMATICA* to calculate the inverses of the matrices in the following examples and show that in each case, $(AB)^{-1} = B^{-1}A^{-1}$.

a. $A = \begin{pmatrix} 2 & -3 \\ -3 & -3 \end{pmatrix}$ and $B = \begin{pmatrix} 0 & -3 \\ -3 & 4 \end{pmatrix}$

b. $A = \begin{pmatrix} -3 & 3 & 1 \\ 3 & 3 & 2 \\ 1 & 2 & 3 \end{pmatrix}$ and $B = \begin{pmatrix} 5 & 3 & 3 \\ -1 & 5 & -1 \\ -2 & -1 & 3 \end{pmatrix}$

5. Use *MATHEMATICA* to show that for each of the following matrices, $(A^T)^{-1} = (A^{-1})^T$.

a. $A = \begin{pmatrix} 2 & -3 \\ -3 & -3 \end{pmatrix}$ b. $A = \begin{pmatrix} 2 & 1 \\ -3 & -2 \end{pmatrix}$ c. $A = \begin{pmatrix} -3 & 3 & 1 \\ 3 & 3 & 2 \\ 1 & 2 & 3 \end{pmatrix}$

6. Show that the matrix

$$A = \begin{pmatrix} 4 & -6 \\ 2 & -3 \end{pmatrix}$$

is not invertible by finding a nontrivial solution of the equation $A\mathbf{x} = \mathbf{0}$.

7. Construct the inverse of the matrix

$$A = \begin{pmatrix} 1 & -3 & -3 \\ -3 & 1 & 3 \\ -3 & 3 & -2 \end{pmatrix}$$

from the calculations that show that A is row equivalent to I_3.

8. Use *MATHEMATICA* to find the inverse of the matrix

$$A = \begin{pmatrix} 1 & a & a^2 \\ 1 & b & b^2 \\ 1 & c & c^2 \end{pmatrix}$$

9. Show that $(A^{-1})^{-1} = A$ and that if $s \neq 0$, then $(sA)^{-1} = \left(\frac{1}{s}\right) A^{-1}$.

ORTHOGONAL MATRICES

Much of the work in Chapter 7 and all of the work in Chapter 9 are based on matrices with the remarkable property that their inverse is simply their transpose.

DEFINITION 2.7 *An invertible real matrix A is **orthogonal** if $A^{-1} = A^T$.*

The reason for this terminology will become clear in Chapter 8 where we will show that the rows and columns of orthogonal matrices are actually *orthogonal vectors* in a precise geometric sense.

EXAMPLE 2.38 ■ **An Orthogonal 2 × 2 Matrix**

Use *MATHEMATICA* to verify that the matrix

$$A = \begin{pmatrix} \frac{1}{\sqrt{2}} & -\frac{1}{\sqrt{2}} \\ \frac{1}{\sqrt{2}} & \frac{1}{\sqrt{2}} \end{pmatrix}$$

is orthogonal.

Solution. We calculate A^{-1} and A^T and compare the results.

In[1]:= A= $\begin{pmatrix} \frac{1}{\sqrt{2}} & -\frac{1}{\sqrt{2}} \\ \frac{1}{\sqrt{2}} & \frac{1}{\sqrt{2}} \end{pmatrix}$; Inverse[A]//MatrixForm

Out[1]//MatrixForm=

$$\begin{pmatrix} \frac{1}{\sqrt{2}} & \frac{1}{\sqrt{2}} \\ -\frac{1}{\sqrt{2}} & \frac{1}{\sqrt{2}} \end{pmatrix}$$

In[2]:= Transpose[A]//MatrixForm

Out[2]//MatrixForm=

$$\begin{pmatrix} \frac{1}{\sqrt{2}} & \frac{1}{\sqrt{2}} \\ -\frac{1}{\sqrt{2}} & \frac{1}{\sqrt{2}} \end{pmatrix}$$

This shows that $A^{-1} = A^T$. Hence the matrix A is orthogonal. ◄►

EXAMPLE 2.39 ■ An Orthogonal 3 × 3 Matrix

Use *MATHEMATICA* to verify that the matrix

$$A = \begin{pmatrix} \cos\theta & -\sin\theta & 0 \\ \sin\theta & \cos\theta & 0 \\ 0 & 0 & 1 \end{pmatrix}$$

is orthogonal.

Solution. We calculate A^{-1} and A^T and compare the results.

In[1]:= A= $\begin{pmatrix} \text{Cos}[\theta] & -\text{Sin}[\theta] & 0 \\ \text{Sin}[\theta] & \text{Cos}[\theta] & 0 \\ 0 & 0 & 1 \end{pmatrix}$

Out[1]= {{Cos[θ],-Sin[θ],0},{Sin[θ],Cos[θ],0},{0,0,1}}

```
In[2]:= Inverse[A]//MatrixForm

Out[2]//MatrixForm=
```

$$\begin{pmatrix} \dfrac{\text{Cos}[\theta]}{\text{Cos}[\theta]^2+\text{Sin}[\theta]^2} & \dfrac{\text{Sin}[\theta]}{\text{Cos}[\theta]^2+\text{Sin}[\theta]^2} & 0 \\ -\dfrac{\text{Sin}[\theta]}{\text{Cos}[\theta]^2+\text{Sin}[\theta]^2} & \dfrac{\text{Cos}[\theta]}{\text{Cos}[\theta]^2+\text{Sin}[\theta]^2} & 0 \\ 0 & 0 & 1 \end{pmatrix}$$

This output is not in its simplest form since $\cos^2\theta + \sin^2\theta = 1$. We therefore simplify the result and compare it with the transpose of A.

```
In[3]:= Simplify[Inverse[A]]==Transpose[A]

Out[3]= True
```

This shows that $A^{-1} = A$. The matrix A is therefore orthogonal. ◄►

MATHEMATICA actually has a built-in function that makes it very easy to test for orthogonality in the case of exact calculations. It tells us whether a specified equation is true or false.

EXAMPLE 2.40   ■ A Boolean Test for Orthogonality

Use the Boolean test definable in *MATHEMATICA* to verify that the matrix

$$A = \begin{pmatrix} \cos\theta & -\sin\theta & 0 \\ \sin\theta & \cos\theta & 0 \\ 0 & 0 & 1 \end{pmatrix}$$

is orthogonal.

Solution. We define A as a list of lists and compare A^{-1} and A^T without inspecting the actual output.

```
In[1]:= A={{Cos[θ],-Sin[θ],0},{Sin[θ],Cos[θ],0},{0,0,1}}

Out[1]= {{Cos[θ],-Sin[θ],0},{Sin[θ],Cos[θ],0},{0,0,1}}
```

```
In[2]:= Simplify[Inverse[A]]==Transpose[A]]

Out[2]= True
```

This shows that the matrix A is orthogonal. ◄►

EXERCISES 2.6

1. Show that the product of two orthogonal matrices is an orthogonal matrix.

2. Show that the matrix $A = \begin{pmatrix} \cos\theta & -\sin\theta \\ \sin\theta & \cos\theta \end{pmatrix}$ is orthogonal for all $0 \le \theta < 2\pi$.

3. Show that the matrix $A = \begin{pmatrix} 0 & 1 \\ 1 & 0 \end{pmatrix}$ is orthogonal.

4. Show that a 2×2 matrix $A = \begin{pmatrix} 1+s & 0 \\ 0 & 1 \end{pmatrix}$ is orthogonal if and only if $s = 0$.

5. Show that each of the following matrices is orthogonal. In part (d), the vector $\mathbf{u}$ is any nonzero vector in $\mathbb{R}^n$ and I is the $n \times n$ identity matrix.

   a. $\begin{pmatrix} \frac{4}{5} & -\frac{3}{5} \\ -\frac{3}{5} & -\frac{4}{5} \end{pmatrix}$ b. $\begin{pmatrix} -1 & 0 \\ 0 & 1 \end{pmatrix}$

   c. $\begin{pmatrix} \frac{15}{19} & -\frac{10}{19} & \frac{6}{19} \\ -\frac{10}{19} & -\frac{6}{19} & \frac{15}{19} \\ \frac{6}{19} & \frac{15}{19} & \frac{10}{19} \end{pmatrix}$ d. $I - \dfrac{2}{\mathbf{u}^T\mathbf{u}}\mathbf{u}\mathbf{u}^T$

LU DECOMPOSITION

An $m \times n$ linear system $A\mathbf{x} = \mathbf{b}$ can often be solved efficiently by LU decomposition—that is, by decomposing the coefficient matrix A into a product LU, where L is a lower-triangular matrix L with 1s on the diagonal and U is a row echelon matrix that is upper triangular if $m = n$. Solving the system $A\mathbf{x} = \mathbf{b}$ is then equivalent to solving two simpler systems $L\mathbf{y} = \mathbf{b}$ and $U\mathbf{x} = \mathbf{y}$. Since L is lower triangular, the system $L\mathbf{y} = \mathbf{b}$ can be solved by **forward substitution**. Moreover, since U is in row echelon form, the system $U\mathbf{x} = \mathbf{y}$ can be solved by **back substitution**.

We will use the built-in **LUDecomposition** function of *MATHEMATICA* to solve LU decomposition problems. Given a real nonsingular matrix A, the command **LUDecomposition[A]** returns a list of three items. The first element is a combination of the upper- and lower-triangular matrices L and U. The second element is a vector specifying rows used for pivoting, and the third element is an estimate of the condition number of A, discussed in Chapter 9. The following example illustrates these ideas.

EXAMPLE 2.41 ■ **Forward and Back Substitution**

Use *LU* decomposition and forward and back substitution to solve the system

$$\begin{pmatrix} 1 & -3 & 2 & -2 \\ 3 & -2 & 0 & -1 \\ 2 & 36 & -28 & 27 \\ 1 & -3 & 22 & 5 \end{pmatrix} \begin{pmatrix} x_1 \\ x_2 \\ x_3 \\ x_4 \end{pmatrix} = \begin{pmatrix} -11 \\ -4 \\ 155 \\ 10 \end{pmatrix}$$

Solution. We begin by finding the *LU* decomposition of the given coefficient matrix. By definition, we need to find a lower-triangular matrix with 1s on the diagonal and an upper-triangular matrix for which

$$\begin{pmatrix} 1 & -3 & 2 & -2 \\ 3 & -2 & 0 & -1 \\ 2 & 36 & -28 & 27 \\ 1 & -3 & 22 & 5 \end{pmatrix} = \begin{pmatrix} 1 & 0 & 0 & 0 \\ a & 1 & 0 & 0 \\ b & c & 1 & 0 \\ d & e & f & 1 \end{pmatrix} \begin{pmatrix} 1 & -3 & 2 & -2 \\ 0 & u & v & w \\ 0 & 0 & x & y \\ 0 & 0 & 0 & z \end{pmatrix}$$

If we multiply out the right-hand side of this equation, we obtain the matrix

$$\begin{pmatrix} 1 & -3 & 2 & -2 \\ a & -3a+u & 2a+v & -2a+w \\ b & -3b+cu & 2b+cv+x & -2b+cw+y \\ d & -3d+eu & 2d+ev+fx & -2d+ew+fy+z \end{pmatrix}$$

By equating entries and solving the resulting equations, we obtain the values $a = 3$, $b = 2$, $c = 6$, $d = 1$, $e = 0$, $f = 5$, $u = 7$, $v = -6$, $w = 5$, $x = 4$, $y = 1$, and $z = 2$. Hence

$$\begin{pmatrix} 1 & -3 & 2 & -2 \\ 3 & -2 & 0 & -1 \\ 2 & 36 & -28 & 27 \\ 1 & -3 & 22 & 5 \end{pmatrix} = LU = \begin{pmatrix} 1 & 0 & 0 & 0 \\ 3 & 1 & 0 & 0 \\ 2 & 6 & 1 & 0 \\ 1 & 0 & 5 & 1 \end{pmatrix} \begin{pmatrix} 1 & -3 & 2 & -2 \\ 0 & 7 & -6 & 5 \\ 0 & 0 & 4 & 1 \\ 0 & 0 & 0 & 2 \end{pmatrix}$$

We now use the matrix L to form the linear system $L\mathbf{y} = \mathbf{b}$, and solve for **y** by forward substitution.

$$\begin{pmatrix} 1 & 0 & 0 & 0 \\ 3 & 1 & 0 & 0 \\ 2 & 6 & 1 & 0 \\ 1 & 0 & 5 & 1 \end{pmatrix} \begin{pmatrix} y_1 \\ y_2 \\ y_3 \\ y_4 \end{pmatrix} = \begin{pmatrix} -11 \\ -4 \\ 155 \\ 10 \end{pmatrix}$$

If we multiply out the left-hand side of this equation, we get

$$\begin{pmatrix} y_1 \\ 3y_1 + y_2 \\ 2y_1 + 6y_2 + y_3 \\ y_1 + 5y_3 + y_4 \end{pmatrix} = \begin{pmatrix} -11 \\ -4 \\ 155 \\ 10 \end{pmatrix}$$

This means that $y_1 = -11$. If we substitute this value in the equation $3y_1 + y_2 = -4$ and solve for y_2, we get $y_2 = 29$. By substituting the values for y_1 and y_2 in the equation $2y_1 + 6y_2 + y_3 = 155$ and solving for y_3, we get $y_3 = 3$. By substituting $y_1 = -11$ and $y_3 = 3$ in the equation $y_1 + 5y_3 + y_4 = 10$ and solving for y_4, we get $y_4 = 6$. Thus

$$\begin{pmatrix} y_1 \\ y_2 \\ y_3 \\ y_4 \end{pmatrix} = \begin{pmatrix} -11 \\ 29 \\ 3 \\ 6 \end{pmatrix}$$

is a solution of the equation $L\mathbf{y} = \mathbf{b}$. For obvious reasons, this process is known as *forward substitution*.

Next we form the system $U\mathbf{x} = \mathbf{y}$, and solve for $\mathbf{x}$.

$$\begin{pmatrix} 1 & -3 & 2 & -2 \\ 0 & 7 & -6 & 5 \\ 0 & 0 & 4 & 1 \\ 0 & 0 & 0 & 2 \end{pmatrix} \begin{pmatrix} x_1 \\ x_2 \\ x_3 \\ x_4 \end{pmatrix} = \begin{pmatrix} -11 \\ 29 \\ 3 \\ 6 \end{pmatrix}$$

If we multiply out the left-hand side of this equation, we get

$$\begin{pmatrix} x_1 - 3x_2 + 2x_3 - 2x_4 \\ 7x_2 - 6x_3 + 5x_4 \\ 4x_3 + x_4 \\ 2x_4 \end{pmatrix} = \begin{pmatrix} -11 \\ 29 \\ 3 \\ 6 \end{pmatrix}$$

As we can see, $x_4 = 6$. If we substitute this value back into the equation $4x_3 + x_4 = 3$, we get a value for x_3. Continuing in this way, we get the vector

$$\begin{pmatrix} x_1 \\ x_2 \\ x_3 \\ x_4 \end{pmatrix} = \begin{pmatrix} 1 \\ 2 \\ 0 \\ 3 \end{pmatrix}$$

as a solution of the equation $U\mathbf{x} = \mathbf{y}$. This process is know as *back substitution*. ◀▶

Example 2.41 shows that we were able to solve the given system in three steps: First we decomposed A into a product LU of a lower-triangular matrix L and an upper-triangular matrix U. We then solved the system $L\mathbf{y} = \mathbf{b}$ for $\mathbf{y}$ by forward substitution, and concluded the process by solving the system $U\mathbf{x} = \mathbf{y}$ for $\mathbf{x}$ by back substitution.

The LU decomposition of a matrix is based on a closer analysis of the process of reducing a matrix to row echelon form. First we observe that the operations $(R_i \rightarrow sR_i)$ are not required since we are initially not interested in converting the pivots of the given matrix to specific values. This means that we need to consider only two cases. The ideal case, requiring only operations of the form $(R_i \rightarrow R_i + sR_j)$, and the mixed case, requiring both $(R_i \rightarrow R_i + sR_j)$ and $(R_i \leftrightarrows R_j)$. We will show that if no permutations of rows $(R_i \leftrightarrows R_j)$ are required, we can decompose A into the form LU. If permutations of rows are required, we can carry them out first, using a suitable permutation matrix P, and then decompose the matrix PA into LU. In both cases, the rows R_j involved in the operations $(R_i \rightarrow R_i + sR_j)$ occur above the rows R_i.

To be able to prove the LU decomposition theorem, we need three facts about triangular matrices. The first is stated in the next theorem.

THEOREM 2.30   *If A is an $n \times n$ row echelon matrix, then A is upper triangular.*

Proof. We prove this theorem by an induction on n.

If $n = 1$, the theorem is obviously true since A is diagonal. This establishes the induction basis.

Suppose, therefore, that $n = k + 1$, and let A_k be the matrix obtained from A by deleting the nth row and the nth column of A. Then

$$A_k = \begin{pmatrix} a_{11} & \cdots & a_{1k} \\ \vdots & \ddots & \vdots \\ a_{k1} & \cdots & a_{kk} \end{pmatrix}$$

The induction hypothesis is that A_k is upper triangular. The induction step consists of showing that this implies that if the $(k + 1) \times (k + 1)$ matrix A is in row echelon form, then A is also upper triangular.

If the $(k + 1)$th row of A is a zero row, then the theorem follows at once since A_k is upper triangular. Let us therefore assume that $a_{(k+1)m} \neq 0$ for some $1 \leq m \leq k + 1$. Since A is a row echelon matrix, we need to consider only two cases: Either $a_{kk} \neq 0$, or $a_{kk} = 0$ and $a_{k(k+1)} \neq 0$. However, the second case is impossible. If $a_{kk} = 0$ and $a_{k(k+1)} \neq 0$, then the nonzero entry $a_{(k+1)m}$ has to occur in the column to the right of the first nonzero entry of the $(k + 1)$th row of A. The entry $a_{(k+1)m}$ therefore has to occur in the column to the right of the last column of A. But there is no such column. Hence A is upper triangular. If $a_{kk} \neq 0$, then any nonzero entry $a_{(k+1)m}$ in the $(k + 1)$th row of A must occur in the $(k + 1)$th column of A. This means that $a_{(k+1)m} = a_{(k+1)(k+1)}$, and A is again upper triangular. ∎

We also need to know that lower- and upper-triangular matrices are closed under multiplication.

THEOREM 2.31   *The product of two $n \times n$ lower-triangular matrices is a lower-triangular matrix.*

Proof. Let $A = (a_{ij})$ and $B = (b_{ij})$ be two $n \times n$ lower-triangular matrices, and let $C = (c_{ij}) = AB$. Then $a_{ij} = b_{ij} = 0$ if $i < j$. Therefore,

$$c_{ij} = a_{i1}b_{1j} + \cdots + a_{in}b_{nj} = a_{ij}b_{jj} + \cdots + a_{in}b_{nj}$$

since $b_{ij} = 0$ for $i < j$. On the other hand, $a_{ij} = 0$ for $i < j$. Therefore,

$$a_{ij}b_{jj} + \cdots + a_{in}b_{nj} = 0$$

Hence $c_{ij} = 0$ for $i < j$. This entails that C is lower triangular. ∎

COROLLARY 2.32   *The product of two $n \times n$ upper-triangular matrices is an upper-triangular matrix.*

Proof. We get this result by reversing the inequalities in the previous proof and adapting the equations as required. ∎

We are now ready to tackle the *LU* decomposition problem. We begin with an example.

EXAMPLE 2.42   ■ An *LU* Decomposition of a Matrix *A*

Find an *LU* decomposition of the matrix

$$A = \begin{pmatrix} 4 & 1 & 3 & 1 & 2 \\ 4 & 2 & 6 & 4 & 5 \\ 8 & 3 & 4 & 1 & 0 \end{pmatrix}$$

Solution. The following calculations yield the desired decomposition.

1.
$$\begin{pmatrix} 1 & 0 & 0 \\ 0 & 1 & 0 \\ -2 & 0 & 1 \end{pmatrix} \begin{pmatrix} 4 & 1 & 3 & 1 & 2 \\ 4 & 2 & 6 & 4 & 5 \\ 8 & 3 & 4 & 1 & 0 \end{pmatrix} = \begin{pmatrix} 4 & 1 & 3 & 1 & 2 \\ 4 & 2 & 6 & 4 & 5 \\ 0 & 1 & -2 & -1 & -4 \end{pmatrix}$$

2.
$$\begin{pmatrix} 1 & 0 & 0 \\ -1 & 1 & 0 \\ 0 & 0 & 1 \end{pmatrix} \begin{pmatrix} 4 & 1 & 3 & 1 & 2 \\ 4 & 2 & 6 & 4 & 5 \\ 0 & 1 & -2 & -1 & -4 \end{pmatrix} = \begin{pmatrix} 4 & 1 & 3 & 1 & 2 \\ 0 & 1 & 3 & 3 & 3 \\ 0 & 1 & -2 & -1 & -4 \end{pmatrix}$$

3.
$$\begin{pmatrix} 1 & 0 & 0 \\ 0 & 1 & 0 \\ 0 & -1 & 1 \end{pmatrix} \begin{pmatrix} 4 & 1 & 3 & 1 & 2 \\ 0 & 1 & 3 & 3 & 3 \\ 0 & 1 & -2 & -1 & -4 \end{pmatrix} = \begin{pmatrix} 4 & 1 & 3 & 1 & 2 \\ 0 & 1 & 3 & 3 & 3 \\ 0 & 0 & -5 & -4 & -7 \end{pmatrix}$$

The matrix

$$U = \begin{pmatrix} 4 & 1 & 3 & 1 & 2 \\ 0 & 1 & 3 & 3 & 3 \\ 0 & 0 & -5 & -4 & -7 \end{pmatrix}$$

is in row echelon form, and the product

$$\begin{pmatrix} 1 & 0 & 0 \\ 0 & 1 & 0 \\ 0 & -1 & 1 \end{pmatrix} \begin{pmatrix} 1 & 0 & 0 \\ -1 & 1 & 0 \\ 0 & 0 & 1 \end{pmatrix} \begin{pmatrix} 1 & 0 & 0 \\ 0 & 1 & 0 \\ -2 & 0 & 1 \end{pmatrix} = \begin{pmatrix} 1 & 0 & 0 \\ -1 & 1 & 0 \\ -1 & -1 & 1 \end{pmatrix}$$

is the required lower-triangular matrix L. As we can see, L has 1s on the diagonal. ◄►

The general decomposition uses similar ideas.

THEOREM 2.33 (*LU* **decomposition theorem**)   *Suppose that A is an m × n matrix that can be reduced to row echelon form without requiring the permutation of rows. Then there exist an m × m lower-triangular matrix L with 1s on the diagonal and an m × n matrix row echelon matrix U such that A = LU.*

Proof. By Corollary 2.12, there exist elementary matrices $E_1, \ldots, E_p$ such that $E_p \cdots E_1 A$ is in row echelon form. Any row echelon matrix satisfies the definition of U. We therefore let $U = E_p \cdots E_1 A$. Since the reduction of A to row echelon form can be achieved without the permutation of rows, we can assume that the required elementary matrices E_k represent operations of the form $(R_i \to R_i + sR_j)$, designed to add multiples sR_j of rows R_j to the rows R_i below. This means that $i > j$ in all cases. Each E_k is therefore a lower-triangular elementary matrix with 1s on the diagonal. We also saw earlier that the inverse of the operation $(R_i \to R_i + sR_j)$ is the operation $(R_i \to R_i - sR_j)$, again with $i > j$. Hence the inverses E_k^{-1} are also lower-triangular matrices with 1s on the diagonal. By Theorem 2.31, the matrix $E_1^{-1} \cdots E_p^{-1}$ is therefore lower triangular and has 1s on the diagonal. If we put $L = E_1^{-1} \cdots E_p^{-1}$, then $A = LU$ is the required decomposition. ∎

The next example shows that some matrices can only be decomposed into the form LU if some of their rows are first permuted.

EXAMPLE 2.43 ■ **An** *LU* **Decomposition of a Matrix** *PA*

Find an LU decomposition of the matrix

$$A = \begin{pmatrix} 0 & 1 & 3 & 1 & 2 \\ 4 & 2 & 6 & 4 & 5 \\ 8 & 3 & 4 & 1 & 0 \end{pmatrix}$$

using a permutation matrix P, if necessary.

Solution. The following steps yield an LU decomposition of A.

1. We first apply ($R_1 \leftrightharpoons R_2$) to A and get

$$
\begin{pmatrix} 0 & 1 & 0 \\ 1 & 0 & 0 \\ 0 & 0 & 1 \end{pmatrix}
\begin{pmatrix} 0 & 1 & 3 & 1 & 2 \\ 4 & 2 & 6 & 4 & 5 \\ 8 & 3 & 4 & 1 & 0 \end{pmatrix}
=
\begin{pmatrix} 4 & 2 & 6 & 4 & 5 \\ 0 & 1 & 3 & 1 & 2 \\ 8 & 3 & 4 & 1 & 0 \end{pmatrix}
$$

2. Then we apply ($R_3 \rightarrow R_3 - 2R_1$) and get

$$
\begin{pmatrix} 1 & 0 & 0 \\ 0 & 1 & 0 \\ -2 & 0 & 1 \end{pmatrix}
\begin{pmatrix} 4 & 2 & 6 & 4 & 5 \\ 0 & 1 & 3 & 1 & 2 \\ 8 & 3 & 4 & 1 & 0 \end{pmatrix}
=
\begin{pmatrix} 4 & 2 & 6 & 4 & 5 \\ 0 & 1 & 3 & 1 & 2 \\ 0 & -1 & -8 & -7 & -10 \end{pmatrix}
$$

3. Next we use ($R_3 \rightarrow R_3 + R_2$) and obtain

$$
\begin{pmatrix} 1 & 0 & 0 \\ 0 & 1 & 0 \\ 0 & 1 & 1 \end{pmatrix}
\begin{pmatrix} 4 & 2 & 6 & 4 & 5 \\ 0 & 1 & 3 & 1 & 2 \\ 0 & -1 & -8 & -7 & -10 \end{pmatrix}
=
\begin{pmatrix} 4 & 2 & 6 & 4 & 5 \\ 0 & 1 & 3 & 1 & 2 \\ 0 & 0 & -5 & -6 & -8 \end{pmatrix}
$$

4. Now we let PA be the matrix

$$
\begin{pmatrix} 0 & 1 & 0 \\ 1 & 0 & 0 \\ 0 & 0 & 1 \end{pmatrix}
\begin{pmatrix} 0 & 1 & 3 & 1 & 2 \\ 4 & 2 & 6 & 4 & 5 \\ 8 & 3 & 4 & 1 & 0 \end{pmatrix}
=
\begin{pmatrix} 4 & 2 & 6 & 4 & 5 \\ 0 & 1 & 3 & 1 & 2 \\ 8 & 3 & 4 & 1 & 0 \end{pmatrix}
$$

The matrices

$$
L = \begin{pmatrix} 1 & 0 & 0 \\ 0 & 1 & 0 \\ 0 & 1 & 1 \end{pmatrix}^{-1}
\begin{pmatrix} 1 & 0 & 0 \\ 0 & 1 & 0 \\ -2 & 0 & 1 \end{pmatrix}^{-1}
=
\begin{pmatrix} 1 & 0 & 0 \\ 0 & 1 & 0 \\ 2 & -1 & 1 \end{pmatrix}
$$

and

$$
U = \begin{pmatrix} 4 & 2 & 6 & 4 & 5 \\ 0 & 1 & 3 & 1 & 2 \\ 0 & 0 & -5 & -6 & -8 \end{pmatrix}
$$

yield an LU decomposition of PA. ◀▶

The use of the interchange ($R_1 \leftrightharpoons R_2$) in this decomposition clearly cannot be avoided. The matrix A itself could not have been decomposed into the form LU without the permutation P. However, we could have applied the row operations in a different order. The following steps also reduce A to row echelon form.

1. First we apply $(R_3 \to R_3 - 2R_2)$:

$$\begin{pmatrix} 1 & 0 & 0 \\ 0 & 1 & 0 \\ 0 & -2 & 1 \end{pmatrix} \begin{pmatrix} 0 & 1 & 3 & 1 & 2 \\ 4 & 2 & 6 & 4 & 5 \\ 8 & 3 & 4 & 1 & 0 \end{pmatrix} = \begin{pmatrix} 0 & 1 & 3 & 1 & 2 \\ 4 & 2 & 6 & 4 & 5 \\ 0 & -1 & -8 & -7 & -10 \end{pmatrix}$$

2. Then we apply $(R_1 \leftrightarrows R_2)$:

$$\begin{pmatrix} 0 & 1 & 0 \\ 1 & 0 & 0 \\ 0 & 0 & 1 \end{pmatrix} \begin{pmatrix} 0 & 1 & 3 & 1 & 2 \\ 4 & 2 & 6 & 4 & 5 \\ 0 & -1 & -8 & -7 & -10 \end{pmatrix} = \begin{pmatrix} 4 & 2 & 6 & 4 & 5 \\ 0 & 1 & 3 & 1 & 2 \\ 0 & -1 & -8 & -7 & -10 \end{pmatrix}$$

3. Finally, we apply $(R_3 \to R_3 + R_2)$:

$$\begin{pmatrix} 1 & 0 & 0 \\ 0 & 1 & 0 \\ 0 & 1 & 1 \end{pmatrix} \begin{pmatrix} 4 & 2 & 6 & 4 & 5 \\ 0 & 1 & 3 & 1 & 2 \\ 0 & -1 & -8 & -7 & -10 \end{pmatrix} = \begin{pmatrix} 4 & 2 & 6 & 4 & 5 \\ 0 & 1 & 3 & 1 & 2 \\ 0 & 0 & -5 & -6 & -8 \end{pmatrix}$$

As we can see, these steps reduce A to the same matrix U. The difference in the two approaches is that in Example 2.43 the permutation $(R_1 \leftrightarrows R_2)$ precedes the operation $(R_3 \to R_3 - 2R_1)$, whereas in the last case, the operation $(R_3 \to R_3 - 2R_2)$ precedes the permutation $(R_1 \leftrightarrows R_2)$.

The next lemma shows that we can always carry out all required permutations of rows before all operations of the form $(R_i \to R_i + sR_j)$.

LEMMA 2.34 *For any $m \times n$ matrix A there exists a permutation matrix P such that the matrix PA can be reduced to row echelon form without the permutation of rows.*

Proof. We can easily convince ourselves that a permutation of rows $(R_i \leftrightarrows R_j)$ has no essential bearing on the use of the operations $(R_k \to R_k + sR_j)$ in the reduction of A to row echelon form. The only difference therefore is that instead of using an operation $(R_k \to R_k + sR_i)$, we may have to use a corresponding operation $(R_k \to R_k + sR_j)$, with $i \neq j$. ∎

THEOREM 2.35 (*PLU decomposition theorem*) *For any $m \times n$ matrix A there exist a permutation matrix P, an $m \times m$ lower-triangular matrix L with 1s on the diagonal, and an $m \times n$ row echelon matrix U such that $PA = LU$.*

Proof. By Lemma 2.34, there exists a permutation matrix P such that PA can be reduced to row echelon form without requiring further permutations of rows. Theorem 2.33 therefore guarantees that there exist suitable matrices L and U for which $PA = LU$. ∎

Let us now consider the special case of an LU decomposition of an invertible matrix. We show that in this case, the decomposition is unique.

THEOREM 2.36 *Suppose that A is an invertible matrix that can be reduced to row echelon form without requiring the permutation of rows. Then there exists a unique lower-triangular matrix L with 1s on the diagonal and a unique upper-triangular matrix U such that $A = LU$.*

Proof. By Theorem 2.33, there exist matrices L and U of the required form for which $A = LU$. It remains to prove the uniqueness of L and U.

Since we took U to be the row echelon form of A, and since elementary row reductions preserve invertibility, the matrix U is invertible. Furthermore, the matrix L is the product of elementary matrices. By Theorem 2.16, each elementary matrix is invertible, and by Corollary 2.28, the product of invertible matrices is invertible. Hence L is invertible.

Now suppose $A = L_1 U_1 = L_2 U_2$. Then

$$
\begin{aligned}
U_1 U_2^{-1} &= \left(L_1^{-1} L_1\right)\left(U_1 U_2^{-1}\right) \\
&= L_1^{-1}(L_1 U_1) U_2^{-1} \\
&= L_1^{-1}(L_2 U_2) U_2^{-1} \\
&= \left(L_1^{-1} L_2\right)\left(U_2 U_2^{-1}\right) \\
&= L_1^{-1} L_2
\end{aligned}
$$

But $U_1 U_2^{-1}$ is upper triangular and $L_1^{-1} L_2$ is lower triangular. Hence there must be a diagonal matrix $L_1^{-1} L_2$ with 1s on the main diagonal. Hence, $U_1 U_2^{-1} = L_1^{-1} L_2 = I$. Therefore, $L_1 = L_2$ and $U_1 = U_2$. ∎

COROLLARY 2.37 *Suppose that A is an invertible matrix and let P be a permutation matrix for which PA is reducible to row echelon form without the permutation of rows. Then there exist a unique lower-triangular matrix L with 1s on the diagonal and a unique upper-triangular matrix U such that $A = LU$.*

Proof. Since A is invertible, Corollary 2.20 tells us that A has n pivots. Since the permutations of rows of A have no effect on the number of pivots of A, the matrix PA has also n pivots. By Corollary 2.20, the matrix PA is therefore invertible. The uniqueness of L and U follows from Theorem 2.36. ∎

MATHEMATICA has a built-in implementation of LU decomposition and LU back substitution that provides an efficient method for solving certain matrix equations. The *MATHEMATICA* version is actually the more general *PLU.* decomposition. If A is an invertible $n \times n$ matrix, the command **LUDecomposition[A]** produces a list $\{M, \pi, c\}$ consisting of a matrix M, a permutation $\pi : \mathbf{n} \to \mathbf{n}$, and a constant c. The matrix M is a combination of the triangular matrices L and U, and the permutation π indicates which permutation matrix

P is used for the decomposition. The constant c is a **condition number** of the matrix A and is of no interest to us at the moment. (Condition numbers are discussed in Chapter 9.) Here is an illustration of these ideas.

If A is the matrix

$$\begin{pmatrix} 1 & 2 & 3 \\ 5 & 5 & 5 \\ 0 & 8 & 0 \end{pmatrix}$$

the command **LUDecomposition[A]** produces the list

$$\{M, \pi, c\} = \{\{\{1, 2, 3\}, \{5, -5, -10\}, \{0, -\frac{8}{5}, -16\}\}, \{1, 2, 3\}, 1\}$$

consisting of the data

$$M = \begin{pmatrix} 1 & 2 & 3 \\ 5 & -5 & -10 \\ 0 & -\frac{8}{5} & -16 \end{pmatrix}, \quad \{1, 2, 3\}, \quad \text{and} \quad 1$$

The list $\{1, 2, 3\}$ denotes the identity permutation of the standard list **3**. It tells us that no permutation is required to extricate the matrices L and U from M. The matrix

$$L = \begin{pmatrix} 1 & 0 & 0 \\ 5 & 1 & 0 \\ 0 & -\frac{8}{5} & 1 \end{pmatrix}$$

consists of the entries of M below the diagonal and has 1s on the diagonal. The matrix

$$U = \begin{pmatrix} 1 & 2 & 3 \\ 0 & -5 & -10 \\ 0 & 0 & -16 \end{pmatrix}$$

on the other hand, consists of the entries of M on and above the diagonal, with zeros below. The matrix M is only a convenient form of combining L and U. It is not obtained from L and U by any algebraic manipulation. A simple calculation shows that

$$\begin{pmatrix} 1 & 2 & 3 \\ 5 & 5 & 5 \\ 0 & 8 & 0 \end{pmatrix} = \begin{pmatrix} 1 & 0 & 0 \\ 5 & 1 & 0 \\ 0 & -\frac{8}{5} & 1 \end{pmatrix} \begin{pmatrix} 1 & 2 & 3 \\ 0 & -5 & -10 \\ 0 & 0 & -16 \end{pmatrix}$$

Next we illustrate a case where *MATHEMATICA* produces an *LU* decomposition involving a nontrivial permutation. Consider the matrix

$$B = \begin{pmatrix} 5 & 0 & 3 \\ 5 & 0 & 5 \\ 0 & 1 & 8 \end{pmatrix}$$

The command **LUDecomposition[B]** yields

$$\{M, \pi, c\} = \{\{\{5, 0, 3\}, \{0, 1, 8\}, \{1, 0, 2\}\}, \{1, 3, 2\}, 1\}$$

This time *MATHEMATICA* is telling us that

$$A = \begin{pmatrix} 5 & 0 & 3 \\ 5 & 0 & 5 \\ 0 & 1 & 8 \end{pmatrix} = \begin{pmatrix} 1 & 0 & 0 \\ 0 & 0 & 1 \\ 0 & 1 & 0 \end{pmatrix} \begin{pmatrix} 1 & 0 & 0 \\ 0 & 1 & 0 \\ 1 & 0 & 1 \end{pmatrix} \begin{pmatrix} 5 & 0 & 3 \\ 0 & 1 & 8 \\ 0 & 0 & 2 \end{pmatrix} = PLU$$

where P is the permutation matrix corresponding to the permutation $\{1, 3, 2\}$, and L and U are obtained in the same way as before.

The fact that *MATHEMATICA* produces a *PLU*. decomposition involving a nontrivial permutation does not mean that for LU decomposition the permutation is actually necessary. Consider the following example. The product

$$\begin{pmatrix} 1 & 4 & 6 & 3 \\ 2 & 1 & 3 & 0 \\ 1 & 8 & 3 & 0 \\ 4 & 5 & 1 & 5 \end{pmatrix} = \begin{pmatrix} 1 & 0 & 0 & 0 \\ 2 & 1 & 0 & 0 \\ 1 & -\frac{4}{7} & 1 & 0 \\ 4 & \frac{11}{7} & \frac{62}{57} & 1 \end{pmatrix} \begin{pmatrix} 1 & 4 & 6 & 3 \\ 0 & -7 & -9 & -6 \\ 0 & 0 & -\frac{57}{7} & -\frac{45}{7} \\ 0 & 0 & 0 & \frac{179}{19} \end{pmatrix}$$

satisfies the definition of an LU decomposition. However, *MATHEMATICA* produces a different answer.

EXAMPLE 2.44 ■ **Using *MATHEMATICA* to Find an *LU* Decomposition**

Use *MATHEMATICA* to find an LU decomposition for the matrix

$$A = \begin{pmatrix} 1 & 4 & 6 & 3 \\ 2 & 1 & 3 & 0 \\ 1 & 8 & 3 & 0 \\ 4 & 5 & 1 & 5 \end{pmatrix}$$

Solution. We begin by defining the matrix A.

$$\text{In[1]:= } A = \begin{pmatrix} 1 & 4 & 6 & 3 \\ 2 & 1 & 3 & 0 \\ 1 & 8 & 3 & 0 \\ 4 & 5 & 1 & 5 \end{pmatrix};$$

Next we apply the **LUDecomposition** function.

```
In[2]:= {M,p,c}=LUDecomposition[A]

Out[2]= {{{1,4,6,3},{1,4,-3,-3},
```

$$\{2,-\tfrac{7}{4},-\tfrac{57}{4},-\tfrac{45}{4}\},\{4,-\tfrac{11}{4},\tfrac{125}{57},\tfrac{179}{19}\}\},\{1,3,2,4\},1\}$$

```
In[3]:= MatrixForm[LUDecomposition[M]]

Out[3]//MatrixForm=
```

$$\begin{pmatrix} 1 & 4 & 6 & 3 \\ 1 & 4 & -3 & -4 \\ 2 & -\frac{7}{4} & -\frac{57}{4} & -\frac{45}{4} \\ 4 & -\frac{11}{4} & \frac{125}{57} & \frac{179}{19} \end{pmatrix}$$

This matrix consists of the combined information in the matrices

$$L = \begin{pmatrix} 1 & 0 & 0 & 0 \\ 1 & 1 & 0 & 0 \\ 2 & -\frac{7}{4} & 1 & 0 \\ 4 & -\frac{11}{4} & \frac{125}{57} & \frac{179}{19} \end{pmatrix} \quad \text{and} \quad U = \begin{pmatrix} 1 & 4 & 6 & 3 \\ 0 & 4 & -3 & -4 \\ 0 & 0 & -\frac{57}{4} & -\frac{45}{4} \\ 0 & 0 & 0 & \frac{179}{19} \end{pmatrix}$$

The permutation $\{1, 3, 2, 4\}$ tells us that to recover the matrix A from L and U, we must multiply LU by the permutation matrix that interchanges rows 2 and 3. This means that

$$\begin{pmatrix} 1 & 0 & 0 & 0 \\ 0 & 0 & 1 & 0 \\ 0 & 1 & 0 & 0 \\ 0 & 0 & 0 & 1 \end{pmatrix} \begin{pmatrix} 1 & 0 & 0 & 0 \\ 1 & 1 & 0 & 0 \\ 2 & -\frac{7}{4} & 1 & 0 \\ 4 & -\frac{11}{4} & \frac{125}{57} & 1 \end{pmatrix} \begin{pmatrix} 1 & 4 & 6 & 3 \\ 0 & 4 & -3 & -3 \\ 0 & 0 & -\frac{57}{4} & -\frac{45}{4} \\ 0 & 0 & 0 & \frac{179}{19} \end{pmatrix} = \begin{pmatrix} 1 & 4 & 6 & 3 \\ 2 & 1 & 3 & 0 \\ 1 & 8 & 3 & 0 \\ 4 & 5 & 1 & 5 \end{pmatrix}$$

is the corresponding *PLU*. decomposition. ◄►

The **LUDecomposition** function in *MATHEMATICA* is designed to go hand in hand with the **LUBackSubstitution** function. The permutations are chosen to optimize calculations involving both functions. The functions are so defined that the joint command

LUBackSubstitution[LUDecomposition[A],b]

produces an efficient solution of the matrix equation $A\mathbf{x} = \mathbf{b}$.

EXAMPLE 2.45 ■ LUBackSubstitution and Matrix Equations

Use the **LUBackSubstitution** function to solve the matrix equation

$$\begin{pmatrix} 1 & 4 & 6 & 3 \\ 2 & 1 & 3 & 0 \\ 1 & 8 & 3 & 0 \\ 4 & 5 & 1 & 5 \end{pmatrix} \begin{pmatrix} w \\ x \\ y \\ z \end{pmatrix} = \begin{pmatrix} 1 \\ 2 \\ 3 \\ 1 \end{pmatrix}$$

Solution. We begin by defining the matrix A.

```
In[1]:= A= ( 1 4 6 3
            2 1 3 0
            1 8 3 0
            4 5 1 5 );
```

Next we apply the **LUDecomposition** and **LUBackSubstitution** functions to A.

```
In[2]:= LUBackSubstitution[LUDecomposition[A],{5,2,-3,1}]
```

$$Out[2]:= \{-\tfrac{8}{179}, -\tfrac{129}{179}, \tfrac{503}{537}, \tfrac{413}{537}\}$$

This shows that the vector

$$\begin{pmatrix} w \\ x \\ y \\ z \end{pmatrix} = \begin{pmatrix} -\frac{8}{179} \\ -\frac{129}{179} \\ \frac{503}{537} \\ \frac{413}{537} \end{pmatrix}$$

is a solution of the given equation. ◄►

If we tried to apply the **LUDecomposition** function to the rectangular matrix

$$\begin{pmatrix} 1 & 2 & 3 \\ 4 & 5 & 6 \end{pmatrix}$$

we would get an error message. The command

LUDecomposition[{{1,2,3},{4,5,6}}],

for example, produces the message

```
LUDecomposition::''matsq'' :   ''Argument {{1,2,3},{4,5,6}}
at position 1 is not a square matrix.''
```

If, on the other hand, we tried to apply the **LUDecomposition** function to the singular matrix

$$\begin{pmatrix} 4 & 9 & -5 \\ 2 & 2 & 0 \\ 7 & 0 & 7 \end{pmatrix}$$

MATHEMATICA would try its best to produce an output. After sending the message

```
LUDecomposition::sing:
''Matrix {{4,9,-5},{2,2,0},{7,0,7}} is singular.''
```

it would produce the output

$$\{\{\{2,2,0\},\{2,5,-5\},\{\tfrac{7}{2},-\tfrac{7}{5},0\}\},\{2,1,3\},1\}$$

Moreover, it is easily seen that

$$\begin{pmatrix} 0 & 1 & 0 \\ 1 & 0 & 0 \\ 0 & 0 & 1 \end{pmatrix}\begin{pmatrix} 1 & 0 & 0 \\ 2 & 1 & 0 \\ \tfrac{7}{2} & -\tfrac{7}{5} & 1 \end{pmatrix}\begin{pmatrix} 2 & 2 & 0 \\ 0 & 5 & -5 \\ 0 & 0 & 0 \end{pmatrix} = \begin{pmatrix} 4 & 9 & -5 \\ 2 & 2 & 0 \\ 7 & 0 & 7 \end{pmatrix}$$

We should mention that in addition to the various built-in techniques for solving linear systems discussed so far, *MATHEMATICA* also has a very general function for solving matrix equations. A consistent linear system of the form $Ax = \mathbf{b}$ can always be solved with the **LinearSolve** function. The command

LinearSolve[A,b]

produces an solution vector. However, one of the drawbacks of **LinearSolve** is the fact that it returns only one of all possible solutions. On the other hand, the function **Solve**, discussed in Chapter 1, returns a general solution. Here is an example of an output produced by the **LinearSolve** function.

EXAMPLE 2.46 ■ Using LinearSolve to Solve a Matrix Equation

Use the **LinearSolve** function of *MATHEMATICA* to find a solution of the equation

$$\begin{pmatrix} 1 & 4 & 6 & 3 \\ 2 & 1 & 3 & 0 \\ 1 & 8 & 3 & 0 \end{pmatrix}\begin{pmatrix} w \\ x \\ y \\ z \end{pmatrix} = \begin{pmatrix} 1 \\ 2 \\ 1 \end{pmatrix}$$

Solution. We begin by defining the matrix A.

$$\text{In[1]:= } A = \begin{pmatrix} 1 & 4 & 6 & 3 \\ 2 & 1 & 3 & 0 \\ 1 & 8 & 3 & 0 \end{pmatrix};$$

Next we apply the **LinearSolve** function to the matrix A.

$$\text{In[2]:= } \textbf{LinearSolve[A, \{1,2,1\}]}$$

$$\text{Out[2]= } \{1,0,0,0\}$$

It follows that the vector

$$\begin{pmatrix} w \\ x \\ y \\ z \end{pmatrix} = \begin{pmatrix} 1 \\ 0 \\ 0 \\ 0 \end{pmatrix}$$

is a solution of the given equation. ◀▶

EXERCISES 2.7

1. Determine which of the following matrices are upper triangular. Explain your answers.

a. $A = \begin{pmatrix} 4 & 1 & 3 & 1 \\ 0 & 1 & 4 & 3 \\ 0 & 0 & 3 & 1 \\ 0 & 0 & 0 & 2 \end{pmatrix}$ b. $A = \begin{pmatrix} 4 & 1 & 3 & 1 \\ 0 & 1 & 4 & 3 \\ 0 & 0 & 3 & 1 \\ 0 & 0 & 0 & 2 \\ 0 & 0 & 0 & 0 \end{pmatrix}$ c. $A = \begin{pmatrix} 1 & 0 & 0 & 0 & 0 \\ 0 & 1 & 0 & 0 & 0 \\ 0 & 0 & 3 & 0 & 0 \\ 0 & 0 & 0 & 0 & 0 \\ 0 & 0 & 0 & 0 & 0 \end{pmatrix}$

2. Take the transpose of the matrices in Exercise 1 and determine which of the resulting matrices are lower triangular. Explain your answers.

3. Generate two random 3×3 matrices A and B and replace all entries above the diagonal in A and B by zero. Show that the product of the resulting matrices is lower triangular.

4. Generate two random 4×4 matrices A and B and replace all entries below the diagonal in A and B by zero. Show that the product of the resulting matrices is upper triangular.

5. Find elementary matrices $E_1, \ldots, E_n$ and an upper-triangular matrix U for the matrix

$$A = \begin{pmatrix} 1 & 2 & 4 & 1 \\ -4 & 1 & 4 & -4 \\ 2 & -2 & -4 & 4 \\ 0 & 0 & 2 & 3 \end{pmatrix}$$

such that $A = E_1^{-1} \cdots E_n^{-1} U$.

6. Use the *LU* decomposition

$$\begin{pmatrix} 5 & 0 \\ -3 & 7 \end{pmatrix} = \begin{pmatrix} 1 & 0 \\ -\frac{3}{5} & 1 \end{pmatrix} \begin{pmatrix} 5 & 0 \\ 0 & 7 \end{pmatrix}$$

and forward and back substitution to solve the linear system

$$\begin{pmatrix} 5 & 0 \\ -3 & 7 \end{pmatrix} \begin{pmatrix} x \\ y \end{pmatrix} = \begin{pmatrix} 3 \\ 4 \end{pmatrix}$$

7. Use the *LU* decomposition

$$\begin{pmatrix} 7 & 5 \\ 1 & 4 \end{pmatrix} = \begin{pmatrix} 1 & 0 \\ \frac{1}{7} & 1 \end{pmatrix} \begin{pmatrix} 7 & 5 \\ 0 & \frac{23}{7} \end{pmatrix}$$

and forward and back substitution to solve the linear system

$$\begin{pmatrix} 7 & 5 \\ 1 & 4 \end{pmatrix} \begin{pmatrix} x \\ y \end{pmatrix} = \begin{pmatrix} 2 \\ 5 \end{pmatrix}$$

8. Use the *LU* decomposition

$$\begin{pmatrix} 5 & 0 & -2 \\ -3 & 7 & 5 \\ 9 & 1 & 4 \end{pmatrix} = \begin{pmatrix} 1 & 0 & 0 \\ -\frac{3}{5} & 1 & 0 \\ \frac{9}{5} & \frac{1}{7} & 1 \end{pmatrix} \begin{pmatrix} 5 & 0 & -2 \\ 0 & 7 & \frac{19}{5} \\ 0 & 0 & \frac{247}{35} \end{pmatrix}$$

and forward and back substitution to solve the linear system

$$\begin{pmatrix} 5 & 0 & -2 \\ -3 & 7 & 5 \\ 9 & 1 & 4 \end{pmatrix} \begin{pmatrix} x \\ y \\ z \end{pmatrix} = \begin{pmatrix} 3 \\ 4 \\ 5 \end{pmatrix}$$

9. Use the *LU* decomposition

$$\begin{pmatrix} 1 & 0 & 2 \\ 4 & 2 & 5 \\ 9 & 1 & 3 \end{pmatrix} = \begin{pmatrix} 1 & 0 & 0 \\ 4 & 1 & 0 \\ 9 & \frac{1}{2} & 1 \end{pmatrix} \begin{pmatrix} 1 & 0 & 2 \\ 0 & 2 & -3 \\ 0 & 0 & -\frac{27}{2} \end{pmatrix}$$

and forward and back substitution to solve the linear system

$$\begin{pmatrix} 1 & 0 & 2 \\ 4 & 2 & 5 \\ 9 & 1 & 3 \end{pmatrix} \begin{pmatrix} x \\ y \\ z \end{pmatrix} = \begin{pmatrix} 2 \\ 5 \\ 3 \end{pmatrix}$$

10. Use elementary row reduction to find an LU decomposition of the following matrices.

a. $\begin{pmatrix} 2 & 2 & 1 \\ 3 & 5 & 0 \end{pmatrix}$
b. $\begin{pmatrix} 2 & 2 & 1 & 0 & 3 \\ 3 & 5 & 0 & 0 & 0 \\ 1 & 3 & 1 & 0 & 2 \end{pmatrix}$
c. $\begin{pmatrix} 5 & 2 & 5 & 3 & 2 & 5 \\ 0 & 5 & 2 & 0 & 3 & 5 \\ 3 & 2 & 1 & 4 & 0 & 4 \\ 3 & 3 & 1 & 5 & 5 & 0 \end{pmatrix}$

11. Use elementary row reduction and suitable permutation matrices P to find an LU decomposition of PA for the following matrices A.

a. $\begin{pmatrix} 0 & 3 \\ 4 & 0 \\ 5 & 5 \end{pmatrix}$
b. $\begin{pmatrix} 0 & 0 & 3 & 5 \\ 1 & 4 & 0 & 4 \end{pmatrix}$
c. $\begin{pmatrix} 0 & 0 & 3 & 1 & 3 \\ 0 & 1 & 5 & 0 & 0 \\ 3 & 0 & 1 & 3 & 0 \\ 3 & 4 & 2 & 3 & 2 \end{pmatrix}$

12. Explore the properties of the **LUDecomposition** and **LUBackSubstitution** functions and compare your findings with the properties of the **LUFactor** function, implemented in the **GaussianElimination** package.

13. Use *MATHEMATICA* to find the LU decomposition and the matrices L and U of the following matrices.

a. $\begin{pmatrix} 3 & 3 & 2 \\ 5 & 5 & 4 \\ 5 & 0 & 2 \end{pmatrix}$
b. $\begin{pmatrix} 1 & 1 & 4 \\ 0 & 4 & 1 \\ 4 & 5 & 4 \end{pmatrix}$
c. $\begin{pmatrix} -4 & 1 & 3 & -2 \\ 2 & -1 & 0 & 5 \\ -2 & 5 & 2 & 1 \\ 2 & -4 & 2 & -5 \end{pmatrix}$

14. Use LU decomposition to solve the following matrix equations $Ax = b_i$.

a. $A = \begin{pmatrix} 1 & 0 & 2 \\ 5 & 2 & 5 \\ 3 & 2 & 5 \end{pmatrix}$ $b_1 = \begin{pmatrix} 1 \\ 2 \\ 3 \end{pmatrix}$ $b_2 = \begin{pmatrix} 0 \\ 5 \\ 4 \end{pmatrix}$ $b_3 = \begin{pmatrix} 6 \\ 7 \\ 1 \end{pmatrix}$

b. $A = \begin{pmatrix} 1 & 5 & 2 & 0 \\ 3 & 5 & 3 & 2 \\ 1 & 4 & 0 & 4 \\ 3 & 3 & 1 & 5 \end{pmatrix}$ $b_1 = \begin{pmatrix} 1 \\ 2 \\ 0 \\ 2 \end{pmatrix}$ $b_2 = \begin{pmatrix} 0 \\ 1 \\ 0 \\ 4 \end{pmatrix}$ $b_3 = \begin{pmatrix} 2 \\ 8 \\ 4 \\ -2 \end{pmatrix}$

15. Use *PLU*. decomposition to solve the following matrix equations $Ax = \mathbf{b}_i$.

a. $A = \begin{pmatrix} 0 & 0 & 2 \\ 5 & 2 & 5 \\ 3 & 2 & 5 \end{pmatrix}$ $\mathbf{b}_1 = \begin{pmatrix} 1 \\ 2 \\ 3 \end{pmatrix}$ $\mathbf{b}_2 = \begin{pmatrix} 0 \\ 5 \\ 4 \end{pmatrix}$ $\mathbf{b}_3 = \begin{pmatrix} 6 \\ 7 \\ 1 \end{pmatrix}$

b. $A = \begin{pmatrix} 0 & 5 & 2 & 0 \\ 3 & 5 & 3 & 2 \\ 1 & 4 & 0 & 4 \\ 3 & 3 & 1 & 5 \end{pmatrix}$ $\mathbf{b}_1 = \begin{pmatrix} 1 \\ 2 \\ 0 \\ 2 \end{pmatrix}$ $\mathbf{b}_2 = \begin{pmatrix} 0 \\ 1 \\ 0 \\ 4 \end{pmatrix}$ $\mathbf{b}_3 = \begin{pmatrix} 2 \\ 8 \\ 4 \\ -2 \end{pmatrix}$

16. Find two distinct LU decompositions $L_1 U_1$ and $L_2 U_2$ of the zero matrix

$$\begin{pmatrix} 0 & 0 & 0 \\ 0 & 0 & 0 \\ 0 & 0 & 0 \end{pmatrix}$$

and explain why this does not contradict Theorem 2.36.

17. Construct an example that illustrates Theorem 2.33.

18. Construct an example that illustrates Theorem 2.35.

19. Use the **LinearSolve** function to find solutions for the equations in Exercise 14.

Complex Matrices

A *complex matrix* is a matrix $A : \mathbf{m} \times \mathbf{n} \rightarrow \mathbb{C}$ whose entries are complex numbers. A *complex polynomial matrix* is a matrix $A : \mathbf{m} \times \mathbf{n} \rightarrow \mathbb{C}[t]$ whose entries are complex polynomials. Since every real number is also a complex number, we can regard all real matrices as complex. Here are some examples of complex matrices that are not real matrices.

EXAMPLE 2.47 ■ **A Complex Matrix**

Find a *MATHEMATICA* representation of the matrix $A : \mathbf{3} \times \mathbf{4} \rightarrow \mathbb{C}$, defined by

$$A[1, 1] = 3, \ A[1, 2] = 4, \ A[1, 3] = \tfrac{1}{2}, \ A[1, 4] = \pi,$$
$$A[2, 1] = 2, \ A[2, 2] = \sqrt{-1} = i, \ A[2, 3] = -i, \ A[2, 4] = 2,$$
$$A[3, 1] = 8, \ A[3, 2] = -\tfrac{6}{7}, \ A[3, 3] = -1 + 2i, \ A[3, 4] = \sqrt{15}$$

Solution. We use the **Create Matrix** palette.

$$\text{In[1]:= } A = \begin{pmatrix} 3 & 4 & \frac{1}{2} & \pi \\ 2 & \sqrt{-1} & -i & 2 \\ 8 & -\frac{6}{7} & -1+2i & \sqrt{15} \end{pmatrix}$$

$\text{Out[1]= } \{\{3,4,\frac{1}{2},\pi\},\{2,i,-i,2\},\{8,-\frac{6}{7},-1+2i,\sqrt{15}\}\}$

As we can see, *MATHEMATICA* represents the given matrix as a list of three lists. ◄►

EXAMPLE 2.48 ■ **A Complex Polynomial Matrix**

Write the matrix $A : \mathbf{3} \times \mathbf{4} \to \mathbb{C}[t]$, defined by

$$A[1,1] = \sqrt{-1}t^4, \ A[1,2] = \sqrt{-1}, \ A[1,3] = t^2 - 1, \ A[1,4] = 2t,$$
$$A[2,1] = 5, \ A[2,2] = t^3, \ A[2,3] = 4\pi, \ A[2,4] = -t^5 + t^3 - 7,$$
$$A[3,1] = 1 + 2\sqrt{-1}, \ A[3,2] = 5, \ A[3,3] = t + 8, \ A[3,4] = \tfrac{1}{3}t + 12$$

as a list of lists and convert the *MATHEMATICA* output to matrix form.

Solution. We first write A as a list of lists $\{r_1, r_2, r_3\}$, where

$$1. \ r_1 = \left\{\sqrt{-1}t^4, \sqrt{-1}, t^2 - 1, 2t\right\}$$
$$2. \ r_2 = \left\{5, t^3, 4\pi, -t^5 + t^3 - 7\right\}$$
$$3. \ r_3 = \left\{1 + 2\sqrt{-1}, 5, t + 8, \tfrac{1}{3}t + 12\right\}$$

In[1]:= A={{√-1t⁴,√-1,t²-1,2t},{5,t³,4π,-t⁵+t³-7},

{1+2√-1,5,t+8,$\frac{1}{3}$t+12}}

Out[1]= {{i t⁴,i,-1+t²,2t},{5,t³,4π,-7+t³-t⁵},

{1+2i,5,8+t,12+$\frac{t}{3}$}}

Then we use the **MatrixForm** function to find the matrix form of A.

In[2]:= **MatrixForm[A]**

Out[2]//MatrixForm=

$$\begin{pmatrix} i\,t^4 & i & -1+t^2 & 2t \\ 5 & t^3 & 4\pi & -7+t^3-t^5 \\ 1+2i & 5 & 8+t & 12+\frac{1}{3}t \end{pmatrix}$$

This output is the required polynomial matrix. ◄►

Hermitian Matrices

If $A = (a_{ij})$ is an $m \times n$ matrix with complex entries, then the *conjugate transpose* A^H of A is the $n \times m$ matrix (b_{ij}) in which $b_{ij} = \overline{a_{ji}}$. For real matrices, the transpose A^T and the conjugate transpose A^H are the same since $\overline{a_{ji}} = a_{ij}$ for all entries of A.

EXAMPLE 2.49 ■ The Conjugate Transpose of a Matrix

Use *MATHEMATICA* to calculate the conjugate transpose of the matrix

$$A = \begin{pmatrix} 1 & 2-4i & 9 & 3 \\ i & 6 & 7+5i & 8 \end{pmatrix}$$

Solution. By combining the **Conjugate** and **Transpose** functions, we can compute A^H in one step.

```
In[1]:= Conjugate[Transpose[{{1,2-4i,9,3},
{i,6,7+5i,8}}]]//MatrixForm

Out[1]//MatrixForm=
```
$$\begin{pmatrix} 1 & -i \\ 2+4i & 6 \\ 9 & 7-5i \\ 3 & 8 \end{pmatrix}$$

This output is the *MATHEMATICA* form of A^H. ◄►

A *Hermitian matrix* is a complex matrix A with the property that $A = A^H$.

EXAMPLE 2.50 ■ A Hermitian Matrix

The matrix

$$A = \begin{pmatrix} a_{11} & a_{12} & a_{13} & a_{14} \\ a_{21} & a_{22} & a_{23} & a_{24} \\ a_{31} & a_{32} & a_{33} & a_{34} \\ a_{41} & a_{42} & a_{43} & a_{44} \end{pmatrix} = \begin{pmatrix} 1 & 2-i & 9 & 3 \\ 2+i & 6 & i & 8 \\ 9 & -i & 4 & 5+7i \\ 3 & 8 & 5-7i & 3 \end{pmatrix}$$

is Hermitian, since $a_{ij} = \overline{a_{ji}}$ for all indices $1 \le i, j \le 4$. ◄►

Unitary Matrices

A *unitary matrix* is a complex matrix A with the property that $A^{-1} = A^H$.

EXAMPLE 2.51 ■ A Unitary Matrix

Use *MATHEMATICA* to show that the matrix

$$A = \begin{pmatrix} \dfrac{1-i}{2} & -\dfrac{1-i}{2} \\ \dfrac{1+i}{2} & \dfrac{1+i}{2} \end{pmatrix}$$

is unitary.

Solution. We use *MATHEMATICA* to create the matrices A^{-1} and A^H and compare the results.

$$\text{In[1]:= A=} \begin{pmatrix} \dfrac{1-i}{2} & -\dfrac{1-i}{2} \\ \dfrac{1+i}{2} & \dfrac{1+i}{2} \end{pmatrix};$$

In[2]:= **MatrixForm[Inverse[A]]**

Out[2]//MatrixForm=

$$\begin{pmatrix} \dfrac{1}{2}+\dfrac{i}{2} & \dfrac{1}{2}-\dfrac{i}{2} \\ -\dfrac{1}{2}-\dfrac{i}{2} & \dfrac{1}{2}-\dfrac{i}{2} \end{pmatrix}$$

In[3]:= **MatrixForm[Conjugate[Transpose[[A]]]]**

Out[3]//MatrixForm=

$$\begin{pmatrix} \dfrac{1}{2}+\dfrac{i}{2} & \dfrac{1}{2}-\dfrac{i}{2} \\ -\dfrac{1}{2}-\dfrac{i}{2} & \dfrac{1}{2}-\dfrac{i}{2} \end{pmatrix}$$

As we can see, the two matrices are the same. Hence A is unitary. ◀▶

EXERCISES 2.8

1. Construct two complex 2×3 matrices A and B, each with 3 real and 3 imaginary entries. Compute the sum $A + B$.

2. Compute the conjugate transposes of the matrices A, B, and $A + B$ in Exercise 1.

3. Construct a complex 3×4 matrix A and a 4×3 matrix B, each with six real and six imaginary entries. Compute the matrices AB and BA.

4. Compute the conjugate transposes of the matrices A, B, AB, and BA in Exercise 3.

5. Construct a 2×2 matrix A whose four entries are complex polynomials and compute the matrix A^2.

6. Find a 2×2 and a 3×3 Hermitian matrix. Use *MATHEMATICA* to verify your answer.

7. Use *MATHEMATICA* to verify that the matrix

$$A = \begin{pmatrix} \frac{1}{\sqrt{2}} & 0 & \frac{i}{\sqrt{2}} \\ 0 & 1 & 0 \\ -\frac{1}{\sqrt{2}} & 0 & \frac{i}{\sqrt{2}} \end{pmatrix}$$

is unitary. Modify A so that the result is a 2×2 unitary matrix.

8. Show that the Pauli spin matrices

a. $\begin{pmatrix} 0 & 1 \\ 1 & 0 \end{pmatrix}$ b. $\begin{pmatrix} 0 & -i \\ i & 0 \end{pmatrix}$ c. $\begin{pmatrix} 1 & 0 \\ 0 & -1 \end{pmatrix}$

used to compute the electron spin in physics are both Hermitian and unitary.

9. Show that the Dirac matrices

a. $\begin{pmatrix} 0 & 0 & 0 & 1 \\ 0 & 0 & 1 & 0 \\ 0 & 1 & 0 & 0 \\ 1 & 0 & 0 & 0 \end{pmatrix}$ b. $\begin{pmatrix} 0 & 0 & 0 & -i \\ 0 & 0 & i & 0 \\ 0 & -i & 0 & 0 \\ i & 0 & 0 & 0 \end{pmatrix}$

c. $\begin{pmatrix} 0 & 0 & 1 & 0 \\ 0 & 0 & 0 & -1 \\ 1 & 0 & 0 & 0 \\ 0 & -1 & 0 & 0 \end{pmatrix}$ d. $\begin{pmatrix} 1 & 0 & 0 & 0 \\ 0 & 1 & 0 & 0 \\ 0 & 0 & -1 & 0 \\ 0 & 0 & 0 & -1 \end{pmatrix}$

also used in physics are both Hermitian and unitary.

APPLICATIONS

Geometry and Invertible Matrices

In this section, we illustrate the geometric properties of invertible 2×2 matrices. We show that these matrices correspond to sequences of contractions, expansions, reflections, and shears.

Let

$$A = \begin{pmatrix} a & b \\ c & d \end{pmatrix}$$

be a 2×2 real invertible matrix. By Corollary 2.19, the matrix A is equal to a product $E_1 \cdots E_n$ of elementary matrices and can therefore be obtained from the 2×2 identity matrix by a sequence of elementary row operations. We now explore the geometric effect of the transformation $\mathbf{x} \to A\mathbf{x}$ on a point $\mathbf{x}$ of $\mathbb{R}^2$.

We recall from Section 1, that the following cases describe the possible types of elementary 2×2 matrices.

1. The operations $(R_1 \to sR_1)$, for $s \neq 0$, yield the elementary matrices

$$\begin{pmatrix} 1 & 0 \\ 0 & 1 \end{pmatrix} (R_1 \to sR_1) \begin{pmatrix} s & 0 \\ 0 & 1 \end{pmatrix} = E_1$$

Left-multiplication by E_1 is called an ***expansion along the x-axis*** if $s > 1$, and a ***contraction along the x-axis*** if $s < 1$. If $s = -1$, then a left-multiplication by E_1 corresponds to a ***reflection about the y-axis***.

2. The operations $(R_2 \to sR_2)$, for $s \neq 0$, yield the elementary matrices

$$\begin{pmatrix} 1 & 0 \\ 0 & 1 \end{pmatrix} (R_2 \to sR_2) \begin{pmatrix} 1 & 0 \\ 0 & s \end{pmatrix} = E_2$$

Left-multiplication by E_2 is called an ***expansion along the y-axis*** if $s > 1$, and a ***contraction along the y-axis*** if $s < 1$. If $s = -1$, then a left-multiplication by E_2 corresponds to a ***reflection about the x-axis***.

3. The operation $(R_1 \rightleftharpoons R_2)$ yields the elementary matrix

$$\begin{pmatrix} 1 & 0 \\ 0 & 1 \end{pmatrix} (R_1 \rightleftharpoons R_2) \begin{pmatrix} 0 & 1 \\ 1 & 0 \end{pmatrix} = E_3$$

Left-multiplication by E_3 is called a ***reflection about the line y=x***.

4. The operations $(R_1 \to R_1 + sR_2)$ yield the elementary matrices

$$\begin{pmatrix} 1 & 0 \\ 0 & 1 \end{pmatrix} (R_1 \to R_1 + sR_2) \begin{pmatrix} 1 & s \\ 0 & 1 \end{pmatrix} = E_4$$

Left-multiplication by E_3 is called a ***shear along the x-axis***.

5. The operations $(R_2 \to R_2 + sR_1)$ yield the elementary matrices

$$\begin{pmatrix} 1 & 0 \\ 0 & 1 \end{pmatrix} (R_2 \to R_2 + sR_1) \begin{pmatrix} 1 & 0 \\ s & 1 \end{pmatrix} = E_5$$

Left-multiplication by E_3 is called a ***shear along the y-axis***.

The operations $(R_1 \rightarrow R_1 + sR_1)$ and $(R_2 \rightarrow R_2 + sR_2)$ are special cases of $(R_i \rightarrow sR_i)$ since $(R_i \rightarrow R_i + sR_i) = (R_i \rightarrow (1 + s)\, R_i)$.

The **AffineMap** function of *MATHEMATICA* provides an efficient tool for calculating the geometric effects of left-multiplication by invertible matrices.

EXAMPLE 2.52   ■ *MATHEMATICA* and 2 × 2 Invertible Matrices

Use *MATHEMATICA* to define the geometric transformations corresponding to left-multiplication by elementary 2 × 2 matrices.

Solution. We begin by loading the **AffineMaps** package.

```
In[1]:= <<ProgrammingInMathematica`AffineMaps`
```

Next we define the transformations determined by the matrices E_1–E_5.

```
In[2]:= m1=AffineMap[{{s,0,0},{0,1,0}}]

Out[2]= - map -
```

```
In[3]:= m2=AffineMap[{{1,0,0},{0,s,0}}]

Out[3]= - map -
```

```
In[4]:= m3=AffineMap[{{0,1,0},{1,0,0}}]

Out[4]= - map -
```

```
In[5]:= m4=AffineMap[{{1,s,0},{0,1,0}}]

Out[5]= - map -
```

```
In[6]:= m5=AffineMap[{{1,0,0},{s,1,0}}]

Out[6]= - map -
```

For any invertible 2×2 matrix $A = E_n \cdots E_1$, the geometric effect of left-multiplying a point (x, y) by A corresponds to the assignment

$$\{x, y\} \rightarrow (m_n \circ \cdots \circ m_1) (\{x, y\})$$

where the m_i are the *MATHEMATICA* transformations determined by the matrices E_i, and where

$$m_n \circ \cdots \circ m_1$$

is the composite transformation of $m_1, \ldots, m_n$. ◄►

The next two examples illustrate the use of the transformations defined in Example 2.52.

EXAMPLE 2.53 ■ Expansions of the Unit Square

Use *MATHEMATICA* to find the rectangle determined by a unit square with vertices $(0, 0)$, $(1, 0)$, $(1, 1)$, and $(0, 1)$, and by an expansion along the x-axis for $s = 2$. Graph the rectangle.

Solution. We first load the **AffineMap** package.

```
In[1]:= <<ProgrammingInMathematica`AffineMaps`
```

Then we define the appropriate *MATHEMATICA* function.

```
In[2]:= ExpansionX := AffineMap[{{1,0,0},{0,1,0}}]
```

Next we calculate the new vertices.

```
In[3]:= Map[ExpansionX,{{0,0},{1,0},{1,1},{0,1}}]

Out[3]= {{0,0},{2,0},{2,1},{0,1}}
```

Finally, we graph the rectangle.

```
In[4]:= Show[Graphics[{GrayLevel[0.4],Rectangle[{0,0},{2,1}]}],
AspectRatio→Automatic,Axes→True]
```

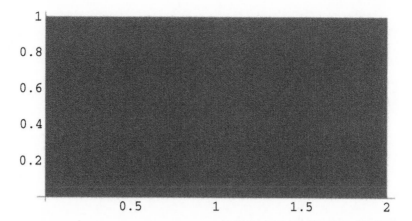

As we can see, the unit square is expanded along the x-axis to a rectangle with a base of length 2 and height 1. ◄►

EXAMPLE 2.54 ■ Four Geometric Transformations

Use *MATHEMATICA* to find the triangle determined by the isosceles triangle with vertices $(0, 0)$, $(4, 0)$, and $(2, 2)$, and by the geometric transformation corresponding to the following operations:

1. A reflection about the line $y = x$,

2. A shear along the x-axis for $s = 3$,

3. A contraction along the y-axis for $s = 1/4$,

4. A reflection about the x-axis.

Solution. We begin by loading the **AffineMaps** package.

```
In[1]:= <<ProgrammingInMathematica`AffineMaps`
```

Next we define the appropriate *MATHEMATICA* functions and compose them.

```
In[2]:= m1 :=AffineMap[{{0,1,0},{1,0,0}}]
```

```
In[3]:= m2 :=AffineMap[{{1,3,0},{0,1,0}}]
```

```
In[4]:= m3 := AffineMap[{{1,0,0},{0,1/4,0}}]
```

```
In[5]:= m4 := AffineMap[{{1,0,0},{0,-1,0}}]
```

```
In[6]  m = Composition[m4,m3,m2,m1]
Out[1]= - map -
```

Now we calculate the new vertices.

```
In[7]:= G={{0,0},{4,0},{2,2}};
```

```
In[8]:= Map[m, G]
Out[8]= {{0,0,{12,-1},{8,-1/2}}}
```

Finally, we graph the triangle.

```
In[9]:= P = Polygon[Map[m,G]];
```

```
In[10]:= Show[Graphics[{GrayLevel[0.4],P}],AspectRatio → .5,
Axes → True]
```

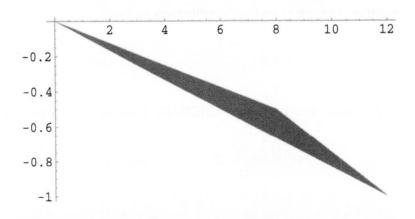

```
Out[10]= - Graphics -
```

As we can see, the vertex $(0, 0)$ remains fixed, the vertex $(4, 0)$ is mapped to $(12, -1)$, and the vertex $(2, 2)$ is mapped to $(8, -\frac{1}{2})$. ◀▶

Reflections about the line $y = x$ are a special case of rotations about the origin. We use the **Rotations** and **Arrow** packages of *MATHEMATICA* to illustrate this fact.

EXAMPLE 2.55 ■ **Reflection About the Line** $y = x$

Let **x** and **y** be the vectors

$$\begin{pmatrix} \frac{1}{2} + \frac{\sqrt{3}}{2} \\ -\frac{1}{2} + \frac{\sqrt{3}}{2} \end{pmatrix} \quad \text{and} \quad \begin{pmatrix} -\frac{1}{2} + \frac{\sqrt{3}}{2} \\ \frac{1}{2} + \frac{\sqrt{3}}{2} \end{pmatrix}$$

respectively. Use *MATHEMATICA* to show that **x** and **y** can be obtained from the vector $(1, 1)$ by rotations about the origin.

Solution. We begin by loading the required graphics packages.

```
In[1]:= <<Geometry`Rotations`;<<Graphics`Arrow`
```

Next we define the rotated vectors.

```
In[2]:= u=Arrow[{0,0},Rotate2D[{1,1},Pi/6]];
```

```
In[3]:= v=Arrow[{0,0},Rotate2D[{1,1},-Pi/6]];
```

We now graph the vectors.

```
In[4]:= Show[Graphics[{u,v,Line[{{0,0},{1,1}}]}],Axes → True,
AspectRatio → Automatic]
```

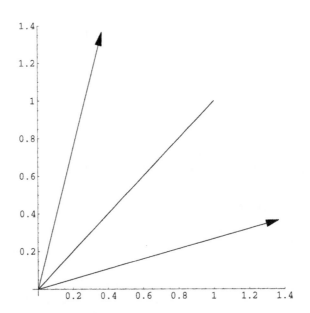

Out[4]= - Graphics -

It remains to show that $\mathbf{u} = \mathbf{x}$ and $\mathbf{v} = \mathbf{y}$.

In[5]:= u[[2]]=={1/2+Sqrt[3]/2,-1/2+Sqrt[3]/2}

Out[5]:= True

In[6]:= v[[2]]=={-1/2+Sqrt[3]/2,1/2+Sqrt[3]/2}

Out[6]:= True

As we can see, the reflections about the line $y = x$ are special rotations. ◄►

The rotation matrices

$$\begin{pmatrix} \cos\theta & -\sin\theta \\ \sin\theta & \cos\theta \end{pmatrix} \quad \text{and} \quad \begin{pmatrix} \cos(-\theta) & -\sin(-\theta) \\ \sin(-\theta) & \cos(-\theta) \end{pmatrix}$$

are inverses of each other. They must therefore be products of 2×2 elementary matrices. We now show how such products can be constructed.

THEOREM 2.38   *For any $\theta \in \mathbb{R}$, the rotation matrix*

$$A = \begin{pmatrix} \cos\theta & -\sin\theta \\ \sin\theta & \cos\theta \end{pmatrix}$$

is a product of elementary matrices.

Proof. Let $a = \cos\theta$ and $b = \sin\theta$. If $\theta = 2k\pi$, then $a = 1$ and $b = 0$. This means that A is the 2×2 identity matrix and is therefore elementary. Suppose now that $\theta \neq 2k\pi$. Then it is clear from trigonometry that $b \neq 0$ and $a^2 + b^2 = 1$. Therefore,

$$
\begin{pmatrix} 1/b & 0 \\ 0 & 1 \end{pmatrix}
\begin{pmatrix} 1 & a \\ 0 & 1 \end{pmatrix}
\begin{pmatrix} -1 & 0 \\ 0 & 1 \end{pmatrix}
\begin{pmatrix} 1 & 0 \\ a & 1 \end{pmatrix}
\begin{pmatrix} 0 & 1 \\ 1 & 0 \end{pmatrix}
\begin{pmatrix} b & 0 \\ 0 & 1 \end{pmatrix}
$$

is a decomposition of A into a product of elementary matrices. ∎

This shows that geometrically, any rotation can be represented by the product of an expansion, a contraction, a reflection about the y-axis, a reflection about the line $y = x$, a shear along the x-axis, and a shear along the y-axis.

EXERCISES 2.9

1. Graph the vectors **x**, A**x**, and B**x**, where

$$
A = \begin{pmatrix} 3 & 0 \\ 0 & 1 \end{pmatrix} \quad
B = \begin{pmatrix} \dfrac{1}{3} & 0 \\ 0 & 1 \end{pmatrix} \quad
\mathbf{x} = \begin{pmatrix} 9 \\ 6 \end{pmatrix}
$$

Explain the connection between the vectors.

2. Graph the vectors **x**, A**x**, and B**x**, where

$$
A = \begin{pmatrix} 1 & 0 \\ 0 & 3 \end{pmatrix} \quad
B = \begin{pmatrix} 1 & 0 \\ 0 & \dfrac{1}{3} \end{pmatrix} \quad
\mathbf{x} = \begin{pmatrix} 9 \\ 6 \end{pmatrix}
$$

Explain the connection between the vectors.

3. Graph the vectors **x**, A**x**, and B**x**, where

$$
A = \begin{pmatrix} 1 & 2 \\ 0 & 1 \end{pmatrix} \quad
B = \begin{pmatrix} 1 & 0 \\ 2 & 1 \end{pmatrix} \quad
\mathbf{x} = \begin{pmatrix} 4 \\ -1 \end{pmatrix}
$$

Explain the connection between the vectors.

4. Graph the vectors **x**, A**x**, and B**x**, where

$$
A = \begin{pmatrix} 0 & 1 \\ 1 & 0 \end{pmatrix} \quad
B = \begin{pmatrix} 0 & -2 \\ -2 & 0 \end{pmatrix} \quad
\mathbf{x} = \begin{pmatrix} 9 \\ 6 \end{pmatrix}
$$

Explain the connection between the vectors.

5. Explain the geometric steps involved in transforming the point

$$
\mathbf{x} = \begin{pmatrix} 2 \\ 3 \end{pmatrix}
$$

to the point $A\mathbf{x}$, where A is the matrix product

$$\begin{pmatrix} 0 & 1 \\ 1 & 0 \end{pmatrix} \begin{pmatrix} \frac{1}{3} & 0 \\ 0 & 1 \end{pmatrix} \begin{pmatrix} 1 & 0 \\ 2 & 1 \end{pmatrix} \begin{pmatrix} 0 & -3 \\ -3 & 0 \end{pmatrix}$$

6. Repeat Exercises 1–5 using the **AffineMap** function of *MATHEMATICA*.

Computer Animation

Computer images are sets of points connected by lines and curves. Much of computer animation is concerned with manipulating these images using matrix multiplication. We have already seen how basic geometric transformations such as expansions, contractions, reflections, rotations, and shears can be achieved through matrix multiplication. However, moving images and three-dimensional objects in space also require translations, which are algebraically nonlinear. The operation $T : \mathbb{R}^2 \to \mathbb{R}^2$ defined by $T(x, y) = (x + h, y + k)$ mapping the origin $(0, 0)$ to the point $(0 + h, 0 + k) = (h, k)$ it is linear if and only if $h = k = 0$. It is therefore impossible to move computer images in the plane by multiplying their points by 2×2 matrices. For the same reason, three-dimensional objects cannot be moved in space by multiplying them by 3×3 matrices.

It is quite remarkable that matrix multiplication can nevertheless be used in a unifying way to reflect, rotate, shear, and translate two-dimensional computer images. The trick consists of embedding two-dimensional images in $\mathbb{R}^3$, using matrix multiplication in $\mathbb{R}^3$, and then converting the new images back into two-dimensional objects.

The embedding that makes this possible is the function $H : \mathbb{R}^2 \to \mathbb{R}^3$ defined by

$$H \begin{pmatrix} x \\ y \end{pmatrix} = \begin{pmatrix} x \\ y \\ 1 \end{pmatrix}$$

The coordinates $(x, y, 1)$ are called the ***homogeneous coordinates*** of the points (x, y). It is important to note that H is not a linear transformation. However, the standard coordinates of points in $\mathbb{R}^2$ can be recovered from their homogeneous coordinates by stripping off the third coordinate. Figure 1 illustrates the relationship between the xy-plane, represented in $\mathbb{R}^3$ by the equation $z = 0$, and its image under H, represented by the equation $z = 1$.

EXAMPLE 2.56 ■ **Translation and Matrix Multiplication**

Find a 3×3 matrix that translates the homogeneous coordinates of the point (x, y) to the homogeneous coordinates of the points $(x + h, y + k)$ by matrix multiplication.

Solution. We show that the matrix

$$A = \begin{pmatrix} 1 & 0 & h \\ 0 & 1 & k \\ 0 & 0 & 1 \end{pmatrix}$$

has the required properties.

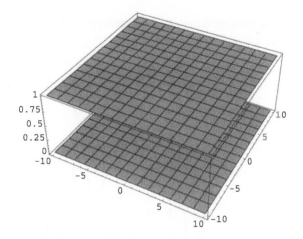

FIGURE 1   The Planes $z = 0$ and $z = 1$ in $\mathbb{R}^3$.

Let (x, y) be any vector in $\mathbb{R}^2$. Then

$$A\left[H\left(\begin{array}{c} x \\ y \end{array}\right)\right] = \left(\begin{array}{ccc} 1 & 0 & h \\ 0 & 1 & k \\ 0 & 0 & 1 \end{array}\right)\left(\begin{array}{c} x \\ y \\ 1 \end{array}\right) = \left(\begin{array}{c} x + h \\ y + k \\ 1 \end{array}\right)$$

We define the inverse of this operation by letting

$$\widehat{H}\left(\begin{array}{c} x + h \\ y + k \\ 1 \end{array}\right) = \left(\begin{array}{c} x + h \\ y + k \end{array}\right)$$

As we can see, the translation of (x, y) to $(x + h, y + k)$ is accomplished in $\mathbb{R}^3$ by multiplying the homogeneous coordinates of (x, y) by the matrix A. ◄►

EXAMPLE 2.57   ■ Translation of a Triangle

Use matrix multiplication to translate the triangle with vertices $(0, 0)$, $(2, 0)$, and $(2, 1)$ to the triangle with vertices $(3, 4)$, $(5, 4)$, and $(5, 5)$.

Solution. We know from Example 2.56 that the matrix

$$A = \left(\begin{array}{ccc} 1 & 0 & 3 \\ 0 & 1 & 4 \\ 0 & 0 & 1 \end{array}\right)$$

is the appropriate translation matrix. If we multiply the homogeneous coordinates of the given points by A, we get

$$\text{a.} \begin{pmatrix} 1 & 0 & 3 \\ 0 & 1 & 4 \\ 0 & 0 & 1 \end{pmatrix} \begin{pmatrix} 0 \\ 0 \\ 1 \end{pmatrix} = \begin{pmatrix} 3 \\ 4 \\ 1 \end{pmatrix}$$

$$\text{b.} \begin{pmatrix} 1 & 0 & 3 \\ 0 & 1 & 4 \\ 0 & 0 & 1 \end{pmatrix} \begin{pmatrix} 2 \\ 0 \\ 1 \end{pmatrix} = \begin{pmatrix} 5 \\ 4 \\ 1 \end{pmatrix}$$

$$\text{c.} \begin{pmatrix} 1 & 0 & 3 \\ 0 & 1 & 4 \\ 0 & 0 & 1 \end{pmatrix} \begin{pmatrix} 2 \\ 1 \\ 1 \end{pmatrix} = \begin{pmatrix} 5 \\ 5 \\ 1 \end{pmatrix}$$

If we apply the function $\widehat{H}$ to these vectors, we get the vertices $(3, 4)$, $(5, 4)$, and $(5, 5)$ of the translated triangle. ◄►

We conclude this section by showing how the **AffineMap** function of *MATHEMATICA* can be used in place of homogeneous coordinates to carry the calculations required for computer animation.

EXAMPLE 2.58 ■ Homogeneous Coordinates and the AffineMap

Show that the **AffineMap** function of *MATHEMATICA* can be used to calculate a sequence of affine transformations.

Solution. Let $T : \mathbb{R}^2 \rightarrow \mathbb{R}^2$ be the affine transformation $T(\mathbf{x}) = A\mathbf{x} + \mathbf{b}$ consisting of a counterclockwise rotation by $\pi/3$, a shear along the x-axis for $s = 3$, an expansion along the y-axis for $s = 2$, and a translation $(x, y) \rightarrow (x + 3, y + 4)$. We show that the representations of T using homogeneous coordinates and using the **AffineMap** function are equivalent.

We know from our previous work that the matrix

$$A = \begin{pmatrix} 1 & 0 & 3 \\ 0 & 1 & 4 \\ 0 & 0 & 1 \end{pmatrix} \begin{pmatrix} \cos \pi/3 & -\sin \pi/3 & 0 \\ \sin \pi/3 & \cos \pi/3 & 0 \\ 0 & 0 & 1 \end{pmatrix} \begin{pmatrix} 1 & 0 & 0 \\ 0 & 2 & 0 \\ 0 & 0 & 1 \end{pmatrix} \begin{pmatrix} 1 & 3 & 0 \\ 0 & 1 & 0 \\ 0 & 0 & 1 \end{pmatrix}$$

$$= \begin{pmatrix} \frac{1}{2} & \frac{3}{2} - \sqrt{3} & 3 \\ \frac{1}{2}\sqrt{3} & \frac{3}{2}\sqrt{3} + 1 & 4 \\ 0 & 0 & 1 \end{pmatrix}$$

represents T relative to homogeneous coordinates. This means that

$$T\begin{pmatrix} x \\ y \end{pmatrix} = \widehat{H}\left[A\begin{pmatrix} x \\ y \\ 1 \end{pmatrix}\right]$$

$$= \widehat{H}\begin{pmatrix} \frac{1}{2}x + \left(\frac{3}{2} - \sqrt{3}\right)y + 3 \\ \frac{1}{2}\sqrt{3}x + \left(\frac{3}{2}\sqrt{3} + 1\right)y + 4 \\ 1 \end{pmatrix}$$

$$= \begin{pmatrix} \frac{1}{2}x + \left(\frac{3}{2} - \sqrt{3}\right)y + 3 \\ \frac{1}{2}\sqrt{3}x + \left(\frac{3}{2}\sqrt{3} + 1\right)y + 4 \end{pmatrix}$$

We now show that we can also use the **AffineMap** function to represent T. We begin by loading the **AffineMaps** package.

```
In[1]:= <<ProgrammingInMathematica`AffineMaps`
```

Next we create the necessary 2×3 matrix. First we form matrix

$$\begin{pmatrix} \cos\pi/3 & -\sin\pi/3 \\ \sin\pi/3 & \cos\pi/3 \end{pmatrix}\begin{pmatrix} 1 & 0 \\ 0 & 2 \end{pmatrix}\begin{pmatrix} 1 & 3 \\ 0 & 1 \end{pmatrix} = \begin{pmatrix} \frac{1}{2} & \frac{3}{2} - \sqrt{3} \\ \frac{1}{2}\sqrt{3} & \frac{3}{2}\sqrt{3} + 1 \end{pmatrix}$$

and then add a third column to represent the translation $(x, y) \to (x + 3, y + 4)$.

```
In[2]:= B= ( 1/2      3/2-√3   3
            1/2√3    3/2√3+1   4 );
```

Using B, we can define the appropriate affine transformation.

```
In[3]:= m := AffineMap[B]
```

We now evaluate m at (x, y).

```
In[4]:= MatrixForm[m[{x,y}]]
```

$$\text{Out[4]} = \begin{pmatrix} 3 + \frac{x}{2} + \left(\frac{3}{2} - \sqrt{3}\right)y \\ 4 + \frac{\sqrt{3}x}{2} + \left(1 + \frac{3\sqrt{3}}{2}\right)y \end{pmatrix}$$

The two obtained vectors are clearly equal. ◀▶

EXERCISES 2.10

1. Find the homogeneous coordinates in $\mathbb{R}^3$ of the points $(3, 5)$, $(-8, 0)$, and $(112, 115)$ in $\mathbb{R}^2$.

2. Find the points in $\mathbb{R}^2$ whose homogeneous coordinates in $\mathbb{R}^3$ are $(4, 5, 1)$, $(-8, -6, 1)$, and $(\pi, -\pi, 1)$.

3. Find the 3×3 matrix A that converts homogeneous coordinates of the image determined by the points $(0, 0)$, $(3, 0)$, $(3, 2)$, and $(0, 2)$ to the homogeneous coordinates of the new image obtained by three consecutive operations:

   a. a shear $\begin{pmatrix} 1 & \frac{1}{3} \\ 0 & 1 \end{pmatrix}$ along the x-axis,

   b. a counterclockwise rotation by $\pi/4$ radians, and

   c. a translation $(x, y) \rightarrow (x + 2, y - 5)$.

4. The homogeneous coordinates of a point (x, y, z) in $\mathbb{R}^3$ are $(x, y, z, 1)$ in $\mathbb{R}^4$. Find the homogeneous coordinates in $\mathbb{R}^4$ of the points $(3, 5, 7)$, $(-8, 0, 1)$, and $(112, 115, 0)$.

5. Find the points in $\mathbb{R}^3$ whose homogeneous coordinates in $\mathbb{R}^4$ are $(4, 5, 2, 1)$, $(\pi, -\pi, 8, 1)$, and $(-8, -6, -1, 1)$.

6. Find the 4×4 translation matrix A that corresponds to the translation of axes $(x, y, z) \rightarrow (x + h, y + k, z + j)$ in $\mathbb{R}^3$. Use homogeneous coordinates and the matrix A to construct an example of a geometric object and its translated image.

7. Show that a translation matrix $\begin{pmatrix} 1 & 0 & h \\ 0 & 1 & k \\ 0 & 0 & 1 \end{pmatrix}$ is orthogonal if and only if $h = k = 0$.

8. Modify Exercises 3 and 6 to be able to use the **AffineMap** function of *MATHEMATICA* to solve the associated geometric problems.

Message Encoding

Matrices and their inverses can be used to encode messages. Here is a simple example.

EXAMPLE 2.59   ■ Encoding a Message

Use *MATHEMATICA* to create a scheme for encoding ASCII text messages.

Solution. We assume that all messages are written in capital letters using the standard English alphabet. We therefore begin by assigning numbers to the required letters. The assignment in Table 2 will do. The symbol ♠ denotes a blank space, and ■ identifies the end of a sentence.

TABLE 2   Numerical Encoding of the English Alphabet.

$(A, 1)$	$(B, 2)$	$(C, 3)$	$(D, 4)$	$(E, 5)$	$(F, 6)$	$(G, 7)$
$(H, 8)$	$(I, 9)$	$(J, 10)$	$(K, 11)$	$(L, 12)$	$(M, 13)$	$(N, 14)$
$(O, 15)$	$(P, 16)$	$(Q, 17)$	$(R, 18)$	$(S, 19)$	$(T, 20)$	$(U, 21)$
$(V, 22)$	$(W, 23)$	$(X, 24)$	$(Y, 25)$	$(Z, 26)$	$(\spadesuit, 27)$	$(\blacksquare, 28)$

Using this numerical encoding, we can replace any string of capital letters, spaces, and punctuations by a unique string of numbers. Thus the string

"THE QUICK BROWN FOX JUMPS OVER THE LAZY DOG"

becomes

$$\begin{pmatrix} 20 & 8 & 5 & 27 & 17 & 21 & 9 & 3 & 11 & 27 \\ 2 & 18 & 15 & 23 & 14 & 27 & 6 & 15 & 24 & 27 \\ 10 & 21 & 13 & 16 & 19 & 27 & 15 & 22 & 5 & 18 \\ 27 & 20 & 8 & 5 & 27 & 12 & 1 & 26 & 25 & 27 \\ 4 & 15 & 7 & 28 & & & & & & \end{pmatrix}$$

To be able to compute with this matrix, we fill the empty cells with zeros. This yields the matrix

$$A = \begin{pmatrix} 20 & 8 & 5 & 27 & 17 & 21 & 9 & 3 & 11 & 27 \\ 2 & 18 & 15 & 23 & 14 & 27 & 6 & 15 & 24 & 27 \\ 10 & 21 & 13 & 16 & 19 & 27 & 15 & 22 & 5 & 18 \\ 27 & 20 & 8 & 5 & 27 & 12 & 1 & 26 & 25 & 27 \\ 4 & 15 & 7 & 28 & 0 & 0 & 0 & 0 & 0 & 0 \end{pmatrix}.$$

The numerical encoding of the message determined by this matrix is achieved by premultiplying A with an invertible matrix B that is only known to the sender and the recipient of the message. Using A and B, we form the encoded message BA. Since A is a 5×10 matrix, B must be a 5×5 matrix. We use *MATHEMATICA* to generate a random 5×5 matrix.

$$\text{In[1]:= } A = \begin{pmatrix} 20 & 8 & 5 & 27 & 17 & 21 & 9 & 3 & 11 & 27 \\ 2 & 18 & 15 & 23 & 14 & 27 & 6 & 15 & 24 & 27 \\ 10 & 21 & 13 & 16 & 19 & 27 & 15 & 22 & 5 & 18 \\ 27 & 20 & 8 & 5 & 27 & 12 & 1 & 26 & 25 & 27 \\ 4 & 15 & 7 & 28 & 0 & 0 & 0 & 0 & 0 & 0 \end{pmatrix};$$

```
In[2]:= B=Array[Random[Integer,{0,9}]&,{5,5}]//MatrixForm
```

Out[2]//MatrixForm=

$$\begin{pmatrix} 1 & 4 & 6 & 3 & 2 \\ 1 & 3 & 0 & 1 & 8 \\ 3 & 0 & 4 & 5 & 1 \\ 5 & 4 & 9 & 0 & 2 \\ 2 & 3 & 7 & 9 & 5 \end{pmatrix}$$

Next we verify that B is invertible. If not, we modify it until we have obtained an invertible matrix. We use *MATHEMATICA* to verify that B is invertible by asking the program to compute the inverse M of B without displaying the result. If B were not invertible, *MATHEMATICA* would produce a message to that effect.

```
In[3]:= M=Inverse[B];
```

Since *MATHEMATICA* produced no message, we know that B is invertible and we can use it to encode A. Of course, we could have asked for an explicit display of M.

Next we calculate the matrix BA.

```
In[4]:= E=B.A//MatrixForm
```

Out[4]//MatrixForm=

$$\begin{pmatrix} 177 & 296 & 181 & 286 & 268 & 327 & 126 & 273 & 212 & 324 \\ 85 & 202 & 114 & 325 & 86 & 114 & 28 & 74 & 108 & 135 \\ 239 & 223 & 114 & 198 & 262 & 231 & 92 & 227 & 178 & 288 \\ 206 & 331 & 216 & 427 & 312 & 456 & 204 & 273 & 196 & 405 \\ 379 & 472 & 253 & 420 & 452 & 420 & 150 & 439 & 354 & 504 \end{pmatrix}$$

The matrix E is the encoded message. To decode it, we must use the matrix M. We recover the original numerical version of the transmitted message by calculating the matrix ME.

```
In[5]:= M.E//MatrixForm
```

Out[5]//MatrixForm=

$$
\begin{pmatrix}
20 & 8 & 5 & 27 & 17 & 21 & 9 & 3 & 11 & 27 \\
2 & 18 & 15 & 23 & 14 & 27 & 6 & 15 & 24 & 27 \\
10 & 21 & 13 & 16 & 19 & 27 & 15 & 22 & 5 & 18 \\
27 & 20 & 8 & 5 & 27 & 12 & 1 & 26 & 25 & 27 \\
4 & 15 & 7 & 28 & 0 & 0 & 0 & 0 & 0 & 0
\end{pmatrix}
$$

The string $\mathbf{r}_1(ME)\ \mathbf{r}_2(ME)\ \mathbf{r}_3(ME)\ \mathbf{r}_4(ME)\ \mathbf{r}_5(ME)$ is the original decoded message. ◀▶

EXERCISES 2.11

1. Use the numerical encoding defined in Example 2.59 and the invertible matrix

$$
B = \begin{pmatrix}
-3 & 5 & 2 & 2 \\
-5 & -3 & 4 & 4 \\
2 & 6 & -4 & 6 \\
4 & 4 & 2 & 5
\end{pmatrix}
$$

to encode the message

"SOME PEOPLE LIKE BASEBALL AND SOME LIKE FOOTBALL."

2. Use the inverse of the matrix B in Exercise 1 to decode the message 3, 15, 14, 7, 18, 1, 20, 21, 12, 1, 20, 9, 15, 14, 19, 28, 27, 25, 15, 21, 27, 8, 1, 22, 5, 27, 4, 15, 14, 5, 27, 9, 20, 28.

3. Make up your own coding scheme using a $6 \times n$ matrix for a suitable n, and encode and decode the message in Exercise 1.

4. Repeat Exercise 3 using a $4 \times n$ matrix for a suitable n.

Leontief Input-Output Models

Economic input-output models are well-known applications of matrix algebra. They are named after the Harvard Professor Wassily Leontief, who used a giant matrix to build a model of the U.S. economy and was awarded a Nobel prize for his work. In the following example, we illustrate the basic idea behind the construction.

EXAMPLE 2.60 ■ **Building a Leontief Economic Input-Output Model**

A small island economy in a Pacific archipelago consists of four industries: agriculture, fishing, manufacturing, and tourism. Each industry can meet its own purchasing needs from these four industries. In addition, the island provides goods and services to some of the neighboring islands. Table 3 describes how the output of the four industries is consumed in a particular year.

TABLE 3   Consumption Vectors.

Supplier/Consumer	Agriculture	Fishing	Manufacturing	Tourism
Agriculture	.10	.20	.10	.40
Fishing	.20	.10	.10	.20
Manufacturing	.30	.40	.30	.30
Tourism	.10	.10	.20	.10

Each column in Table 3 is a ***consumption vector*** $c \in \mathbb{R}^4$, describing the input per unit of output from the other industries. Thus the model involves four consumption vectors:

$$\mathbf{c}_1 = \begin{pmatrix} .10 \\ .20 \\ .30 \\ .10 \end{pmatrix} \quad \mathbf{c}_2 = \begin{pmatrix} .20 \\ .10 \\ .40 \\ .10 \end{pmatrix} \quad \mathbf{c}_3 = \begin{pmatrix} .10 \\ .10 \\ .30 \\ .20 \end{pmatrix} \quad \mathbf{c}_4 = \begin{pmatrix} .40 \\ .20 \\ .30 \\ .10 \end{pmatrix}$$

The consumption of the neighboring islands of the goods and services produced by the four industries is represented by a vector $\mathbf{d} \in \mathbb{R}^4$, called the ***demand vector***. The consumption vectors and the demand vector are related by an equation of the form

$$\mathbf{x} = x_1 \mathbf{c}_1 + x_2 \mathbf{c}_2 + x_3 \mathbf{c}_3 + x_4 \mathbf{c}_4 + \mathbf{d}$$

expressing the input-output relationships between the goods and services produced by the four industries. The vector $\mathbf{x} \in \mathbb{R}^4$ is called the ***production vector*** of the entire economy. In this example, we have

$$\begin{pmatrix} x_1 \\ x_2 \\ x_3 \\ x_4 \end{pmatrix} = x_1 \begin{pmatrix} .10 \\ .20 \\ .30 \\ .10 \end{pmatrix} + x_2 \begin{pmatrix} .20 \\ .10 \\ .40 \\ .10 \end{pmatrix} + x_3 \begin{pmatrix} .10 \\ .10 \\ .30 \\ .20 \end{pmatrix} + x_4 \begin{pmatrix} .40 \\ .20 \\ .30 \\ .10 \end{pmatrix} + \begin{pmatrix} d_1 \\ d_2 \\ d_3 \\ d_4 \end{pmatrix}$$

By multiplying out some of the terms of this equation, we get

$$
\begin{pmatrix} x_1 \\ x_2 \\ x_3 \\ x_4 \end{pmatrix} = \begin{pmatrix} .1x_1 + .2x_2 + .1x_3 + .4x_4 \\ .2x_1 + .1x_2 + .1x_3 + .2x_4 \\ .3x_1 + .4x_2 + .3x_3 + .3x_4 \\ .1x_1 + .1x_2 + .2x_3 + .1x_4 \end{pmatrix} + \begin{pmatrix} d_1 \\ d_2 \\ d_3 \\ d_4 \end{pmatrix}
$$

This equation expresses the fact that the total production is the sum of the *internal* and *external* demand of the economy. The internal demand vector can be expressed as matrix product

$$
\begin{pmatrix} .1x_1 + .2x_2 + .1x_3 + .4x_4 \\ .2x_1 + .1x_2 + .1x_3 + .2x_4 \\ .3x_1 + .4x_2 + .3x_3 + .3x_4 \\ .1x_1 + .1x_2 + .2x_3 + .1x_4 \end{pmatrix} = \begin{pmatrix} .10 & .20 & .10 & .40 \\ .20 & .10 & .10 & .20 \\ .30 & .40 & .30 & .30 \\ .10 & .10 & .20 & .10 \end{pmatrix} \begin{pmatrix} x_1 \\ x_2 \\ x_3 \\ x_4 \end{pmatrix}
$$

The matrix

$$
C = \begin{pmatrix} .10 & .20 & .10 & .40 \\ .20 & .10 & .10 & .20 \\ .30 & .40 & .30 & .30 \\ .10 & .10 & .20 & .10 \end{pmatrix}
$$

is called the **consumption matrix** of the economy, and the equation

$$
\mathbf{x} = C\mathbf{x} + \mathbf{d}
$$

is known as its **Leontief input-output model**. We can solve for **d** and express the external demand of the economy by the equation

$$
\mathbf{x} - C\mathbf{x} = \mathbf{d}
$$

If **d** is known, we can express the total production **x** of the economy required to meet the external demand by

$$
\mathbf{x} = (I - C)^{-1}\mathbf{d}
$$

provided that the matrix $I - C$ is invertible. In the present example, this is the case. We use *MATHEMATICA* to verify this fact.

```
In[1]:= A=IdentityMatrix[4]

Out[1]= {{1,0,0,0},{0,1,0,0},{0,0,1,0},{0,0,0,1}}
```

In[2]:= B={{.10,.20,.10,.40},{.20,.10,.10,.20},

{.30,.40,.30,.30},{.10,.10,.20,.10}}

Out[2]= {{.10,.20,.10,.40},{.20,.10,.10,.20},

{.30,.40,.30,.30},{.10,.10,.20,.10}}

In[3]:= A-B

Out[3]//MatrixForm=

$$\begin{pmatrix} .9 & -.2 & -.1 & -.4 \\ -.2 & .9 & -.1 & -.2 \\ -.3 & -.4 & .7 & -.3 \\ -.1 & -.1 & -.2 & .9 \end{pmatrix}$$

In[4]:= Inverse[A-B]//MatrixForm

Out[4]//MatrixForm=

$$\begin{pmatrix} 1.631 & .783 & .665 & 1.121 \\ .625 & 1.583 & .547 & .812 \\ 1.286 & 1.495 & 2.304 & 1.672 \\ .536 & .595 & .647 & 1.697 \end{pmatrix}$$

Suppose that the external demand for the goods and services of the economy in units of output is given by the column vector

In[5]:= d={{500},{300},{400},{900}}

Out[5]= {{500},{300},{400},{900}}

Then the column vector defined by

In[6]:= Out[4].Out[5]//MatrixForm

Out[6]//MatrixForm=

$$\begin{pmatrix} 2325.3 \\ 1737.0 \\ 3517.9 \\ 2232.6 \end{pmatrix}$$

tabulates the total production of goods and services required to satisfy the demand $\mathbf{d}$. ◄►

EXERCISES 2.12

1. Suppose that

$$C = \begin{pmatrix} .2 & .1 & 0 & .1 \\ .1 & .2 & .2 & .3 \\ .2 & .3 & .2 & .2 \\ 0 & .2 & .2 & .1 \end{pmatrix}$$

   is the consumption matrix of a small economy. Find the consumption vectors.

2. Suppose that $\mathbf{d} = (150, 200, 100, 300)$ and $\mathbf{x} = (x_1, x_2, x_3, x_4)$ are the demand and total production vectors for the economy whose consumption matrix is given in Exercise 1. Find the matrix equation that expresses the assumption that supply equals demand.

3. Find the internal demand vector for the economy in Exercise 2.

4. Find the production vector $\mathbf{x}$ in Example 2.60 if the demand vector $\mathbf{d}$ remains unchanged, but the consumption matrix C is

$$C = \begin{pmatrix} .30 & .20 & .30 & .10 \\ .20 & .30 & .10 & .20 \\ .30 & .10 & .10 & .10 \\ .10 & .40 & .20 & .10 \end{pmatrix}$$

5. Find the demand vector $\mathbf{d}$ in Example 2.60 if the production vector $\mathbf{x}$ remains unchanged, but the consumption matrix C is

$$C = \begin{pmatrix} .10 & .20 & .30 & .15 \\ .10 & .20 & .40 & .10 \\ .30 & .10 & .10 & .10 \\ .10 & .30 & .20 & .10 \end{pmatrix}$$

6. Suppose that an economy consists of three industries, A_1, A_2, and A_3, producing quantities x_1, x_2, and x_3 of some commodities. Suppose further that these companies export 30, 20, and 40 percent of their output, respectively. Find the matrix A for which

$$A \begin{pmatrix} x_1 \\ x_2 \\ x_3 \end{pmatrix} = \begin{pmatrix} 30 \\ 20 \\ 40 \end{pmatrix}$$

when the percentages a_{ij} that company A_i consumes of the output of company A_j are given by the consumption matrix

$$\begin{pmatrix} a_{11} & a_{12} & a_{13} \\ a_{21} & a_{22} & a_{23} \\ a_{31} & a_{32} & a_{33} \end{pmatrix} = \begin{pmatrix} .30 & .15 & .4 \\ .25 & .30 & .1 \\ .20 & .10 & .3 \end{pmatrix}$$

Graph Theory

In this section, we discuss the use of matrices and matrix multiplication to describe the properties of finite graphs. First we deal with undirected graphs, then with directed ones. We motivate the definitions involved by considering the simple network of roads in Figure 2, connecting the cities of Beaconsfield (BFLD), Dorval (DVAL), Lachine (LCHN), and Montreal (MTRL). Table 4 displays the number of direct roads connecting these municipalities.

TABLE 4   Connections.

	BFLD	DVAL	LCHN	MTRL
BFLD	0	1	1	0
DVAL	1	0	2	1
LCHN	1	2	0	2
MTRL	0	1	2	0

In this example, V is the set {BFLD, DVAL, LCHN, MTRL} of *vertices*, and E is a set of *edges* $\{e_1, e_2, e_3, e_4, e_5, e_6, e_7\}$, connected by the function $p : E \to V \times V$ defined by

$$p(e_1) = (\text{LCHN, MTRL}) \quad p(e_2) = (\text{LCHN, MTRL})$$
$$p(e_3) = (\text{BFLD, LCHN}) \quad p(e_4) = (\text{BFLD, DVAL})$$
$$p(e_5) = (\text{DVAL, MTRL}) \quad p(e_6) = (\text{DVAL, LCHN})$$
$$p(e_7) = (\text{DVAL, LCHN})$$

An *undirected graph*

$$G = (V, E, p : E \to V \times V)$$

consists of a set V of vertices, a set E of edges, and a function p that assigns to each edge e a pair $p(e) = (u, v)$ of vertices. We write $e : u \sim v$ or $u \overset{e}{\sim} v$ if $p(e) = (u, v)$.

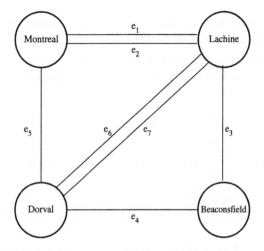

FIGURE 2   A Network of Two-Way Roads.

Two vertices u and v of a graph G are **connected** in G if $(u, v) = p(e)$ for some $e \in E$. A finite sequence $u_1 \overset{e_1}{\sim} u_2 \sim \cdots \sim u_n \overset{e_n}{\sim} u_{n+1}$ is a **path** of length n of G.

We now continue our example. We associate two different matrices with the graph of the given network of roads. The matrix

$$
M = \begin{pmatrix} 0 & 1 & 1 & 0 \\ 1 & 0 & 2 & 1 \\ 1 & 2 & 0 & 2 \\ 0 & 1 & 2 & 0 \end{pmatrix}
$$

records the number of edges connecting the vertices BFLD, DVAL, LCHN, and MTRL. It is symmetric since the edges of G are undirected. We call M the matrix of the graph G. We then form a second matrix

$$
A = \begin{pmatrix} 0 & 1 & 1 & 0 \\ 1 & 0 & 1 & 1 \\ 1 & 1 & 0 & 1 \\ 0 & 1 & 1 & 0 \end{pmatrix}
$$

whose entries are 1 if the graph contains at least one edge connecting two corresponding vertices and 0 otherwise. We call this the adjacency matrix of the graph G.

If $V = \{v_1, \ldots, v_n\}$ is the set of vertices of a graph G, then *the matrix of the graph G* is the $n \times n$ matrix $M = (m_{ij})$ whose entries m_{ij} count the number of edges connecting the vertices v_i and v_j. The *adjacency matrix of the graph G* is the $n \times n$ matrix $A = (a_{ij})$ whose entries a_{ij} are 1 if v_i and v_j are connected by an edge, and are 0 otherwise.

If we examine the given network of roads connecting the four municipalities, we notice that there is no direct road from Beaconsfield to Montreal. This means that the cities Beaconsfield and Montreal are not connected. But as Table 5 shows, we can get from Beaconsfield to Montreal and from Montreal to Beaconsfield via a second or a second and third city.

TABLE 5   Routes from Beaconsfield to Montreal.

Route 1	Route 2
BFLD $\xrightarrow{e_3}$ LCHN $\xrightarrow{e_1}$ MTRL	BFLD $\xrightarrow{e_3}$ LCHN $\xrightarrow{e_2}$ MTRL
Route 3	Route 4
BFLD $\xrightarrow{e_4}$ DVAL $\xrightarrow{e_5}$ MTRL	BFLD $\xrightarrow{e_4}$ DVAL $\xrightarrow{e_7}$ LCHN $\xrightarrow{e_1}$ MTRL

Routes 1 to 4 are *paths* of lengths 2, 2, 2, and 3, respectively, of the graph G. Matrix multiplication allows us to calculate the number of different paths determined by a graph. For example, the matrix

$$M^2 = \begin{pmatrix} 0 & 1 & 1 & 0 \\ 1 & 0 & 2 & 1 \\ 1 & 2 & 0 & 2 \\ 0 & 1 & 2 & 0 \end{pmatrix} \begin{pmatrix} 0 & 1 & 1 & 0 \\ 1 & 0 & 2 & 1 \\ 1 & 2 & 0 & 2 \\ 0 & 1 & 2 & 0 \end{pmatrix} = \begin{pmatrix} 2 & 2 & 2 & 3 \\ 2 & 6 & 3 & 4 \\ 2 & 3 & 9 & 2 \\ 3 & 4 & 2 & 5 \end{pmatrix}$$

shows that there are three paths of length 2 from Beaconsfield to Montreal. If we examine Table 5, we see that these paths correspond precisely to Routes 1, 2, and 3 from Beaconsfield to Montreal.

The matrix

$$M^3 = \begin{pmatrix} 2 & 2 & 2 & 3 \\ 2 & 6 & 3 & 4 \\ 2 & 3 & 9 & 2 \\ 3 & 4 & 2 & 5 \end{pmatrix} \begin{pmatrix} 0 & 1 & 1 & 0 \\ 1 & 0 & 2 & 1 \\ 1 & 2 & 0 & 2 \\ 0 & 1 & 2 & 0 \end{pmatrix} = \begin{pmatrix} 4 & 9 & 12 & 6 \\ 9 & 12 & 22 & 12 \\ 12 & 22 & 12 & 21 \\ 6 & 12 & 21 & 8 \end{pmatrix}$$

on the other hand, shows us that there are six paths of length 3 from Beaconsfield to Montreal. Route 4 is one of them. The number of routes grows quickly as we lengthen the allowable paths.

If there is a path from one vertex of a graph to another, we say that the path in question *connects* the two vertices. Two vertices u and v of a graph G are *pathwise connected* by G if G determines a path from u to v. The following theorem is proved in graph theory and describes the number of connections of a given length determined by a graph.

THEOREM 2.39 *If M_G is the matrix of a graph G and the ijth entry of M_G^p is q, then there are q paths of length p connecting the vertices v_i and v_j in G.* ∎

A question of interest, especially if the graph is large and the number of connections between vertices is not easily gleaned from a table, is whether a given vertex can be reached from any other vertex by some path. A graph G is **strongly connected** if every pair of vertices of G is pathwise connected. If a graph G has n vertices, then G is strongly connected if and only if every pair of vertices of G is connected by a path of length $n - 1$. Using this fact, we have the following beautiful characterization of strongly connected graphs.

THEOREM 2.40 *If A is the adjacency matrix of a graph G, then G is strongly connected if and only if the matrix $B = A + A^2 + \cdots + A^{n-1}$ has no zero entries.* ∎

Let us now turn to directed graphs. Suppose that the four municipalities are trying to reduce the number of road accidents by converting all of their highways into one-way roads. Figure 3 describes one result of this decision.

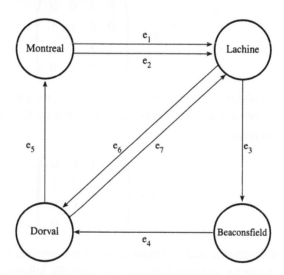

FIGURE 3   A Network of One-Way Roads.

Each edge of the diagram has a unique initial vertex and a unique terminal vertex. This leads us to the following definition.

A **directed graph** is a structure

$$G = (V, E, i : E \to V, t : E \to V)$$

consisting of a set V of vertices, a set E of edges, a function i that assigns to each edge e a vertex $i(e)$ called the **initial vertex** of e, and a function t that assigns to each edge e a vertex $t(e)$ called the **terminal vertex** of e. Two vertices u and v in a directed graph G are **connected** if $i(e) = u$ and $t(e) = v$ for some $e \in E$. We write $e : u \to v$ or $u \overset{e}{\to} v$ if $i(e) = u$ and $t(e) = v$.

Table 6 shows the number of roads connecting the four municipalities.

TABLE 6   Roads.

	BFLD	DVAL	LCHN	MTRL
BFLD	0	1	0	0
DVAL	0	0	1	1
LCHN	1	1	0	0
MTRL	0	0	2	0

The matrix of the graph G defined by this table is

$$M = \begin{pmatrix} 0 & 1 & 0 & 0 \\ 0 & 0 & 1 & 1 \\ 1 & 1 & 0 & 0 \\ 0 & 0 & 2 & 0 \end{pmatrix}$$

This matrix is no longer symmetric, as it was in the case for an undirected graph, since an edge from a vertex u to a vertex v is no longer also an edge from v to u. The adjacency matrix of the given graph is

$$A = \begin{pmatrix} 0 & 1 & 0 & 0 \\ 0 & 0 & 1 & 1 \\ 1 & 1 & 0 & 0 \\ 0 & 0 & 1 & 0 \end{pmatrix}$$

The definitions of a path of length q and of two vertices being pathwise connected are identical to those for undirected graphs. We have the following analogues of the previous theorems.

THEOREM 2.41   *If M is the matrix of a directed graph G and the ijth entry of M^p is q, then there are q paths of length p connecting the vertices v_i and v_j in G.* ■

THEOREM 2.42   *If A is the adjacency matrix of a directed graph G, then G is strongly connected if and only if the matrix $B = A + A^2 + \cdots + A^{n-1}$ has no zero entries.* ■

Here are some additional examples illustrating the preceding definitions.

EXAMPLE 2.61   ■ The Matrix of a Finite Graph

Suppose that the set of vertices of a graph G is $V = \{v_1, v_2, v_3, v_4\}$ and the set of edges of G is $E = \{e_1 : v_1 \sim v_2, e_2 : v_1 \sim v_4, e_3 : v_2 \sim v_3, e_4 : v_2 \sim v_3\}$. Then the matrix

$$M = \begin{pmatrix} 0 & 1 & 0 & 1 \\ 1 & 0 & 2 & 0 \\ 0 & 2 & 0 & 0 \\ 1 & 0 & 0 & 0 \end{pmatrix}$$

represents the graph G. ◄►

EXAMPLE 2.62 ■ A Finite Directed Graph

Suppose that the set of vertices of a graph G is $V = \{v_1, v_2, v_3, v_4\}$ and the set of edges of G is

$$E = \{e_1 : v_1 \rightarrow v_2, e_2 : v_1 \rightarrow v_4, e_3 : v_2 \rightarrow v_3, e_4 : v_3 \rightarrow v_2, e_5 : v_3 \rightarrow v_2\}$$

Then

$$M(G) = \begin{pmatrix} 0 & 1 & 0 & 1 \\ 0 & 0 & 1 & 0 \\ 0 & 2 & 0 & 0 \\ 0 & 0 & 0 & 0 \end{pmatrix} \quad \text{and} \quad A(G) = \begin{pmatrix} 0 & 1 & 0 & 1 \\ 0 & 0 & 1 & 0 \\ 0 & 1 & 0 & 0 \\ 0 & 0 & 0 & 0 \end{pmatrix}$$

are, respectively, the matrix and the adjacency matrix of G. ◄►

EXAMPLE 2.63 ■ Paths of an Undirected Graph

Use *MATHEMATICA* to find the number of paths of length 2 in the graph whose matrix is

$$A = \begin{pmatrix} 0 & 1 & 0 & 1 \\ 1 & 0 & 1 & 0 \\ 0 & 1 & 0 & 0 \\ 1 & 0 & 0 & 0 \end{pmatrix}$$

Solution. First we define the matrix A.

```
In[1]:= A= ( 0 1 0 1
             1 0 1 0
             0 1 0 0
             1 0 0 0 )
Out[1]= {{0,1,0,1},{1,0,1,0},{0,1,0,0},{1,0,0,0}}
```

Next we calculate the powers A^2 and A^3 of A.

In[2]:= **MatrixPower[A,2]//MatrixForm**

Out[2]//MatrixForm=

$$\begin{pmatrix} 2 & 0 & 1 & 0 \\ 0 & 2 & 0 & 1 \\ 1 & 0 & 1 & 0 \\ 0 & 1 & 0 & 1 \end{pmatrix}$$

In[3]:= **MatrixPower[A,3]//MatrixForm**

Out[3]//MatrixForm=

$$\begin{pmatrix} 0 & 3 & 0 & 2 \\ 3 & 0 & 2 & 0 \\ 0 & 2 & 0 & 1 \\ 2 & 0 & 1 & 0 \end{pmatrix}$$

The matrix A^2 shows that there is one path of length 2 from v_3 to v_1, that there are two paths of length 2 from v_1 to v_1, and so on. The path of length 2 from v_3 to v_1 consists of the edges $e_1 : v_1 \sim v_2$ and $e_3 : v_2 \sim v_3$. The two paths of length 2 from v_1 to v_1 consist of the edges $e_1 : v_1 \sim v_2$ and $e_2 : v_1 \sim v_4$. The matrix A^2 also shows that there are three paths of length 3 from v_2 to v_1, but no such paths from v_1 to v_1, and so on. ◄►

EXAMPLE 2.64   ■ Paths of a Directed Graph

Use *MATHEMATICA* to find the number of paths of length 2 in the graph whose matrix is given by

$$A = \begin{pmatrix} 0 & 1 & 0 & 1 \\ 0 & 0 & 1 & 0 \\ 0 & 1 & 0 & 0 \\ 0 & 0 & 0 & 0 \end{pmatrix}$$

Solution. We begin by specifying A. Then we can calculate the powers A^2 and A^3 of A.

In[1]:= **A={{0,1,0,1},{0,0,1,0},{0,1,0,0},{0,0,0,0}}**

Out[1]= {{0,1,0,1},{0,0,1,0},{0,1,0,0},{0,0,0,0}}

```
In[2]:= MatrixPower[A,2]//MatrixForm

Out[2]//MatrixForm=
```

$$\begin{pmatrix} 0 & 0 & 1 & 0 \\ 0 & 1 & 0 & 0 \\ 0 & 0 & 1 & 0 \\ 0 & 0 & 0 & 0 \end{pmatrix}$$

```
In[3]:= MatrixPower[A,3]//MatrixForm

Out[3]//MatrixForm=
```

$$\begin{pmatrix} 0 & 1 & 0 & 0 \\ 0 & 0 & 1 & 0 \\ 0 & 1 & 0 & 0 \\ 0 & 0 & 0 & 0 \end{pmatrix}$$

The matrix A^2 shows that there is one path of length 2 from v_1 to v_3, one path of length 2 from v_2 to v_2, and so on. The path of length 2 from v_2 to v_2 consists of the edges e_4 : $v_3 \rightarrow v_2$ and e_3 : $v_2 \rightarrow v_3$. The two paths of length 2 from v_3 to v_3 consist of the edges e_4 : $v_3 \rightarrow v_2$ and e_3 : $v_2 \rightarrow v_3$. The matrix A^3 shows that there is one path of length 3 from v_2 to v_3, but no such paths from v_1 to v_1, and so on. ◄►

EXERCISES 2.13

1. Find the matrix $M(G)$ and adjacency matrix $A(G)$ of the graph G with vertices $V = \{v_1, v_2, v_3, v_4\}$ and edges $E = \{e_1 : v_1 \sim v_2, e_2 : v_1 \sim v_4, e_3 : v_2 \sim v_3, e_4 : v_2 \sim v_3\}$.

2. Find all paths of lengths 2, 3, and 4 determined by the graph in Exercise 1.

3. Draw an undirected graph G with four vertices v_1, v_2, v_3, and v_4 in which each vertex v_i is connected to each other vertex v_j. Determine the matrix $M(G)$ and the adjacency matrix $A(G)$ of the graph G.

4. Use the powers of the adjacency matrix $A(G)$ in Exercise 3 to find the number of paths of length 3 connecting vertices v_1 and v_2.

5. Four friends, Fred, Bob, Jane, and Sarah, have decided to play a round-robin tennis tournament. In the first match, Fred plays against Bob and Jane against Sarah. In the second match, Fred plays against Jane and Bob against Sarah. In the third match, Fred plays against Sarah and Bob against Jane. Construct three graphs depicting the order of play of

each match, and show that adjacency matrices of the graphs are products of elementary matrices.

6. Draw an undirected graph G that is not pathwise connected and has four vertices. Form the adjacency matrix $A(G)$ of G. Compute the matrix powers $A(G)^2$ and $A(G)^3$, and explain why both matrices have zero entries.

7. An undirected graph is complete if every pair of vertices is connected by exactly one edge. Use the **Combinatorica** package of *MATHEMATICA* to construct a complete graph with five vertices, and show that the graph is strongly connected.

8. Find the matrix $M(G)$ and the adjacency matrix $A(G)$ of the directed graph G determined by the set of vertices $V = \{v_1, v_2, v_3, v_4\}$ and the set of edges

$$E = \{e_1 : v_1 \rightarrow v_2, e_2 : v_1 \rightarrow v_4, e_3 : v_2 \rightarrow v_3, e_4 : v_3 \rightarrow v_2, e_5 : v_3 \rightarrow v_2\}$$

9. Find all paths of lengths 2, 3, and 4 determined by the graph in Exercise 8.

10. Draw a strongly connected directed graph G with four vertices and specify its adjacency matrix $A(G)$. Show that the matrix $A(G) + A(G)^2 + A(G)^3$ has no zero entry.

11. A group of five friends, Stuart, Mary, Tom, Dick, and Julie, are taking the same linear algebra course. They have decided to set up a telephone chain and inform each other of important announcements. They have agreed that Stuart phones Mary, Mary phones Tom, Tom phones Dick, and Dick phones Julie. Show that the graph determined by this communication chain is not strongly connected.

REVIEW

KEY CONCEPTS ▶ Define and discuss each of the following.

Lexicon of Matrices

Diagonal matrix, elementary matrix, elementary permutation matrix, generalized diagonal matrix, Hermitian matrix, identity matrix, invertible matrix, lower-triangular matrix, orthogonal matrix, permutation matrix, rectangular matrix, square matrix, symmetric matrix, unitary matrix, upper-triangular matrix.

Linear Systems

Homogeneous linear system.

Matrix Components

Diagonal entry of a matrix, diagonal of a matrix, entry of a matrix.

Matrix Operations

Conjugate transpose of a matrix, inverse of a matrix, left inverse of a matrix, left-multiplication by a matrix, LU decomposition of a matrix, product of matrices, right inverse of a matrix, right-multiplication by a matrix, row-column product of a matrix, sum of matrices, scalar multiple of a matrix, trace of a matrix, transpose of a matrix.

KEY FACTS ▶ Explain and illustrate each of the following.

1. If A is an $m \times n$ matrix and if B is an $n \times p$ matrix whose columns are $\mathbf{c}_1(B), \ldots, \mathbf{c}_p(B)$, then $AB = (\; A\mathbf{c}_1(B) \quad \cdots \quad A\mathbf{c}_p(B) \;)$.

2. If $\mathbf{a} = (\; a_1 \quad \cdots \quad a_n \;)$ is a $1 \times n$ row vector and $B = (b_{ij})$ is an $n \times p$ matrix, then $\mathbf{a}B = a_1\mathbf{r}_1(B) + \cdots + a_n\mathbf{r}_n(B)$.

3. For all matrices A and B and column vectors $\mathbf{x}$ of compatible dimension, $A(B\mathbf{x}) = (AB)\mathbf{x}$.

4. For all matrices A, B, C of compatible dimension, the following equalities hold: $(AB)C = A(BC)$; $A(B + C) = AB + AC$; $(A + B)C = AC + BC$; for all s, $s(AB) = (sA)B = A(sB)$; $0A = \mathbf{0}$, $A\mathbf{0} = \mathbf{0}$.

5. If $\mathbf{y}$ and $\mathbf{z}$ are solutions of a homogeneous system $A\mathbf{x} = \mathbf{0}$ and a and b are scalars, then $a\mathbf{y} + b\mathbf{z}$ is also a solution of $A\mathbf{x} = \mathbf{0}$.

6. For any $m \times n$ matrix $A = (a_{ij})$ and any $n \times p$ matrix $B = (b_{jk})$, the row-column products $\mathbf{r}_i(A)\mathbf{c}_j(B)$ and $\mathbf{r}_j(B^T)\mathbf{c}_i(A^T)$ are equal.

7. For any two square matrices A and B of compatible dimension, the following equalities hold: $(A^T)^T = A$; for all s, $(sA)^T = sA^T$; $(A + B)^T = A^T + B^T$; $(AB)^T = B^T A^T$.

8. If A and B are two $n \times n$ matrices, then the following equalities hold: trace $A^T =$ trace A; for all s, trace $sA = s$ trace A; trace$(A + B) =$ trace $A +$ trace B; trace $AB =$ trace BA.

9. If the matrix B is the result of an elementary row operation applied to a matrix A, then there exists an elementary matrix E for which $B = EA$.

10. A matrix A is row equivalent to a matrix B if and only if there exist elementary matrices $E_1, \ldots, E_p$ such that $A = E_1 \cdots E_p B$.

11. If B is a right inverse of A and C is a left inverse of A, then $B = C$.

12. If B and C are inverses of A, then $B = C$.

13. If A and B are invertible matrices, then so is AB.

14. Every elementary matrix is invertible.

15. An $n \times n$ matrix A is invertible if and only if it is row equivalent to I_n.

16. An $n \times n$ matrix A is invertible if and only if its reduced row echelon form is I_n.

17. A matrix is invertible if and only if it is a product of elementary matrices.

18. An $n \times n$ matrix A is invertible if and only if it has n pivots.

19. An $n \times n$ matrix A is row equivalent to I_n if and only if the system $A\mathbf{x} = \mathbf{0}$ has only the trivial solution.

20. An $n \times n$ matrix A is invertible if and only if the equation $A\mathbf{x} = \mathbf{0}$ has only the trivial solution.

21. Suppose that A is an $n \times n$ real matrix. Then the linear system $A\mathbf{x} = \mathbf{b}$ has a solution for every $\mathbf{b} \in \mathbb{R}^n$ if and only if A is row equivalent to I_n.

22. An $n \times n$ matrix A is invertible if and only if there exists a matrix B such that $AB = I_n$.

23. A matrix B is the inverse of a matrix A if and only if the matrices $(A \mid I_n)$ and $(I_n \mid B)$ are row equivalent.

24. If A is an invertible matrix, then $(A^{-1})^T$ is the inverse of A^T.

25. An $n \times n$ matrix A is invertible if and only if there exists a matrix B such that $AB = I$.

26. If A and B are invertible, then AB is invertible and $(AB)^{-1} = B^{-1}A^{-1}$.

27. If A and B are real $n \times n$ matrices and AB is invertible, then so are A and B.

28. If A is an $n \times n$ row echelon matrix, then A is upper triangular.

29. The product of two $n \times n$ lower-triangular matrices is a lower-triangular matrix.

30. The product of two $n \times n$ upper-triangular matrices is an upper-triangular matrix.

31. Suppose that A is an $m \times n$ matrix that can be reduced to row echelon form without requiring the permutation of rows. Then there exist an $m \times m$ lower-triangular matrix L with 1s on the diagonal and an $m \times n$ matrix row echelon matrix U such that $A = LU$.

32. For any $m \times n$ matrix A there exists a permutation matrix P such that the matrix PA can be reduced to row echelon form without the permutation of rows.

33. For any $m \times n$ matrix A there exist a permutation matrix P, an $m \times m$ lower-triangular matrix L with 1s on the diagonal, and an $m \times n$ row echelon matrix U such that $PA = LU$.

34. Suppose that A is an invertible matrix that can be reduced to row echelon form without requiring the permutation of rows. Then there exist a unique lower-triangular matrix L with 1s on the diagonal and a unique upper-triangular matrix U such that $A = LU$.

35. Suppose that A is an invertible matrix and let P be a permutation matrix for which PA is reducible to row echelon form without the permutation of rows. Then there exists a unique lower-triangular matrix L with 1s on the diagonal and a unique upper-triangular matrix U such that $A = LU$.

36. For any $\theta \in \mathbb{R}$, the rotation matrix

$$A = \begin{pmatrix} \cos \theta & -\sin \theta \\ \sin \theta & \cos \theta \end{pmatrix}$$

is a product of elementary matrices.

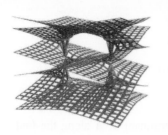

3

DETERMINANTS

With every square matrix we can associate a certain quantity called its ***determinant***. If the matrix is a numerical matrix, the determinant is a number. If the matrix is a polynomial matrix, the determinant is a polynomial. As we study determinants, we will see that they have an amazing variety of mathematical uses. They can be used to show that a real matrix is or is not invertible, to find matrix inverses, to show that two matrices fail to represent the same linear transformation, to compute areas and volumes, to solve systems of linear equations, to find the partial derivatives of functions in several variables, to express relationships between multiple integrals, and so on. In Chapter 6, we will employ determinants of polynomial matrices to study the eigenvalues of square matrices.

THE LAPLACE EXPANSION

The determinant $\det A$ of an $n \times n$ matrix A, for $n \geq 2$, has a very elegant definition by an induction on n, using submatrices of A. The definition is known as the ***Laplace expansion*** of the determinant.

We begin with the case $n = 2$.

DEFINITION 3.1   *For any* 2×2 *matrix* $A = \left(a_{ij}\right)$, $\det A = a_{11}a_{22} - a_{12}a_{21}$.

To define $\det A$ for $n > 2$, we use specific submatrices of A.

DEFINITION 3.2   *The ijth* ***minor*** M_{ij} *of an* $n \times n$ *matrix* A *is the* $(n-1) \times (n-1)$ *submatrix of* A *obtained by deleting the ith row and the jth column of* A.

Next we assume that determinants $\det M_{ij}$ of the minors of A are known, and use them to define the determinant of A itself. We simplify the notation by introducing the following auxiliary concept.

DEFINITION 3.3   *The ijth* ***cofactor*** C_{ij} *of an* $n \times n$ *matrix* A *is* $(-1)^{i+j} \det M_{ij}$.

The formula $(-1)^{i+j}$ assigns $+1$ or -1 to $\det M_{ij}$, depending on whether $i + j$ is even or odd. Using cofactors, we now define the determinant of A.

DEFINITION 3.4 *For any $n \times n$ matrix $A = (a_{ij})$ for which $n > 2$,*

$$\det A = a_{11}C_{11} + \cdots + a_{1n}C_{1n}$$

This formula is called the **Laplace expansion** of the determinant of A along the first row of A. Analogous formulas for the determinant can be defined along any other row of A. The Laplace expansion of the determinant of A along the ith row of A, for example, is $\det A = a_{i1}C_{i1} + \cdots + a_{in}C_{in}$. We could also have defined the determinant as

$$\det A = a_{1j}C_{1j} + a_{2j}C_{2j} + \cdots + a_{nj}C_{nj}$$

using the Laplace expansion along the jth column of A. We will show later that all of these expansions yield the same value.

EXAMPLE 3.1 ■ The Determinant of a 2 × 2 Real Matrix

Compute the determinant of the matrix

$$A = \begin{pmatrix} 3 & 5 \\ 2 & 4 \end{pmatrix}$$

Solution. Let us label the entries of A as

$$\begin{pmatrix} a_{11} & a_{12} \\ a_{21} & a_{22} \end{pmatrix} = \begin{pmatrix} 3 & 5 \\ 2 & 4 \end{pmatrix}$$

Then, by definition,

$$a_{11}a_{22} - a_{12}a_{21} = 3\,(4) - 2\,(5) = 2$$

is the determinant of A. ◄►

EXAMPLE 3.2 ■ The Determinant of a 2 × 2 Polynomial Matrix

Use the definition to show that the determinant of the polynomial matrix

$$A = \begin{pmatrix} t - 3 & 5 \\ 2 & t - 4 \end{pmatrix}$$

is $t^2 - 7t + 2$.

Solution. Once again, we label the entries of A as

$$\begin{pmatrix} a_{11} & a_{12} \\ a_{21} & a_{22} \end{pmatrix} = \begin{pmatrix} t - 3 & 5 \\ 2 & t - 4 \end{pmatrix}$$

It then follows again from the definition that

$$a_{11}a_{22} - a_{12}a_{21} = (t - 3)(t - 4) - 2(5) = t^2 - 7t + 2$$

is the determinant of A. ◄►

EXAMPLE 3.3 ■ **The Determinant of a 3 × 3 Matrix**

Verify that the determinant of the 3 × 3 matrix

$$A = \begin{pmatrix} a_{11} & a_{12} & a_{13} \\ a_{21} & a_{22} & a_{23} \\ a_{31} & a_{32} & a_{33} \end{pmatrix}$$

is the expression $a_{11}a_{22}a_{33} - a_{11}a_{23}a_{32} - a_{12}a_{21}a_{33} + a_{12}a_{23}a_{31} + a_{13}a_{21}a_{32} - a_{13}a_{22}a_{31}$.

Solution. By definition, $\det A = a_{11}C_{11} + a_{12}C_{12} + a_{13}C_{13}$. Moreover, the cofactors C_{ij} of A are

$$C_{11} = (-1)^{1+1} \det \begin{pmatrix} a_{22} & a_{23} \\ a_{32} & a_{33} \end{pmatrix} = +(a_{22}a_{33} - a_{23}a_{32})$$

$$C_{12} = (-1)^{1+2} \det \begin{pmatrix} a_{21} & a_{23} \\ a_{31} & a_{33} \end{pmatrix} = -(a_{21}a_{33} - a_{23}a_{31})$$

$$C_{13} = (-1)^{1+3} \det \begin{pmatrix} a_{21} & a_{22} \\ a_{31} & a_{32} \end{pmatrix} = +(a_{21}a_{32} - a_{22}a_{31})$$

Therefore,

$$\det A = a_{11}(a_{22}a_{33} - a_{23}a_{32}) - a_{12}(a_{21}a_{33} - a_{23}a_{31}) + a_{13}(a_{21}a_{32} - a_{22}a_{21})$$
$$= a_{11}a_{22}a_{33} - a_{11}a_{23}a_{32} - a_{12}a_{21}a_{33} + a_{12}a_{23}a_{31} + a_{13}a_{21}a_{32} - a_{13}a_{22}a_{31}$$

This shows that the determinant of A is equal to the given expression. ◄►

EXAMPLE 3.4 ■ **Using *MATHEMATICA* to Compute the Determinant of a Matrix**

Use *MATHEMATICA* to compute the determinant of the following matrix.

$$A = \begin{pmatrix} t-21 & t^2 & 1 & t^3 \\ 19 & t & 0 & 6 \\ t+1 & 3 & t-19 & 2 \\ 0 & 0 & 8 & t-5 \end{pmatrix}$$

Solution. We first use the matrix template to create A and then apply **Det** to A.

```
In[1]:= A= ( t-21  t²   1     t³  
            19    t    0     6   
            t+1   3    t-19  2   
            0     0    8     t-5 ) ;
```

```
In[2]:= Det[A]

Out[2]= -3309-1453t-962t²-94t³-10t⁴+8t⁵
```

As we can see, the determinant of A is a polynomial in t. ◄►

EXERCISES 3.1

1. Use the definition to calculate the determinants of the following matrices.

a. $A = \begin{pmatrix} -85 & -55 \\ -37 & -35 \end{pmatrix}$
b. $A = \begin{pmatrix} 97 & 50 & 79 \\ 56 & 49 & 63 \\ 57 & -59 & 45 \end{pmatrix}$
c. $A = \begin{pmatrix} 0 & 0 \\ 0 & 0 \end{pmatrix}$

d. $A = \begin{pmatrix} 3 & 9 & 4 & 1 \\ 0 & 8 & 15 & 4 \\ 0 & 0 & 9 & 12 \\ 0 & 0 & 0 & 9 \end{pmatrix}$
e. $A = \begin{pmatrix} 6 & 0 & 0 & 0 \\ 0 & 21 & 0 & 0 \\ 0 & 4 & 3 & 0 \\ 5 & 0 & 0 & 19 \end{pmatrix}$
f. $A = \begin{pmatrix} 1 & 0 & 0 & 0 \\ 0 & 2 & 0 & 0 \\ 0 & 0 & 3 & 0 \\ 0 & 0 & 0 & 2 \end{pmatrix}$

2. Find a 3×3 matrix A with real entries for which

$$\det A \neq a_{11}C_{11} + a_{22}C_{22} + a_{33}C_{33}$$

3. Find a 3×3 matrix A with polynomial entries for which

$$\det A \neq a_{11}C_{11} + a_{22}C_{22} + a_{33}C_{33}$$

4. Find two 3×3 matrices A and B for which $\det(A + B) \neq \det A + \det B$ and $\det 5A \neq 5 \det A$.

5. Use *MATHEMATICA* to calculate the determinant of A, where

$$A = \begin{pmatrix} 54 & -5 & 99 & -61 & -50 \\ -12 & -18 & 31 & -26 & -62 \\ 1 & -47 & -91 & -47 & -61 \\ 41 & -58 & -90 & 53 & -1 \\ 94 & 83 & -86 & 23 & -84 \end{pmatrix}$$

6. Use *MATHEMATICA* to verify that $\det AB = \det BA$ for all 3×3 matrices.

Properties of Determinants

If we take a closer look at the formulas for determinants, we notice that the determinant of an $n \times n$ matrix A is a sum of certain products $e_1 \cdots e_n$ of entries of the matrix. Each product $e_1 \cdots e_n$ consists of n entries. Moreover, each row and column of A is represented in every one of these products exactly once. These specialized products of entries of A are called the *elementary products* of the entries of A.

Elementary products can be described in terms of patterns among the locations ij of the entries a_{ij}. For each row index i, the corresponding column index j can be thought of as the value of a function with input i and output j. The functions involved are called *permutations*.

DEFINITION 3.5   *For any standard set* $\mathbf{n} = \{1, 2, \ldots, n\}$, *a **permutation** of* $\mathbf{n}$ *is a one-one function* $\pi : \mathbf{n} \to \mathbf{n}$.

By a *one-one function* we mean a function π with the property that for all $x, y \in \mathbf{n}$, $\pi(x) = \pi(y)$ implies that $x = y$. The *MATHEMATICA* notation for permutations is very intuitive. If $\mathbf{n} = \mathbf{5}$, for example, the list $\{3, 4, 1, 2, 5\}$ denotes the permutation π defined by $\pi(1) = 3$, $\pi(2) = 4$, $\pi(3) = 1$, $\pi(4) = 2$, and $\pi(5) = 5$.

DEFINITION 3.6   *An **inversion** in a permutation* π *is any pair* $(i, j) \in \mathbf{n} \times \mathbf{n}$ *for which* $i > j$ *and* $\pi(i) < \pi(j)$.

For example, if

$$\pi = \{(1, \pi(1), (2, \pi(2)), (3, \pi(3)), (4, \pi(4))\} = \{(1, 3), (2, 2), (3, 1), (4, 4)\}$$

is the permutation that maps 1 to 3, 2 to 2, 3 to 1, and 4 to 4, then $\pi(1) > \pi(2)$, $\pi(1) > \pi(3)$, and $\pi(2) > \pi(3)$ are the three cases where a larger integer precedes a smaller one. Hence π has three inversions.

DEFINITION 3.7   *A permutation is called **even** if it has an even number of inversions and **odd** otherwise.*

We can use the idea of inversions to attach a sign $\text{sgn}(\pi)$ to a permutation. This sign is called the *parity* of the permutation.

DEFINITION 3.8   *The **parity** of a permutation* π *is* 1 *if* π *has an even number of inversions and is* -1 *otherwise.*

If we look back at the examples of the general determinants given above, we notice that half of the elementary products $e_1 \cdots e_n$ are premultiplied by -1 and the other half by 1. The parity of the permutation π of $\mathbf{n}$ that defines $e_1 \cdots e_n = a_{1\pi(1)} \cdots a_{n\pi(n)}$ determines the sign of $e_1 \cdots e_n$.

The **Combinatorica** package of *MATHEMATICA* makes it easy to calculate permutations and parities. Let us illustrate the method by computing the permutations of the standard lists **2** and **3** and finding their parities.

EXAMPLE 3.5 ■ Using *MATHEMATICA* to Calculate Permutations

Use *MATHEMATICA* to find the permutations of the lists **2** and **3** and find their parities.

Solution. We use the **Combinatorica** package.

```
In[1]:= <<DiscreteMath`Combinatorica`
```

```
In[2]:= Permutations[{1,2}]
```

```
Out[2]= {{1,2},{2,1}}
```

This tells us that $S_2 = \{\{1, 2\}, \{2, 1\}\}$.

```
In[3]:= Inversions[{1,2}]
```

```
Out[3]= 0
```

```
In[4]:= Inversions[{2,1}]
```

```
Out[4]= 1
```

Table 1 records these results.

TABLE 1   Parities for $n = 2$.

Permutations	Parities	sgn
$\sigma_1 = \{1, 2\}$	Even	1
$\sigma_2 = \{2, 1\}$	Odd	−1

```
In[5]:= Permutations[{1,2,3}]
```

```
Out[5]= {{1,2,3},{1,3,2},{2,1,3},{2,3,1},{3,1,2},{3,2,1}}
```

Therefore, $S_3 = \{\{1, 2, 3\}, \{1, 3, 2\}, \{2, 1, 3\}, \{2, 3, 1\}, \{3, 1, 2\}, \{3, 2, 1\}\}$.

```
In[6]:= Inversions[{1,2,3}]
```

```
Out[6]= 0
```

```
In[7]:= Inversions[{1,3,2}]
```

```
Out[7]= 1
```

```
In[8]:= Inversions[{2,1,3}]
```

```
Out[8]= 1
```

```
In[9]:= Inversions[{2,3,1}]
```

```
Out[9]= 2
```

```
In[10]:= Inversions[{3,1,2}]
```

```
Out[10]= 2
```

```
In[11]:= Inversions[{3,2,1}]
```

```
Out[11]= 3
```

Table 2 records the results of these calculations. ◄►

Using permutations and their parities, we can describe determinants without reference to submatrices. An induction on $n \geq 2$ shows that the determinant of an $n \times n$ matrix A is

$$\det A = \sum_{\pi \in S_n} \text{sgn}(\pi) a_{1\pi(1)} \cdots a_{n\pi(n)}$$

Let us illustrate this formula with an example.

TABLE 2   Parities for $n = 3$.

Permutations	Parities	sgn
$\sigma_1 = \{1, 2, 3\}$	Even	1
$\sigma_2 = \{1, 3, 2\}$	Odd	-1
$\sigma_3 = \{2, 1, 3\}$	Odd	-1
$\sigma_4 = \{2, 3, 1\}$	Even	1
$\sigma_5 = \{3, 1, 2\}$	Even	1
$\sigma_6 = \{3, 2, 1\}$	Odd	-1

EXAMPLE 3.6   ■ The Determinant of a 3 × 3 Matrix

Use permutations to find the determinant of a general 3×3 matrix.

Solution. Suppose that

$$A = \begin{pmatrix} a_{11} & a_{12} & a_{13} \\ a_{21} & a_{22} & a_{23} \\ a_{31} & a_{32} & a_{33} \end{pmatrix}$$

is a general 3×3 matrix, and let S_3 be the set of permutations of the standard list **3**. Then

$$\begin{aligned} \det A \quad &= \quad a_{11}a_{22}a_{33} - a_{11}a_{23}a_{32} - a_{12}a_{21}a_{33} \\ &\quad + a_{12}a_{23}a_{31} + a_{13}a_{21}a_{32} - a_{13}a_{22}a_{31} \\ &= \quad \sum_{\pi \in S_3} \text{sgn}(\pi)a_{1\pi(1)}a_{2\pi(2)}a_{3\pi(3)} \end{aligned}$$

Each of the permutations π in S_3 yields a term of $\det A$. ◄►

We now use permutations to prove some of the basic properties of determinants.

THEOREM 3.1   *The determinant $\det I_n$ of the $n \times n$ identity matrix I_n is 1.*

Proof. Since all terms of the form $a_{1\pi(1)} \cdots a_{n\pi(n)}$, except for the term $a_{11} \cdots a_{nn} = 1$, contain 0 as a factor, it is clear that

$$\det I_n = \sum \text{sgn}(\pi)a_{1\pi(1)} \cdots a_{n\pi(n)} = a_{11} \cdots a_{nn}$$

Hence $\det I_n = 1$. ■

THEOREM 3.2   *If A is an $n \times n$ matrix with a zero row, then the determinant of A is 0.*

Proof. Suppose that $A = (a_{ij})$ and that $a_{i1} = \cdots = a_{in} = 0$. Then the terms $a_{1\pi(1)} \cdots a_{i\pi(i)} \cdots a_{n\pi(n)}$ are 0 for all permutations π. Hence $\det A = 0$. ■

Next we explore the effect of elementary row operations on determinants.

THEOREM 3.3 *If B is a matrix resulting from an n × n matrix A by the elementary row operation* $(R_i \rightleftharpoons R_j)$ *and* $i \neq j$, *then* $\det B = -\det A$.

Proof. Suppose that $i \neq j$ and consider the row interchange $(R_i \rightleftharpoons R_j)$. To show that $\det A = -\det B$, we can use the permutation $\tau : \mathbf{n} \to \mathbf{n}$ defined by $\tau(i) = j$, $\tau(j) = i$, and $\tau(x) = x$ otherwise. Then τ is an odd permutation with $\text{sgn}(\tau) = -1$. For any other permutation $\sigma : \mathbf{n} \to \mathbf{n}$, we get

$$\text{sgn}(\tau\sigma) = \text{sgn}(\tau)\,\text{sgn}(\sigma) = -\,\text{sgn}(\sigma)$$

Hence

$$\begin{aligned}
\det B &= \sum \text{sgn}(\sigma)b_{1\sigma(1)} \cdots b_{n\sigma(n)} \\
&= \sum \text{sgn}(\tau\sigma)a_{1(\tau(\sigma(1)))} \cdots a_{n\tau(\sigma(n))} \\
&= -\sum \text{sgn}(\sigma)a_{1\tau(\sigma(1))} \cdots a_{n\tau(\sigma(n))} \\
&= -\det A
\end{aligned}$$

This proves the theorem. ∎

COROLLARY 3.4 *If* $A = (a_{ij})$ *is an n × n matrix containing two identical rows, then the determinant of A is 0.*

Proof. Suppose $\text{row}_i(A) = \text{row}_j(A)$. Let $i \neq j$, and let B be the matrix obtained from A by the elementary row operation $(R_i \rightleftharpoons R_j)$. Then it follows from Theorem 3.3 that $\det B = -\det A$. But $B = A$. Hence $\det A = -\det A$. This can only hold if $\det A = 0$. ∎

THEOREM 3.5 *If B is a matrix resulting from an n × n matrix A by the elementary row operation* $(R_i \to sR_i)$, *then* $\det B = s \det A$.

Proof. Let $B = (b_{ij})$ result from $A = (a_{ij})$ by the application of $(R_i \to sR_i)$. Then

$$\begin{aligned}
\det B &= \sum \text{sgn}(\pi)b_{1\pi(1)} \cdots b_{n\pi(n)} \\
&= \sum \text{sgn}(\pi)a_{1\pi(1)} \cdots sa_{i\pi(i)} \cdots a_{n\pi(n)} \\
&= s \sum \text{sgn}(\pi)a_{1\pi(1)} \cdots a_{i\pi(i)} \cdots a_{n\pi(n)} \\
&= s \det A
\end{aligned}$$

This proves the theorem. ∎

EXAMPLE 3.7 ■ **Determinants and Scalar Multiplication**

Use *MATHEMATICA* to show that if A is the matrix

$$A = \begin{pmatrix} 0 & 5 & -3 & 2 \\ -2 & 2 & 2 & -1 \\ 1 & 4 & 1 & -3 \\ 3 & 5 & 5 & -2 \end{pmatrix}$$

then $\det 9A = 9^4 \det A$.

Solution. We recall that we get the matrix $9A$ by multiplying each entry of A by 9. This is equivalent to multiplying each row by 9. In terms of elementary row operations, $9A$ results from A by the sequence of operations

$$A \ (R_1 \rightarrow 9R_1) \ B \ (R_2 \rightarrow 9R_2) \ C \ (R_3 \rightarrow 9R_3) \ D \ (R_4 \rightarrow 9R_4) \ E$$

Let us use *MATHEMATICA* to compute det A and det E, and compare the results.

$$\text{In[1]:= Det}\begin{bmatrix} \begin{pmatrix} 0 & 5 & -3 & 2 \\ -2 & 2 & 2 & -1 \\ 1 & 4 & 1 & -3 \\ 3 & 5 & 5 & -2 \end{pmatrix} \end{bmatrix}$$

Out[1]= 369

$$\text{In[2]:= Det}\begin{bmatrix} \begin{pmatrix} 9(0) & 9(5) & 9(-3) & 9(2) \\ 9(-2) & 9(2) & 9(2) & 9(-1) \\ 9(1) & 9(4) & 9(1) & 9(-3) \\ 9(3) & 9(5) & 9(5) & 9(-2) \end{pmatrix} \end{bmatrix}$$

Out[2]= 2421009

In[3]:= 2421009/(9^4)

Out[3]= 369

Hence det $9A = 9^4$ det A. ◀▶

EXAMPLE 3.8 ■ **Determinants and the Addition of Rows**

Use *MATHEMATICA* to show that if the matrix

$$A = \begin{pmatrix} 0 & 5 & -3 & 2 \\ -2 & 2 & 2 & -1 \\ 1 & 4 & 1 & -3 \\ 3 & 5 & 5 & -2 \end{pmatrix}$$

is related to the matrix B by $A \ (R_3 \rightarrow R_3 + 7R_2) \ B$, then det $B =$ det A.

Solution. We compute det A and det B and compare the results.

$$\text{In[1]:= Det}\left[A=\begin{pmatrix} 0 & 5 & -3 & 2 \\ -2 & 2 & 2 & -1 \\ 1 & 4 & 1 & -3 \\ 3 & 5 & 5 & -2 \end{pmatrix}\right]$$

Out[1]= 369

$$\text{In[2]:= Det}\left[B=\begin{pmatrix} 0 & 5 & -3 & 2 \\ -2 & 2 & 2 & -1 \\ 1+7\,(-2) & 4+7\,(2) & 1+7\,(2) & -3+7\,(-1) \\ 3 & 5 & 5 & -2 \end{pmatrix}\right]$$

Out[2]= 369

This shows that $\det A = \det B$. ◀▶

THEOREM 3.6 *If B is a matrix resulting from an $n \times n$ matrix A by the elementary row operation $\left(R_i \to R_i + sR_j\right)$, then $\det B = \det A$.*

Proof. Let $B = (b_{ij})$ result from $A = (a_{ij})$ by $R = \left(R_i \to R_i + sR_j\right)$. Then

$$b_{i\pi(i)} = a_{i\pi(i)} + sa_{j\pi(i)}$$

Therefore,

$$\begin{aligned} \det B &= \sum \text{sgn}(\pi)b_{1\pi(1)} \cdots b_{n\pi(n)} \\ &= \sum \text{sgn}(\pi)b_{1\pi(1)} \cdots \left(a_{i\pi(i)} + sa_{j\pi(i)}\right) \cdots b_{n\pi(n)} \\ &= \sum \text{sgn}(\pi)a_{1\pi(1)} \cdots a_{i\pi(i)} \cdots a_{n\pi(n)} \\ &\quad + s \sum \text{sgn}(\pi)a_{1\pi(1)} \cdots a_{j\pi(i)} \cdots a_{j\pi(i)} \cdots a_{n\pi(n)} \end{aligned}$$

Hence

$$\sum \text{sgn}(\pi)a_{1\pi(1)} \cdots a_{i\pi(j)} \cdots a_{i\pi(j)} \cdots a_{n\pi(n)}$$

is the determinant of a matrix whose ith and jth rows are identical. We conclude that

$$\begin{aligned} \det B &= \sum \text{sgn}(\pi) \times a_{1\pi(1)} \cdots a_{i\pi(i)} \cdots a_{n\pi(n)} + 0 \\ &= \sum \text{sgn}(\pi) \times a_{1\pi(1)} \cdots a_{i\pi(i)} \cdots a_{n\pi(n)} \\ &= \det A \end{aligned}$$

This proves the theorem. ∎

COROLLARY 3.7 *If E is an elementary matrix obtained from I_n by $R = (R_i \rightleftharpoons R_j)$ and $i \neq j$, then $\det E = -1$.*

Proof. Since $\det I_n = 1$, Theorem 3.6 implies that $\det E = -\det I_n = -1$. ∎

COROLLARY 3.8   *If E is an elementary matrix obtained from I_n by $R = (R_i \to sR_i)$, then $\det E = s$.*

Proof. Since $\det I_n = 1$, Theorem 3.6 implies that $\det E = s \det I_n = s$. ∎

COROLLARY 3.9   *If E is an elementary matrix obtained from I_n by the operation $R = \left(R_i \to R_i + sR_j\right)$, then $\det E = \det I_n = 1$.*

Proof. This result follows from Theorem 3.6 and the fact that $\det I_n = 1$. ∎

COROLLARY 3.10   *If $A = (a_{ij})$ is an $n \times n$ matrix and E is an $n \times n$ elementary matrix, then $\det EA = \det E \det A$.*

Proof. If $E = I_n$, the result is obvious. In the other cases the result follows from Theorem 3.6 and its corollaries. ∎

We can use the result in the previous section to describe the effect of elementary row operations on determinants.

THEOREM 3.11   *If E is an elementary matrix and $B = EA$ for some $n \times n$ matrix A, then the following statements hold.*

1. If E results from I_n by the operation $(R_i \rightleftharpoons R_j)$ and $i \neq j$, then $\det B = -\det A$.
2. If E results from I_n by the operation $(R_i \to sR_i)$, then $\det B = s \det A$.
3. If E results from I_n by the operation $\left(R_i \to R_i + sR_j\right)$, then $\det B = \det A$.

Proof. Since every elementary row operation corresponds to a left-multiplication by an elementary matrix, the theorem follows from Theorem 3.6 and its corollaries. ∎

THEOREM 3.12   *If two $n \times n$ matrices A and B are row equivalent, then $\det A = 0$ if and only if $\det B = 0$.*

Proof. The matrices A and B are row equivalent if and only if there exist elementary row operations $E_1, \ldots, E_p$ for which $B = E_1 \cdots E_p A$. Using Corollary 3.10, we can conclude that $\det B = s(-1)^r \det A$, where s is the product of nonzero scalars determined by the matrices among $E_1, \ldots, E_p$ corresponding to operations of the form $(R_i \to sR_i)$, and r is the number of matrices among $E_1, \ldots, E_p$ corresponding to operations of the form $(R_i \rightleftharpoons R_j)$. Thus if $\det A = 0$, then $s(-1)^r \det A = 0 = \det B$. Conversely, if $\det B = 0$, then $s(-1)^r \det A = 0$. But since $s(-1)^r \neq 0$, we have $\det A = 0$. ∎

The next theorem provides us with a property of determinant that will be used over and over again in this text.

THEOREM 3.13 (Determinant product theorem)   *If A and B are two $n \times n$ matrices, then $\det AB = \det A \det B$.*

Proof. Suppose that A is not invertible. Then AB is not invertible either since $(AB)^{-1}$ would have to factor into $B^{-1}A^{-1}$. But A^{-1} does not exist. Hence $\det A = \det AB = 0$. Suppose, conversely, that A is invertible. Then it follows from Corollary 2.19 that A is a product $E_1 \cdots E_p$ of elementary matrices. It therefore follows from Corollary 3.10 that

$$\det AB = \det E_1 \cdots E_p B$$
$$= \det E_1 \cdots \det E_p \det B$$
$$= \det A \det B$$

This shows that the determinant of a product is the product of the determinants. ∎

EXAMPLE 3.9   ■ The Determinant of a Matrix Product

Use *MATHEMATICA* to show that the determinant of the product of the matrices

$$A = \begin{pmatrix} 0 & 5 & -3 & 2 \\ -2 & 2 & 2 & -1 \\ 1 & 4 & 1 & -3 \\ 3 & 5 & 5 & -2 \end{pmatrix} \quad \text{and} \quad B = \begin{pmatrix} -9 & 8 & 4 & -5 \\ 10 & -2 & 5 & 7 \\ 7 & -7 & 4 & -8 \\ 2 & 8 & 7 & 8 \end{pmatrix}$$

is equal to the product of the determinants of A and B.

Solution. We use *MATHEMATICA* to compute $\det AB$ and $\det A \det B$ and compare the results.

```
In[1]:= Det[ A = ( 0   5  -3   2 )
                 ( -2  2   2  -1 )
                 ( 1   4   1  -3 )
                 ( 3   5   5  -2 ) ]

Out[1]= 369
```

```
In[2]:= Det[ B = ( -9   8   4  -5 )
                 ( 10  -2   5   7 )
                 ( 7   -7   4  -8 )
                 ( 2    8   7   8 ) ]

Out[2]= 1557
```

```
In[3]:= Det[A] Det[B]

Out[3]= 574533
```

```
In[4]:= Det[A.B]

Out[4]= 574533
```

This shows that det A det B = det AB. ◄►

Determinants provide a useful test for the invertibility of real matrices.

THEOREM 3.14 (Zero determinant theorem)   *A real $n \times n$ matrix A is invertible if and only if* det $A \neq 0$.

Proof. Suppose that A is invertible. If $A = I_n$, then det $A = 1$. If $A \neq I_n$, then we know from Corollary 2.19 that there exist elementary row operations $E_1, \ldots, E_p$ such that

$$E_1 \cdots E_p A = I_n$$

Therefore, det $E_1 \cdots E_p A = $ det $I_n = 1 = s(-1)^r$ det A. It follows that det $A \neq 0$.

Conversely, suppose that A is not invertible. Then the reduced row echelon form B of A must have at least one zero row. Hence every term $b_{1j_1} b_{2j_2} \cdots b_{nj_n}$ in the determinant of B must have a zero factor. Therefore, det $B = 0$. Since det A is a multiple of det B, it follows that det A is also 0. ∎

Since a real number a is invertible if and only if $a \neq 0$, we can reword Theorem 3.14 by saying that "A matrix A is an invertible element of $\mathbb{R}^{n \times n}$ if and only if its determinant det A is an invertible element of $\mathbb{R}$."

Among the many useful applications of this theorem is a description of homogeneous linear systems with nontrivial solutions.

THEOREM 3.15   *A homogeneous linear system $A\mathbf{x} = \mathbf{0}$ with real coefficients has a nonzero solution $\mathbf{x}$ if and only if* det $A = 0$.

Proof. Suppose $A\mathbf{x} = \mathbf{0}$ and det $A \neq 0$. By Theorem 3.14, the matrix A is invertible. We therefore have $A^{-1}A\mathbf{x} = A^{-1}\mathbf{0}$. This means that $\mathbf{x} = \mathbf{0}$. Conversely, suppose that $A\mathbf{x} = \mathbf{0}$ has a nonzero solution $\mathbf{x}$. This is possible only if A is not invertible. Hence det $A = 0$. ∎

THEOREM 3.16   *If $A = (a_{ij})$ is an $n \times n$ triangular matrix, then the determinant of A is the product $a_{11} \cdots a_{nn}$ of the diagonal entries of A.*

Proof. It is sufficient to prove the theorem for upper-triangular matrices. The proof for lower-triangular matrices is analogous. We begin by recalling that

$$\det A = \sum\nolimits_{\pi \in S_n} a_{1\pi(1)} a_{2\pi(2)} \cdots a_{n\pi(n)}$$

If $\pi(i) < i$ for some $i \in \mathbf{n}$, then $a_{i\pi(i)} = 0$ since A is upper triangular. Hence

$$a_{1\pi(1)} a_{2\pi(2)} \cdots a_{n\pi(n)} = 0$$

On the other hand, if $\pi(i) \geq i$ for all $i \in \mathbf{n}$, then $\pi(n) = n$ since n is the largest element of $\mathbf{n}$. Therefore, $\pi(n-1) = n-1$ since π is one-one and since $n-1$ is the largest remaining element of $\mathbf{n}$. Continuing in this way, we conclude that $\pi(i) = i$ for all $i \in \mathbf{n}$. Hence the only nonzero term of the determinant of A is $a_{11} \cdots a_{nn}$. ■

Next we derive a formula for the determinant of a scalar multiple of a matrix.

THEOREM 3.17 *For any $n \times n$ matrix A and any scalar s, $\det sA = s^n \det A$.*

Proof. By the definition of determinants,

$$\begin{aligned}
\det sA &= \sum_{\sigma \in S_n} \text{sgn}(\sigma) \left(sa_{1\sigma(1)}\right) \cdots \left(sa_{n\sigma(n)}\right) \\
&= \sum_{\sigma \in S_n} \text{sgn}(\sigma)s^n \left(a_{1\sigma(1)} \cdots a_{n\sigma(n)}\right) \\
&= s^n \sum_{\sigma \in S_n} \text{sgn}(\sigma) \left(a_{1\sigma(1)} \cdots a_{n\sigma(n)}\right) \\
&= s^n \det A
\end{aligned}$$

This proves the theorem. ■

We conclude this section by establishing a connection between determinants, transposes, and inverses.

THEOREM 3.18 *For any $n \times n$ matrix A, $\det A = \det A^T$.*

Proof. Let $A = (a_{ij})$. Then $A^T = (b_{ij})$, with $b_{ij} = a_{ji}$. The calculations

$$\begin{aligned}
\det A &= \sum_{\sigma \in S_n} \text{sgn}(\sigma)a_{1\sigma(1)} \cdots a_{n\sigma(n)} \\
&= \sum_{\sigma^{-1} \in S_n} \text{sgn}(\sigma^{-1})a_{\sigma^{-1}(1)1} \cdots a_{\sigma^{-1}(n)n} \\
&= \sum_{\sigma \in S_n} \text{sgn}(\sigma)b_{1\sigma(1)} \cdots b_{n\sigma(n)} \\
&= \det A^T
\end{aligned}$$

show that $\det A = \det A^T$. ■

EXAMPLE 3.10 ■ **The Determinant of the Transpose of a Matrix**

Use *MATHEMATICA* to show that the determinant of the matrix

$$A = \begin{pmatrix} -10 & 5 & -4 \\ -8 & -1 & -10 \\ 2 & 3 & -1 \end{pmatrix}$$

is equal to the determinant of A^T.

Solution. We use *MATHEMATICA* to compute $\det A$ and $\det A^T$ and compare the results.

$$\text{In[1]:= Det}\left[A=\begin{pmatrix} -10 & 5 & -4 \\ -8 & -1 & -10 \\ 2 & 3 & -1 \end{pmatrix}\right]$$

Out[1]= -362

In[2]:= Det[Transpose[A]]

Out[2]= -362

This shows that det $A = \det A^T$. ◄►

THEOREM 3.19   *For any invertible matrix A, $\det A^{-1} = 1/\det A$.*

Proof. We know from Theorem 3.1 that $\det I = \det AA^{-1} = 1$, and from Theorem 3.13, $\det AA^{-1} = \det A \det A^{-1}$. Hence $1 = \det A \det A^{-1}$. Since $\det A \neq 0$, we can solve for $\det A^{-1}$ and obtain $1/\det A$. ■

EXAMPLE 3.11   ■ **The Determinant of the Inverse of a Matrix**

Use *MATHEMATICA* to show that the determinant of the inverse of the matrix

$$A = \begin{pmatrix} -1 & -2 & 3 \\ 0 & 6 & 6 \\ 0 & 0 & -7 \end{pmatrix}$$

is the reciprocal of the determinant of A.

Solution. We use *MATHEMATICA* to compute $\det A$ and $\det A^{-1}$ and show that $\det A^{-1} = 1/\det A$.

$$\text{In[1]:= Det}\left[A=\begin{pmatrix} -1 & -2 & 3 \\ 0 & 6 & 6 \\ 0 & 0 & -7 \end{pmatrix}\right]$$

Out[1]= 42

In[2]:= Det[Inverse[A]]

Out[2]= $\frac{1}{42}$

This shows that $\det A^{-1} = 1/\det A$. ◄►

It is important to note that in contrast to matrix products, where the determinant of a product of two matrices is the product of the determinants of the matrices, the determinant of the sum of two matrices is usually not equal to the sum of the two determinants. For example, if

$$A = \begin{pmatrix} 1 & 1 \\ 1 & 1 \end{pmatrix} \quad \text{and} \quad B = \begin{pmatrix} 0 & 2 \\ 2 & 0 \end{pmatrix}$$

then $\det A = 0$ and $\det B = -4$. Therefore, $\det A + \det B = -4$. On the other hand,

$$\det(A + B) = \det \begin{pmatrix} 1 & 3 \\ 3 & 1 \end{pmatrix} = -8$$

Therefore, $\det A + \det B \neq \det(A + B)$.

EXERCISES 3.2

1. Use elementary row operations to simplify the calculation of the determinants of the following matrices, if helpful, and then calculate the determinants.

a. $\begin{pmatrix} 1 & 3 & 2 \\ 0 & 2 & 4 \\ 0 & 0 & 3 \end{pmatrix}$
b. $\begin{pmatrix} 1 & 3 & 2 \\ 0 & 0 & 4 \\ 0 & 0 & 3 \end{pmatrix}$
c. $\begin{pmatrix} 0 & -3 & 2 \\ 4 & 0 & 4 \\ 2 & -1 & 0 \end{pmatrix}$

d. $\begin{pmatrix} 1 & 3 & 2 & 3 \\ 0 & 2 & 4 & 6 \\ 0 & 0 & 0 & 0 \\ 1 & 1 & 1 & 1 \end{pmatrix}$
e. $\begin{pmatrix} 7 & 5 & 0 & 0 \\ 9 & 0 & 57 & -59 \\ 4 & -8 & 0 & 92 \\ 3 & -2 & 7 & 0 \end{pmatrix}$
f. $\begin{pmatrix} 5 & 0 & 0 & 0 \\ 0 & 0 & -3 & 2 \\ 0 & 4 & 0 & 4 \\ 0 & 2 & -1 & 0 \end{pmatrix}$

2. Use *MATHEMATICA* to calculate the determinant of A, where

$$A = \begin{pmatrix} 1 & 0 & 7 & 3 \\ 6 & 8 & 5 & 8 \\ 1 & 9 & 5 & 3 \\ 7 & 0 & 4 & 5 \end{pmatrix}$$

3. Use *MATHEMATICA* to calculate the determinant of A, where

$$A = \begin{pmatrix} 54 & -5 & 99 & -61 & -50 \\ -12 & -18 & 31 & -26 & -62 \\ 1 & -47 & -91 & -47 & -61 \\ 41 & -58 & -90 & 53 & -1 \\ 94 & 83 & -86 & 23 & -84 \end{pmatrix}$$

4. Use the determinant to test the following matrices for invertibility.

a. $\begin{pmatrix} 1 & 1 & 2 \\ 2 & 1 & 3 \\ 3 & 1 & 4 \end{pmatrix}$ b. $\begin{pmatrix} 1 & 1 & 2 \\ 2 & 1 & 3 \\ 3 & 1 & 6 \end{pmatrix}$ c. $\begin{pmatrix} 1 & 1 & 2 & 2 \\ 2 & 1 & 3 & 3 \\ 3 & 1 & 6 & 6 \\ 0 & 7 & 4 & 1 \end{pmatrix}$

5. Use *MATHEMATICA* to generate three random 5×5 matrices with integer entries and verify that their determinants are equal to the determinants of their transposes.

6. Use the **Table** function to generate a general 4×4 matrix A and show that $\det A = \det A^T$.

7. Use *MATHEMATICA* to generate two random 3×3 matrices A and B with real entries and show that $\det AB = \det A \det B$.

8. Use *MATHEMATICA* and the following matrices to show that $\det A^{-1} = 1/\det A$.

a. $A = \begin{pmatrix} 4 & 0 & 0 \\ 0 & 5 & 0 \\ 0 & 0 & 4 \end{pmatrix}$ b. $A = \begin{pmatrix} 4 & 2 & 3 \\ 0 & 5 & 2 \\ 0 & 0 & 4 \end{pmatrix}$ c. $A = \begin{pmatrix} \frac{1}{\sqrt{2}} & \frac{1}{\sqrt{2}} & 0 \\ -\frac{1}{\sqrt{2}} & \frac{1}{\sqrt{2}} & 0 \\ 0 & 0 & 1 \end{pmatrix}$

9. Show for each of the following $n \times n$ matrices that $\det aA = a^n \det A$ for all scalars a.

a. $A = \begin{pmatrix} 0 & 0 & 0 \\ 0 & 0 & 0 \\ 0 & 0 & 0 \end{pmatrix}$ b. $A = \begin{pmatrix} 4 & 2 & 3 \\ -3 & 5 & 2 \\ 7 & 0 & 4 \end{pmatrix}$ d. $A = \begin{pmatrix} \frac{1}{\sqrt{2}} & \frac{1}{\sqrt{2}} \\ -\frac{1}{\sqrt{2}} & \frac{1}{\sqrt{2}} \end{pmatrix}$

10. Prove that for all permutation matrices P, $\det P = \pm 1$.

APPLICATIONS

Cramer's Rule

A well-known procedure for solving linear systems is ***Cramer's rule,*** which uses determinants. If $A\mathbf{x} = \mathbf{b}$ is a linear system and A is an $n \times n$ invertible matrix, then Cramer's rule says that for each $1 \le i \le n$,

$$x_i = \frac{\det A(i/\mathbf{b})}{\det A}$$

where $A(i/\mathbf{b})$ is the matrix obtained from A by replacing the ith column of A by $\mathbf{b}$. We illustrate the use of Cramer's rule in the case of a 3×3 matrix A.

EXAMPLE 3.12 ■ Using Cramer's Rule to Solve a Linear System in Three Variables

Use Cramer's rule to solve the 3×3 linear system

$$\begin{cases} x + y - z = & 4 \\ 2x - y + 2z = & 18 \\ x - y + z = & 5 \end{cases}$$

Solution. The matrix form $A\mathbf{x} = \mathbf{b}$ of the given system is

$$\begin{pmatrix} 1 & 1 & -1 \\ 2 & -1 & 2 \\ 1 & -1 & 1 \end{pmatrix} \begin{pmatrix} x \\ y \\ z \end{pmatrix} = \begin{pmatrix} 4 \\ 18 \\ 5 \end{pmatrix}$$

Since $\det A = 2 \neq 0$, Cramer's rule applies. It tells us that

$$x = \frac{\det A(1/\mathbf{b})}{\det A} = \frac{\det \begin{pmatrix} 4 & 1 & -1 \\ 18 & -1 & 2 \\ 5 & -1 & 1 \end{pmatrix}}{\det \begin{pmatrix} 1 & 1 & -1 \\ 2 & -1 & 2 \\ 1 & -1 & 1 \end{pmatrix}} = \frac{9}{2}$$

and that

$$y = \frac{\det A(2/\mathbf{b})}{\det A} = \frac{\det \begin{pmatrix} 1 & 4 & -1 \\ 2 & 18 & 2 \\ 1 & 5 & 1 \end{pmatrix}}{\det \begin{pmatrix} 1 & 1 & -1 \\ 2 & -1 & 2 \\ 1 & -1 & 1 \end{pmatrix}} = 8$$

and that

$$z = \frac{\det A(3/\mathbf{b})}{\det A} = \frac{\det \begin{pmatrix} 1 & 1 & 4 \\ 2 & -1 & 18 \\ 1 & -1 & 5 \end{pmatrix}}{\det \begin{pmatrix} 1 & 1 & -1 \\ 2 & -1 & 2 \\ 1 & -1 & 1 \end{pmatrix}} = \frac{17}{2}$$

Hence $(x, y, z) = (9/2, 8, 17/2)$ is the solution of the system. ◄►

Cramer's rule provides an excellent tool for elegant definitions in calculus and differential equations. However, its usefulness for solving linear systems is minimized by the fact that, in general, it requires significantly more arithmetic operations than do other methods.

EXERCISES 3.3

1. Convert the following 2×2 linear systems to the matrix form $A\mathbf{x} = \mathbf{b}$. Solve the systems by Cramer's rule if $\det A \neq 0$.

   a. $\begin{cases} 2x + y = 2 \\ x - 3y = 1 \end{cases}$ b. $\begin{cases} 2x + y = 2 \\ -2x - y = 1 \end{cases}$ c. $\begin{cases} 2x + y = 0 \\ x - 3y = 0 \end{cases}$

2. Convert the following 3×3 linear systems to the matrix form $A\mathbf{x} = \mathbf{b}$. Solve the systems by Cramer's rule if $\det A \neq 0$.

   a. $\begin{cases} 2x + y - z = 2 \\ x - 3y + z = 1 \\ 2x + y - 2z = 9 \end{cases}$ b. $\begin{cases} 2x + y - z = 2 \\ z - y + x = 1 \\ 2x - 2y + z = 9 \end{cases}$ c. $\begin{cases} 2x + y - z = 0 \\ x - 3y + z = 0 \\ 2x + y - 4z = 0 \end{cases}$

Adjugates and Inverses

We conclude this section by describing the link between determinants and matrix inverses. For this purpose, we let $A(i/j)$ be the matrix obtained from a matrix A by replacing the ith row of A by its jth row. Our description is based on two simple observations:

1. If $i = j$, then $A(i/j) = A$, and therefore $\det A(i/j) = \det A$.

2. If $i \neq j$, then $A(i/j)$ has two identical rows and by Corollary 3.4, $\det A(i/j) = 0$.

If we use the ith row of $A(i/j)$ to write out the Laplace expansion of $\det A(i/j)$, we get

$$\det A(i/j) = a_{i1}C_{i1} + \cdots + a_{in}C_{in} = \begin{pmatrix} a_{i1} & \cdots & a_{in} \end{pmatrix} \begin{pmatrix} C_{i1} \\ \vdots \\ C_{in} \end{pmatrix}$$

Therefore, $\det A(i/j)$ is the ijth entry of the matrix

$$AC = \begin{pmatrix} a_{11} & \cdots & a_{1n} \\ & \vdots & \\ a_{n1} & \cdots & a_{nn} \end{pmatrix} \begin{pmatrix} C_{11} & \cdots & C_{n1} \\ & \vdots & \\ C_{1n} & \cdots & C_{nn} \end{pmatrix}$$

We call the matrix C the **adjugate** of A and denote it by adj A. We can see that the adjugate is the transpose of the matrix whose entries are the cofactors of A. Moreover, it follows from the stated properties of det $A(i/j)$ that A adj $A = (\det A)I_n$. Therefore,

$$I_n = A\frac{\text{adj } A}{\det A} = AA^{-1}$$

Solving for A^{-1}, we get the formula

$$A^{-1} = \frac{\text{adj } A}{\det A}$$

for the inverse of A, expressed entirely in terms of determinants of submatrices of A.

EXAMPLE 3.13 ■ **Using an Adjugate to Find the Inverse of a Matrix**

Use the adjugate to find the inverse of the matrix

$$A = \begin{pmatrix} -2 & 3 & 4 \\ 4 & -3 & -3 \\ 4 & -1 & 0 \end{pmatrix}$$

Solution. By definition, the entries of the adjugate of A are the cofactors $C_{ij} = (-1)^{i+j} \det M_{ij}$ of A. We must therefore compute $C_{11}, \ldots, C_{33}$.

We leave it as an exercise to verify that $C_{11} = -3$, $C_{12} = -12$, $C_{13} = 8$, $C_{21} = -4$, $C_{22} = -16$, $C_{23} = 10$, $C_{31} = 3$, $C_{32} = 10$, and $C_{33} = -6$. Therefore,

$$\text{adj } A = \begin{pmatrix} -3 & -4 & 3 \\ -12 & -16 & 10 \\ 8 & 10 & -6 \end{pmatrix}$$

Moreover, it is easily seen that $\det A = 2$. By the derived formula,

$$A^{-1} = \frac{1}{2}\begin{pmatrix} -3 & -4 & 3 \\ -12 & -16 & 10 \\ 8 & 10 & -6 \end{pmatrix} = \begin{pmatrix} -\frac{3}{2} & -2 & \frac{3}{2} \\ -6 & -8 & 5 \\ 4 & 5 & -3 \end{pmatrix}$$

is the inverse of A.

Let us use *MATHEMATICA* to verify this fact.

```
In[1]:= A = ( -2  3   4
               4 -3  -3  ) ; B = ( -3/2  -2   3/2
               4 -1   0               -6   -8   5
                                       4    5  -3  ) ;
```

```
In[2]:= A.B==IdentityMatrix[3]
```

Out[2]= True

This confirms that $B = A^{-1}$. ◀▶

EXERCISES 3.4

1. Verify that

$$
\begin{pmatrix} 6 & 3 \\ 4 & 4 \end{pmatrix}^{-1} = \begin{pmatrix} \frac{1}{3} & -\frac{1}{4} \\ -\frac{1}{3} & \frac{1}{2} \end{pmatrix} = \frac{1}{\det A} \begin{pmatrix} C_{11} & C_{21} \\ C_{12} & C_{22} \end{pmatrix}
$$

2. Verify that

$$
\begin{pmatrix} 1 & 0 & 2 \\ 2 & 0 & 3 \\ 4 & 1 & 0 \end{pmatrix}^{-1} = \begin{pmatrix} -3 & 2 & 0 \\ 12 & -8 & 1 \\ 2 & -1 & 0 \end{pmatrix} = \frac{1}{\det A} \begin{pmatrix} C_{11} & C_{21} & C_{31} \\ C_{12} & C_{22} & C_{32} \\ C_{13} & C_{23} & C_{33} \end{pmatrix}
$$

Triangular Matrices

In this section we show that the determinant of a triangular matrix is simply the product of the diagonal entries of the matrix. We use this result to develop strategies for computing determinants.

THEOREM 3.20   *If $A = (a_{ij})$ is an $n \times n$ triangular matrix, then the determinant of A is the product $a_{11} \cdots a_{nn}$ of the diagonal elements of A.*

Proof. We prove the theorem for upper-triangular matrices. The proof for lower-triangular matrices is analogous. By definition,

$$
\det A = \sum_{\sigma \in S_n} \operatorname{sgn}(\sigma) a_{1\sigma(1)} a_{2\sigma(2)} \cdots a_{n\sigma(n)}
$$

is determined by the permutations of $\mathbf{n} = \{1, \ldots, n\}$. If $\sigma(i) < i$ for some $i \in \mathbf{n}$, then $a_{i\sigma(i)} = 0$ since A is upper triangular. Hence $a_{1\sigma(1)} a_{2\sigma(2)} \cdots a_{n\sigma(n)} = 0$. On the other hand, if $\sigma(i) \geq i$ for all $i \in \mathbf{n}$, then $\sigma(n) = n$ since n is largest element of $\mathbf{n}$. Therefore, $\sigma(n-1) = n-1$ since σ is one-one and since $n-1$ is the largest remaining element of $\mathbf{n}$. Continuing in this way, we conclude that $\sigma(i) = i$ for all $i \in \mathbf{n}$. Hence the only nonzero term of the determinant of A is $a_{11} \cdots a_{nn}$. ■

When combined with LU decomposition (discussed in Chapter 2), Theorem 3.20 provides us with an interesting procedure for calculating determinants.

THEOREM 3.21   *For every real square matrix A, there exists an upper-triangular matrix U for which $(-1)^m \det A = \det U$.*

Proof. By Theorem 2.35, there exists a permutation matrix P and triangular matrices L and U for which $PA = LU$. We know from Theorems 3.11 and 3.13 that $\det P = (-1)^m$. Moreover, Theorem 2.35 tells us that L has 1s on the diagonal. By Theorem 3.20, finally, we have $\det L = 1$. Since $\det PA = \det P \det A = \det L \det U$, it therefore follows that $(-1)^m \det A = \det U$. ∎

EXAMPLE 3.14 ■ Determinants and LU Decomposition

Use the **LUDecomposition** function to find the determinant of the matrix

$$A = \begin{pmatrix} 4 & 5 & 0 & 0 \\ 2 & 2 & 2 & 4 \\ 5 & 2 & 5 & 2 \\ 4 & 1 & 3 & 0 \end{pmatrix}$$

Solution. We use the **GaussianElimination** package.

```
In[1]:= <<LinearAlgebra`GaussianElimination`
```

```
In[2]:= A=
```
$$\begin{pmatrix} 4 & 5 & 0 & 0 \\ 2 & 2 & 2 & 4 \\ 5 & 2 & 5 & 2 \\ 4 & 1 & 3 & 0 \end{pmatrix};$$

```
In[3]:= M=LUDecomposition[A]
```
Out[3]= $\{\{2,2,2,4\},\{2,1,-4,-8\},\{\frac{5}{2},-3,-12,-32\},\{2,-3,\frac{13}{12},\frac{8}{3}\},\{2,1,3,4\},1\}$

```
In[4]:= M[[1]]//MatrixForm
```
Out[4]//MatrixForm=
$$\begin{pmatrix} 2 & 2 & 2 & 4 \\ 2 & 1 & -4 & -8 \\ \frac{5}{2} & -3 & -12 & -32 \\ 2 & -3 & \frac{13}{12} & \frac{8}{3} \end{pmatrix}$$

Next we recall from our discussion of the implementation of LU decomposition in *MATH-EMATICA* (Chapter 2) that the matrix

$$\begin{pmatrix} 2 & 2 & 2 & 4 \\ 2 & 1 & -4 & -8 \\ \frac{5}{2} & -3 & -12 & -32 \\ 2 & -3 & \frac{13}{12} & \frac{8}{3} \end{pmatrix}$$

combines the upper- and lower-triangular matrices of the LU decomposition. This means that

$$P\begin{pmatrix} 4 & 5 & 0 & 0 \\ 2 & 2 & 2 & 4 \\ 5 & 2 & 5 & 2 \\ 4 & 1 & 3 & 0 \end{pmatrix} = \begin{pmatrix} 1 & 0 & 0 & 0 \\ 2 & 1 & 0 & 0 \\ \frac{5}{2} & -3 & 1 & 0 \\ 2 & -3 & \frac{13}{12} & 1 \end{pmatrix}\begin{pmatrix} 2 & 2 & 2 & 4 \\ 0 & 1 & -4 & -8 \\ 0 & 0 & -12 & -32 \\ 0 & 0 & 0 & \frac{8}{3} \end{pmatrix}$$

for some permutation matrix P. We can find P by extracting the second row of the matrix M constructed earlier.

In[5]:= **M[[2]]**

Out[5]= {2,1,3,4}

Therefore,

$$P = \begin{pmatrix} 0 & 1 & 0 & 0 \\ 1 & 0 & 0 & 0 \\ 0 & 0 & 1 & 0 \\ 0 & 0 & 0 & 1 \end{pmatrix}$$

is the permutation matrix that interchanges the first and second rows of the identity matrix. We know from Exercises 3.2 that $\det P = -1$. Therefore,

$$\det P \det A = \det \begin{pmatrix} 1 & 0 & 0 & 0 \\ 2 & 1 & 0 & 0 \\ \frac{5}{2} & -3 & 1 & 0 \\ 2 & -3 & \frac{13}{12} & 1 \end{pmatrix} \det \begin{pmatrix} 2 & 2 & 2 & 4 \\ 0 & 1 & -4 & -8 \\ 0 & 0 & -12 & -32 \\ 0 & 0 & 0 & \frac{8}{3} \end{pmatrix}$$

Hence $\det A = (-1)\,(2)\,(1)\,(-12)\,(8/3) = 64.$ ◀▶

The next theorem shows that we can also find the determinant of a matrix by first reducing the matrix to echelon form. In practice, this method is usually an efficient way of calculating a determinant.

THEOREM 3.22   *For every invertible $n \times n$ real matrix A, $\det A = (-1)^m \det B$, where B is a row echelon matrix obtained from A by elementary row operations of types $\left(R_i \rightleftharpoons R_j\right)$, with $i \neq j$, and $(R_i \rightarrow R_i + sR_j)$, and m is the number of row interchanges involved.*

Proof. We know from our work on echelon matrices that every square matrix can be reduced to row echelon form using only operations of types $\left(R_i \rightleftharpoons R_j\right)$ and $(R_i \rightarrow R_i + sR_j)$. Theorem 3.6 tells us that $(R_i \rightarrow R_i + sR_j)$ leaves the determinant of a matrix unchanged, and Corollary 3.7 guarantees that if $i \neq j$, then $\left(R_i \rightleftharpoons R_j\right)$ changes the sign of the determinant. ∎

COROLLARY 3.23   *For every invertible $n \times n$ real matrix A, $\det A = (-1)^m p_1 \cdots p_n$, where $p_1, \ldots, p_n$ are the pivots of A.*

Proof. Since the row echelon matrix B in Theorem 3.22 is upper triangular, its determinant is simply the product of its diagonal entries. It is clear from our work in Chapter 1 that these entries are the pivots of the matrix A. ∎

EXAMPLE 3.15   ■ **Determinants and Pivots**

Use pivots to compute the determinant of the matrix

$$A = \begin{pmatrix} 0 & 5 & 0 \\ 5 & 2 & 5 \\ 2 & 2 & 3 \end{pmatrix}$$

Solution. We first use *MATHEMATICA* to compute the echelon form of A.

```
In[1]:= A=( 0 5 0
           5 2 5
           2 2 3 );
```

```
In[2]:= A=A[[{2,1,3}]]
```

```
Out[2]= {{5,2,5},{0,5,0},{2,2,3}}
```

```
In[3]:= A[[3]]=A[[3]]-2/5A[[1]];A
```

```
Out[3]= {{5,2,5},{0,5,0},{0,6/5,1}}
```

```
In[4]:= A[[3]]=A[[3]]-6/25A[[2]];A
```

Out[4]//MatrixForm=

$$\begin{pmatrix} 5 & 2 & 5 \\ 0 & 5 & 0 \\ 0 & 0 & 1 \end{pmatrix}$$

This shows us that the pivots of A are $5, 5$, and 1. Since we required one interchange of rows to obtain Out[2], we get det $A = (-1)(5)(5)(1) = -25$. ◄►

EXERCISES 3.5

1. Show that

$$\det \begin{pmatrix} 1 & 0 & 0 \\ a & 1 & 0 \\ b & c & 1 \end{pmatrix} = \det \begin{pmatrix} 1 & a & b \\ 0 & 1 & c \\ 0 & 0 & 1 \end{pmatrix} = 1$$

for all a, b, c.

2. Find the determinant of the following matrices by inspection.

$$\text{a. } A = \begin{pmatrix} 4 & 0 & 0 & 1 & 1 \\ 0 & 5 & 0 & -4 & 3 \\ 0 & 0 & 4 & 7 & 2 \\ 0 & 0 & 0 & 0 & 0 \\ 0 & 0 & 0 & 0 & 0 \end{pmatrix} \quad \text{b. } B = \begin{pmatrix} 4 & 0 & 0 & 1 & 1 \\ 0 & 5 & 0 & -4 & 3 \\ 0 & 0 & 4 & 7 & 2 \\ 0 & 0 & 0 & 1 & 0 \\ 0 & 0 & 0 & 0 & 2 \end{pmatrix}$$

3. Let A and B be the matrices in Exercise 2. Use inspection to find the determinant of the matrix AB.

4. Suppose that A is an $n \times n$ diagonal matrix. Find a formula for det A^p in terms of the diagonal entries of A.

5. Find a 3×3 matrix A with nonzero diagonal entries whose trace is equal to the determinant of A.

Geometry and Determinants

The determinant has many beautiful geometric applications. The following examples illustrate some of them.

EXAMPLE 3.16 ■ The Straight Line Determined by Two Points

Use determinants to find an equation for the straight line determined by two distinct points (x_1, y_1) and (x_2, y_2) in $\mathbb{R}^2$.

Solution. The required equation is of the form $ax + by + c = 0$, with a, b, c not all equal to 0. The given points must satisfy this equation, so that we have $ax_1 + by_1 + c = 0$ and $ax_2 + by_2 + c = 0$. These equations determine a homogeneous linear system

$$\begin{cases} ax + by + c = 0 \\ ax_1 + by_1 + c = 0 \\ ax_2 + by_2 + c = 0 \end{cases}$$

whose matrix form $A\mathbf{x} = \mathbf{0}$ is

$$\begin{pmatrix} x & y & 1 \\ x_1 & y_1 & 1 \\ x_2 & y_2 & 1 \end{pmatrix} \begin{pmatrix} a \\ b \\ c \end{pmatrix} = \begin{pmatrix} 0 \\ 0 \\ 0 \end{pmatrix}$$

Since a, b, c are not all 0, the system has a nontrivial solution. It follows from Theorem 3.15 that

$$\det \begin{pmatrix} x & y & 1 \\ x_1 & y_1 & 1 \\ x_2 & y_2 & 1 \end{pmatrix} = 0$$

This means that $ax' + by' + c = 0$ if and only if

$$\det \begin{pmatrix} x' & y' & 1 \\ x_1 & y_1 & 1 \\ x_2 & y_2 & 1 \end{pmatrix} = 0$$

Let us suppose that $(x_1, y_1) = (1, 2)$ and $(x_2, y_2) = (3, 5)$. Then the straight line determined by these points is given by the equation

$$\det \begin{pmatrix} x & y & 1 \\ 1 & 2 & 1 \\ 3 & 5 & 1 \end{pmatrix} = 0$$

Let us use *MATHEMATICA* to extract the desired equation.

```
In[1]:= A= ( x y 1
            1 2 1  ; Det[A]
            3 5 1 )

Out[1]= -1-3x+2y
```

The desired equation is therefore $-3x + 2y - 1 = 0$. ◄►

EXAMPLE 3.17 ■ The Parabola Determined by Three Noncollinear Points

Use determinants to find an equation for the parabola determined by three noncollinear points (x_1, y_1), (x_2, y_2), and (x_3, y_3) in $\mathbb{R}^2$.

Solution. The method is analogous to that used in Example 3.16. Suppose that the desired parabola is of the form $ax^2 + bx + cy + d = 0$, with a, b, c, d not all equal to 0. Then the given points must satisfy this equation. The homogeneous linear system determined by this equation and by the three given points is

$$\begin{cases} ax^2 + bx + cy + d = 0 \\ ax_1^2 + bx_1 + cy_1 + d = 0 \\ ax_2^2 + bx_2 + cy_2 + d = 0 \\ ax_3^2 + bx_3 + cy_3 + d = 0 \end{cases}$$

The matrix form $A\mathbf{x} = \mathbf{0}$ of this equation is

$$\begin{pmatrix} x^2 & x & y & 1 \\ x_1^2 & x_1 & y_1 & 1 \\ x_2^2 & x_2 & y_2 & 1 \\ x_3^2 & x_3 & y_3 & 1 \end{pmatrix} \begin{pmatrix} a \\ b \\ c \\ d \end{pmatrix} = \begin{pmatrix} 0 \\ 0 \\ 0 \\ 0 \end{pmatrix}$$

The determinant equation for the desired parabola is therefore

$$\det \begin{pmatrix} x^2 & x & y & 1 \\ x_1^2 & x_1 & y_1 & 1 \\ x_2^2 & x_2 & y_2 & 1 \\ x_3^2 & x_3 & y_3 & 1 \end{pmatrix} = 0$$

Let us suppose that $(x_1, y_1) = (-1, 1)$, $(x_2, y_2) = (0, 0)$, and $(x_3, y_3) = (1, 1)$. Then the parabola determined by these points is given by

$$\det \begin{pmatrix} x^2 & x & y & 1 \\ 1 & -1 & 1 & 1 \\ 0 & 0 & 0 & 1 \\ 1 & 1 & 1 & 1 \end{pmatrix} = 0$$

Calculations similar to those in Example 3.16 show that the desired equation is $2x^2 - 2y = 0$, which simplifies to $y = x^2$. ◄►

EXAMPLE 3.18 ■ The Circle Determined by Three Noncollinear Points

Use determinants to find an equation for the circle defined by three noncollinear points (x_1, y_1), (x_2, y_2), and (x_3, y_3) in $\mathbb{R}^2$.

Solution. The method is similar to that employed in two previous examples. Suppose that the desired circle is of the form $a(x^2 + y^2) + bx + cy + d = 0$, with a, b, c, d not all equal to 0. Then the given points must satisfy this equation. The homogeneous linear system determined by this equation and the three given points is

$$\begin{cases} a(x^2 + y^2) + bx + cy + d = 0 \\ a(x_1^2 + y_1^2) + bx_1 + cy_1 + d = 0 \\ a(x_2^2 + y_2^2) + bx_2 + cy_2 + d = 0 \\ a(x_3^2 + y_3^2) + bx_3 + cy_3 + d = 0 \end{cases}$$

The matrix form $A\mathbf{x} = \mathbf{0}$ of this equation is

$$\begin{pmatrix} x^2 + y^2 & x & y & 1 \\ x_1^2 + y_1^2 & x_1 & y_1 & 1 \\ x_2^2 + y_2^2 & x_2 & y_2 & 1 \\ x_3^2 + y_3^2 & x_3 & y_3 & 1 \end{pmatrix} \begin{pmatrix} a \\ b \\ c \\ d \end{pmatrix} = \begin{pmatrix} 0 \\ 0 \\ 0 \\ 0 \end{pmatrix}$$

Hence the determinant equation for the desired circle is

$$\det \begin{pmatrix} x^2 + y^2 & x & y & 1 \\ x_1^2 + y_1^2 & x_1 & y_1 & 1 \\ x_2^2 + y_2^2 & x_2 & y_2 & 1 \\ x_3^2 + y_3^2 & x_3 & y_3 & 1 \end{pmatrix} = 0$$

Let us suppose that $(x_1, y_1) = (-1, 1)$, $(x_2, y_2) = (0, 0)$, and $(x_3, y_3) = (1, 1)$. Then the circle defined by these points is $x^2 + (y - 1)^2 = 1$. This equation defines the circle of radius 1 centered at $(0, 1)$, as expected. ◄►

EXAMPLE 3.19 ■ The Area of a Parallelogram

Use determinants to find the area of a parallelogram.

Solution. The area of a parallelogram determined by two vectors is the product of the base and the height of the parallelogram. Let

$$\mathbf{u} = \begin{pmatrix} a \\ b \end{pmatrix} \quad \text{and} \quad \mathbf{v} = \begin{pmatrix} c \\ d \end{pmatrix}$$

be two vectors in $\mathbb{R}^2$, and let θ be the angle between them, with $0 < \theta < \frac{\pi}{2}$. Then

$$\begin{aligned} \text{Area}^2 &= (\text{Base}^2)(\text{Height}^2) \\ &= (a^2 + b^2)(c^2 + d^2) \sin^2 \theta \\ &= (a^2 + b^2)(c^2 + d^2)(1 - \cos^2 \theta) \\ &= (a^2 + b^2)(c^2 + d^2) - (a^2 + b^2)(c^2 + d^2) \cos^2 \theta \end{aligned}$$

$$= (a^2 + b^2)(c^2 + d^2) - (ac + bd)^2$$
$$= (ad - bc)^2$$

Hence the absolute value of the determinant of the matrix

$$\begin{pmatrix} a & c \\ b & d \end{pmatrix}$$

is the area of the parallelogram determined by $\mathbf{u}$ and $\mathbf{v}$. ◀▶

EXAMPLE 3.20 ■ **The Volume of a Parallelepiped**

Use determinants to find the volume of a parallelepiped.

Solution. Suppose that $\mathbf{u} = (u_1, u_2, u_3)$ and $\mathbf{v} = (v_1, v_2, v_3)$ are any two nonzero vectors in $\mathbb{R}^3$ not lying in the same plane, and that

$$\mathbf{i} = \begin{pmatrix} 1 \\ 0 \\ 0 \end{pmatrix} \quad \mathbf{j} = \begin{pmatrix} 0 \\ 1 \\ 0 \end{pmatrix} \quad \mathbf{k} = \begin{pmatrix} 0 \\ 0 \\ 1 \end{pmatrix}$$

Then the determinant of the *symbolic* matrix

$$\begin{pmatrix} \mathbf{i} & \mathbf{j} & \mathbf{k} \\ u_1 & u_2 & u_3 \\ v_1 & v_2 & v_3 \end{pmatrix}$$

is the vector

$$\det \begin{pmatrix} u_2 & u_3 \\ v_2 & v_3 \end{pmatrix} \mathbf{i} - \det \begin{pmatrix} u_1 & u_3 \\ v_1 & v_3 \end{pmatrix} \mathbf{j} + \det \begin{pmatrix} u_1 & u_2 \\ v_1 & v_2 \end{pmatrix} \mathbf{k}$$

We denote this vector by $\mathbf{u} \times \mathbf{v}$ and call it the *cross product* of $\mathbf{u}$ and $\mathbf{v}$. A simple matrix calculation shows that

$$\mathbf{u}^T (\mathbf{u} \times \mathbf{v}) = \mathbf{v}^T (\mathbf{u} \times \mathbf{v}) = 0$$

In Chapter 8, we will show that the vector $\mathbf{u} \times \mathbf{v}$ is **orthogonal** to $\mathbf{u}$ and $\mathbf{v}$ in the sense of Euclidean geometry. This implies that

$$\|\mathbf{u} \times \mathbf{v}\|^2 = \left[\det \begin{pmatrix} u_2 & u_3 \\ v_2 & v_3 \end{pmatrix} \right]^2 + \left[\det \begin{pmatrix} u_1 & u_3 \\ v_1 & v_3 \end{pmatrix} \right]^2 + \left[\det \begin{pmatrix} u_1 & u_2 \\ v_1 & v_2 \end{pmatrix} \right]^2$$

is the square of the area of the parallelogram determined by $\mathbf{u}$ and $\mathbf{v}$.

If $\mathbf{w} = (w_1, w_2, w_3)$ is a third vector in $\mathbb{R}^3$ not lying in the plane determined by $\mathbf{u}$ and $\mathbf{v}$, then the volume of the parallelepiped determined by $\mathbf{u}$, $\mathbf{v}$, and $\mathbf{w}$ is given by the formula

$$\text{Volume} = (\text{Base})(\text{Height}) = \|\mathbf{u} \times \mathbf{v}\| \, \|\mathbf{w}\| \, |\cos \theta|$$

where θ is the angle between the vectors $\mathbf{u} \times \mathbf{v}$ and $\mathbf{w}$, and where

$$\|\mathbf{w}\|^2 = w_1^2 + w_2^2 + w_3^2$$

Hence the absolute value of the determinant of the matrix

$$\begin{pmatrix} w_1 & w_2 & w_3 \\ u_1 & u_2 & u_3 \\ v_1 & v_2 & v_3 \end{pmatrix}$$

is the volume of the parallelepiped determined by the vectors $\mathbf{u}$, $\mathbf{v}$, and $\mathbf{w}$. ◄►

EXERCISES 3.6

1. Use determinants to find equations for the straight lines containing the following points.

    a. $(3, 4)$ and $(2, 7)$ b. $(3, 4)$ and $(2, -7)$

    c. $(3, 4)$ and $(-2, 7)$ d. $(3, 4)$ and $(-2, -7)$

2. Use determinants to find equations for the parabolas containing the following points.

    a. $(3, 4)$, $(2, 7)$, and $(9, 4)$ b. $(3, 4)$, $(2, 7)$, and $(9, -4)$

    c. $(3, 4)$, $(2, 7)$, and $(-9, 4)$ d. $(3, 4)$, $(2, 7)$, and $(-9, -4)$

3. Use determinants to find equations for the circles containing the points in Exercise 2.

4. Use determinants to find the areas of the parallelograms determined by the following points.

    a. $(3, 4)$, $(2, 7)$, and $(9, 4)$ b. $(3, 4)$, $(2, 7)$, and $(-7, 6)$

    c. $(3, 4)$, $(2, 7)$, and $(-2, 2)$ d. $(3, 4)$, $(2, 7)$, and $(-12, -15)$

5. Use the fact that

$$\det \begin{pmatrix} a & c \\ b & d \end{pmatrix} = \det \begin{pmatrix} a & c & 0 \\ b & d & 0 \\ 1 & 1 & 1 \end{pmatrix}$$

to show that the area of a parallelogram in $\mathbb{R}^2$ is preserved by the homogeneous coordinate embedding of $\mathbb{R}^2$ in $\mathbb{R}^3$.

Calculus and Determinants

Determinants can be used in calculus to solve a variety of problems that include the computation of partial derivatives, problems involving multiple integrals, the identification of maxima and minima, and the proof that a set of functions is linearly independent.

Jacobians

Suppose that $F(u, v, x, y) = 0$ and $G(u, v, x, y) = 0$ are two differentiable functions in some region of $\mathbb{R}^4$ and that u and v are differentiable functions of x and y. Determinants and Cramer's rule can be used to find the partial derivatives u_x, v_x, u_y, and v_y.

We proceed as follows. Since $F(u, v, x, y) = G(u, v, x, y) = 0$, the differentials dF and dG are 0. This means that

$$dF = F_u du + F_v dv + F_x dx + F_y dy = 0$$
$$dG = G_u du + G_v dv + G_x dx + G_y dy = 0$$

Furthermore, $du = u_x dx + u_y dy$ and $dv = v_x dx + v_y dy$ since u and v are differentiable functions of x and y. Therefore,

$$
\begin{aligned}
dF &= F_u du + F_v dv + F_x dx + F_y dy = 0 \\
&= F_u \left(u_x dx + u_y dy\right) + F_v \left(v_x dx + v_y dy\right) + F_x dx + F_y dy = 0 \\
&= (F_u u_x + F_v v_x + F_x)\, dx + \left(F_u u_y + F_v v_y + F_y\right) dy = 0
\end{aligned}
$$

and

$$
\begin{aligned}
dG &= G_u du + G_v dv + G_x dx + G_y dy = 0 \\
&= G_u \left(u_x dx + u_y dy\right) + G_v \left(v_x dx + v_y dy\right) + G_x dx + G_y dy = 0 \\
&= (G_u u_x + G_v v_x + G_x)\, dx + \left(G_u u_y + G_v v_y + G_y\right) dy = 0
\end{aligned}
$$

Since x and y are independent variables, the coefficients of dx and dy in these equations must be zero. Hence we obtain the two linear systems

$$
\begin{cases}
F_u u_x + F_v v_x &= -F_x \\
G_u u_x + G_v v_x &= -G_x
\end{cases}
\quad \text{and} \quad
\begin{cases}
F_u u_y + F_v v_y &= -F_y \\
G_u u_y + G_v v_y &= -G_y
\end{cases}
$$

By Cramer's rule, the formulas

$$
u_x = -\frac{\det\begin{pmatrix} F_x & F_v \\ G_x & G_v \end{pmatrix}}{\det\begin{pmatrix} F_u & F_v \\ G_u & G_v \end{pmatrix}}
\quad \text{and} \quad
v_x = -\frac{\det\begin{pmatrix} F_u & F_x \\ G_u & G_x \end{pmatrix}}{\det\begin{pmatrix} F_u & F_v \\ G_u & G_v \end{pmatrix}}
$$

and

$$
u_y = -\frac{\det\begin{pmatrix} F_y & F_v \\ G_y & G_v \end{pmatrix}}{\det\begin{pmatrix} F_u & F_v \\ G_u & G_v \end{pmatrix}}
\quad \text{and} \quad
v_y = -\frac{\det\begin{pmatrix} F_u & F_y \\ G_u & G_y \end{pmatrix}}{\det\begin{pmatrix} F_u & F_v \\ G_u & G_v \end{pmatrix}}
$$

are valid at all points (x, y) at which $(F_u G_v - F_v G_u) \neq 0$. The function $(F_u G_v - F_v G_u)$ is usually denoted by

$$ J \left(\frac{F, G}{u, v} \right) \quad \text{or} \quad \frac{\partial(F, G)}{\partial(u, v)} $$

and is called the *Jacobian* of F and G with respect to u and v. The determinants

$$ J \left(\frac{F, G}{x, v} \right) = \det \begin{pmatrix} F_x & F_v \\ G_x & G_v \end{pmatrix} \quad J \left(\frac{F, G}{y, v} \right) = \det \begin{pmatrix} F_y & F_v \\ G_y & G_v \end{pmatrix} $$

$$ J \left(\frac{F, G}{u, x} \right) = \det \begin{pmatrix} F_u & F_x \\ G_u & G_x \end{pmatrix} \quad J \left(\frac{F, G}{u, y} \right) = \det \begin{pmatrix} F_u & F_y \\ G_u & G_y \end{pmatrix} $$

are interpreted analogously. As such, they are the Jacobians of F and G with respect to x and v, y and v, u and x, and u and y, respectively.

Jacobians and Cramer's rule can be used to develop analogous formulas for partial derivatives involving other combinations variables and functions.

EXAMPLE 3.21 ■ The Jacobian of the Polar Coordinate Transformation

Find the Jacobian $J \left(\frac{F, G}{r, \theta} \right)$ of the transformation

$$ \begin{cases} F(r, \theta) = r \cos \theta \\ G(r, \theta) = r \sin \theta \end{cases} $$

known as the *polar coordinate transformation*.

Solution. By definition,

$$ J \left(\frac{F, G}{r, \theta} \right) = \det \begin{pmatrix} \frac{\partial}{\partial r} (r \cos \theta) & \frac{\partial}{\partial \theta} (r \cos \theta) \\ \frac{\partial}{\partial r} (r \sin \theta) & \frac{\partial}{\partial \theta} (r \sin \theta) \end{pmatrix} $$

$$ = \begin{pmatrix} \cos \theta & -r \sin \theta \\ \sin \theta & r \cos \theta \end{pmatrix} $$

$$ = r $$

is the Jacobian of the given transformation. ◀▶

EXAMPLE 3.22 ■ Using Jacobians to Calculate Partial Derivatives

Use *MATHEMATICA* and Jacobians to find the partial derivative u_x, u_y, v_x and v_y of the functions u and v defined by the system

$$ \begin{cases} F(u, v, x, y) & = \quad x^2 + y^3 x + u^2 + v^3 = 0 \\ G(u, v, x, y) & = \quad x^2 + 3xy + u^4 - v^2 = 0 \end{cases} $$

assuming that u and v are differentiable functions of x and y.

Solution. We use the method described in Example 3, and begin by representing the functions F and G in *MATHEMATICA*.

```
In[1]:= F:=x^2+y^3x+u^2+v^3
```

```
In[2]:= G:=x^2+3yx+u^4-v^2
```

Next we define the five required Jacobians.

```
In[3]:= J[F,G,x,v]:=Det[{{D[F,x],D[F,v]},{D[G,x],D[G,v]}}]
```

```
In[4]:=J[F,G,u,v]:=Det[{{D[F,u],D[F,v]},{D[G,u],D[G,v]}}]
```

```
In[5]:=J[F,G,u,x]:=Det[{{D[F,u],D[F,x]},{D[G,u],D[G,x]}}]
```

```
In[6]:=J[F,G,y,v]:=Det[{{D[F,y],D[F,v]},{D[G,y],D[G,v]}}]
```

```
In[7]:=J[F,G,u,y]:=Det[{{D[F,u],D[F,y]},{D[G,u],D[G,y]}}]
```

Now we combine the Jacobians and compute the partial derivatives. As was shown in Example 3, we can use Cramer's rule.

```
In[8]:= u[x]=Simplify[-J[F,G,x,v]/J[F,G,u,v]]
```
$$\text{Out[8]}= -\frac{(4+6v)x+9vy+2y^3}{4(u+3u^3v)}$$

```
In[9]:= v[x]=Simplify[-J[F,G,u,x]/J[F,G,u,v]]
```
$$\text{Out[9]}= \frac{2x-4u^2x+3y-2u^2y^3}{2v+6u^2v^2}$$

In[10]:= u[y]=Simplify[-J[F,G,y,v]/J[F,G,u,v]]

Out[10]= $-\dfrac{3x(3v+2y^2)}{4(u+3u^3v)}$

In[11]:= v[y]=Simplify[-J[F,G,u,y]/J[F,G,u,v]]

Out[11]= $\dfrac{3x-6u^2xy^2}{2v+6u^2v^2}$

The functions produced in steps 8–11 are u_x, v_x, u_y, and v_y, respectively. ◀▶

Jacobian determinants also play a role in the theory of multiple integration. Suppose, for example, that $f : \mathbb{R}^2 \to \mathbb{R}$ is an integrable function on some domain D, that $f(x, y) = f(\varphi(u, v), \psi(u, v))$ on D, that φ and ψ have partial derivatives on D, and that $T : \mathbb{R}^2 \to \mathbb{R}^2$ is an appropriate transformation of the xy-plane into the uv-plane given by $T(u, v) = (\varphi(u, v), \psi(u, v))$. If $E = T^{-1}(D)$ is the set of all (u, v) in the uv-plane for which $T(u, v) \in D$, then

$$\int\int_D f(x, y)dx\,dy = \int\int_E f(\varphi(u, v), \psi(u, v)) \left| \det J\left(\frac{\varphi, \psi}{u, v}\right) \right| du\,dv$$

The number

$$\left| \det J\left(\frac{\varphi, \psi}{u, v}\right) \right|$$

is the ratio of the areas of the domains D and E.

Hessians

Determinants can be used to classify critical points of differentiable functions. Suppose, for example, that $f : \mathbb{R}^2 \to \mathbb{R}$ is a function with continuous second partial derivatives f_{xx}, f_{xy}, f_{yx}, and f_{yy}. Let H_f be the matrix

$$H_f = \begin{pmatrix} f_{xx} & f_{xy} \\ f_{yx} & f_{yy} \end{pmatrix}$$

H_f is called a **Hessian matrix**. The determinant $D = \det H_f(\mathbf{v})$ of H_f is called the **discriminant** of f. We know from calculus that if $\mathbf{v} = (x, y) \in \mathbb{R}^2$ is a critical point of f. In other words, if $f_x(\mathbf{v}) = f_y(\mathbf{v}) = 0$, and $D > 0$, then $f(\mathbf{v})$ is a local minimum of f if $f_{xx}(\mathbf{v}) > 0$, and a local maximum if $f_{xx}(\mathbf{v}) < 0$. If $D < 0$, then $\mathbf{v}$ is a saddle point of f. If $D = 0$, the test fails.

EXAMPLE 3.23 ■ **A Minimax Problem**

Show that if an object with a specific volume is to be packed in a box with minimum surface area, the box must be a cube.

Solution. Let V be the volume of a cubic solid, and let x, y, and z be the length, width, and height of the box B enclosing V. Then the surface area of B is $S(x, y, z) = 2(xy + xz + yz)$ and $V = xyz$, with $x, y, z > 0$. Hence

$$z = \frac{V}{xy} \quad \text{and} \quad S(x, y, z) = 2\left(xy + \frac{V}{y} + \frac{V}{x}\right)$$

The critical points of S are therefore the solutions of the systems

$$\begin{cases} S_x = 2y - \dfrac{2V}{x^2} = 0 \\[3mm] S_y = 2x - \dfrac{2V}{y^2} = 0 \end{cases}$$

It follows that $xy^2 = V = x^2y$, so that $x^2y - xy^2 = xy(x - y) = 0$. Since x and y are nonzero, we must have $x = y$. Therefore,

$$2x - \frac{2V}{x^2} = 0$$

This implies that $V = x^3$.

Let us use the discriminant to verify that V is a minimum. We first compute the second derivatives $S_{xy} = 2$, $S_{yx} = 2$, $S_{xx} = 4(V/x^3)$, and $S_{yy} = 4(V/y^3)$ and form the Hessian matrix

$$H_S = \begin{pmatrix} S_{xx} & S_{xy} \\ S_{yx} & S_{yy} \end{pmatrix} = \begin{pmatrix} 4\left(\dfrac{V}{x^3}\right) & 2 \\[3mm] 2 & 4\left(\dfrac{V}{y^3}\right) \end{pmatrix}$$

Its determinant at $x = y$ is

$$D = \det \begin{pmatrix} 4\left(\dfrac{V}{x^3}\right) & 2 \\[3mm] 2 & 4\left(\dfrac{V}{x^3}\right) \end{pmatrix} = 4\left(\frac{4V^2 - x^6}{x^6}\right)$$

The fact that $V = x^3$ implies that

$$D = 4\left(\frac{4x^6 - x^6}{x^6}\right) = 12 > 0$$

Hence S has a minimum at $x = y = z$. ◀▶

Wronskians

Matrices whose entries are partial derivatives are useful for describing the solutions of differential equations. Suppose that $S = \{f_1, \ldots, f_n\}$ is a set of functions $f_i : \mathbb{R} \to \mathbb{R}$ $(1 \le i \le n)$ on the real line, each differentiable $(n-1)$ times, and that the functions $f_i^{(m)} : \mathbb{R} \to \mathbb{R}$ are their mth derivatives. Then the determinant $W(f_1, \ldots, f_n)$ of the matrix of functions

$$
\begin{pmatrix}
f_1 & \cdots & f_n \\
& \vdots & \\
f_1^{(n-1)} & \cdots & f_n^{(n-1)}
\end{pmatrix}
$$

can be used to test whether the set S is **linearly independent**, where S is linearly independent when

$$a_1 f_1 + \cdots + a_n f_n = 0$$

if and only if $a_1 = \cdots = a_n = 0$. The determinant $W(f_1, \ldots, f_n)$ is called the **Wronskian** of S. It is proved in calculus that S is linearly independent if and only if there exists a point $x \in \mathbb{R}$ for which $W(f_1, \ldots, f_n)(x) \neq 0$.

EXAMPLE 3.24 ■ A Wronskian Determinant

Use Wronskian determinants to show that the set of functions

$$S = \left\{ f_1 = x^5, \ f_2 = e^x, \ f_3 = e^{2x} \right\}$$

is linearly independent.

Solution. By definition, the Wronskian of S is

$$
W(f_1, f_2, f_3) = \det \begin{pmatrix}
x^5 & e^x & e^{2x} \\
5x^4 & e^x & 2e^{2x} \\
20x^3 & e^x & 4e^{2x}
\end{pmatrix} = 2x^5 e^x e^{2x} - 15x^4 e^x e^{2x} + 20x^3 e^x e^{2x}
$$

Let $x = 1$. Then $W(f_1, f_2, f_3)(1) = 7e^3$. Since $7e^3$ is not equal to 0, the set S is linearly independent. ◄►

EXERCISES 3.7

1. Suppose that the variables u and v are defined implicitly as functions of x. Use Jacobians to find their first partial derivatives, where

$$F(u, v, x) = x^2 + 2uv - 1 = 0 \quad \text{and} \quad G(u, v, x) = x^3 - u^3 + v^3 - 1 = 0$$

2. Suppose that the variables u and v are defined implicitly as functions of the remaining variables. Use Jacobians to find their first partial derivatives, where $F(u, v, x, y) = u + e^v - x - y = 0$ and $G(u, v, x, y) = e^u + v - x + y = 0$.

3. Suppose that the variables u, v, and z are defined implicitly as functions of x and y. Use Jacobians to find their first partial derivatives with respect to x and y, where $F(x, y, z, u, v) = u + z + y - x + v = 0$ and $G(u, v, w, x, y) = uv + vz - y^2 + x^2 = 0$.

4. Use Hessians to classify the critical points of the following functions.

$$\text{a. } f(x, y) = x^3 - 6x^2 + 7x + 4y - 5$$
$$\text{b. } f(x, y) = x^4 + y^2 + 8x + 9y$$
$$\text{c. } f(x, y) = x^3 - 6x^2 + 7xy + 4y - 5$$

5. Find x, y, z for which $xyz = 1$ and $x^2 + y^2 + z^2$ is minimal.

6. Find the dimensions of a rectangular box with an open top whose volume equals 100 cubic feet and whose surface area is minimal.

7. Use the Wronskian to test the following sets of functions for linear independence.

a. $S_1 = \{e^x, e^{2x}, e^{3x}\}$ b. $S_2 = \{x, x^2, \sin x\}$ c. $S_3 = \{1, (x - 2), (x - 2)^3\}$

REVIEW

KEY CONCEPTS ▶ Define and discuss each of the following.

Square Matrices

Adjugate of a square matrix, cofactors of a square matrix, elementary product of square matrix entries, minors of a square matrix.

Permutations

Elementary product of matrix entries, even permutation, inversion of the elements of a permutation, odd permutation, parity of a permutation.

Determinant Functions

Laplace expansion of a determinant, permutation form of a determinant.

Linear Systems

Cramer's rule.

KEY FACTS ▶ Explain and illustrate each of the following.

1. The determinant $\det I_n$ of the $n \times n$ identity matrix I_n is 1.

2. If A is an $n \times n$ matrix with a zero row, then the determinant of A is 0.

3. If B is a matrix resulting from an $n \times n$ matrix A by the elementary row operation $\left(R_i \rightleftharpoons R_j \right)$ and $i \neq j$, then $\det B = -\det A$.

4. If $A = (a_{ij})$ is an $n \times n$ matrix containing two identical rows, then the determinant of A is 0.

5. If B is a matrix resulting from an $n \times n$ matrix A by the elementary row operation $(R_i \rightarrow s R_i)$, then $\det B = s \det A$.

6. If B is a matrix resulting from an $n \times n$ matrix A by the elementary row operation $\left(R_i \rightarrow R_i + s R_j \right)$, then $\det B = \det A$.

7. If E is an elementary matrix obtained from I_n by $R = (R_i \rightleftharpoons R_j)$ and $i \neq j$, then $\det E = -1$.

8. If E is an elementary matrix obtained from I_n by $R = (R_i \rightarrow s R_i)$, then $\det E = s$.

9. If E is an elementary matrix obtained from I_n by $R = \left(R_i \rightarrow R_i + s R_j \right)$, then $\det E = \det I_n = 1$.

10. If $A = (a_{ij})$ is an $n \times n$ matrix and E is an $n \times n$ elementary matrix, then $\det EA = \det E \det A$.

11. If E is an elementary matrix and $B = EA$ for some $n \times n$ matrix A, then the following statements hold.

    a. $\det B = -\det A$ if E results from I_n by the operation $(R_i \rightleftharpoons R_j)$, provided that $i \neq j$.

    b. $\det B = s \det A$ if E results from I_n by the operation $(R_i \rightarrow sR_i)$.

    c. $\det B = \det A$ if E results from I_n by the operation $\left(R_i \rightarrow R_i + sR_j\right)$.

12. If two $n \times n$ matrices A and B are row equivalent, then $\det A = 0$ if and only if $\det B = 0$.

13. If A and B are two $n \times n$ matrices, then $\det AB = \det A \det B$.

14. A real $n \times n$ matrix A is invertible if and only if $\det A \neq 0$.

15. A homogeneous linear system $A\mathbf{x} = \mathbf{0}$ with real coefficients has a nonzero solution $\mathbf{x}$ if and only if $\det A = 0$.

16. If $A = (a_{ij})$ is an $n \times n$ triangular matrix, then the determinant of A is the product $a_{11} \cdots a_{nn}$ of the diagonal entries of A.

17. For any $n \times n$ matrix A and any scalar s, $\det sA = s^n \det A$.

18. For any matrix A, $\det A = \det A^T$.

19. For any invertible matrix A, $\det A^{-1} = 1/\det A$.

20. If $A = (a_{ij})$ is an $n \times n$ triangular matrix, then the determinant of A is the product $a_{11} \cdots a_{nn}$ of the diagonal elements of A.

21. For every real square matrix A, there exists an upper-triangular matrix U for which $(-1)^m \det A = \det U$.

22. For every invertible $n \times n$ real matrix A, $\det A = (-1)^m \det B$, where B is a row echelon matrix obtained from A by elementary row operations of types $\left(R_i \rightleftharpoons R_j\right)$, with $i \neq j$, and $(R_i \rightarrow R_i + sR_j)$, and m is the number of row interchanges involved.

23. For every invertible $n \times n$ real matrix A, $\det A = (-1)^m \, p_1 \cdots p_n$, where $p_1, \ldots, p_n$ are the pivots of A.

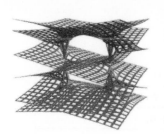

4

VECTOR SPACES

What do real numbers, matrices, continuous functions, solutions of differential equations, polynomials, and many other mathematical objects have in common? The answer is that they can all be *added* to produce new objects of the same kind. They can also be *multiplied by constants* to produce objects of the same kind. For example, if a and b are real numbers, then so are $a + b$ and ab. If $\mathbf{x}$ and $\mathbf{y}$ are two column vectors with real entries and have the same number of rows, then $\mathbf{x} + \mathbf{y}$ is a column vector with the same number of rows and so is the multiple $a\mathbf{x}$, for any real number a. If A and B are two $m \times n$ matrices with real entries, then so are $A + B$ and aA, for any real number a.

In linear algebra, these different kinds of objects are referred to as *vectors* belonging to different *vector spaces*. Thus vectors can be real numbers, continuous functions, matrices, and many other objects. The constants used to multiply vectors to produce new vectors are called *scalars*. In this text, scalars will be either real or complex numbers. More specifically, they will be real numbers unless otherwise specified.

REAL VECTOR SPACES

In this chapter, we lay the foundation for the study of *real vector spaces*. The vectors will vary: They will be columns of real numbers, polynomials, matrices, and other objects. The scalars, however, will always be the real numbers. We examine the spaces, their bases, and their subspaces. The pivotal concept for this endeavor will be that of a *linearly independent* set of vectors. The sets $\mathbb{R}^n$ of real column vectors of height n will play an essential role in guiding our intuition.

DEFINITION 4.1   *A real **vector space** V consists of a set of objects called **vectors**, a distinguished vector $\mathbf{0}$ called the **zero vector**, and the set $\mathbb{R}$ of real numbers, whose elements are called **scalars**. The space V is equipped with two operations called **vector addition**, denoted by $\mathbf{u} + \mathbf{v}$, and **scalar multiplication**, denoted by $a\mathbf{v}$. Table 1 lists the eight axioms satisfied by these operations.*

The first four axioms express the facts that vector addition is associative and commutative, that adding the zero vector to a vector leaves the vector unchanged, and that each vector has an additive inverse. The next three axioms describe how matrix addition and scalar multiplication

TABLE 1   Real Vector Space Axioms.

1.	Vector Addition	For all $\mathbf{u}, \mathbf{v}, \mathbf{w} \in V$, $\mathbf{u} + (\mathbf{v} + \mathbf{w}) = (\mathbf{u} + \mathbf{v}) + \mathbf{w}$.
2.	Vector Addition	For all $\mathbf{u}, \mathbf{v} \in V$, $\mathbf{u} + \mathbf{v} = \mathbf{v} + \mathbf{u}$.
3.	Vector Addition	For all $\mathbf{u} \in V$, $\mathbf{u} + \mathbf{0} = \mathbf{u}$.
4.	Vector Addition	For all $\mathbf{u} \in V$ there exists a unique $-\mathbf{u} \in V$ for which $\mathbf{u} + (-\mathbf{u}) = \mathbf{0}$.
5.	Scalar Multiplication	For all $a \in \mathbb{R}$ and all $\mathbf{u}, \mathbf{v} \in V$, $a(\mathbf{u} + \mathbf{v}) = a\mathbf{u} + a\mathbf{v}$.
6.	Scalar Multiplication	For all $a, b \in \mathbb{R}$ and all $\mathbf{u} \in V$, $(a + b)\mathbf{u} = a\mathbf{u} + b\mathbf{u}$.
7.	Scalar Multiplication	For all $a, b \in \mathbb{R}$ and all $\mathbf{u} \in V$, $(ab)\mathbf{u} = a(b\mathbf{u})$.
8.	Scalar Multiplication	For all $a \in \mathbb{R}$ and all $\mathbf{u} \in V$, $1\mathbf{u} = \mathbf{u}$.

interact. The last axiom indicates that multiplying a vector by the real number 1 leaves the vector unchanged.

The statement that vector addition and scalar multiplication are *operations* on V is equivalent to saying that if $\mathbf{u}$ and $\mathbf{v}$ are any two vectors in V and a and b are any two scalars, then $a\mathbf{u} + b\mathbf{v} \in V$. We express this fact by saying that V is *closed under vector addition and scalar multiplication*. We follow the usual practice and write $\mathbf{u} + (-\mathbf{v})$ as $\mathbf{u} - \mathbf{v}$ and refer to this operation as the *subtraction* of vectors.

If we replace the set $\mathbb{R}$ of real numbers by the set $\mathbb{C}$ of complex numbers in Definition 4.1, we get a *complex vector space*. More general vector spaces can be defined by varying the set of scalars of the spaces.

In all vector spaces, the zero vector and the zero scalar are connected by the following basic property.

THEOREM 4.1   *For any scalar a and any vector $\mathbf{v}$, it holds that $\mathbf{0} = a\mathbf{0} = 0\mathbf{v}$.*

Proof. Since $\mathbf{0} = \mathbf{0} + \mathbf{0}$, it follows that $a\mathbf{0} = a\mathbf{0} + a\mathbf{0}$ for all scalars a. By subtracting $a\mathbf{0}$ from both sides of this equation, we get $\mathbf{0} = a\mathbf{0} - a\mathbf{0} = a\mathbf{0} + a\mathbf{0} - a\mathbf{0} = a\mathbf{0}$. Moreover, since $0 = 0 + 0$, we have $0\mathbf{v} = 0\mathbf{v} + 0\mathbf{v}$ for all vectors $\mathbf{v}$. By subtracting the vector $0\mathbf{v}$ from both sides of this equation, we get $\mathbf{0} = 0\mathbf{v} - 0\mathbf{v} = 0\mathbf{v} + 0\mathbf{v} - 0\mathbf{v} = 0\mathbf{v}$. ∎

We can strengthen this theorem by noting that the following is true.

THEOREM 4.2   *For any scalar a and any vector $\mathbf{v}$, it holds that $a\mathbf{v} = \mathbf{0}$ if and only if $a = 0$ or $\mathbf{v} = \mathbf{0}$.*

Proof. In view of Theorem 4.1, it suffices to prove that if $a\mathbf{v} = \mathbf{0}$ and $a \neq 0$, then $\mathbf{v} = \mathbf{0}$. Since $a \neq 0$, there exists a nonzero inverse a^{-1} such that

$$\mathbf{0} = a^{-1}\mathbf{0} = a^{-1}a\mathbf{v} = 1\mathbf{v} = \mathbf{v}$$

The theorem therefore follows. ∎

EXERCISES 4.1

1. Show that in any vector space, $(-a)\mathbf{u} = a(-\mathbf{u}) = -(a\mathbf{u})$ for all scalars a and all vectors u.

2. Show that in any vector space, $(a + b)(\mathbf{u} + \mathbf{v}) = a\mathbf{u} + b\mathbf{u} + a\mathbf{v} + b\mathbf{v}$ for all scalars a, b and all vectors $\mathbf{u}, \mathbf{v}$.

3. Show that the set of points $(x, y) \in \mathbb{R} \times \mathbb{R}$ with $(x, y) + (x', y') = (x + x', y + y')$ and $a(x, y) = (ax, ay)$ is a vector space.

4. Define vector space operations on the set of all points $(x, y, z) \in \mathbb{R} \times \mathbb{R} \times \mathbb{R}$ analogous to the operations in Exercise 3 and show that the result is a real vector space.

5. Use the operations as defined in Exercises 3 and 4 to show that the following sets contain the zero vector and are closed under vector addition and scalar multiplication.

    a. The set of all points on the line $y = 3x$

    b. The set of all points on the plane $x + 2y + 3z = 0$

    c. The set $V = \{0\}$ whose only vector is the number 0

    d. The set of all real upper-triangular $n \times n$ matrices

    e. The set of all real diagonal $n \times n$ matrices

    f. The set of all real symmetric $n \times n$ matrices

6. Explain why the set of points on the line $y = 3x + 5$ is not a vector space.

7. Let $V = \{a_0 + a_1 t + a_2 t^5 + t^3 : a_0, a_1, a_2 \in \mathbb{R}\}$. Explain why V is not a vector space.

8. Explain why the set of all 2×2 matrices with rational entries is not a real vector space.

We group the basic vector spaces studied in this text into four kinds: coordinate spaces, matrix spaces, polynomial spaces, and others.

To show that the following examples are real vector spaces, we must show that each specified set contains the zero vector; is closed under vector addition and scalar multiplication; and that the set, with the given operations, satisfies the vector space axioms. In all but one case, we leave the routine verification of the axioms as exercises.

Coordinate Spaces

EXAMPLE 4.1   ■ The Space $\mathbb{R}$

If $n = 1$, we can think of $\mathbb{R}^1$ as $\mathbb{R}$ by identifying the column vectors (x) with the real numbers x. The usual addition and multiplication of real numbers determine a vector space structure on $\mathbb{R}$. ◀▶

EXAMPLE 4.2 ■ **The Space** $\mathbb{R}^n$

The space $\mathbb{R}^n$ of all $n \times 1$ column vectors $A : \mathbf{n} \times \mathbf{1} \to \mathbb{R}$, equipped with matrix addition and the scalar multiplication of matrices, is a real vector space. ◀▶

The advantage of thinking of the elements of $\mathbb{R}^n$ as column vectors is that it provides an optimal notation for describing the two-dimensional link between the vector space operations. For example, by the commutativity of the multiplication of real numbers, the combination of vectors

$$3 \begin{pmatrix} 5 \\ 6 \end{pmatrix} + 2 \begin{pmatrix} 8 \\ 7 \end{pmatrix} = \begin{pmatrix} 3(5) \\ 3(6) \end{pmatrix} + \begin{pmatrix} 2(8) \\ 2(7) \end{pmatrix} = \begin{pmatrix} 3(5) + 2(8) \\ 3(6) + 2(7) \end{pmatrix}$$

can be expressed as the matrix product

$$\begin{pmatrix} 5 & 8 \\ 6 & 7 \end{pmatrix} \begin{pmatrix} 3 \\ 2 \end{pmatrix} = \begin{pmatrix} 5(3) + 8(2) \\ 6(3) + 7(2) \end{pmatrix}$$

The identity

$$\begin{pmatrix} 3(5) + 2(8) \\ 3(6) + 2(7) \end{pmatrix} = \begin{pmatrix} 5(3) + 8(2) \\ 6(3) + 7(2) \end{pmatrix}$$

holds because $3(5) = 5(3)$, $2(8) = 8(2)$, $3(6) = 6(3)$, and $2(7) = 7(2)$.

EXAMPLE 4.3 ■ **Verifying the Vector Space Axioms for** $\mathbb{R}^3$

Use *MATHEMATICA* to verify that the addition of vectors and the multiplication of vectors by scalars in $\mathbb{R}^3$ satisfy the axioms of a vector space.

Solution. We must verify eight axioms.

```
In[1]:= {a₁,a₂,a₃}+({b₁,b₂,b₃}+{c₁,c₂,c₃})==
({a₁,a₂,a₃}+{b₁,b₂,b₃})+{c₁,c₂,c₃}

Out[1]= True
```

```
In[2]:= {a₁,a₂,a₃}+{b₁,b₂,b₃}=={b₁,b₂,b₃}+{a₁,a₂,a₃}

Out[2]= True
```

```
In[3]:= {a₁,a₂,a₃}+{0,0,0}=={a₁,a₂,a₃}

Out[3]= True
```

```
In[4]:= {a₁,a₂,a₃}+{-a₁,-a₂,-a₃}=={0,0,0}

Out[4]= True
```

```
In[5]:= Simplify[a({a₁,a₂,a₃}+{b₁,b₂,b₃})]==

Simplify[a{a₁,a₂,a₃}+a{b₁,b₂,b₃}]

Out[5]= True
```

```
In[6]:= Simplify[(a+b){a₁,a₂,a₃}]==

Simplify[a{a₁,a₂,a₃}+b{a₁,a₂,a₃}]

Out[6]= True
```

```
In[7]:= Simplify[(a b){a₁,a₂,a₃}]==Simplify[a(b{a₁,a₂,a₃})]

Out[7]= True
```

```
In[8]:= Simplify[1{a₁,a₂,a₃}]=={a₁,a₂,a₃}

Out[8]= True
```

This shows that $\mathbb{R}^3$ satisfies the axioms of a vector space. In these calculations, it is tacitly assumed that the constants a, b, a_i, b_i, and c_i are real numbers. ◄►

EXERCISES 4.2

1. Use *MATHEMATICA* to verify that the vector space axioms hold for $\mathbb{R}_2[t]$.

2. Show that the vector space axioms hold for $\mathbb{R}^n$.

Matrix Spaces

EXAMPLE 4.4 ■ The Space $\mathbb{R}^{m \times n}$

The space $\mathbb{R}^{m \times n}$ consists of all $m \times n$ matrices A with real entries, equipped with matrix addition and the scalar multiplication of matrices as vector space operations. We refer to this space as the space of real $m \times n$ matrices. ◄►

EXERCISES 4.3

1. Show that the vector space axioms hold for $\mathbb{R}^{2\times 3}$.

2. Explain the difference between the vectors in the following real vector spaces.

    a. The spaces $\mathbb{R}$ and $\mathbb{R}^{1\times 1}$

    b. The spaces $\mathbb{R}^3$, $\mathbb{R}^{3\times 1}$, and $\mathbb{R}^{1\times 3}$

    c. The spaces $\mathbb{R} \times \mathbb{R}$ and $\mathbb{R}^2$

    d. The spaces $\mathbb{R}^{2\times 3}$ and $\mathbb{R}^{3\times 2}$

    e. The spaces $\mathbb{R}^{2\times 3}$ and $\mathbb{R}^6$

3. Show that the set of real 3×2 matrices with nonnegative entries is not a real vector space.

Polynomial Spaces

EXAMPLE 4.5   ■ The Space $\mathbb{R}[t]$

If $p(t) = a_0 + a_1 t + \cdots + a_n t^n$ and $q(t) = b_0 + b_1 t + \cdots + b_n t^n$ are two polynomials in $\mathbb{R}[t]$, then we can think of them as vectors. The definitions

$$
\begin{aligned}
p(t) + q(t) &= \quad (a_0 + b_0) + (a_1 + b_1)t + \cdots + (a_n + b_n)t^n \\
ap(t) &= \quad aa_0 + aa_1 t + \cdots + aa_n t^n \\
\mathbf{0} &= \quad 0
\end{aligned}
$$

turn $\mathbb{R}[t]$ into a real vector space. If $p(t) = a_0 + a_1 t + \cdots + a_n t^n$ and $q(t) = b_0 + b_1 t + \cdots + b_m t^m$, and $m \neq n$, we simply add the required number of zero terms to the polynomial of lower degree so we can apply the definition of $p(t) + q(t)$. For example, if $p(t) = 1 + t$ and $q(t) = 7 - 3t^2$, we put $p(t) = 1 + t + 0t^2$ and $q(t) = 7 + 0t - 3t^2$ and form the sum $p(t) + q(t) = 8 + t - 3t^2$. ◀▶

EXERCISES 4.4

1. Describe in detail the vector space operations for all polynomials in $\mathbb{R}[t]$ of degree 3 or less.

2. Show that the vector space axioms hold for $\mathbb{R}[t]$.

3. Show that the set of all polynomials in $\mathbb{R}[t]$ of degree 4 or less is closed under vector addition and scalar multiplication.

4. Show that the set of all polynomials in $\mathbb{R}[t]$ of degree 4 is not closed under vector addition and not closed under scalar multiplication.

Other Spaces

EXAMPLE 4.6 ■ **The Space $\mathbb{R}_n[t]$**

The set of real polynomials of degree n or less can be made into a vector space by using the vector space operations on $\mathbb{R}[t]$. Since the sum of two polynomials of degree n may be a polynomial of degree less than n, it is essential to include in the vectors of this space all polynomials of degree less than n. ◄►

EXAMPLE 4.7 ■ **The Space $\mathbb{R}^{\mathbb{N}}$**

The set $\mathbb{R}^{\mathbb{N}}$ of real-valued infinite sequences $(x_1, \ldots, x_n, \ldots)$ can be turned into a vector space by defining

$$(v_1, \ldots, v_n, \ldots) + (w_1, \ldots, w_n, \ldots) = (v_1 + w_1, \ldots, v_n + w_n, \ldots)$$
$$a(v_1, \ldots, v_n, \ldots) = (av_1, \ldots, av_n, \ldots)$$
$$\mathbf{0} = (0, \ldots, 0, \ldots)$$

In this example, it would not be helpful to try to visualize the vectors as columns since such columns would have to be of infinite height. ◄►

EXAMPLE 4.8 ■ **The Space $\mathbb{R}^{\infty}$**

The set $\mathbb{R}^{\infty}$ of all infinite sequences $(x_1, \ldots, x_n, \ldots)$ of real numbers for which

$$|x_1|^2 + \cdots + |x_n|^2 + \cdots < \infty$$

is a real vector space. Vector addition and scalar multiplication are defined as in the previous example. ◄►

EXAMPLE 4.9 ■ **The Solution Space of $A\mathbf{x} = \mathbf{0}$**

The set of solutions of a homogeneous linear system with real coefficients forms a vector space. We know from Theorems 1.4 and 2.7 that this set contains the zero vector and that if $\mathbf{x}$ and $\mathbf{y}$ are solutions and a and b are two scalars, then $a\mathbf{x} + b\mathbf{y}$ is a solution. It is therefore a real vector space. ◄►

EXAMPLE 4.10 ■ **The Solution Space of $y'(t) + ay(t) = 0$**

Suppose that $y(t)$ is a differentiable function $\mathbb{R} \to \mathbb{R}$, that $y'(t)$ is the derivative of $y(t)$, and that $a \in \mathbb{R}$. Then it is shown in calculus that

$$V = \left\{ y(t) : y(t) = ce^{-at} \text{ and } c \in \mathbb{R} \right\}$$

is the set of solutions of the equation $y'(t) + ay(t) = 0$. This set is closed under addition and the multiplication by scalars. The constant function $y(t) = 0e^{-at} = 0$ is the $\mathbf{0}$ vector. Therefore, V is a real vector space. ◄►

EXAMPLE 4.11 ■ **The Solution Space of $y''(t) + p(t)y'(t) + q(t)y(t) = 0$**

Suppose that $p(t)$ and $q(t)$ are two continuous functions on some interval (r, s), that $y(t)$ is a twice-differentiable function on (r, s), and that $y'(t)$ and $y''(t)$ are the first and second derivatives of $y(t)$, respectively. Suppose further than $y_1(t)$ and $y_2(t)$ are two solutions of the given equation on (r, s). Then it can be shown in calculus that the set

$$V = \{ay_1(t) + by_2(t) : a, b \in \mathbb{R}\}$$

is the solution space of the equation if and only if the Wronskian determinant

$$W(y_1(t), y_2(t)) = \det \begin{pmatrix} y_1(t) & y_2(t) \\ y_1'(t) & y_2'(t) \end{pmatrix}$$

is nonzero for some $t \in (r, s)$. The space V is a real vector space. The constant function $f(t) = 0$ is the $\mathbf{0}$ vector of the space. ◄►

EXAMPLE 4.12   ■ The Space $D^\infty(\mathbb{R}, \mathbb{R})$

Show that the set $D^\infty(\mathbb{R}, \mathbb{R})$ of infinitely differentiable functions $f : \mathbb{R} \to \mathbb{R}$ forms a real vector space.

Solution. Suppose that f and g belong to $D^\infty(\mathbb{R}, \mathbb{R})$. Then

$$\frac{d}{dt}(af(t) + bg(t)) = a\frac{d}{dt}(f(t)) + b\frac{d}{dt}(g(t))$$

Therefore, $D^\infty(\mathbb{R}, \mathbb{R})$ is closed under vector addition and scalar multiplication. Moreover, the zero function $f(t) = 0$ for all $t \in \mathbb{R}$ belongs to $D^\infty(\mathbb{R}, \mathbb{R})$. Therefore $D^\infty(\mathbb{R}, \mathbb{R})$ is a real vector space. ◄►

EXAMPLE 4.13   ■ The Zero Space

If $\mathbf{0}$ is the zero vector of a real vector space V, then the set $\{\mathbf{0}\}$ is a real vector space. Since $a\mathbf{0} + b\mathbf{0} = \mathbf{0}$ for all real numbers a and b, it is clear that $\{\mathbf{0}\}$ contains the zero vector and is closed under vector addition and scalar multiplication. Hence it is a real vector space. ◄►

What the zero vector in the space $\{\mathbf{0}\}$ looks like depends on the ambient space V. As Table 2 shows, the zero vector of $\mathbb{R}^3$, for example, consists of a column vector of height 3 whose coordinates are all zero, whereas the zero vector of the space $\mathbb{R}^{3\times2}$ of 3×2 real matrices is the 3×2 matrix all of whose entries are zero.

TABLE 2   Zero Vectors.

$\mathbb{R}^3$	$\mathbb{R}^{3\times2}$
$\mathbf{0} = \begin{pmatrix} 0 \\ 0 \\ 0 \end{pmatrix}$	$\mathbf{0} = \begin{pmatrix} 0 & 0 \\ 0 & 0 \\ 0 & 0 \end{pmatrix}$

Nevertheless, we usually speak of the *zero space* since it is always clear from the context which zero vector is involved.

EXERCISES 4.5

1. Show that the set of real polynomials of the form $a_0 + a_1 t + a_2 t^2$ is a real vector space.

2. Show that $\mathbb{R}^N$ is a real vector space.

3. Explain the difference between the vector space $\mathbb{R}^N$ and the vector space $\mathbb{R}^\infty$.

4. Show that the set of all solutions of the linear system

$$\begin{cases} 3x + 7y - z = 0 \\ x + y + z = 0 \end{cases}$$

   is a real vector space.

5. Show that the set of all solutions of the linear system

$$\begin{cases} 3x + 7y - z = 0 \\ x + y + z = 0 \\ 2x - y + 3z = 0 \end{cases}$$

   is a real vector space.

6. Show that the space $D^\infty(\mathbb{R}, \mathbb{R})$ is a real vector space.

7. Show that the solution space of the differential equation $y'(t) + 3y(t) = 0$ is a real vector space.

8. Show that the solution space of the differential equation $y^{(2)}(t) + 3y^{(1)}(t) + 6y(t) = 0$ is a real vector space.

9. Describe the zero vectors in the spaces $\mathbb{R} \times \mathbb{R}$, $\mathbb{R}_4[t]$, $\mathbb{R}^{2 \times 3}$, $\mathbb{R}^6$, $D^\infty(\mathbb{R}, \mathbb{R})$, and $\mathbb{R}^N$.

BASES AND DIMENSION

Although infinite, a vector space may have the property that all of its vectors can be built up from a fixed set of finitely many of its vectors using vector addition and scalar multiplication. This is similar to a language being based on a finite alphabet in which the infinite set of words of the language can be written. It is this feature that will allow us to use matrices to study the properties of certain functions on real vector spaces. It will be essential for our work that all sets of vectors used in this context are *ordered*. The ordered lists *tea* and *eat*, for example, are different *words* determined by the unordered set of letters {a, e, t}. Similarly, the lists

$$\mathcal{B} = \{\mathbf{x}_2, \mathbf{x}_3, \mathbf{x}_1\} = \left\{ \begin{pmatrix} 5 \\ 7 \end{pmatrix}, \begin{pmatrix} 3 \\ 8 \end{pmatrix}, \begin{pmatrix} 2 \\ 2 \end{pmatrix} \right\}$$

and

$$C = \{\mathbf{x}_1, \mathbf{x}_3, \mathbf{x}_2\} = \left\{ \begin{pmatrix} 2 \\ 2 \end{pmatrix}, \begin{pmatrix} 3 \\ 8 \end{pmatrix}, \begin{pmatrix} 5 \\ 7 \end{pmatrix} \right\}$$

are different **ordered sets of vectors**, determined by the vectors $\mathbf{x}_1$, $\mathbf{x}_2$, and $\mathbf{x}_3$. However, since all sets of vectors in this text are assumed to be ordered, we simplify the notation and terminology and speak of **sets of vectors** when we should really speak of ordered sets of vectors.

Linear Combinations

Since much of the work that follows is based on bases and linear combinations, we motivate these concepts with an example.

EXAMPLE 4.14 ■ Linear Combinations of Vectors

On CRT computer monitors, color is usually represented as a vector in **RGB** space, where **R** is the red component, **G** is the green component, and **B** is the blue component. **RGB** space is a subset of the vector space $\mathbb{R}^3$ obtained from the **basis**

$$\begin{aligned}
\mathbf{R} &= (1, 0, 0) \quad \text{(pure red)} \\
\mathbf{G} &= (0, 1, 0) \quad \text{(pure green)} \\
\mathbf{B} &= (0, 0, 1) \quad \text{(pure blue)}
\end{aligned}$$

by taking all **linear combinations** of **R**, **G**, and **B**, using nonnegative scalars. In the context of color, these basis vectors are referred to as primaries. We know that any color C in **RGB** space can be represented as a linear combination of red, green, and blue:

$$C = \alpha_1 \mathbf{R} + \alpha_2 \mathbf{G} + \alpha_3 \mathbf{B}$$

Here are some examples of colors in **RGB** space:

$$\begin{aligned}
(0, 0, 0) &\quad \text{black} \\
(1, 1, 1) &\quad \text{white} \\
(1, 0, 1) &\quad \text{magenta} \\
(0, 1, 1) &\quad \text{cyan} \\
(1, 1, 0) &\quad \text{yellow}
\end{aligned}$$

Along with red, green, and blue, the above colors define the eight corners of a unit cube known as the **RGB** *color cube*. The main diagonal of the cube, which extends from white to black, is called the *grayscale*, because colors that fall on this diagonal are gray, or colorless. This line is defined by the weights $\alpha_1 = .299$, $\alpha_2 = .587$, and $\alpha_3 = .114$. The process of converting an arbitrary color to grayscale is the following. Consider an arbitrary color

$C = (C_x, C_y, C_z)$. The grayscale representation C_G of C is given by

$$C_G = \begin{pmatrix} .299C_x + .587C_y + .114C_z \\ .299C_x + .587C_y + .114C_z \\ .299C_x + .587C_y + .114C_z \end{pmatrix}$$

Each of its components is a *linear combination* of the colors C_x, C_y, and C_z. Observe that all **R**, **G**, and **B** components are now identical. ◀▶

It is interesting to note that the green weight is higher than the red and blue weights. Much of image processing is done in grayscale because images appear in high contrast when displayed in this way. The greater importance of green in the grayscale formula seems to have a historical explanation. Apparently, the human eye sees greater contrast in green than in the other primaries because as predators in prehistoric times, humans needed to see prey that were well-camouflaged in green forests. Experiments have shown that if we can take an arbitrary image and look at its red, green, and blue channels individually, our eyes will see the most contrast when looking at the green channel and the least contrast in the blue channel. This explains the low weight attached to blue in the grayscale formula.

The idea of a linear combination of vectors is fundamental to the study of vector spaces.

DEFINITION 4.2 *Let $v_1, \ldots, v_n$ be vectors in a vector space V and let $a_1, \ldots, a_n$ be corresponding scalars. Then the vector $a_1 v_1 + \cdots + a_n v_n$ is a **linear combination** of the given vectors and scalars.*

EXAMPLE 4.15 ■ A Linear Combination of Column Vectors

The vectors in $\mathbb{R}^n$ are columns of real numbers. We can therefore think of a linear combination $a_1 v_1 + \cdots + a_n v_n$ of vectors in $\mathbb{R}^n$ as a matrix product

$$a_1 v_1 + \cdots + a_n v_n = M \begin{pmatrix} a_1 \\ \vdots \\ a_n \end{pmatrix}$$

where $M = (v_1 \cdots v_n)$ is the matrix whose columns are the vectors v_i. ◀▶

DEFINITION 4.3 *The **span** of a nonempty subset S of vectors of a vector space V is the set of all linear combinations in V that can be formed with the vectors in S. If S is empty, then the span of S in V is the set $\{0\}$, containing only the zero vector of V.*

We write span (S) for the span of S. If $S = \{x_1, \ldots, x_n\}$ is a finite set of vectors, we also write span $\{x_1, \ldots, x_n\}$ for the span of S. It is important to notice that the span of a set S of vectors depends on the set of scalars of V.

EXAMPLE 4.16 ■ **The Span of a Set of Column Vectors**

Show that the set

$$S = \left\{ \begin{pmatrix} 1 \\ 2 \\ 3 \end{pmatrix}, \begin{pmatrix} 0 \\ 4 \\ 4 \end{pmatrix}, \begin{pmatrix} 1 \\ 2 \\ 0 \end{pmatrix}, \begin{pmatrix} 3 \\ 2 \\ 1 \end{pmatrix} \right\}$$

spans $\mathbb{R}^3$.

Solution. We use *MATHEMATICA* to show that every vector in $\mathbb{R}^3$ can be written as a linear combination of vectors in S. Since there are infinitely many such possible linear combinations, these sums are not unique. We begin by forming a general linear combination of the form

$$a \begin{pmatrix} 1 \\ 2 \\ 3 \end{pmatrix} + b \begin{pmatrix} 0 \\ 4 \\ 4 \end{pmatrix} + c \begin{pmatrix} 1 \\ 2 \\ 0 \end{pmatrix} + d \begin{pmatrix} 3 \\ 2 \\ 1 \end{pmatrix} = \begin{pmatrix} x \\ y \\ z \end{pmatrix}$$

We then evaluate the left-hand side of this equation.

$$\text{In[1]:= } a \begin{pmatrix} 1 \\ 2 \\ 3 \end{pmatrix} + b \begin{pmatrix} 0 \\ 4 \\ 4 \end{pmatrix} + c \begin{pmatrix} 1 \\ 2 \\ 0 \end{pmatrix} + d \begin{pmatrix} 3 \\ 2 \\ 1 \end{pmatrix}$$

Out[1]= {{a+c+3d},{2a+4b+2c+2d},{3a+4b+d}}

We therefore solve the equation

$$\begin{pmatrix} a + c + 3d \\ 2a + 4b + 2c + 2d \\ 3a + 4b + d \end{pmatrix} = \begin{pmatrix} x \\ y \\ z \end{pmatrix}$$

For any given x, y, z, we have a system of three equations in the four unknowns a, b, c, and d. If we choose a, b, and c as basic variables and d as a free variable, *MATHEMATICA* finds a, b, and c in terms of d.

In[2]:= **Solve[{a+c+3d==x, 2a+4b+2c+2d==y, 3a+4b+d==z},{a,b,c}]**

Out[2]={{a → $\frac{1}{3}$(-5d+2x-y+z), b → $\frac{1}{4}$(4d-2x+y), c → $\frac{1}{3}$(-4d+x+y-z)}}

Since there are infinitely many $d \in \mathbb{R}$, we have infinitely many different linear combinations of the given vectors spanning $\mathbb{R}^3$. ◄►

DEFINITION 4.4   *A subset S of a vector space V is **linearly dependent** if there exist vectors $v_1, \ldots, v_n \in S$ and scalars $a_1, \ldots, a_n$, not all 0, such that*

$$a_1 v_1 + \cdots + a_n v_n = 0$$

DEFINITION 4.5   *A subset S of a vector space V is **linearly independent** if it is not linearly dependent.*

The definition of linear independence is equivalent to the statement that the zero vector can be written in only one way as a linear combination of vectors in S. If

$$a_1 v_1 + \cdots + a_n v_n = 0 \quad \text{and} \quad b_1 v_1 + \cdots + b_n v_n = 0$$

for example, then

$$(a_1 - b_1) v_1 + \cdots + (a_n - b_n) v_n = 0 v_1 + \cdots + 0 v_n = 0$$

If S is linearly independent, then $a_i - b_i = 0$ for all $i \in \mathbf{n}$. Hence the representation of 0 is unique. Conversely, if the representation is unique, then it follows that $a_i - b_i = 0$ for all $i \in \mathbf{n}$, and S is therefore linearly independent.

EXAMPLE 4.17   ■ Linearly Independent Column Vectors

Use *MATHEMATICA* to show that the set

$$S = \left\{ \begin{pmatrix} 1 \\ 2 \\ 3 \end{pmatrix}, \begin{pmatrix} 0 \\ 4 \\ 4 \end{pmatrix}, \begin{pmatrix} 1 \\ 2 \\ 0 \end{pmatrix} \right\}$$

is linearly independent.

Solution. First we express the zero vector as a linear combination of the vectors in S :

$$a \begin{pmatrix} 1 \\ 2 \\ 3 \end{pmatrix} + b \begin{pmatrix} 0 \\ 4 \\ 4 \end{pmatrix} + c \begin{pmatrix} 1 \\ 2 \\ 0 \end{pmatrix} = \begin{pmatrix} 0 \\ 0 \\ 0 \end{pmatrix}$$

We then evaluate the left-hand side of this equation.

```
In[1]:= a ( 1 )   + b ( 0 )   + c ( 1 )
          ( 2 )       ( 4 )       ( 2 )
          ( 3 )       ( 4 )       ( 0 )

Out[1]={{a+c,2a+4b+2c,3a+4b}}
```

We therefore solve the equation

$$\begin{pmatrix} a + c \\ 2a + 4b + 2c \\ 3a + 4b \end{pmatrix} = \begin{pmatrix} 0 \\ 0 \\ 0 \end{pmatrix}$$

```
In[2]:= Solve[{a+c==0,2a+4b+2c==0,3a+4b==0},{a,b,c}]

Out[2]= {{a → 0, b → 0, c → 0}}
```

This shows that the linear combination of the vectors in S is $\mathbf{0}$ if and only if $a = b = c = 0$. The set S is therefore linearly independent. ◄►

The idea behind linear independence is that a set of vectors $\mathbf{v}_1, \ldots, \mathbf{v}_n$ should be linearly independent if no vector $\mathbf{v}_i$ in this set can be written as a linear combination of the other vectors $\mathbf{v}_j$. If $a_1\mathbf{v}_1 = a_2\mathbf{v}_2 + \cdots + a_n\mathbf{v}_n$, for example, and $a_1 \neq 0$, then at least one of the coefficients $a_2, \ldots, a_n$ must be nonzero. Hence $a_1\mathbf{v}_1 + a_2\mathbf{v}_2 + \cdots + a_n\mathbf{v}_n = \mathbf{0}$ for some nonzero coefficients a_i. Here are some important facts about linearly dependent and linearly independent sets.

1. The empty set $\emptyset$ is linearly independent. For suppose that $\emptyset$ is linearly dependent. Then there exists a vector $\mathbf{x} \in \emptyset$ and a nonzero scalar a for which $a\mathbf{x} = \mathbf{0}$. This contradicts the fact that $\emptyset$ is empty. The linear independence of $\emptyset$ will be useful later. It will allow us to assign the dimension zero to the zero space.

2. Any set $\{\mathbf{v}\}$ containing a single nonzero vector is linearly independent since $a\mathbf{v} = \mathbf{0}$ only if $a = 0$.

3. The set $\{\mathbf{0}\}$ is linearly dependent since $a\mathbf{0} = \mathbf{0}$ for all scalars a.

4. A set $\{\mathbf{v}, \mathbf{w}\}$ containing two nonzero vectors is linearly independent if $\mathbf{v}$ is not a scalar multiple of $\mathbf{w}$ since $a\mathbf{v} + b\mathbf{w} = \mathbf{0}$ is equivalent to saying that $a\mathbf{v} = -b\mathbf{w}$. If $a \neq 0$, then we can solve this equation for $\mathbf{v}$ and write $\mathbf{v} = -\frac{b}{a}\mathbf{w}$.

5. All vectors in a nonempty linearly independent set S are nonzero. For suppose $\mathbf{v} \in S$ and $\mathbf{v} = \mathbf{0}$. Then $a\mathbf{v} = \mathbf{0}$ for all scalars a, and S is therefore linearly dependent.

Matrix Test for Linear Independence

Corollary 2.22 provides an easy test for the linear independence of a set $\{\mathbf{x}_1, \ldots, \mathbf{x}_n\}$ of column vectors in $\mathbb{R}^n$.

THEOREM 4.3 (Matrix test for linear independence)   *A set $\{\mathbf{x}_1, \ldots, \mathbf{x}_n\}$ of column vectors of height n is linearly independent if and only if the matrix $A = (\mathbf{x}_1 \ \cdots \ \mathbf{x}_n)$ is invertible.*

Proof. If we combine the vectors $\mathbf{x}_1, \ldots, \mathbf{x}_n$ into an $n \times n$ matrix $A = (\mathbf{x}_1 \ \cdots \ \mathbf{x}_n)$ and use the vector $\mathbf{x} = (a_1, \ldots, a_n)$ to form the homogeneous equation $A\mathbf{x} = \mathbf{0}$, then Corollary 2.22 tells us that A is invertible if and only if the equation $A\mathbf{x} = \mathbf{0}$ has only the trivial solution. Since

$$A\mathbf{x} = (\mathbf{x}_1 \ \cdots \ \mathbf{x}_n) \begin{pmatrix} a_1 \\ \vdots \\ a_n \end{pmatrix} = a_1\mathbf{x}_1 + \cdots + a_n\mathbf{x}_n$$

this statement is equivalent to saying that $A\mathbf{x} = \mathbf{0}$ has only the trivial solution if and only if the vectors $\mathbf{x}_1, \ldots, \mathbf{x}_n$ are linearly independent. ∎

COROLLARY 4.4 *A set $\{\mathbf{x}_1, \ldots, \mathbf{x}_n\}$ of column vectors of height n is linearly independent if and only if the matrix $A = (\mathbf{x}_1 \ \cdots \ \mathbf{x}_n)$ has n pivots.*

Proof. By Corollary 2.20, A is invertible if and only if it has n pivots. ∎

As noted, we refer to Theorem 4.3 as the *matrix test for linear independence*. The theorem also provides us with an infinite set of linearly independent subsets of $\mathbb{R}^n$ since the columns of invertible real $n \times n$ matrices are linearly independent sets.

EXAMPLE 4.18 ■ A Linearly Independent Subset of $\mathbb{R}^4$

Use *MATHEMATICA* and Theorem 4.3 to try to generate a linearly independent subset $\{\mathbf{x}_1, \mathbf{x}_2, \mathbf{x}_3, \mathbf{x}_4\}$ of $\mathbb{R}^4$.

Solution. We use *MATHEMATICA* to generate a random 4×4 matrix A and test it for invertibility. If it is invertible, we use its columns as the required vectors. If A is not invertible, we stop since we have no guarantee that *MATHEMATICA* will produce an invertible matrix in a finite number of attempts.

```
In[1]:= MatrixForm[A=Array[Random[Integer,{0,5}]&,{4,4}]]

Out[1]//MatrixForm=
```

$$\begin{pmatrix} 5 & 4 & 3 & 4 \\ 0 & 1 & 4 & 2 \\ 2 & 0 & 4 & 4 \\ 2 & 1 & 4 & 1 \end{pmatrix}$$

Next we ask *MATHEMATICA* to invert the matrix A, without requesting an explicit display of A^{-1}. If A is not invertible, *MATHEMATICA* will respond to our command with a message indicating that A is not invertible.

```
In[2]:= Inverse[A];
```

Since *MATHEMATICA* carried out the required operation without producing a output, we know that A is invertible. Hence its columns are linearly independent. Therefore, the set

$$\left\{ \mathbf{x}_1 = \begin{pmatrix} 5 \\ 0 \\ 2 \\ 2 \end{pmatrix}, \mathbf{x}_2 = \begin{pmatrix} 4 \\ 1 \\ 0 \\ 1 \end{pmatrix}, \mathbf{x}_3 = \begin{pmatrix} 3 \\ 4 \\ 4 \\ 4 \end{pmatrix}, \mathbf{x}_4 = \begin{pmatrix} 4 \\ 2 \\ 4 \\ 1 \end{pmatrix} \right\}$$

is the required set of vectors. ◄►

So far, we have defined linear dependence and independence as a property of sets. However, the following theorem shows that we can also speak of the linear dependence and independence as a property of vectors.

THEOREM 4.5   *A nonempty set $S = \{\mathbf{v}_1, \ldots, \mathbf{v}_n\}$ is linearly dependent if and only if one of the vectors $\mathbf{v}_i$ can be written as a linear combination of the remaining vectors.*

Proof. Suppose that S is linearly dependent. Then there exist scalars $a_1, \ldots, a_n$ not all 0, for which $a_1\mathbf{v}_1 + \cdots + a_n\mathbf{v}_n = \mathbf{0}$. By suitably renumbering the vectors in S, we may assume that $a_1 \neq 0$. Therefore, $a_1\mathbf{v}_1 = -a_2\mathbf{v}_2 - \cdots - a_n\mathbf{v}_n$. Hence $\mathbf{v}_1$ can be written as a linear combination of $\mathbf{v}_2, \ldots, \mathbf{v}_n$ using the coefficients $-a_2/a_1$ to $-a_n/a_1$.

Conversely, suppose that $\mathbf{v}_1 = b_2\mathbf{v}_2 + \cdots + b_n\mathbf{v}_n$. Then

$$1\mathbf{v}_1 - b_2\mathbf{v}_2 - \cdots - b_n\mathbf{v}_n = a_1\mathbf{v}_1 + a_2\mathbf{v}_2 + \cdots + a_n\mathbf{v}_n = \mathbf{0}$$

and not all a_i are 0. Hence S is linearly dependent. ∎

COROLLARY 4.6   *A nonempty set $S = \{\mathbf{v}_1, \ldots, \mathbf{v}_n\}$ is linearly independent if and only if none of the vectors $\mathbf{v}_i$ can be written as a linear combination of the remaining vectors.*

Proof. This fact is easily established by rewriting the proof of Theorem 4.5, using the definition of linear independence in place of linear dependence. ∎

Using Theorem 4.5 and its corollary, we say that a vector $\mathbf{x} \in V$ is ***linearly independent*** of the vectors $\mathbf{v}_1, \ldots, \mathbf{v}_n$ if $\mathbf{x}$ is not expressible as a linear combination of $\mathbf{v}_1, \ldots, \mathbf{v}_n$. Equivalently, we say that a vector $\mathbf{x} \in V$ is ***linearly independent*** of a set S of vectors if $\mathbf{x}$ is not in the span of S. The idea of ***linearly dependent vectors*** is analogous.

EXAMPLE 4.19   ■ **Linearly Independent Vectors**
Show that the vector $\mathbf{x} = (1, 2, 3)$ is linearly independent of the set

$$S = \text{span} \left\{ \begin{pmatrix} 1 \\ 2 \\ 0 \end{pmatrix}, \begin{pmatrix} 0 \\ 3 \\ 4 \end{pmatrix} \right\}$$

Solution. Suppose that

$$\begin{pmatrix} 1 \\ 2 \\ 3 \end{pmatrix} = a \begin{pmatrix} 1 \\ 2 \\ 0 \end{pmatrix} + b \begin{pmatrix} 0 \\ 3 \\ 4 \end{pmatrix} = \begin{pmatrix} a \\ 2a + 3b \\ 4b \end{pmatrix}$$

Then $3 = 4b$, $a = 1$, and $2 = 2a + 3b$. Hence we must have both $b = 3/4$ and $b = 0$. This obviously is impossible. ◄▶

EXERCISES 4.6

1. Explain in what sense the lists of ingredients and their quantities on food product labels can be thought of as linear combinations.

2. Use the given vectors and scalars to construct all possible linear combinations of the given vectors.

   a. Scalars: 3, −8, 5; vectors: (3, 5), (−8, 2)

   b. Scalars: π, π^2, 0; vectors: $\begin{pmatrix} 1 & 2 \\ 3 & 4 \end{pmatrix}$, $\begin{pmatrix} 1 & 2 \\ 3 & 4 \end{pmatrix}$, $\begin{pmatrix} 1 & 2 \\ 3 & 4 \end{pmatrix}$

3. Describe the spans of the vectors in Exercise 2.

4. Describe the span of the set of polynomials $S = \{t - 7, t^3, t^5, t^6 + 1\}$.

5. Test the following sets of vectors for linear dependence.

   a. $S = \left\{ \begin{pmatrix} 3 \\ 5 \end{pmatrix}, \begin{pmatrix} -8 \\ 2 \end{pmatrix} \right\}$

   b. $S = \left\{ \begin{pmatrix} 3 \\ 5 \end{pmatrix}, \begin{pmatrix} -8 \\ 2 \end{pmatrix}, \begin{pmatrix} 0 \\ 1 \end{pmatrix} \right\}$

   c. $S = \left\{ \begin{pmatrix} 3 \\ 5 \\ 4 \end{pmatrix}, \begin{pmatrix} -8 \\ 2 \\ 3 \end{pmatrix}, \begin{pmatrix} 0 \\ 1 \\ 1 \end{pmatrix} \right\}$

   d. $S = \left\{ \begin{pmatrix} 3 \\ 5 \\ 4 \end{pmatrix}, \begin{pmatrix} -8 \\ 2 \\ 3 \end{pmatrix}, \begin{pmatrix} 0 \\ 1 \\ 1 \end{pmatrix}, \begin{pmatrix} 6 \\ 6 \\ 6 \end{pmatrix} \right\}$

6. Use linear independence to test the following matrices for invertibility.

$$
a. \begin{pmatrix} 3 & -8 & 0 \\ 5 & 0 & 1 \\ 4 & 3 & 6 \end{pmatrix} \quad
b. \begin{pmatrix} 3 & -8 & 1 \\ 0 & 0 & 0 \\ 4 & 3 & 1 \end{pmatrix} \quad
c. \begin{pmatrix} 3 & -8 & -5 \\ 5 & 0 & 5 \\ 4 & 3 & 7 \end{pmatrix}
$$

7. Use pivots to test the columns of the following matrices for linear independence.

$$
a. \begin{pmatrix} 3 & -8 & 0 \\ 5 & 0 & 1 \\ 4 & 3 & 6 \end{pmatrix} \quad
b. \begin{pmatrix} 3 & -8 & 1 & 2 & 0 \\ 0 & 0 & 0 & 1 & 0 \\ 4 & 3 & 1 & 0 & 5 \\ 1 & 4 & 0 & 0 & 0 \\ 2 & 5 & 0 & 0 & 1 \end{pmatrix} \quad
c. \begin{pmatrix} 3 & -8 & -5 \\ 5 & 0 & 5 \\ 4 & 3 & 7 \end{pmatrix}
$$

Bases

We now show that all vectors in a vector space can be written as unique linear combinations of special sets of vectors.

DEFINITION 4.6 *A **basis** for a vector space V is any linearly independent ordered subset B of V that spans V.*

It is perhaps surprising that every vector space has a basis since, in the general case, the proof of this fact requires a powerful axiom of set theory. It is less surprising to find that if B and C are two finite bases for the same vector space, then B and C have the same number of elements. We will use this fact later to define the dimension of a vector space.

EXAMPLE 4.20 ■ **A Basis for** $\mathbb{R}^3$

Use *MATHEMATICA* to show that the set of vectors

$$
B = \left\{ x_1 = \begin{pmatrix} 1 \\ 2 \\ 3 \end{pmatrix}, x_2 = \begin{pmatrix} 0 \\ 5 \\ 2 \end{pmatrix}, x_3 = \begin{pmatrix} 1 \\ 2 \\ 0 \end{pmatrix} \right\}
$$

is a basis for $\mathbb{R}^3$.

Solution. We first show that B spans $\mathbb{R}^3$. We show that for any vector $x = (x, y, z)$ there exist scalars a, b, c such that

$$
a x_1 + b x_2 + c x_3 = x
$$

The scalars a, b, c are determined by the scalars x, y, z. The following linear system expresses their relationship.

$$
a \begin{pmatrix} 1 \\ 2 \\ 3 \end{pmatrix} + b \begin{pmatrix} 0 \\ 5 \\ 2 \end{pmatrix} + c \begin{pmatrix} 1 \\ 2 \\ 0 \end{pmatrix} = \begin{pmatrix} x \\ y \\ z \end{pmatrix}
$$

To find a, b, and c, we solve the equation

$$\begin{pmatrix} a + c \\ 2a + 5b + 2c \\ 3a + 2b \end{pmatrix} = \begin{pmatrix} x \\ y \\ z \end{pmatrix}$$

```
In[1]:= Solve[{a+c==x,2a+5b+2c==y,3a+2b==z},{a,b,c}]

Out[1]= {{a→ 1/15 (4x-2y+5z),b→ 1/5 (-2x+y),c→ 1/15 (11x+2y-5z)}}
```

Since a, b, c depend only on x, y, z, the set $\mathcal{B}$ spans $\mathbb{R}^3$. It remains to verify that $\mathcal{B}$ is linearly independent. We do so by showing that if $x = y = z = 0$, then $a = b = c = 0$.

```
In[2]:= Solve[{a+c==0,2a+5b+2c==0,3a+2b==0}]

Out[2]= {{a→0,b→0,c→0}}
```

Therefore, $\mathcal{B}$ is a basis for $\mathbb{R}^3$. ◄►

EXAMPLE 4.21 ■ The Basis of the Zero Space

By definition, the span of the empty set $\emptyset$ of vectors of a vector space is the space $\{\mathbf{0}\}$. As we pointed out earlier, the set $\emptyset$ is linearly independent. Hence it satisfies both parts of the definition of a basis. ◄►

Finite bases have an easy description in terms of linear combinations.

THEOREM 4.7 (Unique combination theorem) *A subset $\mathcal{B} = \{\mathbf{x}_1, \ldots, \mathbf{x}_n\}$ of a vector space V is a basis for V if and only if every vector $\mathbf{x} \in V$ can be written as a unique linear combination $a_1\mathbf{x}_1 + \cdots + a_n\mathbf{x}_n$.*

Proof. Suppose that $\mathcal{B}$ is a basis and that

$$\mathbf{x} = a_1\mathbf{x}_1 + \cdots + a_n\mathbf{x}_n = b_1\mathbf{x}_1 + \cdots + b_n\mathbf{x}_n$$

Then

$$\mathbf{x} - \mathbf{x} = \mathbf{0} = (a_1 - b_1)\mathbf{x}_1 + \cdots + (a_n - b_n)\mathbf{x}_n$$

Since the basis vectors $\mathbf{x}_i$ are linearly independent, it follows that $a_i - b_i = 0$ for all i. Hence the linear combination is unique.

Conversely, suppose that every $\mathbf{x} \in V$ can be written as a unique linear combination $a_1\mathbf{x}_1 + \cdots + a_n\mathbf{x}_n$. Then $\mathcal{B}$ obviously spans V.

Now suppose that $a_1\mathbf{x}_1 + \cdots + a_n\mathbf{x}_n = \mathbf{0}$. Then

$$a_1\mathbf{x}_1 + \cdots + a_n\mathbf{x}_n = 0\mathbf{x}_1 + \cdots + 0\mathbf{x} = \mathbf{0}$$

Since the representation of $\mathbf{0}$ as a linear combination of vectors in $\mathcal{B}$ is unique, it follows that $a_1 = \cdots = a_n = 0$. Hence $\mathcal{B}$ is linearly independent. Therefore, the set $\mathcal{B}$ is a basis for V. ■

EXERCISES 4.7

1. Describe all possible bases of the space $\mathbb{R}$ of real numbers.

2. Use invertible matrices to construct two different bases for the space $\mathbb{R}^4$.

3. Show that the set

$$B = \left\{ \begin{pmatrix} 1 & 2 \\ 3 & 4 \end{pmatrix}, \begin{pmatrix} 2 & 0 \\ 0 & 5 \end{pmatrix}, \begin{pmatrix} 0 & 2 \\ 5 & 0 \end{pmatrix}, \begin{pmatrix} 1 & 1 \\ 1 & 1 \end{pmatrix} \right\}$$

is a basis for the space $\mathbb{R}^{2\times2}$ of 2×2 matrices.

4. The fact that the linear system

$$\begin{cases} a + 2b - c = 0 \\ 2a + 4b + 2c + d = 0 \\ 3a + 7b + 5c + d = 0 \\ 4a + 5b + c + 3d = 0 \end{cases}$$

has only the trivial solution ensures that the set

$$B = \left\{ \begin{pmatrix} 1 & 2 \\ 3 & 4 \end{pmatrix}, \begin{pmatrix} 2 & 4 \\ 7 & 5 \end{pmatrix}, \begin{pmatrix} -1 & 2 \\ 5 & 1 \end{pmatrix}, \begin{pmatrix} 0 & 1 \\ 1 & 3 \end{pmatrix} \right\}$$

is a basis for $\mathbb{R}^{2\times2}$. Explain why.

5. Find two different bases for the space $\mathbb{R}_5[t]$ of polynomials of degree 5 or less.

6. Find a basis for the solution space of the homogeneous linear system

$$\begin{pmatrix} 2 & 1 & 4 & 2 \\ 3 & 7 & 4 & 5 \\ 4 & 8 & 4 & 1 \end{pmatrix} \begin{pmatrix} w \\ x \\ y \\ z \end{pmatrix} = \begin{pmatrix} 0 \\ 0 \\ 0 \end{pmatrix}$$

7. Find a basis for the solution space of the differential equation $y'(t) - y(t) = 0$.

Standard Bases

Among the bases of some vector spaces are bases that are particularly simple. The vectors involved are often built from the scalars 0 and 1, but may also have other special properties. These bases are known as **standard bases**. Except for the order in which the vectors are listed, standard bases are unique.

We can describe the standard bases of coordinate spaces and matrix spaces using the functions $\delta : \mathbf{m} \times \mathbf{n} \to \{0, 1\}$, defined by

$$\delta_{ij} = \begin{cases} 1 & \text{if} \quad i = j \\ 0 & \text{if} \quad i \neq j \end{cases}$$

These functions are known as the **Kronecker delta** functions and are built directly into *MATHEMATICA*. The input **KroneckerDelta[i,j]** produces the output 1 if $i = j$ and 0 otherwise. The input **KroneckerDelta[i]** produces 1 if $i = 0$ and 0 otherwise.

EXAMPLE 4.22 ■ The Standard Basis of $\mathbb{R}^2$

The standard basis of the coordinate space $\mathbb{R}^2$ can be defined using the Kronecker delta function $\delta : \mathbf{2} \times \mathbf{2} \to \{0, 1\}$. The set $\mathcal{E}$ consisting of the two vectors $\mathbf{e}_1 = (\delta_{11}, \delta_{12}) = (1, 0)$ and $\mathbf{e}_2 = (\delta_{21}, \delta_{22}) = (0, 1)$ is the standard basis of $\mathbb{R}^2$. ◄►

EXAMPLE 4.23 ■ The Standard Basis of $\mathbb{R}^3$

The standard basis $\mathcal{E}$ of $\mathbb{R}^3$ consists of the three vectors $\mathbf{e}_1 = (\delta_{11}, \delta_{12}, \delta_{13}) = (1, 0, 0)$, $\mathbf{e}_2 = (\delta_{21}, \delta_{22}, \delta_{23}) = (0, 1, 0)$, and $\mathbf{e}_3 = (\delta_{31}, \delta_{32}, \delta_{33}) = (0, 0, 1)$.

We can combine the **Table** and **KroneckerDelta** functions to generate these vectors in *MATHEMATICA*.

```
In[1]:=e1=Table[KroneckerDelta[n],{n,0,2}]//MatrixForm

Out[1]//MatrixForm=
                    ⎛ 1 ⎞
                    ⎜ 0 ⎟
                    ⎝ 0 ⎠
```

```
In[2]:=e2=Table[KroneckerDelta[n],{n,-1,1}]//MatrixForm

Out[2]//MatrixForm=
                    ⎛ 0 ⎞
                    ⎜ 1 ⎟
                    ⎝ 0 ⎠
```

```
In[3]:=e3=Table[KroneckerDelta[n],{n,-2,0}]//MatrixForm
Out[3]//MatrixForm=
```

$$\begin{pmatrix} 0 \\ 0 \\ 1 \end{pmatrix}$$

As we can see, $\mathbf{e}_1$, $\mathbf{e}_2$, and $\mathbf{e}_3$ are the required vectors. ◄►

In vector geometry, the vectors $\mathbf{e}_1$, $\mathbf{e}_2$, $\mathbf{e}_3$ in $\mathbb{R}^3$ are often denoted by $\mathbf{i}$, $\mathbf{j}$, and $\mathbf{k}$. Figure 1 shows how these vectors are represented geometrically.

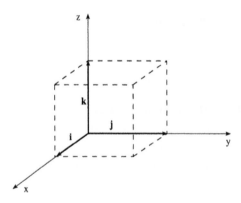

FIGURE 1   The Standard Basis of $\mathbb{R}^3$.

EXAMPLE 4.24 ■ The Standard Basis of $\mathbb{R}^{2 \times 2}$

The standard basis of the matrix space $\mathbb{R}^{2 \times 2}$ can also be defined using the Kronecker delta function $\delta : \mathbf{4} \times \mathbf{4} \to \{0, 1\}$. The set

$$\mathcal{E} = \left\{ \begin{pmatrix} \delta_{11} & \delta_{12} \\ \delta_{13} & \delta_{14} \end{pmatrix}, \begin{pmatrix} \delta_{21} & \delta_{22} \\ \delta_{23} & \delta_{24} \end{pmatrix}, \begin{pmatrix} \delta_{31} & \delta_{32} \\ \delta_{33} & \delta_{34} \end{pmatrix}, \begin{pmatrix} \delta_{41} & \delta_{42} \\ \delta_{43} & \delta_{44} \end{pmatrix} \right\}$$

is the standard basis of $\mathbb{R}^{2 \times 2}$. ◄►

EXAMPLE 4.25 ■ The Standard Basis of $\mathbb{R}^n$

The standard basis of the coordinate space $\mathbb{R}^n$ is defined using the Kronecker delta function $\delta : \mathbf{n} \times \mathbf{n} \to \{0, 1\}$. It consists of the vectors $\mathbf{e}_i = (\delta_{i1}, \ldots, \delta_{in})$, for $1 \leq i \leq n$. ◄►

EXAMPLE 4.26 ■ The Standard Basis of $\mathbb{R}_n[t]$

The set $\mathcal{E} = \{1, t, \ldots, t^n\}$ is the standard basis of the polynomial space $\mathbb{R}_n[t]$. ◄►

EXAMPLE 4.27 ■ The Standard Basis of $\mathbb{R}[t]$

The set $\mathcal{E} = \{1, t, \ldots, t^n, \ldots\}$ is the standard basis of the polynomial space $\mathbb{R}[t]$. It is infinite since it contains the polynomials t^n for all $n \in \mathbb{N}$. ◄►

EXAMPLE 4.28 ■ The Standard Basis of a Homogeneous System

Consider the homogeneous system $A\mathbf{x} = \mathbf{0}$, determined by the linear equations

$$\begin{cases} x_1 + 3x_2 - x_5 + x_6 = 0 \\ 2x_1 + x_3 - x_4 + 4x_5 = 0 \\ x_2 + x_3 + x_4 - x_6 = 0 \end{cases}$$

We use the reduced row echelon form of the augmented matrix

$$(A \mid \mathbf{0}) = \begin{pmatrix} 1 & 3 & 0 & 0 & -1 & 1 & 0 \\ 2 & 0 & 1 & -1 & 4 & 0 & 0 \\ 0 & 1 & 1 & 1 & 0 & -1 & 0 \end{pmatrix}$$

of $A\mathbf{x} = \mathbf{0}$ to define the standard basis of the solution space of $A\mathbf{x} = \mathbf{0}$.

We expedite the calculations by using *MATHEMATICA* to row-reduce the matrix $(A \mid \mathbf{0})$.

```
In[1]:=(A|0)={{1,3,0,0,-1,1,0},{2,0,1,-1,4,0,0},
{0,1,1,1,0,-1,0}};
```

```
In[2]:=B=RowReduce[(A|0)]//MatrixForm

Out[2]//MatrixForm=
```

$$\begin{pmatrix} 1 & 0 & 0 & -\frac{6}{7} & \frac{11}{7} & \frac{4}{7} & 0 \\ 0 & 1 & 0 & \frac{2}{7} & -\frac{6}{7} & \frac{1}{7} & 0 \\ 0 & 0 & 1 & \frac{5}{7} & \frac{6}{7} & -\frac{8}{7} & 0 \end{pmatrix}$$

We can tell from the matrix B that the variables of the system $A\mathbf{x} = \mathbf{0}$ are related by the equations

$$\begin{cases} x_1 = \frac{6}{7}x_4 - \frac{11}{7}x_5 - \frac{4}{7}x_6 \\ x_2 = -\frac{2}{7}x_4 + \frac{6}{7}x_5 - \frac{1}{7}x_6 \\ x_3 = -\frac{5}{7}x_4 - \frac{6}{7}x_5 + \frac{8}{7}x_6 \end{cases}$$

The solutions of $A\mathbf{x} = \mathbf{0}$ are therefore all vectors of the form

$$
\begin{pmatrix}
\frac{6}{7}x_4 - \frac{11}{7}x_5 - \frac{4}{7}x_6 \\
-\frac{2}{7}x_4 + \frac{6}{7}x_5 - \frac{1}{7}x_6 \\
-\frac{5}{7}x_4 - \frac{6}{7}x_5 + \frac{8}{7}x_6 \\
x_4 \\
x_5 \\
x_6
\end{pmatrix}
= x_4
\begin{pmatrix}
\frac{6}{7} \\
-\frac{2}{7} \\
-\frac{5}{7} \\
1 \\
0 \\
0
\end{pmatrix}
+ x_5
\begin{pmatrix}
-\frac{11}{7} \\
\frac{6}{7} \\
-\frac{6}{7} \\
0 \\
1 \\
0
\end{pmatrix}
+ x_6
\begin{pmatrix}
-\frac{4}{7} \\
-\frac{1}{7} \\
\frac{8}{7} \\
0 \\
0 \\
1
\end{pmatrix}
$$

written as unique linear combinations of the set

$$
\mathcal{E} = \left\{
\begin{pmatrix}
\frac{6}{7} \\
-\frac{2}{7} \\
-\frac{5}{7} \\
1 \\
0 \\
0
\end{pmatrix},
\begin{pmatrix}
-\frac{11}{7} \\
\frac{6}{7} \\
-\frac{6}{7} \\
0 \\
1 \\
0
\end{pmatrix},
\begin{pmatrix}
-\frac{4}{7} \\
-\frac{1}{7} \\
\frac{8}{7} \\
0 \\
0 \\
1
\end{pmatrix}
\right\}
$$

By construction, the set $\mathcal{E}$ spans the solution space of $A\mathbf{x} = \mathbf{0}$. It is also clear that $\mathcal{E}$ is linearly independent. Hence $\mathcal{E}$ is a basis for the solution space of the given homogeneous linear system. Since this basis is uniquely determined by the reduced row echelon form of $(A \mid \mathbf{0})$, we call it a standard basis. ◄►

EXERCISES 4.8

1. Use the Kronecker delta function to generate the standard basis of $\mathbb{R}^4$.

2. Use the Kronecker delta function to generate the standard basis of $\mathbb{R}^{3\times3}$.

3. Use the Kronecker delta function to generate the standard basis of $\mathbb{R}^{2\times3}$.

4. Discuss the idea of a standard basis for the following spaces.

   a. The space $\mathbb{R}$ of real numbers, considered as a real vector space

   b. The space of real upper-triangular 3×3 matrices

   c. The space of real diagonal 4×4 matrices

   d. The space of real symmetric 4×4 matrices

5. Define the standard basis of the solution space of the homogeneous linear system

$$
\begin{cases}
2w + x + 4y + 2z = 0 \\
3w + 7x + 4y + 5z = 0
\end{cases}
$$

6. Discuss the relevance of the **KroneckerDelta** function to the construction of standard bases with *MATHEMATICA*.

Bases and Linear Systems

If $\mathbf{x}$ is any vector in the vector space $\mathbb{R}^n$ and $\mathcal{B} = \{\mathbf{v}_1, \ldots, \mathbf{v}_n\}$ is a basis for $\mathbb{R}^n$, then $\mathbf{x}$ can be written uniquely in the form $\mathbf{x} = a_1\mathbf{v}_1 + \cdots + a_n\mathbf{v}_n$ in the basis $\mathcal{B}$. The coefficients $a_1, \ldots, a_n$ can be found by solving a system of linear equations.

Consider the basis

$$
\mathcal{B} = \left\{ \mathbf{v}_1 = \begin{pmatrix} 1 \\ 0 \\ 2 \\ 3 \end{pmatrix}, \mathbf{v}_2 = \begin{pmatrix} 1 \\ 0 \\ 0 \\ 1 \end{pmatrix}, \mathbf{v}_3 = \begin{pmatrix} 0 \\ 3 \\ -1 \\ 0 \end{pmatrix}, \mathbf{v}_4 = \begin{pmatrix} 1 \\ 1 \\ 1 \\ 1 \end{pmatrix} \right\}
$$

for $\mathbb{R}^4$, and let $\mathbf{x} = (1, 2, 3, 4) \in \mathbb{R}^4$. We show how the vector $\mathbf{x}$ is written as a linear combination of the vectors in $\mathcal{B}$.

Since $\mathcal{B}$ is a basis, there exist scalars a, b, c, d for which

$$
\begin{pmatrix} 1 \\ 2 \\ 3 \\ 4 \end{pmatrix} = a \begin{pmatrix} 1 \\ 0 \\ 2 \\ 3 \end{pmatrix} + b \begin{pmatrix} 1 \\ 0 \\ 0 \\ 1 \end{pmatrix} + c \begin{pmatrix} 0 \\ 3 \\ -1 \\ 0 \end{pmatrix} + d \begin{pmatrix} 1 \\ 1 \\ 1 \\ 1 \end{pmatrix}
$$

For the purpose of illustration, we use *MATHEMATICA* to find the coefficients a, b, c, d by solving the linear system

$$
\begin{cases} a + b + d = 1 \\ 3c + d = 2 \\ 2a - c + d = 3 \\ 3a + b + d = 4 \end{cases}
$$

In[1]:= Solve[{a+b+d==1,3c+d==2,2a-c+d==3,3a+b+d==4},
{a,b,c,d}]

Out[1]= {{a→ $\frac{3}{2}$,b→-1,c→ $\frac{1}{2}$,d→ $\frac{1}{2}$}}

Therefore, $a = \frac{3}{2}$, $b = -1$, $c = \frac{1}{2}$, and $d = \frac{1}{2}$ are the unique scalars that allow us to represent the vector $\mathbf{x}$ in the basis $\mathcal{B}$. The expression

$$
\frac{3}{2} \begin{pmatrix} 1 \\ 0 \\ 2 \\ 3 \end{pmatrix} - 1 \begin{pmatrix} 1 \\ 0 \\ 0 \\ 1 \end{pmatrix} + \frac{1}{2} \begin{pmatrix} 0 \\ 3 \\ -1 \\ 0 \end{pmatrix} + \frac{1}{2} \begin{pmatrix} 1 \\ 1 \\ 1 \\ 1 \end{pmatrix}
$$

is the actual linear combination.

We formulate these observations as a theorem.

THEOREM 4.8 *If $\mathcal{B} = \{\mathbf{v}_1, \ldots, \mathbf{v}_n\}$ is any basis for $\mathbb{R}^n$ and $\mathbf{x}$ is any vector in $\mathbb{R}^n$, then the coefficients $a_1, \ldots, a_n$ in the unique linear combination $\mathbf{x} = a_1\mathbf{v}_1 + \cdots + a_n\mathbf{v}_n$ can be found by solving a system of linear equations.*

Proof. Let $i \in \mathbf{n}$ and suppose that the vectors $\mathbf{v}_i = (v_{i1}, \ldots, v_{in})$ form a basis for $\mathbb{R}^n$. Suppose further that $\mathbf{x} = (x_1, \ldots, x_n)$ is an arbitrary vector in $\mathbb{R}^n$. Then

$$
\begin{pmatrix} x_1 \\ \vdots \\ x_n \end{pmatrix} = a_1 \begin{pmatrix} v_{11} \\ \vdots \\ v_{1n} \end{pmatrix} + \cdots + a_n \begin{pmatrix} v_{n1} \\ \vdots \\ v_{nn} \end{pmatrix}
$$

$$
= \begin{pmatrix} a_1 v_{11} \\ \vdots \\ a_1 v_{1n} \end{pmatrix} + \cdots + \begin{pmatrix} a_n v_{n1} \\ \vdots \\ a_n v_{nn} \end{pmatrix}
$$

and

$$
\begin{pmatrix} a_1 v_{11} \\ \vdots \\ a_1 v_{1n} \end{pmatrix} + \cdots + \begin{pmatrix} a_n v_{n1} \\ \vdots \\ a_n v_{nn} \end{pmatrix} = \begin{pmatrix} a_1 v_{11} + \cdots + a_n v_{n1} \\ \vdots \\ a_1 v_{1n} + \cdots + a_n v_{nn} \end{pmatrix}
$$

Therefore,

$$
\begin{pmatrix} a_1 v_{11} + \cdots + a_n v_{n1} \\ \vdots \\ a_1 v_{1n} + \cdots + a_n v_{nn} \end{pmatrix} = \begin{pmatrix} x_1 \\ \vdots \\ x_n \end{pmatrix}
$$

This shows that the linear system

$$
\begin{cases} a_1 v_{11} + \cdots + a_n v_{n1} = x_1 \\ \quad\quad\quad \vdots \\ a_1 v_{1n} + \cdots + a_n v_{nn} = x_n \end{cases}
$$

determines the coefficients $a_1, \ldots, a_n$. ■

EXERCISES 4.9

1. Use the basis

$$
\mathcal{B} = \left\{ \begin{pmatrix} 4 \\ 5 \\ -9 \end{pmatrix}, \begin{pmatrix} 4 \\ -5 \\ 4 \end{pmatrix}, \begin{pmatrix} 2 \\ 9 \\ 0 \end{pmatrix} \right\}
$$

and the vectors

$$\mathbf{b}_1 = \begin{pmatrix} 3 \\ -2 \\ 1 \end{pmatrix}, \mathbf{b}_2 = \begin{pmatrix} 7 \\ 5 \\ 3 \end{pmatrix}, \mathbf{b}_3 = \begin{pmatrix} 1 \\ 1 \\ 1 \end{pmatrix}$$

to create three linear systems $A\mathbf{x} = \mathbf{b}_1$, $A\mathbf{x} = \mathbf{b}_2$, and $A\mathbf{x} = \mathbf{b}_3$. Find the solutions of the systems and use the fact that $\mathcal{B}$ is a basis to explain why the solutions are unique.

2. Use the fact that the linear system

$$\begin{cases} 6w + 2x - 6y + 6z = 3 \\ 2w - 3x + 5y + 2z = 2 \\ 2w - 5x - 3y + 4z = 0 \\ 4w + 2x + 6y - 4z = 1 \end{cases}$$

has a unique solution to explain why the columns of the coefficient matrix of the system form a basis for $\mathbb{R}^4$.

3. Construct and solve the linear system that determines the coefficients a_1, a_2, a_3, and a_4 for which

$$\begin{pmatrix} 1 \\ 2 \\ 3 \\ 4 \end{pmatrix} = a_1 \begin{pmatrix} 8 \\ 3 \\ -1 \\ -5 \end{pmatrix} + a_2 \begin{pmatrix} 4 \\ -5 \\ 0 \\ -2 \end{pmatrix} + a_3 \begin{pmatrix} -5 \\ 8 \\ 3 \\ -1 \end{pmatrix} + a_4 \begin{pmatrix} -5 \\ 5 \\ -4 \\ -9 \end{pmatrix}$$

and explain why the coefficients a_1, a_2, a_3, and a_4 are unique.

Dimension

A vector space is *finite-dimensional* if it has a finite basis. A space that is not finite-dimensional is called *infinite-dimensional*.

All coordinate spaces $\mathbb{R}^n$ and all matrix spaces $\mathbb{R}^{m \times n}$ are finite-dimensional. So are the polynomial spaces $\mathbb{R}_n[t]$. The polynomial space $\mathbb{R}[t]$, on the other hand, is infinite-dimensional. So are the sequence spaces $\mathbb{R}^N$ and $\mathbb{R}^\infty$. However, most spaces studied in this text are finite-dimensional.

The main theorem in this section establishes that all bases of a finite-dimensional vector space V have the same number of elements. We can use this property to define the *dimension* of V.

THEOREM 4.9   *Let V be a vector space spanned by the vectors $\mathbf{v}_1, \ldots, \mathbf{v}_m$. Then any linearly independent set of vectors in V is finite and contains no more than m elements.*

Proof. It is sufficient to show that any set containing more than m elements is linearly dependent. Let S be such a set, containing n distinct vectors $\mathbf{w}_1, \ldots, \mathbf{w}_n$, where $n > m$. Since the vectors $\mathbf{v}_1, \ldots, \mathbf{v}_m$ span V, there exist scalars a_{ij} such that $\mathbf{w}_j = \sum_{i \in \mathbf{m}} a_{ij} \mathbf{v}_i$. Moreover, for any other scalars $b_1, \ldots, b_n$, we have

$$b_1 \mathbf{w}_1 + \cdots + b_n \mathbf{w}_n = \sum_{j \in \mathbf{n}} b_i \left(\sum_{i \in \mathbf{m}} a_{ij} \mathbf{v}_i \right)$$
$$= \sum_{i \in \mathbf{m}} \left(\sum_{j \in \mathbf{n}} b_i a_{ij} \right) \mathbf{v}_i$$

Suppose that $\sum_{i \in \mathbf{m}} \left(\sum_{j \in \mathbf{n}} b_i a_{ij} \right) \mathbf{v}_i = 0$. Since the $\mathbf{v}_i$ are linearly independent, we get the homogeneous linear system

$$\sum_{j \in \mathbf{n}} b_i a_{ij} = 0 \quad (1 \le i \le m)$$

Since $n > m$, Theorem 1.5 guarantees the existence of a nontrivial solution $(b_1, \ldots, b_n)$. Therefore S is linearly dependent. ∎

THEOREM 4.10 (Dimension theorem) *If $\mathcal{B}$ and $\mathcal{C}$ are two bases of a finite-dimensional vector space V, then $\mathcal{B}$ and $\mathcal{C}$ have the same number of elements.*

Proof. Suppose that $\mathcal{B} = \{\mathbf{x}_1, \ldots, \mathbf{x}_n\}$ and $\mathcal{C} = \{\mathbf{y}_1, \ldots, \mathbf{y}_m\}$ are two bases for V. Since $\mathcal{B}$ spans V, Theorem 4.9 implies that the linearly independent set $\mathcal{C}$ has at most n elements. Therefore, $m \le n$. On the other hand, $\mathcal{C}$ is also a basis and spans V. Hence the linearly independent set $\mathcal{B}$ has at most m elements, so that $n \le m$. If follows that $n = m$. ∎

This theorem tells us that the following definition makes sense.

DEFINITION 4.7 *If V is a vector space with a finite basis $\mathcal{B}$, then the number of elements of $\mathcal{B}$ is the **dimension** of V.*

EXAMPLE 4.29 ■ **The Dimension of the Coordinate Space $\mathbb{R}^n$**
    The space $\mathbb{R}^n$ has dimension n since its standard basis $\mathcal{E} = \{\mathbf{e}_1, \ldots, \mathbf{e}_n\}$, for example, has n elements. ◄►

EXAMPLE 4.30 ■ **The Dimension of the Matrix Space $\mathbb{R}^{m \times n}$**
    The space $\mathbb{R}^{m \times n}$ of $m \times n$ matrices over $\mathbb{R}$ has dimension nm since its standard basis, for example, has mn elements. ◄►

Contracting Spanning Sets
If a set S of vectors of a finite-dimensional vector space V spans V, then there are procedures for extracting a basis $\mathcal{B}$ for V from S.

EXAMPLE 4.31 ■ **Extracting a Basis from a Finite Spanning Set**
    Use *MATHEMATICA* and the fact that $\mathbb{R}^3$ has dimension 3 to extract a basis for $\mathbb{R}^3$ from the spanning set

$$S = \left\{ \mathbf{v}_1 = \begin{pmatrix} 1 \\ 2 \\ 1 \end{pmatrix}, \mathbf{v}_2 = \begin{pmatrix} 4 \\ 4 \\ 4 \end{pmatrix}, \mathbf{v}_3 = \begin{pmatrix} 1 \\ 0 \\ 1 \end{pmatrix}, \mathbf{v}_4 = \begin{pmatrix} 2 \\ 4 \\ 2 \end{pmatrix}, \mathbf{v}_5 = \begin{pmatrix} 0 \\ 1 \\ 1 \end{pmatrix} \right\}$$

Solution. Since the dimension of $\mathbb{R}^3$ is 3, any linearly independent subset of S that spans $\mathbb{R}^3$ must be among the sets $S_1 = \{\mathbf{v}_1, \mathbf{v}_2, \mathbf{v}_3\}$, $S_2 = \{\mathbf{v}_1, \mathbf{v}_2, \mathbf{v}_4\}$, $S_3 = \{\mathbf{v}_1, \mathbf{v}_2, \mathbf{v}_5\}$, $S_4 = \{\mathbf{v}_2, \mathbf{v}_3, \mathbf{v}_4\}$, $S_5 = \{\mathbf{v}_2, \mathbf{v}_3, \mathbf{v}_5\}$, or $S_6 = \{\mathbf{v}_3, \mathbf{v}_4, \mathbf{v}_5\}$.

To test these sets for linear independence, we use Corollary 4.32 proved later in the chapter. The corollary guarantees that a set S_i of three 3×1 column vectors is linearly independent if and only if the associated 3×3 matrix has an inverse.

The calculation

```
In[1]:= Inverse[{{1,2,1},{4,4,4},{1,0,1}}]

Inverse::  sing:

Matrix {{1,2,1},{4,4,4},{1,0,1}} is singular.

Out[1]= Inverse[{{1,2,1},{4,4,4},{1,0,1}}]
```

shows that the matrix associated with S_1 is singular. Hence S_1 is linearly dependent.

We therefore test one of the other sets. The calculation

```
In[2]:= Inverse[{{1,2,1},{4,4,4},{0,1,1}}]

Out[2]= {{0,1/4,-1},{1,-1/4,0},{-1,1/4,1}}
```

shows that the matrix associated with S_3 is invertible. Hence S_3 is linearly independent and is therefore a basis for $\mathbb{R}^3$. ◀▶

EXAMPLE 4.32   ■ Extracting a Basis from an Infinite Spanning Set

Let n be a positive integer, and let S be the infinite set of vectors in $\mathbb{R}^2$ of the form $(n, n+1)$. Find a subset of S that is a basis for $\mathbb{R}^2$.

Solution. We know that $\mathbb{R}^2$ has dimension 2 and we can easily verify that the set

$$\mathcal{B} = \left\{ \begin{pmatrix} 1 \\ 1 \end{pmatrix}, \begin{pmatrix} 5 \\ 4 \end{pmatrix} \right\}$$

for example, is a basis for $\mathbb{R}^2$. Moreover, trial and error shows that we can write the vectors in S as linear combinations of vectors in $\mathcal{B}$:

$$\text{a.} \begin{pmatrix} 1 \\ 1 \end{pmatrix} = \begin{pmatrix} 2 \\ 3 \end{pmatrix} - \begin{pmatrix} 1 \\ 2 \end{pmatrix} \quad \text{b.} \begin{pmatrix} 5 \\ 4 \end{pmatrix} = \begin{pmatrix} 5 \\ 6 \end{pmatrix} - 2 \begin{pmatrix} 0 \\ 1 \end{pmatrix}$$

This shows that

$$\begin{pmatrix} 1 \\ 1 \end{pmatrix} \in \text{span} \left\{ \begin{pmatrix} 1 \\ 2 \end{pmatrix}, \begin{pmatrix} 2 \\ 3 \end{pmatrix} \right\} \quad \text{and} \quad \begin{pmatrix} 5 \\ 4 \end{pmatrix} \in \text{span} \left\{ \begin{pmatrix} 0 \\ 1 \end{pmatrix}, \begin{pmatrix} 5 \\ 6 \end{pmatrix} \right\}$$

Therefore, the finite subset

$$S' = \left\{ \mathbf{v}_1 = \begin{pmatrix} 0 \\ 1 \end{pmatrix}, \mathbf{v}_2 = \begin{pmatrix} 2 \\ 3 \end{pmatrix}, \mathbf{v}_3 = \begin{pmatrix} 1 \\ 2 \end{pmatrix}, \mathbf{v}_4 = \begin{pmatrix} 5 \\ 6 \end{pmatrix} \right\}$$

of S spans $\mathbb{R}^2$. It remains for us to extract a linearly independent subset $\mathcal{C}$ from the set S'.

We construct $\mathcal{C}$ step by step. We begin with $\mathcal{C}_1 = \{\mathbf{v}_1\}$. Since this set contains a single nonzero vector, it is linearly independent. Next we check $\mathcal{C}_2 = \{\mathbf{v}_1, \mathbf{v}_2\}$. Suppose that

$$a\mathbf{v}_1 + b\mathbf{v}_2 = a \begin{pmatrix} 0 \\ 1 \end{pmatrix} + b \begin{pmatrix} 1 \\ 2 \end{pmatrix} = \begin{pmatrix} b \\ a + 2b \end{pmatrix} = \begin{pmatrix} 0 \\ 0 \end{pmatrix}$$

We can tell immediately that $a = b = 0$. Hence $\mathcal{C}_2$ is linearly independent and we have the desired basis. ◀▶

In the proof of the next theorem, we explain how a basis can be extracted from a given spanning set.

THEOREM 4.11 (Basis contraction theorem)   *Suppose that S is a subset of a finite-dimensional vector space V that spans V. Then S has a linearly independent subset S' that is a basis for V.*

Proof. Since V is finite-dimensional, it has a finite basis $\mathcal{B} = \{\mathbf{v}_1, \ldots, \mathbf{v}_n\}$. Since $\text{span}(S) = V$, we can write each $\mathbf{v}_i \in \mathcal{B}$ as a finite linear combination of vectors in S. The vector $\mathbf{v}_1$ is a linear combination of vectors in a finite set $S_1 \subseteq S$, the vector $\mathbf{v}_2$ is a linear combination of vectors in a finite set $S_2 \subseteq S$, and so on. Continuing in this way we see that the vector $\mathbf{v}_n$ is a linear combination of vectors in a finite set $S_n \subseteq S$.

Let $S_0 = S_1 \cup S_2 \cup \cdots \cup S_n$. Since S_0 is a finite union of finite sets, S_0 is finite. By construction, it spans V. We can therefore construct a basis S' from S_0 by eliminating systematically from left to right all vectors that are linearly dependent on the preceding ones. ∎

Expanding Linearly Independent Sets

Bases are optimal subsets of vector spaces. They are large enough to span a space and yet small enough to be linearly independent. Theorem 4.11 explains how to contract a spanning set to a minimal spanning set. In the next theorem we describe a procedure for extending a linearly independent set to a maximally linearly independent set. We begin with an example.

EXAMPLE 4.33 ■ **Extending a Linearly Independent Set to a Basis**

Use *MATHEMATICA* to extend the linearly independent set

$$S = \left\{ \mathbf{v}_1 = \begin{pmatrix} 1 \\ 0 \\ 3 \end{pmatrix}, \mathbf{v}_2 = \begin{pmatrix} 0 \\ 0 \\ 1 \end{pmatrix} \right\}$$

to a basis for $\mathbb{R}^3$.

Solution. The idea behind the construction is to take a basis $\mathcal{B} = \{\mathbf{w}_1, \mathbf{w}_2, \mathbf{w}_3\}$ for $\mathbb{R}^3$ and append one of the vectors from $\mathcal{B}$ to S. Since $\mathcal{B}$ is a basis and therefore spans $\mathbb{R}^3$, and since S is linearly independent, there must be a vector in S that is linearly independent of $\mathbf{v}_1$ and $\mathbf{v}_2$.

Let $\mathcal{B}$ be the standard basis

$$\left\{ \mathbf{w}_1 = \begin{pmatrix} 1 \\ 0 \\ 0 \end{pmatrix}, \mathbf{w}_2 = \begin{pmatrix} 0 \\ 1 \\ 0 \end{pmatrix}, \mathbf{w}_3 = \begin{pmatrix} 0 \\ 0 \\ 1 \end{pmatrix} \right\}$$

for $\mathbb{R}^3$. We test for linear independence as in Example 4.31. The calculation

```
In[1]:= Inverse[{{1,0,3},{0,0,1},{1,0,0}}]

Inverse::  sing:

Out[1]= Inverse[{{1,0,3},{0,0,1},{1,0,0}}]
```

shows that $\mathbf{w}_1$ is linearly dependent on $\mathbf{v}_1$ and $\mathbf{v}_2$. We therefore discard it. On the other hand, the calculation

```
In[2]:= Inverse[{{1,0,3},{0,0,1},{0,1,0}}]

Out[2]= {{1,-3,0},{0,0,1},{0,1,0}}
```

shows that the matrix

$$\begin{pmatrix} 1 & 0 & 3 \\ 0 & 0 & 1 \\ 0 & 1 & 0 \end{pmatrix}$$

is invertible. The vector $\mathbf{w}_2$ is therefore linearly independent of $\mathbf{v}_1$ and $\mathbf{v}_2$. Hence the set $\mathcal{C} = \{\mathbf{v}_1, \mathbf{v}_2, \mathbf{w}_2\}$ is a basis for $\mathbb{R}^3$. ◀▶

THEOREM 4.12 (Basis extension theorem) *Any linearly independent set of vectors S of a finite-dimensional vector space V can be extended to a basis for V.*

Proof. Suppose that dim $V = n$, and that $S = \{\mathbf{w}_1, \ldots, \mathbf{w}_p\} \subseteq V$ is linearly independent, with $p \leq n$. If $p = n$, then S spans V and is the basis for V. If $p < n$, then S does not span V. We therefore take an arbitrary basis $\mathcal{B} = \{\mathbf{v}_1, \ldots, \mathbf{v}_n\}$ and use vectors from $\mathcal{B}$ not in S to extend S to a basis. We form a sequence of linearly independent sets

$$S_1 = S \cup \{\mathbf{y}_1\}, \quad S_2 = S_1 \cup \{\mathbf{y}_2\}, \quad \ldots, \quad S_q = S_{q-1} \cup \{\mathbf{y}_q\}$$

by appending to each set already constructed, the first vector $\mathbf{y}_i \in S$ (from left to right) that is linearly independent of the vectors in the previously constructed set. We continue until we have a set S_q that spans V. Since V is finite-dimensional, this process must end after finitely many steps. By construction, S_q is a basis for V. ∎

EXERCISES 4.10

1. Extend the set $S = \left\{ \begin{pmatrix} 1 \\ 2 \end{pmatrix} \right\}$ to a basis for $\mathbb{R}^2$.

2. Extend the set $S = \left\{ \begin{pmatrix} 1 \\ 2 \\ 3 \end{pmatrix} \right\}$ to a basis for $\mathbb{R}^3$.

3. Extend the set $S = \left\{ \begin{pmatrix} 1 \\ 2 \\ 3 \\ 4 \end{pmatrix}, \begin{pmatrix} 5 \\ 6 \\ 7 \\ 8 \end{pmatrix} \right\}$ to a basis for $\mathbb{R}^4$.

4. Select a linearly independent subset S_0 from the set

$$S = \left\{ \begin{pmatrix} 1 \\ 2 \\ 3 \end{pmatrix}, \begin{pmatrix} 2 \\ 4 \\ 6 \end{pmatrix}, \begin{pmatrix} -1 \\ -2 \\ -3 \end{pmatrix} \right\}$$

and extend S_0 to a basis for $\mathbb{R}^3$.

5. Select a linearly independent subset S_0 from the set

$$S = \left\{ \begin{pmatrix} 1 \\ 2 \\ 3 \end{pmatrix}, \begin{pmatrix} 4 \\ 5 \\ 6 \end{pmatrix}, \begin{pmatrix} -10 \\ -11 \\ -12 \end{pmatrix} \right\}$$

and extend S_0 to a basis for $\mathbb{R}^3$.

6. Contract the set $S = \left\{ \begin{pmatrix} 1 \\ 2 \end{pmatrix}, \begin{pmatrix} 8 \\ 16 \end{pmatrix}, \begin{pmatrix} 2 \\ 1 \end{pmatrix} \right\}$ to a basis for $\mathbb{R}^2$.

7. Contract the set $S = \left\{ \begin{pmatrix} 1 \\ 2 \\ 3 \end{pmatrix}, \begin{pmatrix} 1 \\ 0 \\ 1 \end{pmatrix}, \begin{pmatrix} -8 \\ 5 \\ 7 \end{pmatrix}, \begin{pmatrix} 4 \\ 4 \\ 4 \end{pmatrix} \right\}$ to a basis for $\mathbb{R}^3$.

8. Extend the set $S = \{t, t^2\}$ to a basis for $\mathbb{R}_2[t]$.

9. Extend the set $S = \{1, t^2\}$ to a basis for $\mathbb{R}_4[t]$.

10. Contract the set $S = \{1, t, t^5, t^2, t + t^2, t^3, t^4\}$ to a basis for $\mathbb{R}_5[t]$.

11. Contract the set $S = \{1, t, t^5, 3t^2, t + t^2, t^3, 8t^4\}$ to a basis for $\mathbb{R}_5[t]$.

12. Let

$$S_1 = \left\{ \begin{pmatrix} 1 \\ 2 \\ 3 \end{pmatrix}, \begin{pmatrix} 2 \\ 2 \\ 2 \end{pmatrix} \right\} \quad \text{and} \quad S_2 = \left\{ \begin{pmatrix} 3 \\ 6 \\ 9 \end{pmatrix}, \begin{pmatrix} 2 \\ 4 \\ 6 \end{pmatrix}, \begin{pmatrix} 1 \\ 0 \\ 0 \end{pmatrix} \right\}$$

be two sets of vectors of $\mathbb{R}^3$. The linearly independent set S_1 is too small to span $\mathbb{R}^3$, and the set S_2 is not linearly independent. Find a linearly independent subset of $S_1 \cup S_2$ that spans $\mathbb{R}^3$.

13. Show that the set $\mathcal{B} = \left\{ \begin{pmatrix} 1 \\ 0 \\ 1 \end{pmatrix}, \begin{pmatrix} 0 \\ 2 \\ 0 \end{pmatrix}, \begin{pmatrix} 0 \\ 0 \\ 5 \end{pmatrix} \right\}$ is a basis for $\mathbb{R}^3$.

14. Show that the set

$$S = \left\{ \begin{pmatrix} 1 \\ 2 \\ 1 \end{pmatrix}, \begin{pmatrix} 4 \\ 4 \\ 4 \end{pmatrix}, \begin{pmatrix} 1 \\ 0 \\ 1 \end{pmatrix}, \begin{pmatrix} 2 \\ 4 \\ 2 \end{pmatrix}, \begin{pmatrix} 0 \\ 1 \\ 0 \end{pmatrix} \right\}$$

does not span $\mathbb{R}^3$.

15. Verify that the set

$$S = \left\{ \begin{pmatrix} 1 \\ 2 \\ 1 \end{pmatrix}, \begin{pmatrix} 4 \\ 4 \\ 4 \end{pmatrix}, \begin{pmatrix} 1 \\ 0 \\ 1 \end{pmatrix}, \begin{pmatrix} 2 \\ 4 \\ 2 \end{pmatrix}, \begin{pmatrix} 0 \\ 1 \\ 1 \end{pmatrix} \right\}$$

spans $\mathbb{R}^3$.

16. Extend the set $S = \left\{ \begin{pmatrix} 1 & 0 \\ 1 & 2 \end{pmatrix} \right\}$ to a basis for the space $\mathbb{R}^{2 \times 2}$ of real 2×2 matrices.

17. Let $\mathcal{S}$ be the set

$$\left\{ \begin{pmatrix} 1 & 0 \\ 1 & 2 \end{pmatrix}, \begin{pmatrix} 0 & 0 \\ 0 & 0 \end{pmatrix}, \begin{pmatrix} 1 & 0 \\ 0 & 1 \end{pmatrix}, \begin{pmatrix} 0 & 0 \\ 1 & 0 \end{pmatrix}, \begin{pmatrix} 0 & 1 \\ 0 & 0 \end{pmatrix}, \begin{pmatrix} 1 & 0 \\ 0 & 0 \end{pmatrix} \right\}$$

Verify that $\mathcal{S}$ spans the space $\mathbb{R}^{2\times 2}$ and contract $\mathcal{S}$ to a basis for $\mathbb{R}^{2\times 2}$.

18. Extend the set $\mathcal{S} = \left\{ \begin{pmatrix} 1 & 2 \\ 3 & 4 \end{pmatrix}, \begin{pmatrix} 0 & 0 \\ 1 & 0 \end{pmatrix}, \begin{pmatrix} 1 & 1 \\ 0 & 0 \end{pmatrix} \right\}$ to a basis for $\mathbb{R}^{2\times 2}$.

19. Show that the space of real numbers is a vector space of dimension 1.

20. Prove that if V is an n-dimensional vector space and $\mathcal{S} = \{\mathbf{x}_1, \ldots, \mathbf{x}_n\}$ is a subset of V that spans V, then $\mathcal{S}$ is a basis for V.

21. Prove that if V is an n-dimensional vector space and $\mathcal{S} = \{\mathbf{x}_1, \ldots, \mathbf{x}_n\}$ is a linearly independent subset of V, then $\mathcal{S}$ is a basis for V.

Bases and Invertible Matrices

The coordinate spaces $\mathbb{R}^n$ have an abundant supply of bases determined by invertible matrices.

THEOREM 4.13 (Bases and invertible matrix theorem)   *Every invertible real $n \times n$ matrix determines a basis for $\mathbb{R}^n$.*

Proof. Let $A = [\mathbf{x}_1 \; \cdots \; \mathbf{x}_n]$ be an invertible real $n \times n$ matrix with columns $\mathbf{x}_i$. Then Corollary 2.20 tells us that A has n pivots. By Corollary 4.4, the set of column vectors $\mathcal{B} = \{\mathbf{x}_1, \ldots, \mathbf{x}_n\}$ is linearly independent. Since the dimension of $\mathbb{R}^n$ is n, it is clear that $\mathcal{B}$ also spans $\mathbb{R}^n$. ∎

EXAMPLE 4.34 ■ A Basis Built from an Invertible Matrix

Use *MATHEMATICA* and invertible matrices to construct a basis for $\mathbb{R}^4$.

Solution. We begin by generating random 4×4 matrices until we have found an invertible one.

```
In[1]:= A=Array[Random[Integer,{-6,6}]&,{4,4}]

Out[1]= {{1,-5,0,3},{4,0,5,1},{-6,6,6,3},{0,1,2,3}}
```

If we apply the **Inverse** function to the matrix A, *MATHEMATICA* produces an actual matrix. This shows that A is invertible. Hence we can use it to construct a basis. To be able to read off the columns of A, we write A in matrix form.

```
In[2]:= MatrixForm[A]

Out[2]//MatrixForm=
```

$$\begin{pmatrix} 1 & -5 & 0 & 3 \\ 4 & 0 & 5 & 1 \\ -6 & 6 & 6 & 3 \\ 0 & 1 & 2 & 3 \end{pmatrix}$$

It follows that the set

$$\mathcal{B} = \left\{ \begin{pmatrix} 1 \\ 4 \\ -6 \\ 0 \end{pmatrix}, \begin{pmatrix} -5 \\ 0 \\ 6 \\ 1 \end{pmatrix}, \begin{pmatrix} 0 \\ 5 \\ 6 \\ 2 \end{pmatrix}, \begin{pmatrix} 3 \\ 1 \\ 3 \\ 3 \end{pmatrix} \right\}$$

is a basis. ◄►

EXERCISES 4.11

1. Use the **Array** and **Random** functions of *MATHEMATICA* to generate an invertible 4×4 real matrix A. It may take more than one attempt to obtain an invertible matrix. Show that the columns of A and the columns of A^{-1} are bases for $\mathbb{R}^4$.

2. Verify that the columns of the matrix

$$A = \begin{pmatrix} -8 & -5 & 7 & 7 & -9 \\ -5 & 4 & -9 & -8 & 4 \\ 7 & -9 & 5 & 3 & 0 \\ 7 & -8 & 3 & 5 & -1 \\ -9 & 4 & 0 & -1 & -5 \end{pmatrix}$$

form a basis for $\mathbb{R}^5$ by showing that A is invertible.

3. Verify that the matrix

$$A = \begin{pmatrix} 19 & 15 & 21 & 28 \\ 0 & 34 & 3 & 35 \\ 0 & 0 & 8 & 9 \\ 0 & 0 & 0 & 35 \end{pmatrix}$$

is invertible by showing that the columns of A form a basis for $\mathbb{R}^4$.

Coordinate Vectors

To link linear algebra with matrix algebra, we must find a way of representing vectors numerically. This can be done by choosing a basis and writing each vector as a linear combination of basis vectors. The coefficients arising in this way are the required scalars. Table 3 lists the required components of the coordinate vector construction.

TABLE 3   Coordinate Vectors.

Objects	Explanations
$\mathcal{B} = \{\mathbf{v}_1, \ldots, \mathbf{v}_n\}$	Basis for V
$\mathbf{x}$	Vector in V
$\mathbf{x} = a_1\mathbf{v}_1 + \cdots + a_n\mathbf{v}_n$	Linear combination in V
$[\mathbf{x}]_{\mathcal{B}} = \begin{pmatrix} a_1 \\ \vdots \\ a_n \end{pmatrix}$	Coordinate vector in $\mathbb{R}^n$

DEFINITION 4.8   *The **coordinate vector** of a vector $\mathbf{x}$ in an n-dimensional real vector space V relative to a basis $\mathcal{B} = \{\mathbf{v}_1, \ldots, \mathbf{v}_n\}$ for V is the unique column vector*

$$[\mathbf{x}]_{\mathcal{B}} = \begin{pmatrix} a_1 \\ \vdots \\ a_n \end{pmatrix} \in \mathbb{R}^n$$

for which $\mathbf{x} = a_1\mathbf{v}_1 + \cdots + a_n\mathbf{v}_n$.

Since the coordinate vector $[\mathbf{x}]_{\mathcal{B}}$ of a vector $\mathbf{x}$ is unique, we can recover $\mathbf{x}$ from $[\mathbf{x}]_{\mathcal{B}}$ in an easy way. Each column vector

$$\mathbf{y} = \begin{pmatrix} b_1 \\ \vdots \\ b_n \end{pmatrix} \in \mathbb{R}^n$$

determines a unique linear combination $\hat{\mathbf{y}}_{\mathcal{B}} = b_1\mathbf{v}_1 + \cdots + b_n\mathbf{v}_n \in V$. Figure 2 illustrates the link between coordinate vectors and linear combinations in $\mathbb{R}^2$.

The function $\mathbf{x} \to [\mathbf{x}]_{\mathcal{B}}$ that assigns to each vector $\mathbf{x} \in V$ the coordinate vector $[\mathbf{x}]_{\mathcal{B}} \in \mathbb{R}^n$ is the **coordinate vector function** from V to $\mathbb{R}^n$ determined by the basis $\mathcal{B}$ for V. Conversely, the function $\mathbf{y} \to \hat{\mathbf{y}}_{\mathcal{B}}$ that assigns to each column vector $\mathbf{y} \in \mathbb{R}^n$ the linear combination $\hat{\mathbf{y}}_{\mathcal{B}} \in V$ is the **linear combination function** from $\mathbb{R}^n$ to V determined by the basis $\mathcal{B}$ for V.

THEOREM 4.14 (Coordinate function theorem)   *The functions $\mathbf{x} \to [\mathbf{x}]_{\mathcal{B}}$ and $\mathbf{y} \to \hat{\mathbf{y}}_{\mathcal{B}}$ are inverses of each other.*

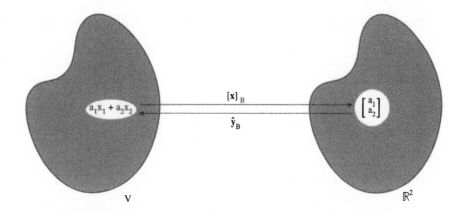

FIGURE 2   Coordinate Vector and Linear Combination.

Proof. The calculations required to prove this theorem are easy. By definition,

$$\mathbf{x} = a_1\mathbf{v}_1 + \cdots + a_n\mathbf{v}_n$$

$$\rightarrow \begin{pmatrix} a_1 \\ \vdots \\ a_n \end{pmatrix} = [\mathbf{x}]_\mathcal{B} = \mathbf{y}$$

$$\rightarrow \hat{\mathbf{y}}_\mathcal{B} = a_1\mathbf{v}_1 + \cdots + a_n\mathbf{v}_n$$

and

$$\mathbf{y} = \begin{pmatrix} a_1 \\ \vdots \\ a_n \end{pmatrix}$$

$$\rightarrow \hat{\mathbf{y}}_\mathcal{B} = a_1\mathbf{v}_1 + \cdots + a_n\mathbf{v}_n$$

$$\rightarrow [a_1\mathbf{v}_1 + \cdots + a_n\mathbf{v}_n]_\mathcal{B}$$

$$= \mathbf{y}$$

Therefore, the coordinate vector functions are inverses of each other. ∎

The coordinate vector and linear combination functions are useful because they are inverses of each other and because they preserve vector addition and scalar multiplication in the sense of the following theorem.

THEOREM 4.15   *The coordinate vector function* $\mathbf{x} \rightarrow [\mathbf{x}]_\mathcal{B}$ *from V to* $\mathbb{R}^n$ *has the property that* $[\mathbf{x} + \mathbf{y}]_\mathcal{B} = [\mathbf{x}]_\mathcal{B} + [\mathbf{y}]_\mathcal{B}$ *and that* $[a\mathbf{x}]_\mathcal{B} = a[\mathbf{x}]_\mathcal{B}$.

Proof. Let $\mathbf{x} = a_1\mathbf{v}_1 + \cdots + a_n\mathbf{v}_n$ and $\mathbf{y} = b_1\mathbf{v}_1 + \cdots + b_n\mathbf{v}_n$ be two vectors in V and let

$$\mathbf{x} + \mathbf{y} = (a_1 + b_1)\mathbf{v}_1 + \cdots + (a_n + b_n)\mathbf{v}_n$$

be their sum. Then

$$[\mathbf{x} + \mathbf{y}]_{\mathcal{B}} = \begin{pmatrix} a_1 + b_1 \\ \vdots \\ a_n + b_n \end{pmatrix} = \begin{pmatrix} a_1 \\ \vdots \\ a_n \end{pmatrix} + \begin{pmatrix} b_1 \\ \vdots \\ b_n \end{pmatrix} = [\mathbf{x}]_{\mathcal{B}} + [\mathbf{y}]_{\mathcal{B}}$$

Moreover, for any scalar a,

$$a\mathbf{x} = a(a_1\mathbf{v}_1 + \cdots + a_n\mathbf{v}_n) = (aa_1)\mathbf{v}_1 + \cdots + (aa_n)\mathbf{v}_n$$

so that

$$[a\mathbf{x}]_{\mathcal{B}} = \begin{pmatrix} aa_1 \\ \vdots \\ aa_n \end{pmatrix} = a \begin{pmatrix} a_1 \\ \vdots \\ a_n \end{pmatrix} = a[\mathbf{x}]_{\mathcal{B}}$$

This proves the theorem. ■

COROLLARY 4.16   *The linear combination function $\mathbf{y} \rightarrow \widehat{\mathbf{y}}_B$ from $\mathbb{R}^n$ to V has the property that $\widehat{\mathbf{x} + \mathbf{y}}_B = \widehat{\mathbf{x}}_B + \widehat{\mathbf{y}}_B$ and that $\widehat{a\mathbf{x}}_B = a\widehat{\mathbf{x}}_B$.*

Proof. By reversing the conversion steps in the proof of Theorem 4.15, we obtain the stated equalities. ■

A coordinate vector function is an example of a ***linear transformation***. Its key property is the fact that it "preserves" linear combinations of vectors.

The obvious question to ask is what connection exists between the coordinate vectors $[\mathbf{x}]_{\mathcal{B}}$ and $[\mathbf{x}]_{\mathcal{C}}$ of a given vector $\mathbf{x}$ relative to two different bases $\mathcal{B}$ and $\mathcal{C}$ for V.

The answer turns out to be quite simple and uses the idea of coordinate conversion. Table 4 lists the components required for the construction of a coordinate conversion matrix.

THEOREM 4.17 (Coordinate conversion theorem)   *Let V be an n-dimensional real vector space, let $\mathcal{B}$ and $\mathcal{C}$ be two bases for V, and let $\mathbf{x}$ be a vector in V with coordinate vectors $[\mathbf{x}]_{\mathcal{B}}$ and $[\mathbf{x}]_{\mathcal{C}}$. Then there exists an invertible matrix P for which $P[\mathbf{x}]_{\mathcal{B}} = [\mathbf{x}]_{\mathcal{C}}$ and $P^{-1}[\mathbf{x}]_{\mathcal{C}} = [\mathbf{x}]_{\mathcal{B}}$.*

Proof. Let $\mathcal{B} = \{\mathbf{v}_1, \ldots, \mathbf{v}_n\}$ and $\mathcal{C} = \{\mathbf{w}_1, \ldots, \mathbf{w}_n\}$, and suppose that $\mathbf{x} = a_1\mathbf{v}_1 + \cdots + a_n\mathbf{v}_n$ is any vector in V, written as a linear combination in the basis $\mathcal{B}$. From the fact that the coordinate vector function $\mathbf{x} \rightarrow [\mathbf{x}]_{\mathcal{C}}$ preserves vector addition and scalar multiplication, it follows that

$$[\mathbf{x}]_{\mathcal{C}} = [a_1\mathbf{v}_1 + \cdots + a_n\mathbf{v}_n]_{\mathcal{C}}$$
$$= a_1[\mathbf{v}_1]_{\mathcal{C}} + \cdots + a_n[\mathbf{v}_n]_{\mathcal{C}}$$

TABLE 4   Coordinate Conversion Matrix.

Objects	Explanations
$\mathcal{B} = \{\mathbf{v}_1, \ldots, \mathbf{v}_n\}$	Basis for $\mathbb{R}^n$
$\mathcal{C} = \{\mathbf{w}_1, \ldots, \mathbf{w}_n\}$	Basis for $\mathbb{R}^n$
$[\mathbf{v}_1]_{\mathcal{C}}, \ldots, [\mathbf{v}_n]_{\mathcal{C}}$	Coordinate vectors in $\mathbb{R}^n$
$P = \Big([\mathbf{v}_1]_{\mathcal{C}} \quad \cdots \quad [\mathbf{v}_n]_{\mathcal{C}} \Big)$	Conversion matrix from $\mathcal{B}$ to $\mathcal{C}$
$P[\mathbf{x}]_{\mathcal{B}} = [\mathbf{x}]_{\mathcal{C}}$	Action of P
$[\mathbf{w}_1]_{\mathcal{B}}, \ldots, [\mathbf{w}_n]_{\mathcal{B}}$	Coordinate vectors in $\mathbb{R}^n$
$P^{-1} = \Big([\mathbf{w}_1]_{\mathcal{B}} \quad \cdots \quad [\mathbf{w}_n]_{\mathcal{B}} \Big)$	Conversion matrix from $\mathcal{C}$ to $\mathcal{B}$
$P^{-1}[\mathbf{y}]_{\mathcal{B}} = [\mathbf{y}]_{\mathcal{C}}$	Action of P^{-1}

$$= \Big([\mathbf{v}_1]_{\mathcal{C}} \quad \cdots \quad [\mathbf{v}_n]_{\mathcal{C}} \Big) \begin{pmatrix} a_1 \\ \vdots \\ a_n \end{pmatrix}$$

$$= \Big([\mathbf{v}_1]_{\mathcal{C}} \quad \cdots \quad [\mathbf{v}_n]_{\mathcal{C}} \Big) [\mathbf{x}]_{\mathcal{B}}$$

$$= P[\mathbf{x}]_{\mathcal{B}}$$

Conversely, let $\mathbf{y} = b_1\mathbf{w}_1 + \cdots + b_n\mathbf{w}_n$ be any vector in V, written as a linear combination in the basis $\mathcal{C}$. From the fact that the coordinate vector function $\mathbf{y} \rightarrow [\mathbf{y}]_{\mathcal{B}}$ preserves vector addition and scalar multiplication, it follows that

$$[\mathbf{y}]_{\mathcal{B}} = [b_1\mathbf{w}_1 + \cdots + b_n\mathbf{w}_n]_{\mathcal{B}}$$

$$= b_1[\mathbf{w}_1]_{\mathcal{B}} + \cdots + b_n[\mathbf{w}_n]_{\mathcal{B}}$$

$$= \Big([\mathbf{w}_1]_{\mathcal{B}} \quad \cdots \quad [\mathbf{w}_n]_{\mathcal{B}} \Big) \begin{pmatrix} b_1 \\ \vdots \\ b_n \end{pmatrix}$$

$$= \Big([\mathbf{w}_1]_{\mathcal{B}} \quad \cdots \quad [\mathbf{w}_n]_{\mathcal{B}} \Big) [\mathbf{y}]_{\mathcal{C}}$$

$$= Q[\mathbf{y}]_{\mathcal{C}}$$

This means that $Q = P^{-1}$. ∎

DEFINITION 4.9   *The $n \times n$ matrices P and P^{-1} with the property that $P[\mathbf{x}]_{\mathcal{B}} = [\mathbf{x}]_{\mathcal{C}}$ and $P^{-1}[\mathbf{x}]_{\mathcal{C}} = [\mathbf{x}]_{\mathcal{B}}$ are the **coordinate conversion matrices** from $\mathcal{B}$ to $\mathcal{C}$, and from $\mathcal{C}$ to $\mathcal{B}$, respectively.*

Every coordinate space $\mathbb{R}^n$ has an unlimited supply of such matrices.

THEOREM 4.18 *Every invertible real $n \times n$ matrix is a coordinate conversion matrix for $\mathbb{R}^n$.*

Proof. Let $A = (\mathbf{x}_1 \ \cdots \ \mathbf{x}_n)$ be an invertible real $n \times n$ matrix with columns $\mathbf{x}_i$ and let $\mathbf{x}$ be any vector in $\mathbb{R}^n$. By Theorem 4.13, the associated set of vectors $\mathcal{B} = \{\mathbf{x}_1, \ldots, \mathbf{x}_n\}$ is a basis for $\mathbb{R}^n$. Therefore, there exist unique scalars $a_1, \ldots, a_n$ for which $\mathbf{x} = a_1\mathbf{x}_1 + \cdots + a_n\mathbf{x}_n$. This means that

$$\mathbf{x} = (\mathbf{x}_1 \ \cdots \ \mathbf{x}_n) \begin{pmatrix} a_1 \\ \vdots \\ a_n \end{pmatrix} = A[\mathbf{x}]_\mathcal{B}$$

Since A is invertible, $[\mathbf{x}]_\mathcal{B} = A^{-1}\mathbf{x}$. Therefore, A is a coordinate conversion matrix from $\mathcal{B}$ to the standard basis for $\mathbb{R}^n$. ∎

EXAMPLE 4.35 ■ A Coordinate Conversion in $\mathbb{R}^2$

Discuss the coordinate conversion on $\mathbb{R}^2$ determined by the invertible matrix

$$A = \begin{pmatrix} 2 & 1 \\ 1 & 3 \end{pmatrix}$$

Solution. Since the determinant of A is not zero, A is an invertible matrix. By Theorem 4.18, the matrix A therefore determines a coordinate conversion from the basis

$$\mathcal{B} = \left\{ \mathbf{u} = \begin{pmatrix} 2 \\ 1 \end{pmatrix}, \mathbf{v} = \begin{pmatrix} 1 \\ 3 \end{pmatrix} \right\}$$

to the standard basis

$$\mathcal{E} = \left\{ \mathbf{x} = \begin{pmatrix} 1 \\ 0 \end{pmatrix}, \mathbf{y} = \begin{pmatrix} 0 \\ 1 \end{pmatrix} \right\}$$

of $\mathbb{R}^2$. Figure 3 illustrates the link between the coordinate vectors

$$\begin{pmatrix} 3 \\ 4 \end{pmatrix}_\mathcal{E} \quad \text{and} \quad \begin{pmatrix} 1 \\ 1 \end{pmatrix}_\mathcal{B}$$

created by the matrix A.

We can see geometrically that

$$\begin{pmatrix} 1 \\ 1 \end{pmatrix}_\mathcal{B} = \mathbf{u} + \mathbf{v} = \begin{pmatrix} 2 \\ 1 \end{pmatrix}_\mathcal{E} + \begin{pmatrix} 1 \\ 3 \end{pmatrix}_\mathcal{E} = \begin{pmatrix} 3 \\ 4 \end{pmatrix}_\mathcal{E}$$

The matrix equation

$$\begin{pmatrix} 2 & 1 \\ 1 & 3 \end{pmatrix} \begin{pmatrix} 1 \\ 1 \end{pmatrix} = \begin{pmatrix} 3 \\ 4 \end{pmatrix}$$

confirms algebraically that this relationship holds. ◀▶

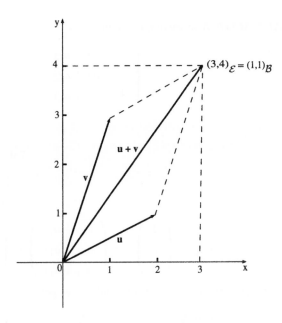

FIGURE 3   A Coordinate Conversion.

EXAMPLE 4.36   ■ A Coordinate Conversion in $\mathbb{R}^3$

Use *MATHEMATICA* to convert the basis vectors in

$$\mathcal{B} = \left\{ \mathbf{v}_1 = \begin{pmatrix} 1 \\ 2 \\ 3 \end{pmatrix}, \mathbf{v}_2 = \begin{pmatrix} 0 \\ 1 \\ 0 \end{pmatrix}, \mathbf{v}_3 = \begin{pmatrix} 1 \\ 0 \\ 1 \end{pmatrix} \right\}$$

to coordinate vectors in the basis

$$\mathcal{C} = \left\{ \mathbf{w}_1 = \begin{pmatrix} 0 \\ 1 \\ 1 \end{pmatrix}, \mathbf{w}_2 = \begin{pmatrix} 1 \\ 1 \\ 0 \end{pmatrix}, \mathbf{w}_3 = \begin{pmatrix} 1 \\ 0 \\ 1 \end{pmatrix} \right\}$$

and use the result to create a coordinate conversion matrix from $\mathcal{B}$ to $\mathcal{C}$.

Solution. We begin by computing the coordinate vector $[\mathbf{v}_1]_{\mathcal{C}}$. Since

$$\begin{pmatrix} 1 \\ 2 \\ 3 \end{pmatrix} = a_{11} \begin{pmatrix} 0 \\ 1 \\ 1 \end{pmatrix} + a_{12} \begin{pmatrix} 1 \\ 1 \\ 0 \end{pmatrix} + a_{13} \begin{pmatrix} 1 \\ 0 \\ 1 \end{pmatrix} = \begin{pmatrix} a_{12} + a_{13} \\ a_{11} + a_{12} \\ a_{11} + a_{13} \end{pmatrix}$$

we use *MATHEMATICA* to solve the system

$$\begin{cases} 1 = a_{12} + a_{13} \\ 2 = a_{11} + a_{12} \\ 3 = a_{11} + a_{13} \end{cases}$$

$\text{In[1]:= } \textbf{Solve[\{a}_{12}\textbf{+a}_{13}\textbf{==1,a}_{11}\textbf{+a}_{12}\textbf{==2,a}_{11}\textbf{+a}_{13}\textbf{==3\},\{a}_{11}\textbf{,a}_{12}\textbf{,a}_{13}\textbf{\}]}$

$\text{Out[1]= } \{\{a_{13} \to 1, a_{11} \to 2, a_{12} \to 0\}\}$

Therefore,

$$[\mathbf{v}_1]_{\mathcal{C}} = \begin{pmatrix} 1 \\ 2 \\ 3 \end{pmatrix}_{\mathcal{C}} = \begin{pmatrix} a_{12} + a_{13} \\ a_{11} + a_{12} \\ a_{11} + a_{13} \end{pmatrix} = \begin{pmatrix} 2 \\ 0 \\ 1 \end{pmatrix}$$

By repeating the previous steps for $\mathbf{v}_2$ and $\mathbf{v}_3$, we get

$$[\mathbf{v}_2]_{\mathcal{C}} = \begin{pmatrix} 0 \\ 1 \\ 0 \end{pmatrix}_{\mathcal{C}} = a_{21} \begin{pmatrix} 0 \\ 1 \\ 1 \end{pmatrix} + a_{22} \begin{pmatrix} 1 \\ 1 \\ 0 \end{pmatrix} + a_{23} \begin{pmatrix} 1 \\ 0 \\ 1 \end{pmatrix}$$

$$= \begin{pmatrix} a_{22} + a_{23} \\ a_{21} + a_{22} \\ a_{21} + a_{23} \end{pmatrix} = \begin{pmatrix} \frac{1}{2} \\ \frac{1}{2} \\ -\frac{1}{2} \end{pmatrix}$$

and

$$[\mathbf{v}_3]_{\mathcal{C}} = \begin{pmatrix} 1 \\ 0 \\ 1 \end{pmatrix}_{\mathcal{C}} = a_{31} \begin{pmatrix} 0 \\ 1 \\ 1 \end{pmatrix} + a_{32} \begin{pmatrix} 1 \\ 1 \\ 0 \end{pmatrix} + a_{33} \begin{pmatrix} 1 \\ 0 \\ 1 \end{pmatrix}$$

$$= \begin{pmatrix} a_{32} + a_{33} \\ a_{31} + a_{32} \\ a_{31} + a_{33} \end{pmatrix} = \begin{pmatrix} 0 \\ 0 \\ 1 \end{pmatrix}$$

Using these results, we can build the coordinate conversion matrix P from $\mathcal{B}$ to $\mathcal{C}$.

$$P = \begin{pmatrix} [\mathbf{v}_1]_{\mathcal{C}} & [\mathbf{v}_2]_{\mathcal{C}} & [\mathbf{v}_3]_{\mathcal{C}} \end{pmatrix} = \begin{pmatrix} 2 & \frac{1}{2} & 0 \\ 0 & \frac{1}{2} & 0 \\ 1 & -\frac{1}{2} & 1 \end{pmatrix}$$

As expected, we get

$$P[\mathbf{v}_1]_{\mathcal{B}} = \begin{pmatrix} 2 & \frac{1}{2} & 0 \\ 0 & \frac{1}{2} & 0 \\ 1 & -\frac{1}{2} & 1 \end{pmatrix} \begin{pmatrix} 1 \\ 0 \\ 0 \end{pmatrix} = \begin{pmatrix} 2 \\ 0 \\ 1 \end{pmatrix} = [\mathbf{v}_1]_{\mathcal{C}}$$

$$P[v_2]_\mathcal{B} = \begin{pmatrix} 2 & \frac{1}{2} & 0 \\ 0 & \frac{1}{2} & 0 \\ 1 & -\frac{1}{2} & 1 \end{pmatrix} \begin{pmatrix} 0 \\ 1 \\ 0 \end{pmatrix} = \begin{pmatrix} \frac{1}{2} \\ \frac{1}{2} \\ -\frac{1}{2} \end{pmatrix} = [v_2]_\mathcal{C}$$

$$P[v_3]_\mathcal{B} = \begin{pmatrix} 2 & \frac{1}{2} & 0 \\ 0 & \frac{1}{2} & 0 \\ 1 & -\frac{1}{2} & 1 \end{pmatrix} \begin{pmatrix} 0 \\ 0 \\ 1 \end{pmatrix} = \begin{pmatrix} 0 \\ 0 \\ 1 \end{pmatrix} = [v_3]_\mathcal{C}$$

Next we use P^{-1}, the coordinate conversion matrix from $\mathcal{C}$ to $\mathcal{B}$, to reverse these calculations.

In[2]:= **MatrixForm**$\left[\text{**Inverse**}\left[\left(\begin{matrix} 2 & \frac{1}{2} & 0 \\ 0 & \frac{1}{2} & 0 \\ 1 & -\frac{1}{2} & 1 \end{matrix}\right)\right]\right]$

Out[2]//MatrixForm=

$$\begin{pmatrix} \frac{1}{2} & -\frac{1}{2} & 0 \\ 0 & 2 & 0 \\ -\frac{1}{2} & \frac{3}{2} & 1 \end{pmatrix}$$

As expected, we get

$$P^{-1}[v_1]_\mathcal{C} = \begin{pmatrix} \frac{1}{2} & -\frac{1}{2} & 0 \\ 0 & 2 & 0 \\ -\frac{1}{2} & \frac{3}{2} & 1 \end{pmatrix} \begin{pmatrix} 2 \\ 0 \\ 1 \end{pmatrix} = \begin{pmatrix} 1 \\ 0 \\ 0 \end{pmatrix} = [v_1]_\mathcal{B}$$

$$P^{-1}[v_2]_\mathcal{C} = \begin{pmatrix} \frac{1}{2} & -\frac{1}{2} & 0 \\ 0 & 2 & 0 \\ -\frac{1}{2} & \frac{3}{2} & 1 \end{pmatrix} \begin{pmatrix} \frac{1}{2} \\ \frac{1}{2} \\ -\frac{1}{2} \end{pmatrix} = \begin{pmatrix} 0 \\ 1 \\ 0 \end{pmatrix} = [v_2]_\mathcal{B}$$

$$P^{-1}[v_3]_\mathcal{C} = \begin{pmatrix} \frac{1}{2} & -\frac{1}{2} & 0 \\ 0 & 2 & 0 \\ -\frac{1}{2} & \frac{3}{2} & 1 \end{pmatrix} \begin{pmatrix} 0 \\ 0 \\ 1 \end{pmatrix} = \begin{pmatrix} 0 \\ 0 \\ 1 \end{pmatrix} = [v_3]_\mathcal{B}$$

This completes the coordinate changes. ◄►

EXERCISES 4.12

1. Find the coordinate vectors in $\mathbb{R}^4$ of the polynomials $3 + 7t - t^2$, $4 + t^3$, $21t^2$, $1 + t + t^2$, and $3t^3 + 2t$ in $\mathbb{R}_3[t]$ in the standard bases.

2. Use the standard bases to define a coordinate vector function $[x]_\mathcal{B}$ from the space of real 2×3 matrices $\mathbb{R}^{2\times 3}$ to $\mathbb{R}^6$ and find the associated coordinate vectors of the following matrices.

a. $\begin{pmatrix} 4 & 1 & 5 \\ 4 & 3 & -4 \end{pmatrix}$ b. $\begin{pmatrix} 0 & 0 & 0 \\ 0 & 1 & 0 \end{pmatrix}$ c. $\begin{pmatrix} 1 & -1 & 1 \\ -1 & \frac{3}{5} & \pi \end{pmatrix}$

3. Let $\mathcal{E}$ be the standard basis for $\mathbb{R}^3$ and let

$$ \mathcal{B} = \left\{ \begin{pmatrix} 0 \\ 0 \\ 1 \end{pmatrix}, \begin{pmatrix} 0 \\ 1 \\ 1 \end{pmatrix}, \begin{pmatrix} 1 \\ 1 \\ 1 \end{pmatrix} \right\} $$

be a second basis.

a. Find the coordinate conversion matrix that converts coordinate vectors in the basis $\mathcal{E}$ to coordinate vectors in the basis $\mathcal{B}$.

b. Convert the coordinate vectors

$$ \begin{pmatrix} 1 \\ 2 \\ 3 \end{pmatrix}, \begin{pmatrix} -1 \\ 1 \\ -21 \end{pmatrix}, \begin{pmatrix} 2 \\ 2 \\ 2 \end{pmatrix}, \begin{pmatrix} 0 \\ 0 \\ 0 \end{pmatrix}, \begin{pmatrix} 3 \\ -7 \\ 1 \end{pmatrix} $$

from vectors in the standard basis to vectors in the basis $\mathcal{B}$.

c. Find the coordinate conversion matrix that converts coordinate vectors in the basis $\mathcal{B}$ to coordinate vectors in the basis $\mathcal{E}$.

d. Convert the coordinate vectors

$$ \begin{pmatrix} 1 \\ 2 \\ 3 \end{pmatrix}, \begin{pmatrix} -1 \\ 1 \\ -21 \end{pmatrix}, \begin{pmatrix} 2 \\ 2 \\ 2 \end{pmatrix}, \begin{pmatrix} 0 \\ 0 \\ 0 \end{pmatrix}, \begin{pmatrix} 3 \\ -7 \\ 1 \end{pmatrix} $$

from vectors in the basis $\mathcal{B}$ to vectors in the basis $\mathcal{E}$.

4. Explain why there is no coordinate conversion matrix from the standard basis for the polynomial space $\mathbb{R}[t]$ to any coordinate space $\mathbb{R}^n$.

5. Find the basis $\mathcal{B}$ for $\mathbb{R}^3$ for which

$$ \begin{pmatrix} 2 & 0 & 0 \\ 0 & 0 & 2 \\ 0 & 2 & 2 \end{pmatrix} $$

is the coordinate conversion matrix from the standard basis $\mathcal{E}$ to $\mathcal{B}$.

6. Use *MATHEMATICA* to find the coordinate conversion matrices for the bases $\mathcal{B} = \{\mathbf{v}_1 = (1, 2), \mathbf{v}_2 = (3, 4)\}$ and $\mathcal{C} = \{\mathbf{w}_1 = (1, 0), \mathbf{w}_2 = (0, 1)\}$ for $\mathbb{R}^2$.

7. Let $A = (a_{ij})$ be an invertible 3×3 real matrix, let $\mathcal{B}$ be the basis for $\mathbb{R}^3$ determined by A, and let $\mathcal{E}$ be the standard basis of $\mathbb{R}^3$. Show that

$$\begin{pmatrix} a_{11} + a_{12} + a_{13} \\ a_{21} + a_{22} + a_{23} \\ a_{31} + a_{32} + a_{33} \end{pmatrix}_{\mathcal{E}} = \begin{pmatrix} 1 \\ 1 \\ 1 \end{pmatrix}_{\mathcal{B}}$$

Explain this result geometrically.

SUBSPACES

Certain subsets of a vector space are vector spaces in their own right, with the same operations as those of the ambient space. For example, the coordinate planes

$$\mathcal{S}_{xy} = \left\{ \begin{pmatrix} x \\ y \\ 0 \end{pmatrix} : x, y \in \mathbb{R} \right\}, \mathcal{S}_{xz} = \left\{ \begin{pmatrix} x \\ 0 \\ z \end{pmatrix} : x, z \in \mathbb{R} \right\}, \mathcal{S}_{yz} = \left\{ \begin{pmatrix} 0 \\ y \\ z \end{pmatrix} : y, z \in \mathbb{R} \right\}$$

shown in Figure 4 are subspaces of $\mathbb{R}^3$.

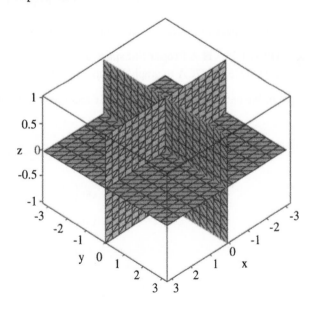

FIGURE 4   The Coordinate-Plane Subspaces of $\mathbb{R}^3$.

The vector space operations of these spaces are those of $\mathbb{R}^3$, restricted to $\mathcal{S}_{xy}$, $\mathcal{S}_{xz}$, and $\mathcal{S}_{xy}$. The operations on $\mathcal{S}_{xy}$, for example, are

$$\begin{pmatrix} x_1 \\ y_1 \\ 0 \end{pmatrix} + \begin{pmatrix} x_2 \\ y_2 \\ 0 \end{pmatrix} = \begin{pmatrix} x_1 + x_2 \\ y_1 + y_2 \\ 0 \end{pmatrix} \quad \text{and} \quad a \begin{pmatrix} x \\ y \\ 0 \end{pmatrix} = \begin{pmatrix} ax \\ ay \\ 0 \end{pmatrix}$$

Since the set S_{xy} is a proper subset of $\mathbb{R}^3$, it does not contain all vectors of $\mathbb{R}^3$. Nevertheless, the sum of two vectors $\mathbf{x} + \mathbf{y}$ in S_{xy} is the same as the sum of these vectors in $\mathbb{R}^3$. Moreover, any scalar multiple $a\mathbf{x}$ of a vector $\mathbf{x}$ in S_{xy} is the same as the scalar multiple $a\mathbf{x}$ in $\mathbb{R}^3$. The set S_{xy} is an example of a *subspace* of $\mathbb{R}^3$.

DEFINITION 4.10   *A nonempty subset W of a vector space V is a **subspace** of V if for all vectors* $\mathbf{x}$ *and* $\mathbf{y}$ *in W and all scalars a and b, the linear combination* $a\mathbf{x} + b\mathbf{y}$ *belongs to W.*

We note that the fact that subspaces are nonempty implies that they contain the zero vector of V. If $\mathbf{x}$ is any vector in W, $0\mathbf{x} + 0\mathbf{x} = \mathbf{0}$ belongs to W. Moreover, subspaces are vector spaces in their own right. First of all, it is implicit in the definition that W and V have the same set of scalars. Secondly, the vector space operation $a\mathbf{x}$ and $\mathbf{x} + \mathbf{y}$ of W are inherited from V. They therefore satisfy the vector space axioms in W since they satisfy these axioms in V.

EXAMPLE 4.37   ■ **The Zero Subspace**

For any vector space V, the space $\{\mathbf{0}\} \subseteq V$ is a subspace of V. ◄►

EXAMPLE 4.38   ■ **The Ambient Space**

It is obvious from the definition that every vector space is a subspace of itself. ◄►

Every subspace of a vector space V other than V itself is called a ***proper subspace***.

EXAMPLE 4.39   ■ **A Proper Subspace of** $\mathbb{R}^2$

Use *MATHEMATICA* to graph the subspace of $\mathbb{R}^2$ determined by the equation $y = 17x$.

Solution. We enter the expression $y = 17x$ and invoke the **Plot** function.

```
In[1]:= Plot[y=17x,{x,-10,10}]
```

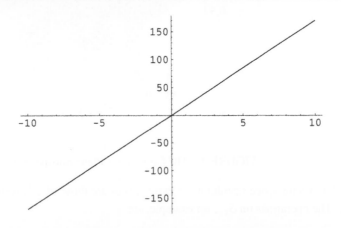

As we can see, the space W consists of a line through the origin of $\mathbb{R}^2$. ◄►

Here is a general method for constructing subspaces.

THEOREM 4.19 (Subspace theorem)   *The span of a subset W of a vector space V is a subspace of V.*

Proof. Suppose $W = \emptyset$ and that $\text{span}(W) = \{\mathbf{0}\}$. Then $\text{span}(W)$ is the zero subspace of V. Now suppose $W \neq \emptyset$, that $\mathbf{x}$ and $\mathbf{y}$ belong to $\text{span}(W)$, and that a and b are two scalars. Since $\text{span}(W)$ is closed under linear combinations, it follows that $0\mathbf{x} + 0\mathbf{y} = \mathbf{0} \in \text{span}(W)$ and $a\mathbf{x} + b\mathbf{y} \in \text{span}(W)$. Hence $\text{span}(W)$ is a subspace of V. ∎

Another method for constructing subspaces is to intersect them.

THEOREM 4.20 (Intersection theorem)   *If U and W are subspaces of a vector space V, then so is $U \cap W$.*

Proof. First we observe that since U and W are subspaces, $\mathbf{0} \in U \cap W$. Suppose that in addition, $\mathbf{x}, \mathbf{y} \in U \cap W$. Then $\mathbf{x}, \mathbf{y} \in U$ and $\mathbf{x}, \mathbf{y} \in W$. Since U and W are subspaces, $\mathbf{x} + \mathbf{y} \in U$ and for any scalar a, $a\mathbf{x} \in U$. Similarly $\mathbf{x} + \mathbf{y} \in W$ and $a\mathbf{x} \in W$. Hence $\mathbf{x} + \mathbf{y} \in U \cap W$ and $a\mathbf{x} \in U \cap W$ for all scalars a. ∎

If $U \cap W = \{\mathbf{0}\}$, we say that U and W are ***disjoint subspaces***. For example, the space $U = \{(x, 0) : x \in \mathbb{R}\}$ and $W = U = \{(0, y) : y \in \mathbb{R}\}$ are disjoint subspaces of $\mathbb{R}^2$.

Here is an example of the intersection of two subspaces of $\mathbb{R}^3$.

EXAMPLE 4.40 ■ **The Intersection of Two Subspaces**

Use *MATHEMATICA* to plot the two subspaces

$$U = \left\{(x, y, z) \in \mathbb{R}^3 : z = 3x + 17y\right\} \quad \text{and} \quad V = \left\{(x, y, z) \in \mathbb{R}^3 : z = x - y\right\}$$

of $\mathbb{R}^3$. Explain why U and V are not disjoint.

Solution. We use the **Plot3D** and **Show** functions. The commands

   a.   `p=Plot3D[z=3x+17y,{x,-10,10},{y,-10,10}]`
   b.   `q=Plot3D[z=x-y,{x,-10,10},{y,-10,10}]`
   c.   `Show[p,q]`

produce the graphs in Figures 5, 6, and 7. The graph in Figure 5 illustrates the subspace determined by the equation $z = 3x + 17y$.

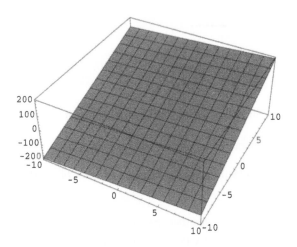

FIGURE 5   The Subspace U of $\mathbb{R}^3$.

The graph in Figure 6 illustrates the subspace determined by the equation $z = x - y$.

Figure 7, finally, depicts the subspace

$$L = U \cap V = \left\{ (x, y, z) \in \mathbb{R}^3 : 3x + 17y = x - y \right\} = \left\{ (x, -\tfrac{1}{9}x, z) : x, z \in \mathbb{R} \right\}$$

It consists of the line determined by the intersection of U and V.

As we can see, the solutions of the linear system

$$\begin{cases} x + 17y - z = 0 \\ x - y - z = 0 \end{cases}$$

form the line corresponding to the intersection of the spaces U and V. ◄►

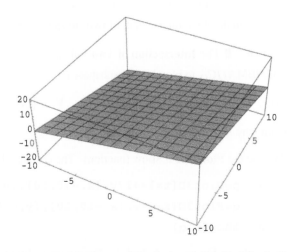

FIGURE 6   The Subspace V of $\mathbb{R}^3$.

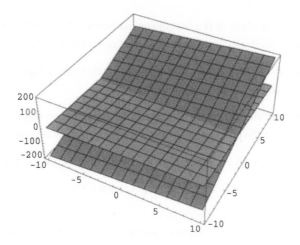

FIGURE 7   The Subspace $U \cap V$ of $\mathbb{R}^3$.

EXAMPLE 4.41 ■ The Subspaces of $\mathbb{R}^3$

Since all subspaces of $\mathbb{R}^3$ must contain the zero vector, it is clear from geometry that the only possible subspaces are lines and planes passing through the origin **0**, together with the zero subspace and $\mathbb{R}^3$ itself. ◄►

The situation where a subspace W of a finite-dimensional vector space V is all of V has a very simple description in terms of dimensions.

THEOREM 4.21   *If W is a subspace of a finite-dimensional vector space V, and W and V have the same dimensions, then $W = V$.*

Proof. We prove this theorem by deriving a contradiction from the assumption that $W \neq V$.

Suppose that span$(W) \neq V$ and that $\mathcal{B} = \{\mathbf{w}_1, \ldots, \mathbf{w}_n\}$ is a basis for W. Then there exists a vector $\mathbf{v} \in V$ not in W that is linearly independent of the vectors in $\mathcal{B}$. Therefore, the set $\mathcal{B}' = \{\mathbf{w}_1, \ldots, \mathbf{w}_n, \mathbf{v}\}$, consisting of $n + 1$ vectors, is linearly independent subset of V. This contradicts the fact that W and V have the same dimensions. ∎

EXERCISES 4.13

1. Show that the proper subspaces of $\mathbb{R}^2$ are the zero subspace and all lines through the origin.

2. Give a geometric description of the proper subspaces of $\mathbb{R}^3$.

3. Find bases for the following subspaces of $\mathbb{R}^3$.

      a. The plane $2x - y + 3z = 0$ b. $W = \{(a, a - b, b) : a, b \in \mathbb{R}\}$

      c. The line $x = t, y = -2t, z = 3t$ d. $(2x - y + 3z = 0) \cap (x + y - z = 0)$

4. Prove that if $A\mathbf{x} = 0$ is a homogeneous linear system with real coefficients consisting of m equations in n unknowns, then the set of solution vectors is a subspace of $\mathbb{R}^n$.

5. Determine whether the set of real polynomials of even degree is a subspace of the polynomial space $\mathbb{R}[t]$. Prove your assertion.

6. Determine whether the set of real polynomials of even degree less than or equal to 10 is a subspace of the polynomial space $\mathbb{R}_{15}[t]$. Prove your assertion.

7. Prove that the set of symmetric real $n \times n$ matrices is a subspace of $\mathbb{R}^{n \times n}$.

8. Determine whether the set of orthogonal $n \times n$ matrices is a subspace of $\mathbb{R}^{n \times n}$. Prove your assertion.

9. Determine whether the set of invertible $n \times n$ matrices is a subspace of $\mathbb{R}^{n \times n}$. Prove your assertion.

10. Show by means of an example that the union of two subspaces of $\mathbb{R}^3$ is not necessarily a subspace of $\mathbb{R}^3$.

Fundamental Subspaces

Let A be an $m \times n$ matrix over $\mathbb{R}$. Then A determines four important vector spaces, known as the *fundamental subspaces determined by* A. They are the column space, the null space, the row space, and the left null space. Two of the spaces are subspaces of $\mathbb{R}^m$ and two of them are subspaces of $\mathbb{R}^n$. It left as an exercise to show that the defined sets of vectors are subspaces.

DEFINITION 4.11   *The **column space** of a real $m \times n$ matrix A is the subspace* Col A *of* $\mathbb{R}^m$ *spanned by the columns of A.*

The dimension of the column space is called the **column rank** of A. It counts the number of linearly independent columns of A.

DEFINITION 4.12   *The **null space** of a real $m \times n$ matrix A is the subspace* Nul A *of* $\mathbb{R}^n$ *consisting of all vectors* $\mathbf{x} \in \mathbb{R}^n$ *for which* $A\mathbf{x} = \mathbf{0}$.

The fact that Nul A is a subspace of $\mathbb{R}^n$ was essentially shown in Example 4.9. The dimension of Nul A is called the **nullity** of A.

Although the emphasis in this book is on column spaces, we sometimes need the space Col A^T whose columns are the rows of A.

DEFINITION 4.13   *The **row space** of a real $m \times n$ matrix A is the subspace* Col A^T *of* $\mathbb{R}^n$ *spanned by the columns of A^T.*

We sometimes write Row A for Col A^T to highlight the that the vectors in the Col A^T are obtained by transposing the rows of A. The dimension of the row space of A is called the **row rank** of A. It counts the number of linearly independent rows of A.

DEFINITION 4.14   *The **left null space** of a real $m \times n$ matrix A is the subspace $\text{Nul } A^T$ of $\mathbb{R}^m$ consisting of all vectors $\mathbf{y} \in \mathbb{R}^n$ for which $A^T \mathbf{y} = \mathbf{0}$.*

These spaces can be described in terms of interesting bases. Alternative descriptions, based on the singular value decomposition of a matrix, are given in Chapter 9. The geometric relationships between the fundamental subspaces are discussed in Chapter 7.

The Column Space

Column spaces have a straightforward description in terms of pivot columns.

THEOREM 4.22 (Column space basis theorem)   *The pivot columns of an $m \times n$ matrix A over $\mathbb{R}$ form a basis for $\text{Col } A$.*

Proof. Let B be the reduced row echelon form of A. By Theorem 1.6, B and A have the same linear dependence relations. Since the pivot columns of B are linearly independent, it follows that the pivot columns of A are linearly independent. Similarly, since the nonpivot columns of B are linear combinations of the pivot columns of B, the nonpivot columns of A are linear combinations of the nonpivot columns of A. Hence the pivot columns of A form a basis for $\text{Col } A$. ■

EXAMPLE 4.42   ■ A Basis for a Column Space

Use *MATHEMATICA* to find a basis for the column space of the matrix

$$A = \begin{pmatrix} 3 & 1 & 0 & 2 & 4 \\ 1 & 1 & 0 & 0 & 2 \\ 5 & 2 & 0 & 3 & 7 \end{pmatrix}$$

Solution. According to Theorem 4.22, the pivot columns of A form a basis for $\text{Col } A$. We can find these columns by converting A^T to reduced row echelon form.

In[1]:= **Transpose** $\left[A = \begin{pmatrix} 3 & 1 & 0 & 2 & 4 \\ 1 & 1 & 0 & 0 & 2 \\ 5 & 2 & 0 & 3 & 7 \end{pmatrix} \right]$

Out[1]= {{3,1,5},{1,1,2},{0,0,0},{2,0,3},{4,2,7}}

In[2]:= **RowReduce[Transpose[A]]**

Out[2]= {{1,0,$\frac{3}{2}$},{0,1,$\frac{1}{2}$},{0,0,0},{0,0,0},{0,0,0}}

The first two rows of the last output are the pivot columns of A. We extract them with the **Take** function.

```
In[3]:= Take[Out[2],2]

Out[3]= {{1,0,3/2},{0,1,1/2}}
```

Thus the set

$$B = \left\{ \begin{pmatrix} 1 \\ 0 \\ \frac{3}{2} \end{pmatrix}, \begin{pmatrix} 0 \\ 1 \\ \frac{1}{2} \end{pmatrix} \right\}$$

is a basis for the column space of A. ◄►

The Null Space

Using the concept of free variables discussed in Chapter 1, we can give an easy description of a basis for Nul A. The idea is to produce the standard basis of the solution space of the system $A\mathbf{x} = \mathbf{0}$, as described in Example 4.28.

THEOREM 4.23 (Null space basis theorem)   *The set of column vectors obtained by writing the general solution of the system $A\mathbf{x} = \mathbf{0}$ as a linear combination of the free variables of A is a basis for* Nul A.

Proof. If we convert the matrix A to reduced row echelon form and write the general solution of the system as a linear combination of the free variables of the system $A\mathbf{x} = \mathbf{0}$, then it follows from the discussion in Example 4.28 that the set of column vectors arising in this way is a basis for Nul A. ■

EXAMPLE 4.43   ■ A Basis for a Null Space

Use *MATHEMATICA* to find a basis for the null space of the matrix

$$A = \begin{pmatrix} 3 & 1 & 0 & 2 & 4 \\ 1 & 1 & 0 & 0 & 2 \\ 5 & 2 & 0 & 3 & 7 \end{pmatrix}$$

Solution. We begin by finding the general solution of the system $A\mathbf{x} = \mathbf{0}$.

```
In[1]:= A=( 3  1  0  2  4
            1  1  0  0  2
            5  2  0  3  7 );
```

```
In[2]:= x={{a},{b},{c},{d},{e}};
```

```
In[3]:= Solve[A.x=={{0},{0},{0}}]

Solve::svars:Equations may not give solutions for all
''solve'' variables.

Out[3]= {{a→-d-e,b→d-e}}
```

This tells us that the general solution **x** of $A\mathbf{x} = \mathbf{0}$ is

$$\begin{pmatrix} -d-e \\ d-e \\ c \\ d \\ e \end{pmatrix} = \begin{pmatrix} -d \\ d \\ c \\ d \\ e \end{pmatrix} + \begin{pmatrix} -e \\ -e \\ c \\ d \\ e \end{pmatrix} = c\begin{pmatrix} 0 \\ 0 \\ 1 \\ 0 \\ 0 \end{pmatrix} + d\begin{pmatrix} -1 \\ 1 \\ 0 \\ 1 \\ 0 \end{pmatrix} + e\begin{pmatrix} -1 \\ -1 \\ 0 \\ 0 \\ 1 \end{pmatrix}$$

The *MATHEMATICA* message refers to the fact that in the last output, the variable c is not accounted for. By Theorem 4.23, the set of vectors

$$\mathcal{B} = \left\{ \begin{pmatrix} 0 \\ 0 \\ 1 \\ 0 \\ 0 \end{pmatrix}, \begin{pmatrix} -1 \\ 1 \\ 0 \\ 1 \\ 0 \end{pmatrix}, \begin{pmatrix} -1 \\ -1 \\ 0 \\ 0 \\ 1 \end{pmatrix} \right\}$$

is a basis for the null space of A. We can use the **NullSpace** function to confirm this fact.

```
In[4]:= NullSpace[A]

Out[4]= {{-1,-1,0,0,1},{-1,1,0,1,0},{0,0,1,0,0}}
```

As we can see, the *MATHEMATICA* output coincides with the basis $\mathcal{B}$. ◀▶

The Row Space

The next theorem provides an easy description of a basis for the row space of A. The proof of the theorem uses the following lemma.

LEMMA 4.24 *The nonzero rows of a matrix in reduced row echelon form are linearly independent.*

Proof. Let A be an $m \times n$ matrix in reduced row echelon form and suppose that $\mathbf{r}_1 = (a_1 \cdots a_n)$ and $\mathbf{r}_2 = (b_1 \cdots b_n)$ are two distinct nonzero rows of A, with $\mathbf{r}_1$ occurring above $\mathbf{r}_2$. Assume that $a_1 = 1$ is a pivot of A. Since A is in reduced row echelon form, $b_1 = 0$. Hence, $\mathbf{r}_1$ and $\mathbf{r}_2$ must be linearly independent. Otherwise, either $\mathbf{r}_2 = c\mathbf{r}_1$ or $\mathbf{r}_1 = d\mathbf{r}_2$ for a nonzero c or a nonzero d. However, the equation $\mathbf{r}_2 = c\mathbf{r}_1$ then forces b_1 to be nonzero. This contradicts the fact that A is in reduced row echelon form. For the same reason, it is impossible for $\mathbf{r}_1$ to be a multiple of $\mathbf{r}_2$. If some initial terms of $\mathbf{r}_1$ are zero, a similar argument shows that $\mathbf{r}_1$ and $\mathbf{r}_2$ cannot be multiples of each other. ∎

We use this lemma to construct a basis for the row space of A.

THEOREM 4.25 (Row space basis theorem)   *The nonzero rows of the reduced row echelon form B of an $m \times n$ matrix A are a basis for $\operatorname{Col} A^T$.*

Proof. Suppose that A is a matrix over $\mathbb{R}$. By Theorem 1.6, the matrices A and B have the same linear dependence relations. Therefore, the nonzero rows of B and the nonzero rows of A span the same subspace of $\mathbb{R}^n$. But by Lemma 4.24, the nonzero rows of B are linearly independent. They therefore form a basis for the row space of B, and hence also for the row space of A. ∎

Theorem 4.25 also provides us with an alternative description of a basis for the column space of A.

COROLLARY 4.26   *The nonzero columns of the reduced row echelon form B of an $m \times n$ matrix A are a basis for $\operatorname{Col} A$.*

Proof. In terms of columns, Theorem 4.25 says that the nonzero columns of the reduced row echelon form B of an $m \times n$ matrix A^T are a basis for $\operatorname{Col} A^T$. If we replace A by A^T and use the fact that $A^{TT} = A$, we get the column version of the theorem. ∎

EXAMPLE 4.44   ■ A Basis for a Row Space

Use *MATHEMATICA* to find a basis for the row space of the matrix

$$A = \begin{pmatrix} 3 & 1 & 0 & 2 & 4 \\ 1 & 1 & 0 & 0 & 2 \\ 5 & 2 & 0 & 3 & 7 \end{pmatrix}$$

Solution. We begin by finding the reduced row echelon form B of the matrix A.

In[1]:= A=$\begin{pmatrix} 3 & 1 & 0 & 2 & 4 \\ 1 & 1 & 0 & 0 & 2 \\ 5 & 2 & 0 & 3 & 7 \end{pmatrix}$; MatrixForm[B=RowReduce[A]]

Out[1]//MatrixForm=

$$\begin{pmatrix} 1 & 0 & 0 & 1 & 1 \\ 0 & 1 & 0 & -1 & 1 \\ 0 & 0 & 0 & 0 & 0 \end{pmatrix}$$

In[2]:= MatrixForm[Transpose[B]]

Out[2]//MatrixForm=

$$\begin{pmatrix} 1 & 0 & 0 \\ 0 & 1 & 0 \\ 0 & 0 & 0 \\ 1 & -1 & 0 \\ 1 & 1 & 0 \end{pmatrix}$$

According to Theorem 4.25, the set

$$\mathcal{B} = \left\{ \begin{pmatrix} 1 \\ 0 \\ 0 \\ 1 \\ 1 \end{pmatrix}, \begin{pmatrix} 0 \\ 1 \\ 0 \\ -1 \\ 1 \end{pmatrix} \right\}$$

determined by the nonzero columns of B^T is a basis for Col A^T. ◄►

The Left Null Space

Next we describe the construction of a basis for a left null space. For this purpose we use the LU decomposition of rectangular matrices discussed in Chapter 2.

THEOREM 4.27 (Left null space basis theorem)   *If A is an $m \times n$ matrix of rank k, and if $PA = LU$ is the permutation-free LU decomposition of A, then the last $m - k$ rows of $L^{-1}P$ are a basis for* Nul A^T.

Proof. Suppose that $PA = LU$ is the permutation-free LU decomposition of PA. We know from Theorems 2.15 and 2.16 that L is invertible. Therefore, $L^{-1}PA = U$. Since U is upper triangular, all of its zero rows occur at the bottom and are produced by the row-column products $\mathbf{r}_i(L^{-1}P)\mathbf{c}_j(A)$, for $k < i \leq m$. Hence the rows $\mathbf{r}_i(L^{-1}P)$ form a basis for the left null space of A. ∎

EXAMPLE 4.45 ■ A Basis for a Left Null Space

Use *MATHEMATICA* and Theorem 4.27 to find a basis for the left null space of the matrix

$$A = \begin{pmatrix} -1 & -2 & 39 & 8 \\ 0 & 0 & 0 & 0 \\ 0 & -7 & 2 & 0 \\ 0 & 1 & 2 & 12 \end{pmatrix}$$

Solution. We begin by noting that the product

$$EPA = \begin{pmatrix} 1 & 0 & 0 & 0 \\ 0 & 1 & 0 & 0 \\ 0 & 7 & 1 & 0 \\ 0 & 0 & 0 & 1 \end{pmatrix} \begin{pmatrix} 1 & 0 & 0 & 0 \\ 0 & 0 & 0 & 1 \\ 0 & 0 & 1 & 0 \\ 0 & 1 & 0 & 0 \end{pmatrix} \begin{pmatrix} -1 & -2 & 39 & 8 \\ 0 & 0 & 0 & 0 \\ 0 & -7 & 2 & 0 \\ 0 & 1 & 2 & 12 \end{pmatrix}$$

$$= \begin{pmatrix} 1 & 0 & 0 & 0 \\ 0 & 0 & 0 & 1 \\ 0 & 0 & 1 & 7 \\ 0 & 1 & 0 & 0 \end{pmatrix} \begin{pmatrix} -1 & -2 & 39 & 8 \\ 0 & 0 & 0 & 0 \\ 0 & -7 & 2 & 0 \\ 0 & 1 & 2 & 12 \end{pmatrix} = \begin{pmatrix} -1 & -2 & 39 & 8 \\ 0 & 1 & 2 & 12 \\ 0 & 0 & 16 & 84 \\ 0 & 0 & 0 & 0 \end{pmatrix} = U$$

is upper triangular. This tells us that in the multiplication of EP and A, the row vector $(0\ 1\ 0\ 0)$ of EP produces the zero row $(0\ 0\ 0\ 0)$ of U. By Theorem 4.27, the set

$$\mathcal{B} = \left\{ \mathbf{x}_0 = \begin{pmatrix} 0 \\ 1 \\ 0 \\ 0 \end{pmatrix} \right\}$$

is therefore a basis for Nul A^T. We can verify directly that $A^T\mathbf{x}_0 = \mathbf{0}$.

In[1]:= A = $\begin{pmatrix} -1 & -2 & 39 & 8 \\ 0 & 0 & 0 & 0 \\ 0 & -7 & 2 & 0 \\ 0 & 1 & 2 & 12 \end{pmatrix}$; Transpose[A]

Out[1]= {{-1,0,0,0},{2,0,-7,1},{39,0,2,2},{8,0,0,12}}

```
In[2]:= Transpose[A].{0,1,0,0}=={0,0,0,0}

Out[2]= True
```

Hence **x** is in the null space of A^T. We can use the **NullSpace** function to corroborate our result.

```
In[3]:= NullSpace[Transpose[A]]

Out[3]= {{0,1,0,0}}
```

This result confirms that, as expected, the set $\mathcal{B}$ is a basis for the left null space of A. ◄►

If A is a rectangular matrix, we cannot use the LU decomposition function. However, we can still find a basis for the left null space of A by using the **NullSpace** function.

EXAMPLE 4.46   ■ A Basis for the Left Null Space of a Rectangular Matrix
Use *MATHEMATICA* to find a basis for the left null space of the matrix

$$A = \begin{pmatrix} 3 & 1 & 0 & 2 & 4 \\ 1 & 1 & 0 & 0 & 2 \\ 5 & 2 & 0 & 3 & 7 \end{pmatrix}$$

Solution. We begin by using the **NullSpace** function to find a basis. Then we verify that we can find the same basis by using the procedure described in Theorem 4.27.

```
In[1]:= A=Transpose[( 3 1 0 2 4
                       1 1 0 0 2
                       5 2 0 3 7 )];
```

```
In[2]:= NullSpace[A]

Out[2]= {{-3,-1,2}}
```

MATHEMATICA is telling us that the set

$$\mathcal{B} = \left\{ \begin{pmatrix} -3 \\ -1 \\ 2 \end{pmatrix} \right\}$$

is a basis for Nul A^T. ◄►

Rank and Fundamental Subspaces

It is clear that if A is an $m \times n$ matrix over $\mathbb{R}$ and $m \neq n$, then Col A and Col A^T are subspaces of column spaces of different dimensions. It is remarkable, therefore, that they nevertheless have the same dimension.

THEOREM 4.28 (Rank theorem)   *The spaces* Col A *and* Col A^T *have the same dimension.*

Proof. Suppose that A has m rows and n columns. We show that dim Col $A \leq$ dim Row A and that dim Row $A \leq$ dim Col A. Let us denote the m rows of A by $\mathbf{r}_1, \ldots, \mathbf{r}_m$ and assume that dim Row $A = k$. Suppose further that $\mathcal{B} = \{\mathbf{v}_1, \ldots, \mathbf{v}_k\}$ is a basis for Row A. Since Row $A = $ Col $A^T \subseteq \mathbb{R}^n$, every basis vector is of the form $\mathbf{v}_i = (v_{i1} \cdots v_{in})$. We can therefore write the rows of A as linear combinations of these basis vectors:

$$
\begin{cases}
\mathbf{r}_1 &= a_{11}\mathbf{v}_1 + \cdots + a_{1k}\mathbf{v}_k \\
&\vdots \\
\mathbf{r}_i &= a_{i1}\mathbf{v}_1 + \cdots + a_{ik}\mathbf{v}_k \\
&\vdots \\
\mathbf{r}_{m1} &= a_{m1}\mathbf{v}_1 + \cdots + a_{mk}\mathbf{v}_k
\end{cases}
$$

If we take jth components on each side of these equations, we get

$$
\begin{cases}
a_{1j} &= a_{11}v_{1j} + \cdots + a_{1k}v_{kj} \\
&\vdots \\
a_{ij} &= a_{i1}v_{1j} + \cdots + a_{ik}v_{kj} \\
&\vdots \\
a_{mj} &= a_{m1}v_{1j} + \cdots + a_{mk}v_{kj}
\end{cases}
$$

In vector notation, this means that

$$
\begin{pmatrix} a_{1j} \\ \vdots \\ a_{ij} \\ \vdots \\ a_{mj} \end{pmatrix} = v_{1j} \begin{pmatrix} a_{11} \\ \vdots \\ a_{i1} \\ \vdots \\ a_{m1} \end{pmatrix} + \cdots + v_{mj} \begin{pmatrix} a_{1k} \\ \vdots \\ a_{ik} \\ \vdots \\ a_{mk} \end{pmatrix}
$$

for all $j \in \mathbf{n}$. Thus every column of A can be written as a linear combination of the k vectors

$$
\mathbf{w}_1 = (a_{11}, \ldots, a_{i1}, \ldots, a_{m1}), \ldots, \mathbf{w}_k = (a_{1k}, \ldots, a_{ik}, \ldots, a_{mk})
$$

Therefore, dim Col $A \leq k = $ dim Row A. Applied to A^T, the same calculations can be used to prove that dim Col $A^T \leq$ dim Row A^T. But since Col $A^T = $ Row A and Row $A^T = $ Col A, the last inequality shows that $\dim \text{Row } A \leq \dim \text{Col } A$. Hence the spaces $\text{Col } A$ and $\text{Row } A$ have the same dimension. ■

This theorem justifies the following definition.

DEFINITION 4.15   *The **rank** of an $m \times n$ matrix A is the common dimension $\text{rank}(A)$ of the column space and the row space of A.*

An $n \times n$ matrix is said to be of **full rank** if it has rank n. Otherwise, it is said to be **rank deficient**.

Although we cannot use a built-in *MATHEMATICA* function to compute the rank of a matrix, we can use the conclusion of the next theorem to define one.

THEOREM 4.29   *For any real $m \times n$ matrix A, the sum of the rank and nullity of A is n.*

Proof. Let A be a real $m \times n$ matrix. Then A has n columns and the homogeneous linear system $A\mathbf{x} = \mathbf{0}$ therefore consists of m equations in n variables. By Theorem 4.23, the number of free variables of A determines the nullity of A, and by Theorem 4.22, the number of basic variables of A determines the rank of A. It follows that $\text{nullity}(A) + \text{rank}(A) = n$. ■

As Theorems 4.28 and 4.29 show, the rank of a matrix uniquely determines the dimensions of the fundamental subspaces. Table 5 lists the dimensions of these subspaces for an $m \times n$ matrix A of rank r.

TABLE 5   Subspace Dimensions.

Fundamental Subspaces	Dimensions
Col A	r
Col A^T	r
Nul A	$n - r$
Nul A^T	$m - r$

We use these facts to define *MATHEMATICA* functions that compute the rank of a matrix, as well as a basis for the row space of that matrix.

EXAMPLE 4.47   ■ **The Rank of a Matrix**

Use *MATHEMATICA* to compute the rank of the matrix

$$A = \begin{pmatrix} 3 & 1 & 0 & 2 & 4 \\ 1 & 1 & 0 & 0 & 2 \\ 5 & 2 & 0 & 3 & 7 \end{pmatrix}$$

Solution. We know from Example 4.43 that the nullity of A is 3. Since A has five columns, its rank is 2. We therefore first define a function that computes the column rank of A. Then we define a function that computes the rank as the column rank.

```
In[1]:= ColumnRank[A_]:=Length[Transpose[A]]-
Length[NullSpace[A]]
```

```
In[2]:= Rank[A_]:=ColumnRank[A]
```

Now we use the **Rank** function.

```
In[3]:= A=⎛ 3  1  0  2  4 ⎞ ; Rank[A]
        ⎜ 1  1  0  0  2 ⎟
        ⎝ 5  2  0  3  7 ⎠

Out[3]= 2
```

As expected, the rank of *A* is 2. ◄►

EXAMPLE 4.48 ■ **The Row Space of a Matrix**

Use *MATHEMATICA* to compute the row space of the matrix

$$A = \begin{pmatrix} 3 & 1 & 0 & 2 & 4 \\ 1 & 1 & 0 & 0 & 2 \\ 5 & 2 & 0 & 3 & 7 \end{pmatrix}$$

Solution. We use the **Rank** function defined in Example 4.47, together with the **RowSpace** function.

```
In[1]:= RowRank[A_]:=Length[A]-Length[NullSpace[Transpose[A]]]
```

```
In[2]:= Rank[A_]:=RowRank[A_]
```

```
In[3]:= RowSpace[A_]:=Take[RowReduce[A],Rank[A]]
```

We now use the **RowSpace** function find a basis for the row space of *A*.

$$\text{In[4]:= } \mathbf{A} = \begin{pmatrix} 3 & 1 & 0 & 2 & 4 \\ 1 & 1 & 0 & 0 & 2 \\ 5 & 2 & 0 & 3 & 7 \end{pmatrix} \text{; RowSpace[A]}$$

$$\text{Out[4]= } \{\{1,0,0,1,1\},\{0,1,0,-1,1\}\}$$

Our calculation shows that the set

$$\mathcal{B} = \left\{ \begin{pmatrix} 1 \\ 0 \\ 0 \\ 1 \\ 1 \end{pmatrix}, \begin{pmatrix} 0 \\ 1 \\ 0 \\ -1 \\ 1 \end{pmatrix} \right\}$$

is a basis for the row space of A. ◄►

Invertibility and Fundamental Subspaces

There are several useful connections between the invertibility of a matrix and the fundamental subspaces it determines.

THEOREM 4.30   *A real $n \times n$ matrix A is invertible if and only if $\text{Col } A = \mathbb{R}^n$.*

Proof. By Corollary 2.22, the matrix A is invertible if and only if the system $A\mathbf{x} = \mathbf{0}$ has only the trivial solution. If we think of the matrix $A = (A_1 \cdots A_n)$ as a vector whose components are the columns A_i, then the equation $A\mathbf{x} = \mathbf{0}$ is equivalent to the linear combination

$$x_1 A_1 + \cdots + x_n A_n = \mathbf{0}$$

We must therefore show that the A_i are linearly independent. Since the dimension of $\mathbb{R}^n$ is n, it follows that $\text{span}(A_1, \ldots, A_n) = \mathbb{R}^n$. The columns A_i therefore form a basis for $\mathbb{R}^n$. Moreover, they are linearly independent if and only if $x_1 = \cdots = x_n = 0$. This is so if and only if $\mathbf{x} = \mathbf{0}$. It follows that A is invertible if and only if $A\mathbf{x} = \mathbf{0}$ has only the trivial solution. ∎

Having proved that the columns of A are linearly independent and span $\mathbb{R}^n$, we have also established the following corollary.

COROLLARY 4.31   *A real $n \times n$ matrix A is invertible if and only if it has rank n.*

Proof. Since $\dim \mathbb{R}^n = n$ and $\text{Col } A = \mathbb{R}^n$, the result follows from Theorem 4.30. ∎

The next corollary is a restatement of Theorem 4.3. The new proof is less elementary since it refers to bases, column spaces, and dimensions.

COROLLARY 4.32   *A real $n \times n$ matrix A is invertible if and only if the columns of A are linearly independent.*

Proof. This corollary follows from Corollary 4.31. The dimension of the column space of an $n \times n$ matrix is n if and only if the columns form a basis for A. This holds if and only if the columns of A are linearly independent. ■

EXAMPLE 4.49   ■ A Basis for $\mathbb{R}^3$

Use *MATHEMATICA* to find an invertible 3×3 matrix, and use the columns of this matrix to construct a basis for $\mathbb{R}^3$.

Solution. We use the **Array** and **Random** functions to find an invertible matrix.

```
In[1]:= A=Array[Random[Integer,{-5,5}]&,{3,3}]

Out[1]= {{-1,3,-4},{5,0,-1},{5,5,-1}}
```

The calculation

```
In[2]:= MatrixForm[Inverse[A]]

Out[2]//MatrixForm=
```

$$\begin{pmatrix} -\frac{1}{3} & \frac{1}{3} & -\frac{1}{2} \\ -\frac{1}{3} & \frac{5}{6} & -\frac{5}{4} \\ -1 & \frac{3}{2} & -\frac{11}{4} \end{pmatrix}$$

shows that A has an inverse. We can therefore use the columns of A and form the basis

$$\mathcal{B} = \left\{ \begin{pmatrix} -1 \\ 3 \\ -4 \end{pmatrix}, \begin{pmatrix} 5 \\ 0 \\ -1 \end{pmatrix}, \begin{pmatrix} 5 \\ 5 \\ -1 \end{pmatrix} \right\}$$

for $\mathbb{R}^3$. Since A is invertible if and only if A^{-1} is invertible, we could also have taken the columns of A^{-1} to construct a basis for $\mathbb{R}^3$. ◄►

EXAMPLE 4.50   ■ Matrices and Linear Dependence

Use *MATHEMATICA* to show that the set

$$\mathcal{S} = \left\{ \begin{pmatrix} -3 \\ -3 \\ 3 \end{pmatrix}, \begin{pmatrix} 5 \\ 8 \\ -8 \end{pmatrix}, \begin{pmatrix} 17 \\ 17 \\ -17 \end{pmatrix} \right\}$$

is linearly dependent and therefore fails to form a basis for $\mathbb{R}^3$.

Solution. We begin by constructing the required matrix.

```
In[1]:= A={{-3,-3,3},{5,8,-8},{17,17,-17}}
Out[1]= {{-3,-3,3},{5,8,-8},{17,17,-17}}
```

Next we try to compute the inverse of A.

```
In[2]:= Inverse[A]

Inverse::sing:  Matrix {{-3,-3,3},{5,8,-8},{17,17,-17}} is
singular.

Out[2]= Inverse[{{-3,-3,3},{5,8,-8},{17,17,-17}}]
```

MATHEMATICA is telling us that A is not invertible. By Corollary 4.32, the columns of S are therefore linearly dependent and cannot form a basis for $\mathbb{R}^3$. ◄►

By reversing the role of the rows and columns of A, we get the analogous result for the row space of A.

COROLLARY 4.33 *A real $n \times n$ matrix A is invertible if and only if the dimension of Col A^T is n.*

Proof. By Theorem 4.28, the row rank of A is the same as the column rank of A. Hence the result follows from Corollary 4.32. ∎

COROLLARY 4.34 *A real $n \times n$ matrix A is invertible if and only if Nul $A = \{\mathbf{0}\}$.*

Proof. By definition, Nul $A = \{\mathbf{x} \in \mathbb{R}^n \mid A\mathbf{x} = \mathbf{0}\}$, and by Corollary 2.22, A is invertible if and only if the equation $A\mathbf{x} = \mathbf{0}$ has only the trivial solution. Thus A is invertible if and only if Nul $A = \{\mathbf{0}\}$. ∎

EXERCISES 4.14

1. Show that for any real $m \times n$ matrix A, the spaces Col A, Nul A, Col A^T, and Nul A^T are subspaces and therefore satisfy the axioms of a real vector space.

2. Show that the function

   ```
   ColumnRank[A_]:=Length[Transpose[A]]-Length[NullSpace[A]]
   ```

   calculates the column rank of a matrix.

3. Show that the function

   RowRank[A_]:=Length[A]-Length[NullSpace[Transpose[A]]]

   calculates the row rank of a matrix.

4. Explain why the function

   Rank[A_]:=RowRank[A]

   calculates the rank of a matrix.

5. Show that the function

   RowSpace[A_]:=Take[RowReduce[A],Rank[A]]

   calculates a basis for the row space of a matrix.

6. Show that the function

   ColumnSpace[A_]:=RowSpace[Transpose[A]]

   calculates a basis for the column space of a matrix.

7. Use *MATHEMATICA* to find the column space, row space, null space, and left null space of the matrix

$$A = \begin{pmatrix} 1 & 2 & 4 \\ 0 & 0 & 1 \\ 3 & 6 & 12 \end{pmatrix}$$

   and give a geometric description of these subspaces of $\mathbb{R}^3$.

8. Use *MATHEMATICA* to find the column space, row space, null space, and left null space of the matrix

$$A = \begin{pmatrix} 1 & 2 & 4 \\ 0 & 0 & 1 \end{pmatrix}$$

   and give a geometric description of these subspaces of $\mathbb{R}^2$ and $\mathbb{R}^3$.

9. Use *MATHEMATICA* to find the column space, row space, null space, and left null space of the matrix

$$A = \begin{pmatrix} 1 & 2 & 4 & 3 \\ 0 & 0 & 1 & 3 \end{pmatrix}$$

10. Show that the intersection of the column space of the matrix

$$A = \begin{pmatrix} 1 & 2 & 4 \\ 0 & 0 & 1 \\ 3 & 6 & 12 \end{pmatrix}$$

and the null space of the matrix

$$B = \begin{pmatrix} 1 & 0 & 1 \\ 1 & 0 & 1 \\ 1 & 0 & 1 \\ 0 & 0 & 0 \end{pmatrix}$$

is a subspace of $\mathbb{R}^3$.

11. Construct 3×4 matrices of rank 0, 1, 2, and 3.

12. Construct 5×5 matrices of rank 0, 1, 2, 3, 4, and 5.

13. Show that the rank of a matrix is equal to the rank of the transpose of the matrix.

14. Prove that Gaussian elimination preserves rank.

15. Prove that the rank of a matrix is equal to the number of nonzero rows of the reduced row echelon form of the matrix.

16. Prove that the rank of a matrix is equal to the number of pivots of the row echelon form of the matrix.

17. Use *MATHEMATICA* to generate 10 random 8×8 matrices and compute their rank. Compute the probability, based on your experiment, that a random 8×8 matrix is of full rank.

18. Let A be an $m \times n$ matrix, B be an $m \times m$ invertible matrix, and C be an $n \times n$ invertible matrix. Show that the matrices A, BA, and AC have the same rank.

19. Let A be an $m \times n$ matrix and B be an $n \times p$ matrix. Show that the rank of AB is the smaller of the rank of A and the rank of B.

20. Prove that a real $n \times n$ matrix is invertible if and only its nullity is zero.

Direct Sums

Contrary to intersections, the union $U \cup W$ of two subspaces U and W of a vector space V may not be a subspace of V since it may not contain all linear combinations of the vectors in U and W. Let

$$U = \left\{ \begin{pmatrix} a \\ 0 \end{pmatrix} : a \in \mathbb{R} \right\} \quad \text{and} \quad W = \left\{ \begin{pmatrix} 0 \\ b \end{pmatrix} : b \in \mathbb{R} \right\}$$

be two subspaces of $\mathbb{R}^2$. Then the vector $(1, 0)$ belongs to U, the vector $(0, 2)$ belongs to W, but the vector

$$\begin{pmatrix} 1 \\ 0 \end{pmatrix} + \begin{pmatrix} 0 \\ 2 \end{pmatrix} = \begin{pmatrix} 1 \\ 2 \end{pmatrix}$$

does not belong to $U \cup W$. Therefore, $U \cup W$ is not a subspace of V. To construct a subspace from U and V, we need to form the span of $U \cup W$. We call this set the **sum** of U and W. The space

$$U + W = \{a\mathbf{x} + b\mathbf{y} \mid \mathbf{x} \in U, \ \mathbf{y} \in V, \ \text{and } a, b \in \mathbb{R}\}$$

is constructed from U and W by building all possible linear combinations of vectors $a\mathbf{x} + b\mathbf{y}$. It is therefore closed under vector addition and scalar multiplication. By Theorem 4.19, $U + W$ is a subspace of V.

THEOREM 4.35 *Let U and W be subspaces of a finite-dimensional vector space V. Then* $\dim(U + W) = \dim U + \dim W - \dim (U \cap W)$.

Proof. Let $\dim U = p$ and $\dim V = q$. The idea of the proof is to take a basis for the common part $U \cap V$ of U and V containing $r \geq 0$ elements, and to extend this set to a basis for U and a basis for V. This process gives rise to three sets of vectors: the basis for $U \cap V$, the set of p additional vectors required for a basis for U (not in V), and the set of q additional vectors required for a basis for V (not in U). The theorem is proved by showing that these sets are linearly independent and that they span $U + V$.

Let $\mathcal{B} = \{\mathbf{v}_1, \ldots, \mathbf{v}_n\}$ be a basis for $U \cap V$, $\mathcal{C} = \{\mathbf{v}_1, \ldots, \mathbf{x}_r, \mathbf{w}_1, \ldots, \mathbf{w}_p\}$ be a basis for U, and $\mathcal{D} = \{\mathbf{v}_1, \ldots, \mathbf{x}_r, \mathbf{u}_1, \ldots, \mathbf{u}_q\}$ be a basis for V. We show that $\mathcal{E} = \mathcal{B} \cup \mathcal{C} \cup \mathcal{D}$ is a basis for $U + V$. Since the fact that $\text{span}(\mathcal{E}) = U + V$ is obvious, it remains to show that $\mathcal{E}$ is linearly independent.

Suppose that

$$a_1\mathbf{v}_1 + \cdots + a_n\mathbf{v}_n + b_1\mathbf{w}_1 + \cdots + b_p\mathbf{w}_p + c_1\mathbf{u}_1 + \cdots + c_q\mathbf{u}_q = 0$$

This vector is made up of three distinct parts:

1. $a_1\mathbf{v}_1 + \cdots + a_n\mathbf{v}_n \in U \cap V$
2. $b_1\mathbf{w}_1 + \cdots + b_p\mathbf{w}_p \in U$
3. $c_1\mathbf{u}_1 + \cdots + c_q\mathbf{u}_q \in V$

The equation can therefore be rewritten as

$$c_1\mathbf{u}_1 + \cdots + c_q\mathbf{u}_q = -(a_1\mathbf{v}_1 + \cdots + a_n\mathbf{v}_n + b_1\mathbf{w}_1 + \cdots + b_p\mathbf{w}_p)$$

This implies that $c_1\mathbf{u}_1 + \cdots + c_q\mathbf{u}_q \in U$. Therefore,

$$c_1\mathbf{u}_1 + \cdots + c_q\mathbf{u}_q = d_1\mathbf{v}_1 + \cdots + d_n\mathbf{v}_n$$

for some $d_1, \ldots, d_n$.

We rewrite the last equation and get

$$c_1\mathbf{u}_1 + \cdots + c_q\mathbf{u}_q - (d_1\mathbf{v}_1 + \cdots + d_n\mathbf{v}_n) = 0$$

Since $\mathcal{D}$ is linearly independent, we have $c_1 = \cdots = c_q = d_1 = \cdots = d_n = 0$. In other words, $c_1 = \cdots = c_q = 0$. A similar calculation shows that $b_1 = \cdots = b_p = 0$. Therefore,

$$\dim(U + V) = n + p + q$$
$$= (n + p) + (n + q) - n$$
$$= \dim U + \dim V - \dim(U \cap V)$$

This proves the theorem. ■

COROLLARY 4.36 *If U and W are disjoint subspaces of V, then*

$$\dim(U + W) = \dim U + \dim W$$

Proof. Since U and W are disjoint, $\dim(U \cap W) = 0$. By Theorem 4.35,

$$\dim(U + W) = \dim U + \dim W - \dim(U \cap W)$$
$$= \dim U + \dim W - 0$$
$$= \dim U + \dim W$$

as required. ■

One of the standard techniques for studying a complex mathematical object is to break it down into simpler parts, study the simpler parts, and then reassemble the object in a transparent way. In the case of vector spaces, the simpler parts are frequently disjoint subspaces.

DEFINITION 4.16 *If U and W are disjoint subspaces of a vector space V, then $U + W$ is the **direct sum** of U and W, denoted by $U \oplus W$.*

This definition extends to any finite number of subspaces. If $U_1, \ldots, U_n$ are pairwise disjoint subspaces of a vector space V, then the space

$$U_1 + \cdots + U_n = U_1 \oplus \cdots \oplus U_n$$

is the direct sum of $U_1, \ldots, U_n$. Direct sums of subspaces are an important tool for analyzing the structure of vector spaces. They rely on the following fundamental property.

THEOREM 4.37 (Direct sum theorem) *If $V = U \oplus W$, then every vector $\mathbf{x} \in V$ can be written as a unique sum $\mathbf{u} + \mathbf{w}$, with $\mathbf{u} \in U$ and $\mathbf{w} \in W$.*

Proof. Suppose that $\mathbf{x} \in V$ can be written as $\mathbf{u}_1 + \mathbf{w}_1$ and $\mathbf{u}_2 + \mathbf{w}_2$, with $\mathbf{u}_1, \mathbf{u}_2 \in U$ and $\mathbf{w}_1, \mathbf{w}_2 \in W$. Then $\mathbf{u}_1 + \mathbf{w}_1 = \mathbf{u}_2 + \mathbf{w}_2$, so that $\mathbf{u}_1 - \mathbf{u}_2 = \mathbf{w}_2 - \mathbf{w}_1$. But since $\mathbf{u}_1 - \mathbf{u}_2 \in U$ and $\mathbf{w}_2 - \mathbf{w}_1 \in W$, it follows that $\mathbf{u}_1 - \mathbf{u}_2 = \mathbf{w}_2 - \mathbf{w}_1 \in U \cap W$. Since $U \cap W = \{\mathbf{0}\}$, we have $\mathbf{u}_1 - \mathbf{u}_2 = \mathbf{w}_2 - \mathbf{w}_1 = \mathbf{0}$. Therefore, $\mathbf{u}_1 = \mathbf{u}_2$ and $\mathbf{w}_2 = \mathbf{w}_1$. ■

The following theorem expresses one of the basic properties of direct sums.

THEOREM 4.38 (Basis union theorem)   *If U and W are disjoint subspaces of a vector space V with bases B and C, then $V = U \oplus W$ if and only if $B \cup C$ is a basis for V.*

Proof. Suppose that $V = U \oplus W$ and that $B = \{\mathbf{u}_1, \ldots, \mathbf{u}_p\}$ and $C = \{\mathbf{w}_1, \ldots, \mathbf{w}_q\}$. Then for any $\mathbf{x} \in V$, $\mathbf{x} = \mathbf{u} + \mathbf{w}$ with $\mathbf{u} \in U$ and $\mathbf{w} \in W$. Since B is a basis for U and C is a basis for W, $\mathbf{u}$ is a linear combination of vectors in B and $\mathbf{w}$ is a linear combination of vectors in C. Hence $\mathbf{x}$ is a linear combination of vectors in $B \cup C$. Therefore, $B \cup C$ spans V. It remains to show that $B \cup C$ is linearly independent.

Suppose that $a_1\mathbf{u}_1 + \cdots + a_p\mathbf{u}_p + b_1\mathbf{w}_1 + \cdots + b_q\mathbf{w}_q = \mathbf{0}$ and that

$$a_1\mathbf{u}_1 + \cdots + a_p\mathbf{u}_p \in U \quad \text{and} \quad b_1\mathbf{w}_1 + \cdots + b_q\mathbf{w}_q \in W$$

Since the sum $U \oplus W$ is direct, the zero vector $\mathbf{0} \in V$ is the unique sum of $\mathbf{0} \in U$ and $\mathbf{0} \in W$. It therefore follows from the unique decomposition property of direct sums that $a_1\mathbf{u}_1 + \cdots + a_p\mathbf{u}_p = \mathbf{0}$ and $b_1\mathbf{w}_1 + \cdots + b_q\mathbf{w}_q = \mathbf{0}$. Since B and C are bases, we get $a_1 = \cdots = a_p = b_1 = \cdots = b_q = 0$. This shows that $B \cup C$ is linearly independent.

Conversely, suppose that $B \cup C$ is a basis for V. Then for any $\mathbf{x} \in V$,

$$\mathbf{x} = a_1\mathbf{u}_1 + \cdots + a_p\mathbf{u}_p + b_1\mathbf{w}_1 + \cdots + b_q\mathbf{w}_q = \mathbf{u} + \mathbf{w}$$

with

$$\mathbf{u} = a_1\mathbf{u}_1 + \cdots + a_p\mathbf{u}_p \in U \quad \text{and} \quad \mathbf{w} = b_1\mathbf{w}_1 + \cdots + b_q\mathbf{w}_q \in W$$

Since B and C are bases, the linear combinations $a_1\mathbf{u}_1 + \cdots + a_p\mathbf{u}_p$ and $b_1\mathbf{w}_1 + \cdots + b_q\mathbf{w}_q$ are unique. Hence the linear combination $a_1\mathbf{u}_1 + \cdots + a_p\mathbf{u}_p + b_1\mathbf{w}_1 + \cdots + b_q\mathbf{w}_q$ is unique. Therefore, $V = U \oplus W$. ∎

With appropriate notational changes we can extend Theorem 4.38 to any finite number of subspaces.

COROLLARY 4.39   *If $U_1, \ldots, U_r$ are disjoint subspaces of a vector space V and $B_1, \ldots, B_r$ are bases for these spaces, then $V = U_1 \oplus \cdots \oplus U_r$ if and only if $B_1 \cup \cdots \cup B_r$ is a basis for V.*

Proof. This corollary follows from Theorem 4.38 by induction on r. ∎

EXERCISES 4.15

1. Let U be the set of all solutions of the equation $y = 3x$ and W the set of all solutions of the equation $y = -7x$ in $\mathbb{R}^2$. Show that $\mathbb{R}^2 = U \oplus W$.

2. Express $\mathbb{R}^3$ as a direct sum of subspaces in three different ways.

3. Show that $\mathbb{R}^3$ is the direct sum of the row space and the null space of the matrix

$$\begin{pmatrix} 1 & 0 & 1 \\ 1 & 0 & 1 \\ 1 & 0 & 1 \\ 0 & 0 & 0 \end{pmatrix}$$

4. Show that $\mathbb{R}^4$ is the direct sum of the column space and the left null space of the matrix

$$\begin{pmatrix} 1 & 0 & 1 \\ 1 & 0 & 1 \\ 1 & 0 & 1 \\ 0 & 0 & 0 \end{pmatrix}$$

5. Let $\mathcal{B} = \{\mathbf{v}_1, \ldots, \mathbf{v}_r, \mathbf{v}_{r+1}, \ldots, \mathbf{v}_n\}$ be a basis for a vector space V, and let

$$U = \text{span}\{\mathbf{v}_1, \ldots, \mathbf{v}_r\}$$

Show that if $V = U \oplus W$, then $\mathcal{C} = \{\mathbf{v}_{r+1}, \ldots, \mathbf{v}_n\}$ is a basis for W.

COMPLEX VECTOR SPACES

At various times in our work we will need the complex analogues of **coordinate spaces**, **matrix spaces**, and **polynomial spaces**.

EXAMPLE 4.51   ■ The Space $\mathbb{C}^n$

The space $\mathbb{C}^n$ of all $n \times 1$ column vectors $A : \mathbf{n} \times \mathbf{1} \to \mathbb{C}$, equipped with matrix addition and the scalar multiplication of matrices, is a complex vector space. ◀▶

EXAMPLE 4.52   ■ The Space $\mathbb{C}^{m \times n}$

The space $\mathbb{C}^{m \times n}$ consists of all $m \times n$ matrices A with complex entries, equipped with matrix addition and the scalar multiplication of matrices as vector space operations. We refer to this space as the space of complex $m \times n$ matrices. ◀▶

EXAMPLE 4.53   ■ The Space $\mathbb{C}[t]$

If $p(t) = a_0 + a_1 t + \cdots + a_n t^n$ and $q(t) = b_0 + b_1 t + \cdots + b_n t^n$ are two polynomials in $\mathbb{C}[t]$, then we can think of them as vectors. The definitions

$$p(t) + q(t) = (a_0 + b_0) + (a_1 + b_1)t + \cdots + (a_n + b_n)t^n$$
$$ap(t) = aa_0 + aa_1 t + \cdots + aa_n t^n$$
$$\mathbf{0} = 0$$

turn $\mathbb{C}[t]$ into a real vector space. ◀▶

EXERCISES 4.16

1. Verify that the set $\mathbb{C}^2$ of complex column vectors of height 2 satisfies the axioms for a complex vector space.

2. Show that the set $\mathbb{C}$ of complex numbers, together with the set $\mathbb{R}$ of real numbers as scalars, satisfies the axioms of a real vector space.

3. Show that the set $\mathbb{C}^2$ of complex column vectors of height 2, together with the set $\mathbb{R}$ of real numbers as scalars, satisfies the axioms of a real vector space.

4. Describe the set of linear combinations $a\mathbf{x} + b\mathbf{y}$ in $\mathbb{C}^2$ determined by the set

$$S = \left\{ \begin{pmatrix} 1 \\ 0 \end{pmatrix}, \begin{pmatrix} 0 \\ 1 \end{pmatrix} \right\}$$

and all $a, b \in \mathbb{C}$.

5. Describe the set of linear combinations $a\mathbf{x} + b\mathbf{y}$ in $\mathbb{C}^2$ determined by the set

$$S = \left\{ \begin{pmatrix} 1 \\ 0 \end{pmatrix}, \begin{pmatrix} 0 \\ 1 \end{pmatrix} \right\}$$

and all $a, b \in \mathbb{R}$.

6. Describe the set of linear combinations $a\mathbf{x} + b\mathbf{y}$ in $\mathbb{C}^2$ determined by the set

$$S = \left\{ \begin{pmatrix} 1 \\ 0 \end{pmatrix}, \begin{pmatrix} 0 \\ i \end{pmatrix}, \begin{pmatrix} 0 \\ 1 \end{pmatrix}, \begin{pmatrix} i \\ 0 \end{pmatrix} \right\}$$

and all $a, b, c, d \in \mathbb{R}$.

7. Describe the set of linear combinations $a\mathbf{x} + b\mathbf{y}$ in $\mathbb{C}^2$ determined by the set

$$S = \left\{ \begin{pmatrix} 1 \\ 0 \end{pmatrix}, \begin{pmatrix} 0 \\ i \end{pmatrix}, \begin{pmatrix} 0 \\ 1 \end{pmatrix}, \begin{pmatrix} i \\ 0 \end{pmatrix} \right\}$$

and all $a, b, c, d \in \mathbb{C}$.

8. Explain why the set

$$S = \left\{ \begin{pmatrix} 1 \\ 0 \end{pmatrix}, \begin{pmatrix} 0 \\ i \end{pmatrix}, \begin{pmatrix} 0 \\ 1 \end{pmatrix}, \begin{pmatrix} i \\ 0 \end{pmatrix} \right\}$$

is linearly independent in $\mathbb{C}^2$ when $\mathbb{C}^2$ is considered as a real vector space, but linearly dependent when $\mathbb{C}^2$ is considered as a complex vector space.

9. Find a basis for the space $\mathbb{C}_3[t]$ of complex polynomials of degree 3 or less, considered as a complex vector space.

10. Find a basis for the space $\mathbb{C}[t]$ of complex polynomials, considered as a complex vector space.

REVIEW

KEY CONCEPTS ▶ Define and discuss each of the following.

Bases
Finite basis, linear independence of a basis, span of a basis, standard basis.

Linear Transformations
Coordinate vector function, linear combination function.

Sets of Vectors
Basis for a vector space, linearly dependent vectors, linearly independent vectors, standard basis of a vector space.

Subspaces
Column space, fundamental subspace, left null space, null space, row space, zero space.

Vector Operations
Linear combination of vectors, scalar multiplication, subtraction of vectors, vector addition.

Vector Space Properties
Dimension of a vector space, finite-dimensional space, infinite-dimensional space.

Vector Space Operations
Direct sum of vector spaces, subspace of a vector space, sum.

Vector Spaces
Complex vector space, coordinate space, matrix space, polynomial space, real vector space.

Vectors
Coordinate vector, linear combination of vectors, scalar multiple of a vector, zero vector.

KEY FACTS ▶ Explain and illustrate each of the following.

1. For any scalar a and any vector $\mathbf{v}$, it holds that $\mathbf{0} = a\mathbf{0} = 0\mathbf{v}$.

2. For any scalar a and any vector $\mathbf{v}$, it holds that $a\mathbf{v} = \mathbf{0}$ if and only if $a = 0$ or $\mathbf{v} = \mathbf{0}$.

3. A set $\{\mathbf{x}_1, \ldots, \mathbf{x}_n\}$ of column vectors of height n is linearly independent if and only if the matrix $A = (\mathbf{x}_1 \ \cdots \ \mathbf{x}_n)$ is invertible.

4. A set $\{\mathbf{x}_1, \ldots, \mathbf{x}_n\}$ of column vectors of height n is linearly independent if and only if the matrix $A = (\mathbf{x}_1 \ \cdots \ \mathbf{x}_n)$ has n pivots.

5. A nonempty set $S = \{\mathbf{v}_1, \ldots, \mathbf{v}_n\}$ is linearly dependent if and only if one of the vectors $\mathbf{v}_i$ can be written as a linear combination of the remaining vectors.

6. A nonempty set $S = \{\mathbf{v}_1, \ldots, \mathbf{v}_n\}$ is linearly independent if and only if none of the vectors $\mathbf{v}_i$ can be written as a linear combination of the remaining vectors.

7. A subset $B = \{\mathbf{x}_1, \ldots, \mathbf{x}_n\}$ of a vector space V is a basis for V if and only if every vector $\mathbf{x} \in V$ can be written as a unique linear combination $a_1\mathbf{x}_1 + \cdots + a_n\mathbf{x}_n$.

8. If $B = \{\mathbf{v}_1, \ldots, \mathbf{v}_n\}$ is any basis for $\mathbb{R}^n$ and $\mathbf{x}$ is any vector in $\mathbb{R}^n$, then the coefficients $a_1, \ldots, a_n$ in the unique linear combination $\mathbf{x} = a_1\mathbf{v}_1 + \cdots + a_n\mathbf{v}_n$ can be found by solving a system of linear equations.

9. Let V be a vector space spanned by the vectors $\mathbf{v}_1, \ldots, \mathbf{v}_m$. Then any linearly independent set of vectors in V is finite and contains no more than m elements.

10. If B and C are two bases of a finite-dimensional vector space V, then B and C have the same number of elements.

11. Suppose that S is a subset of a finite-dimensional vector space V that spans V. Then S has a linearly independent subset S' that is a basis for V.

12. Any linearly independent set of vectors S of a finite-dimensional vector space V can be extended to a basis for V.

13. Every invertible real $n \times n$ matrix determines a basis for $\mathbb{R}^n$.

14. The functions $\mathbf{x} \rightarrow [\mathbf{x}]_B$ and $\mathbf{y} \rightarrow \hat{\mathbf{y}}_B$ are inverses of each other.

15. The coordinate vector function $\mathbf{x} \rightarrow [\mathbf{x}]_B$ from V to $\mathbb{R}^n$ has the property that $[\mathbf{x} + \mathbf{y}]_B = [\mathbf{x}]_B + [\mathbf{y}]_B$ and that $[a\mathbf{x}]_B = a[\mathbf{x}]_B$.

16. The linear combination function $\mathbf{y} \rightarrow \widehat{\mathbf{y}}_B$ from $\mathbb{R}^n$ to V has the property that $\widehat{\mathbf{x} + \mathbf{y}}_B = \widehat{\mathbf{x}}_B + \widehat{\mathbf{y}}_B$ and that $\widehat{a\mathbf{x}}_B = a\widehat{\mathbf{x}}_B$.

17. Let V be an n-dimensional vector space, let B and C be two bases for V, and let $\mathbf{x}$ be a vector in V with coordinate vectors $[\mathbf{x}]_B$ and $[\mathbf{x}]_C$. Then there exists an invertible matrix P for which $P[\mathbf{x}]_B = [\mathbf{x}]_C$ and $P^{-1}[\mathbf{x}]_C = [\mathbf{x}]_B$.

18. Every invertible real $n \times n$ matrix is a coordinate conversion matrix for $\mathbb{R}^n$.

19. The span of a subset W of a vector space V is a subspace of V.

20. If U and W are subspaces of a vector space V, then so is $U \cap W$.

21. If W is a subspace of a finite-dimensional vector space V, and W and V have the same dimension, then $W = V$.

22. The pivot columns of an $m \times n$ matrix A over $\mathbb{R}$ form a basis for Col A.

23. The set of column vectors obtained by writing the general solution of the system $A\mathbf{x} = \mathbf{0}$ as a linear combination of the free variables of A is a basis for Nul A.

24. The nonzero rows of matrix in reduced row echelon form are linearly independent.

25. The nonzero rows of the reduced row echelon form B of an $m \times n$ matrix A are a basis for Col A^T.

26. The nonzero columns of the reduced row echelon form B of an $m \times n$ matrix A are a basis for Col A.

27. If A is an $m \times n$ matrix of rank k, and if $PA = LU$ is the permutation-free LU decomposition of A, then the last $m - k$ rows of $L^{-1}P$ are a basis for Nul A^T.

28. The spaces Col A and Col A^T have the same dimension.

29. For any real $m \times n$ matrix A, the sum of the rank and nullity of A is n.

30. A real $n \times n$ matrix A is invertible if and only if Col $A = \mathbb{R}^n$.

31. Let U and W be subspaces of a finite-dimensional vector space V. Then $\dim(U + W) = \dim U + \dim W - \dim(U \cap W)$.

32. A real $n \times n$ matrix A is invertible if and only if it has rank n.

33. A real $n \times n$ matrix A is invertible if and only if the columns of A are linearly independent.

34. A real $n \times n$ matrix A is invertible if and only if the dimension of Col A^T is n.

35. A real $n \times n$ matrix A is invertible if and only if Nul $A = \{\mathbf{0}\}$.

36. If U and W are disjoint subspaces of V, then

$$\dim(U + W) = \dim U + \dim W$$

37. If $V = U \oplus W$, then every vector $\mathbf{x} \in V$ can be written as a unique sum $\mathbf{u} + \mathbf{w}$, with $\mathbf{u} \in U$ and $\mathbf{w} \in W$.

38. If U and W are disjoint subspaces of a vector space V with bases $\mathcal{B}$ and $\mathcal{C}$, then $V = U \oplus W$ if and only if $\mathcal{B} \cup \mathcal{C}$ is a basis for V.

39. If $U_1, \ldots, U_r$ are disjoint subspaces of a vector space V and $\mathcal{B}_1, \ldots, \mathcal{B}_r$ are bases for these spaces, then $V = U_1 \oplus \cdots \oplus U_r$ if and only if $\mathcal{B}_1 \cup \cdots \cup \mathcal{B}_r$ is a basis for V.

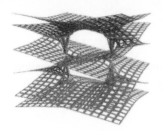

5

LINEAR TRANSFORMATIONS

LINEAR TRANSFORMATIONS

In mathematics, we study both individual objects and their interactions. In linear algebra, the objects are vector spaces. The interactions between vector spaces have various names, such as linear maps, vector space homomorphisms, and linear transformations; in this text we call them *linear transformations*. From one point of view, linear algebra deals almost exclusively with the study of the properties of linear transformations, which map the vectors in one space—called the *domain* of the function—to another space—called the *codomain* of the function (sometimes also called the *range*). To be linear, these functions must *preserve linear combinations*. Figure 1 illustrates the connection between the domain and codomain of a linear transformation.

DEFINITION 5.1 *If V and W are two real vector spaces, a **linear transformation** from V to W is a function $T : V \to W$ with the property that*

$$
\begin{aligned}
T(\mathbf{x} + \mathbf{y}) &= T(\mathbf{x}) + T(\mathbf{y}) \\
T(a\mathbf{x}) &= aT(\mathbf{x})
\end{aligned}
$$

for all vectors $\mathbf{x}, \mathbf{y} \in V$ and all scalars $a \in \mathbb{R}$.

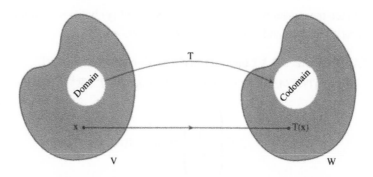

FIGURE 1   A Linear Transformation from V to W.

We can combine the two conditions defining a linear transformation into a single equation by saying that a function $T : V \to W$ is a linear transformation if

$$T(a\mathbf{x} + b\mathbf{y}) = aT(\mathbf{x}) + bT(\mathbf{y})$$

for all $\mathbf{x}, \mathbf{y} \in V$ and all $a, b \in \mathbb{R}$. This equation expresses the idea that T *preserves linear combinations*: the linear combination $a\mathbf{x} + b\mathbf{y} \in V$ is mapped to the linear combination $aT(\mathbf{x}) + bT(\mathbf{y}) \in W$, with the same scalars.

Instead of saying that T is a linear transformation, we often simply will say that T is linear.

EXAMPLE 5.1   ■ A Linear Transformation from $\mathbb{R}^{2 \times 3}$ to $\mathbb{R}^2$

Use *MATHEMATICA* to show that the function $T : \mathbb{R}^{2 \times 3} \to \mathbb{R}^2$ defined by

$$T \begin{pmatrix} a_{11} & a_{12} & a_{13} \\ a_{21} & a_{22} & a_{23} \end{pmatrix} = \begin{pmatrix} a_{11} \\ a_{21} \end{pmatrix}$$

is a linear transformation.

Solution. We begin by defining the function T.

```
In[1]:= T[v_]:={v[[1,1]],v[[2,1]]}
```

Next we specify two general 2×3 matrices $\mathbf{x}$ and $\mathbf{y}$.

```
In[2]:= x=Array[p,{2,3}]; y=Array[q,{2,3}];
```

Finally, we verify that $T(a\mathbf{x} + b\mathbf{y}) = aT(\mathbf{x}) + bT(\mathbf{y})$.

```
In[3]:= T[ax+by]==aT[x]+bT[y]

Out[3]= True
```

Since the equation $T(a\mathbf{x} + b\mathbf{y}) = aT(\mathbf{x}) + bT(\mathbf{y})$ holds for arbitrary scalars a and b and arbitrary vectors $\mathbf{x}$ and $\mathbf{y}$, the function T is linear. ◄►

EXAMPLE 5.2   ■ A Linear Transformation from $\mathbb{R}^4$ to $\mathbb{R}[t]$

Use *MATHEMATICA* to show that the function $T : \mathbb{R}^4 \to \mathbb{R}[t]$ given by

$$T \begin{pmatrix} a_0 \\ a_1 \\ a_2 \\ a_3 \end{pmatrix} = a_0 + a_1 t + a_2 t^2 + a_3 t^3$$

is a linear transformation.

Solution. We recall that in *MATHEMATICA*, the entries of a general column vector **x** of height 4 are $x[[1, 1]]$, $x[[2, 1]]$, $x[[3, 1]]$, and $x[[4, 1]]$. We therefore define the function T taking column vectors to polynomials in t as follows:

```
In[1]:= T[x_]:=x[[1,1]]+x[[2,1]]t+x[[3,1]]t²+x[[4,1]]t³
```

Next we create two arbitrary column vectors.

```
In[2]:= A=Array[p,{4,1}]; B=Array[2,{4,1}];
```

We conclude by verifying that $T(a\mathbf{x} + b\mathbf{y}) = aT(\mathbf{x}) + bT(\mathbf{y})$.

```
In[3]:= Expand[T[aA+bB]]==Expand[aT[A]+bT[B]]

Out[3]= True
```

This shows that $T(a\mathbf{x} + b\mathbf{y}) = aT(\mathbf{x}) + bT(\mathbf{y})$. Hence T is linear. ◄►

EXAMPLE 5.3   ■ A Linear Transformation from $\mathbb{R}^4$ to $\mathbb{R}_3[t]$

Let $T : \mathbb{R}^4 \to \mathbb{R}_3[t]$ be the function defined by the same rule as the linear transformation in Example 5.2. Then the preceding calculations still apply and show that T is a linear transformation. The only difference between the linear transformations in the two examples is that the codomain of T in Example 5.2 is infinite-dimensional, whereas the codomain of T in the present example has dimension 4. ◄►

EXAMPLE 5.4   ■ A Linear Transformation from $\mathbb{R}_3[t]$ to $\mathbb{R}^4$

Use *MATHEMATICA* to show that the function $T : \mathbb{R}_3[t] \to \mathbb{R}^4$ given by

$$T(a_0 + a_1 t + a_2 t^2 + a_3 t^3) = \begin{pmatrix} a_0 \\ a_1 \\ a_2 \\ a_3 \end{pmatrix}$$

is a linear transformation.

Solution. We begin by defining the function T.

```
In[1]:= T[a_+b_t+c_t²+d_t³]:={{a},{b},{c},{d}}
```

We prove that T is linear by showing that it preserves linear combinations.

```
In[2]:= p=a1+a2 t+a3 t²+a4 t³;
```

```
In[3]:= q=b1+b2 t+b3 t²+b4 t³;
```

```
In[4]:= T[Collect[a p+b q,t]]==T[Expand[a p]]+T[Expand[b q]]

Out[4]= True
```

By manipulating the various polynomials involved in the calculation into a form to which the function T could be applied, we were able to show that $T(a\mathbf{x} + b\mathbf{y}) = aT(\mathbf{x}) + bT(\mathbf{y})$. Hence T is linear. ◄►

EXAMPLE 5.5 ■ **A Linear Transformation from $\mathbb{R}^{2\times 2}$ to $\mathbb{R}$**

Use *MATHEMATICA* to show that the trace function $T : \mathbb{R}^{2\times 2} \to \mathbb{R}$ defined by

$$T\begin{pmatrix} a & b \\ c & d \end{pmatrix} = a + d$$

is a linear transformation.

Solution. We first define T.

```
In[1]:= T[{{a_,b_},{c_,d_}}]:=a+d
```

Next we build two general 2×2 matrices.

```
In[2]:= A=Array[a,{2,2}];B=Array[b,{2,2}];
```

Now we compare $aT[A] + bT[B]$ and $T[aA + bB]$.

```
In[3]:= T[aA+bB]==Expand[aT[A]+bT[B]]

Out[3]= True
```

The last calculation shows that $aT(\mathbf{x}) + bT(\mathbf{y}) = T(a\mathbf{x} + b\mathbf{y})$ for all vectors $\mathbf{x}$ and $\mathbf{y}$ and all scalars a and b. Hence T is linear. ◄►

Matrix Transformations

Among the linear transformations are the transformations between coordinate spaces whose functional rules are determined by matrix multiplication.

DEFINITION 5.2   *A **matrix transformation** is a function $T : \mathbb{R}^n \to \mathbb{R}^m$ for which there exists a real $m \times n$ matrix A with the property that $T(\mathbf{x}) = A\mathbf{x}$.*

Since T depends on a matrix A, we denote T by T_A.

THEOREM 5.1   *Every matrix transformation is a linear transformation.*

Proof. The linearity of T_A follows from the linearity of matrix multiplication:

$$T_A(a\mathbf{x} + b\mathbf{y}) = A(a\mathbf{x} + b\mathbf{y})$$
$$= aA\mathbf{x} + bA\mathbf{y}$$
$$= aT_A(\mathbf{x}) + bT_A(\mathbf{y})$$

By Definition 5.1, T_A is a linear transformation. ∎

We will see later that this theorem has a converse. All linear transformations on finite-dimensional vector spaces are matrix transformations. However, the matrix representation of a linear transformation $T : \mathbb{R}^m \to \mathbb{R}^n$ is virtually never unique. It varies with the bases for $\mathbb{R}^m$ and $\mathbb{R}^n$ chosen to represent T. In fact, much of our work will consist of developing strategies for finding suitable bases in which a linear transformation has a simple matrix representation. By "simple," we often mean diagonal.

EXAMPLE 5.6   ■ A Matrix Transformation from $\mathbb{R}^2$ to $\mathbb{R}^3$

Use *MATHEMATICA* to show that the function $T_A : \mathbb{R}^2 \to \mathbb{R}^3$ defined by

$$T_A \begin{pmatrix} x \\ y \end{pmatrix} = \begin{pmatrix} 1 & 2 \\ 3 & 4 \\ 5 & 6 \end{pmatrix} \begin{pmatrix} x \\ y \end{pmatrix} = \begin{pmatrix} x + 2y \\ 3x + 4y \\ 5x + 6y \end{pmatrix}$$

is linear.

Solution. We must show that

$$\begin{pmatrix} 1 & 2 \\ 3 & 4 \\ 5 & 6 \end{pmatrix} \left[a \begin{pmatrix} a_1 \\ a_2 \end{pmatrix} + b \begin{pmatrix} b_1 \\ b_2 \end{pmatrix} \right] = a \begin{pmatrix} 1 & 2 \\ 3 & 4 \\ 5 & 6 \end{pmatrix} \begin{pmatrix} a_1 \\ a_2 \end{pmatrix} + b \begin{pmatrix} 1 & 2 \\ 3 & 4 \\ 5 & 6 \end{pmatrix} \begin{pmatrix} b_1 \\ b_2 \end{pmatrix}$$

```
In[1]:= A={{1,2},{3,4},{5,6}};
```

```
In[2]:= x={{a1},{a2}};
```

```
In[3]:= y={{b1},{b2}};
```

```
In[4]:= Expand[A.(ax+by)]==Expand[aA.x+bA.y]

Out[4]= True
```

This shows that $A(a\mathbf{x} + b\mathbf{y}) = aA\mathbf{x} + bA\mathbf{y}$. Hence T_A is linear. ◄►

EXAMPLE 5.7   ■ **A Matrix Transformation from $\mathbb{R}^3$ to $\mathbb{R}^2$**

Use *MATHEMATICA* to show that the function $T_A : \mathbb{R}^3 \to \mathbb{R}^2$ defined by

$$T_A \begin{pmatrix} x \\ y \end{pmatrix} = \begin{pmatrix} 1 & 3 & 5 \\ 2 & 4 & 6 \end{pmatrix} \begin{pmatrix} x \\ y \\ z \end{pmatrix} = \begin{pmatrix} x + 3y + 5z \\ 2x + 4y + 6z \end{pmatrix}$$

is linear.

Solution. We show that the vector

$$\begin{pmatrix} 1 & 3 & 5 \\ 2 & 4 & 6 \end{pmatrix} \left[a \begin{pmatrix} a_1 \\ a_2 \\ a_3 \end{pmatrix} + b \begin{pmatrix} b_1 \\ b_2 \\ b_3 \end{pmatrix} \right]$$

is equal to the vector

$$a \begin{pmatrix} 1 & 3 & 5 \\ 2 & 4 & 6 \end{pmatrix} \begin{pmatrix} a_1 \\ a_2 \\ a_3 \end{pmatrix} + b \begin{pmatrix} 1 & 3 & 5 \\ 2 & 4 & 6 \end{pmatrix} \begin{pmatrix} b_1 \\ b_2 \\ b_3 \end{pmatrix}$$

We simplify the *MATHEMATICA* notation by defining

$$\begin{pmatrix} a_1 \\ a_2 \\ a_3 \end{pmatrix} \quad \text{and} \quad \begin{pmatrix} b_1 \\ b_2 \\ b_3 \end{pmatrix}$$

as *MATHEMATICA* vectors and not, as in the previous example, as *MATHEMATICA* column vectors. Often *MATHEMATICA* uses the context to interpret the shape of vectors.

```
In[1]:= A={{1,3,5},{2,4,6}};
```

```
In[2]:= x={a1,a2,a3};
```

```
In[3]:= y={b1,b2,b3};
```

Next we verify that $A(a\mathbf{x} + b\mathbf{y}) = aA\mathbf{x} + bA\mathbf{y}$.

```
In[4]:= Expand[A.(ax+by)]==Expand[aA.x+bA.y]
Out[4]= True
```

This equality shows that $A(a\mathbf{x} + b\mathbf{y}) = aA\mathbf{x} + bA\mathbf{y}$. Hence T_A is linear. ◄►

Examples 5.6 and 5.7 illustrate the fact that with every matrix transformation $T_A : \mathbb{R}^n \rightarrow \mathbb{R}^m$ we can associate a linear transformation $T_{(A^T)} : \mathbb{R}^m \rightarrow \mathbb{R}^n$ defined by

$$T_{(A^T)}(\mathbf{x}) = A^T\mathbf{x}$$

The transformation $T_{(A^T)}$ is the **transpose** of T_A.

One of the basic properties of a linear transformation from a vector space V to a vector space W is the fact that it maps the zero vector $\mathbf{0}_V$ of V to the zero vector $\mathbf{0}_W$ of W.

THEOREM 5.2 *If $T : V \rightarrow W$ is a linear transformation, then $T(\mathbf{0}_V) = \mathbf{0}_W$.*

Proof. Recall from Theorem 4.1 that $a\mathbf{0} = \mathbf{0}$ in all vector spaces and for any scalar a. Therefore $T(\mathbf{0}_V) = T(0\mathbf{0}_V) = 0T(\mathbf{0}_V) = \mathbf{0}_W$, where $\mathbf{0}_V$ is the zero vector of V and $\mathbf{0}_W$ is the zero vector of W. ∎

This property of linear transformations prevents certain functions whose graphs are linear from being linear transformations. For example, the translation function $T : \mathbb{R}^2 \rightarrow \mathbb{R}^2$ defined by

$$T\begin{pmatrix} x \\ y \end{pmatrix} = \begin{pmatrix} x + h \\ y + k \end{pmatrix}$$

is not a linear transformation if either h or k is not zero. The failure of translations to be linear has given rise to a separate subject called **affine geometry**, which deals with transformations $T : \mathbb{R}^n \rightarrow \mathbb{R}^m$ of the form $T(\mathbf{x}) = A\mathbf{x} + \mathbf{b}$, determined by $m \times n$ matrices A and fixed translation vectors $\mathbf{b} \in \mathbb{R}^m$. As indicated in Chapter 2, these transformations are known as **affine transformations**. The **AffineMap** function of *MATHEMATICA* provides a useful tool for calculating the values of affine transformations of the plane.

EXAMPLE 5.8   ■ **An Affine Transformation** $T : \mathbb{R}^2 \rightarrow \mathbb{R}^2$

Use *MATHEMATICA* to calculate the values of the affine transformation $T : \mathbb{R}^2 \rightarrow \mathbb{R}^2$ defined by

$$A = \begin{pmatrix} 1 & 2 \\ 3 & 4 \end{pmatrix} \quad \text{and} \quad \mathbf{b} = \begin{pmatrix} 5 \\ 6 \end{pmatrix}$$

Solution. We form the augmented matrix

$$\begin{pmatrix} 1 & 2 & 5 \\ 3 & 4 & 6 \end{pmatrix}$$

and use the **AffineMap** function.

```
In[1]:= <<ProgrammingInMathematica`AffineMaps`
```

```
In[2]:= B= ( 1  2  5 )
           ( 3  4  6 ) ;
```

```
In[3]:= T:=AffineMap[B]
```

```
In[4]:= T[{x,y}]
Out[4]= {5+x+2y, 6+3x+4y}
```

As we can see, **T[{x,y}]** corresponds to the vector

$$\begin{pmatrix} 1 & 2 \\ 3 & 4 \end{pmatrix} \begin{pmatrix} x \\ y \end{pmatrix} + \begin{pmatrix} 5 \\ 6 \end{pmatrix}$$

obtained by forming the vector $A\mathbf{x}$ and then adding the vector $\mathbf{b}$ to $A\mathbf{x}$. ◀▶

Bases and Linear Transformations

In this section, we show that every basis $\mathcal{B} = \{\mathbf{x}_1, \mathbf{x}_2, \ldots, \mathbf{x}_n\}$ for a real vector space V and every list $\{\mathbf{y}_1, \mathbf{y}_2, \ldots, \mathbf{y}_n\}$ of n not necessarily distinct vectors in a vector space W determine a unique linear transformation $T : V \rightarrow W$. We express this fact by saying *that every linear transformation is uniquely determined by its values on a basis.* We will see later that if W is an m-dimensional real vector space, then this property implies that, relative to fixed bases for $\mathbb{R}^n$ and $\mathbb{R}^m$, every linear transformation $T : V \to W$ corresponds to a unique matrix transformation $T_A : \mathbb{R}^n \to \mathbb{R}^m$. The matrix A is a real $m \times n$ matrix. The transformation T is therefore defined by mn scalars. This makes linear transformations very special functions. Although $\mathbb{R}^n$ and $\mathbb{R}^m$ consist of infinitely many vectors and T usually has infinitely many distinct values, all of them are determined by a finite amount of information, namely the entries of the matrix A. It is this remarkable property that allows us to use matrix algebra to study linear transformations.

THEOREM 5.3 (Linear transformation theorem)   *If $\mathcal{B} = \{\mathbf{x}_1, \mathbf{x}_2, \ldots, \mathbf{x}_n\}$ is a basis for a vector space V and if $\mathbf{y}_1, \mathbf{y}_2, \ldots, \mathbf{y}_n$ is any list of vectors in a vector space W, then there exists a unique linear transformation $T : V \to W$ for which $T(\mathbf{x}_1) = \mathbf{y}_1, \ldots, T(\mathbf{x}_n) = \mathbf{y}_n$.*

Proof. We first observe that since $\mathcal{B}$ is a basis, every vector $\mathbf{x} \in V$ is a unique linear combination $a_1\mathbf{x}_1 + \cdots + a_n\mathbf{x}_n$. Since $T : V \to W$ is linear, we have

$$T(\mathbf{x}) = T(a_1\mathbf{x}_1 + \cdots + a_n\mathbf{x}_n) = a_1 T(\mathbf{x}_1) + \cdots + a_n T(\mathbf{x}_n)$$

Suppose now that $T(\mathbf{x}_1) = \mathbf{y}_1, \ldots, T(\mathbf{x}_n) = \mathbf{y}_n$. Then

$$a_1 T(\mathbf{x}_1) + \cdots + a_n T(\mathbf{x}_n) = a_1\mathbf{y}_1 + \cdots + a_n\mathbf{y}_n = T(\mathbf{x})$$

Since the coefficients $a_1, \ldots, a_n$ uniquely determine both $\mathbf{x}$ and $T(\mathbf{x})$, the transformation T is unique. ∎

EXAMPLE 5.9   ■ **Defining a Linear Transformation $T : \mathbb{R}^3 \to \mathbb{R}^4$ on a Basis**

Show that the linear transformation $T : \mathbb{R}^3 \to \mathbb{R}^4$ defined by

$$T \begin{pmatrix} a_1 \\ a_2 \\ a_3 \end{pmatrix} = \begin{pmatrix} a_1 + a_2 \\ a_2 \\ a_3 \\ a_1 - a_3 \end{pmatrix}$$

is uniquely determined by its values on the standard basis $\mathcal{E}$ of $\mathbb{R}^3$.

Solution. We can see by direct calculation that the values of T on the standard basis

$$\mathcal{E} = \{\mathbf{e}_1 = (1, 0, 0), \mathbf{e}_2 = (0, 1, 0), \mathbf{e}_3 = (0, 0, 1)\}$$

are

$$T(\mathbf{e}_1) = (1, 0, 0, 1), \ T(\mathbf{e}_2) = (1, 1, 0, 0), \ T(\mathbf{e}_3) = (0, 0, 1, -1)$$

If $\mathbf{x} = a_1\mathbf{e}_1 + a_2\mathbf{e}_2 + a_3\mathbf{e}_3$ is any vector in $\mathbb{R}^3$, written as a linear combination in the basis $\mathcal{E}$, then the linearity of T implies that

$$T(\mathbf{x}) = T(a_1\mathbf{e}_1 + a_2\mathbf{e}_2 + a_3\mathbf{e}_3) = a_1 T(\mathbf{e}_1) + a_2 T(\mathbf{e}_2) + a_3 T(\mathbf{e}_3)$$

This shows that the three vectors $T(\mathbf{e}_1)$, $T(\mathbf{e}_2)$, $T(\mathbf{e}_3)$ and the three scalars a_1, a_2, a_3 contain all the information required to compute the vector $T(\mathbf{x})$ in the basis $\mathcal{E}$. ◄►

EXAMPLE 5.10 ■ **Using *MATHEMATICA* to Define a Linear Transformation**

Use the **Function** command of *MATHEMATICA* to define the linear transformation $T : \mathbb{R}^5 \to \mathbb{R}^4$ given by $T(u, v, x, y, z) = (2x - 7z, z, u - v, y)$, and calculate the vector $T(\mathbf{x}) = T(1, 2, 3, 4, 5)$.

Solution. We define the function T by specifying its input and corresponding output.

```
In[1]= T:=Function[{x},{2x[[3]]-7x[[5]],x[[5]],
x[[1]]-x[[2]],x[[4]]}]
```

Next we specify the vector $\mathbf{x} = (1, 2, 3, 4, 5)$.

```
In[2]= x={1,2,3,4,5};
```

We can now calculate $T(\mathbf{x})$.

```
In[3]= T[x]
Out[3]= {-29,5,-1,4}
```

This shows that $T(\mathbf{x}) = (-29, 5, -1, 4) \in \mathbb{R}^4$. ◄►

EXERCISES 5.1

1. Show that for all $\theta \in \mathbb{R}$, the matrix

$$A = \begin{pmatrix} \cos \theta & -\sin \theta \\ \sin \theta & \cos \theta \end{pmatrix}$$

determines a linear transformation $T_\theta : \mathbb{R}^2 \to \mathbb{R}^2$.

2. Explain why the translation function $T : \mathbb{R}^2 \to \mathbb{R}^2$ defined by

$$T \begin{pmatrix} x \\ y \end{pmatrix} = \begin{pmatrix} x + h \\ y + k \end{pmatrix}$$

is almost never linear. When is it linear?

3. Explain why the following functions $T : \mathbb{R}^1 \to \mathbb{R}^1$ are not linear transformations.

a. $T(x) = x + 2$ b. $T(x) = \cos x$ c. $T(x) = x^2$

d. $T(x) = 3$ e. $T(x) = \exp x$ f. $T(x) = \log x$

4. Determine which of the following functions $T : \mathbb{R}^1 \to \mathbb{R}^2$ are linear transformations.

a. $T(x) = \begin{pmatrix} x \\ x \end{pmatrix}$ b. $T(x) = \begin{pmatrix} x \\ 2x \end{pmatrix}$ c. $T(x) = \begin{pmatrix} x \\ 0 \end{pmatrix}$

d. $T(x) = \begin{pmatrix} 0 \\ x \end{pmatrix}$ e. $T(x) = \begin{pmatrix} 0 \\ 0 \end{pmatrix}$ f. $T(x) = 3 \begin{pmatrix} x \\ x \end{pmatrix}$

5. Determine which of the following functions $T : \mathbb{R}^2 \to \mathbb{R}^1$ are linear transformations.

a. $T \begin{pmatrix} x \\ y \end{pmatrix} = x$ b. $T \begin{pmatrix} x \\ y \end{pmatrix} = y$ c. $T \begin{pmatrix} x \\ y \end{pmatrix} = 0$

d. $T \begin{pmatrix} x \\ y \end{pmatrix} = xy$ e. $T \begin{pmatrix} x \\ y \end{pmatrix} = 7$ f. $T \begin{pmatrix} x \\ y \end{pmatrix} = x + y$

6. Determine which of the following functions $T : \mathbb{R}^2 \to \mathbb{R}^2$ are linear transformations.

a. $T \begin{pmatrix} x \\ y \end{pmatrix} = \begin{pmatrix} 42x + 71y \\ 18x + 44y \end{pmatrix}$ b. $T \begin{pmatrix} x \\ y \end{pmatrix} = \begin{pmatrix} x \\ y^2 \end{pmatrix}$

c. $T \begin{pmatrix} x \\ y \end{pmatrix} = \begin{pmatrix} \cos x \\ y \end{pmatrix}$ d. $T \begin{pmatrix} x \\ y \end{pmatrix} = 3 \begin{pmatrix} x \\ y \end{pmatrix} - 5 \begin{pmatrix} y \\ x \end{pmatrix}$

7. Determine which of the following functions are linear transformations.

a. $T : \mathbb{R}^{n \times n} \to \mathbb{R}$ defined by $T(A) = \det A$

b. $T : \mathbb{R}^{n \times n} \to \mathbb{R}$ defined by $T(A) = \text{trace}(A)$

c. $T : \mathbb{R}^{n \times n} \to \mathbb{R}^n$ defined by $T(A) = \text{diag}(A)$

d. $T : \mathbb{R}^{m \times n} \to \mathbb{R}^{n \times m}$ defined by $T(A) = A^T$

e. $T : \mathbb{R}^{m \times n} \to \mathbb{R}^{n \times m}$ defined by $T(A) = -A^T$

8. Explain why the function $T(A) = A^{-1}$ is not a linear transformation.

9. Find the images of the standard basis vectors of the matrix transformation $T_A : \mathbb{R}^3 \to \mathbb{R}^4$ determined by the transpose of the matrix

$$\begin{pmatrix} 7 & 17 & 38 & 27 \\ 0 & 22 & 7 & 34 \\ 0 & 0 & 9 & 25 \end{pmatrix}$$

10. Explain why the distributive law $a(b+c) = ab+ac$ guarantees that the conversion formula $f(x) = 5280x$ from miles to feet is a linear transformation $f : \mathbb{R} \to \mathbb{R}$.

11. Explain why the conversion formula $F = \frac{9}{5}C + 32$ from degrees Celsius to degrees Fahrenheit is an affine transformation that is not linear.

12. Describe the images of the vertices $(-1, 0)$, $(1, 0)$, $(0, 2)$ of an isosceles triangle under the affine transformation

$$
T\begin{pmatrix} x \\ y \end{pmatrix} = \begin{pmatrix} 3 & 0 \\ 0 & 5 \end{pmatrix}\begin{pmatrix} x \\ y \end{pmatrix} + \begin{pmatrix} 4 \\ 5 \end{pmatrix}
$$

13. Let $T_1, T_2 : U \to V$ be two linear transformations and let $T_1 + T_2 : U \to V$ be their **sum**, defined by

$$
(T_1 + T_2)(\mathbf{x}) = T_1(\mathbf{x}) + T_2(\mathbf{x}).
$$

Show that $T_1 + T_2$ is a linear transformation.

14. Let $T : U \to V$ be a linear transformation, let s be any scalar, and let $sT : U \to V$ be a **scalar multiple** of T, defined by

$$
(sT)(\mathbf{x}) = s(T(\mathbf{x})).
$$

Show that sT is a linear transformation.

15. Use *MATHEMATICA* to verify that the function $f : \mathbb{R}^2 \to \mathbb{R}$ defined by

$$
f\begin{pmatrix} x \\ y \end{pmatrix} = 3x + 4y
$$

is linear.

16. Verify that for any fixed vector $\begin{pmatrix} x_0 \\ y_0 \end{pmatrix}$, the function $f : \mathbb{R}^2 \to \mathbb{R}$ defined by

$$
f\begin{pmatrix} x \\ y \end{pmatrix} = x_0 x + y_0 y
$$

is linear.

17. Verify that the function $f(x, y, z) = 5x + 2y - 7z$ is a linear transformation $f : \mathbb{R}^3 \to \mathbb{R}$. Use *MATHEMATICA* to confirm this fact.

18. Verify that for any fixed vector $\mathbf{x}_0 \in \mathbb{R}^n$ the function $f : \mathbb{R}^n \to \mathbb{R}$ defined by $f(\mathbf{y}) = \mathbf{x}_0^T \mathbf{y}$ is a linear transformation. Use *MATHEMATICA* to confirm this fact for $n = 5$.

19. Verify that the function $f : \mathbb{R}[t] \to \mathbb{R}$ defined by $f(p(t)) = p(0)$ is a linear transformation. Write a *MATHEMATICA* routine for calculating the values of f.

20. Use the **AffineMap** function of *MATHEMATICA* to find the images $A\mathbf{x}+\mathbf{b}$ of the vertices $(0, 0)$, $(1, 0)$, $(1, 1)$, and $(0, 1)$ of the unit square for the following affine transformations.

   a. $A = \begin{pmatrix} 2 & 3 \\ 1 & 6 \end{pmatrix}$ and $\mathbf{b} = \begin{pmatrix} -3 \\ 4 \end{pmatrix}$

   b. $A = \begin{pmatrix} 0 & 1 \\ -1 & 0 \end{pmatrix}$ and $\mathbf{b} = \begin{pmatrix} 5 \\ 9 \end{pmatrix}$

21. Repeat Exercise 12 using the **AffineMap** function of *MATHEMATICA*.

22. Show that a *MATHEMATICA* function **AffineMap[A]** defined by a real 2×3 matrix A determines a linear $T : \mathbb{R}^2 \to \mathbb{R}^2$ if and only if the third column of A is a zero column.

Invertible Transformations

In many applications, we require linear transformations that are *invertible*. In geometry, for example, we often require linear transformations that preserve length, distance, and angles. This will happen only if the transformations are invertible. A linear transformation $T : V \to W$ is ***invertible*** if it is paired with another linear transformation $S : W \to V$ for which $S(T(\mathbf{v})) = \mathbf{v}$ for all $\mathbf{v} \in V$ and $T(S(\mathbf{w})) = \mathbf{w}$ for all $\mathbf{w} \in W$.

In this section, we discuss the basic properties of invertible linear transformations. For reasons that will become clear later, we also refer to invertible linear transformations as ***isomorphisms***. The prefix iso comes from Greek and means "equal," and morphism also comes from Greek and means "form." Isomorphisms between vector spaces are functions between two mathematical objects of the same form preserving that form. We will prove that for every n-dimensional real vector space V there exists an isomorphism from V to the coordinate space $\mathbb{R}^n$. This fact is fundamental for the link between linear and matrix algebra.

The definition of invertible linear transformations is based on *identity transformations* and the *composition* of linear transformations.

DEFINITION 5.3   *The **identity transformation** $I_V : V \to V$ on a vector space V is the function defined by $I_V(\mathbf{x}) = \mathbf{x}$ for all $\mathbf{x} \in V$.*

Identity transformations are linear since $I_V(a\mathbf{x} + b\mathbf{y}) = a\mathbf{x} + b\mathbf{y} = aI_V(\mathbf{x}) + bI_V(\mathbf{y})$ for all vectors $\mathbf{x}$ and $\mathbf{y}$ and all scalars a and b. We write I in place of I_V if V is understood from the context.

DEFINITION 5.4   *The **composition** of two linear transformations $T : U \to V$ and $S : V \to W$ is the function $S \circ T : U \to W$, defined by*

$$(S \circ T)(\mathbf{x}) = S(T(\mathbf{x}))$$

Composite linear transformations $S \circ T$ are linear since

$$(S \circ T)(a\mathbf{x} + b\mathbf{y}) = S(T(a\mathbf{x} + b\mathbf{y}))$$
$$= S(aT(\mathbf{x}) + bT(\mathbf{y}))$$
$$= a(S(T(\mathbf{x}))) + b(S(T(\mathbf{y})))$$
$$= a(S \circ T)(\mathbf{x}) + b(S \circ T)(\mathbf{y})$$

for all vectors $\mathbf{x}$ and $\mathbf{y}$ and all scalars a and b. We write ST in place of $S \circ T$ when the intended meaning is clear from the context.

We use identity transformations and composition to define the inverse of a linear transformation.

DEFINITION 5.5 *If $T : U \to V$ and $S : V \to U$ are linear transformations with the property that $S \circ T = I_U$ and $T \circ S = I_V$, then S is an **inverse** of T and T is an **inverse** of S.*

It is easy to see that the inverse of a linear transformation T is uniquely determined by T. We can therefore speak of **the inverse** of T. We write T^{-1} for the inverse of T.

DEFINITION 5.6 *If $T : U \to V$ is a linear transformation that has an inverse, T is **invertible**.*

It is an easy exercise to prove that the composition of two invertible linear transformations (in other words, of two isomorphisms) is an isomorphism.

The invertibility of a linear transformation can be described in terms of two basic mapping properties.

1. A linear transformation $T : V \to W$ is ***one-one*** if $T(\mathbf{x}) = T(\mathbf{y})$ implies that $\mathbf{x} = \mathbf{y}$. This is equivalent to saying that T is one-one if $\mathbf{x} \neq \mathbf{y}$ implies that $T(\mathbf{x}) \neq T(\mathbf{y})$. Figure 2 illustrates the idea of a one-one linear transformation. We note that this definition generalizes the definition of one-one functions used in Chapter 3 to define permutations of standard lists.

2. A linear transformation $T : V \to W$ is ***onto*** if for every vector $\mathbf{w} \in W$ there exists an $\mathbf{x} \in V$ such that $T(\mathbf{x}) = \mathbf{w}$. In other words, a linear transformation T is *not* onto if there exists a vector $\mathbf{w} \in W$ that is not in the image of T. Figure 3 shows graphically an onto linear transformation.

We now show that invertible linear transformations can be described as linear transformations that are both one-one and onto.

THEOREM 5.4 *If $T : V \to W$ is an invertible linear transformation, then T is one-one.*

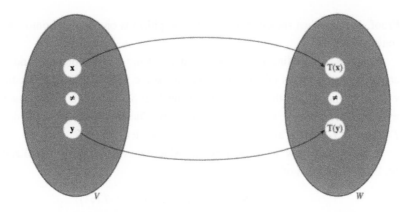

FIGURE 2   A One-One Transformation.

Proof. We prove this theorem by showing that if $T(\mathbf{x}) = T(\mathbf{y})$, then $\mathbf{x}$ must be equal to $\mathbf{y}$. Suppose that T is invertible and that S is its inverse. If $T(\mathbf{x}) = T(\mathbf{y})$, then $S(T(\mathbf{x})) = S(T(\mathbf{y}))$. But since $S(T(\mathbf{x})) = (S \circ T)(\mathbf{x}) = I_V(\mathbf{x}) = \mathbf{x}$ and $S(T(\mathbf{y})) = (S \circ T)(\mathbf{y}) = I_V(\mathbf{y}) = \mathbf{y}$, it follows that $\mathbf{x} = \mathbf{y}$. ∎

THEOREM 5.5   *If $T : V \rightarrow W$ is an invertible linear transformation, then T is onto.*

Proof. Suppose that T is invertible and that S is its inverse. If $\mathbf{y} \in W$, then $S(\mathbf{y}) \in V$ and $T(S(\mathbf{y})) = (T \circ S)(\mathbf{y}) = I_W(\mathbf{y}) = \mathbf{y}$. Therefore, $\mathbf{x} = S(\mathbf{y})$ is a vector in V whose value in W is $\mathbf{y}$. ∎

We now prove the converse of Theorems 5.4 and 5.5.

THEOREM 5.6   *A linear transformation $T : V \rightarrow W$ is an isomorphism if and only if T is one-one and onto.*

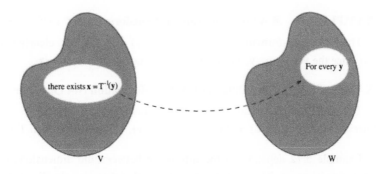

FIGURE 3   An Onto Transformation.

Proof. We have just shown that every isomorphism is one-one and onto. We now prove the converse.

Suppose that the linear transformation $T : V \rightarrow W$ is one-one and onto. Since T is onto, every $\mathbf{w} \in W$ is of the form $T(\mathbf{v})$ for some $\mathbf{v} \in V$. Since T is one-one, the vector $\mathbf{v} \in V$ is unique. We therefore define a function $S : W \rightarrow V$ that associates with each $\mathbf{w} \in W$ the unique $\mathbf{v} \in V$ for which $T(\mathbf{v}) = \mathbf{w}$. Then it follows from the definition of S that $T(S(\mathbf{w})) = T(\mathbf{v}) = \mathbf{w}$ and $S(T(\mathbf{v})) = S(\mathbf{w}) = \mathbf{v}$. Hence $S = T^{-1}$.

It remains to show that S is linear. Consider two vectors $\mathbf{w}_1 = T(\mathbf{v}_1)$ and $\mathbf{w}_2 = T(\mathbf{v}_2)$ in W. By the definition of S,

$$
\begin{aligned}
S(\mathbf{w}_1 + \mathbf{w}_2) &= S(T(\mathbf{v}_1) + T(\mathbf{v}_2)) \\
&= S(T(\mathbf{v}_1 + T\mathbf{v}_2)) \\
&= (S \circ T)(\mathbf{v}_1 + \mathbf{v}_2) \\
&= \mathbf{v}_1 + \mathbf{v}_2 \\
&= S(\mathbf{w}_1) + S(\mathbf{w}_2)
\end{aligned}
$$

Moreover, for any scalar a,

$$
\begin{aligned}
S(a\mathbf{w}_1) &= S(aT(\mathbf{v}_1)) \\
&= S(T(a\mathbf{v}_1)) \\
&= (S \circ T)(a\mathbf{v}_1) \\
&= a\mathbf{v}_1 \\
&= aS(\mathbf{w}_1)
\end{aligned}
$$

Hence S is a linear transformation. ∎

Using Theorem 5.6 and elementary properties of functions, we can easily show that a composition of isomorphisms is an isomorphism.

Next we show examples of linear transformations that are one-one but not onto, and onto but not one-one.

EXAMPLE 5.11   ■ A One-One Linear Transformation $T : \mathbb{R}^2 \to \mathbb{R}^3$ Not Onto

The linear transformation defined by $T(x, y) = (x, y, 0)$ is clearly one-one. But it is not onto since the vector $(x, y, 1)$, for example, is not a value of T. ◀▶

EXAMPLE 5.12   ■ An Onto Linear Transformation Not One-One

The linear transformation $T : \mathbb{R}^3 \to \mathbb{R}^2$ defined by $T(x, y, z) = (x, y)$ is clearly onto. However, since $T(x, y, 1) = T(x, y, 2) = (x, y)$, the transformation T is not one-one. ◀▶

Example 5.12 depends on the difference between the dimensions of the domains and codomains involved. The next example shows that there are also linear transformations with identical domains and codomains that are one-one but not onto.

EXAMPLE 5.13 ■ **A One-One Linear Transformation $T : \mathbb{R}[t] \to \mathbb{R}[t]$ Not Onto**

Show that the linear transformation $T : \mathbb{R}[t] \to \mathbb{R}[t]$ defined by $T(p(t)) = tp(t)$ is one-one but not onto.

Solution. The calculation

$$T(ap(t) + bq(t)) = t(ap(t) + bq(t))$$
$$= atp(t) + btq(t)$$
$$= aT(p(t)) + bT(q(t))$$

shows that T is linear. In particular, $T(\mathbf{0}) = t\mathbf{0} = \mathbf{0}$. Moreover, if $T(p(t)) = T(q(t))$, then $tp(t) = tq(t)$. This implies that

$$tp(t) - tq(t) = t(p(t) - q(t))$$
$$= p(t) - q(t)$$
$$= 0$$

Therefore, T is one-one. On the other hand, no nonzero constant polynomial is a value of T. Hence the transformation T is not onto. ◀▶

Later we will show that a linear transformation $T : V \to W$ cannot be one-one but not onto if the vector spaces V and W have the same finite dimension. In Theorem 5.27 we will prove that if V is finite-dimensional, then every one-one linear transformation from V to V is also onto, and every onto linear transformation from V to V is also one-one.

EXAMPLE 5.14 ■ **An Isomorphism Between $\mathbb{R}_3[t]$ and $\mathbb{R}^4$**

Construct an isomorphism between the matrix space $\mathbb{R}_3[t]$ and the coordinate space $\mathbb{R}^4$.

Solution. Let $T_1 : \mathbb{R}^4 \to \mathbb{R}_3[t]$ be the linear transformation defined in Example 5.2 and $T_2 : \mathbb{R}_3[t] \to \mathbb{R}^4$ be the linear transformations defined in Example 5.4. It is easy to check that $I = T_2 \circ T_1 : \mathbb{R}^4 \to \mathbb{R}^4$ is the identity transformation on $\mathbb{R}^4$ and that $I = T_1 \circ T_2 : \mathbb{R}_3[t] \to \mathbb{R}_3[t]$ is the identity transformation on $\mathbb{R}_3[t]$. Therefore, T_1 is an isomorphism between $\mathbb{R}_3[t]$ and $\mathbb{R}^4$, and so is its inverse T_2. ◀▶

EXAMPLE 5.15 ■ **An Isomorphism Between $\mathbb{R}^{2 \times 2}$ and $\mathbb{R}^4$**

Construct an isomorphism between the matrix space $\mathbb{R}^{2 \times 2}$ and the coordinate space $\mathbb{R}^4$.

Solution. Let

$$A = \begin{pmatrix} a_{11} & a_{12} \\ a_{21} & a_{22} \end{pmatrix}$$

be an arbitrary matrix in $\mathbb{R}^{2\times 2}$. Then the function $T : \mathbb{R}^{2\times 2} \to \mathbb{R}^4$ defined by

$$T \begin{pmatrix} a_{11} & a_{12} \\ a_{21} & a_{22} \end{pmatrix} = \begin{pmatrix} a_{11} \\ a_{12} \\ a_{21} \\ a_{22} \end{pmatrix}$$

is an isomorphism between $\mathbb{R}^{2\times 2}$ and $\mathbb{R}^4$. ◀▶

EXAMPLE 5.16 ■ An Isomorphism Between $\mathbb{R}^{2\times 3}$ and $\mathbb{R}_5[t]$

Construct an isomorphism between the matrix space $\mathbb{R}^{2\times 3}$ and the polynomial space $\mathbb{R}_5[t]$.

Solution. Let

$$A = \begin{pmatrix} a_{11} & a_{12} & a_{13} \\ a_{21} & a_{22} & a_{23} \end{pmatrix}$$

be an arbitrary matrix in $\mathbb{R}^{2\times 3}$. Then the function $T : \mathbb{R}^{2\times 3} \to \mathbb{R}_5[t]$ defined by

$$T \begin{pmatrix} a_{11} & a_{12} & a_{13} \\ a_{21} & a_{22} & a_{23} \end{pmatrix} = a_{11} + a_{12}t + a_{13}t^2 + a_{21}t^3 + a_{22}t^4 + a_{23}t^5$$

is an isomorphism between $\mathbb{R}^{2\times 3}$ and $\mathbb{R}_5[t]$. ◀▶

Several of the examples of linear transformations given so far are defined in terms of matrices. This suggests that matrices may be useful for defining linear transformations more generally, and this is actually the case. All real vector spaces of the same dimension d are isomorphic to the coordinate space $\mathbb{R}^d$. We can therefore study a linear transformation $T : V \to W$ from an n-dimensional vector space V to an m-dimensional vector space W by studying a corresponding linear transformation $T' : \mathbb{R}^n \to \mathbb{R}^m$. After working with T', we can go back to V and W using given isomorphisms. Moreover, we will see later that the linear transformation T' can be represented by an $m \times n$ matrix A with the property that $T'(\mathbf{x}) = A\mathbf{x}$. This means that we will be able to find the values of T' by matrix multiplication.

THEOREM 5.7 (Coordinate space isomorphism theorem) *For any n-dimensional real vector space V and any basis B for V, the coordinate vector function $[-]_B : V \to \mathbb{R}^n$ is an isomorphism.*

Proof. In Theorem 4.15 and its Corollary 4.16, we proved that $[-]_B$ is an invertible linear transformation. ■

COROLLARY 5.8   *If two real vector spaces have the same dimension, they are isomorphic.*

Proof. Suppose that V and W have dimension n. By Theorem 5.7, there exists an isomorphism $T : V \to \mathbb{R}^n$ and an isomorphism $S : W \to \mathbb{R}^n$. It is easy to check that the linear transformation $S^{-1} \circ T : V \to W$ is an isomorphism. ∎

COROLLARY 5.9   *The matrix space $\mathbb{R}^{m \times n}$ is isomorphic to the coordinate space $\mathbb{R}^{mn}$.*

Proof. Since the dimension of both $\mathbb{R}^{m \times n}$ and $\mathbb{R}^{mn}$ is mn, the result follows from Corollary 5.8. ∎

COROLLARY 5.10   *The polynomial space $\mathbb{R}_n[t]$ is isomorphic to the coordinate space $\mathbb{R}^{n+1}$.*

Proof. Since $\mathbb{R}_n[t]$ and $\mathbb{R}^{n+1}$ are both of dimension $n + 1$, the result follows again from Corollary 5.8. ∎

EXERCISES 5.2

1. Construct two linear transformations S and T for which $ST \neq TS$.

2. Find two linear transformations S and T for which $ST = TS$.

3. Show that if $S : U \to V$ and $T : V \to W$ are two isomorphisms, then $T \circ S$ is an isomorphism.

4. Show by means of an example that an isomorphism is one-one and onto.

5. Show by means of an example that a one-one, onto linear transformation is invertible.

6. Prove that the inverse of a linear transformation is unique.

7. Prove or refute the claim that if T is a one-one linear transformation $\mathbb{R}^n \to \mathbb{R}^m$, then $n = m$.

8. Prove or refute the claim that if T is an onto linear transformation $\mathbb{R}^n \to \mathbb{R}^m$, then $n = m$.

9. Show by means of examples that a linear transformation fails to be invertible if it is not one-one or not onto.

10. Prove or refute the claim that if T is an isomorphism $\mathbb{R}^n \to \mathbb{R}^m$, then $n = m$.

11. Prove that if T is a one-one linear transformation $\mathbb{R}^n \to \mathbb{R}^m$, then $\mathbb{R}^n$ is isomorphic to a subspace of $\mathbb{R}^m$.

12. Show that the functions defined in Examples 5.15 and 5.16 are isomorphisms.

13. Show that the equations

$$T_1 \circ (T_2 + T_3) = T_1 \circ T_2 + T_1 \circ T_3$$
$$(T_1 + T_2) \circ T_3 = T_1 \circ T_3 + T_2 \circ T_3$$
$$s(T_1 \circ T_2) = (sT_1) \circ T_2 = T_1 \circ (sT_2)$$

hold for all linear transformations T_1, T_2, T_3 and all scalars s.

14. Let V and W be two real vector spaces. Show that the set $\text{Hom}(V, W)$ of all linear transformations from V to W is a real vector space.

15. Show that the set $\text{Hom}(V, V)$ of all linear transformations from V to V is closed under sums, scalar multiples, and composition. Vector spaces with this structure satisfying the equations in Exercise 11 are called **linear algebras**.

16. Use *MATHEMATICA* to verify that the sets $\text{Hom}(\mathbb{R}, \mathbb{R})$ and $\text{Hom}(\mathbb{R}^2, \mathbb{R}^2)$ are linear algebras.

17. It follows from Exercise 14 that for any real vector space V, the set $\text{Hom}(V, \mathbb{R})$ is a real vector space. It is called the **dual space** of V and is denoted by V^*. The elements of V^* are called **linear functionals**. Show that if $V = \mathbb{R}^2$, then the projection functions $p_1(x_1, x_2) = x_1$ and $p_2(x_1, x_2) = x_2$ are linear functionals.

18. Let $\varphi : V \to V^{**}$ be the linear transformation $\varphi(\mathbf{v}) : V^* \to \mathbb{R}$ defined by $\varphi(\mathbf{v})(f) = f(\mathbf{v})$. Verify that φ is linear. Suppose that $\varphi(\mathbf{v}) = 0$ only if $\mathbf{v} = 0$. Show that this implies that φ is an isomorphism.

19. Let $\mathcal{B} = \{\mathbf{v}_1, \mathbf{v}_2, \mathbf{v}_3\}$ be a basis for a real vector space V and associate with each $\mathbf{v}_i$ a linear functional $\varphi_i : V \to \mathbb{R}$ defined by

$$\varphi_i(\mathbf{v}_j) = \begin{cases} 1 & \text{if } i = j \\ 0 & \text{if } i \neq j \end{cases}$$

Show that the set $\mathcal{C} = \{\varphi_1, \varphi_2, \varphi_3\}$ is a basis for V^*. (*Hint:* Let $f : V \to \mathbb{R}$ be any linear functional determined by its three values $f(\mathbf{v}_i) = a_i \in \mathbb{R}$ on the basis $\mathcal{B}$. Show that $f = a_1\varphi_1 + a_2\varphi_2 + a_3\varphi_3$. Show that this equation holds for the basis vector $\mathbf{v}_i$.)

20. Suppose that $T : \mathbb{R}^3 \to \mathbb{R}^2$ is a linear transformation. Show that the associated function $T^* : \text{Hom}(\mathbb{R}^2, \mathbb{R}) \to \text{Hom}(\mathbb{R}^3, \mathbb{R})$ defined by $T^*(f) = f \circ T$ is a linear transformation.

21. In communication theory, a *signal* is sometimes defined to be a function

$$f(t) = \frac{1}{2}a_0 + \sum_{k \in \mathbf{n}} b_k \cos 2\pi kt + \sum_{k \in \mathbf{n}} c_k \sin 2\pi kt$$

with $a_0, b_k, c_k \in \mathbb{R}$. Prove that the set of all signals $f(t)$ is isomorphic to $\mathbb{R}^{2n+1}$.

Singular Transformations

A linear transformation $T : V \to W$ is **singular** if there exists a nonzero vector $\mathbf{x} \in V$ for which $T(\mathbf{x}) = 0$. A linear transformation that is not singular is called **nonsingular**. Singular transformations cannot be one-one since $T(\mathbf{0}) = T(\mathbf{x}) = 0$ for some nonzero vector $\mathbf{x}$. Therefore, singular transformations are not invertible. On the other hand, they can certainly be onto. Let $W = \{\mathbf{0}\}$ be a zero space and let V contain nonzero vectors. Since $T(\mathbf{0}) = \mathbf{0}$, T is onto. But there also exists a nonzero vector $\mathbf{x} \in V$ for which $T(\mathbf{x}) = \mathbf{0}$. Hence T is singular.

Here is an example of a singular linear transformation, called a **projection**. It projects the space $\mathbb{R}^3$ onto the xz-plane in $\mathbb{R}^3$.

EXAMPLE 5.17 ■ **A Projection Function**

Show that the function $T : \mathbb{R}^3 \to \mathbb{R}^3$ defined by $T(x, y, z) = (x, 0, z)$ is a singular linear transformation.

Solution. First we note that

$$
T\left[a \begin{pmatrix} a_1 \\ a_2 \\ a_3 \end{pmatrix} + b \begin{pmatrix} b_1 \\ b_2 \\ b_3 \end{pmatrix} \right] = T \begin{pmatrix} aa_1 + bb_1 \\ aa_2 + bb_2 \\ aa_3 + bb_3 \end{pmatrix} = \begin{pmatrix} aa_1 + bb_1 \\ 0 \\ aa_3 + bb_3 \end{pmatrix}
$$

and

$$
\begin{pmatrix} aa_1 + bb_1 \\ 0 \\ aa_3 + bb_3 \end{pmatrix} = \begin{pmatrix} aa_1 \\ 0 \\ aa_3 \end{pmatrix} + \begin{pmatrix} bb_1 \\ 0 \\ bb_3 \end{pmatrix} = a \begin{pmatrix} a_1 \\ 0 \\ a_3 \end{pmatrix} + b \begin{pmatrix} b_1 \\ 0 \\ b_3 \end{pmatrix}
$$

Therefore,

$$
T\left[a \begin{pmatrix} a_1 \\ a_2 \\ a_3 \end{pmatrix} + b \begin{pmatrix} b_1 \\ b_2 \\ b_3 \end{pmatrix} \right] = aT \begin{pmatrix} a_1 \\ a_2 \\ a_3 \end{pmatrix} + bT \begin{pmatrix} b_1 \\ b_2 \\ b_3 \end{pmatrix}
$$

This shows that T is linear. In addition, T is singular since $T(0, y, 0) = (0, 0, 0)$ for any $y \in \mathbb{R}$. ◄►

Projections will be discussed more fully in Chapter 8.

EXERCISES 5.3

1. Show that a linear transformation $T : \mathbb{R}^n \to \mathbb{R}^n$ is singular if and only if there exists an $n \times n$ singular matrix A for which $T(\mathbf{x}) = A\mathbf{x}$.

2. Show that the linear transformation T_A and T_B determined by the matrices

$$A = \begin{pmatrix} 1 & 0 & 0 \\ 2 & 5 & 0 \\ 3 & 0 & 0 \end{pmatrix} \quad \text{and} \quad B = \begin{pmatrix} 0 & 0 & 1 \\ 0 & 0 & 7 \\ 0 & 0 & 4 \end{pmatrix}$$

are singular. Does this imply that $T_{(A+B)}$ is singular?

3. Show that the identity transformation $I : \mathbb{R}^n \to \mathbb{R}^n$ is a sum of singular transformations.

4. Prove or disprove the claim that the composition of two singular transformations is singular.

5. Prove or disprove the claim that the composition of a singular and a nonsingular transformation is singular.

6. Show that if A is a square matrix and the matrix transformation T_A is nonsingular, then the matrix transformation determined by A^n is nonsingular for all n.

7. Let A be a matrix obtained from the identity matrix I_3 by replacing one or two of its diagonal entries by 0. The matrix A is called a ***projection matrix***. Show that projection matrices are singular and explain in what sense they determine matrix transformations that project $\mathbb{R}^3$ onto subspaces of $\mathbb{R}^3$.

MATRICES OF LINEAR TRANSFORMATIONS

We know from Theorem 5.1 that every real $m \times n$ matrix A determines a linear transformation $T_A : \mathbb{R}^n \to \mathbb{R}^m$. Now we show that the converse is also true. Every linear transformation $T : \mathbb{R}^n \to \mathbb{R}^m$ is a matrix transformation T_A for some A. Table 1 summarizes the components of the construction.

TABLE 1   Matrix of a Linear Transformation.

Objects	Explanations
$T : \mathbb{R}^n \to \mathbb{R}^m$	Linear transformation
$\mathcal{B} = \{\mathbf{x}_1, \ldots, \mathbf{x}_n\}$	Basis for $\mathbb{R}^n$
$\mathcal{C} = \{\mathbf{y}_1, \ldots, \mathbf{y}_m\}$	Basis for $\mathbb{R}^m$
$T(\mathbf{x}_1), \ldots, T(\mathbf{x}_n)$	Image vectors in $\mathbb{R}^m$
$[T(\mathbf{x}_1)]_{\mathcal{C}}, \ldots, [T(\mathbf{x}_n)]_{\mathcal{C}}$	Coordinate vectors in $\mathbb{R}^m$
$[T]_{\mathcal{C}}^{\mathcal{B}} = \left([T(\mathbf{x}_1)]_{\mathcal{C}} \quad \cdots \quad [T(\mathbf{x}_n)]_{\mathcal{C}} \right)$	Matrix of T

The matrix A depends on a basis $\mathcal{B}$ for $\mathbb{R}^n$ and a basis $\mathcal{C}$ for $\mathbb{R}^m$. We build this dependence into our notation by writing $[T]_{\mathcal{C}}^{\mathcal{B}}$ for the matrix A. The columns of $[T]_{\mathcal{C}}^{\mathcal{B}}$ are the coordinate vectors $[T(\mathbf{x})]_\mathcal{C}$ of the images $T(\mathbf{x})$ in the basis $\mathcal{C}$ of the basis vectors $\mathbf{x} \in \mathcal{B}$. If $\mathcal{B} = \mathcal{C}$, we write $[T]_\mathcal{B}$ for $[T]_\mathcal{C}^\mathcal{B}$.

DEFINITION 5.7 *If $T : \mathbb{R}^n \to \mathbb{R}^m$ is a linear transformation and $\mathcal{B}$ is a basis for $\mathbb{R}^n$ and $\mathcal{C}$ a basis for $\mathbb{R}^m$, then the matrix $[T]_\mathcal{C}^\mathcal{B}$ is the **matrix of T in the bases $\mathcal{B}$ and $\mathcal{C}$**.*

If $\mathcal{B}$ and $\mathcal{C}$ are standard bases, we call $[T]_\mathcal{C}^\mathcal{B}$ the **standard matrix of T**. We say that the matrix $[T]_\mathcal{C}^\mathcal{B}$ **represents** T in the bases $\mathcal{B}$ and $\mathcal{C}$.

EXAMPLE 5.18 ■ The Matrix of a Linear Transformation

Let $T : \mathbb{R}^3 \to \mathbb{R}^2$ be the linear transformation defined by

$$T\begin{pmatrix} x \\ y \\ z \end{pmatrix} = \begin{pmatrix} 3x + 2y - 4z \\ x - 5y + 3z \end{pmatrix}$$

Use *MATHEMATICA* to find the matrix $[T]_\mathcal{C}^\mathcal{B}$ of T in the bases

$$\mathcal{B} = \left\{ \mathbf{x}_1 = \begin{pmatrix} 1 \\ 1 \\ 1 \end{pmatrix}, \mathbf{x}_2 = \begin{pmatrix} 1 \\ 1 \\ 0 \end{pmatrix}, \mathbf{x}_3 = \begin{pmatrix} 1 \\ 0 \\ 0 \end{pmatrix} \right\}$$

and

$$\mathcal{C} = \left\{ \mathbf{y}_1 = \begin{pmatrix} 1 \\ 3 \end{pmatrix}, \mathbf{y}_2 = \begin{pmatrix} 2 \\ 5 \end{pmatrix} \right\}$$

Solution. By definition,

$$[T]_\mathcal{C}^\mathcal{B} = \left([T(\mathbf{x}_1)]_\mathcal{C} \quad [T(\mathbf{x}_2)]_\mathcal{C} \quad [T(\mathbf{x}_3)]_\mathcal{C} \right)$$

We therefore begin by computing the image vectors $T(\mathbf{x}_1)$, $T(\mathbf{x}_2)$, and $T(\mathbf{x}_3)$.

```
In[1]:= T[{x_,y_,z_}]:={3x+2y-4z,x-5y+3z}
```

```
In[2]:= x1={1,1,1};x2={1,1,0};x3={1,0,0};
```

```
In[3]:= T[x1]

Out[3]= {1,-1}
```

```
In[4]:= T[x2]
Out[4]= {5,-4}
```

```
In[5]:= T[x3]
Out[5]= {3,1}
```

Next we need to find the coordinate vectors of $T(\mathbf{x}_1)$, $T(\mathbf{x}_2)$, and $T(\mathbf{x}_3)$ in the basis $\mathcal{C}$. We therefore need to find scalars a, b, c, d, e, and f satisfying the following equations.

$$\text{a.} \quad \left[\begin{pmatrix} 1 \\ -1 \end{pmatrix}\right]_{\mathcal{C}} = a\begin{pmatrix} 1 \\ 2 \end{pmatrix} + b\begin{pmatrix} 2 \\ 5 \end{pmatrix} = \begin{pmatrix} a+2b \\ 2a+5b \end{pmatrix}$$

$$\text{b.} \quad \left[\begin{pmatrix} 5 \\ -4 \end{pmatrix}\right]_{\mathcal{C}} = c\begin{pmatrix} 1 \\ 2 \end{pmatrix} + d\begin{pmatrix} 2 \\ 5 \end{pmatrix} = \begin{pmatrix} c+2d \\ 2c+5d \end{pmatrix}$$

$$\text{c.} \quad \left[\begin{pmatrix} 3 \\ 1 \end{pmatrix}\right]_{\mathcal{C}} = e\begin{pmatrix} 1 \\ 2 \end{pmatrix} + f\begin{pmatrix} 2 \\ 5 \end{pmatrix} = \begin{pmatrix} e+2f \\ 2e+5f \end{pmatrix}$$

To find these scalars, we solve the following linear systems.

$$\text{a.} \begin{cases} 1 = a+2b \\ -1 = 3a+5b \end{cases} \quad \text{b.} \begin{cases} 5 = c+2d \\ -4 = 3c+5d \end{cases} \quad \text{c.} \begin{cases} 3 = e+2f \\ 1 = 3e+5f \end{cases}$$

```
In[8]:= Solve[{1==a+2b,-1==3a+5b}]
Out[8]= {{a→-7,b→4}}
```

```
In[9]:= Solve[{5==c+2d,-4==3c+5d}]
Out[9]= {{c→-33,d→19}}
```

```
In[10]:= Solve[{3==e+2f,1==3e+5f}]
Out[10]= {{e→-13,f→8}}
```

Hence

$$[T]_{\mathcal{C}}^{\mathcal{B}} = \begin{pmatrix} [T(\mathbf{x}_1)]_{\mathcal{C}} & [T(\mathbf{x}_2)]_{\mathcal{C}} & [T(\mathbf{x}_3)]_{\mathcal{C}} \end{pmatrix} = \begin{pmatrix} -7 & -33 & -13 \\ 4 & 19 & 8 \end{pmatrix}$$

is the required matrix. ◄►

If C is the standard basis $\mathcal{E}$ for $\mathbb{R}^2$, the last few steps in Example 5.18 are redundant. In this case, the matrix $[T]_{\mathcal{E}}^{\mathcal{B}}$ is simply

$$\left(\ [T(\mathbf{x}_1)]_{\mathcal{E}} \quad [T(\mathbf{x}_2)]_{\mathcal{E}} \quad [T(\mathbf{x}_3)]_{\mathcal{E}} \ \right) = \begin{pmatrix} 1 & 5 & 3 \\ -1 & -4 & 1 \end{pmatrix}$$

We have already pointed out that the purpose of representing linear transformations by matrices is so that we can find their values using matrix multiplication. We now show how this works.

THEOREM 5.11 (Matrix representation theorem)   *If T is a linear transformation from $\mathbb{R}^n$ to $\mathbb{R}^m$, and if $\mathcal{B}$ is a basis for $\mathbb{R}^n$ and C is a basis for $\mathbb{R}^m$, then $[T(\mathbf{x})]_C = [T]_C^{\mathcal{B}} [\mathbf{x}]_{\mathcal{B}}$ for all $\mathbf{x} \in \mathbb{R}^n$.*

Proof. Suppose that $\mathcal{B} = \{\mathbf{x}_1, \ldots, \mathbf{x}_n\}$ and that $\mathbf{x} = a_1\mathbf{x}_1 + \cdots + a_n\mathbf{x}_n$ is any vector in $\mathbb{R}^n$. Then Theorem 4.15 implies that

$$\begin{aligned} [T(\mathbf{x})]_C &= [T(a_1\mathbf{x}_1 + \cdots + a_n\mathbf{x}_n)]_C \\ &= a_1[T(\mathbf{x}_1)]_C + \cdots + a_n[T(\mathbf{x}_n)]_C \\ &= ([T(\mathbf{x}_1)]_C \ \cdots \ [T(\mathbf{x}_n)]_C) \begin{pmatrix} a_1 \\ \vdots \\ a_n \end{pmatrix} \\ &= [T]_C^{\mathcal{B}} [\mathbf{x}]_{\mathcal{B}} \end{aligned}$$

This shows that the image vector $[T(\mathbf{x})]_C$ can be computed by matrix multiplication. ■

Relative to suitable bases $\mathcal{B}$ and C, the matrix $[T]_C^{\mathcal{B}}$ of a linear transformation T may have a particularly simple form. It may, for example, be diagonal, triangular, or symmetric.

EXAMPLE 5.19   ■ A Linear Transformation Represented by a Symmetric Matrix

Let $\mathcal{E}$ be the standard basis for $\mathbb{R}^3$, and let $T : \mathbb{R}^3 \to \mathbb{R}^3$ be a linear transformation defined by

$$\begin{cases} T(\mathbf{e}_1) = a_1\mathbf{e}_1 + a_2\mathbf{e}_2 + a_3\mathbf{e}_3 \\ T(\mathbf{e}_2) = a_2\mathbf{e}_1 + a_4\mathbf{e}_2 + a_5\mathbf{e}_3 \\ T(\mathbf{e}_3) = a_3\mathbf{e}_1 + a_5\mathbf{e}_2 + a_6\mathbf{e}_3 \end{cases}$$

Then the standard matrix

$$[T]_{\mathcal{E}}^{\mathcal{E}} = \begin{pmatrix} a_1 & a_2 & a_3 \\ a_2 & a_4 & a_5 \\ a_3 & a_5 & a_6 \end{pmatrix}$$

of T is symmetric. ◄►

EXAMPLE 5.20 ■ **Coordinate Conversion Matrices and the Identity Transformations**

Show that relative to two bases $\mathcal{B}$ and $\mathcal{C}$ for $\mathbb{R}^n$, the matrix $[I]_{\mathcal{C}}^{\mathcal{B}}$ of the identity transformation $I : \mathbb{R}^n \to \mathbb{R}^n$ is the coordinate conversion matrix from $\mathcal{B}$ to $\mathcal{C}$, and the matrix $[I]_{\mathcal{B}}^{\mathcal{C}}$ is the coordinate conversion matrix from $\mathcal{C}$ to $\mathcal{B}$.

Solution. Let $\mathcal{B} = \{\mathbf{x}_1, \ldots, \mathbf{x}_n\}$ and $\mathcal{C} = \{\mathbf{y}_1, \ldots, \mathbf{y}_n\}$ be two bases for $\mathbb{R}^n$. Then it is clear from the definition of a coordinate conversion matrix in Chapter 3 that

$$[I]_{\mathcal{C}}^{\mathcal{B}} = ([I(\mathbf{x}_1)]_{\mathcal{C}} \quad \cdots \quad [I(\mathbf{x}_n)]_{\mathcal{C}}) = ([\mathbf{x}_1]_{\mathcal{C}} \quad \cdots \quad [\mathbf{x}_n]_{\mathcal{C}})$$

is the coordinate conversion matrix from $\mathcal{B}$ to $\mathcal{C}$, and that

$$[I]_{\mathcal{B}}^{\mathcal{C}} = ([I(\mathbf{y}_1)]_{\mathcal{B}} \quad \cdots \quad [I(\mathbf{y}_n)]_{\mathcal{B}}) = ([\mathbf{y}_1]_{\mathcal{B}} \quad \cdots \quad [\mathbf{y}_n]_{\mathcal{B}})$$

is the coordinate conversion matrix from $\mathcal{C}$ to $\mathcal{B}$. ◀▶

EXAMPLE 5.21 ■ **A Coordinate Conversion**

Calculate the coordinate conversion matrices $[I]_{\mathcal{C}}^{\mathcal{B}}$ and $[I]_{\mathcal{B}}^{\mathcal{C}}$ for the bases

$$\mathcal{B} = \left\{ \begin{pmatrix} 0 \\ 4 \\ 2 \end{pmatrix}, \begin{pmatrix} 3 \\ 0 \\ 0 \end{pmatrix}, \begin{pmatrix} 1 \\ 3 \\ 4 \end{pmatrix} \right\} \quad \text{and} \quad \mathcal{C} = \left\{ \begin{pmatrix} 0 \\ 0 \\ 2 \end{pmatrix}, \begin{pmatrix} 3 \\ 0 \\ 0 \end{pmatrix}, \begin{pmatrix} 1 \\ 1 \\ 1 \end{pmatrix} \right\}$$

for $\mathbb{R}^3$ and show that $[I]_{\mathcal{C}}^{\mathcal{B}} [\mathbf{x}]_{\mathcal{B}} = [\mathbf{x}]_{\mathcal{C}}$ and $[I]_{\mathcal{B}}^{\mathcal{C}} [\mathbf{x}]_{\mathcal{C}} = [\mathbf{x}]_{\mathcal{B}}$ for $\mathbf{x} = (1, 2, 3)$.

Solution. It is clear from Example 5.20 that

$$[I]_{\mathcal{C}}^{\mathcal{B}} = \left(\left[\begin{pmatrix} 0 \\ 4 \\ 2 \end{pmatrix} \right]_{\mathcal{C}} \quad \left[\begin{pmatrix} 3 \\ 0 \\ 0 \end{pmatrix} \right]_{\mathcal{C}} \quad \left[\begin{pmatrix} 1 \\ 3 \\ 4 \end{pmatrix} \right]_{\mathcal{C}} \right) = \begin{pmatrix} -1 & 0 & \frac{1}{2} \\ -\frac{4}{3} & 1 & -\frac{2}{3} \\ 4 & 0 & 3 \end{pmatrix}$$

and

$$[I]_{\mathcal{B}}^{\mathcal{C}} = \left(\left[\begin{pmatrix} 0 \\ 0 \\ 2 \end{pmatrix} \right]_{\mathcal{B}} \quad \left[\begin{pmatrix} 3 \\ 0 \\ 0 \end{pmatrix} \right]_{\mathcal{B}} \quad \left[\begin{pmatrix} 1 \\ 1 \\ 1 \end{pmatrix} \right]_{\mathcal{B}} \right) = \begin{pmatrix} -\frac{3}{5} & 0 & \frac{1}{10} \\ -\frac{4}{15} & 1 & \frac{4}{15} \\ \frac{4}{5} & 0 & \frac{1}{5} \end{pmatrix}$$

Moreover,

$$[\mathbf{x}]_{\mathcal{B}} = \begin{pmatrix} -\frac{1}{10} \\ \frac{1}{15} \\ \frac{4}{5} \end{pmatrix} \quad \text{and} \quad [\mathbf{x}]_{\mathcal{C}} = \begin{pmatrix} \frac{1}{2} \\ -\frac{1}{3} \\ 2 \end{pmatrix}$$

Therefore,

$$[I]_\mathcal{C}^\mathcal{B} [\mathbf{x}]_\mathcal{B} = \begin{pmatrix} -\frac{3}{5} & 0 & \frac{1}{10} \\ -\frac{4}{15} & 1 & \frac{4}{15} \\ \frac{4}{5} & 0 & \frac{1}{5} \end{pmatrix} \begin{pmatrix} \frac{1}{2} \\ -\frac{1}{3} \\ 2 \end{pmatrix} = \begin{pmatrix} -\frac{1}{10} \\ \frac{1}{15} \\ \frac{4}{5} \end{pmatrix} = [\mathbf{x}]_\mathcal{C}$$

and

$$[I]_\mathcal{B}^\mathcal{C} [\mathbf{x}]_\mathcal{C} = \begin{pmatrix} -1 & 0 & \frac{1}{2} \\ -\frac{4}{3} & 1 & -\frac{2}{3} \\ 4 & 0 & 3 \end{pmatrix} \begin{pmatrix} -\frac{1}{10} \\ \frac{1}{15} \\ \frac{4}{5} \end{pmatrix} = \begin{pmatrix} \frac{1}{2} \\ -\frac{1}{3} \\ 2 \end{pmatrix} = [\mathbf{x}]_\mathcal{B}$$

It is easy to show that these equations hold for any vector $\mathbf{x} \in \mathbb{R}^3$. Hence $[I]_\mathcal{C}^\mathcal{B}$ is the coordinate conversion matrix from $\mathcal{B}$ to $\mathcal{C}$ and $[I]_\mathcal{B}^\mathcal{C}$ is the coordinate conversion matrix from $\mathcal{C}$ to $\mathcal{B}$. ◄►

EXAMPLE 5.22   ■ A Matrix of an Identity Transformation

Show that in the standard basis $\mathcal{E} = \{\mathbf{e}_1, \ldots, \mathbf{e}_n\}$ for $\mathbb{R}^n$, the identity transformation $I : \mathbb{R}^n \to \mathbb{R}^n$ is represented by the identity matrix I_n.

Solution. By Theorem 5.11, it holds that $[I]_\mathcal{E}^\mathcal{E} = ([I(\mathbf{e}_1)]_\mathcal{E} \cdots [I(\mathbf{e}_n)]_\mathcal{E})$. But $[I(\mathbf{e}_i)]_\mathcal{E} = \mathbf{e}_i$ for all $i \in \mathbf{n}$. Therefore, $[I]_\mathcal{E}^\mathcal{E} = (\mathbf{e}_1 \cdots \mathbf{e}_n) = I_n$. This shows that the identity matrix represents the identity transformation. ◄►

EXERCISES 5.4

1. Find the standard matrix of the following linear transformations.

   a. $T : \mathbb{R}^2 \to \mathbb{R}^2$ defined by $T(1, 0) = (4, 2)$ and $T(0, 1) = (8, 3)$

   b. $T : \mathbb{R}^3 \to \mathbb{R}^3$ defined by $T(1, 0, 0) = (4, 2, 1)$, $T(0, 1, 0) = (1, 1, 1)$, and $T(0, 0, 1) = (8, 3, 4)$

   c. $T : \mathbb{R}^2 \to \mathbb{R}^2$ defined by $T(1, 1) = (4, 2)$ and $T(2, -1) = (8, 3)$

   d. $T : \mathbb{R}^3 \to \mathbb{R}^3$ defined by $T(1, 0, -5) = (4, 2, 1)$, $T(0, 0, 3) = (1, 1, 1)$, and $T(5, -3, 0) = (8, 3, 4)$

2. Let $\mathcal{B}$ be the basis of $\mathbb{R}^2$ determined by the invertible matrix

$$\begin{pmatrix} 1 & 3 \\ -2 & 0 \end{pmatrix}$$

and let $\mathcal{E}$ be the standard basis of $\mathbb{R}^2$. Find the matrix $[T]_\mathcal{E}^\mathcal{B}$ of the linear transformation

$$T\begin{pmatrix} x \\ y \end{pmatrix} = \begin{pmatrix} x + y \\ x \end{pmatrix}$$

3. Find the matrix $[T]_{\mathcal{B}}^{\mathcal{E}}$ of the linear transformation in Exercise 2.

4. Find the matrix $[T]_{\mathcal{B}}^{\mathcal{B}}$ of the linear transformation in Exercise 2.

5. Let $\mathcal{B}$ be the basis of $\mathbb{R}^3$ determined by the invertible matrix

$$\begin{pmatrix} -1 & -2 & 2 \\ 0 & 1 & 2 \\ 0 & 0 & 4 \end{pmatrix}$$

and let $\mathcal{E}$ be the standard matrix of $\mathbb{R}^3$. Find the matrix $[T]_{\mathcal{E}}^{\mathcal{B}}$ of the linear transformation $T(x, y, z) = (x + y, z, y)$.

6. Find the matrix $[T]_{\mathcal{B}}^{\mathcal{E}}$ of the linear transformation in Exercise 5.

7. Find the matrix $[T]_{\mathcal{B}}^{\mathcal{B}}$ of the linear transformation in Exercise 5.

8. Let $\mathcal{B}$ and $\mathcal{C}$ be two bases for $\mathbb{R}^3$ determined by the invertible matrices

$$\begin{pmatrix} 5 & 1 & -3 \\ 0 & 5 & 3 \\ 0 & 0 & -2 \end{pmatrix} \quad \text{and} \quad \begin{pmatrix} 0 & 4 & 4 \\ 5 & 0 & 0 \\ 0 & 0 & 4 \end{pmatrix}$$

Find the matrices $[I]_{\mathcal{C}}^{\mathcal{B}}$ and $[I]_{\mathcal{B}}^{\mathcal{C}}$ of the identity transformation $I : \mathbb{R}^3 \to \mathbb{R}^3$.

9. Let $D_t : \mathbb{R}_3[t] \to \mathbb{R}_2[t]$ be the differentiation operation

$$D_t(a_0 + a_1 t + a_2 t^2 + a_3 t^3) = a_1 + 2a_2 t + 3a_3 t^2$$

and $I_t : \mathbb{R}_2[t] \to \mathbb{R}_3[t]$ the integration operation

$$I_t(a_0 + a_1 t + a_2 t^2) = a_0 t + \frac{1}{2} a_1 t^2 + \frac{1}{3} a_2 t^3$$

Find the standard matrices $A = [D_t]$ and $B = [I_t]$ with respect to the standard bases, and show that

$$AB = \begin{pmatrix} 1 & 0 & 0 \\ 0 & 1 & 0 \\ 0 & 0 & 1 \end{pmatrix} \quad \text{and} \quad BA = \begin{pmatrix} 0 & 0 & 0 & 0 \\ 0 & 1 & 0 & 0 \\ 0 & 0 & 1 & 0 \\ 0 & 0 & 0 & 1 \end{pmatrix}$$

10. Prove or disprove the following statements.

   a. If T_1 and T_2 are two linear transformations $\mathbb{R}^n \to \mathbb{R}^n$ and $[T_1]_{\mathcal{B}}^{\mathcal{B}} = [T_2]_{\mathcal{C}}^{\mathcal{C}}$ in two bases $\mathcal{B}$ and $\mathcal{C}$ for $\mathbb{R}^n$, then $T_1 = T_2$.

   b. If T_1 and T_2 are two linear transformations $\mathbb{R}^n \to \mathbb{R}^n$ and $[T_1]_{\mathcal{B}}^{\mathcal{B}} = [T_2]_{\mathcal{B}}^{\mathcal{B}}$ in some basis $\mathcal{B}$ for $\mathbb{R}^n$, then $T_1 = T_2$.

c. If T is a linear transformation $\mathbb{R}^n \rightarrow \mathbb{R}^n$ and $[T]_{\mathcal{B}}^{\mathcal{B}} = [T]_{\mathcal{C}}^{\mathcal{C}}$ in two bases $\mathcal{B}$ and $\mathcal{C}$ for $\mathbb{R}^n$, then $\mathcal{B} = \mathcal{C}$.

11. Use *MATHEMATICA* to find the coordinate conversion matrices $[I]_{\mathcal{C}}^{\mathcal{B}}$ and $[I]_{\mathcal{B}}^{\mathcal{C}}$ for the bases $\mathcal{B}$ and $\mathcal{C}$ for $\mathbb{R}^4$ determined by the columns of the following invertible matrices.

$$\begin{pmatrix} 3 & 5 & 5 & 2 \\ 5 & 2 & -2 & 0 \\ 5 & -2 & 6 & 4 \\ 2 & 0 & 4 & -3 \end{pmatrix} \quad \text{and} \quad \begin{pmatrix} 3 & 6 & 5 & 6 \\ 4 & 6 & 7 & 7 \\ 3 & 6 & 6 & 6 \\ 7 & 7 & 2 & 7 \end{pmatrix}$$

Composition of Linear Transformations and Matrix Multiplication

We now explain the connection between matrix multiplication and the composition of matrix transformations.

THEOREM 5.12 (Composition theorem)   *Suppose that $T_1 : \mathbb{R}^n \rightarrow \mathbb{R}^m$ and $T_2 : \mathbb{R}^m \rightarrow \mathbb{R}^p$ are two linear transformations and that the matrices $[T_1]_{\mathcal{C}}^{\mathcal{B}}$ and $[T_2]_{\mathcal{D}}^{\mathcal{C}}$ represent the transformation T_1 and T_2 in the bases $\mathcal{B}$, $\mathcal{C}$, and $\mathcal{D}$. Then*

$$[T_2 \circ T_1]_{\mathcal{D}}^{\mathcal{B}} = [T_2]_{\mathcal{D}}^{\mathcal{C}}[T_1]_{\mathcal{C}}^{\mathcal{B}}$$

Proof. Let $\mathbf{x}$ be any vector in $\mathbb{R}^n$. Then we know from Theorem 5.11 that

$$\begin{aligned} [T_2 \circ T_1]_{\mathcal{D}}^{\mathcal{B}}[\mathbf{x}]_{\mathcal{B}} &= [(T_2 \circ T_1)(\mathbf{x})]_{\mathcal{D}} \\ &= [T_2(T_1(\mathbf{x}))]_{\mathcal{D}} \\ &= [T_2]_{\mathcal{D}}^{\mathcal{C}}[T_1(\mathbf{x})]_{\mathcal{C}} \\ &= [T_2]_{\mathcal{D}}^{\mathcal{C}}([T_1]_{\mathcal{C}}^{\mathcal{B}}[\mathbf{x}]_{\mathcal{B}}) \\ &= ([T_2]_{\mathcal{D}}^{\mathcal{C}}[T_1]_{\mathcal{C}}^{\mathcal{B}})[\mathbf{x}]_{\mathcal{B}} \end{aligned}$$

for all $\mathbf{x} \in \mathbb{R}^n$. Therefore, $[T_2]_{\mathcal{D}}^{\mathcal{C}}[T_1]_{\mathcal{C}}^{\mathcal{B}}$ is the matrix of $T_2 \circ T_1$ in the basis $\mathcal{B}$ for $\mathbb{R}^n$ and $\mathcal{D}$ for $\mathbb{R}^p$. ∎

EXAMPLE 5.23   ■ **Composition of Linear Transformations and Matrix Multiplication**

Illustrate by an example that matrix multiplication corresponds to the composition of matrix transformations.

Solution. Let $T_1 : \mathbb{R}^3 \rightarrow \mathbb{R}^2$ and $T_2 : \mathbb{R}^2 \rightarrow \mathbb{R}^2$ be the transformations defined by

$$T_1 \begin{pmatrix} x \\ y \\ z \end{pmatrix} = \begin{pmatrix} 3x + 2y - 4z \\ x - 5y + 3z \end{pmatrix} \quad \text{and} \quad T_2 \begin{pmatrix} u \\ v \end{pmatrix} = \begin{pmatrix} 4u - 2v \\ 2u + v \end{pmatrix}$$

and let A be the standard matrix of T_1 and B be the standard matrix of T_2. Then

$$A = \left(T_1 \begin{pmatrix} 1 \\ 0 \\ 0 \end{pmatrix} \ T_1 \begin{pmatrix} 0 \\ 1 \\ 0 \end{pmatrix} \ T_1 \begin{pmatrix} 0 \\ 0 \\ 1 \end{pmatrix} \right) = \begin{pmatrix} 3 & 2 & -4 \\ 1 & -5 & 3 \end{pmatrix}$$

and

$$B = \left(T_2 \begin{pmatrix} 1 \\ 0 \end{pmatrix} \ T_2 \begin{pmatrix} 0 \\ 1 \end{pmatrix} \right) = \begin{pmatrix} 4 & -2 \\ 2 & 1 \end{pmatrix}$$

We show that $T_B \circ T_A = T_{BA}$.

1. The composite transformation $T_B \circ T_A$, applied to an arbitrary vector $\mathbf{x} \in \mathbb{R}^3$, yields

$$(T_B \circ T_A) \begin{pmatrix} x \\ y \\ z \end{pmatrix} = T_B T_A \begin{pmatrix} x \\ y \\ z \end{pmatrix}$$

$$= \begin{pmatrix} 4 & -2 \\ 2 & 1 \end{pmatrix} \begin{pmatrix} 3 & 2 & -4 \\ 1 & -5 & 3 \end{pmatrix} \begin{pmatrix} x \\ y \\ z \end{pmatrix}$$

$$= \begin{pmatrix} 4 & -2 \\ 2 & 1 \end{pmatrix} \begin{pmatrix} 3x + 2y - 4z \\ x - 5y + 3z \end{pmatrix}$$

$$= \begin{pmatrix} 10x + 18y - 22z \\ 7x - y - 5z \end{pmatrix}$$

2. Next we examine the matrix transformation $T_{BA} : \mathbb{R}^3 \to \mathbb{R}^2$. Since

$$BA = \begin{pmatrix} 4 & -2 \\ 2 & 1 \end{pmatrix} \begin{pmatrix} 3 & 2 & -4 \\ 1 & -5 & 3 \end{pmatrix} = \begin{pmatrix} 10 & 18 & -22 \\ 7 & -1 & -5 \end{pmatrix}$$

we have

$$T_{BA} \begin{pmatrix} x \\ y \\ z \end{pmatrix} = \begin{pmatrix} 10 & 18 & -22 \\ 7 & -1 & -5 \end{pmatrix} \begin{pmatrix} x \\ y \\ z \end{pmatrix}$$

$$= \begin{pmatrix} 10x + 18y - 22z \\ 7x - y - 5z \end{pmatrix}$$

Hence $(T_B \circ T_A)(\mathbf{x}) = T_{BA}(\mathbf{x})$ for all $\mathbf{x} \in \mathbb{R}^3$. ◄►

EXERCISES 5.5

1. Let $T_A : \mathbb{R}^2 \to \mathbb{R}^2$ be a linear transformation defined by

$$T_A \begin{pmatrix} x \\ y \end{pmatrix} = A \begin{pmatrix} x \\ y \end{pmatrix} = \begin{pmatrix} 4 & 8 \\ 2 & 3 \end{pmatrix} \begin{pmatrix} x \\ y \end{pmatrix}$$

and $T_B : \mathbb{R}^2 \to \mathbb{R}^2$ be a linear transformation defined by

$$T_B \begin{pmatrix} x \\ y \end{pmatrix} = B \begin{pmatrix} x \\ y \end{pmatrix} = \begin{pmatrix} 6 & 2 \\ 0 & 1 \end{pmatrix} \begin{pmatrix} x \\ y \end{pmatrix}$$

Show that $T_B \circ T_A = T_{BA}$ and $T_A \circ T_B = T_{AB}$. Use *MATHEMATICA* to verify your calculations.

2. Let $T_A : \mathbb{R}^3 \to \mathbb{R}^3$ be a linear transformation defined by

$$T_A \begin{pmatrix} x \\ y \\ z \end{pmatrix} = A \begin{pmatrix} x \\ y \\ z \end{pmatrix} = \begin{pmatrix} 4 & 8 & 2 \\ 2 & 3 & 6 \\ 0 & 0 & 0 \end{pmatrix} \begin{pmatrix} x \\ y \\ z \end{pmatrix}$$

and $T_B : \mathbb{R}^3 \to \mathbb{R}^3$ be a linear transformation defined by

$$T_B \begin{pmatrix} x \\ y \\ z \end{pmatrix} = B \begin{pmatrix} x \\ y \\ z \end{pmatrix} = \begin{pmatrix} 6 & 2 & 0 \\ 0 & 1 & 0 \\ 1 & 3 & 0 \end{pmatrix} \begin{pmatrix} x \\ y \\ z \end{pmatrix}$$

Show that $T_B \circ T_A = T_{BA}$ and $T_A \circ T_B = T_{AB}$. Use *MATHEMATICA* to verify your calculations.

3. Suppose the matrix product $A = PLU$ is

$$\begin{pmatrix} 0 & 2 & 5 \\ 1 & 5 & 2 \\ 3 & 2 & 1 \end{pmatrix} = \begin{pmatrix} 0 & 1 & 0 \\ 1 & 0 & 0 \\ 0 & 0 & 1 \end{pmatrix} \begin{pmatrix} 1 & 0 & 0 \\ 0 & 1 & 0 \\ 3 & -\frac{13}{2} & 1 \end{pmatrix} \begin{pmatrix} 1 & 5 & 2 \\ 0 & 2 & 5 \\ 0 & 0 & \frac{55}{2} \end{pmatrix}$$

Show that the linear transformation T_A is equal to the composition $T_P \circ T_L \circ T_U$. Use *MATHEMATICA* to verify your calculations.

4. Decompose the invertible matrix

$$A = \begin{pmatrix} 1 & 2 \\ 7 & 8 \end{pmatrix}$$

into a product of elementary matrices $E_1, \ldots, E_n$ and explain the geometric effect of the matrix transformations $T_{E_1}, \ldots, T_{E_n}$.

5. Describe the geometric effect of the matrix transformations E_A, E_B, and E_P, where

$$AB = \begin{pmatrix} 1 & 0 & 0 \\ 0 & 0 & 1 \\ 0 & 1 & 0 \end{pmatrix} \begin{pmatrix} 0 & 0 & 1 \\ 0 & 1 & 0 \\ 1 & 0 & 0 \end{pmatrix} = \begin{pmatrix} 0 & 0 & 1 \\ 1 & 0 & 0 \\ 0 & 1 & 0 \end{pmatrix} = P$$

is a product of elementary permutation matrices.

The Inverse of a Linear Transformation and Matrix Inversion

It is easy to show that the inverse of a linear transformation is represented by the inverse of the matrix of the transformation.

THEOREM 5.13   *If $T : \mathbb{R}^n \to \mathbb{R}^n$ is an invertible linear transformation represented by a matrix A, then $\left[T^{-1}\right] = A^{-1}$.*

Proof. By Theorem 5.11, the transformation $T^{-1} : \mathbb{R}^n \to \mathbb{R}^n$ is represented by some matrix B. It follows from Theorem 5.12 that

$$\left[T^{-1} \circ T\right] = BA = \left[T \circ T^{-1}\right] = AB = [I] = I_n$$

Hence $B = A^{-1}$. ∎

EXAMPLE 5.24   ■ **The Matrix of an Inverse Transformation**

Show by an example that the matrix of the inverse of a linear transformation is the inverse of the matrix of the transformation.

Solution. Let $T : \mathbb{R}^2 \to \mathbb{R}^2$ be the linear transformation defined by

$$T(x, y) = (x + y, x - y)$$

It is easy to check that inverse of T is the linear transformation

$$S(x, y) = \left(\frac{1}{2}(x + y), \frac{1}{2}(x - y) \right)$$

In the $\mathcal{E}$ standard basis of $\mathbb{R}^2$, the matrices of T and S are

$$[T]_{\mathcal{E}} = \begin{pmatrix} 1 & 1 \\ 1 & -1 \end{pmatrix} \quad \text{and} \quad [S]_{\mathcal{E}} = \begin{pmatrix} \frac{1}{2} & \frac{1}{2} \\ \frac{1}{2} & -\frac{1}{2} \end{pmatrix}$$

Since

$$\begin{pmatrix} 1 & 1 \\ 1 & -1 \end{pmatrix} \begin{pmatrix} \frac{1}{2} & \frac{1}{2} \\ \frac{1}{2} & -\frac{1}{2} \end{pmatrix} = \begin{pmatrix} 1 & 0 \\ 0 & 1 \end{pmatrix}$$

it is clear that $\left[T^{-1}\right]_{\mathcal{E}} = [T]_{\mathcal{E}}^{-1}$. ◀▶

EXERCISES 5.6

1. Prove that if $T_A : \mathbb{R}^n \rightarrow \mathbb{R}^n$ is a matrix transformation and $T_{A^{-1}} : \mathbb{R}^n \rightarrow \mathbb{R}^n$ is the matrix transformation determined by A^{-1}, then $T_{A^{-1}} \circ T_A$ and $T_A \circ T_{A^{-1}}$ are the identity transformations on $\mathbb{R}^n$.

2. Let $T_A : \mathbb{R}^2 \rightarrow \mathbb{R}^2$ be the linear transformation defined by

$$T_A \begin{pmatrix} x \\ y \end{pmatrix} = A \begin{pmatrix} x \\ y \end{pmatrix} = \begin{pmatrix} 4 & 8 \\ 2 & 3 \end{pmatrix} \begin{pmatrix} x \\ y \end{pmatrix}$$

Find the inverse of T_A.

3. Let $T_A : \mathbb{R}^3 \rightarrow \mathbb{R}^3$ be the linear transformation defined by

$$T_A \begin{pmatrix} x \\ y \\ z \end{pmatrix} = A \begin{pmatrix} x \\ y \\ z \end{pmatrix} = \begin{pmatrix} 4 & 8 & 2 \\ 2 & 3 & 6 \\ 0 & 0 & 0 \end{pmatrix} \begin{pmatrix} x \\ y \\ z \end{pmatrix}$$

Show that T_A does not have an inverse.

4. Prove that if A is an orthogonal $n \times n$ matrix, then $T_A \circ T_{A^T}$ is the identity transformation on $\mathbb{R}^n$.

IMAGES AND KERNELS

Every linear transformation $T : V \rightarrow W$ determines two special subspaces: the subspace of W consisting of the values $T(\mathbf{x})$ of T, and the subspace of V consisting of all vectors $\mathbf{x} \in V$ mapped to $\mathbf{0} \in W$ by T.

DEFINITION 5.8   *The **image** im T of a linear transformation $T : V \rightarrow W$ is the set of all vectors $\mathbf{y} \in W$ with the property that $\mathbf{y} = T(\mathbf{x})$ for some $\mathbf{x} \in V$.*

Figure 4 illustrates an image.

THEOREM 5.14 (Image theorem)   *The image of a linear transformation $T : V \rightarrow W$ is a subspace of W.*

Proof. Since $T(0_V) = 0_W$, the image of T is nonempty. We need to show that it is also closed under linear combinations. Let $\mathbf{u}$ and $\mathbf{v}$ be two vectors in the image of T, and let a and b be any two scalars. Then $\mathbf{u} = T(\mathbf{x})$ and $\mathbf{v} = T(\mathbf{y})$ for some vectors $\mathbf{x}$ and $\mathbf{y}$ in V. By the linearity of T,

$$aT(\mathbf{x}) + bT(\mathbf{y}) = T(a\mathbf{x} + b\mathbf{y})$$

Since $T(a\mathbf{x} + b\mathbf{y}) \in \operatorname{im} T$, we can conclude that $aT(\mathbf{x}) + bT(\mathbf{y}) \in \operatorname{im} T$. The image of T is therefore closed under linear combinations. ∎

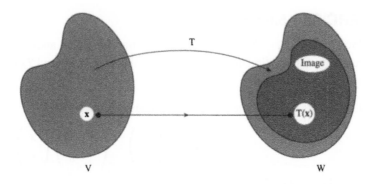

FIGURE 4   Image of a Linear Transformation.

If T is a matrix transformation, then the image of T has a familiar form.

THEOREM 5.15   *If A is a real $m \times n$ matrix and $T_A : \mathbb{R}^n \to \mathbb{R}^m$ is the associated matrix transformation, then* $\operatorname{im} T_A = \operatorname{Col} A$.

Proof. Suppose that

$$A = \begin{pmatrix} a_{11} & \cdots & a_{1n} \\ & \vdots & \\ a_{m1} & \cdots & a_{mn} \end{pmatrix}$$

and that

$$\mathbf{x} = \begin{pmatrix} x_1 \\ \vdots \\ x_n \end{pmatrix} \in \mathbb{R}^n$$

Then the vector

$$\begin{aligned} A\mathbf{x} &= \begin{pmatrix} a_{11} & \cdots & a_{1n} \\ & \vdots & \\ a_{m1} & \cdots & a_{mn} \end{pmatrix} \begin{pmatrix} x_1 \\ \vdots \\ x_n \end{pmatrix} \\ &= \begin{pmatrix} a_{11}x_1 + \cdots + a_{1n}x_n \\ \vdots \\ a_{m1}x_1 + \cdots + a_{mn}x_n \end{pmatrix} \\ &= x_1 \begin{pmatrix} a_{11} \\ \vdots \\ a_{m1} \end{pmatrix} + \cdots + x_n \begin{pmatrix} a_{1n} \\ \vdots \\ a_{mn} \end{pmatrix} \end{aligned}$$

is a linear combination of the columns of A. ∎

EXAMPLE 5.25 ■ An Image as a Column Space

Use *MATHEMATICA* to find the image of the matrix transformation $T_A : \mathbb{R}^3 \to \mathbb{R}^2$ determined by the matrix

$$A = \begin{pmatrix} 0 & 0 & 2 \\ 0 & 1 & 0 \end{pmatrix}$$

Solution. By Theorem 5.15, the image of T_A is identical with the column space of A. Since *MATHEMATICA* does not have a built-in function for computing the column space, we compute the row space of A^T instead. We know from Theorem 4.25 that the nonzero rows of the reduced row echelon form A^T are a basis for Col A.

```
In[1]:= A= ( 0  0  2 );
         ( 0  1  0 )
```

```
In[2]:= MatrixForm[RowReduce[Transpose[A]]]

Out[2]//MatrixForm=
         ( 1  0 )
         ( 0  1 )
         ( 0  0 )
```

This shows that the set

$$\left\{ \begin{pmatrix} 1 \\ 0 \end{pmatrix}, \begin{pmatrix} 0 \\ 1 \end{pmatrix} \right\}$$

is a basis for the image of T_A. ◄►

We can get a better understanding of the nature of the image of a linear transformation by considering how the image relates to the two defining properties of a basis.

THEOREM 5.16 *If $T : V \to W$ is a linear transformation and if $S = \{\mathbf{x}_1, \ldots, \mathbf{x}_n\}$ spans V, then $\{T(\mathbf{x}_1), \ldots, T(\mathbf{x}_n)\}$ spans the image of T.*

Proof. Suppose that $\mathbf{y} \in \operatorname{im} T$. Then $\mathbf{y} = T(\mathbf{x})$ for some $\mathbf{x} \in V$. Since S spans V, there exist scalars $a_1, \ldots, a_n$ for which $\mathbf{x} = a_1\mathbf{x}_1 + \cdots + a_n\mathbf{x}_n$. Since T is linear,

$$T(\mathbf{x}) = T(a_1\mathbf{x}_1 + \cdots + a_n\mathbf{x}_n) = a_1 T(\mathbf{x}_1) + \cdots + a_n T(\mathbf{x})$$

Therefore, $\mathbf{y}$ is a linear combination of the vectors $T(\mathbf{x}_1), \ldots, T(\mathbf{x}_n)$. ■

THEOREM 5.17 *If $T : V \to W$ is a one-one linear transformation and if the set of values $\{T(\mathbf{x}_1), \ldots, T(\mathbf{x}_n)\}$ of T spans the image of T, then $\{\mathbf{x}_1, \ldots, \mathbf{x}_n\}$ spans V.*

Proof. Let $\mathbf{x} \in V$. Then $T(\mathbf{x}) \in \operatorname{im} T$. Since $\{T(\mathbf{x}_1), \ldots, T(\mathbf{x}_n)\}$ spans the image of T, it follows that

$$T(\mathbf{x}) = a_1 T(\mathbf{x}_1) + \cdots + a_n T(\mathbf{x}_n) = T(a_1 \mathbf{x}_1 + \cdots + a_n \mathbf{x}_n)$$

Since T is one-one, $\mathbf{x} = a_1 \mathbf{x}_1 + \cdots + a_n \mathbf{x}_n$. ∎

THEOREM 5.18 *If $T : V \to W$ is a one-one linear transformation and the vectors $\mathbf{x}_1, \ldots, \mathbf{x}_n$ are linearly independent in V, then the vectors $T(\mathbf{x}_1), \ldots, T(\mathbf{x}_n)$ are linearly independent in W.*

Proof. Let $a_1 T(\mathbf{x}_1) + \cdots + a_n T(\mathbf{x}_n) = \mathbf{0}$. By the linearity of T, we have

$$T(a_1 \mathbf{x}_1 + \cdots + a_n \mathbf{x}_n) = \mathbf{0} = T(\mathbf{0})$$

Since T is one-one, it follows that $a_1 \mathbf{x}_1 + \cdots + a_n \mathbf{x}_n = \mathbf{0}$. The linear independence of $\mathbf{x}_1, \ldots, \mathbf{x}_n$ guarantees that $a_1 = \cdots = a_n = 0$. ∎

THEOREM 5.19 *If $T : V \to W$ is a linear transformation and the vectors $T(\mathbf{x}_1), \ldots, T(\mathbf{x}_n)$ are linearly independent in W, then the vectors $\mathbf{x}_1, \ldots, \mathbf{x}_n$ are linearly independent in V.*

Proof. Let $a_1 \mathbf{x}_1 + \cdots + a_n \mathbf{x}_n = \mathbf{0}$. Then

$$\begin{aligned}
T(\mathbf{0}) &= T(a_1 \mathbf{x}_1 + \cdots + a_n \mathbf{x}_n) \\
&= a_1 T(\mathbf{x}_1) + \cdots + a_n T(\mathbf{x}_n) \\
&= \mathbf{0}
\end{aligned}$$

The linear independence of $T(\mathbf{x}_1), \ldots, T(\mathbf{x}_n)$ implies that $a_1 = \cdots = a_n = 0$. ∎

As mentioned earlier, a linear transformation from V to W also determines an important subspace of V.

DEFINITION 5.9 *The **kernel** $\ker T$ of a linear transformation $T : V \to W$ is the set of all vectors $\mathbf{x} \in V$ with the property that $T(\mathbf{x}) = \mathbf{0}$.*

Figure 5 illustrates a kernel.

THEOREM 5.20 (Kernel theorem) *The kernel of a linear transformation $T : V \to W$ is a subspace of V.*

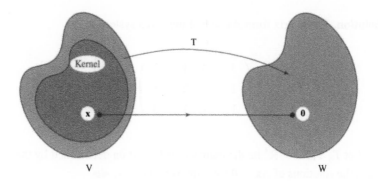

T

Kernel

x

0

V

W

FIGURE 5   Kernel of a Linear Transformation.

Proof. Since $T(\mathbf{0}) = \mathbf{0}$, it follows that $\mathbf{0} \in \ker T$. Furthermore, the linearity of T implies that for all $\mathbf{x}_1, \mathbf{x}_2 \in \ker T$, and all scalars a and b,

$$T(a\mathbf{x}_1 + b\mathbf{x}_2) = aT(\mathbf{x}_1) + bT(\mathbf{x}_2) = a\mathbf{0} + b\mathbf{0} = \mathbf{0}$$

Hence $a\mathbf{x}_1 + b\mathbf{x}_2 \in \ker T$. Therefore the kernel of T is nonempty and is closed under linear combinations. This means that it is a subspace of V. ∎

THEOREM 5.21   *The kernel of a matrix transformation $T_A : \mathbb{R}^n \to \mathbb{R}^m$ is the null space of A.*

Proof. We recall that the null space of the matrix A is the set of all vectors $\mathbf{x} \in \mathbb{R}^n$ for which $A\mathbf{x} = \mathbf{0}$. Moreover, the kernel of T_A consists of all vectors $\mathbf{x} \in \mathbb{R}^n$ for which $T_A(\mathbf{x}) = \mathbf{0}$. Since $T_A(\mathbf{x}) = A\mathbf{x}$, it follows that the kernel of T_A and the null space of A consist of precisely the same vectors. ∎

EXAMPLE 5.26 ■ The Kernel of a Linear Transformation

Let $T : \mathbb{R}^3 \to \mathbb{R}^3$ be the **projection** defined by $T(x, y, z) = (x, y, 0)$. Then $T(x, y, z) = (0, 0, 0)$ if and only if $(x, y, z) = (0, 0, z)$. Therefore, the set of vectors $\{(0, 0, z) : z \in \mathbb{R}\}$ is the kernel of T. ◄►

EXAMPLE 5.27 ■ Kernels and Homogeneous Systems

Show that the set of solutions of the homogeneous linear system

$$\begin{cases} 3x + 4y + w + z = 0 \\ x - y + 2w - z = 0 \\ x + y + w = 0 \end{cases}$$

is the kernel of a linear transformation.

Solution. The matrix form $A\mathbf{x} = \mathbf{0}$ of the given system is

$$
\begin{pmatrix}
3 & 4 & 1 & 1 \\
1 & -1 & 2 & -1 \\
1 & 1 & 1 & 0
\end{pmatrix}
\begin{pmatrix}
x \\
y \\
w \\
z
\end{pmatrix}
=
\begin{pmatrix}
0 \\
0 \\
0
\end{pmatrix}
$$

Let $T_A : \mathbb{R}^4 \to \mathbb{R}^3$ be the matrix transformation determined by the coefficient matrix A. Then the solutions of $A\mathbf{x} = \mathbf{0}$ form the kernel of T_A. ◀▶

Whether or not a linear transformation is one-one can be expressed elegantly in terms of its kernel.

THEOREM 5.22   *A linear transformation T is one-one if and only if* $\ker T = \{\mathbf{0}\}$.

Proof. Suppose that T is one-one and that $\mathbf{x} \in \ker T$. Then $T(\mathbf{x}) = \mathbf{0} = T(\mathbf{0})$. Since T is one-one, we have $\mathbf{x} = \mathbf{0}$. On the other hand, if $\ker T = \{\mathbf{0}\}$ and $T(\mathbf{x}) = T(\mathbf{y})$ for some $\mathbf{x}, \mathbf{y} \in V$, then $T(\mathbf{x}) - T(\mathbf{y}) = T(\mathbf{x} - \mathbf{y}) = \mathbf{0}$, so that $\mathbf{x} - \mathbf{y} \in \ker T$. This can only be the case if $\mathbf{x} - \mathbf{y} = \mathbf{0}$. Therefore, $\mathbf{x} = \mathbf{y}$. ■

EXERCISES 5.7

1. Use *MATHEMATICA* to find the image and kernel of the matrix transformations determined by the following matrices.

a. $\begin{pmatrix} 1 & 0 & -2 & 6 \\ 0 & 5 & 7 & 1 \end{pmatrix}$
   b. $\begin{pmatrix} 1 & 0 \\ 0 & 5 \\ -2 & 7 \\ 6 & 1 \end{pmatrix}$
   c. $\begin{pmatrix} 0 & -8 & 3 \\ 8 & 2 & 77 \\ 3 & -7 & 9 \end{pmatrix}$

d. $\begin{pmatrix} 1 & 0 & 0 \\ 0 & 3 & 0 \\ 0 & 0 & 7 \\ 0 & 0 & 0 \end{pmatrix}$
   e. $\begin{pmatrix} 1 & 0 & 0 & 0 \\ 0 & 3 & 0 & 0 \\ 0 & 0 & 0 & 0 \end{pmatrix}$
   f. $\begin{pmatrix} 1 & 0 & 0 & 9 \\ 0 & 0 & 0 & 0 \\ 0 & 0 & 0 & 0 \\ 0 & 0 & 0 & 0 \end{pmatrix}$

2. Find a linear transformation $T : \mathbb{R}^2 \to \mathbb{R}^2$ whose image in the standard basis is spanned by set

$$
S = \left\{ \begin{pmatrix} -8 \\ 2 \end{pmatrix}, \begin{pmatrix} 3 \\ 77 \end{pmatrix} \right\}
$$

3. Find a linear transformation $T : \mathbb{R}^3 \to \mathbb{R}^3$ whose image in the standard basis is spanned by set

$$S = \left\{ \begin{pmatrix} -8 \\ 2 \\ 3 \end{pmatrix}, \begin{pmatrix} 3 \\ 77 \\ 5 \end{pmatrix} \right\}$$

4. Suppose that $T : \mathbb{R}^3 \to \mathbb{R}^2$ is a linear transformation and $T(5, 0, 1) = (-4, 7)$, $T(2, 8, 0) = (1, 1)$, and $T(4, 5, 8) = (0, 5)$. Show that the set

$$S = \left\{ \begin{pmatrix} -4 \\ 7 \end{pmatrix}, \begin{pmatrix} 1 \\ 1 \end{pmatrix}, \begin{pmatrix} 0 \\ 5 \end{pmatrix} \right\}$$

spans the image of T.

5. Let $T : \mathbb{R}^3 \to \mathbb{R}^2$ be the linear transformation

$$T \begin{pmatrix} x \\ y \\ z \end{pmatrix} = \begin{pmatrix} 2x + 4y - z \\ -x + y - 8z \end{pmatrix}$$

Show that the kernel of T is the solution space of the homogeneous linear system

$$\begin{cases} 2x + 4y - z = 0 \\ -x + y - 8z = 0 \end{cases}$$

and find a basis for the kernel of T.

6. Describe the image and kernel of the differentiation operation $D : \mathbb{R}_4[t] \to \mathbb{R}_3[t]$ defined by

$$D_t(a_0 + a_1 t + a_2 t^2 + a_3 t^3 + a_4 t^4) = a_1 + 2a_2 t + 3a_3 t^2 + 4a_4 t^3$$

7. Describe the image and kernel of the integration operation $I_t : \mathbb{R}_3[t] \to \mathbb{R}[t]$ defined by

$$I_t(a_0 + a_1 t + a_2 t^2 + a_3 t^3) = a_0 t + \frac{1}{2} a_1 t^2 + \frac{1}{3} a_2 t^3 + \frac{1}{4} a_3 t^4$$

8. Describe the kernel of the differentiation operation $D : D^\infty(\mathbb{R}, \mathbb{R}) \to D^\infty(\mathbb{R}, \mathbb{R})$ defined by $D(f) = f'$. Show that ker D is the set of constant functions.

9. Describe the kernel of the differentiation operation $D^2 : D^\infty(\mathbb{R}, \mathbb{R}) \to D^\infty(\mathbb{R}, \mathbb{R})$ defined by $D(f) = f''$.

10. Describe the basis for the kernel of the matrix transformation T_A determined by the matrix

$$A = \begin{pmatrix} 1 & 0 \\ 0 & -1 \end{pmatrix}$$

11. Describe the kernel and image of the linear transformation $T : \mathbb{R}^{2 \times 3} \rightarrow \mathbb{R}^6$ defined by

$$T \begin{pmatrix} a & b & c \\ d & e & f \end{pmatrix} = (a, b, c, 0, 0, 0)$$

12. Show that the set $S = \left\{ \begin{pmatrix} 0 & 0 & 0 \\ 0 & 0 & 0 \end{pmatrix} \right\}$ is the kernel of the linear transformation $T : \mathbb{R}^{2 \times 3} \rightarrow$ $\mathbb{R}^6$ defined by

$$T \begin{pmatrix} a & b & c \\ d & e & f \end{pmatrix} = (d, e, f, a, b, c)$$

and explain why this means that the image of T is $\mathbb{R}^6$. Conclude that T is an isomorphism.

13. Suppose $\mathbf{x} \in \ker T_A$. Show that $T_A(\mathbf{x} + \mathbf{x}_0) = \mathbf{b}$ if and only if $T_A(\mathbf{x}_0) = \mathbf{b}$.

Rank and Nullity

Images and kernels of linear transformations are important for reconstructing vector spaces from their constituent parts. Since many calculations with images and kernels involve the dimensions of these spaces, these dimensions have been given special names.

DEFINITION 5.10   *The **rank** of a linear transformation T, denoted by rank T, is the dimension of the image of T.*

DEFINITION 5.11   *The **nullity** of T, denoted by nullity T, is the dimension of the kernel of T.*

One of the important connections between the rank and the nullity of a linear transformation $T : V \rightarrow W$ is the fact that if V is finite-dimensional, then dim $V =$ nullity $T +$ rank T. This fact is proved in Theorem 5.25.

THEOREM 5.23   *The rank of a linear transformation $T_A : \mathbb{R}^n \rightarrow \mathbb{R}^m$ is the dimension of the column space of A.*

Proof. By Theorem 5.15, the image of T_A is the column space of A. The rank of T_A therefore coincides with the dimension of the column space of A. ∎

THEOREM 5.24   *The nullity of a linear transformation $T_A : \mathbb{R}^n \rightarrow \mathbb{R}^m$ is the dimension of the null space of A.*

Proof. By definition, $T_A(\mathbf{x}) = A\mathbf{x}$ for all $\mathbf{x} \in \mathbb{R}^n$, and the null space of A consists of all $\mathbf{x} \in \mathbb{R}^n$ for which $A\mathbf{x} = \mathbf{0}$. The nullity of T_A therefore coincides with the dimension of the null space of A. ∎

EXAMPLE 5.28 ■ **The Rank and Nullity of a Linear Transformation**

Use *MATHEMATICA* to find the rank and nullity of the linear transformation $T : \mathbb{R}^5 \to \mathbb{R}^4$ defined by

$$T \begin{pmatrix} u \\ v \\ x \\ y \\ z \end{pmatrix} = \begin{pmatrix} x - y + z + u \\ x + 2y - u \\ x + y + 3v - 5z \\ v \end{pmatrix}$$

Solution. We begin by loading two *MATHEMATICA* functions, discussed earlier, that compute the rank and nullity of an arbitrary $m \times n$ matrix. It is shown in the exercises that these functions calculate the rank and nullity of a matrix.

```
In[1]:= Rank[A_]:=Length[Transpose[A]]-Length[NullSpace[A]]
```

```
In[2]:= Nullity[A_]:=Length[NullSpace[A]]
```

Next we find a matrix representation of T. In the standard bases of $\mathbb{R}^4$ and $\mathbb{R}^5$, the transformation T is represented by the matrix

$$A = \begin{pmatrix} 1 & 0 & 1 & -1 & 1 \\ -1 & 0 & 1 & 2 & 0 \\ 0 & 3 & 1 & 1 & -5 \\ 0 & 1 & 0 & 0 & 0 \end{pmatrix}$$

It remains to represent the matrix A in *MATHEMATICA* and to compute its rank and nullity.

```
In[3]:= A= ( 1   0   1  -1   1
            -1   0   1   2   0
             0   3   1   1  -5
             0   1   0   0   0 );
```

```
In[4]:= Nullity[A]

Out[4]= 1
```

```
In[5]:= Rank[A]

Out[5]= 4
```

This shows that the rank and nullity of T are both 2. ◀▶

In this example, rank T + nullity $T = 4 + 1 = 5 = \dim \mathbb{R}^5$. The corollary of the next theorem shows that this relationship between the rank and nullity of a linear transformation is no accident.

THEOREM 5.25   *Suppose that $\mathcal{B} = \{x_1, \ldots, x_p\}$ is a basis for the kernel of a linear transformation $T : V \to W$ and that $\mathcal{C} = \{x_1, \ldots, x_p, x_{p+1}, \ldots, x_n\}$ is a basis for V. Then the set $\mathcal{D} = \{T(x_{p+1}), \ldots, T(x_n)\}$ is a basis for the image of T.*

Proof. If $\ker T = \{0\}$, then its basis is $\emptyset$. Since T is nonsingular, it preserves linear independence. The image of any basis for V, which is the complement of $\emptyset$, is therefore a basis for im T.

Suppose that $\ker T \neq \{0\}$. If $\mathcal{C} = \{x_1, \ldots, x_p, x_{p+1}, \ldots, x_n\}$ is a basis for V, and $\mathcal{B} = \{x_1, \ldots, x_p\}$ is a basis for $\ker T$, then

$$T(\mathcal{C}) = \{T(x_1), \ldots, T(x_p), T(x_{p+1}), \ldots, T(x_n)\} = \{0, T(x_{p+1}), \ldots, T(x_n)\}$$

This means that the set $\mathcal{D} = \{T(x_{p+1}), \ldots, T(x_n)\}$ spans the image of T. It remains to show that D is linearly independent.

Suppose that $a_{p+1} T(x_{p+1}) + \cdots + a_n T(x_n) = 0$ for some scalars $a_{p+1}, \ldots, a_n$. Then the linearity of T implies that $T(a_{p+1} x_{p+1} + \cdots + a_n x_n) = 0$. This means that

$$a_{p+1} x_{p+1} + \cdots + a_n x_n \in \ker T$$

so that $a_{p+1} x_{p+1} + \cdots + a_n x_n = a_1 x_1 + \cdots + a_p x_p$ for some scalars $a_1, \ldots, a_p$. Hence

$$a_1 x_1 + \cdots + a_p x_p - a_{p+1} x_{p+1} - \cdots - a_n x_n = 0$$

Since $\mathcal{B}$ is a basis, it follows that $a_i = 0$ for all $i \in \mathbf{n}$. ∎

COROLLARY 5.26   *If $T : V \to W$ is a linear transformation and V is finite-dimensional, then $\dim V = $ nullity $T + $ rank T.*

Proof. Theorem 5.25 shows that $\dim V = p + (n - p) = n$, where $p = \dim(\ker T)$ and $n - p = \dim(\operatorname{im} T)$. Since $\dim(\ker T)$ is the nullity of T and $\dim(\operatorname{im} T)$ is the rank of T. Therefore, $\dim V = $ nullity $T + $ rank T. ∎

We can use Corollary 5.26 to characterize isomorphisms between finite-dimensional vector spaces.

THEOREM 5.27 (Isomorphism theorem)   *For any linear transformation* $T : V \rightarrow V$ *on a finite-dimensional vector space* V, *the following statements are equivalent.*

1. T is nonsingular.
2. T is one-one.
3. T is onto.
4. T is an isomorphism.

Proof. Theorem 5.22 shows that (1) is equivalent to (2). Theorem 5.6 shows that (4) implies (2) and (3) and that (2) and (3) together imply (4). It therefore remains to show that if V is finite-dimensional, then (2) and (3) are equivalent.

Suppose, therefore, that the dimension of V is finite. Then it follows from Corollary 4.36 that

$$\dim V = \dim(\ker T) + \dim(\operatorname{im} T)$$

Since the kernel of T and the image of T are disjoint, $\dim V = \dim(\operatorname{im} T)$ if and only if $\dim(\ker T) = 0$. In other words, $\dim V = \dim(\operatorname{im} T)$ if and only if T is one-one. By Theorem 4.21 this is so if and only if $V = \operatorname{im} T$. ∎

EXERCISES 5.8

1. Compute the rank and nullity of the matrix transformations determined by the matrices below and show that in each case, $\operatorname{rank}(T) + \operatorname{nullity}(T)$ is equal to the dimension of the domain of T.

   a. $\begin{pmatrix} 1 & 0 & -2 & 6 \\ 0 & 5 & 7 & 1 \end{pmatrix}$ b. $\begin{pmatrix} 1 & 0 & 0 \\ 0 & 5 & 0 \\ -2 & 7 & 0 \\ 6 & 1 & 0 \end{pmatrix}$ c. $\begin{pmatrix} 0 & 0 & 0 \\ 8 & 2 & 77 \\ 3 & -7 & 9 \end{pmatrix}$

   d. $\begin{pmatrix} 1 & 0 & 0 \\ 0 & 3 & 0 \\ 0 & 0 & 7 \\ 0 & 0 & 0 \end{pmatrix}$ e. $\begin{pmatrix} 1 & 0 & 0 & 0 \\ 0 & 3 & 0 & 0 \\ 0 & 0 & 0 & 0 \end{pmatrix}$ f. $\begin{pmatrix} 1 & 0 & 0 & 9 \\ 0 & 0 & 0 & 0 \\ 0 & 0 & 0 & 0 \\ 0 & 0 & 0 & 0 \end{pmatrix}$

2. Use *MATHEMATICA* to show that if

   $$A = \begin{pmatrix} 5 & 1 & 0 & 0 & 1 & 1 & 2 \\ 3 & 0 & 0 & 0 & 0 & 3 & 0 \\ 8 & 1 & 0 & 0 & 1 & 4 & 2 \\ 0 & 0 & 0 & 0 & 0 & 0 & 0 \\ 3 & 0 & 0 & 0 & 1 & 0 & 2 \end{pmatrix}$$

then the nullity of the matrix transformation $T_A : \mathbb{R}^7 \rightarrow \mathbb{R}^5$ is 4. Use *MATHEMATICA* to find a basis $\mathcal{B} = \{x_1, x_2, x_3, x_4\}$ for ker T_A. Extend $\mathcal{B}$ to a basis

$$\mathcal{C} = \{x_1, x_2, x_3, x_4, x_5, x_6, x_7\}$$

for $\mathbb{R}^7$ and show that the set $\{T_A(x_5), T_A(x_6), T_A(x_7)\}$ is a basis for im T_A.

3. Show that a matrix transformation $T_A : \mathbb{R}^n \rightarrow \mathbb{R}^n$ is an isomorphism if and only if its rank is n.

4. Show that a matrix transformation $T_A : \mathbb{R}^n \rightarrow \mathbb{R}^n$ is an isomorphism if and only if its nullity is 0.

5. Show that the function **Rank[A_]:=Length[Transpose[A]]** **-Length[NullSpace[A]]** calculates the rank of any $m \times n$ matrix A.

6. Show that the function **Nullity[A_]:=Length[NullSpace[A]]** calculates the nullity of any $m \times n$ matrix A.

Areas and Linear Transformations

We know from Example 3.19 that the area of the parallelogram determined by two vectors

$$\mathbf{u} = \begin{pmatrix} u_1 \\ u_2 \end{pmatrix} \quad \text{and} \quad \mathbf{v} = \begin{pmatrix} v_1 \\ v_2 \end{pmatrix}$$

is the absolute value of the determinant of the matrix

$$P = \begin{pmatrix} u_1 & v_1 \\ u_2 & v_2 \end{pmatrix}$$

We now explore the change in the area of the parallelogram if the vectors $\mathbf{u}$ and $\mathbf{v}$ are modified by a matrix transformation.

EXAMPLE 5.29 ■ **The Area of the Image of a Parallelogram**

Suppose that $T_A : \mathbb{R}^2 \rightarrow \mathbb{R}^2$ is a linear transformation defined by the matrix

$$A = \begin{pmatrix} a & c \\ b & d \end{pmatrix}$$

and that the columns of the matrix

$$P = \begin{pmatrix} u_1 & v_1 \\ u_2 & v_2 \end{pmatrix}$$

are two vectors determining a parallelogram in $\mathbb{R}^2$. Show that the area of the image of the parallelogram under T_A is det A det P.

Solution. Consider the parallelogram determined by the vectors $A\mathbf{u}$ and $A\mathbf{v}$. We know from Theorem 2.1 that the matrix $(\ A\mathbf{u}\quad A\mathbf{v}\ )$ whose columns are $A\mathbf{u}$ and $A\mathbf{v}$ equals AP. Therefore, the area of the parallelogram determined by $A\mathbf{u}$ and $A\mathbf{v}$ is the absolute value of

$$\det(\ A\mathbf{u}\quad A\mathbf{v}\ ) = \det AP = \det A \det P$$

This shows that the area of the original parallelogram is magnified by the absolute value of the determinant of the matrix A. ◀▶

In Example 5.29 we assumed that the parallelogram was determined by two vectors starting at the origin. If this is not the case, we can use homogeneous coordinates to find the area.

EXAMPLE 5.30 ■ **Homogeneous Coordinates and the Area of an Image**

Let P be the parallelogram with vertices $(2, 3)$, $(10, -4)$, $(0, -9)$, and $(-8, -2)$. Use homogeneous coordinates to find the area of the image of P under the matrix transformation $T_A : \mathbb{R}^2 \to \mathbb{R}^2$ determined by

$$A = \begin{pmatrix} 3 & 1 \\ 0 & -2 \end{pmatrix}$$

Solution. An easy calculation shows that P is determined by the vertices $(2, 3)$, $(10, -4)$, and $(-8, -2)$. Let

$$H = \begin{pmatrix} 2 & 10 & -8 \\ 3 & -4 & -2 \\ 1 & 1 & 1 \end{pmatrix}$$

be the matrix of the homogeneous coordinates of the given vertices. We translate H so that the coordinates of one of the vertices are the homogeneous coordinates of the origin of $\mathbb{R}^2$.

We have three choices and opt for the translation that moves $(2, 3)$ to $(0, 0)$.

$$A_1 = \begin{pmatrix} 1 & 0 & -2 \\ 0 & 1 & -3 \\ 0 & 0 & 1 \end{pmatrix} \begin{pmatrix} 2 & 10 & -8 \\ 3 & -4 & -2 \\ 1 & 1 & 1 \end{pmatrix} = \begin{pmatrix} 0 & 8 & -10 \\ 0 & -7 & -5 \\ 1 & 1 & 1 \end{pmatrix}$$

The last two columns of A_1 are the homogeneous coordinates of the new vectors determining the translated parallelogram.

Next we find the homogeneous coordinates of the images of the new vectors.

$$A_2 = \begin{pmatrix} 3 & 1 & 0 \\ 0 & -2 & 0 \\ 0 & 0 & 1 \end{pmatrix} \begin{pmatrix} 0 & 8 & -10 \\ 0 & -7 & -5 \\ 1 & 1 & 1 \end{pmatrix} = \begin{pmatrix} 0 & 17 & -35 \\ 0 & 14 & 10 \\ 1 & 1 & 1 \end{pmatrix}$$

It is easy to check that the area of the image of P is the absolute value of the determinant of the matrix A_2.

$$\text{Area of im } P = \det \begin{pmatrix} 0 & 17 & -35 \\ 0 & 14 & 10 \\ 1 & 1 & 1 \end{pmatrix} = 660$$

This confirms what we know from Example 5.30. The areas of P and im P are connected by the determinant of A :

$$\det \begin{pmatrix} 17 & -35 \\ 14 & 10 \end{pmatrix} = 660 = \det \begin{pmatrix} 3 & 1 \\ 0 & -2 \end{pmatrix} \det \begin{pmatrix} 8 & -10 \\ -7 & -5 \end{pmatrix} = (-6)(-110)$$

As we can see, the magnification factor is the absolute value of the determinant of A. ◀▶

EXERCISES 5.9

1. Find the area of the image of the triangle determined by the vectors $(2, 3)$ and $(5, -1)$ under the matrix transformation

$$T \begin{pmatrix} x \\ y \end{pmatrix} = \begin{pmatrix} x + y \\ y \end{pmatrix}$$

2. Find the area of the image of the parallelogram determined by the vertices

$$\begin{pmatrix} -2\sqrt{2} \\ 5\sqrt{2} \end{pmatrix}, \quad \begin{pmatrix} 0 \\ 7\sqrt{2} \end{pmatrix}, \quad \begin{pmatrix} -\sqrt{2} \\ 8\sqrt{2} \end{pmatrix}, \quad \begin{pmatrix} -3\sqrt{2} \\ 6\sqrt{2} \end{pmatrix}$$

under the linear transformation

$$T \begin{pmatrix} x \\ y \end{pmatrix} = \begin{pmatrix} x - y \\ y \end{pmatrix}$$

3. Use *MATHEMATICA* to plot the quadrilateral determined by the points $(1, 2)$, $(5, -3)$, $(7, 8)$, and $(12, 2)$ and find the area of its image under the matrix transformation

$$T \begin{pmatrix} x \\ y \end{pmatrix} = \begin{pmatrix} x - y \\ x + y \end{pmatrix}$$

4. Construct a matrix transformation $T_A : \mathbb{R}^3 \to \mathbb{R}^3$ that does not preserve linear independence. Justify your construction by exhibiting two linearly independent vectors x_1 and x_2 for which $T_A(x_1)$ and $T_A(x_2)$ are linearly dependent.

5. Use *MATHEMATICA* to find the image of the matrix transformation $T_A : \mathbb{R}^3 \rightarrow \mathbb{R}^4$ determined by the matrix

$$A = \begin{pmatrix} 1 & 0 & 2 \\ 0 & 1 & 0 \\ -6 & 0 & 1 \\ 0 & 0 & 0 \end{pmatrix}$$

SIMILARITY

We saw earlier that there are many different ways of representing a linear transformation $T : \mathbb{R}^n \rightarrow \mathbb{R}^m$ by a matrix. Different choices of bases yield different matrices. What is the connection between these matrices? If $n = m$, the answer is quite simple. Two matrices A and B represent the same transformation $T : \mathbb{R}^n \rightarrow \mathbb{R}^n$ if there exists a coordinate conversion matrix P for which $B = PAP^{-1}$. Before proving this fact, we give a name to this connection between A and B.

DEFINITION 5.12 *Two matrices A and B are **similar** if there exists an invertible matrix P such that $B = PAP^{-1}$.*

THEOREM 5.28 (Matrix similarity theorem) *Two square matrices A and B represent the same linear transformation if and only if they are similar.*

Proof. Suppose that $\mathcal{B}$ and $\mathcal{C}$ are two bases for V, that $A = [T]_{\mathcal{C}}$ and $B = [T]_{\mathcal{B}}$, and that P is the coordinate conversion matrix from $\mathcal{B}$ to $\mathcal{C}$. Then we know from Example 5.20 and Theorem 5.12 that

$$\begin{aligned}
P^{-1}AP[\mathbf{x}]_{\mathcal{B}} &= [I]_{\mathcal{B}}^{\mathcal{C}} [T]_{\mathcal{C}} [I]_{\mathcal{C}}^{\mathcal{B}} [\mathbf{x}]_{\mathcal{B}} \\
&= [I]_{\mathcal{B}}^{\mathcal{C}} [T]_{\mathcal{C}} [\mathbf{x}]_{\mathcal{C}} \\
&= [I]_{\mathcal{B}}^{\mathcal{C}} [T(\mathbf{x})]_{\mathcal{C}} \\
&= [T(\mathbf{x})]_{\mathcal{B}} \\
&= [T]_{\mathcal{B}} [\mathbf{x}]_{\mathcal{B}} \\
&= B[\mathbf{x}]_{\mathcal{B}}
\end{aligned}$$

Since this equation holds for all $\mathbf{x} \in V$, it follows that $P^{-1}AP = B$.

Conversely, suppose that $A = PBP^{-1}$ and that $A = [T_1]_{\mathcal{B}}$ and $B = [T_2]_{\mathcal{C}}$. Then

$$\begin{aligned}
[T_1(\mathbf{x})]_{\mathcal{B}} &= [T_1]_{\mathcal{B}} [\mathbf{x}]_{\mathcal{B}} \\
&= [I]_{\mathcal{C}}^{\mathcal{B}} [T_2]_{\mathcal{C}} [I]_{\mathcal{B}}^{\mathcal{C}} [\mathbf{x}]_{\mathcal{B}} \\
&= [I]_{\mathcal{C}}^{\mathcal{B}} [T_2]_{\mathcal{C}} [\mathbf{x}]_{\mathcal{C}} \\
&= [I]_{\mathcal{C}}^{\mathcal{B}} [T_2(\mathbf{x})]_{\mathcal{C}} \\
&= [T_2(\mathbf{x})]_{\mathcal{B}}
\end{aligned}$$

Therefore, $[T_1]_B[\mathbf{x}]_B = [T_2]_B[\mathbf{x}]_B$ for all $\mathbf{x} \in V$. Hence $T_1 = T_2$. ■

The next two theorems provide useful tests for showing that two square matrices are *not* similar.

THEOREM 5.29 (Trace test)   *If A and B are similar, then* trace $A =$ trace B.

Proof. Suppose that A and B are similar. Then $A = P^{-1}BP$. We know from Theorem 2.10 that trace $DE =$ trace ED. Therefore

$$\begin{aligned}
\text{trace } A &= \text{trace } P^{-1}BP \\
&= \text{trace } P^{-1}(BP) \\
&= \text{trace}(BP)P^{-1} \\
&= \text{trace } BPP^{-1} \\
&= \text{trace } B
\end{aligned}$$

This proves the theorem. ■

This theorem shows that if trace $A \neq$ trace B, then A and B are not similar. However, two nonsimilar matrices may still have the same trace.

EXAMPLE 5.31 ■ **Two Matrices with a Different Trace**

Use the trace test to show that the matrices

$$A = \begin{pmatrix} 3 & 2 \\ 1 & 2 \end{pmatrix} \quad \text{and} \quad B = \begin{pmatrix} 1 & 2 \\ -1 & 2 \end{pmatrix}$$

are not similar.

Solution. The trace of A is $3 + 2 = 5$ and that of B is $1 + 2 = 3$. Therefore, A and B are not similar since they have a different trace. ◄►

THEOREM 5.30 (Determinant test)   *If A and B are similar, then* det $A =$ det B.

Proof. Suppose that A and B are similar. Then $A = P^{-1}BP$. We know from Theorem 3.13 that det $DEF =$ det D det E det F. Therefore,

$$\begin{aligned}
\det A &= \det P^{-1}BP \\
&= \det P^{-1} \det B \det P \\
&= \det P^{-1} \det P \det B \\
&= \det B
\end{aligned}$$

This proves the theorem. ■

This theorem shows that if det $A \neq$ det B, then A and B are not similar. Again the test may be nonconclusive: Two nonsimilar matrices may have the same determinant.

EXAMPLE 5.32   ■ Two Matrices with a Different Determinant

Use *MATHEMATICA* and the determinant test to show that the matrices

$$A = \begin{pmatrix} 3 & 2 \\ 1 & 2 \end{pmatrix} \quad \text{and} \quad B = \begin{pmatrix} 1 & 2 \\ -1 & 1 \end{pmatrix}$$

are not similar.

Solution. We begin by defining A and B.

```
In[1]:= A= ( 3  2 )  ; B= ( 1  2 )  ;
         ( 1  2 )      (-1  1 )
```

We then use the **Det** function.

```
In[2]:= Det [A]

Out[2]= 4
```

```
In[3]:= Det [B]

Out[3]= 3
```

Since det $A \neq$ det B, the matrices A and B are not similar. ◄►

Neither the trace test nor the determinant test allows us to conclude that two matrices are similar. The matrices

$$\begin{pmatrix} 1 & 1 & 0 \\ 3 & 3 & 2 \\ 2 & 2 & 3 \end{pmatrix} \quad \text{and} \quad \begin{pmatrix} 1 & 1 & 0 \\ 3 & 3 & 2 \\ 0 & 0 & 3 \end{pmatrix}$$

for example, have the same trace and the same determinant. However, in Example 6.11 we will show that they are not similar.

Furthermore, neither the equality of the trace of two matrices nor the equality of their determinant allows us to conclude that two square matrices are similar. They provide necessary but not sufficient conditions for similarity. Sufficient conditions also exist, but their study is beyond the scope of this book.

EXERCISES 5.10

1. Use the trace test to show that the following three matrices are not similar.

$$
\text{a. } A = \begin{pmatrix} 1 & 0 & -2 \\ 0 & 5 & 7 \\ 0 & 0 & 0 \end{pmatrix} \quad
\text{b. } A = \begin{pmatrix} 0 & 0 & 0 \\ 0 & 0 & 0 \\ 0 & 0 & 0 \end{pmatrix} \quad
\text{c. } A = \begin{pmatrix} 5 & -8 & 3 \\ 4 & 2 & 7 \\ 4 & -5 & 9 \end{pmatrix}
$$

2. Use the determinant test to show that the following three matrices are not similar.

$$
\text{a. } A = \begin{pmatrix} 1 & 0 & -2 \\ 0 & 3 & 7 \\ 0 & 0 & 2 \end{pmatrix} \quad
\text{b. } A = \begin{pmatrix} 6 & 0 & 0 \\ 0 & 0 & 0 \\ 0 & 0 & 0 \end{pmatrix} \quad
\text{c. } A = \begin{pmatrix} 2 & -8 & 3 \\ 4 & 2 & 7 \\ 4 & -5 & 2 \end{pmatrix}
$$

3. Show that the $n \times n$ zero matrix is not similar to any nonzero $n \times n$ matrix.

4. Find all matrices that are similar to the $n \times n$ identity matrix.

5. Show that if a matrix A is similar to a matrix B and s is any scalar, then sA is similar to sB.

6. Find two similar 2×2 matrices A and B, and two similar 2×2 matrices C and D, with the property that $A + C$ is not similar to $B + D$, and AC is not similar to BD.

7. Show that the matrices

$$
\begin{pmatrix} 3 & 0 & 0 \\ 0 & 2 & 0 \\ 0 & 0 & 2 \end{pmatrix} \quad \text{and} \quad \begin{pmatrix} 2 & 0 & 0 \\ 0 & 3 & 0 \\ 0 & 0 & 2 \end{pmatrix}
$$

are similar.

8. Explain why two matrices with the same trace but different determinants are not similar. Use this test to construct two nonsimilar 2×2 matrices.

9. Explain why two real $n \times n$ matrices AB and BA are not necessarily similar.

10. Find two real 3×3 matrices A and B for which the matrices AB and BA are not similar.

REVIEW

KEY CONCEPTS ▶ Define and discuss each of the following.

Components of a Linear Transformation

Codomain of a linear transformation, domain of a linear transformation, image of a linear transformation, kernel of a linear transformation, range of a linear transformation.

Matrices and Linear Transformations

Matrix of a transformation, matrix transformation, similar matrices and matrix transformations, standard matrix of a linear transformation.

Operations on Linear Transformations

Addition of linear transformations, composition of linear transformations, inversion of a transformation, scalar multiplication of a linear transformation, transposition of a linear transformation.

Properties of a Linear Transformation

Invertible transformation, isomorphism, nullity of a linear transformation, one-one linear transformation, onto linear transformation, singular transformation.

Vectors and Linear Transformations

Coordinate vector function, linear combination function.

KEY FACTS ▶ Explain and illustrate each of the following.

1. Every matrix transformation is a linear transformation.

2. If $T : V \to W$ is a linear transformation, then $T(\mathbf{0}_V) = \mathbf{0}_W$.

3. If $\mathcal{B} = \{\mathbf{x}_1, \mathbf{x}_2, \ldots, \mathbf{x}_n\}$ is a basis for a vector space V and if $\mathbf{y}_1, \mathbf{y}_2, \ldots, \mathbf{y}_n$ is any list of vectors in a vector space W, then there exists a unique linear transformation $T : V \to W$ for which $T(\mathbf{x}_1) = \mathbf{y}_1, \ldots, T(\mathbf{x}_n) = \mathbf{y}_n$.

4. If $T : V \to W$ is an invertible linear transformation, then T is one-one.

5. If $T : V \to W$ is an invertible linear transformation, then T is onto.

6. A linear transformation $T : V \to W$ is an isomorphism if and only if T is one-one and onto.

7. For any basis $\mathcal{B}$ of a finite-dimensional real vector space V, the coordinate vector function $[-]_{\mathcal{B}} : V \to \mathbb{R}^n$ is an isomorphism.

8. If two real vector spaces have the same dimension, they are isomorphic.

9. The matrix space $\mathbb{R}^{m \times n}$ is isomorphic to the coordinate space $\mathbb{R}^{mn}$.

10. The polynomial space $\mathbb{R}_n[t]$ is isomorphic to the coordinate space $\mathbb{R}^{n+1}$.

11. If T is a linear transformation from $\mathbb{R}^n$ to $\mathbb{R}^m$ and if $\mathcal{B}$ is a basis for $\mathbb{R}^n$ and $\mathcal{C}$ is a basis for $\mathbb{R}^m$, then $[T(\mathbf{x})]_{\mathcal{C}} = [T]_{\mathcal{C}}^{\mathcal{B}} [\mathbf{x}]_{\mathcal{B}}$ for all $\mathbf{x} \in \mathbb{R}^n$.

12. Suppose that $T_1 : \mathbb{R}^n \to \mathbb{R}^m$ and $T_2 : \mathbb{R}^m \to \mathbb{R}^p$ are two linear transformations and that the matrices $[T_1]_{\mathcal{C}}^{\mathcal{B}}$ and $[T_2]_{\mathcal{D}}^{\mathcal{C}}$ represent the transformation T_1 and T_2 in the bases $\mathcal{B}, \mathcal{C}$, and $\mathcal{D}$. Then

$$[T_2 \circ T_1]_{\mathcal{D}}^{\mathcal{B}} = [T_2]_{\mathcal{D}}^{\mathcal{C}} [T_1]_{\mathcal{C}}^{\mathcal{B}}$$

13. If $T : \mathbb{R}^n \to \mathbb{R}^n$ is an invertible linear transformation represented by a matrix A, then $\left[T^{-1}\right] = A^{-1}$.

14. The image of a linear transformation $T : V \to W$ is a subspace of W.

15. If A is a real $m \times n$ matrix and $T_A : \mathbb{R}^n \to \mathbb{R}^m$ is the associated matrix transformation, then $\text{im}(T_A) = \text{Col } A$.

16. If $T : V \to W$ is a linear transformation and if $\mathcal{S} = \{\mathbf{x}_1, \ldots, \mathbf{x}_n\}$ spans V, then the set $\{T(\mathbf{x}_1), \ldots, T(\mathbf{x}_n)\}$ spans the image of T.

17. If $T : V \to W$ is a one-one linear transformation and if $\{T(\mathbf{x}_1), \ldots, T(\mathbf{x}_n)\}$ spans the image of T, then $\{\mathbf{x}_1, \ldots, \mathbf{x}_n\}$ spans V.

18. If $T : V \to W$ is a one-one linear transformation and the vectors $\mathbf{x}_1, \ldots, \mathbf{x}_n$ are linearly independent in V, then the vectors $T(\mathbf{x}_1), \ldots, T(\mathbf{x}_n)$ are linearly independent in W.

19. If $T : V \to W$ is a linear transformation and the vectors $T(\mathbf{x}_1), \ldots, T(\mathbf{x}_n)$ are linearly independent in W, then the vectors $\mathbf{x}_1, \ldots, \mathbf{x}_n$ are linearly independent in V.

20. The kernel of a linear transformation $T : V \to W$ is a subspace of V.

21. The kernel of a matrix transformation $T_A : \mathbb{R}^n \to \mathbb{R}^m$ is the null space of A.

22. A linear transformation T is one-one if and only if $\ker T = \{\mathbf{0}\}$.

23. The rank of a linear transformation $T_A : \mathbb{R}^n \to \mathbb{R}^m$ is the dimension of the column space of A.

24. The nullity of a linear transformation $T_A : \mathbb{R}^n \to \mathbb{R}^m$ is the dimension of the null space of A.

25. Suppose that $\mathcal{B} = \{\mathbf{x}_1, \ldots, \mathbf{x}_p\}$ is a basis for the kernel of a linear transformation $T : V \to W$ and that $\mathcal{C} = \{\mathbf{x}_1, \ldots, \mathbf{x}_p, \mathbf{x}_{p+1}, \ldots, \mathbf{x}_n\}$ is a basis for V. Then the set $\mathcal{D} = \{T(\mathbf{x}_{p+1}), \ldots, T(\mathbf{x}_n)\}$ is a basis for the image of T.

26. If $T : V \rightarrow W$ is a linear transformation and V is finite-dimensional, then $\dim V =$ nullity $T + $ rank T.

27. For any linear transformation $T : V \rightarrow V$ on a finite-dimensional vector space V, the following statements are equivalent:

    a. T is nonsingular

    b. T is one-one

    c. T is onto

    d. T is an isomorphism

28. Two square matrices A and B represent the same linear transformation if and only if they are similar.

29. If A and B are similar, then trace $A = $ trace B.

30. If A and B are similar, then $\det A = \det B$.

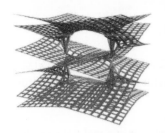

6

EIGENVALUES AND EIGENVECTORS

INTRODUCTION

In many applications involving a linear transformation $T : V \to V$ it is useful to find nonzero vectors $\mathbf{x}$ that are parallel to their images $T(\mathbf{x})$. By parallel vectors we mean vectors that are scalar multiples of each other. We are interested in finding scalars λ and nonzero vectors $\mathbf{x}$ for which $T(\mathbf{x}) = \lambda\mathbf{x}$. The scalar λ is called an *eigenvalue* and the associated vector $\mathbf{x}$ an *eigenvector* of T.

Our goal in this chapter is to describe how eigenvalues and eigenvectors can be found and to discuss conditions under which T has enough linearly independent eigenvectors to form a basis for V. Not every linear transformation has eigenvalues and eigenvectors. If V is finite-dimensional, we can reduce the problem of finding eigenvalues and eigenvectors for T to the problem of finding eigenvalues and eigenvectors for any matrix A of T.

THEOREM 6.1 *If $T : V \to V$ is a linear transformation on a finite-dimensional vector space V and $A = [T]_{\mathcal{B}}^{\mathcal{B}}$ is the matrix of T in some basis $\mathcal{B}$, then $T(\mathbf{x}) = \lambda\mathbf{x}$ if and only if $A[\mathbf{x}]_{\mathcal{B}} = \lambda[\mathbf{x}]_{\mathcal{B}}$.*

Proof. Suppose that $T(\mathbf{x}) = \lambda\mathbf{x}$. Then $[T(\mathbf{x})]_{\mathcal{B}} = [\lambda\mathbf{x}]_{\mathcal{B}}$. Since $A = [T]_{\mathcal{B}}^{\mathcal{B}}$, we know from the construction of A that $[T(\mathbf{x})]_{\mathcal{B}} = A[\mathbf{x}]_{\mathcal{B}}$. We also know from the linearity of the coordinate vector function, that $[\lambda\mathbf{x}]_{\mathcal{B}} = \lambda[\mathbf{x}]_{\mathcal{B}}$. Hence $A[\mathbf{x}]_{\mathcal{B}} = \lambda[\mathbf{x}]_{\mathcal{B}}$. By reversing these arguments, we see that if $A[\mathbf{x}]_{\mathcal{B}} = \lambda[\mathbf{x}]_{\mathcal{B}}$, then $T(\mathbf{x}) = \lambda\mathbf{x}$. ∎

It is an important consequence of this theorem that the eigenvalue λ of T is independent of the basis $\mathcal{B}$ involved in the construction of $[T]_{\mathcal{B}}^{\mathcal{B}}$. Therefore, λ is an eigenvalue of T if and only if λ is an eigenvalue of T_A for any $A = [T]_{\mathcal{B}}^{\mathcal{B}}$, and $\mathbf{x}$ is an eigenvector of T belonging to λ if and only if $\mathbf{x}$ is an eigenvector of T_A for any $A = [T]_{\mathcal{B}}^{\mathcal{B}}$. It therefore makes sense to define eigenvalues and eigenvectors for square matrices.

DEFINITION 6.1 *A real number λ is an **eigenvalue** of a real $n \times n$ matrix A if there exists a nonzero column vector $\mathbf{x} \in \mathbb{R}^n$ for which $A\mathbf{x} = \lambda\mathbf{x}$.*

Since every matrix A determines a matrix transformation T_A for which $A\mathbf{x} = T_A(\mathbf{x})$, Theorem 6.1 implies that similar matrices have the same eigenvalues. We will later confirm this fact using properties of determinants.

DEFINITION 6.2   *A nonzero column vector* $\mathbf{x} \in \mathbb{R}^n$ *is an **eigenvector** of a real* $n \times n$ *matrix* A *if there exists a real number* λ *for which* $A\mathbf{x} = \lambda\mathbf{x}$.

We say that an eigenvector $\mathbf{x}$ ***belongs to*** an eigenvalue λ, or that $\mathbf{x}$ *is associated with* λ. It may happen that an eigenvector $\mathbf{x}$ belongs to the eigenvalue $\lambda = 0$. Thus eigenvalues can be zero, but eigenvectors cannot.

In the language of eigenvalues and eigenvectors, Theorem 6.1 says that if $A = [T]_B^B$ is the matrix of a linear transformation T, then λ is an eigenvalue of T if and only if it is an eigenvalue of A, and $\mathbf{x}$ is an eigenvector of T if and only if $[\mathbf{x}]_B$ is an eigenvector of A. Let us consider some examples.

EXAMPLE 6.1   ■ The Eigenvalues and Eigenvectors of an Identity Transformation

Let $T : V \rightarrow V$ be the identity transformation. Then $T(\mathbf{x}) = \mathbf{x} = 1\mathbf{x}$ for all $\mathbf{x} \in V$. Therefore, 1 is an eigenvalue of T and any nonzero vector $\mathbf{x} \in V$ is an eigenvector belonging to the eigenvalue 1. ◄►

EXAMPLE 6.2   ■ The Eigenvalues and Eigenvectors of a Zero Transformation

Let $T : V \rightarrow V$ be the zero transformation. Then $T(\mathbf{x}) = \mathbf{0} = 0\mathbf{x}$ for all $\mathbf{x} \in V$. Therefore, 0 is an eigenvalue of T and any nonzero vector $\mathbf{x} \in V$ is an eigenvector belonging to the eigenvalue 0. ◄►

The next examples are more interesting. They show that a linear transformation $T : \mathbb{R}^n \rightarrow \mathbb{R}^n$ may have any number of distinct eigenvalues. However, in no case can T have more than n such values.

EXAMPLE 6.3   ■ A Linear Transformation with Three Eigenvalues

Let $T : \mathbb{R}^3 \rightarrow \mathbb{R}^3$ be the transformation defined by $T(x, y, z) = (x, 2y, 3z)$. Then 1, 2, and 3 are eigenvalues of T since

$$T(x, 0, 0) = (x, 0, 0) = 1(x, 0, 0)$$
$$T(0, y, 0) = (0, 2y, 0) = 2(0, y, 0)$$
$$T(0, 0, z) = (0, 0, 3z) = 3(0, 0, z)$$

for all $x, y, z \in \mathbb{R}$. ◄►

If we modify this example and let

$$T_2(x, y, z) = (x, 2y, 2z) \quad \text{and} \quad T_3(x, y, z) = (2x, 2y, 2z)$$

then T_2 is an example of a linear transformation $\mathbb{R}^3 \to \mathbb{R}^3$ with two distinct eigenvalues, and T_3 is an example of a linear transformation $\mathbb{R}^3 \to \mathbb{R}^3$ with only one eigenvalue.

If the number of distinct eigenvalues of a linear transformation $T : \mathbb{R}^n \to \mathbb{R}^n$ is less than n, it is often a challenge to find several linearly independent eigenvectors belonging to the same eigenvalue. We will in fact see it is not always possible.

DEFINITION 6.3   *An $n \times n$ matrix is **diagonalizable** if it has n linearly independent eigenvectors.*

This will mean that there exists a diagonal matrix D whose diagonal entries are eigenvalues of A, and an invertible matrix P whose columns are eigenvectors of A, for which $A = PDP^{-1}$. This definition applies both to real and to complex matrices. We call this the ***eigenvector-eigenvalue decomposition*** of A. In Chapter 9, we will study techniques for dealing with matrices that fail to have such decompositions.

DEFINITION 6.4   *A linear transformation $T : V \to V$ is **diagonalizable** if there exists a basis $\mathcal{B}$ for which the matrix $A = [T]_{\mathcal{B}}^{\mathcal{B}}$ is diagonal.*

EXAMPLE 6.4   ■ A Diagonalizable Transformation

Find the eigenvalues and eigenvectors of the linear transformation $T_A : \mathbb{R}^2 \to \mathbb{R}^2$ defined by

$$T_A \begin{pmatrix} x \\ y \end{pmatrix} = \begin{pmatrix} 3 & 0 \\ 0 & 2 \end{pmatrix} \begin{pmatrix} x \\ y \end{pmatrix}$$

and discuss the geometric properties of T_A.

Solution. Since

$$\begin{pmatrix} 3 & 0 \\ 0 & 2 \end{pmatrix} \begin{pmatrix} x \\ y \end{pmatrix} = \begin{pmatrix} 3x \\ 2y \end{pmatrix} = 3 \begin{pmatrix} x \\ 0 \end{pmatrix} + 2 \begin{pmatrix} 0 \\ y \end{pmatrix}$$

it follows that $T_A(x, 0) = 3(x, 0)$ and $T_A(0, y) = 2(0, y)$ for all $x, y \in \mathbb{R}$. Hence $\lambda_1 = 3$ and $\lambda_2 = 2$ are eigenvalues of T_A and $(x, 0)$ and $(0, y)$ are their associated eigenvectors, provided x and y are not zero.

We can illustrate the geometric properties of T_A by considering its effect on the triangle determined by the points $(0, 0)$, $(1, 0)$, and $(0, 1)$. Since $T(0, 0) = (0, 0)$, $T(1, 0) = (3, 0)$, and $T(0, 1) = (0, 2)$, the transformation T stretches the triangle three units along the x-axis and two units along the y-axis. Figure 1 shows the two corresponding triangles. ◂▸

MATHEMATICA has a built-in **Eigenvalues** function that can be used to find eigenvalues of square matrices.

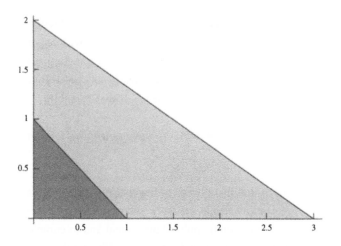

FIGURE 1   The Geometric Effect of T_A.

EXAMPLE 6.5   ■ A Linear Transformation with Two Eigenvalues

Find the eigenvalues of the linear transformation $T : \mathbb{R}^2 \to \mathbb{R}^2$ defined by

$$T\begin{pmatrix} x \\ y \end{pmatrix} = \begin{pmatrix} x + y \\ x - y \end{pmatrix}$$

Solution. We must find a λ for which

$$\begin{pmatrix} x + y \\ x - y \end{pmatrix} = \lambda \begin{pmatrix} x \\ y \end{pmatrix} = \begin{pmatrix} \lambda x \\ \lambda y \end{pmatrix}$$

We begin by representing T in the standard basis $\mathcal{E} = \{\mathbf{e}_1, \mathbf{e}_2\}$ of $\mathbb{R}^2$.

$$[T]_{\mathcal{E}} = ([T(\mathbf{e}_1)]_{\mathcal{E}} \ \ [T(\mathbf{e}_2)]_{\mathcal{E}}) = \begin{pmatrix} 1 & 1 \\ 1 & -1 \end{pmatrix}$$

Then we use *MATHEMATICA* to find the eigenvalues and associated eigenvectors of $[T]_{\mathcal{E}}$.

```
In[1]:= Eigenvalues[ A= ( 1   1
                          1  -1 ) ]

Out[1]= { -√2̄ , √2̄ }
```

We show that $-\sqrt{2}$ is an eigenvalue by finding an associated eigenvector. The verification for $\sqrt{2}$ is analogous. If $-\sqrt{2}$ is an eigenvalue of A, then

$$\begin{pmatrix} x+y \\ x-y \end{pmatrix} = -\sqrt{2}\begin{pmatrix} x \\ y \end{pmatrix} = \begin{pmatrix} -\sqrt{2}x \\ -\sqrt{2}y \end{pmatrix}$$

This means that

$$\begin{cases} x+y = -\sqrt{2}x \\ x-y = -\sqrt{2}y \end{cases}$$

If this system has a nontrivial solution, then $-\sqrt{2}$ is an eigenvalue of A.

```
In[2]:= Solve[{x+y= =-√2x,x-y= =-√2y},{x,y}]

Solve::svars:

Equations may not give solutions for all ''solve'' vari-
ables.

Out[2]= {{x→ (1-√2y)}}
```

The message preceding the rule $\{\{x \to (1-\sqrt{2}y)\}\}$ indicates that this rule may not give all solutions. However, for our purposes, one solution will suffice. Let $y = 1$, then the rule tells us that we should let $x = 1 - \sqrt{2}$. The calculation

$$T\begin{pmatrix} 1-\sqrt{2} \\ 1 \end{pmatrix} = \begin{pmatrix} \left(1-\sqrt{2}\right)+1 \\ \left(1-\sqrt{2}\right)-1 \end{pmatrix} = -\sqrt{2}\begin{pmatrix} 1-\sqrt{2} \\ 1 \end{pmatrix}$$

shows that, indeed, the vector $(x, y) = (1 - \sqrt{2}, 1)$ is an eigenvector belonging to the eigenvalue $-\sqrt{2}$. A similar calculation shows that the vector $(x, y) = (1+\sqrt{2}, 1)$ is an eigenvector belonging to $\sqrt{2}$. Hence the matrix A has the two real eigenvalues $-\sqrt{2}$ and $\sqrt{2}$.

We could have used the more comprehensive **Eigensystem** function to find both the eigenvalues and two associated eigenvectors of the linear transformation directly.

```
In[3]:= Eigensystem[A]

Out[3]= {{-√2,√2},{{1-√2,1},{{1+√2,1}}}
```

In this output, the component $\{-\sqrt{2}, \sqrt{2}\}$ lists the eigenvalues of A, and the component $\{\{1-\sqrt{2}, 1\}, \{\{1+\sqrt{2}, 1\}\}$ lists two associated eigenvectors. ◄►

The next example shows that a linear transformation may have no eigenvalues at all. It also explains geometrically why this is the case.

EXAMPLE 6.6 ■ A Linear Transformation Without Eigenvalues

Find the real eigenvalues of the linear transformation $T : \mathbb{R}^2 \to \mathbb{R}^2$ defined by

$$T \begin{pmatrix} x \\ y \end{pmatrix} = \begin{pmatrix} 0 & -1 \\ 1 & 0 \end{pmatrix} \begin{pmatrix} x \\ y \end{pmatrix}$$

if they exist.

Solution. Since

$$T \begin{pmatrix} x \\ y \end{pmatrix} = \begin{pmatrix} 0 & -1 \\ 1 & 0 \end{pmatrix} \begin{pmatrix} x \\ y \end{pmatrix} = \begin{pmatrix} -y \\ x \end{pmatrix}$$

it is clear that T rotates the points in $\mathbb{R}^2$ counterclockwise by $90°$. Therefore,

$$T \begin{pmatrix} x \\ y \end{pmatrix} = \lambda \begin{pmatrix} x \\ y \end{pmatrix} \quad \text{if and only if} \quad \begin{pmatrix} -y \\ x \end{pmatrix} = \lambda \begin{pmatrix} x \\ y \end{pmatrix} = \begin{pmatrix} \lambda x \\ \lambda y \end{pmatrix}$$

This means that $-y = \lambda x$ and $x = \lambda y$. These equations hold only if $(\lambda^2 + 1)y = 0$ and $(\lambda^2 + 1)x = 0$. Since the equation $\lambda^2 + 1 = 0$ has no solution in the real numbers, x and y must be 0. This shows that there are no nonzero vectors $\mathbf{x}$ for which $T(\mathbf{x}) = \lambda \mathbf{x}$. Therefore T has no eigenvalues. ◄►

The existence or nonexistence of eigenvalues and eigenvectors of linear transformations depends both on the geometric properties of the transformations and on the spaces on which they act. If we were to expand the domain and codomain of the matrix transformation $T_A :$ $\mathbb{R}^2 \to \mathbb{R}^2$ in Example 6.6 to the complex vector space $\mathbb{C}^2$, then T_A would have eigenvalues and eigenvectors. The complex roots of the equation $\lambda^2 + 1 = 0$ are $\lambda = i$ and $\lambda = -i$, and

$$T \begin{pmatrix} y \\ iy \end{pmatrix} = \begin{pmatrix} 0 & -1 \\ 1 & 0 \end{pmatrix} \begin{pmatrix} y \\ iy \end{pmatrix} = \begin{pmatrix} -iy \\ y \end{pmatrix} = -i \begin{pmatrix} y \\ iy \end{pmatrix}$$

Hence, all vectors of the form (y, iy) are eigenvectors belonging to the eigenvalue $-i$. Similarly, all vectors of the form (iy, y) are eigenvectors belonging to the eigenvalue i.

EXERCISES 6.1

1. Sketch the geometric properties of the following linear transformations $T : \mathbb{R}^2 \to \mathbb{R}^2$.

   a. $T(x, y) = (3x, 4y)$ b. $T(x, y) = (-3x, 4y)$
   c. $T(x, y) = (3x, -4y)$ d. $T(x, y) = (-3x, -4y)$

2. Sketch the mapping properties of the following linear transformations $T : \mathbb{R}^3 \to \mathbb{R}^3$.

   a. $T(x, y, z) = (3x, 4y, 5z)$ b. $T(x, y, z) = (-3x, 4y, 5z)$
   c. $T(x, y, z) = (-3x, -4y, 5z)$ d. $T(x, y, z) = (3x, 4y, -5z)$

3. Let $T : \mathbb{R}^2 \to \mathbb{R}^2$ be a linear transformation whose values on the standard basis of $\mathbb{R}^2$ are $T(e_1) = 5e_1$ and $T(e_2) = -2e_2$. Explain why 5 and -2 are eigenvalues of T, and find two eigenvectors for each eigenvalue.

4. Let $T : \mathbb{R}^3 \to \mathbb{R}^3$ be a linear transformation whose values on the standard basis of $\mathbb{R}^3$ are $T(e_1) = 5e_1$, $T(e_2) = -2e_2$, and $T(e_3) = 7e_3$. Explain why 5, -2, and 7 are eigenvalues of T, and find two eigenvectors for each eigenvalue.

5. Show that -5, -4, and 3 are eigenvalues of the matrix transformation $T_A : \mathbb{R}^3 \to \mathbb{R}^3$ determined by the matrix

$$A = \begin{pmatrix} 0 & 0 & 60 \\ 1 & 0 & 7 \\ 0 & 1 & -6 \end{pmatrix}$$

and find two eigenvectors for each eigenvalue.

6. Use *MATHEMATICA* to find the real eigenvalues of the following matrices, if they exist.

a. $\begin{pmatrix} 1 & 0 \\ 0 & 0 \end{pmatrix}$ b. $\begin{pmatrix} 0 & -1 \\ 1 & 0 \end{pmatrix}$ c. $\begin{pmatrix} 0 & 1 \\ -1 & 1 \end{pmatrix}$

d. $\begin{pmatrix} 1 & 0 & -2 \\ 0 & 5 & 7 \\ 0 & 0 & 0 \end{pmatrix}$ e. $\begin{pmatrix} 0 & -1 & 0 \\ 1 & 0 & 0 \\ 0 & 0 & 1 \end{pmatrix}$ f. $\begin{pmatrix} 2 & 0 & 0 & 0 \\ 0 & 3 & 0 & 0 \\ 0 & 0 & 2 & 0 \\ 0 & 0 & 0 & 5 \end{pmatrix}$

g. $\begin{pmatrix} 1 & 0 & 0 & 0 \\ 0 & 1 & 0 & 0 \\ 0 & 0 & 1 & 0 \\ 0 & 0 & 0 & 1 \end{pmatrix}$ h. $\begin{pmatrix} 6.0 & 3 & 6 & 5 \\ 3 & 4 & 4 & 3 \\ 6 & 4 & 2 & 2 \\ 5 & 3 & 2 & 4 \end{pmatrix}$ i. $\begin{pmatrix} 1 & 0 & 0 & 9 \\ 0 & 1 & 7 & 0 \\ 0 & 7 & 1 & 0 \\ 9 & 0 & 0 & 1 \end{pmatrix}$

7. Use *MATHEMATICA* to generate a random 3×3 invertible matrix P and its inverse P^{-1}. Show that -5, -4, and 3 are eigenvalues of the matrix transformation determined by PAP^{-1}, where A is the matrix in Exercise 5.

8. Use *MATHEMATICA* to find the eigenvalues of the matrix

$$A = \begin{pmatrix} 0 & 0 & 0 & 0 & 120 \\ 1 & 0 & 0 & 0 & -274 \\ 0 & 1 & 0 & 0 & 225 \\ 0 & 0 & 1 & 0 & -85 \\ 0 & 0 & 0 & 1 & 15 \end{pmatrix}$$

Show that A has enough linearly independent eigenvectors to span $\mathbb{R}^5$.

9. Explain the connection between the eigenvectors of the matrix transformation $T_A : \mathbb{R}^3 \to \mathbb{R}^3$ determined by the matrix

$$A = \begin{pmatrix} 0 & 0 & 6 \\ 1 & 0 & -11 \\ 0 & 1 & 6 \end{pmatrix}$$

and the solutions of the homogeneous linear systems $(A - I)\mathbf{x} = 0$, $(A - 2I)\mathbf{x} = 0$, and $(A - 3I)\mathbf{x} = 0$.

10. Let $T_P : \mathbb{R}^{2 \times 2} \to \mathbb{R}^{2 \times 2}$ be the similarity transformation defined by the matrix

$$P = \begin{pmatrix} 24 & 17 \\ 9 & 6 \end{pmatrix}$$

Compare the eigenvalues of the matrices A and $T_P(A)$ for the case

$$A = \begin{pmatrix} 1 & 2 \\ 3 & 4 \end{pmatrix}$$

11. Let $T_P : \mathbb{R}^{3 \times 3} \to \mathbb{R}^{3 \times 3}$ be the similarity transformation defined by the matrix

$$P = \begin{pmatrix} 3 & 0 & 8 \\ 0 & 1 & 3 \\ 8 & 3 & 0 \end{pmatrix}$$

Compare the eigenvalues of the matrices A and $T_P(A)$ for the case

$$A = \begin{pmatrix} 1 & 2 & 0 \\ 3 & 4 & 0 \\ 0 & 0 & 1 \end{pmatrix}$$

12. Show that if λ is an eigenvalue of a linear transformation $T : V \to V$, then λ^2 is an eigenvalue of T^2.

13. Show by an example that the existence of an eigenvalue for $T^2 : V \to V$ does not imply the existence of an eigenvalue for $T : V \to V$.

14. Explain why every set of eigenvectors belonging to an eigenvalue of a real matrix is infinite, but why every set of linearly independent eigenvectors belonging to an eigenvalue is finite.

15. Show that the linear transformation T^2 obtained from T in Exercise 3 by composing T with itself has a real eigenvalue.

16. Show that the matrices A and A^T have the same eigenvalues.

17. Show that the symmetric matrices AA^T and A^TA have the same eigenvalues.

18. The eigenvalues λ of the matrix $A = \begin{pmatrix} 2 & 5 & 6 \\ 0 & 3 & 7 \\ 0 & 0 & 4 \end{pmatrix}$ are 2, 3, and 4. Show that for each λ, the matrix $(A - \lambda I)$ is singular.

19. Use *MATHEMATICA* to generate a random 3×3 invertible matrix P and show that if $\lambda = 1, \frac{5}{2} + \frac{1}{2}\sqrt{33}$, or $\frac{5}{2} - \frac{1}{2}\sqrt{33}$, and A is the matrix in Exercise 11, then the matrices $\left(PAP^{-1} - \lambda I\right)$ are singular.

20. Let $T_2 \circ T_1 : V \to V$ be a composite linear transformation, and suppose that $T_1(\mathbf{x}) = \lambda \mathbf{x}$ and $T_2(\mathbf{x}) = \mu \mathbf{x}$ for some nonzero vector $\mathbf{x} \in V$. Show that $\lambda \mu$ is an eigenvalue of $T_2 \circ T_1$.

21. Verify that the matrix transformation $T_A : \mathbb{C}^3 \to \mathbb{C}^3$, defined on the complex vector space $\mathbb{C}^3$ by the matrix

$$A = \begin{pmatrix} 0 & 0 & 2 \\ 1 & 0 & 1 \\ 0 & 1 & 1 \end{pmatrix}$$

has the eigenvalues $2, -\frac{1}{2} + \frac{1}{2}i\sqrt{3}$, and $-\frac{1}{2} - \frac{1}{2}i\sqrt{3}$. Find an eigenvector for each of the three eigenvalues.

CHARACTERISTIC POLYNOMIALS

In this section we show that eigenvalues of matrices are *roots of polynomials*. Suppose that A is a real $n \times n$ matrix and that there exists a nonzero column vector $\mathbf{x} \in \mathbb{R}^n$ for which $A\mathbf{x} = \lambda \mathbf{x}$. Using the identity matrix I, we can rewrite this equation as $A\mathbf{x} = \lambda I \mathbf{x}$. Therefore, $A\mathbf{x} - \lambda I \mathbf{x} = (A - \lambda I)\mathbf{x} = \mathbf{0}$. Finding an eigenvector for A is therefore equivalent to finding a nontrivial solution of the homogeneous system $(A - \lambda I)\mathbf{x} = \mathbf{0}$. We know from our work on linear systems that this is possible only if the real coefficient matrix $(A - \lambda I)$ is singular. We also know that this happens precisely when the determinant $(A - \lambda I)$ is 0. This leads us to real polynomials. For any variable t, the determinant of the matrix $(A - tI)$ is a polynomial $c_A(t) \in \mathbb{R}[t]$, and the required scalars λ will be among the roots of $c_A(t)$.

DEFINITION 6.5   *The matrix $(A - tI)$ is called the **characteristic matrix** of A in the variable t, and the polynomial $c_A(t)$ is called the **characteristic polynomial** of A in the variable t.*

EXAMPLE 6.7   ■ **A Characteristic Polynomial**

Use *MATHEMATICA* to find the characteristic polynomial $c_A(t)$ of the matrix

$$A = \begin{pmatrix} 1 & 1 & 0 \\ 3 & 3 & 2 \\ 2 & 2 & 3 \end{pmatrix}$$

Solution. First we form the characteristic matrix $(A - tI)$.

```
In[1]:= M= ( 1 1 0 )   ( 1 0 0 )
          ( 3 3 2 ) -t ( 0 1 0 )
          ( 2 2 3 )   ( 0 0 1 )
Out[1]= {{1-t,1,0},{3,3-t,2},{2,2,3-t}}
```

Then we compute the determinant of M.

```
In[2]:= Det[M]
Out[2]= -8t+7t²-t³
```

Thus the characteristic polynomial $c_A(t) = -8t + t^2 - t^3$. ◄►

EXAMPLE 6.8   ■ **The Characteristic Polynomial of a Random Matrix**

Use the **CharacteristicPolynomial** function to compute the characteristic polynomial of a random real 6×6 matrix.

Solution. We begin by generating a random 6×6 matrix.

```
In[1]:= A=Array[Random[Integer,{-10,10}]&,{6,6}]
Out[1]= {{9,10,-6,-7,1,8},{-3,3,-3,8,-7,-1},{7,-5,6,1,-5,1},
        {4,-1,5,5,9,-5},{-4,-6,-6,-1,10,4},{9,1,5,1,2,-4}}
```

```
In[2]:= CharacteristicPolynomial[A,t]
Out[2]= 611829+65170t-27965t²+1145t³+237t⁴-29t⁵+t⁶
```

Thus the polynomial $c_A(t)$ is the characteristic polynomial of A.   ◄►

The characteristic polynomials of certain matrices are easy to calculate.

EXAMPLE 6.9 ■ **The Characteristic Polynomial of a Triangular Matrix**

Find the characteristic polynomial $c_A(t)$ of the matrix

$$A = \begin{pmatrix} 6 & 7 & 4 & 5 \\ 0 & 3 & 7 & 5 \\ 0 & 0 & 3 & 3 \\ 0 & 0 & 0 & 5 \end{pmatrix}$$

Solution. By definition, the characteristic polynomial $c_A(t)$ is the determinant of the characteristic matrix $(A - tI)$. Since A is upper triangular, the matrix

$$(A - tI) = \begin{pmatrix} 6 & 7 & 4 & 5 \\ 0 & 3 & 7 & 5 \\ 0 & 0 & 3 & 3 \\ 0 & 0 & 0 & 5 \end{pmatrix} - t \begin{pmatrix} 1 & 0 & 0 & 0 \\ 0 & 1 & 0 & 0 \\ 0 & 0 & 1 & 0 \\ 0 & 0 & 0 & 1 \end{pmatrix} = \begin{pmatrix} 6-t & 7 & 4 & 5 \\ 0 & 3-t & 7 & 5 \\ 0 & 0 & 3-t & 3 \\ 0 & 0 & 0 & 5-t \end{pmatrix}$$

is also upper triangular. It therefore follows from the property of determinants that

$$\det(A - tI) = (6 - t)(3 - t)(3 - t)(5 - t)$$

Collecting identical terms, we get $c_A(t) = (6 - t)(3 - t)^2(5 - t)$. ◀▶

The polynomial matrix version of Theorem 3.20 shows that this example is typical for all triangular matrices: The characteristic polynomial of a triangular matrix A is the product of the diagonal entries of the characteristic matrix of A.

The following theorem establishes the fundamental connection between eigenvalues and the roots of characteristic polynomials. From a theoretical point of view, the theorem provides a basic tool for identifying the eigenvalues of a matrix. In practice, however, the theorem is not particularly helpful: A remarkable reality of mathematics is that there is no general algorithm for finding the roots of polynomials of degree five or higher. In practical applications, approximation techniques are used for calculating eigenvalues.

THEOREM 6.2 (Eigenvalue theorem) *For every scalar λ and real square matrix A, the following statements are equivalent.*

1. The scalar λ is an eigenvalue of A.

2. The matrix $(A - \lambda I)$ is singular.

3. The scalar λ is a root of the characteristic polynomial $c_A(t)$ of A.

4. The set $\{\mathbf{x} : A\mathbf{x} = \lambda\mathbf{x}\}$ is the solution space of the system $(A - \lambda I)\mathbf{x} = 0$.

Proof. We prove the theorem by showing that the first statement implies the second, the second implies the third, the third implies the fourth, and the fourth implies the first.

1. (1) implies (2). Suppose that λ is an eigenvalue of A. Then $A\mathbf{x} = \lambda\mathbf{x}$ for some nonzero vector $\mathbf{x}$, so that $A\mathbf{x} - \lambda\mathbf{x} = A\mathbf{x} - \lambda I\mathbf{x} = (A - \lambda I)\mathbf{x} = 0$. Hence it follows from Theorem 3.15 that $(A - \lambda I)$ is singular.

2. (2) implies (3). We know from Theorem 3.14 that if $A - \lambda I$ is singular, then $\det(A - \lambda I) = c_A(\lambda) = 0$. Hence λ is a root of the characteristic polynomial of A.

3. (3) implies (4). Suppose that λ is a root of the characteristic polynomial of A. Then $c_A(\lambda) = \det(A - \lambda I) = 0$. This tells us that λ is an eigenvalue of A, and the set of all vectors $\mathbf{x}$ for which $A\mathbf{x} = \lambda\mathbf{x}$ is the solution space of the system $(A - \lambda I)\mathbf{x} = 0$.

4. (4) implies (1). We close the circle by noting that if $\mathbf{x}$ is a nonzero vector for which $(A - \lambda I)\mathbf{x} = 0$, then λ is an eigenvalue of A. ∎

COROLLARY 6.3 *Suppose that $T : V \to V$ is a linear transformation on a real finite-dimensional vector space V and that $c_A(t)$ is the characteristic polynomial of the matrix $A = [T]_{\mathcal{B}}^{\mathcal{B}}$. Then every eigenvalue λ of T is a real root $c_A(t)$.*

Proof. The corollary follows immediately from Theorem 6.1. ∎

Characteristic Polynomials of Similar Matrices

Since similar matrices represent the same linear transformation in different bases, we would expect that similar matrices have the same characteristic polynomial. The next theorem shows that this is in fact the case.

THEOREM 6.4 (Characteristic polynomial test) *If A and B are similar matrices, then $c_A(t) = c_B(t)$.*

Proof. Suppose that A and B are two similar matrices. Then there exists an invertible matrix P such that $B = P^{-1}AP$. Hence

$$
\begin{aligned}
B - tI &= P^{-1}AP - tI \\
&= P^{-1}AP - P^{-1}tIP \\
&= P^{-1}(A - tI)P
\end{aligned}
$$

This means that

$$c_B(t) = \det(B - tI) = \det P^{-1}(A - tI)P = \det P^{-1}c_A(t)\det P = c_A(t)$$

The two characteristic polynomials are therefore identical. ∎

EXAMPLE 6.10   ■ The Characteristic Polynomials of A and PAP^{-1}

Use *MATHEMATICA* to show that if A and P are two random 2×2 matrices and P is invertible, then A and PAP^{-1} have the same characteristic polynomial.

Solution. We begin by generating a random matrix.

```
In[1]:= A=Array[Random[Integer,{50,100}]&,{2,2}]

Out[1]= {{54,60},{64,66}}
```

```
In[2]:= CharacteristicPolynomial[A,t]

Out[2]= -276-120t+t²
```

For P, we need an invertible matrix. We try our luck by generating another random matrix and test P for invertibility.

```
In[3]:= P=Array[Random[Integer,{-5,5}]&,{2,2}]

Out[3]= {{1,3},{1,1}}
```

```
In[4]:= Q=Inverse[P]

Out[4]= {{-½,³⁄₂},{½,-½}}
```

```
In[5]:= CharacteristicPolynomial[P.A.Q]

Out[5]= -276-120t+t²
```

This shows that the matrices A and PAP^{-1} have the same characteristic polynomial. ◀▶

EXAMPLE 6.11 ■ **The Characteristic Polynomials of Two Nonsimilar 3 × 3 Matrices**

Use characteristic polynomials to show that the matrices

$$A = \begin{pmatrix} 1 & 1 & 0 \\ 3 & 3 & 2 \\ 2 & 2 & 3 \end{pmatrix} \quad \text{and} \quad B = \begin{pmatrix} 1 & 1 & 0 \\ 3 & 3 & 2 \\ 0 & 0 & 3 \end{pmatrix}$$

are not similar.

Solution. In Example 6.10, we showed that the characteristic polynomial $c_A(t)$ of A is $-8t + t^2 - t^3$. By Theorem 6.4, the matrices A and B are not similar if the characteristic polynomial $c_B(t)$ is different from $c_A(t)$. We therefore compute $c_B(t)$.

```
In[1]:= M={{1,1,0},{3,3,2},{0,0,3}}-t*IdentityMatrix[3]

Out[1]= {{1-t,1,0},{3,3-t,2},{0,0,3-t}}
```

Next we compute the determinant of M.

```
In[2]:= Det[M]

Out[2]= -12t+7t²-t³
```

As we can see, $c_B(t) = -12t + 7t^2 - t^3$. Since $c_B(t) \neq c_A(t)$, the matrices A and B are not similar. ◄►

It is easy to check that in Example 6.11, $\det A = \det B$ and trace A = trace B. This shows that these equalities are necessary conditions for the similarity of A and B, but they are not sufficient.

EXERCISES 6.2

1. Find the characteristic matrices in the variable t of the following matrices.

a. $\begin{pmatrix} 1 & 2 \\ 0 & 0 \end{pmatrix}$
   b. $\begin{pmatrix} 0 & -1 \\ 1 & 0 \end{pmatrix}$
   c. $\begin{pmatrix} 0 & 1 \\ -1 & 1 \end{pmatrix}$

d. $\begin{pmatrix} 1 & 0 & -2 \\ 0 & 5 & 7 \\ 0 & 0 & 0 \end{pmatrix}$
   e. $\begin{pmatrix} 0 & -1 & 0 \\ 1 & 0 & 0 \\ 0 & 0 & 1 \end{pmatrix}$
   f. $\begin{pmatrix} 2 & 0 & 0 & 0 \\ 0 & 3 & 0 & 0 \\ 0 & 0 & 2 & 0 \\ 0 & 0 & 0 & 5 \end{pmatrix}$

g. $\begin{pmatrix} 1 & 0 & 0 & 0 \\ 0 & 1 & 0 & 0 \\ 0 & 0 & 1 & 0 \\ 0 & 0 & 0 & 1 \end{pmatrix}$
   h. $\begin{pmatrix} 6 & 3 & 6 & 5 \\ 3 & 4 & 4 & 3 \\ 6 & 4 & 2 & 2 \\ 5 & 3 & 2 & 4 \end{pmatrix}$
   i. $\begin{pmatrix} 1 & 0 & 0 & 9 \\ 0 & 1 & 7 & 0 \\ 0 & 7 & 1 & 0 \\ 9 & 0 & 0 & 1 \end{pmatrix}$

2. Use *MATHEMATICA* to find the characteristic polynomials in the variable t of the matrices in Exercise 1.

3. For each matrix A in Exercise 1, show that $c_A(A)$ is the zero matrix, where $c_A(A)$ is obtained from the characteristic polynomial $c_A(t)$ by replacing the variable t by the matrix A and replacing the constant term a_0 of $c_A(t)$ by $a_0 I$.

4. Find two linearly independent nontrivial solutions of the homogeneous system $A\mathbf{x} = \mathbf{0}$ defined by

$$\begin{pmatrix} 5 & 2 & 7 \\ 0 & 0 & 0 \\ 0 & 0 & 0 \end{pmatrix} \begin{pmatrix} x \\ y \\ z \end{pmatrix} = \begin{pmatrix} 0 \\ 0 \\ 0 \end{pmatrix}$$

and show that they are eigenvectors of A.

5. Use the fact that the matrix

$$A = \begin{pmatrix} 5 & 2 & 7 \\ 0 & 2 & 1 \\ 0 & 0 & 2 \end{pmatrix}$$

has three distinct eigenvalues to explain why the homogeneous system $A\mathbf{x} = \mathbf{0}$ cannot have two linearly independent nontrivial solutions.

6. Use characteristic polynomials to find the eigenvalues of the following linear transformations.

   a. $T(x, y) = (3x + 4y, x - y)$

   b. $T(x, y, z) = (3x + y - z, x + y, x + z)$

7. Prove that if A is a matrix (a_{ij}), in which $a_{i1} = 0$ if $i > 1$, and B is the submatrix of A obtained by deleting the first row and first column of A, then the characteristic polynomial $c_A(t)$ of A is $(a_{11} - t)c_B(t)$.

8. Construct an example of two 3×3 matrices having the same eigenvalues but different sets of eigenvectors.

9. Use *MATHEMATICA* to find the eigenvectors of the matrices

$$A = \begin{pmatrix} 8 & 4 & -5 & -5 \\ 3 & -5 & 8 & 5 \\ 0 & 0 & 0 & 0 \\ 0 & 0 & 0 & 0 \end{pmatrix} \quad \text{and} \quad A^T = \begin{pmatrix} 8 & 3 & 0 & 0 \\ 4 & -5 & 0 & 0 \\ -5 & 8 & 0 & 0 \\ -5 & 5 & 0 & 0 \end{pmatrix}$$

and compare your results.

10. Let A and B be two $n \times n$ matrices. Show that the matrices AB and BA have the same eigenvalues.

11. The intermediate value theorem of calculus says that if f is a continuous function on a closed interval $[a, b]$ and if y is any number between $f(a)$ and $f(b)$, then there exists a number x between a and b such that $f(x) = y$. Use this theorem to prove that every real

3×3 matrix A has at least one real eigenvalue. (*Hint:* If $p(t) = a_0 + a_1 t + a_2 t^2 + t^3$ is the characteristic polynomial of A, then for any nonzero real number r, we have

$$p(r) = a_0 + a_1 r + a_2 r^2 + r^3 = r^3 \left(\frac{a_0}{r^3} + \frac{a_1}{r^2} + \frac{a_2}{r} + 1 \right) = r^3 q(r)$$

If we take $|r|$ sufficiently large, then $q(r)$ is close to 1. This means that $q(r) > 0$ for some r. Let $r_1 < 0$ and $|r_1|$ be large enough to ensure that $q(r_1) > 0$, and let $r_2 > 0$ be large enough to ensure that $q(r_2) > 0$. Then the intermediate value theorem tells us that there exists a number r between r_1 and r_2 for which $q(r) = 0$.)

The Cayley-Hamilton Theorem

One of the conceptually beautiful properties of characteristic polynomials is that they are satisfied by the matrices that determine them. This result is known as the Cayley-Hamilton theorem.

THEOREM 6.5 (Cayley-Hamilton theorem) *Every matrix satisfies its characteristic polynomial.*

Proof. Let A be any $n \times n$ matrix. We must prove that

$$c_A(A) = A^n + a_{n-1} A^{n-1} + \cdots + a_1 A + a_0 I = 0$$

where $c_A(t) = t^n + a_{n-1} t^{n-1} + \cdots + a_1 t + a_0$ is the characteristic polynomial of A. For this purpose we form the characteristic matrix $(A - tI)$, and let $A(t)$ be the adjugate of $A - tI$. The elements of $A(t)$ are the cofactors of the matrix $(A - tI)$, that is to say, they are polynomials in t of degree $\leq n - 1$. We now write $A(t)$ as a sum of the form $B_{n-1} t^{n-1} + \cdots + B_1 t + B_0$, where B_i are $n \times n$ matrices not containing t. We know from the properties of adjugates, established earlier, that $(A - tI)A(t) = \det(A - tI)I$. Therefore,

$$(A - tI)(B_{n-1} t^{n-1} + \cdots + B_1 t + B_0) = (t^n + a_{n-1} t^{n-1} + \cdots + a_1 t + a_0)I$$

If we expand and regroup the left-hand side of the above equation, we get

$$B(t) = -B_{n-1} t^n + (AB_{n-1} - B_{n-2})t^{n-1} + \cdots + (AB_1 - B_0)t + AB_0$$

By comparing coefficients, we conclude that

$$-B_{n-1} = I$$
$$AB_{n-1} - B_{n-2} = a_{n-1} I$$
$$AB_{n-2} - B_{n-3} = a_{n-2} I$$
$$\vdots$$
$$AB_1 - B_0 = a_1 I$$
$$AB_0 = a_0 I$$

If we multiply these equations from the top down by descending powers of A, in other words, by A^n, A^{n-1}, ..., A, I, we get

$$
\begin{aligned}
-A^n B_{n-1} &= A^n \\
A^n B_{n-1} - A^{n-1} B_{n-2} &= a_{n-1} A^{n-1} \\
A^{n-1} B_{n-2} - A^{n-2} B_{n-3} &= a_{n-2} A^{n-2} \\
&\ \ \vdots \\
A^3 B_2 - A^2 B_1 &= a_2 A^2 \\
A^2 B_1 - A B_0 &= a_1 A \\
A B_0 &= a_0 I
\end{aligned}
$$

By adding these equations, we get $0 = A_n + a_{n-1} A_{n-1} + \cdots + a_1 A + a_0 I$, as required. ∎

EXAMPLE 6.12   ■ **An Instance of the Proof of the Cayley-Hamilton Theorem**

Use *MATHEMATICA* to illustrate the proof of the Cayley-Hamilton theorem in the case of a 2×2 matrix.

Solution. First we create a random 2×2 matrix.

```
In[1]:= A=Array[Random[Integer,{-100,100}]&,{2,2}]

Out[1]= {{-85,-55},{-35,97}}
```

Next we load the **Combinatorica** package to access the **Cofactor** function.

```
In[2]:= <<DiscreteMath`Combinatorica`
```

Then we build up the left-hand side of the equation.

```
In[3]:= A1=t*IdentityMatrix[2]-A

Out[3]= {{85+t,55},{35,-97+t}}
```

Now we calculate the adjugate of the characteristic matrix **A1**.

```
In[4]:= C11=Cofactor[A1,{1,1}];
```

```
In[5]:= C12=Cofactor[A1,{1,2}];
```

```
In[6]:= C21=Cofactor[A1,{2,1}];
```

```
In[7]:= C22=Cofactor[A1,{2,2}];
```

```
In[8]:= Adjugate[A1]={{C11,C21},{C12,C22}}

Out[8]= {{-97+t,-55},{-35,85+t}}
```

We now multiply the characteristic matrix **A1** and its adjugate.

```
In[9]:= LHS=A1.Adjugate[A1]

Out[9]= {{-1925+(-97+t)(85+t),0},{0,-1925+(-97+t)(85+t)}}
```

```
In[10]:= Expand[LHS]

Out[10]= {{-10170-12t+t², 0},{0,-10170-12t+t²}}
```

Finally, we build up the right-hand side of the equation.

```
In[11]:= Det[A1]*IdentityMatrix[2]

Out[11]= {{-10170-12t+t², 0},{0,-10170-12t+t²}}
```

Since the last two outputs are equal, the equation holds. ◄►

EXAMPLE 6.13 ■ **The Zero Matrix Determined by a Characteristic Polynomial**
Use *MATHEMATICA* to illustrate the Cayley-Hamilton theorem.

Solution. Let A be the matrix

$$\begin{pmatrix} 1 & 2 \\ 3 & 2 \end{pmatrix}$$

Then

$$t\begin{pmatrix} 1 & 0 \\ 0 & 1 \end{pmatrix} - \begin{pmatrix} 1 & 2 \\ 3 & 2 \end{pmatrix} = \begin{pmatrix} t-1 & -2 \\ -3 & t-2 \end{pmatrix}$$

is the characteristic matrix of A. We begin by computing its determinant.

```
In[1]:= Det[(  t-1   -2  )]
          [(  -3    t-2 )]

Out[1]= -4-3t+t²
```

The Cayley-Hamilton theorem asserts that $-4I_2 - 3A + A^2$ is the zero matrix.

```
In[2]:= A={{1,2},{3,2}}

Out[2]= {{1,2},{3,2}}
```

```
In[3]:= -4*IdentityMatrix[2]-3*A+MatrixPower[A,2]==
{{0,0},{0,0}}

Out[3]= True
```

As we can see, the constructed matrix is the zero matrix. ◄►

EXERCISES 6.3

1. Use *MATHEMATICA* to show that the following matrices satisfy their characteristic polynomials.

$$a. \ A = \begin{pmatrix} 1 & 2 \\ 3 & 4 \end{pmatrix} \quad b. \ A = \begin{pmatrix} 97 & 50 \\ 79 & 56 \end{pmatrix} \quad c. \ A = \begin{pmatrix} 0 & 2 & 1 \\ 1 & 0 & 0 \\ -1 & 0 & 3 \end{pmatrix}$$

2. The characteristic polynomial of the matrix

$$\begin{pmatrix} 0 & 2 & 1 \\ 1 & 0 & 0 \\ -1 & 0 & 3 \end{pmatrix}$$

is $c_A(t) = 6 - t - 3t^2 + t^3$. Moreover, $c_A(t)$ is the product of the polynomials $p_1(t) = (t-2)$ and $p_2(t) = (t^2 - t - 3)$. Show that neither $p_1(A)$ nor $p_2(A)$ is the zero matrix. Why does the Cayley-Hamilton theorem nevertheless guarantee that $p_1(A) \ p_2(A)$ is the zero matrix?

3. Use the Cayley-Hamilton theorem to find the inverse of the matrix

$$A = \begin{pmatrix} -86 & 23 & -84 & 19 & -50 \\ 88 & -53 & 85 & 49 & 78 \\ 17 & 72 & -99 & -85 & -86 \\ 30 & 80 & 72 & 66 & -29 \\ -91 & -53 & -19 & -47 & 68 \end{pmatrix}$$

4. Show that the following statements about real $n \times n$ matrices are false.

   a. If $c_A(t)$ is the characteristic polynomial of A, then $c_A(A^2) = \mathbf{0}$.

   b. If $p(t)$ is a polynomial with real coefficients and $p(A^2) = \mathbf{0}$, then $p(A) = \mathbf{0}$.

5. Use the Cayley-Hamilton theorem to prove that the inverse of an invertible $n \times n$ matrix A is a linear combination of $A, A^2, \ldots, A^{n-1}$.

Computing Characteristic Polynomials

Although the definition of characteristic polynomials is conceptually elegant, it is also computationally unmanageable. For general matrices, the number of terms of $\det(A - tI)$ grows too rapidly as the dimension of A increases to make a direct computation of $c_A(t)$ manually or mechanically feasible. We must therefore look for simpler matrices B for which $c_B(t)$ can easily be computed, and for which $c_B(t) = c_A(t)$. Theorem 6.4 provides the clue. If B is a triangular matrix, then Example 6.9 shows that the characteristic polynomial of B is the product of the diagonal entries of the matrix $(B - tI)$. It can therefore be determined immediately from $(B - tI)$. If B is similar to A, then $c_B(t) = c_A(t)$. Unfortunately, not every square matrix is similar to a triangular matrix.

EXAMPLE 6.14   ■ A Real Matrix Not Similar to a Triangular Matrix

Show that the real matrix

$$A = \begin{pmatrix} 0 & -1 \\ 1 & 0 \end{pmatrix}$$

is not similar to any real triangular matrix B.

Solution. Let

$$B = \begin{pmatrix} a & b \\ 0 & c \end{pmatrix}$$

be a general upper-triangular matrix. If A and B are similar, then it follows from Theorems 5.29 and 5.30 that $\det A = \det B$ and trace $A = $ trace B. Since $\det A = 1$ and $\det B = ac$, we must have $ac = 1$. Since trace $A = 0$ and trace $B = a + c$, we must have $a + c = 0$. If we combine these two results, it follows that $-c^2$ must equal 1. But this is impossible for any real number c. Hence A is not similar to B for any real numbers a, b, c. ◄►

The question arises whether we can find a more general substitute for triangular matrices that can be used to compute characteristic polynomials in reasonable time. It turns out that there are such matrices. A **Hessenberg matrix** is a matrix $H = (h_{ij})$ for which $h_{ij} = 0$ if $i > j + 1$. This means that H is almost triangular in the sense that all entries of H below the "subdiagonal" are 0. Moreover, there exist similarity transformations $M \to QMQ^{-1}$, known as **Householder transformations**, for converting A to H. The characteristic polynomial $c_H(t)$ can be found easily either by applying the definition or by using a number of special algorithms devised for this purpose.

EXAMPLE 6.15   ■ A Hessenberg Matrix

Explain why the matrix

$$A = \begin{pmatrix} 3 & 4 & 2 & 1 & 5 \\ 7 & 1 & 5 & 1 & 1 \\ 0 & 2 & 6 & 2 & 7 \\ 0 & 0 & 1 & 7 & 3 \\ 0 & 0 & 0 & 0 & 3 \end{pmatrix}$$

is a Hessenberg matrix.

Solution. If $i > j + 1$, then $h_{ij} \in \{h_{31}, h_{41}, h_{42}, h_{51}, h_{52}, h_{53}\}$. We can see from an inspection of A that $h_{31} = h_{41} = h_{42} = h_{51} = h_{52} = h_{53} = 0$, as required. ◄►

A real $n \times n$ **Householder matrix** Q is a matrix of the form

$$I - \frac{2}{\mathbf{u}^T\mathbf{u}}\mathbf{u}\mathbf{u}^T$$

determined by any nonzero vector $\mathbf{u} \in \mathbb{R}^n$. It is quite remarkable that each matrix constructed in this way is both symmetric and orthogonal: $Q = Q^T = Q^{-1}$. To see that Q is symmetric, we note that

$$\left(I - \frac{2}{\mathbf{u}^T\mathbf{u}}\mathbf{u}\mathbf{u}^T\right)^T = I^T - \frac{2}{\mathbf{u}^T\mathbf{u}}\left(\mathbf{u}\mathbf{u}^T\right)^T = I - \frac{2}{\mathbf{u}^T\mathbf{u}}\mathbf{u}^{TT}\mathbf{u}^T = I - \frac{2}{\mathbf{u}^T\mathbf{u}}\mathbf{u}\mathbf{u}^T$$

To see that Q is orthogonal, we observe that

$$\begin{aligned} QQ^T &= \left(I - \frac{2}{\mathbf{u}^T\mathbf{u}}\mathbf{u}\mathbf{u}^T\right)\left(I - \frac{2}{\mathbf{u}^T\mathbf{u}}\mathbf{u}\mathbf{u}^T\right)^T \\ &= \left(I - \frac{2}{\mathbf{u}^T\mathbf{u}}\mathbf{u}\mathbf{u}^T\right)\left(I^T - \frac{2}{\mathbf{u}^T\mathbf{u}}\left(\mathbf{u}\mathbf{u}^T\right)^T\right) \\ &= \left(I - \frac{2}{\mathbf{u}^T\mathbf{u}}\mathbf{u}\mathbf{u}^T\right)\left(I - \frac{2}{\mathbf{u}^T\mathbf{u}}\mathbf{u}\mathbf{u}^T\right) \\ &= I - \frac{4}{\mathbf{u}^T\mathbf{u}}\mathbf{u}\mathbf{u}^T + \frac{4}{(\mathbf{u}^T\mathbf{u})(\mathbf{u}^T\mathbf{u})}\left(\mathbf{u}\mathbf{u}^T\right)\left(\mathbf{u}\mathbf{u}^T\right) \\ &= I - \frac{4}{\mathbf{u}^T\mathbf{u}}\mathbf{u}\mathbf{u}^T + \frac{4}{\mathbf{u}^T\mathbf{u}}\mathbf{u}\mathbf{u}^T = I \end{aligned}$$

If we choose the vector $\mathbf{u}$ in Q wisely, we can use a sequence of Householder transformations to reduce any real $n \times n$ matrix A to a Hessenberg matrix B. Here is how this can be done. Suppose that

$$
\mathbf{x} = \begin{pmatrix} x_1 \\ \vdots \\ x_{k-1} \\ x_k \\ x_{k+1} \\ \vdots \\ x_n \end{pmatrix} \quad \text{and} \quad \mathbf{u} = \begin{pmatrix} 0 \\ \vdots \\ 0 \\ x_k - a \\ x_{k+1} \\ \vdots \\ x_n \end{pmatrix}
$$

are two vectors in $\mathbb{R}^n$, and that a is either $\sqrt{x_k^2 + \cdots + x_n^2}$ or $-\sqrt{x_k^2 + \cdots + x_n^2}$, chosen so that $ax_k \leq 0$. If Q is the Householder matrix determined by the specified vector $\mathbf{u}$, then

$$
Q\mathbf{x} = \begin{pmatrix} x_1 \\ \vdots \\ x_{k-1} \\ a \\ 0 \\ \vdots \\ 0 \end{pmatrix}
$$

For any column of an arbitrary $n \times n$ real matrix A, we can use this recipe to find the appropriate Householder transformations that convert A to a similar Hessenberg matrix H.

EXAMPLE 6.16 ■ A Householder Conversion of a Matrix to Hessenberg Form

Use an appropriate Householder transformation to convert the matrix

$$
A = \begin{pmatrix} 7 & 3 & 1 & 0 \\ 2 & 5 & 4 & 5 \\ 0 & 4 & 6 & 6 \\ 0 & 3 & 5 & 5 \end{pmatrix}
$$

to a Hessenberg matrix H, and use H to find the characteristic polynomial of A.

Solution. We must find a Householder matrix Q such that $QAQ = H = (h_{ij})$ and $h_{31} = h_{41} = h_{42} = 0$. Since h_{31} and h_{41} are already 0, it suffices to find a matrix Q converting h_{42} to 0. Consider the second column vector

$$
\mathbf{x} = \begin{pmatrix} x_1 \\ x_2 \\ x_3 \\ x_4 \end{pmatrix} = \begin{pmatrix} 3 \\ 5 \\ 4 \\ 3 \end{pmatrix}
$$

By the definition of $\mathbf{u}$, we have $x_4 = x_{k+1}$ and $k = 3$. Therefore,

$$a = \pm\sqrt{x_3^2 + x_4^2} = \pm\sqrt{4^2 + 3^2} = \pm 5$$

Since $x_3 - a$ must be less than or equal to 0, we take a to be -5. This means that

$$\mathbf{u} = \begin{pmatrix} 0 \\ 0 \\ x_3 - a \\ x_4 \end{pmatrix} = \begin{pmatrix} 0 \\ 0 \\ 4 - (-5) \\ 3 \end{pmatrix} = \begin{pmatrix} 0 \\ 0 \\ 9 \\ 3 \end{pmatrix}$$

is the vector required to construct the matrix $Q = I - b(\mathbf{u}\mathbf{u}^T)$, with

$$b = 2/(\mathbf{u}^T \mathbf{u}) = \frac{2}{90} = \frac{1}{45}$$

Thus

$$Q = I - b\mathbf{u}\mathbf{u}^T = \begin{pmatrix} 1 & 0 & 0 & 0 \\ 0 & 1 & 0 & 0 \\ 0 & 0 & 1 & 0 \\ 0 & 0 & 0 & 1 \end{pmatrix} - \frac{1}{45}\begin{pmatrix} 0 \\ 0 \\ 9 \\ 3 \end{pmatrix}\begin{pmatrix} 0 \\ 0 \\ 9 \\ 3 \end{pmatrix}^T = \begin{pmatrix} 1 & 0 & 0 & 0 \\ 0 & 1 & 0 & 0 \\ 0 & 0 & -\frac{4}{5} & -\frac{3}{5} \\ 0 & 0 & -\frac{3}{5} & \frac{4}{5} \end{pmatrix}$$

The product

$$QAQ = \begin{pmatrix} 1 & 0 & 0 & 0 \\ 0 & 1 & 0 & 0 \\ 0 & 0 & -\frac{4}{5} & -\frac{3}{5} \\ 0 & 0 & -\frac{3}{5} & \frac{4}{5} \end{pmatrix}\begin{pmatrix} 7 & 3 & 1 & 0 \\ 2 & 5 & 4 & 5 \\ 0 & 4 & 6 & 6 \\ 0 & 3 & 5 & 5 \end{pmatrix}\begin{pmatrix} 1 & 0 & 0 & 0 \\ 0 & 1 & 0 & 0 \\ 0 & 0 & -\frac{4}{5} & -\frac{3}{5} \\ 0 & 0 & -\frac{3}{5} & \frac{4}{5} \end{pmatrix}$$

$$= \begin{pmatrix} 7 & 3 & -\frac{4}{5} & -\frac{3}{5} \\ 2 & 5 & -\frac{31}{5} & \frac{8}{5} \\ 0 & -5 & \frac{273}{25} & -\frac{39}{25} \\ 0 & 0 & -\frac{14}{25} & \frac{2}{25} \end{pmatrix}$$

is the required Hessenberg matrix H. If we use the Laplace expansion formula for $\det H$ along the first column, we get

$$c_H(t) = 18 - 112t + 130t^2 - 23t^3 + t^4$$

By Theorem 6.4, $c_A(t) = c_H(t)$ is the characteristic polynomial of the matrix A. ◄►

EXERCISES 6.4

1. For $n = 3, 4$, and 5, create three $n \times n$ Hessenberg matrices.

2. For $n = 3, 4$, and 5, create three $n \times n$ Householder matrices and use *MATHEMATICA* to show that each matrix Q constructed in this way is both symmetric and orthogonal.

3. Convert the matrix

$$A = \begin{pmatrix} 0 & 0 & 0 & 0 & 200 \\ 1 & 0 & 0 & 0 & -220 \\ 0 & 1 & 0 & 0 & 38 \\ 0 & 0 & 1 & 0 & 23 \\ 0 & 0 & 0 & 1 & -4 \end{pmatrix}$$

to a Hessenberg matrix H, and use H to find the characteristic polynomial of A.

EIGENSPACES

Each eigenvalue λ of a linear transformation $T : V \to V$ determines a subspace $E_\lambda = \{\mathbf{x} : T(\mathbf{x}) = \lambda\mathbf{x}\}$ of V called the *eigenspace* of T associated with λ. If T is a matrix transformation $T_A : \mathbb{R}^n \to \mathbb{R}^n$ for some real $n \times n$ matrix A, then we call E_λ an *eigenspace* of A.

THEOREM 6.6 (Eigenspace theorem)   *For every eigenvalue λ of a linear transformation $T : V \to V$, the set $E_\lambda = \{\mathbf{x} : T(\mathbf{x}) = \lambda\mathbf{x}\}$ is a subspace of V.*

Proof. Since $T(\mathbf{0}) = 0 = \lambda\mathbf{0}$, the zero vector belongs to E_λ. We show that E_λ is also closed under linear combinations. Suppose that $\mathbf{x}$ and $\mathbf{y}$ belong to E_λ. Then

$$T(a\mathbf{x} + b\mathbf{y}) = T(a\mathbf{x}) + T(b\mathbf{y})$$
$$= aT(\mathbf{x}) + bT(\mathbf{y})$$
$$= a(\lambda\mathbf{x}) + b(\lambda\mathbf{y})$$
$$= \lambda(a\mathbf{x} + b\mathbf{y})$$

Hence $a\mathbf{x} + b\mathbf{y} \in E_\lambda$. Therefore, E_λ is a subspace of V. ∎

EXAMPLE 6.17   ■ The Eigenspaces of a Matrix

Describe the eigenspaces of the matrix

$$A = \begin{pmatrix} 5 & 0 & 0 & 0 \\ 0 & 5 & 0 & 0 \\ 0 & 0 & 5 & 0 \\ 0 & 0 & 0 & 7 \end{pmatrix}$$

Solution. For any vector $\mathbf{x} \in \mathbb{R}^4$,

$$A\mathbf{x} = \begin{pmatrix} 5 & 0 & 0 & 0 \\ 0 & 5 & 0 & 0 \\ 0 & 0 & 5 & 0 \\ 0 & 0 & 0 & 7 \end{pmatrix} \begin{pmatrix} w \\ x \\ y \\ z \end{pmatrix} = \begin{pmatrix} 5w \\ 5x \\ 5y \\ 7z \end{pmatrix}$$

If $z = 0$, then

$$A\mathbf{x} = \begin{pmatrix} 5 & 0 & 0 & 0 \\ 0 & 5 & 0 & 0 \\ 0 & 0 & 5 & 0 \\ 0 & 0 & 0 & 7 \end{pmatrix} \begin{pmatrix} w \\ x \\ y \\ 0 \end{pmatrix} = \begin{pmatrix} 5w \\ 5x \\ 5y \\ 0 \end{pmatrix} = 5 \begin{pmatrix} w \\ x \\ y \\ 0 \end{pmatrix} = 5\mathbf{x}$$

This means that 5 is an eigenvalue of A, and the eigenspace $E_5 = \{\mathbf{x} : A\mathbf{x} = 5\mathbf{x}\}$ consists of all vectors $\mathbf{x}$ in which $z = 0$. On the other hand, if $w = x = y = 0$, then

$$A\mathbf{x} = \begin{pmatrix} 5 & 0 & 0 & 0 \\ 0 & 5 & 0 & 0 \\ 0 & 0 & 5 & 0 \\ 0 & 0 & 0 & 7 \end{pmatrix} \begin{pmatrix} 0 \\ 0 \\ 0 \\ z \end{pmatrix} = \begin{pmatrix} 0 \\ 0 \\ 0 \\ 7z \end{pmatrix} = 7 \begin{pmatrix} 0 \\ 0 \\ 0 \\ z \end{pmatrix} = 7\mathbf{x}$$

This means that 7 is also an eigenvalue of A. The eigenspace $E_7 = \{\mathbf{x} : A\mathbf{x} = 7\mathbf{x}\}$ consists of all vectors $\mathbf{x}$ in which $w = x = y = 0$.

Since the characteristic polynomial of A is easily seen to be $c_A(t) = (t - 5)^3(t - 7)$, we can conclude that A has no other eigenvalues and therefore no other eigenspaces. ◄►

The number of repetitions of the factors $(t - 5)$ and $(t - 7)$ in the characteristic polynomial $c_A(t) = (t - 5)^3(t - 7)$ in Example 6.17 is called the *algebraic multiplicity* of the eigenvalues 5 and 7. Thus the eigenvalue 5 has algebraic multiplicity 3 and the eigenvalue 7 has multiplicity 1. For any $n \times n$ matrix A, the algebraic multiplicity of an eigenvalue λ of A is the largest integer m for which the characteristic polynomial $c_A(t)$ factors into the form $c_A(t) = (t - \lambda)^m p(t)$. In this particular case, these multiplicities coincide with the dimensions of the eigenspaces. These dimensions are called the *geometric multiplicities*. In the example, the two multiplicities coincide because the matrix is diagonal. In general, however, the algebraic multiplicity of an eigenvalue is often greater than its geometric multiplicity.

THEOREM 6.7 *The algebraic multiplicity of an eigenvalue is always greater than or equal to its geometric multiplicity.*

Proof. Suppose that λ is an eigenvalue of a linear transformation T on an n-dimensional vector space V and that the geometric multiplicity of λ is m. Then, by definition, the subspace E_λ has a basis $\mathcal{B} = \{\mathbf{x}_1, \ldots, \mathbf{x}_m\}$ consisting of eigenvectors of T. By Theorem 4.12, we can extend $\mathcal{B}$ to a basis $\mathcal{C} = \{x_1, \ldots, x_m, x_{m+1}, \ldots, x_n\}$ of V. If we represent T in the basis $\mathcal{C}$, we get a matrix

$$A = ([T(x_1)]_{\mathcal{C}} \quad \cdots \quad [T(x_m)]_{\mathcal{C}} \quad [T(x_{m+1})]_{\mathcal{C}} \quad \cdots \quad [T(x_n)]_{\mathcal{C}})$$

of the form

$$\begin{pmatrix} \lambda I & P \\ 0 & Q \end{pmatrix}$$

consisting of the $m \times m$ diagonal block λI, the zero block 0, and the blocks P and Q. This means that the characteristic matrix $(A - tI)$ of A is upper triangular. Therefore,

$$c_A(t) = \det(\lambda I_m - tI)\det(Q - tI) = (t - \lambda)^m p(t)$$

Hence the characteristic polynomial $(t - \lambda)^m$ of λI is a multiple of $c_A(t)$. If $p(t)$ contains the factor $(t - \lambda)$, then m is less than the algebraic multiplicity of λ. On the other hand, if $p(t)$ does not contain the factor $(t - \lambda)$, then m equals the algebraic multiplicity of λ. ∎

EXAMPLE 6.18 ■ A Maximally Linearly Independent Set of Eigenvectors

Find all eigenvalues and a maximum set S of linearly independent eigenvectors for the matrix

$$A = \begin{pmatrix} 5 & 6 \\ 3 & -2 \end{pmatrix}$$

Solution. We use *MATHEMATICA* to find the characteristic polynomial of A.

```
In[1]:= Eigenvalues[A={{5,6},{3,-2}}]

Out[1]= {-4,7}
```

Thus the eigenvalues of A are $\lambda_1 = 7$ and $\lambda_2 = -4$. To find the eigenvectors corresponding to λ_1, we form the matrix

$$(A - \lambda_1 I) = \begin{pmatrix} 5 & 6 \\ 3 & -2 \end{pmatrix} - \begin{pmatrix} 7 & 0 \\ 0 & 7 \end{pmatrix} = \begin{pmatrix} -2 & 6 \\ 3 & -9 \end{pmatrix}$$

The eigenvectors belonging to λ_1 must therefore satisfy the equation

$$\begin{pmatrix} -2 & 6 \\ 3 & -9 \end{pmatrix}\begin{pmatrix} x \\ y \end{pmatrix} = \begin{pmatrix} -2x + 6y \\ 3x - 9y \end{pmatrix} = \begin{pmatrix} 0 \\ 0 \end{pmatrix}$$

We solve the resulting homogeneous linear system.

```
In[2]:= Solve[{-2x+6y==0,3x-9y==0},{x}]

Out[2]= {{x→3y}}
```

This means that $x - 3y = 0$. Therefore $\mathbf{x}_1 = (3, 1)$ is an eigenvector of λ_1. All other eigenvectors belonging to λ_1 are multiples of $\mathbf{x}_1$. To find the eigenvectors corresponding to λ_2, we form the matrix equation

$$(A - \lambda_2 I) = \begin{pmatrix} 5 & 6 \\ 3 & -2 \end{pmatrix} - \begin{pmatrix} -4 & 0 \\ 0 & -4 \end{pmatrix} = \begin{pmatrix} 9 & 6 \\ 3 & 2 \end{pmatrix}$$

The eigenvectors belonging to λ_2 must therefore satisfy the matrix equation

$$\begin{pmatrix} 9 & 6 \\ 3 & 2 \end{pmatrix} \begin{pmatrix} x \\ y \end{pmatrix} = \begin{pmatrix} 9x + 6y \\ 3x + 2y \end{pmatrix} = \begin{pmatrix} 0 \\ 0 \end{pmatrix}$$

We solve the resulting homogeneous linear system.

```
In[3]:= Solve[{9x+6y==0,3x+2y==0},{x}]

Out[3]= {{x→ -2y/3}}
```

If we let $y = -3$, then $x = 2$. Therefore, the vector $\mathbf{x}_2 = (2, -3)$, for example, is an eigenvector of λ_2. All other eigenvectors belonging to λ_2 are multiples of $\mathbf{x}_2$.

We know from Chapter 3 that any maximally linearly independent subset of $\mathbb{R}^2$ can have at most two elements. If we can show that the vectors $\mathbf{x}_1$ and $\mathbf{x}_2$ are linearly independent, then $S = \{\mathbf{x}_1, \mathbf{x}_2\}$ is a maximally linearly independent set of eigenvectors of A. Suppose, therefore, that

$$a\mathbf{x}_1 + b\mathbf{x}_2 = a \begin{pmatrix} 3 \\ 1 \end{pmatrix} + b \begin{pmatrix} 2 \\ -3 \end{pmatrix} = \begin{pmatrix} 3a + 2b \\ a - 3b \end{pmatrix} = \begin{pmatrix} 0 \\ 0 \end{pmatrix}$$

Then $a = 3b$, so that $9b + 2b = 11b = 0$. Hence $b = 0$. This implies that a is also 0. Therefore, $\mathbf{x}_1$ and $\mathbf{x}_2$ are linearly independent. ◄►

EXAMPLE 6.19 ■ The Algebraic Multiplicities of Eigenvalues

Use *MATHEMATICA* to find the algebraic multiplicities of the eigenvalues of the matrix

$$A = \begin{pmatrix} 0 & 0 & 0 & -40 \\ 1 & 0 & 0 & 68 \\ 0 & 1 & 0 & -42 \\ 0 & 0 & 1 & 11 \end{pmatrix}$$

Solution. First we compute the characteristic polynomial of A.

$$\text{In[1]:= } A = \begin{pmatrix} 0 & 0 & 0 & -40 \\ 1 & 0 & 0 & 68 \\ 0 & 1 & 0 & -42 \\ 0 & 0 & 1 & 11 \end{pmatrix};$$

```
In[2]:= p=CharacteristicPolynomial[A,t]

Out[2]= 40-68t+42t²-11t³+t⁴
```

```
In[3]:= Roots[p==0,t]

Out[3]= t==2 || t==2 || t==2 || t==5
```

Thus A has the eigenvalues 2 and 5. The algebraic multiplicity of the eigenvalue 2 is 3, and that of the eigenvalue 5 is 1. ◄►

It is elegant to express the eigenvalues of matrices as the roots of characteristic polynomials. However, a remarkable discovery made by Galois in the 19th century, shows that it is impossible to find an algorithm for computing roots of arbitrary polynomials of degree greater than 4. For polynomials of degree 5 or higher, there are no variants of the familiar quadratic formula

$$\frac{-b \pm \sqrt{b^2 - 4ac}}{2a}$$

used to find the roots of the quadratic equation $at^2 + bt + c$. This means that in applications involving large matrices, we must use numerical techniques and usually work with approximations of eigenvalues. For example, if we generate a random matrix

```
In[1]:= A=Array[Random[Integer,{-10,10}]&,{5,5}]

Out[1]= {{-10,-5,-6,-7,10},{-4,-1,8,-6,-4},{-4,1,3,5,-5},
        {6,3,-8,6,-6},{-4,8,-5,0,-3}}
```

and ask *MATHEMATICA* to compute the eigenvalues of A, either directly using the **Eigenvalues** function, or indirectly, using the **CharacteristicPolynomial** and **Roots** functions, we get a symbolic output from which approximations of roots have to be calculated.

```
In[2]:= Eigenvalues[A]

Out[2]= {Root[81672+8330#1+514#1^2+30#1^3+5#1^4+#1^5&,1],
Root[81672+8330#1+514#1^2+30#1^3+5#1^4+#1^5&,2],
Root[81672+8330#1+514#1^2+30#1^3+5#1^4+#1^5&,3],
Root[81672+8330#1+514#1^2+30#1^3+5#1^4+#1^5&,4],
Root[81672+8330#1+514#1^2+30#1^3+5#1^4+#1^5&,5]}
```

```
In[3]:= p=CharacteristicPolynomial[A,t]

Out[3]= -81672-8330t-514t^2-30t^3-5t^4-t^5
```

```
In[4]:= Roots[p==0,t]

Out[4]= {Root[81672+8330#1+514#1^2+30#1^3+5#1^4+#1^5&,1],
Root[81672+8330#1+514#1^2+30#1^3+5#1^4+#1^5&,2],
Root[81672+8330#1+514#1^2+30#1^3+5#1^4+#1^5&,3],
Root[81672+8330#1+514#1^2+30#1^3+5#1^4+#1^5&,4],
Root[81672+8330#1+514#1^2+30#1^3+5#1^4+#1^5&,5]}
```

MATHEMATICA is telling us that the eigenvalues of the matrix A are among the roots of the characteristic polynomial. The symbolic name of the ith root of this polynomial is

$$\text{Root}[81672+8330\#1+514\#1^2+30\#1^3+5\#1^4+\#1^5\&,i]$$

We can find approximations of the eigenvalues of the matrix A with *MATHEMATICA* by designating the entries of A as *real* numbers.

```
In[5]:= Eigenvalues[{{-10,-5,-6,-7,10.0},{-4,-1,8,-6,-4},
{-4,1,3,5,-5},{6,3,-8,6,-6},{-4,8,-5,0,-3}}]

Out[5]= {6.70707+8.02912i,6.70707-8.02912i,
-4.70257+7.79187i,-4.70257-7.79187i,-9.00901}
```

As we can see, the matrix A has one real eigenvalue near -9.00901. The complex roots of the characteristic polynomial do not count as eigenvalues.

The **Eigensystem** function provides a quick way to compute the eigenspaces of a particular matrix. It produces both the eigenvalues of the matrix and for each eigenvalue, it calculates a maximally linearly independent set of eigenvectors.

EXAMPLE 6.20 ■ **The Eigenspaces of a Diagonal Matrix**

Use *MATHEMATICA* to find the eigenspaces of the matrix

$$A = \begin{pmatrix} 5 & 0 & 0 & 0 & 0 \\ 0 & 5 & 0 & 0 & 0 \\ 0 & 0 & 7 & 0 & 0 \\ 0 & 0 & 0 & 7 & 0 \\ 0 & 0 & 0 & 0 & 7 \end{pmatrix}$$

Solution. Since the matrix A is diagonal, we know that its eigenvalues are its diagonal entries 5 and 7. We use the **Eigensystem** function to produce two bases for the associated eigenspaces.

```
In[1]:= A = ⎛ 5  0  0  0  0 ⎞
          ⎜ 0  5  0  0  0 ⎟
          ⎜ 0  0  7  0  0 ⎟ ;
          ⎜ 0  0  0  7  0 ⎟
          ⎝ 0  0  0  0  7 ⎠
```

```
In[2]:= Eigensystem[A]
Out[2]= {{5,5,7,7,7},{{0,1,0,0,0},{1,0,0,0,0},
{0,0,0,0,1},{0,0,0,1,0},{0,0,1,0,0}}}
```

For each of the five occurrences of eigenvalues of the matrix A, *MATHEMATICA* has produced a matching eigenvector. Table 1 summarizes the results.

TABLE 1   Eigenvalues and Eigenvectors.

Eigenvalues	Eigenvectors
5	{0,1,0,0,0},{1,0,0,0,0}
7	{0,0,0,0,1},{0,0,0,1,0},{0,0,1,0,0}

As we can see, the eigenspace E_5 of the eigenvalue 5 has a basis consisting of two vectors, and the eigenspace E_7 of the eigenvalue 7 has a basis consisting of three vectors. ◄►

EXAMPLE 6.21   ■ The Eigenspaces of a Nondiagonalizable Matrix

Use *MATHEMATICA* to find the eigenspaces of the matrix

$$A = \begin{pmatrix} 2 & 1 & 4 & 5 & 0 \\ 0 & 2 & 3 & 0 & 9 \\ 0 & 0 & 3 & 6 & 0 \\ 0 & 0 & 0 & 3 & 0 \\ 0 & 0 & 0 & 0 & 3 \end{pmatrix}$$

Solution. Since the matrix A is upper triangular, we know that its eigenvalues are its diagonal entries 2 and 3. We use the **Eigensystem** function to produce bases for the associated eigenspaces.

```
In[1]:= A= ( 2 1 4 5 0
            0 2 3 0 9
            0 0 3 6 0
            0 0 0 3 0
            0 0 0 0 3 ) ;
```

```
In[2]:= Eigensystem[A]
Out[2]= {{2,2,3,3,3},{{1,0,0,0,0},{0,0,0,0,0},
{9,9,0,0,1},{7,3,1,0,0},{0,0,0,0,0}}}
```

For each of the five occurrences of eigenvalues of the matrix A, *MATHEMATICA* has produced a matching eigenvector. However, the vector associated with the second occurrence of 2 is zero and so is the vector associated with the third occurrence of 3. Since eigenvectors must be nonzero, we have the following match-up of eigenvalues and eigenvectors, shown in Table 2.

TABLE 2   Eigenvalues and Eigenvectors.

Eigenvalues	Eigenvectors
2	{1,0,0,0,0}
3	{9,9,0,0,1},{7,3,1,0,0}

As we can see, the eigenspace E_2 of the eigenvalue 2 has a basis consisting of one vector, and the eigenspace E_3 of the eigenvalue 3 has a basis consisting of two vectors. This tells us that the matrix A does not have five linearly independent eigenvectors and is therefore not diagonalizable. ◄►

The **Eigensystem** function also works for matrices without real eigenvalues. The command **Eigensystem[A={{0,-1},{1,0}}],** for example, yields

$$\{\{-i,i\},\{\{-i,1\},\{i,1\}\}\}$$

This tells us that the rotation matrix A has no real eigenvalues. Considered as a complex matrix, however, it has the eigenvalues $\pm i$, with associated eigenvectors $(-i, 1)$ and $(i, 1)$.

EXERCISES 6.5

1. Use *MATHEMATICA* to find bases for the eigenspaces of the following matrices, if they exist. Use only real eigenvalues and bases for real eigenspaces.

a. $\begin{pmatrix} 1 & 2 \\ 0 & 0 \end{pmatrix}$ b. $\begin{pmatrix} 1 & -1 \\ 1 & 1 \end{pmatrix}$ c. $\begin{pmatrix} 1 & 2 \\ 3 & 2 \end{pmatrix}$

g. $\begin{pmatrix} 0 & 0 & 0 & -225 \\ 1 & 0 & 0 & -60 \\ 0 & 1 & 0 & 26 \\ 0 & 0 & 1 & 4 \end{pmatrix}$ h. $\begin{pmatrix} 6 & 3 & 6 & 5 \\ 3 & 4 & 4.0 & 3 \\ 6 & 4 & 2 & 2 \\ 5 & 3 & 2 & 4 \end{pmatrix}$ i. $\begin{pmatrix} 1 & 0 & 0 & 9 \\ 0 & 1 & 7 & 0 \\ 0 & 7 & 1 & 0 \\ 9 & 0 & 0 & 1 \end{pmatrix}$

2. Let λ be an eigenvalue of an $n \times n$ real matrix A, and suppose that the matrix $(A - \lambda I)$ has rank r. Find the dimension of the eigenspace E_λ.

3. Suppose that λ is an eigenvalue of a matrix A and that the vector $\mathbf{x}$ belongs to the eigenspace E_λ. Show that for all $n \geq 2$, the vectors $A^n \mathbf{x}$ belong to E_λ.

4. Show that a nondiagonal 3×3 real matrix cannot have an eigenspace E_λ of dimension 3.

5. Construct a nondiagonal 3×3 real matrix whose eigenspaces E_λ and $E\mu$ have dimensions 1 and 2.

6. Show that for each eigenvalue λ of a real matrix A, the eigenspace E_λ is $\text{Nul}(A - \lambda I)$.

7. Show that if the geometric multiplicity of an eigenvalue λ of an $n \times n$ real matrix A is n, then $A = \lambda I_n$.

8. Show that if E_λ and E_μ are eigenspaces of a linear transformation $T : V \rightarrow V$ and $\lambda \neq \mu$, then E_λ and E_μ are disjoint.

Invertibility and Eigenvalues

The invertibility of a matrix is intimately connected to the nature of its eigenvalues.

THEOREM 6.8   *A real square matrix A is invertible if and only if the number 0 is not an eigenvalue of A.*

Proof. Suppose that A is invertible and that λ is an eigenvalue of A. Then there exists a nonzero vector $\mathbf{x}$ such that $A\mathbf{x} = \lambda\mathbf{x}$. Therefore,

$$A^{-1}A(\mathbf{x}) = \mathbf{x} = A^{-1}(\lambda\mathbf{x}) = \lambda A^{-1}(\mathbf{x})$$

This can only be so if $\lambda \neq 0$. Conversely, suppose that there exists a nonzero vector $\mathbf{x}$ for which $A\mathbf{x} = 0\mathbf{x} = \mathbf{0}$. Then the invertibility of A would lead to the contradiction that

$$A^{-1}A\mathbf{x} = \mathbf{x} = A^{-1}(0\mathbf{x}) = 0A^{-1}(\mathbf{x}) = \mathbf{0}$$

This proves the equivalence. ∎

EXAMPLE 6.22   ■ An Invertible Matrix and Its Eigenvalues

Show that the matrix

$$A = \begin{pmatrix} 1 & 0 & 0 \\ 3 & 2 & 2 \\ 0 & 0 & 3 \end{pmatrix}$$

is invertible by showing that 0 is not an eigenvalue of A.

Solution. Let us begin by computing the determinant of A.

```
In[1]:= Det[A={{1,0,0},{3,2,2},{0,0,3}}]
Out[1]= 6
```

By Theorem 3.14, A is invertible since its determinant is nonzero. Let us show that 0 is not an eigenvalue of A.

```
In[2]:= p=CharacteristicPolynomial[A,t]
Out[2]= 6-11t+6t²-t³
```

```
In[3]:= Roots[p,t]

Out[3]= t== 1 || t==2 || t==3
```

Therefore, 0 is not an eigenvalue of A. ◄►

We can use this theorem to relate the eigenvalues of T^{-1} to those of T.

THEOREM 6.9 (Inverse eigenvalue theorem) *If λ is an eigenvalue of an invertible linear transformation T on a real vector space V, then λ^{-1} is an eigenvalue of T^{-1}.*

Proof. Suppose that λ is an eigenvalue of T. Then there exists a nonzero vector $\mathbf{x} \in V$ for which $T(\mathbf{x}) = \lambda\mathbf{x}$. This means that $T^{-1}(T(\mathbf{x})) = \mathbf{x} = T^{-1}(\lambda\mathbf{x}) = \lambda T^{-1}(\mathbf{x})$. Hence $\lambda \neq 0$, and we can rewrite the equation $\mathbf{x} = \lambda T^{-1}(\mathbf{x})$ as $\lambda^{-1}\mathbf{x} = T^{-1}(\mathbf{x})$. This shows that λ^{-1} is an eigenvalue of T^{-1}. ∎

EXAMPLE 6.23   ■ **The Eigenvalues of the Inverse of a Linear Transformation**

Use *MATHEMATICA* to show that the eigenvalues of the inverse of the matrix transformation defined by

$$
T_A\begin{pmatrix} x \\ y \\ z \end{pmatrix} = \begin{pmatrix} 1 & 0 & 0 \\ 3 & 2 & 2 \\ 0 & 0 & 3 \end{pmatrix}\begin{pmatrix} x \\ y \\ z \end{pmatrix} = \begin{pmatrix} x \\ 3x + 2y + 2z \\ 3z \end{pmatrix}
$$

are the reciprocals of the eigenvalues of T.

Solution. By Corollary 6.3, the eigenvalues of T_A are the roots of the characteristic polynomial of T_A. By definition, the characteristic polynomial $c_T(t)$ of T is $c_A(t)$, and we saw in Example 6.22 that the roots of $c_A(t)$ are 1, 2, and 3. By Theorem 5.13, the transformation T^{-1} is represented by A^{-1}. Hence its eigenvalues are those of A^{-1}.

Let us find the eigenvalues of this matrix.

```
In[1]:= Eigenvalues[Inverse[A={{1,0,0},{3,2,2},{0,0,3}}]]

Out[1]= {1, 1/2, 1/3}
```

This shows that the eigenvalues of T^{-1} are the reciprocals of the eigenvalues of T_A. ◄►

EXERCISES 6.6

1. Show that the following matrices are invertible by showing that 0 is not an eigenvalue.

a. $\begin{pmatrix} 1 & 2 \\ 3 & 0 \end{pmatrix}$
b. $\begin{pmatrix} 4 & 0 & 0 \\ 0 & 4 & 0 \\ 0 & 0 & 3 \end{pmatrix}$
c. $\begin{pmatrix} 1 & 2 & 1 & 2 \\ 3 & 0 & 2 & 2 \\ 1 & 0 & 1 & 2 \\ 2 & 2 & 2 & 2 \end{pmatrix}$

d. $\begin{pmatrix} 0 & -1 \\ 1 & 0 \end{pmatrix}$
e. $\begin{pmatrix} 1 & 2 & 1 \\ 3 & 0 & 2 \\ 1 & 0 & 1 \end{pmatrix}$
f. $\begin{pmatrix} 1 & 2 & 1 & 2 & 0 \\ 3 & 0 & 2 & 2 & 0 \\ 1 & 0 & 1 & 2 & 0 \\ 2 & 2 & 2 & 2 & 0 \\ 0 & 0 & 0 & 0 & 3 \end{pmatrix}$

2. Find the eigenvalues of the inverses of the following matrices without computing the inverse matrices.

a. $\begin{pmatrix} 3 & 0 \\ 4 & -4 \end{pmatrix}$
b. $\begin{pmatrix} -4 & -3 \\ -3 & -1 \end{pmatrix}$
c. $\begin{pmatrix} 2 & 0 & 0 & 0 \\ 1 & 2 & 0 & 0 \\ 0 & \frac{3}{2} & 3 & 0 \\ 0 & 0 & 1 & 4 \end{pmatrix}$

d. $\begin{pmatrix} 7 & 7 \\ 0 & -7 \end{pmatrix}$
e. $\begin{pmatrix} 8 & 4 & -8 \\ 0 & 7 & 2 \\ 0 & 0 & 8 \end{pmatrix}$
f. $\begin{pmatrix} 0 & 0 & 0 & 0 & 2250 \\ 1 & 0 & 0 & 0 & -2850 \\ 0 & \frac{3}{2} & 0 & 0 & 2130 \\ 0 & 0 & 1 & 0 & -522 \\ 0 & 0 & 0 & \frac{1}{3} & 21 \end{pmatrix}$

Linearly Independent Eigenvectors

In many applications we require linearly independent eigenvectors. It is therefore fortunate that eigenvalues belonging to distinct eigenvalues are always linearly independent. This fact is proved by mathematical induction. The following example explains the idea behind the proof.

EXAMPLE 6.24   ■ Linearly Independent Eigenvectors

Suppose x_1, x_2, x_3, and x_4 are eigenvectors belonging to the distinct nonzero eigenvalues λ_1, λ_2, λ_3, and λ_4 of a linear transformation $T : V \to V$, and suppose that x_1, x_2, and x_3 are linearly independent. Then x_1, x_2, x_3, and x_4 are also linearly independent. The following calculation proves this fact. Let $a\mathbf{x}_1 + b\mathbf{x}_2 + c\mathbf{x}_3 + d\mathbf{x}_4 = \mathbf{0}$. Then

$$T(a\mathbf{x}_1 + b\mathbf{x}_2 + c\mathbf{x}_3 + d\mathbf{x}_4) = aT(\mathbf{x}_1) + bT(\mathbf{x}_2) + cT(\mathbf{x}_3) + dT(\mathbf{x}_4)$$
$$= a\lambda_1\mathbf{x}_1 + b\lambda_2\mathbf{x}_2 + c\lambda_3\mathbf{x}_3 + d\lambda_4\mathbf{x}_4$$
$$= \mathbf{0}$$

Moreover, $a\lambda_4\mathbf{x}_1 + b\lambda_4\mathbf{x}_2 + c\lambda_4\mathbf{x}_3 + d\lambda_4\mathbf{x}_4 = \mathbf{0}$. Therefore,

$$a(\lambda_1 - \lambda_4)\mathbf{x}_1 + b(\lambda_2 - \lambda_4)\mathbf{x}_2 + c(\lambda_3 - \lambda_4)\mathbf{x}_3 + d(\lambda_4 - \lambda_4)\mathbf{x}_4 = \mathbf{0}$$

Since $d(\lambda_4 - \lambda_4)\mathbf{x}_4 = \mathbf{0}$, the preceding equation reduces to

$$a(\lambda_1 - \lambda_4)\mathbf{x}_1 + b(\lambda_2 - \lambda_4)\mathbf{x}_2 + c(\lambda_3 - \lambda_4)\mathbf{x}_3 + 0 = \mathbf{0}$$

Since the vectors $\mathbf{x}_1$, $\mathbf{x}_2$, and $\mathbf{x}_3$ are linearly independent, the scalars $a(\lambda_1 - \lambda_4)$, $b(\lambda_2 - \lambda_4)$, and $c(\lambda_3 - \lambda_4)$ must be zero. Furthermore, the eigenvalues λ_1, λ_2, λ_3, and λ_4 are distinct, so that the scalars $(\lambda_1 - \lambda_4)$, $(\lambda_2 - \lambda_4)$, and $(\lambda_3 - \lambda_4)$ are all nonzero. Therefore a, b, and c must be zero. This implies that $d\mathbf{x}_4 = \mathbf{0}$. However, since $\mathbf{x}_4$ is an eigenvector and is therefore nonzero, the scalar d must be 0. ◄►

The proof of the next theorem is modeled after this calculation, somewhat disguised in a general induction.

THEOREM 6.10 (Eigenvector theorem)   *Eigenvectors belonging to distinct eigenvalues of a linear transformation $T : V \to V$ are linearly independent.*

Proof. Let $\lambda_1, \ldots, \lambda_n$ be a sequence of distinct eigenvalues of T, and let $\mathbf{x}_1, \ldots, \mathbf{x}_n$ be a sequence of corresponding eigenvectors. We must show that if $a_1\mathbf{x}_1 + \cdots + a_n\mathbf{x}_n = 0$, then $a_1 = \cdots = a_n = 0$. The proof is by induction on n.

Let $n = 1$ and $a_1\mathbf{x}_1 = 0$. Since $\mathbf{x}_1$ is an eigenvector and is therefore nonzero, the scalar a_1 must be zero.

Now suppose that $n > 1$, assume that the theorem holds for $(n - 1)$, and consider the equation $a_1\mathbf{x}_1 + \cdots + a_n\mathbf{x}_n = 0$. We first observe that

$$T(a_1\mathbf{x}_1 + \cdots + a_n\mathbf{x}_n) = a_1T(\mathbf{x}_1) + \cdots + a_nT(\mathbf{x}_n)$$
$$= a_1\lambda_1\mathbf{x}_1 + \cdots + a_n\lambda_n\mathbf{x}_n$$
$$= \mathbf{0}$$

Therefore $a_1\lambda_1\mathbf{x}_1 + \cdots + a_n\lambda_n\mathbf{x}_n = \mathbf{0}$. Next we multiply the zero vector $a_1\mathbf{x}_1 + \cdots + a_n\mathbf{x}_n$ by the last eigenvalue λ_n and note that

$$a_1\lambda_n\mathbf{x}_1 + \cdots + a_{n-1}\lambda_n\mathbf{x}_{n-1} + a_n\lambda_n\mathbf{x}_n = \mathbf{0}$$

We now subtract the last two equations and get

$$a_1(\lambda_1 - \lambda_n)\mathbf{x}_1 + \cdots + a_{n-1}(\lambda_{n-1} - \lambda_n)\mathbf{x}_{n-1} + a_n(\lambda_n - \lambda_n)\mathbf{x}_n = \mathbf{0}$$

Therefore,

$$a_1(\lambda_1 - \lambda_n)\mathbf{x}_1 + \cdots + a_{n-1}(\lambda_{n-1} - \lambda_n)\mathbf{x}_{n-1} = \mathbf{0}$$

By the induction hypothesis, the vectors $\mathbf{x}_1, \ldots, \mathbf{x}_{n-1}$ are linearly independent. Therefore, the scalars $a_i(\lambda_i - \lambda_n)$ must all be zero for $1 \le i \le n-1$. However, since the given eigenvalues are assumed to be distinct, $\lambda_i - \lambda_n \ne 0$ for $1 \le i \le n-1$. Hence the scalars $a_1, \ldots, a_{n-1}$ must be zero. This in turn implies that

$$\mathbf{0} = a_1\mathbf{x}_1 + \cdots + a_{n-1}\mathbf{x}_{n-1} + a_n\mathbf{x}_n = \mathbf{0} + a_n\mathbf{x}_n = a_n\mathbf{x}_n$$

Since $\mathbf{x}_n$ is an eigenvector and is therefore nonzero, we must have $a_n = 0$. ∎

EXERCISES 6.7

1. Find an eigenvector for each eigenvalue of the following matrices.

   a. $\begin{pmatrix} 1 & 0 \\ 0 & 0 \end{pmatrix}$ b. $\begin{pmatrix} 0 & 1 \\ 1 & 0 \end{pmatrix}$ c. $\begin{pmatrix} 1 & 1 \\ 0 & 1 \end{pmatrix}$

   d. $\begin{pmatrix} 1 & 0 & -2 & 6 \\ 0 & 5 & 7 & 1 \\ 0 & 0 & 0 & 1 \\ 0 & 0 & 0 & 1 \end{pmatrix}$ e. $\begin{pmatrix} 0 & 0 & 0 & 0 \\ 0 & 0 & 0 & 0 \\ 0 & 0 & 0 & 0 \\ 0 & 0 & 0 & 0 \end{pmatrix}$ f. $\begin{pmatrix} 1 & 0 & 0 & 9 \\ 0 & 1 & 7 & 0 \\ 0 & 7 & 1 & 0 \\ 9 & 0 & 0 & 1 \end{pmatrix}$

2. Show that the eigenvectors of the matrices in Exercise 1 associated with distinct eigenvalues are linearly independent.

3. Show that if $\mathbf{x}$ is an eigenvector associated with an eigenvalue λ and a is a nonzero scalar, then the vector $a\mathbf{x}$ is an eigenvector associated with λ.

4. Show that if $\mathbf{x}_1$ and $\mathbf{x}_2$ are eigenvectors associated with the same eigenvalue λ, and a and b are two scalars for which $a\mathbf{x}_1 + b\mathbf{x}_2 \ne \mathbf{0}$, then the linear combination $a\mathbf{x}_1 + b\mathbf{x}_2$ is an eigenvector associated with λ.

5. Prove or disprove the statement that if $\mathbf{x}_1$ and $\mathbf{x}_2$ are two eigenvectors of a real matrix A, associated with two different eigenvalues λ_1 and λ_2, then the product $\mathbf{x}_1^T \mathbf{x}_2$ is zero.

6. Prove or disprove the statement that if two eigenvectors $\mathbf{x}$ and $\mathbf{y}$ are linearly independent, then they are associated with distinct eigenvalues.

7. Show that if a real $n \times n$ matrix A has n distinct eigenvalues, then the eigenvectors associated with these eigenvalues form a basis for $\mathbb{R}^n$.

DIAGONALIZING SQUARE MATRICES

At the beginning of this chapter, we discussed some of the conceptual and computational benefits of diagonal matrices. Some of these benefits transfer to nondiagonal matrices, as long as the matrices involved are similar to diagonal ones. Table 3 lists the elements involved in using similarity to link diagonal and nondiagonal matrices.

TABLE 3   Eigenvector-Eigenvalue Decomposition.

Objects	Explanations
A	Real $n \times n$ matrix
$\{\lambda_1, \ldots, \lambda_n\}$	Eigenvalues of A in $\mathbb{R}$
$\{\mathbf{x}_1, \ldots, \mathbf{x}_n\}$	Eigenvectors in $\mathbb{R}^n$ belonging to $\lambda_1, \ldots, \lambda_n$
$P = (\mathbf{x}_1 \ \cdots \ \mathbf{x}_n)$	Real $n \times n$ matrix whose columns are $\mathbf{x}_1, \ldots, \mathbf{x}_n$
$D = (\lambda_i \delta_{ij})$	Real diagonal matrix whose diagonal entries are λ_i
PDP^{-1}	Diagonal decomposition of A

In this section, we will show that a real matrix A is **diagonalizable** if and only if there exists a real diagonal matrix D and a real invertible matrix P for which $A = PDP^{-1}$. If the matrix A is diagonalizable, we say that D and P **diagonalize** A. To stress the nature of the matrices P and D involved in a diagonalization, we call a product PDP^{-1} an **eigenvector-eigenvalue decomposition** of A.

If D and P diagonalize the matrix A, then the diagonal entries of D are the eigenvalues of A, and the columns of P are eigenvectors belonging to the matching eigenvalues. If $A = PDP^{-1}$, then $AP = PD$. Therefore,

$$
\begin{aligned}
AP &= A\,(\mathbf{x}_1 \ \cdots \ \mathbf{x}_n) \\
&= (A\mathbf{x}_1 \ \cdots \ A\mathbf{x}_n) \\
&= PD \\
&= P\,\mathrm{diag}(\lambda_1, \ldots, \lambda_n) \\
&= (\lambda\mathbf{x}_1 \ \cdots \ \lambda_n\mathbf{x}_n)
\end{aligned}
$$

The next theorem explains why a matrix is diagonalizable only if has enough linearly independent eigenvectors.

THEOREM 6.11 (Diagonalizability theorem)   *An $n \times n$ matrix A is similar to a diagonal matrix D if and only if A has n linearly independent eigenvectors.*

Proof. Suppose that A is diagonalizable and that $A = PDP^{-1}$. Then by right-multiplication we have $AP = PD$. Hence

$$
(A\mathbf{x}_1 \ A\mathbf{x}_2 \ \cdots \ A\mathbf{x}_n) = (\lambda_1\mathbf{x}_1 \ \lambda_2\mathbf{x}_2 \ \cdots \ \lambda_n\mathbf{x}_n)
$$

By equating columns, we get

$$Ax_1 = \lambda_1 x_1, \ Ax_2 = \lambda_2 x_2, \ldots, Ax_n = \lambda_n x_n$$

Since P is invertible, the column vectors $x_1, x_2, \ldots, x_n$ must be linearly independent. They must also be nonzero. Therefore, $\lambda_1, \lambda_2, \ldots, \lambda_n$ are eigenvalues and $x_1, x_2, \ldots, x_n$ are corresponding eigenvectors.

Suppose, conversely, that $x_1, x_2, \ldots, x_n$ are linearly independent eigenvectors. We can use them as the columns of an invertible matrix P, and let $D = \text{diag}(\lambda_1, \ldots, \lambda_n)$, where $\lambda_1, \lambda_2, \ldots, \lambda_n$ are the eigenvalues associated with $x_1, x_2, \ldots, x_n$. It follows that $AP = PD$. Since the eigenvectors are linearly independent, the matrix P is invertible. We therefore get $A = PDP^{-1}$. ■

EXAMPLE 6.25 ■ A Diagonalizable 3 × 3 Matrix

Use *MATHEMATICA* to verify that the matrices

$$D = \begin{pmatrix} 3 & 0 & 0 \\ 0 & 3 & 0 \\ 0 & 0 & 5 \end{pmatrix} \quad \text{and} \quad P = \begin{pmatrix} 1 & 1 & 1 \\ -1 & 0 & 2 \\ 0 & 1 & 1 \end{pmatrix}$$

diagonalize the matrix

$$A = \begin{pmatrix} 4 & 1 & -1 \\ 2 & 5 & -2 \\ 1 & 1 & 2 \end{pmatrix}$$

Solution. We begin by representing A, P, and D.

```
In[1]:= A= ( 4  1  -1 )     ( 1  1  1 )          ( 3  0  0 )
          ( 2  5  -2 ) ; P= ( -1 0  2 ) ; Diag=  ( 0  3  0 ) ;
          ( 1  1   2 )     ( 0  1  1 )          ( 0  0  5 )
```

Next we show that $A = PDP^{-1}$.

```
In[2]:= A= =P.Diag.Inverse[P]
Out[2]= True
```

This shows that the matrices D and P diagonalize the matrix A. ◄►

EXAMPLE 6.26 ■ **A Diagonalizable Linear Transformation**

Let $T : \mathbb{R}^2 \to \mathbb{R}^2$ be the linear transformation defined by

$$T(x, y) = \left(\begin{array}{c} 6x - y \\ 3x + 2y \end{array} \right)$$

Use *MATHEMATICA* to find an eigenvector-eigenvalue decomposition for the matrix $[T]_{\mathcal{E}}$.

Solution. Let $A = [T]_{\mathcal{E}}$. It is easy to verify that

$$A = \left(\begin{array}{cc} 6 & -1 \\ 3 & 2 \end{array} \right)$$

We use the **Eigensystem** function to find an eigenvector-eigenvalue decomposition of A. This provides us with both the eigenvalues of A and with corresponding eigenvectors at the same time.

```
In[1]:= Eigensystem[A={{6,-1},{3,2}}]

Out[1]= {{3,5},{{1,3},{1,1}}}
```

This tells us that $\{3, 5\}$ is the set of eigenvalues of A, that $(1, 3)$ is an eigenvector belonging to the eigenvalue 3, and $(1, 1)$ is an eigenvector belonging to the eigenvalue 5. It follows that the matrices

$$D = \left(\begin{array}{cc} 3 & 0 \\ 0 & 5 \end{array} \right) \quad \text{and} \quad P = \left(\begin{array}{cc} 1 & 1 \\ 3 & 1 \end{array} \right)$$

yield an eigenvector-eigenvalue decomposition of A.

We now verify that $A = PDP^{-1}$.

```
In[2]:= Diag={{3,0},{0,5}}; P={{1,1},{3,1}};
```

```
In[3]:= A==P.Diag.Inverse[P]

Out[3]= True
```

Therefore, $A = PDP^{-1}$. ◄►

EXAMPLE 6.27 ■ **A Diagonalizable Matrix with Repeated Eigenvalues**

Use *MATHEMATICA* to diagonalize the matrix

$$A = \left(\begin{array}{ccc} 4 & 1 & -1 \\ 2 & 5 & -2 \\ 1 & 1 & 2 \end{array} \right)$$

Solution. We begin by computing the eigenvalues and corresponding eigenvectors of A.

In[1]:= **Eigenvalues**$\left[A = \begin{pmatrix} 4 & 1 & -1 \\ 2 & 5 & -2 \\ 1 & 1 & 2 \end{pmatrix} \right]$

Out[1]= $\{3, 3, 5\}$

In[2]:= **Solve**$[A.\{a,b,c\}==3\{a,b,c\},\{a\}]$

Out[2]= $\{\{a \rightarrow -b+c\}\}$

In[3]:= **Solve**$[A.\{a,b,c\}==5\{a,b,c\},\{a,b\}]$

Out[3]= $\{\{a \rightarrow c, b \rightarrow 2c\}\}$

To diagonalize the matrix A, we need three linearly independent eigenvectors: one vector for the eigenvalue 5 and two vectors for the eigenvalue 3. First we find an eigenvector for $\lambda_1 = 5$. The matrix equation

$$A - 5 \begin{pmatrix} 1 & 0 & 0 \\ 0 & 1 & 0 \\ 0 & 0 & 1 \end{pmatrix} = \begin{pmatrix} 4 & 1 & -1 \\ 2 & 5 & -2 \\ 1 & 1 & 2 \end{pmatrix} - 5 \begin{pmatrix} 1 & 0 & 0 \\ 0 & 1 & 0 \\ 0 & 0 & 1 \end{pmatrix}$$

$$= \begin{pmatrix} -1 & 1 & -1 \\ 2 & 0 & -2 \\ 1 & 1 & -3 \end{pmatrix}$$

determines the homogeneous system

$$\begin{cases} -x + y - z = 0 \\ 2x - 2z = 0 \\ x + y - 3z = 0 \end{cases}$$

which simplifies to

$$\begin{cases} x - z = 0 \\ y - 2z = 0 \end{cases}$$

and yields the vector $\mathbf{x}_1 = (1, 2, 1)$ if $z = 1$.

Next we find two linearly independent eigenvectors for $\lambda_2 = 3$. The matrix equation

$$A - 3 \begin{pmatrix} 1 & 0 & 0 \\ 0 & 1 & 0 \\ 0 & 0 & 1 \end{pmatrix} = \begin{pmatrix} 4 & 1 & -1 \\ 2 & 5 & -2 \\ 1 & 1 & 2 \end{pmatrix} - 3 \begin{pmatrix} 1 & 0 & 0 \\ 0 & 1 & 0 \\ 0 & 0 & 1 \end{pmatrix} = \begin{pmatrix} 1 & 1 & -1 \\ 2 & 2 & -2 \\ 1 & 1 & -1 \end{pmatrix}$$

determines the homogeneous system $x + y - z = 0$. If we let $z = 0$ or 1, and assign 1 to x, we get two solutions: $\mathbf{x}_2 = (1, -1, 0)$ and $\mathbf{x}_3 = (1, 0, 1)$. It is easy to check that $\mathbf{x}_2$ and $\mathbf{x}_3$ are linearly independent. By Theorem 6.10, both vectors are linearly independent of $\mathbf{x}_1$. Therefore, $\mathcal{B} = \{\mathbf{x}_1, \mathbf{x}_2, \mathbf{x}_3\}$ is a basis for $\mathbb{R}^3$ consisting of eigenvectors of A. Let

$$D = \begin{pmatrix} 5 & 0 & 0 \\ 0 & 3 & 0 \\ 0 & 0 & 3 \end{pmatrix} \quad \text{and} \quad P = \begin{pmatrix} 1 & 1 & 1 \\ 2 & -1 & 0 \\ 1 & 0 & 1 \end{pmatrix}$$

Then $A = PDP^{-1}$. ◄►

The next example shows that not all square matrices have enough linearly independent eigenvectors to be diagonalizable.

EXAMPLE 6.28 ■ A Nondiagonalizable Matrix

Use *MATHEMATICA* to show that the matrix

$$A = \begin{pmatrix} -3 & 1 & -1 \\ -7 & 5 & -1 \\ -6 & 6 & -2 \end{pmatrix}$$

is not diagonalizable.

Solution. Let us first find the eigenvalues of A.

```
In[1]:= Eigenvalues[ A=( -3  1  -1
                          -7  5  -1
                          -6  6  -2 )]

Out[1]= {-2,-2,4}
```

As we can see, the matrix A has the two eigenvalues $\lambda_1 = -2$ and $\lambda_2 = 4$. From

$$\left[\begin{pmatrix} -3 & 1 & -1 \\ -7 & 5 & -1 \\ -6 & 6 & -2 \end{pmatrix} + 2 \begin{pmatrix} 1 & 0 & 0 \\ 0 & 1 & 0 \\ 0 & 0 & 1 \end{pmatrix} \right] \begin{pmatrix} x \\ y \\ z \end{pmatrix} = \begin{pmatrix} 1 & -1 & 1 \\ 7 & -7 & 1 \\ 6 & -6 & 0 \end{pmatrix} \begin{pmatrix} x \\ y \\ z \end{pmatrix}$$

we get the homogeneous linear system

$$\begin{cases} x - y + z = 0 \\ 7x - 7y + z = 0 \\ 6x - 6y = 0 \end{cases}$$

We use *MATHEMATICA* to solve this system. Let us solve the system for y and z.

```
In[2]:= Solve[{x-y+z==0,7x-7y+z==0,6x-6y==0},{y,z}]

Out[2]= {{y→x,z→0}}
```

This shows that x is the only free variable of the system. Therefore we cannot find two linearly independent eigenvectors belonging to the eigenvalue 2. Hence the matrix A is not diagonalizable. ◄►

EXERCISES 6.8

All eigenvalues in the following exercises must be real numbers and all eigenvectors must have real entries.

1. Show that the following real matrices have two linearly independent eigenvectors.

  a. $\begin{pmatrix} 1 & 1 \\ 1 & 1 \end{pmatrix}$ b. $\begin{pmatrix} 1 & 1 \\ 0 & 3 \end{pmatrix}$ c. $\begin{pmatrix} 1 & 5 \\ 5 & 3 \end{pmatrix}$ d. $\begin{pmatrix} -5 & 5 \\ 5 & 3 \end{pmatrix}$

2. Show that the following real matrices are not diagonalizable by showing that they fail to have two linearly independent eigenvectors.

  a. $\begin{pmatrix} 2 & 1 \\ 0 & 2 \end{pmatrix}$ b. $\begin{pmatrix} 1 & 1 \\ 0 & 1 \end{pmatrix}$ c. $\begin{pmatrix} 1 & 0 \\ -1 & 1 \end{pmatrix}$ d. $\begin{pmatrix} 0 & -1 \\ 1 & 0 \end{pmatrix}$

3. Show that the following real matrices are diagonalizable by showing that they have three linearly independent eigenvectors.

  a. $\begin{pmatrix} 2 & 0 & 0 \\ 0 & 1 & 0 \\ -1 & \frac{1}{2} & 1 \end{pmatrix}$ b. $\begin{pmatrix} 2 & 0 & 0 \\ 5 & 0 & 5 \\ 0 & \frac{1}{2} & 1 \end{pmatrix}$ c. $\begin{pmatrix} 2 & -\frac{1}{2} & 0 \\ 0 & 1 & 0 \\ -1 & \frac{1}{2} & 1 \end{pmatrix}$

4. Show that the following real matrices are not diagonalizable by showing that they do not have three linearly independent eigenvectors.

  a. $\begin{pmatrix} 2 & 3 & 4 \\ 0 & 1 & 5 \\ 0 & 0 & 1 \end{pmatrix}$ b. $\begin{pmatrix} 2 & -\frac{1}{2} & 0 \\ 0 & 1 & 0 \\ -1 & 0 & 1 \end{pmatrix}$ c. $\begin{pmatrix} 2 & 3 & 4 \\ 0 & 2 & 5 \\ 0 & 0 & 2 \end{pmatrix}$

5. Show that not every invertible matrix is diagonalizable and that not every diagonalizable matrix is invertible.

6. Show that each of the following real matrices has four linearly independent eigenvectors.

a.
$$\begin{pmatrix} 0 & 0 & 0 & 0 \\ 0 & 0 & 0 & 0 \\ 0 & 0 & 0 & 0 \\ 0 & 0 & 0 & 0 \end{pmatrix}$$
b.
$$\begin{pmatrix} 3 & 0 & 0 & 0 \\ 0 & 3 & 0 & 0 \\ 0 & 0 & 3 & 0 \\ 0 & 0 & 0 & 3 \end{pmatrix}$$
c.
$$\begin{pmatrix} 2 & 0 & 0 & 0 \\ -2 & 3 & 0 & 0 \\ 0 & 0 & 3 & 0 \\ -1 & 0 & 0 & 3 \end{pmatrix}$$

7. Find all real numbers a for which the following matrices are diagonalizable.

a.
$$\begin{pmatrix} 1 & 1 \\ a & 1 \end{pmatrix}$$
b.
$$\begin{pmatrix} 1 & a \\ 0 & 3 \end{pmatrix}$$
c.
$$\begin{pmatrix} 2 & 0 & 0 \\ 0 & 2 & 0 \\ a & 0 & 2 \end{pmatrix}$$
d.
$$\begin{pmatrix} 2 & 0 & a \\ 0 & 2 & 0 \\ 1 & 0 & 2 \end{pmatrix}$$

8. Use diagonalization to compute A^{12}, where

$$A = \begin{pmatrix} 1 & 0 & 3 & 0 & 0 \\ 0 & 1 & 0 & 0 & 0 \\ 0 & 0 & 0 & 0 & 0 \\ 0 & 0 & 2 & 3 & 0 \\ 1 & 0 & 0 & 0 & 3 \end{pmatrix}$$

9. Use Theorem 6.11 to show that if a real $n \times n$ matrix A has an eigenspace E_λ of dimension n, then A is diagonal.

10. Show that if A is diagonalized by an $n \times n$ matrix D, then the determinant of A is the product $\lambda_1 \lambda_2 \cdots \lambda_n$, where λ_i are the diagonal entries of D.

11. Every polynomial of the form $p(t) = t^n + a_{n-1}t^{n-1} + \cdots + a_1 t + a_0 \in \mathbb{R}[t]$ has an associated $n \times n$ real matrix

$$A = \begin{pmatrix} 0 & 0 & \cdots & 0 & -a_0 \\ 1 & 0 & \cdots & 0 & -a_1 \\ 0 & 1 & \cdots & 0 & -a_2 \\ \vdots & \vdots & \vdots & \vdots & \vdots \\ 0 & 0 & \cdots & 1 & -a_{n-1} \end{pmatrix}$$

called the *companion matrix* of $p(t)$. If $p(t)$ has n distinct real roots, then A is diagonalizable. Construct the companion matrix of $p(t) = (t - 2)(t - 3)(t - 4)$, and find a diagonal matrix D and an invertible matrix P that diagonalize A.

12. Repeat Exercise 11 for polynomials of degree 4 with the following real roots.

a. $\lambda_1 = 1; \lambda_2 = 2; \lambda_3 = 3; \lambda_4 = 4$ b. $\lambda_1 = 3; \lambda_2 = 5; \lambda_3 = -3; \lambda_4 = 7$

c. $\lambda_1 = 0; \lambda_2 = 2; \lambda_3 = 4; \lambda_4 = 5$ d. $\lambda_1 = 5; \lambda_2 = 6; \lambda_3 = -5; \lambda_4 = 12$

APPLICATIONS

Calculus

Let V be the real vector space of functions $f : \mathbb{R} \to \mathbb{R}$ spanned by the set $S = \{\sin x, \cos x\}$, and let D_x be the differentiation operator on V. Since $D_x(af + bg) = aD_x f(x) + bD_x g(x)$ preserves linear combinations, it is a linear transformation on V. We show that D_x has no real eigenvalues.

EXAMPLE 6.29   ■ The Eigenvalues of D_x

Use *MATHEMATICA* to show that D_x has no real eigenvalues.

Solution. First we find the values of D_x on the basis vectors.

```
In[1]:= D[Sin[x],x]

Out[1]= Cos[x]
```

```
In[2]:= D[Cos[x],x]

Out[2]= -Sin[x]
```

This means that

$$D_x \sin x = \cos x$$
$$= 0 \sin x + 1 \cos x$$

and

$$D_x \cos x = -\sin x$$
$$= -1 \sin x + 0 \cos x$$

Therefore,

$$D_x(a \sin x + b \cos x) = aD_x \sin x + bD_x \cos x$$
$$= a \cos x - b \sin x$$
$$= -b \sin x + a \cos x$$

On the other hand, if $\mathbf{x} = a \sin x + b \cos x$ is an eigenvector of D_x, then $\mathbf{x} \neq 0$, and

$$D_x(a \sin x + b \cos x) = \lambda(a \sin x + b \cos x)$$
$$= (\lambda a) \sin x + (\lambda b) \cos x$$

Therefore, $(\lambda a) \sin x + (\lambda b) \cos x = -b \sin x + a \cos x$. This means that $\lambda a = -b$ and $\lambda b = a$. Since either a or b must be nonzero, these equations can hold only if $\lambda^2 = -1$. Hence D_x has no real eigenvalues. ◄►

Differential Equations

Suppose that $\mathbf{y}' = A\mathbf{y}$ denotes the system

$$
\begin{cases}
y_1' &= a_{11}y_1 + \cdots + a_{1n}y_n \\
&\vdots \\
y_n' &= a_{n1}y_1 + \cdots + a_{nn}y_n
\end{cases}
$$

of linear differential equations, with y_i varying over t, and let $\mathbf{y} = P\mathbf{z}$ for some invertible matrix P. Then $P\mathbf{z}' = \mathbf{y}'$. Therefore,

$$
\mathbf{z}' = P^{-1}\mathbf{y}' = P^{-1}A\mathbf{y} = P^{-1}AP\mathbf{z}
$$

If A is diagonalizable, we can choose P to be a matrix whose columns are eigenvectors of A. This means that $P^{-1}AP$ is a diagonal matrix D whose diagonal entries are the eigenvalues of A. Therefore,

$$
\mathbf{z}' = \begin{pmatrix} z_1(t)' \\ \vdots \\ z_n(t)' \end{pmatrix} = \mathrm{diag}(\lambda_1, \ldots, \lambda_n) \begin{pmatrix} z_1(t) \\ \vdots \\ z_n(t) \end{pmatrix} = \begin{pmatrix} \lambda_1 z_1(t) \\ \vdots \\ \lambda_n z_n(t) \end{pmatrix}
$$

Moreover, we know from calculus that the general solution of the differential equation $z_i'(t) = \lambda_i z_i(t)$ is $z_i(t) = a_1 e^{\lambda_i t}$, for an arbitrary scalar a_i. Therefore,

$$
\mathbf{y} = \begin{pmatrix} y_1(t) \\ \vdots \\ y_n(t) \end{pmatrix} = P\mathbf{z} = P \begin{pmatrix} a_1 e^{\lambda_i t} \\ \vdots \\ a_n e^{\lambda_n t} \end{pmatrix}
$$

We apply this technique to solve two types of differential equations. In Example 6.30, we show how a higher-order homogeneous differential equation can be converted to a system of first-order differential equations and solved by diagonalization. In Example 6.31, we show how a problem in population dynamics can be solved by diagonalization.

EXAMPLE 6.30 ■ A Third-Order Differential Equation

Solve the homogeneous third-order differential equation

$$
y''' - 2y'' - y' + 2y = 0
$$

by reducing the equation to a system of first-order equations.

Solution. We introduce two new variables u and v connected by $y' = u$ and $u' = v$. Using these variables, we rewrite the given equation as

$$y''' = 2y'' + y' - 2y = 2v + u - 2y$$

We now write this equation as the linear system

$$\begin{cases} y' = u \\ u' = v \\ v' = -2y + u + 2v \end{cases}$$

and convert this system to matrix form:

$$\begin{pmatrix} y' \\ u' \\ v' \end{pmatrix} = \begin{pmatrix} 0 & 1 & 0 \\ 0 & 0 & 1 \\ -2 & 1 & 2 \end{pmatrix} \begin{pmatrix} y \\ u \\ v \end{pmatrix}$$

Next we use *MATHEMATICA* to diagonalize the coefficient matrix.

$$\text{In[1]:= } \mathbf{Eigensystem}\left[A = \begin{pmatrix} 0 & 1 & 0 \\ 0 & 0 & 1 \\ -2 & 1 & 2 \end{pmatrix} \right]$$

Out[1]= {{-1,1,2},{{1,-1,1},{1,1,1},{1,2,4}}}

Therefore, the eigenvalues of A are -1, 1, and 2 and the vectors $\mathbf{x}_1 = (-1, 1, -1)$, $\mathbf{x}_2 = (1, 1, 1)$, and $\mathbf{x}_3 = (1, 2, 4)$ are corresponding eigenvectors. We use this information to construct an eigenvector-eigenvalue decomposition PDP^{-1} of A.

Let

$$P = \begin{pmatrix} -1 & 1 & 1 \\ 1 & 1 & 2 \\ -1 & 1 & 4 \end{pmatrix} \quad \text{and} \quad D = \begin{pmatrix} -1 & 0 & 0 \\ 0 & 1 & 0 \\ 0 & 0 & 2 \end{pmatrix}$$

Then it is clear from our introductory remarks that

$$\mathbf{z}(t)' = P^{-1}AP\mathbf{z}(t) = D\mathbf{z}(t)$$

In other words,

$$\begin{pmatrix} z_1(t)' \\ z_2(t)' \\ z_3(t)' \end{pmatrix} = \begin{pmatrix} -1 & 0 & 0 \\ 0 & 1 & 0 \\ 0 & 0 & 2 \end{pmatrix} \begin{pmatrix} z_1(t) \\ z_2(t) \\ z_3(t) \end{pmatrix} = \begin{pmatrix} -z_1(t) \\ z_2(t) \\ 2z_3(t) \end{pmatrix}$$

We know from calculus that

$$\begin{pmatrix} -z_1(t) \\ z_2(t) \\ 2z_3(t) \end{pmatrix} = \begin{pmatrix} ae^{-t} \\ be^t \\ ce^{2t} \end{pmatrix}$$

Therefore,

$$\mathbf{y}(t) = \begin{pmatrix} y(t) \\ u(t) \\ v(t) \end{pmatrix} = \begin{pmatrix} -1 & 1 & 1 \\ 1 & 1 & 2 \\ -1 & 1 & 4 \end{pmatrix} \begin{pmatrix} ae^{-t} \\ be^t \\ ce^{2t} \end{pmatrix} = \begin{pmatrix} -ae^{-t} + be^t + ce^{2t} \\ ae^{-t} + be^t + 2ce^{2t} \\ -ae^{-t} + be^t + 4ce^{2t} \end{pmatrix}$$

We conclude by using *MATHEMATICA* to verify that the calculated functions satisfy the given differential equation.

```
In[2]:= y=-a Exp[-t]+b Exp[t]+c Exp[2t];
```

```
In[3]:= D[y,{t,3}]==Expand[2 D[y,{t,2}]+D[y,{t,1}]-2y]

Out[3]= True
```

As we can see, the equation holds. ◄►

In population biology, the rates of change of certain populations are sometimes given by linear combinations of other populations. If the growth rate of each species, measured over time, is a function of the size of the different species involved, the resulting model is a system of first-order differential equations. We now show how the growth rates of the individual species can be calculated when the coefficient matrix of the system is diagonalizable.

EXAMPLE 6.31   ■ Biological Rates of Change

Use *MATHEMATICA* to calculate the rates of change of the populations $y_1(t)$ and $y_2(t)$ of two interacting species determined by the system

$$\begin{cases} y_1' = 5y_1 + 2y_2 \\ y_2' = 3y_1 + 4y_2 \end{cases}$$

if $y_1(0) = 500$ and $y_2(0) = 700$.

Solution. First we write the given system of differential equations in matrix form.

$$\begin{pmatrix} y_1' \\ y_2' \end{pmatrix} = \begin{pmatrix} 5 & 2 \\ 3 & 4 \end{pmatrix} \begin{pmatrix} y_1 \\ y_2 \end{pmatrix}$$

Next we diagonalize the coefficient matrix and the matrices P and D.

In[1]:= A=$\begin{pmatrix} 5 & 2 \\ 3 & 4 \end{pmatrix}$;

In[2]:= **Eigensystem[A]**

Out[2]= {{2,7},{{-2,3},{1,1}}}

Hence the eigenvalues of A are 2 and 7, and two corresponding eigenvectors are $\{-2, 3\}$ and $\{1, 1\}$. We therefore form the matrices

$$P = \begin{pmatrix} -2 & 1 \\ 3 & 1 \end{pmatrix} \quad \text{and} \quad D = \begin{pmatrix} 2 & 0 \\ 0 & 7 \end{pmatrix}$$

We know from the introduction to this section that

$$\begin{pmatrix} z_1(t)' \\ z_2(t)' \end{pmatrix} = \begin{pmatrix} 2 & 0 \\ 0 & 7 \end{pmatrix} \begin{pmatrix} z_1(t) \\ z_2(t) \end{pmatrix} = \begin{pmatrix} 2z_1t \\ 7z_2t \end{pmatrix}$$

and

$$\begin{pmatrix} y_1(t) \\ y_2(t) \end{pmatrix} = \begin{pmatrix} -2 & 1 \\ 3 & 1 \end{pmatrix} \begin{pmatrix} ae^{2t} \\ be^{7t} \end{pmatrix} = \begin{pmatrix} -2ae^{2t} + be^{7t} \\ 3ae^{2t} + be^{7t} \end{pmatrix}$$

We now use the size of the populations at time $t = 0$ to calculate the scalars a and b.

$$\begin{cases} y_1(0) = -2a + b = 500 \\ y_2(0) = 3a + b = 700 \end{cases}$$

In[3]:= **Solve[{-2a+b==500,3a+b==700},{a,b}]**

Out[3]= {{a→40,b→580}}

Hence $y_1(t) = -80e^{2t} + 580e^{7t}$ and $y_2(t) = 120e^{2t} + 580e^{7t}$ are the rate-of-change functions for the two species. ◀▶

EXERCISES 6.9

1. Suppose that $D : D^\infty(\mathbb{R}, \mathbb{R}) \to D^\infty(\mathbb{R}, \mathbb{R})$ is the differentiation operation. Then λ is an eigenvalue of D if and only if $D(f) = f' = \lambda f$. The eigenvectors of D belonging to λ are therefore the solutions of the differential equation $y' - \lambda y = 0$. Use this information to show that every real number λ is an eigenvalue of D and find the eigenvectors belonging to λ.

2. Find the eigenvalues and eigenvectors of the differentiation operation $D : \mathbb{R}_2[t] \to \mathbb{R}_2[t]$. Explain why D is not diagonalizable.

3. Diagonalize the linear transformation $T : \mathbb{R}_3[t] \to \mathbb{R}_3[t]$ defined by $T(p(t)) = p(0)$.

4. Suppose that $D : D^\infty(\mathbb{R}, \mathbb{R}) \to D^\infty(\mathbb{R}, \mathbb{R})$ is the differentiation operation. Find the eigenvalues and eigenvectors of each of the following linear transformations.

a. $D^2 : D^\infty(\mathbb{R}, \mathbb{R}) \to D^\infty(\mathbb{R}, \mathbb{R})$ b. $D^2 + D : D^\infty(\mathbb{R}, \mathbb{R}) \to D^\infty(\mathbb{R}, \mathbb{R})$

5. Suppose that $D : D^\infty(\mathbb{R}, \mathbb{R}) \to D^\infty(\mathbb{R}, \mathbb{R})$ is the differentiation operation. Show that a vector $f \in D^\infty(\mathbb{R}, \mathbb{R})$ may be an eigenvector of $D^2 : D^\infty(\mathbb{R}, \mathbb{R}) \to D^\infty(\mathbb{R}, \mathbb{R})$ without being an eigenvector of D.

6. Use the isomorphism $\varphi(a \cos x + b \sin x) = (a, b)$ and the fact that the matrix

$$\begin{pmatrix} 0 & -1 \\ 1 & 0 \end{pmatrix}$$

has no real eigenvalues to show that the differentiation operator in Example 6.29 has no real eigenvalues.

7. Find the rate-of-change functions for an ecosystem made up of three species whose populations are related by

$$\begin{pmatrix} y_1(t)' \\ y_3(t)' \\ y_3(t)' \end{pmatrix} = \begin{pmatrix} 9 & 0 & 2 \\ 6 & 5 & 3 \\ -4 & 0 & 3 \end{pmatrix} \begin{pmatrix} y_1(t) \\ y_3(t) \\ y_3(t) \end{pmatrix}$$

if $y_1(0) = y_2(0) = y_3(0) = 400$.

Discrete Dynamical Systems

Sequences of vectors generated by a vector and a matrix are frequently used to model physical processes. In this section, we discuss some of the mathematical ideas involved. Although the range of applications is more general, we limit ourselves to the diagonalizable case.

DEFINITION 6.6   *Let x_0 be a fixed vector in $\mathbb{R}^n$ and let A be a diagonalizable $n \times n$ real matrix. Then the sequence of vectors*

$$x_0, Ax_0, A^2x_0, \ldots, A^nx_0, \ldots$$

*is a **discrete dynamical system**.*

We can rewrite the system in terms a family of equations $x_{k+1} = Ax_k$, with $k \in \mathbb{N}$ ranging over discrete intervals of time such as days, months, years, and so on. Hence the system is called discrete. It is called dynamical because at each stage $k + 1$, the system models an evolving process determined only by its condition at stage k and by the matrix A. If $\mathbf{x}_0$ is an eigenvector A belonging to an eigenvalue λ, then the system

$$\mathbf{x}_0, A\mathbf{x}_0, A^2\mathbf{x}_0, \ldots, A^n\mathbf{x}_0, \ldots$$

becomes

$$\mathbf{x}_0, \lambda\mathbf{x}_0, \lambda^2\mathbf{x}_0, \ldots, \lambda^n\mathbf{x}_0, \ldots$$

The long-term behavior of the system therefore can be analyzed in terms of the limit of $\lambda^n\mathbf{x}_0$, as $n \to \infty$, if such a limit exists.

EXAMPLE 6.32   ■ A Discrete Dynamical System

Describe the long-term behavior of the discrete dynamical system determined by

$$A = \begin{pmatrix} \frac{9}{10} & \frac{5}{100} \\ \frac{1}{10} & \frac{95}{100} \end{pmatrix} \quad \text{and} \quad \mathbf{x}_0 = \begin{pmatrix} 30 \\ 20 \end{pmatrix}$$

Solution. We begin by showing that A has two linearly independent eigenvectors. We can then decompose the vector $\mathbf{x}_0$ into a linear combination of eigenvectors and compute the required limits. The calculation

$$\text{In[1]:= Eigensystem}\left[A = \begin{pmatrix} \frac{9}{10} & \frac{5}{100} \\ \frac{1}{10} & \frac{95}{100} \end{pmatrix}\right]$$

$$\text{Out[1]= } \left\{\left\{\tfrac{17}{20}, 1\right\}, \left\{\{-1, 1\}, \left\{\tfrac{1}{2}, 1\right\}\right\}\right\}$$

shows that $17/20$ and 1 are the eigenvalues of A and that the vectors $\mathbf{x}_1 = (-1, 1)$ and $\mathbf{x}_2 = (1/2, 1)$ are eigenvectors of A belonging to the distinct eigenvalues $17/20$ and 1. They are therefore linearly independent and form a basis for $\mathbb{R}^2$. Hence there exist unique scalars a and b for which

$$\begin{pmatrix} 30 \\ 20 \end{pmatrix} = a \begin{pmatrix} -1 \\ 1 \end{pmatrix} + b \begin{pmatrix} \frac{1}{2} \\ 1 \end{pmatrix}$$

Now we find the scalars a and b.

$$\text{In[2]:= Solve[\{-a+b/2==30, a+b==20\}, \{a, b\}]}$$

$$\text{Out[2]= } \left\{\left\{a \to -\tfrac{40}{3}, b \to \tfrac{100}{3}\right\}\right\}$$

Therefore, $a = -40/3$ and $b = 100/3$. This means that long-term behavior of the system $\mathbf{x}_{k+1} = A\mathbf{x}_k$ is described by the following limit:

$$L = \lim_{n \to \infty} \left[\begin{pmatrix} \frac{9}{10} & \frac{5}{100} \\ \frac{1}{10} & \frac{95}{100} \end{pmatrix}^n \begin{pmatrix} 30 \\ 20 \end{pmatrix} \right]$$

$$= \lim_{n \to \infty} \left[\begin{pmatrix} \frac{9}{10} & \frac{5}{100} \\ \frac{1}{10} & \frac{95}{100} \end{pmatrix}^n \left(-\frac{40}{3} \begin{pmatrix} -1 \\ 1 \end{pmatrix} + \frac{100}{3} \begin{pmatrix} \frac{1}{2} \\ 1 \end{pmatrix} \right) \right]$$

$$= \lim_{n \to \infty} \left[-\frac{40}{3} \begin{pmatrix} \frac{9}{10} & \frac{5}{100} \\ \frac{1}{10} & \frac{95}{100} \end{pmatrix}^n \begin{pmatrix} -1 \\ 1 \end{pmatrix} + \frac{100}{3} \begin{pmatrix} \frac{9}{10} & \frac{5}{100} \\ \frac{1}{10} & \frac{95}{100} \end{pmatrix}^n \begin{pmatrix} \frac{1}{2} \\ 1 \end{pmatrix} \right]$$

$$= -\frac{40}{3} \lim_{n \to \infty} \left(\frac{17}{20} \right)^n \begin{pmatrix} -1 \\ 1 \end{pmatrix} + \frac{100}{3} \lim_{n \to \infty} (1)^n \begin{pmatrix} \frac{1}{2} \\ 1 \end{pmatrix}$$

$$= \frac{100}{3} \begin{pmatrix} \frac{1}{2} \\ 1 \end{pmatrix} = \begin{pmatrix} \frac{50}{3} \\ \frac{100}{3} \end{pmatrix}$$

Hence the limiting vector L of the system is $(50/3, 100/3)$. ◀▶

EXERCISES 6.10

1. Use *MATHEMATICA* to create a diagonalizable nondiagonal 3×3 matrix A. (Hint: Use companion matrices.) Construct the dynamical system $\mathbf{x}_{k+1} = A\mathbf{x}_k$ and compute $\mathbf{x}_{10}$ for an arbitrary initial vector $\mathbf{x}_1$.

2. Repeat Exercise 1 for 4×4 matrices with the following choices of eigenvalues:

   a. $\lambda_1 = 1; \lambda_2 = 2; \lambda_3 = 3; \lambda_4 = 4$ b. $\lambda_1 = 3; \lambda_2 = 3; \lambda_3 = -3; \lambda_4 = 7$
   c. $\lambda_1 = 0; \lambda_2 = 0; \lambda_3 = 4; \lambda_4 = 5$ d. $\lambda_1 = 5; \lambda_2 = 5; \lambda_3 = 5; \lambda_4 = 5$

3. Explain why the compound interest formula $P_{n+1} = 1.08 P_n$ for investing a fixed amount P at 8% per year for $n + 1$ years determines a dynamical system.

4. Find the dynamical system $\mathbf{x}_{n+1} = A\mathbf{x}_n$ that yields the value of four investments certificates after $n + 1$ years if the first certificate is invested at 6%, the second at 4.5%, the third at 8%, and the fourth certificate at 5%.

5. The sequence $1, 1, 2, 3, 5, 8, \ldots$ of Fibonacci numbers is defined by the formula $f_{i+2} = f_{i+1} + f_i$. Show that the dynamical system $\mathbf{x}_{i+1} = A\mathbf{x}_i$ determined by

$$A = \begin{pmatrix} 1 & 1 \\ 1 & 0 \end{pmatrix} \quad \text{and} \quad \mathbf{x} = \begin{pmatrix} 1 \\ 1 \end{pmatrix}$$

generates the sequence of Fibonacci numbers.

6. A population of 1000 gulls on a secluded island is being attacked by a communicable disease. The disease is not hereditary. Moreover, a gull becomes immune to the disease if it has survived the first month of infection. Each month, one quarter of the healthy

gulls contract the disease and one fifth of them die after a month. Nevertheless, the gull population increases by 10% each month, after births and deaths have been taken into account. Find the dynamical system $\mathbf{x}_{n+1} = A\mathbf{x}_n$ that describes the growth of the gull population and calculate the population after twelve months.

Markov Chains

Frequently a discrete dynamical system is a sequence of probability vectors determined by a stochastic matrix. A **probability vector** is a vector $\mathbf{u} = (u_1, \ldots, u_n) \in \mathbb{R}^n$ with the property that $0 \le u_1, \ldots, u_n$ and $u_1 + \cdots + u_n = 1$, and a **stochastic matrix** is an $n \times n$ matrix $A = (\mathbf{u}_1 \cdots \mathbf{u}_n)$ whose columns are probability vectors. The matrix A is called the **transition matrix** of the chain. A stochastic matrix A is **regular** if there exists an $n \in \mathbb{N}$ for which all entries of A^n are nonzero.

EXAMPLE 6.33 ■ **Probability Vectors**

Show that the vectors $\mathbf{x}_1 = (0, 1, 0)$, $\mathbf{x}_2 = (\frac{1}{4}, \frac{1}{4}, \frac{1}{4}, \frac{1}{4})$, $\mathbf{x}_3 = (\frac{1}{3}, 0, \frac{2}{3})$, and $\mathbf{x}_4 = (\frac{1}{4}, \frac{3}{8}, \frac{1}{8}, \frac{1}{4})$ are probability vectors.

Solution. Since the entries of all vectors are nonnegative, and since the sums of the entries of each column are 1, each vector is a probability vector. ◄►

EXAMPLE 6.34 ■ **Stochastic Matrices**

Show that the following matrices are stochastic.

a. $\begin{pmatrix} 1 & 0 & 0 \\ 0 & 1 & 0 \\ 0 & 0 & 1 \end{pmatrix}$ b. $\begin{pmatrix} \frac{1}{3} & 0 & 0 \\ \frac{1}{3} & \frac{1}{2} & 0 \\ \frac{1}{3} & \frac{1}{2} & 1 \end{pmatrix}$

c. $\begin{pmatrix} \frac{1}{4} & \frac{1}{3} & \frac{1}{3} \\ \frac{1}{2} & \frac{1}{3} & \frac{1}{3} \\ \frac{1}{4} & \frac{1}{3} & \frac{1}{3} \end{pmatrix}$ d. $\begin{pmatrix} \frac{1}{2} & \frac{1}{2} & \frac{1}{3} \\ \frac{1}{2} & 0 & \frac{1}{3} \\ 0 & \frac{1}{2} & \frac{1}{3} \end{pmatrix}$

are stochastic.

Solution. Since each of the four matrices is square and since the columns of all four matrices are probability vectors, the matrices are stochastic. ◄►

EXAMPLE 6.35 ■ **Regular Stochastic Matrices**

Determine which of the stochastic matrices in Example 6.34 are regular.

Solution. Since the powers of a diagonal matrix are diagonal and therefore contain non-diagonal zero entries, the identity matrix A_1 is not regular. Moreover, the product of two lower-triangular matrices is lower triangular. Hence the powers of A_2 contain zero entries above the diagonal. The matrix A_2 is therefore not regular. The matrix A_3 is regular since it contains no zero entries. A direct calculation shows that

$$A_4^2 = \begin{pmatrix} \frac{1}{2} & \frac{1}{2} & \frac{1}{3} \\ \frac{1}{2} & 0 & \frac{1}{3} \\ 0 & \frac{1}{2} & \frac{1}{3} \end{pmatrix} \begin{pmatrix} \frac{1}{2} & \frac{1}{2} & \frac{1}{3} \\ \frac{1}{2} & 0 & \frac{1}{3} \\ 0 & \frac{1}{2} & \frac{1}{3} \end{pmatrix} = \begin{pmatrix} \frac{1}{2} & \frac{5}{12} & \frac{4}{9} \\ \frac{1}{4} & \frac{5}{12} & \frac{5}{18} \\ \frac{1}{4} & \frac{1}{6} & \frac{5}{18} \end{pmatrix}$$

Hence A_4 is regular. ◄►

A **steady-state vector** of a stochastic matrix A is a probability vector $\mathbf{s}$ for which $A\mathbf{s} = \mathbf{s}$. It can be shown that every stochastic matrix has a steady-state vector.

EXAMPLE 6.36 ■ Steady-State Vectors

Find steady-state vectors for the stochastic matrices in Example 6.34.

Solution. We show that 1 is an eigenvalue of each matrix, and for each matrix we compute an eigenvector associated with the eigenvalue 1.

The result is obvious for the identity matrix A_1. Moreover, every probability vector in $\mathbb{R}^3$ is an eigenvector belonging to 1 and is therefore a steady-state vector.

Since diagonal entries of a lower-triangular matrix are the eigenvalues of the matrix and since 1 is a diagonal entry of A, we know immediately that A has the eigenvalue 1. It therefore remains to find a probability vector in the eigenspace of 1. Such a vector must satisfy the equation

$$\begin{pmatrix} \frac{1}{3} & 0 & 0 \\ \frac{1}{3} & \frac{1}{2} & 0 \\ \frac{1}{3} & \frac{1}{2} & 1 \end{pmatrix} \begin{pmatrix} x \\ y \\ z \end{pmatrix} = \begin{pmatrix} \frac{1}{3}x \\ \frac{1}{3}x + \frac{1}{2}y \\ \frac{1}{3}x + \frac{1}{2}y + z \end{pmatrix} = \begin{pmatrix} x \\ y \\ z \end{pmatrix}$$

We can tell by inspection that $x = 0$, $y = 0$, and $z = 1$ are the coordinates of the required probability vector.

It follows from an exercise that the end of this section that 1 is also an eigenvalue of A_3 and that $\mathbf{x}_1 = (4/13, 5/13, 4/13)$ is the required steady-state vector.

It will be shown in Exercise 6 at the end of this section that 1 is also an eigenvalue of A_4 and that $x_1 = (6/13, 4/13, 3/13)$ is the required steady-state vector. ◄►

DEFINITION 6.7 *A **Markov chain** is a sequence of probability vectors*

$$\mathbf{x}_1, A\mathbf{x}_1, A^2\mathbf{x}_1 \ldots, A^n\mathbf{x}_1, \ldots$$

determined by a stochastic matrix A and an initial probability vector $\mathbf{x}_1$.

A steady-state vector of the matrix A is a steady-state vector of the chain. A Markov chain is *regular* if there exists an n for which all entries of the matrix power A^n are nonzero. The next theorem explains why regularity is important.

THEOREM 6.12 *If A is a regular stochastic matrix, then the powers of A approach a stochastic matrix $P = (\mathbf{x} \cdots \mathbf{x})$ with identical columns and positive entries.* ∎

This theorem has the following important corollary.

COROLLARY 6.13 *If $\mathbf{x}_0$ is any initial probability vector and $(\mathbf{x}_0, A\mathbf{x}_0, \ldots, A^n\mathbf{x}_0, \ldots)$ is a regular Markov chain for which the powers of A approach the stochastic matrix $P = (\mathbf{x} \cdots \mathbf{x})$, then $A\mathbf{x} = \mathbf{x}$ and $\lim_{n \to \infty} (A^n\mathbf{x}_0) = \mathbf{x}$.* ∎

Many practical problems can be solved using regular Markov chains. One of the basic examples of a Markov chain deals with population movements.

EXAMPLE 6.37 ■ Markov Chains

Assume that the population of an island is divided into two distinct groups, city dwellers and country dwellers. Every year, 10% of the city dwellers move to the country, and 5% of the country dwellers move to the city. If the total population in the year 2000 is 50,000, with 30,000 living in the city and 20,000 living in the country, how many people will live in the city and how many will live in the country between the years 2000 and 2003?

Solution. To solve this problem, we note that the percentage changes in the two populations can be expressed by the linear system

$$\begin{cases} C = .90x + .05y \\ S = .10x + .95y \end{cases}$$

where C is the percentage of the population living in the city and S is the percentage of population living in the country at the end of one year. The variables x and y range over the city and country population percentages at the beginning of a year. Since initially 60% of the population live in the city and 40% live in the country, the initial probability vector of the chain is $\mathbf{x}_1 = (.6, .4)$ and we have the Markov chain $\mathbf{x}_1, A\mathbf{x}_1, A^2\mathbf{x}_1, A^3\mathbf{x}_1, \ldots$, determined by the matrix

$$A = \begin{pmatrix} .90 & .05 \\ .10 & .95 \end{pmatrix}$$

The first three vectors of the chain,

$$A\mathbf{x}_1 = \begin{pmatrix} .56 \\ .44 \end{pmatrix}, A^2\mathbf{x}_1 = \begin{pmatrix} .526 \\ .474 \end{pmatrix}, A^3\mathbf{x}_1 = \begin{pmatrix} .4971 \\ .5029 \end{pmatrix}$$

determine the population distributions of the island at the beginning of 2001, 2002, and 2003, respectively. Table 4 lists the actual distributions, measured in thousands.

Since the matrix A is regular, we can also calculate the long-term population distribution of the island. We first calculate the steady-state $\mathbf{s}$ of A.

TABLE 4   Population Distribution.

	2000	2001	2002	2003
City	30	28	26.3	24.855
Country	20	22	23.7	25.145

In[1]:= **Eigenvalues** $\left[A = \begin{pmatrix} .90 & .05 \\ .10 & .95 \end{pmatrix} \right]$

Out[1]= {1., 0.85}

This shows that the scalar 1 is an eigenvalue of A. Now let us find an eigenvector belonging to 1.

In[2]:= **Eigenvectors** $\left[A = \begin{pmatrix} .90 & .05 \\ .10 & .95 \end{pmatrix} \right]$

Out[2]= {{-0.447214, -0.894427}, {-0.707107, 0.707107}}

Therefore,

$$\mathbf{q} = \begin{pmatrix} -.447214 \\ -.894427 \end{pmatrix}$$

is an eigenvector belonging to the eigenvalue 1. But it is obviously not a probability vector. The desired steady-state vector **s** will be a scalar multiple of $a\mathbf{q}$. An easy calculation shows that

$$a\mathbf{q} = -.745\,36 \begin{pmatrix} -.447214 \\ -.894427 \end{pmatrix} = \begin{pmatrix} .333\,34 \\ .666\,67 \end{pmatrix} = \mathbf{s}$$

is the required probability vector **s**. It follows that in the long run, one-third of the island's population will live in the city and two-thirds will live in the country.

Let us calculate the population of the island for the year 2103 and estimate how close to its final state it will have come at that time.

In[1]:= **x=MatrixPower[{{.90,.05},{.10,.95}},103].{.6,.4}**

Out[1]= {0.333333, 0.666667}

Therefore,

$$\mathbf{x}_{104} = \begin{pmatrix} .333333 \\ .666667 \end{pmatrix}$$

We can see that at the beginning of the year 2103, the population of the island will have essentially reached its final distribution. ◄►

EXERCISES 6.11

1. Use the **Random** function to generate two probabilities p and q and form the matrix

$$A = \begin{pmatrix} 1-p & q \\ p & 1-q \end{pmatrix}$$

Show that 1 is an eigenvalue of A and find a steady-state vector of A.

2. Prove that for all probabilities p and q, the vector (q, p) is a steady-state vector of the matrix

$$A = \begin{pmatrix} 1-p & q \\ p & 1-q \end{pmatrix}$$

3. Determine which of the following matrices are stochastic.

a. $\begin{pmatrix} .3 & .8 \\ .7 & .2 \end{pmatrix}$ b. $\begin{pmatrix} 0 & 1 \\ 1 & 0 \end{pmatrix}$

c. $\begin{pmatrix} .1 & .5 & 0 \\ .1 & .2 & 1 \\ .8 & .3 & 0 \end{pmatrix}$ d. $\begin{pmatrix} .3 & .5 & .2 \\ .3 & .2 & .2 \\ .4 & .3 & .5 \end{pmatrix}$

4. Use *MATHEMATICA* to find the steady-state vectors of the following stochastic matrices.

a. $\begin{pmatrix} .5 & .3 \\ .5 & .7 \end{pmatrix}$ b. $\begin{pmatrix} .4 & .6 \\ .6 & .4 \end{pmatrix}$ c. $\begin{pmatrix} .1 & .5 & .2 \\ .1 & .2 & .5 \\ .8 & .3 & .3 \end{pmatrix}$

5. Use *MATHEMATICA* to show that 1 is an eigenvalue of the stochastic matrix

$$A = \begin{pmatrix} \frac{1}{4} & \frac{1}{3} & \frac{1}{3} \\ \frac{1}{2} & \frac{1}{3} & \frac{1}{3} \\ \frac{1}{4} & \frac{1}{3} & \frac{1}{3} \end{pmatrix}$$

and find a steady-state vector for A.

6. Use *MATHEMATICA* to show that 1 is an eigenvalue of the stochastic matrix

$$A = \begin{pmatrix} \frac{1}{2} & \frac{1}{2} & \frac{1}{3} \\ \frac{1}{2} & 0 & \frac{1}{3} \\ 0 & \frac{1}{2} & \frac{1}{3} \end{pmatrix}$$

and find a steady-state vector for A.

7. Use *MATHEMATICA* to determine which of the following stochastic matrices are regular.

a. $\begin{pmatrix} 1 & .3 & .7 & 0 \\ .1 & .2 & 0 & .8 \\ 0 & .5 & 0 & .2 \\ .8 & 0 & .3 & 0 \end{pmatrix}$
b. $\begin{pmatrix} 0 & 1 & .4 & .2 \\ 1 & 0 & .2 & .5 \\ 0 & 0 & .2 & .1 \\ 0 & 0 & .2 & .3 \end{pmatrix}$

c. $\begin{pmatrix} .1 & .5 & 0 \\ .1 & 0 & 1 \\ .8 & .5 & 0 \end{pmatrix}$
d. $\begin{pmatrix} .3 & .5 & .2 \\ .3 & .2 & .2 \\ .4 & .3 & .6 \end{pmatrix}$

8. A student has the following study habits. If he studies one evening, he is 60% certain of not studying the next evening. Moreover, the probability that he studies two nights in a row is also 60%. How often does he study in the long run?

9. A businesswoman is the manager of branches A, B, and C of a large corporation. She never visits the same branch on consecutive days. If she visits branch A one day, she visits branch B the next day. If she visits either branch B or C that day, then the next day she is twice as likely to visit branch A as to visit branch B or C. How often does she visit each branch in the long run?

10. Suppose that a businessman manages three branches of a large corporation. The matrix

$$A = \begin{pmatrix} 0 & \frac{1}{2} & \frac{1}{2} \\ 1 & 0 & \frac{1}{2} \\ 0 & \frac{1}{2} & 0 \end{pmatrix}.$$

describes the probabilities of his visiting patterns. Describe these patterns in words and calculate how often he visits each branch in the long run.

11. Suppose that a large circle is drawn on the floor and that five points are marked off on the circle in counterclockwise order. Imagine that a girl performs a random walk on the circle by moving from point to point. Let p be the probability that she moves clockwise from one point to the next, and $q = 1 - p$ the probability that she moves counterclockwise. Calculate the transition matrix of this Markov chain.

12. A car rental company in Montreal has three locations: Dorval Airport, Old Montreal, and the Dominion Square. Of the cars rented at Dorval Airport, 60% are returned to Dorval Airport, 10% are returned to Old Montreal, and 30% are returned to Dominion Square. Of the cars rented in Old Montreal, 30% are returned to Old Montreal, 50% are returned to Dorval Airport, and 20% are returned to Dominion Square. Of the cars rented in Dominion Square, 30% are returned to Dominion Square, 10% are returned to Old Montreal, and 60% are returned to Dorval Airport. Construct the transition matrix for this Markov chain and determine the distribution of the cars among the three locations in the long run.

REVIEW

KEY CONCEPTS ▶ Define and discuss each of the following.

Eigenspaces
Eigenvector basis.

Eigenvalues
Algebraic multiplicity, geometric multiplicity.

Eigenvectors
Eigenvector belonging to an eigenvalue.

Linear Transformations
Diagonalizable transformation, Householder transformation.

Matrices
Characteristic matrix, diagonalizable matrix, Hessenberg matrix.

Polynomials
Characteristic polynomial, root of a characteristic polynomial.

KEY FACTS ▶ Explain and illustrate each of the following.

1. If $T : V \to V$ is a linear transformation on a finite-dimensional vector space V and $A = [T]_{\mathcal{B}}^{\mathcal{B}}$ is the matrix of T in some basis $\mathcal{B}$, then $T(\mathbf{x}) = \lambda\mathbf{x}$ if and only if $A[\mathbf{x}]_{\mathcal{B}} = \lambda[\mathbf{x}]_{\mathcal{B}}$.

2. For any scalar λ and any real square matrix A, the following are equivalent: (a) The scalar λ is an eigenvalue of A. (b) The matrix $(A - \lambda I)$ is singular. (c) The scalar λ is a root of the characteristic polynomial $c_A(t)$ of A. (d) The set $\{\mathbf{x} : A\mathbf{x} = \lambda\mathbf{x}\}$ is the solution space of the system $(A - \lambda I)\mathbf{x} = 0$.

3. Suppose that $T : V \to V$ is a linear transformation on a real finite-dimensional vector space V and that $c_A(t)$ is the characteristic polynomial of the matrix $A = [T]_{\mathcal{B}}^{\mathcal{B}}$. Then every eigenvalue λ of T is a real root $c_A(t)$.

4. If A and B are similar matrices, then $c_A(t) = c_B(t)$.

5. Every matrix satisfies its characteristic polynomial.

6. For every eigenvalue λ of a linear transformation $T : V \to V$, the set $E_\lambda = \{\mathbf{x} : T(\mathbf{x}) = \lambda\mathbf{x}\}$ is a subspace of V.

7. The algebraic multiplicity of an eigenvalue is always greater than or equal to its geometric multiplicity.

8. A real square matrix A is invertible if and only if the number 0 is not an eigenvalue of A.

9. If λ is an eigenvalue of an invertible linear transformation T on a real vector space V, then λ^{-1} is an eigenvalue of T^{-1}.

10. Eigenvectors belonging to distinct eigenvalues of a linear transformation $T : V \rightarrow V$ are linearly independent.

11. An $n \times n$ matrix A is similar to a diagonal matrix D if and only if A has n linearly independent eigenvectors.

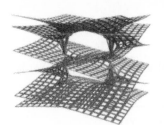

7

NORMS AND INNER PRODUCTS

In this chapter, we lay the foundation for doing geometry in vector spaces. The underlying spaces will be the coordinate spaces. Most ideas developed in this chapter are motivated by our understanding of Euclidean geometry in $\mathbb{R}^2$ and $\mathbb{R}^3$.

The three basic geometric concepts needed in our work are length, distance, and angle. Let us recall how these concepts are defined in ordinary geometry.

1. **Length.** Let $\mathbf{x} \in \mathbb{R}^2$ be the vector determined by the line segment from the origin $(0, 0)$ to the point (x_1, x_2). The length of $\mathbf{x}$ is measured by the Pythagorean formula

$$\|\mathbf{x}\| = \sqrt{x_1^2 + x_2^2}$$

Since the length of a vector is always a nonnegative real number, the assignment $\mathbf{x} \to \|\mathbf{x}\|$ is a function from $\mathbb{R}^2$ to the set $[0, \infty) = \{x \in \mathbb{R} : 0 \le x\}$ of nonnegative real numbers.

2. **Distance.** If $\mathbf{y} \in \mathbb{R}^2$ is a second vector determined by the line segment from $(0, 0)$ to the point (y_1, y_2), then the distance between $\mathbf{x}$ and $\mathbf{y}$ is measured by the formula

$$d(\mathbf{x}, \mathbf{y}) = \|\mathbf{x} - \mathbf{y}\| = \sqrt{(x_1 - y_1)^2 + (x_2 - y_2)^2}$$

The measure $d(\mathbf{x}, \mathbf{y}) = \|\mathbf{x} - \mathbf{y}\|$ is a function from $\mathbb{R}^2 \times \mathbb{R}^2$ to $[0, \infty)$.

3. **Angles.** Angles are defined indirectly through their cosines. The cosine of the angle θ between $\mathbf{x}$ and $\mathbf{y}$ is given by the formula

$$\cos \theta = \frac{\mathbf{x}^T \mathbf{y}}{\|\mathbf{x}\| \, \|\mathbf{y}\|}$$

The assignment $(\mathbf{x}, \mathbf{y}) \to \cos \theta$ is a function from $\mathbb{R}^2 \times \mathbb{R}^2$ to $\mathbb{R}$.

Analogous formulas are used to measure length, distance, and angles in $\mathbb{R}^3$.

1. The length of a vector $\mathbf{x}$ from the origin $(0, 0, 0)$ to the point (x_1, x_2, x_3) is measured by the formula

$$\|\mathbf{x}\| = \sqrt{x_1^2 + x_2^2 + x_3^2}$$

2. The distance between $\mathbf{x}$ and a second vector $\mathbf{y}$ from $(0, 0, 0)$ to the point (y_1, y_2, y_3) is measured by the formula

$$\|\mathbf{x} - \mathbf{y}\| = \sqrt{(x_1 - y_1)^2 + (x_2 - y_2)^2 + (x_3 - y_3)^2}$$

3. The cosine of the angle θ between $\mathbf{x}$ and $\mathbf{y}$ is measured by the formula

$$\cos\theta = \frac{\mathbf{x}^T \mathbf{y}}{\|\mathbf{x}\| \, \|\mathbf{y}\|}$$

The scalar $\mathbf{x}^T\mathbf{y}$ in the cosine formulas is of course the familiar *dot product* $\mathbf{x} \cdot \mathbf{y}$ of $\mathbf{x}$ and $\mathbf{y}$. One of the objectives of this chapter is to generalize the dot product and thereby expand the applicability of geometric ideas.

A remarkable fact about the dot product is that length, distance, and angles can all be defined in terms of it. Consider two points $\mathbf{x} = (x_1, x_2)$ and $\mathbf{y} = (y_1, y_2)$ in $\mathbb{R}^2$. The products $\mathbf{x} \cdot \mathbf{x}$ and $\mathbf{y} \cdot \mathbf{y}$ are equal to $x_1 x_1 + x_2 x_2 = x_1^2 + x_2^2$ and $y_1 y_1 + y_2 y_2 = y_1^2 + y_2^2$. This means that the length of $\mathbf{x}$ is $\sqrt{\mathbf{x} \cdot \mathbf{x}}$ and that of $\mathbf{y}$ is $\sqrt{\mathbf{y} \cdot \mathbf{y}}$. The cosine

$$\frac{\mathbf{x} \cdot \mathbf{y}}{\|\mathbf{x}\| \, \|\mathbf{y}\|} = \frac{\mathbf{x} \cdot \mathbf{y}}{\sqrt{\mathbf{x} \cdot \mathbf{x}} \sqrt{\mathbf{y} \cdot \mathbf{y}}}$$

of the angle between $\mathbf{x}$ and $\mathbf{y}$ is therefore completely determined by the dot product. Moreover, the distance $d(\mathbf{x}, \mathbf{y})$ between $\mathbf{x}$ and $\mathbf{y}$ is simply the square root of the dot product

$$(\mathbf{x} - \mathbf{y}) \cdot (\mathbf{x} - \mathbf{y})$$

In spite of its beauty and simplicity, the dot product fails to provide the right geometry for certain applications. Fortunately, it can be generalized and other geometries can be developed. We will show that the dot product is just one of many examples of so-called *inner products* on real vector spaces. All of them can be used to generalize the ideas of length, distance, and angles. On the other hand, it is also possible to define length and distance for vectors independently of inner products. We therefore begin this chapter by exploring these concepts in this more general context. At the same time, we begin the study of the quantitative effect of matrix transformations. We show that we can assign different measures $\|A\|$ to a matrix A that give us a good deal of information about how the transformation T_A changes the form of geometric objects. We will describe four such matrix measures, three of which are based on analogous measures for row and column vectors.

NORMS

Our goal in this section is to define the size for vectors and matrices. We refer to these measures collectively as *norms*.

Vector norms

We begin with the definition of vector norms.

DEFINITION 7.1   *A **vector norm** is a function* $\mathbf{x} \to \|\mathbf{x}\|$ *on a vector space V that assigns a nonnegative real number* $\|\mathbf{x}\|$ *to every vector* $\mathbf{x} \in V$ *and has the following properties.*

$$
\begin{array}{rlcl}
1. & \|\mathbf{x}\| & > & 0 \text{ if } \mathbf{x} \neq \mathbf{0} \\
2. & \|\mathbf{x}\| & = & 0 \text{ if } \mathbf{x} = \mathbf{0} \\
3. & \|a\mathbf{x}\| & = & |a|\,\|\mathbf{x}\| \\
4. & \|\mathbf{x} + \mathbf{y}\| & \leq & \|\mathbf{x}\| + \|\mathbf{y}\|
\end{array}
$$

The inequality $\|\mathbf{x} + \mathbf{y}\| \leq \|\mathbf{x}\| + \|\mathbf{y}\|$ is known as the **triangle inequality**. A **normed vector space** is a vector space equipped with a vector norm.

The following are basic examples of norms on finite-dimensional real vector spaces.

EXAMPLE 7.1   ■ The One-Norm on $\mathbb{R}^n$

The function defined by $\|\mathbf{x}\|_1 = |x_1| + \cdots + |x_n|$ on the vector $\mathbf{x} = (x_1, \ldots, x_n) \in \mathbb{R}^n$ is a vector norm on $\mathbb{R}^n$. The absolute values $|x_i|$ are zero if $x_i = 0$ and positive otherwise. Hence $\|\mathbf{x}\|_1$ satisfies the first and second axioms. The properties of the absolute value function also imply that

$$
\begin{aligned}
\|a\mathbf{x}\|_1 & = |ax_1| + \cdots + |ax_n| \\
& = |a|\,(|x_1| + \cdots + |x_n|) \\
& = |a|\,\|\mathbf{x}\|_1
\end{aligned}
$$

Therefore, the function $\|\mathbf{x}\|_1$ satisfies the third axiom. It remains to verify the triangle inequality. Consider any two vectors $\mathbf{x} = (x_1, \ldots, x_n)$ and $\mathbf{y} = (y_1, \ldots, y_n)$. By the definition of the function $\|\cdot\|_1$ and by the triangle inequality for absolute values,

$$
\begin{aligned}
\|\mathbf{x} + \mathbf{y}\|_1 & = |x_1 + y_1| + \cdots + |x_n + y_n| \\
& \leq |x_1| + \cdots + |x_n| + |y_1| + \cdots + |y_n| \\
& = \|\mathbf{x}\|_1 + \|\mathbf{y}\|_1
\end{aligned}
$$

Therefore, the function $\mathbf{x} \to \|\mathbf{x}\|_1$ is a vector norm on $\mathbb{R}^n$. ◄►

EXAMPLE 7.2   ■ The Two-Norm on $\mathbb{R}^n$

Let $\mathbf{x} = (x_1, \ldots, x_n)$ be a vector in $\mathbb{R}^n$ and let $\mathbf{x} \cdot \mathbf{x} = x_1^2 + \cdots + x_n^2$ be the dot product of $\mathbf{x}$ with itself. Then the function $\|\mathbf{x}\|_2 = \sqrt{\mathbf{x} \cdot \mathbf{x}}$ is a vector norm on $\mathbb{R}^n$. The first two axioms are satisfied since we are taking the square root and since $x_1^2 + \cdots + x_n^2 = 0$ is zero if and only if all coordinates x_i are zero. Moreover, since $a\mathbf{x} = (ax_1, \ldots, ax_n)$, we have

$$
\begin{aligned}
\|a\mathbf{x}\|_2 & = \sqrt{(ax_1)^2 + \cdots + (ax_n)^2} \\
& = \sqrt{a^2 x_1^2 + \cdots + a^2 x_n^2}
\end{aligned}
$$

$$= |a| \sqrt{x_1^2 + \cdots + x_n^2}$$
$$= |a| \, \|\mathbf{x}\|_2$$

Hence the third axiom is satisfied. Theorem 7.6 guarantees that the triangle inequality also holds. ◄►

The norm $\|\mathbf{x}\|_2$ is also called the *Euclidean norm* on $\mathbb{R}^n$, and the space $\mathbb{R}^n$, equipped with the Euclidean norm, is called the *Euclidean n-space*.

EXAMPLE 7.3 ■ **The Infinity-Norm on $\mathbb{R}^n$**

Let $\mathbf{x} = (x_1, \dots, x_n)$ be a vector in $\mathbb{R}^n$. Then the function $\|\mathbf{x}\|_\infty = \max(|x_1|, \dots, |x_n|)$ is a vector norm on $\mathbb{R}^n$. The properties of the absolute value function guarantee that the function $\|\mathbf{x}\|_\infty$ satisfies the first two axioms. Moreover, $a\mathbf{x} = (ax_1, \dots, ax_n)$. Therefore,

$$\|a\mathbf{x}\|_\infty = \max(|ax_1|, \dots, |ax_n|)$$
$$= \max(|a| \, |x_1|, \dots, |a| \, |x_n|)$$
$$= |a| \max(|x_1|, \dots, |x_n|)$$
$$= |a| \, \|\mathbf{x}\|_\infty$$

Hence the third axiom holds. In addition, it follows from the triangle inequality for absolute values and the properties of the max function that for all vectors $\mathbf{x} = (x_1, \dots, x_n)$ and $\mathbf{y} = (y_1, \dots, y_n)$ in $\mathbb{R}^n$,

$$\max(|x_1 + y_1|, \dots, |x_n + y_n|) \le \max(|x_1| + |y_1|, \dots, |x_n| + |y_n|)$$
$$\le \max(|x_1|, \dots, |x_n|) + \max(|y_1|, \dots, |y_n|)$$

Hence the triangle inequality $\|\mathbf{x} + \mathbf{y}\|_\infty \le \|\mathbf{x}\|_\infty + \|\mathbf{y}\|_\infty$ holds. ◄►

We can get an idea of the geometric properties of these norms by examining the shapes of the unit spheres determined by them. A *unit vector* is any vector $\mathbf{x}$ for which $\|\mathbf{x}\| = 1$, and the *unit sphere* determined by a norm is the set of all unit vectors of that space. In the two-dimensional case, we usually refer to a unit sphere as a *unit circle*.

EXAMPLE 7.4 ■ **Three Unit Circles in $\mathbb{R}^2$**

Draw the unit circles $\{\mathbf{x} : \|\mathbf{x}\|_1 = 1\}$, $\{\mathbf{x} : \|\mathbf{x}\|_2 = 1\}$, and $\{\mathbf{x} : \|\mathbf{x}\|_\infty = 1\}$ in $\mathbb{R}^2$.

Solution. We use *MATHEMATICA* to plot the sets for which

$$|x| + |y| = 1 \quad x^2 + y^2 = 1 \quad \max(|x|, |y|) = 1$$

```
In[1]:= <<Graphics`ImplicitPlot`
```

```
In[2]:= ImplicitPlot[{x²+y²==1, Abs[x]+Abs[y]==1,
Max[Abs[x],Abs[y]]==1}, {x,-1,1}, {y,-1,1}]
```

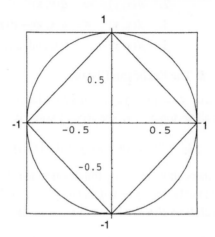

```
Out[2]= - Graphics -
```

EXERCISES 7.1

1. Compute the one-norm, two-norm, and infinity-norm of the following vectors in $\mathbb{R}^2$.

   a. $\mathbf{x} = (3, 4)$ b. $\mathbf{x} = (0, 0)$ c. $\mathbf{x} = (-3, 4)$
   d. $\mathbf{x} = (1, 0)$ e. $\mathbf{x} = (-9, 11)$ f. $\mathbf{x} = (-9, -11)$

2. Compute the one-norm, two-norm, and infinity-norm of the following vectors in $\mathbb{R}^3$.

   a. $\mathbf{x} = (3, 4, 5)$ b. $\mathbf{x} = (0, 0, 0)$ c. $\mathbf{x} = (-3, 4, 5)$
   d. $\mathbf{x} = (1, 0, 1)$ e. $\mathbf{x} = (0, 0, 1)$ f. $\mathbf{x} = (-9, -11, -5)$

Distance

If V is a vector space equipped with a vector norm $\mathbf{x} \rightarrow \|\mathbf{x}\|$ and if $\mathbf{x}$ and $\mathbf{y}$ are two vectors in V, then the function

$$d(\mathbf{x}, \mathbf{y}) = \|\mathbf{x} - \mathbf{y}\|$$

defines the distance between $\mathbf{x}$ and $\mathbf{y}$ in the given vector norm. This function is an example of a whole family of functions satisfying the following definition.

DEFINITION 7.2   *A **distance function** on a vector space V is a function that assigns a nonnegative real number $d(\mathbf{x}, \mathbf{y})$ to every pair $(\mathbf{x}, \mathbf{y})$ of vectors, and has the following properties.*

$$
\begin{aligned}
1. \quad & d(\mathbf{x}, \mathbf{y}) \; \geq \; 0 \\
2. \quad & d(\mathbf{x}, \mathbf{y}) \; = \; d(\mathbf{y}, \mathbf{x}) \\
3. \quad & d(\mathbf{x}, \mathbf{y}) \; \leq \; d(\mathbf{x}, \mathbf{z}) + d(\mathbf{z}, \mathbf{y}) \\
4. \quad & d(\mathbf{x}, \mathbf{y}) \; = \; 0 \text{ if and only if } \mathbf{x} = \mathbf{y}
\end{aligned}
$$

EXAMPLE 7.5   ■ **The Euclidean Distance**

The function $d(\mathbf{x}, \mathbf{y}) = \|\mathbf{x} - \mathbf{y}\|_2$ is a distance function. The first axiom is satisfied since $\|\mathbf{x} - \mathbf{y}\|_2 \geq 0$ for all vectors $\mathbf{x} - \mathbf{y}$. The second axiom is satisfied since

$$
\|\mathbf{x} - \mathbf{y}\|_2 = \|(-1)(\mathbf{y} - \mathbf{x})\|_2 = |(-1)| \, \|(\mathbf{y} - \mathbf{x})\|_2 = \|(\mathbf{y} - \mathbf{x})\|_2
$$

The triangle inequality for $\|-\|_2$ implies that the third axiom is satisfied. Moreover, $\|\mathbf{x} - \mathbf{y}\|_2 = 0$ if and only if $\mathbf{x} - \mathbf{y} = \mathbf{0}$ if and only if $\mathbf{x} = \mathbf{y}$. Hence $\|-\|_2$ is a distance function. It is called the ***Euclidean distance*** between $\mathbf{x}$ and $\mathbf{y}$. ◄►

EXAMPLE 7.6   ■ **The Euclidean Distance in $\mathbb{R}^2$**

Use *MATHEMATICA* to find the Euclidean distance between the vectors $\mathbf{x} = (1, 2)$ and $\mathbf{y} = (3, 4)$.

Solution. By definition, $d(\mathbf{x}, \mathbf{y}) = \|\mathbf{x} - \mathbf{y}\|_2 = \sqrt{(1 - 3)^2 + (2 - 4)^2}$.

```
In[1]:= d[x,y]=√(1-3)²+(2-4)²

Out[1]= 2√2
```

Thus the Euclidean distance between $\mathbf{x}$ and $\mathbf{y}$ is $2\sqrt{2}$. ◄►

EXAMPLE 7.7   ■ **The One-Norm Distance Between Two Vectors in $\mathbb{R}^2$**

Use *MATHEMATICA* to find the one-norm distance between the vectors $\mathbf{x} = (1, 2)$ and $\mathbf{y} = (3, 4)$ in $\mathbb{R}^2$.

Solution. By definition, $d(\mathbf{x}, \mathbf{y}) = \|\mathbf{x} - \mathbf{y}\|_1 = |1 - 3| + |2 - 4|$.

```
In[1]:= d[x,y]=Abs[1-3]+Abs[2-4]

Out[1]= 4
```

Thus the distance between $\mathbf{x}$ and $\mathbf{y}$ in the one-norm is 4. ◄►

EXAMPLE 7.8 ■ **The Infinity-Norm Distance Between Two Vectors in $\mathbb{R}^2$**

Use *MATHEMATICA* to find the distance between the vectors $\mathbf{x} = (1, 2)$ and $\mathbf{y} = (3, 4)$ in the infinity-norm.

Solution. By definition, $d(\mathbf{x}, \mathbf{y}) = \|\mathbf{x} - \mathbf{y}\|_\infty = \max(|1 - 3|, |2 - 4|)$.

```
In[1]:= d[x,y]=Max[Abs[1-3],Abs[2-4]]

Out[1]= 2
```

Thus the distance between $\mathbf{x}$ and $\mathbf{y}$ in the infinity-norm is 2. ◄►

EXERCISES 7.2

1. Find the one-norm distance between the following pairs of vectors in $\mathbb{R}^2$.

   a. $\mathbf{x} = (4, 5)$ and $\mathbf{y} = (1, 2)$ b. $\mathbf{x} = (4, 5)$ and $\mathbf{y} = (-1, 2)$
   c. $\mathbf{x} = (4, 5)$ and $\mathbf{y} = (0, -2)$ d. $\mathbf{x} = (3, 0)$ and $\mathbf{y} = (4, 5)$
   e. $\mathbf{x} = (1, 2)$ and $\mathbf{y} = (0, 0)$ f. $\mathbf{x} = (1, 0)$ and $\mathbf{y} = (0, 1)$

2. Find the two-norm distance between the pairs of vectors in Exercise 1.

3. Find the infinity-norm distance between the pairs of vectors in Exercise 1.

4. Find the two-norm distance between the following pairs of vectors in $\mathbb{R}^3$.

   a. $\mathbf{x} = (1, 2, 3)$ and $\mathbf{y} = (4, 5, 7)$ b. $\mathbf{x} = (4, 5, 7)$ and $\mathbf{y} = (-1, 2, 0)$
   c. $\mathbf{x} = (4, 5, 7)$ and $\mathbf{y} = (0, -2, -3)$ d. $\mathbf{x} = (1, 0, 0)$ and $\mathbf{y} = (4, 5, 7)$
   e. $\mathbf{x} = (1, 2, 3)$ and $\mathbf{y} = (0, 0, 0)$ f. $\mathbf{x} = (1, 0, 0)$ and $\mathbf{y} = (0, 0, 1)$

5. Find the infinity-norm distance between the pairs of vectors in Exercise 4.

6. Show that the function

$$d(\mathbf{x}, \mathbf{y}) = \begin{cases} 1 & \text{if } \mathbf{x} \neq \mathbf{y} \\ 0 & \text{if } \mathbf{x} = \mathbf{y} \end{cases}$$

   satisfies the axioms of a distance function.

7. Show that the attempt to use the distance function $d(\mathbf{x}, \mathbf{y})$ in Exercise 6 to define a norm by putting $\|\mathbf{x}\| = d(\mathbf{x}, \mathbf{0})$ fails.

8. In his special relativity theory, Einstein defined a dependence between space, $\mathbb{R}^3$, and time, $\mathbb{R}$, by combining the two quantities in the vector space $\mathbb{R}^4$. The theory is based on the star

function $\star$, defined by $\mathbf{x} \star \mathbf{y} = -xx' - yy' - zz' + tt'$ for any two points $\mathbf{x} = (x, y, z, t)$ and $\mathbf{y} = (x', y', z', t')$ in $\mathbb{R}^4$. Let

$$\|\mathbf{x}\| = \sqrt{|\mathbf{x} \star \mathbf{x}|} \quad \text{and} \quad d(\mathbf{x}, \mathbf{y}) = \|\mathbf{x} - \mathbf{y}\|$$

Show that $d(\mathbf{x}, \mathbf{y})$ is a distance function, but that $\|\mathbf{x}\|$ fails to be a vector norm.

Matrix Norms

Using the one-, two-, and infinity-norms for vectors, we now develop three measures for the size of an $n \times n$ matrix. These measures will provide us with estimates of the ratios $\|A\mathbf{x}\| / \|\mathbf{x}\|$ of the norms of the image vectors $A\mathbf{x}$ and the norms of the vectors $\mathbf{x}$ in a way that is independent of $\mathbf{x}$. Some general properties of vector norm will be helpful. To prove them, we require a fact established in advanced texts: for each matrix A and each nonzero vector $\mathbf{x}$, the set of real numbers $\left\{ \frac{\|A\mathbf{x}\|}{\|\mathbf{x}\|} : \|\mathbf{x}\| \neq 0 \right\}$ has a maximum.

THEOREM 7.1   *Let A be a real $n \times n$ matrix, and let $\mathbf{x} \in \mathbb{R}^n$. Then*

$$\max \left\{ \frac{\|A\mathbf{x}\|}{\|\mathbf{x}\|} : \|\mathbf{x}\| \neq 0 \right\} = \max \left\{ \|A\mathbf{x}\| : \|\mathbf{x}\| = 1 \right\}$$

Proof.

$$\max \left\{ \frac{\|A\mathbf{x}\|}{\|\mathbf{x}\|} : \|\mathbf{x}\| \neq 0 \right\} = \max \left\{ \frac{1}{\|\mathbf{x}\|} \|A\mathbf{x}\| : \|\mathbf{x}\| \neq 0 \right\}$$

$$= \max \left\{ \left\| \frac{1}{\|\mathbf{x}\|} (A\mathbf{x}) \right\| : \|\mathbf{x}\| \neq 0 \right\}$$

$$= \max \left\{ \left\| A \frac{\mathbf{x}}{\|\mathbf{x}\|} \right\| : \|\mathbf{x}\| \neq 0 \right\}$$

$$= \max \left\{ \|A\mathbf{x}\| : \left\| \frac{\mathbf{x}}{\|\mathbf{x}\|} \right\| \neq 0 \right\}$$

$$= \max \left\{ \|A\mathbf{x}\| : \|\mathbf{x}\| = 1 \right\}$$

This proves the theorem. ∎

A geometric interpretation of this result tells us that

$$\max \left\{ \frac{\|A\mathbf{x}\|}{\|\mathbf{x}\|} : \|\mathbf{x}\| \neq \mathbf{0} \right\}$$

is the maximum stretching factor obtained when multiplying the vectors on the unit sphere by the matrix A.

Since the composition of matrix transformations, and therefore matrix multiplication, plays a major role in linear algebra, it is reasonable to assume that any measure of the size of a matrix should be stable for products. The size $\|AB\|$ of the product of two matrices should be no larger than the product $\|A\| \|B\|$ of the size $\|A\|$ of A and the size $\|B\|$ of B. This consideration leads us to the definition of a matrix norm.

DEFINITION 7.3   *A **matrix norm** on a space V of square matrices is a function $A \to \|A\|$ that assigns to each matrix A a nonnegative real number and satisfies the following properties.*

1. $\|A\| \geq 0$
2. $\|A\| = 0$ only if $A = \mathbf{0}$
3. $\|cA\| = |c|\,\|A\|$
4. $\|A + B\| \leq \|A\| + \|B\|$
5. $\|AB\| \leq \|A\|\,\|B\|$

It is not unexpected that for any vector norm $\|A\mathbf{x}\|$, the function $\max\{\|A\mathbf{x}\| : \|\mathbf{x}\| = 1\}$ is a matrix norm.

THEOREM 7.2 (Matrix norm theorem)   *The function $\|A\| = \max\left\{\frac{\|A\mathbf{x}\|}{\|\mathbf{x}\|} : \|\mathbf{x}\| \neq \mathbf{0}\right\} = \max\{\|A\mathbf{x}\| : \|\mathbf{x}\| = 1\}$ is a matrix norm.*

Proof. Since $\|A\mathbf{x}\|$ is a vector norm, $\|A\mathbf{x}\| \geq 0$. Moreover, since $\|\mathbf{x}\| = 1$, the quantity $\|A\mathbf{x}\| = 0$ only if $A\mathbf{x} = \mathbf{0}$. Hence the first two axioms are satisfied. The third axiom is satisfied since

$$\|cA\| = \max\{\|(cA)\,\mathbf{x}\| : \|\mathbf{x}\| = 1\}$$
$$= \max\{\|c\,(A\mathbf{x})\| : \|\mathbf{x}\| = 1\}$$
$$= \max\{|c|\,\|A\mathbf{x}\| : \|\mathbf{x}\| = 1\}$$
$$= |c|\max\{\|A\mathbf{x}\| : \|\mathbf{x}\| = 1\}$$
$$= |c|\,\|A\|$$

Furthermore, $\|A + B\| = \max\{\|(A + B)\,\mathbf{x}\| : \|\mathbf{x}\| = 1\}$. Let $\mathbf{y}$ be a vector that causes the right-hand side to attain a maximum. Then

$$\|A + B\| = \|(A + B)\,\mathbf{y}\|$$
$$= \|A\mathbf{y} + B\mathbf{y}\|$$
$$\leq \|A\mathbf{y}\| + \|B\mathbf{y}\|$$
$$\leq \|A\|\,\|\mathbf{y}\| + \|B\|\,\|\mathbf{y}\|$$
$$= \|A\| + \|B\|$$

Hence the fourth axiom is satisfied. Finally, consider the quantity $\|AB\| = \max\{\|(AB)\,\mathbf{x}\| = 1\}$. Suppose that $\|(AB)\,\mathbf{x}\|$ is a maximum at some vector $\mathbf{y}$. Then

$$\|AB\| = \|(AB)\,\mathbf{y}\|$$
$$= \|A\,(B\mathbf{y})\|$$
$$\leq \|A\|\,\|B\mathbf{y}\|$$
$$\leq \|A\|\,\|B\|\,\|\mathbf{y}\| = \|A\|\,\|B\|$$

Therefore, the function $\max\{\|A\mathbf{x}\| : \|\mathbf{x}\| = 1\}$ is a matrix norm. ∎

EXAMPLE 7.9   ■ **Stretching a Unit Circle in $\mathbb{R}^2$**

Use *MATHEMATICA* to graph the set $W = \left\{ A\mathbf{x} \in \mathbb{R}^2 : \|\mathbf{x}\|_2 = 1 \right\}$ for

$$A = \begin{pmatrix} 2 & 0 \\ 0 & 3 \end{pmatrix}$$

Solution. Let $\mathbf{x} = (x, y)$ be a point in $\mathbb{R}^2$ whose Euclidean vector norm $\|\mathbf{x}\|_2 = \sqrt{x^2 + y^2}$ is 1. Then $x^2 + y^2 = 1$. Furthermore, the image $A\mathbf{x}$ of $\mathbf{x}$ is

$$\begin{pmatrix} 2 & 0 \\ 0 & 3 \end{pmatrix} \begin{pmatrix} x \\ y \end{pmatrix} = \begin{pmatrix} 2x \\ 3y \end{pmatrix}$$

This means that

$$W = \left\{ \begin{pmatrix} 2x \\ 3y \end{pmatrix} : x^2 + y^2 = 1 \right\} = \left\{ \begin{pmatrix} x \\ y \end{pmatrix} : 9x^2 + 4y^2 = 36 \right\}$$

We use *MATHEMATICA* to plot W.

```
In[1]:= <<Graphics`ImplicitPlot`
```

```
In[2]:= ImplicitPlot[9x²+4y²==36,{x,-2,2}]
```

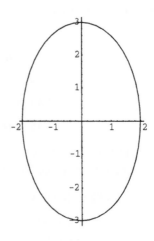

```
Out[2]= - Graphics -
```

As we can see, the set W forms an ellipse. The longest vector in W is 3 units long. Thus

$$\|A\| = \left\| \begin{pmatrix} 2 & 0 \\ 0 & 3 \end{pmatrix} \right\| = 3$$

In this case, the norm of A corresponds to the maximum stretch of the unit circle in the direction of the y-axis. ◄►

We now define the four matrix norms that are directly linked to the three vector norms discussed earlier. They are called the one-norm $\|A\|_1$, the two-norm $\|A\|_2$, the infinity-norm $\|A\|_\infty$, and the Frobenius norm $\|A\|_F$. We omit the verification that these functions are indeed matrix norms.

To be able to compare the vector and matrix forms of these norms, we treat column vectors of height n as $n \times 1$ matrices. Suppose that

$$A = \begin{pmatrix} a_{11} & a_{12} \\ a_{21} & a_{22} \end{pmatrix}$$

is a given 2×2 matrix and that $\mathbf{x}$ is the first column of A. In Table 1 we compare the one-, two-, and infinity-norms for $\mathbf{x}$ and the corresponding matrix norms for A.

TABLE 1   Vector and Matrix Norms.

Norms	Vectors in $\mathbb{R}^{2 \times 1}$	Matrices in $\mathbb{R}^{2 \times 2}$
$\|A\|_1$	$\|a_{11}\| + \|a_{21}\|$	$\max\{\|a_{11}\| + \|a_{21}\|, \|a_{12}\| + \|a_{22}\|\}$
$\|A\|_2$	$\left(a_{11}^2 + a_{21}^2\right)^{1/2}$	$\max\left\{\sqrt{\lambda_1}, \sqrt{\lambda_2}\right\}$, where λ_i is an eigenvalue of $A^T A$
$\|A\|_\infty$	$\max\{\|a_{11}\|, \|a_{21}\|\}$	$\max\{\|a_{11}\| + \|a_{12}\|, \|a_{21}\| + \|a_{22}\|\}$
$\|A\|_F$	$\left(a_{11}^2 + a_{21}^2\right)^{1/2}$	$\left(a_{11}^2 + a_{12}^2 + a_{21}^2 + a_{22}^2\right)^{1/2}$

Each matrix norm $\|A\|_i$ is defined as the maximum of the vector norms $\|A\mathbf{x}\|_i$ on the unit sphere $\{\mathbf{x} : \|\mathbf{x}\|_i = 1\}$. The scalars $\sigma_i = \sqrt{\lambda_i}$ are called the *singular values* of A.

1. The *one-norm* $\|A\|_1$ of A is the maximum of the absolute value sums of the columns of A. If A is a one-column vector, then $\|A\|_1$ corresponds to the one-norm of the vector A.

2. The *two-norm* $\|A\|_2$ of A is the largest singular value of A. Chapter 9 deals in depth with such values.

3. The *infinity-norm* $\|A\|_\infty$ of A is the maximum of the absolute value sums of the rows of A. If A is a one-column matrix, then $\|A\|_\infty$ corresponds to the infinity-norm of the vector A.

4. The *Frobenius norm* $\|A\|_F$ of A is the square root of the sum of the squares of the entries of A. It is also known as the *two-norm*. If A is a one-column vector, then counterpart of $\|A\|_F$ is the two-norm of A. It is also equal to the square root of the trace of $A^T A$.

The general rules for computing these matrix norms are analogous to those described for the 2×2 case.

EXAMPLE 7.10 ■ The Two-Norm of a Matrix

Use *MATHEMATICA* to compute the two-norm of the matrix

$$A = \begin{pmatrix} 1 & 3 \\ 0 & 2 \end{pmatrix}$$

Solution. We use *MATHEMATICA* to compute the eigenvalues of $A^T A$.

$\text{In[1]:= A=} \begin{pmatrix} 1 & 3 \\ 0 & 2 \end{pmatrix}$

$\text{Out[1]= } \{\{1,3\},\{0,2\}\}$

$\text{In[2]:= Eigenvalues[Transpose[A].A]}$

$\text{Out[2]= } \{7-3\sqrt{5}, 7+3\sqrt{5}\}$

This means that the singular values of A are $\sigma_1 = \sqrt{\lambda_1} = \sqrt{7 + 3\sqrt{5}}$ and $\sigma_2 = \sqrt{\lambda_2} = \sqrt{7 - 3\sqrt{5}}$. Since $\sigma_1 > \sigma_2$, the function $\|A\|_2 = \sigma_1$ is the two-norm of A. ◄►

EXAMPLE 7.11 ■ The One-Norm and the Infinity-Norm of a Matrix

Compute $\|A\|_1$ and $\|A\|_\infty$ for the matrix

$$A = \begin{pmatrix} 3 & -1 & 0 \\ 2 & 0 & 1 \\ -3 & 5 & 1 \end{pmatrix}$$

Solution. It is clear from our discussion that

$$\|A\|_1 = \max\{|3| + |2| + |-3|, |-1| + |0| + |5|, |0| + |1| + |1|\} = \max\{8, 6, 2\} = 8$$

and

$$\|A\|_\infty = \max\{|3| + |-1| + |0|, |2| + |0| + |1|, |-3| + |5| + |1|\} = \max\{4, 3, 9\} = 9$$

as expected. ◄►

EXAMPLE 7.12 The Frobenius Norm of a Matrix

Use *MATHEMATICA* to find the Frobenius norm $\|A\|_F$ of the matrix

$$A = \begin{pmatrix} 9 & -3 \\ -2 & 1 \end{pmatrix}$$

Solution. We first compute the trace of $A^T A$.

```
In[1]:= A=  ( 9  -3 );
          ( -2  1 )
```

Next we compute the trace of $A^T A$.

```
In[2]:= Tr[Transpose[A].A]

Out[2]= 95
```

Therefore, $\|A\|_F = \sqrt{95}$ is the Frobenius norm of A. ◄►

EXERCISES 7.3

1. Let $S_1 = \{x_1 = (2, 3), x_2 = (5, -6), x_3 = (1, 1), x_4 = (0, 0), x_5 = (9, -2)\}$ be a set of vectors in $\mathbb{R}^2$ and let

$$A = \begin{pmatrix} 3 & 4 \\ -1 & 2 \end{pmatrix}$$

Compute max $\{\|Ax\|_1 : x \in S_1\}$ in $\mathbb{R}^2$.

2. Repeat Exercise 1 using the two-norm.

3. Repeat Exercise 1 using the infinity-norm.

4. Find the one-norm, two-norm, infinity-norm, and the Frobenius norm of the following matrices.

a. $\begin{pmatrix} 3 & 4 \\ -1 & 2 \end{pmatrix}$ b. $\begin{pmatrix} 3 & 0 \\ 0 & 2 \end{pmatrix}$ c. $\begin{pmatrix} 1 & 0 \\ 0 & 1 \end{pmatrix}$

d. $\begin{pmatrix} 1 & -1 & 0 \\ 0 & 1 & 0 \\ 0 & 0 & 1 \end{pmatrix}$ e. $\begin{pmatrix} -3 & 2 & 0 \\ 2 & -3 & 0 \\ 0 & 0 & 4 \end{pmatrix}$ f. $\begin{pmatrix} -3 & 2 & 4 \\ 2 & -3 & 1 \\ 1 & 0 & 4 \end{pmatrix}$

5. Show that the Frobenius norm of the matrix $A = \begin{pmatrix} a & 2 \\ 3 & 4 \end{pmatrix}$ is less than the Frobenius

norm of the matrix $B = \begin{pmatrix} b & 2 \\ 3 & 4 \end{pmatrix}$ if and only if $|a| < |b|$.

6. Use the Frobenius norm to compute the distance between the matrices

$$\begin{pmatrix} 1.4 & 2 \\ 3 & 4 \end{pmatrix} \quad \text{and} \quad \begin{pmatrix} 1.13 & 2 \\ 3 & 4 \end{pmatrix}$$

7. Show that the idea of matrix norms can be generalized from square matrices to rectangular ones and give one example each.

REAL INNER PRODUCTS

We now begin the study of inner products and show how they can be used to link the concepts of length, distance, and angles. In the case of matrices, the trace will play a key role in the specification of an important inner product.

DEFINITION 7.4   *An **inner product** on a real vector space V is a real-valued function* $(\mathbf{x}, \mathbf{y}) \to \langle \mathbf{x}, \mathbf{y} \rangle$, *satisfying the following four axioms.*

1. $\langle a\mathbf{x} + b\mathbf{y}, \mathbf{z} \rangle = a \langle \mathbf{x}, \mathbf{z} \rangle + b \langle \mathbf{y}, \mathbf{z} \rangle$
2. $\langle \mathbf{x}, \mathbf{y} \rangle = \langle \mathbf{y}, \mathbf{x} \rangle$
3. $\langle \mathbf{x}, \mathbf{x} \rangle \geq 0$
4. $\langle \mathbf{x}, \mathbf{x} \rangle = 0$ if and only if $\mathbf{x} = \mathbf{0}$

A **real inner product space** is a real vector space V equipped with a real inner product $(\mathbf{x}, \mathbf{y}) \to \langle \mathbf{x}, \mathbf{y} \rangle$.

Inner products are **bilinear** because they are linear in each of their two variables. The linearity $\langle a\mathbf{x} + b\mathbf{y}, \mathbf{z} \rangle = a \langle \mathbf{x}, \mathbf{z} \rangle + b \langle \mathbf{y}, \mathbf{z} \rangle$ in the first variable is part of the definition. The linearity in the second variable follows from their **symmetry**. The fact that $\langle \mathbf{x}, \mathbf{y} \rangle = \langle \mathbf{y}, \mathbf{x} \rangle$ entails that

$$\langle \mathbf{u}, a\mathbf{v} + b\mathbf{w} \rangle = \langle a\mathbf{v} + b\mathbf{w}, \mathbf{u} \rangle = a\langle \mathbf{v}, \mathbf{u} \rangle + b\langle \mathbf{w}, \mathbf{u} \rangle = a\langle \mathbf{u}, \mathbf{v} \rangle + b\langle \mathbf{u}, \mathbf{w} \rangle$$

Hence $\langle \mathbf{x}, \mathbf{y} \rangle$ is linear in $\mathbf{y}$.

EXAMPLE 7.13   ■ **The Standard Inner Product on $\mathbb{R}^n$**

Show that the function

$$\mathbf{x}^T \mathbf{y} = (x_1 \cdots x_n) \begin{pmatrix} y_1 \\ \vdots \\ y_n \end{pmatrix} = x_1 y_1 + \cdots + x_n y_n$$

is an inner product on $\mathbb{R}^n$.

Solution. Let $\mathbf{x}$ and $\mathbf{y}$ be two vectors in $\mathbb{R}^n$. Then

$$
\begin{aligned}
\langle a\mathbf{x} + b\mathbf{y}, \mathbf{z} \rangle &= (a\mathbf{x} + b\mathbf{y})^T \mathbf{z} \\
&= \left((a\mathbf{x})^T + (b\mathbf{y})^T \right) \mathbf{z} \\
&= \left(a\mathbf{x}^T + b\mathbf{y}^T \right) \mathbf{z} \\
&= a\mathbf{x}^T \mathbf{z} + b\mathbf{y}^T \mathbf{z} \\
&= a\langle \mathbf{x}, \mathbf{z} \rangle + b\langle \mathbf{y}, \mathbf{z} \rangle
\end{aligned}
$$

Therefore, $\mathbf{x}^T \mathbf{y}$ is linear in the first variable. The calculation

$$
\begin{aligned}
\mathbf{x}^T \mathbf{y} &= x_1 y_1 + \cdots + x_n y_n \\
&= y_1 x_1 + \cdots + y_n x_n \\
&= \mathbf{y}^T \mathbf{x} = (\mathbf{x}^T \mathbf{y})^T
\end{aligned}
$$

shows that $\mathbf{x}^T \mathbf{y}$ is also symmetric. Moreover, $\langle \mathbf{x}, \mathbf{x} \rangle = \mathbf{x}^T \mathbf{x} = x_1^2 + \cdots + x_n^2 \geq 0$ for all $\mathbf{x}$, and is 0 if and only if $x_i = 0$ for all i. Hence the function $\mathbf{x}^T \mathbf{y}$ is an inner product on $\mathbb{R}^n$. ◄►

The inner product in this example has various names. In addition to being called the *dot product*, it is also known as the *standard inner product* and the *Euclidean inner product*. In *MATHEMATICA*, it is called the dot product.

EXAMPLE 7.14   ■ A Nonstandard Inner Product on $\mathbb{R}^2$

Show that the function $\langle \mathbf{x}, \mathbf{y} \rangle = 2x_1 y_1 + 3x_2 y_2$ is an inner product on $\mathbb{R}^2$, where $\mathbf{x} = (x_1, x_2)$ and $\mathbf{y} = (y_1, y_2)$.

Solution. Let $\mathbf{z} = (z_1, z_2)$ be an arbitrary third vector in $\mathbb{R}^2$, and consider the arbitrary functional value $\langle a\mathbf{x} + b\mathbf{y}, \mathbf{z} \rangle$, with

$$
a\mathbf{x} + b\mathbf{y} = \begin{pmatrix} ax_1 \\ ax_2 \end{pmatrix} + \begin{pmatrix} by_1 \\ by_2 \end{pmatrix} = \begin{pmatrix} ax_1 + by_1 \\ ax_2 + by_2 \end{pmatrix}
$$

By the laws of arithmetic,

$$
\begin{aligned}
\langle a\mathbf{x} + b\mathbf{y}, \mathbf{z} \rangle &= 2(ax_1 + by_1)z_1 + 3(ax_2 + by_2)z_2 \\
&= 2ax_1 z_1 + 3ax_2 z_2 + 2by_1 z_1 + 3by_2 z_2 \\
&= a\langle \mathbf{x}, \mathbf{z} \rangle + b\langle \mathbf{y}, \mathbf{z} \rangle
\end{aligned}
$$

Hence $\langle \mathbf{x}, \mathbf{y} \rangle$ is linear in $\mathbf{x}$. The symmetry of $\langle \mathbf{x}, \mathbf{y} \rangle$ follows from the fact that

$$
\langle \mathbf{x}, \mathbf{y} \rangle = 2x_1 y_1 + 3x_2 y_2 = 2y_1 x_1 + 3y_2 x_2 = \langle \mathbf{y}, \mathbf{x} \rangle
$$

Moreover, $\langle \mathbf{x}, \mathbf{x} \rangle = 2x_1 x_1 + 3x_2 x_2 \geq 0$, and is 0 if and only if $x_1 = x_2 = 0$. This shows that the function $\langle \mathbf{x}, \mathbf{y} \rangle$ is an inner product. ◄►

The next example is an important inner product for matrix spaces. Earlier we showed that the trace of the matrix $A^T A$ defines a norm for A. We now show that the trace actually defines an inner product.

EXAMPLE 7.15   ■ The Trace as an Inner Product

Let $A = (a_{ij})$ and $B = (b_{ij})$ be two real $m \times n$ matrices. Then the function

$$\langle A, B \rangle = \text{trace } B^T A$$

is an inner product on $\mathbb{R}^{m \times n}$. It is easy to check that

$$\langle A, B \rangle = \sum_{i \in m} \sum_{j \in n} a_{ij} b_{ij}$$

is the sum of the products of the diagonal entries of $B^T A$. Therefore $\langle A, A \rangle \geq 0$ and is 0 if and only if $A = \mathbf{0}$. Moreover, the distributive and commutative laws $a(b + c) = ab + ac$ and $ab = ba$ for multiplication and addition of real numbers show that $\langle A, B \rangle$ is linear in A and symmetric. Hence $\langle A, B \rangle$ is an inner product. ◀▶

EXAMPLE 7.16   ■ The Trace Inner Product on $\mathbb{R}^{2 \times 3}$

Use *MATHEMATICA* to calculate the inner product trace $B^T A$ of the matrices

$$A = \begin{pmatrix} a_{11} & a_{12} & a_{13} \\ a_{21} & a_{22} & a_{23} \end{pmatrix} \quad \text{and} \quad B = \begin{pmatrix} b_{11} & b_{12} & b_{13} \\ b_{21} & b_{22} & b_{23} \end{pmatrix}$$

Solution. We begin by defining the matrices A and B.

```
In[1]:= A=Array[a,{2,3}]

Out[1]= {{a[1,1],a[1,2],a[1,3]},{a[2,1],a[2,2],a[2,3]}}
```

```
In[2]:= B=Array[b,{2,3}]

Out[2]= {{b[1,1],b[1,2],b[1,3]},{b[2,1],b[2,2],b[2,3]}}
```

Next we compute the transpose of B.

```
In[3]:= Transpose[B]

Out[3]= {{b[1,1],b[2,1]},{b[1,2],b[2,2]},{b[1,3],b[2,3]}}
```

We can now compute the required trace.

```
In[4]:= Tr[Transpose[B].A]

Out[4]= a[1,1]b[1,1]+a[1,2]b[1,2]+a[1,3]b[1,3]

+a[2,1]b[2,1]+a[2,2]b[2,2]+a[2,3]b[2,3]
```

This shows that $\langle A, B \rangle$ is $\sum_{i \in 2} \sum_{j \in 3} a_{ij} b_{ij}$. ◀▶

Every nonzero vector **x** in an inner product space can be converted to a unit vector by dividing it by its norm $\|\mathbf{x}\|$. This process is called ***normalization***. Unit vectors are therefore also called ***normal vectors***.

THEOREM 7.3 (Normalization theorem)   *For every nonzero vector* **x** *in an inner product space* V, *the vector* $\mathbf{y} = \mathbf{x}/\|\mathbf{x}\|$ *is a unit vector.*

Proof. By definition,

$$\|\mathbf{y}\|^2 = \langle \mathbf{y}, \mathbf{y} \rangle = \left\langle \frac{\mathbf{x}}{\|\mathbf{x}\|}, \frac{\mathbf{x}}{\|\mathbf{x}\|} \right\rangle = \frac{1}{\|\mathbf{x}\|} \left\langle \mathbf{x}, \frac{\mathbf{x}}{\|\mathbf{x}\|} \right\rangle = \frac{1}{\|\mathbf{x}\|^2} \langle \mathbf{x}, \mathbf{x} \rangle = \frac{\langle \mathbf{x}, \mathbf{x} \rangle}{\langle \mathbf{x}, \mathbf{x} \rangle} = 1$$

Hence **y** is a unit vector. ∎

The next theorem relates the size of the values of inner products to the size of the vectors involved. It is known as the ***Cauchy-Schwarz inequality***. Its proof uses all four of the defining properties of inner products. We let $\|\mathbf{x}\|$ be the scalar $\sqrt{\langle \mathbf{x}, \mathbf{x} \rangle}$ determined by $\langle \mathbf{x}, \mathbf{y} \rangle$.

THEOREM 7.4 (Cauchy-Schwarz inequality)   *If* $(\mathbf{x}, \mathbf{y}) \to \langle \mathbf{x}, \mathbf{y} \rangle$ *is an inner product on a real vector space* V, *then* $|\langle \mathbf{x}, \mathbf{y} \rangle| \leq \|\mathbf{x}\| \|\mathbf{y}\|$ *for all* $\mathbf{x}, \mathbf{y} \in V$.

Proof. Suppose first that either **x** or **y** is 0. Then either $\|\mathbf{x}\| = 0$ or $\|\mathbf{y}\| = 0$. It follows that $\langle \mathbf{x}, \mathbf{y} \rangle = 0$. Hence $0 = -\|\mathbf{x}\| \|\mathbf{y}\| = \langle \mathbf{x}, \mathbf{y} \rangle = \|\mathbf{x}\| \|\mathbf{y}\| = 0$. This means that $|\langle \mathbf{x}, \mathbf{y} \rangle| = \|\mathbf{x}\| \|\mathbf{y}\|$ and the inequality holds.

Suppose next that neither **x** nor **y** is 0. It follows that $\|\mathbf{x}\| > 0$ and $\|\mathbf{y}\| > 0$. Since $\langle \mathbf{z}, \mathbf{z} \rangle \geq 0$ for all $\mathbf{z} \in V$, the bilinearity and symmetry of inner products therefore implies that

$$\left\langle \frac{\mathbf{x}}{\|\mathbf{x}\|} + \frac{\mathbf{y}}{\|\mathbf{y}\|}, \frac{\mathbf{x}}{\|\mathbf{x}\|} + \frac{\mathbf{y}}{\|\mathbf{y}\|} \right\rangle = \left\langle \frac{\mathbf{x}}{\|\mathbf{x}\|}, \frac{\mathbf{x}}{\|\mathbf{x}\|} \right\rangle + 2 \left\langle \frac{\mathbf{x}}{\|\mathbf{x}\|}, \frac{\mathbf{y}}{\|\mathbf{y}\|} \right\rangle + \left\langle \frac{\mathbf{y}}{\|\mathbf{y}\|}, \frac{\mathbf{y}}{\|\mathbf{y}\|} \right\rangle$$

$$= \frac{1}{\|\mathbf{x}\|^2} \langle \mathbf{x}, \mathbf{x} \rangle + \frac{2}{\|\mathbf{x}\| \|\mathbf{y}\|} \langle \mathbf{x}, \mathbf{y} \rangle + \frac{1}{\|\mathbf{y}\|^2} \langle \mathbf{y}, \mathbf{y} \rangle$$

$$= \frac{1}{\|\mathbf{x}\|^2} \|\mathbf{x}\|^2 + \frac{2}{\|\mathbf{x}\| \|\mathbf{y}\|} \langle \mathbf{x}, \mathbf{y} \rangle + \frac{1}{\|\mathbf{y}\|^2} \|\mathbf{y}\|^2$$

$$= 1 + \frac{2}{\|\mathbf{x}\| \|\mathbf{y}\|} \langle \mathbf{x}, \mathbf{y} \rangle + 1 \geq 0$$

Hence $-\|\mathbf{x}\| \|\mathbf{y}\| \leq \langle \mathbf{x}, \mathbf{y} \rangle$. Similarly,

$$
\left\langle \frac{\mathbf{x}}{\|\mathbf{x}\|} - \frac{\mathbf{y}}{\|\mathbf{y}\|}, \frac{\mathbf{x}}{\|\mathbf{x}\|} - \frac{\mathbf{y}}{\|\mathbf{y}\|} \right\rangle = \left\langle \frac{\mathbf{x}}{\|\mathbf{x}\|}, \frac{\mathbf{x}}{\|\mathbf{x}\|} \right\rangle - 2 \left\langle \frac{\mathbf{x}}{\|\mathbf{x}\|}, \frac{\mathbf{y}}{\|\mathbf{y}\|} \right\rangle + \left\langle \frac{\mathbf{y}}{\|\mathbf{y}\|}, \frac{\mathbf{y}}{\|\mathbf{y}\|} \right\rangle
$$

$$
= \frac{1}{\|\mathbf{x}\|^2} \langle \mathbf{x}, \mathbf{x} \rangle - \frac{2}{\|\mathbf{x}\| \|\mathbf{y}\|} \langle \mathbf{x}, \mathbf{y} \rangle + \frac{1}{\|\mathbf{y}\|^2} \langle \mathbf{y}, \mathbf{y} \rangle
$$

$$
= \frac{1}{\|\mathbf{x}\|^2} \|\mathbf{x}\|^2 - \frac{2}{\|\mathbf{x}\| \|\mathbf{y}\|} \langle \mathbf{x}, \mathbf{y} \rangle + \frac{1}{\|\mathbf{y}\|^2} \|\mathbf{y}\|^2
$$

$$
= 1 - \frac{2}{\|\mathbf{x}\| \|\mathbf{y}\|} \langle \mathbf{x}, \mathbf{y} \rangle + 1 \geq 0
$$

Hence $\langle \mathbf{x}, \mathbf{y} \rangle \leq \|\mathbf{x}\| \|\mathbf{y}\|$. By combining the two inequalities, we obtain the theorem. ■

EXERCISES 7.4

1. Use the one-norm, two-norm, and infinity-norm to normalize the following vectors in $\mathbb{R}^2$.

   a. $\mathbf{x} = (3, 4)$ b. $\mathbf{x} = (0, 0)$ c. $\mathbf{x} = (-3, 4)$

   d. $\mathbf{x} = (1, 0)$ e. $\mathbf{x} = (-9, 11)$ f. $\mathbf{x} = (-9, -11)$

2. Use the one-norm, two-norm, and infinity-norm to normalize the following vectors in $\mathbb{R}^3$.

   a. $\mathbf{x} = (3, 4, 5)$ b. $\mathbf{x} = (0, 0, 0)$ c. $\mathbf{x} = (-3, 4, 5)$

   d. $\mathbf{x} = (1, 0, 1)$ e. $\mathbf{x} = (0, 0, 1)$ f. $\mathbf{x} = (-9, -11, -5)$

3. Calculate the inner products $\mathbf{x}^T \mathbf{y}$ for the following pairs of vectors.

   a. $\mathbf{x} = (1, 2)$ and $\mathbf{y} = (4, 5)$ b. $\mathbf{x} = (-1, 2)$ and $\mathbf{y} = (4, 5)$

   c. $\mathbf{x} = (0, -2)$ and $\mathbf{y} = (4, 5)$ d. $\mathbf{x} = (1, 2)$ and $\mathbf{y} = (0, 0)$

4. Calculate the inner products $\langle \mathbf{x}, \mathbf{y} \rangle = 5x_1 y_1 + 7x_2 y_2$ of the pairs of vectors $\mathbf{x} = (x_1, x_2)$ and $\mathbf{y} = (y_1, y_2)$ in Exercise 1.

5. Use *MATHEMATICA* to calculate $\mathbf{x}^T \mathbf{y}$ for the following pairs of vectors.

   a. $\mathbf{x} = (1, 2, 3)$ and $\mathbf{y} = (4, 5, 7)$ b. $\mathbf{x} = (-1, 2, 0)$ and $\mathbf{y} = (4, 5, 7)$

   c. $\mathbf{x} = (0, -2, -3)$ and $\mathbf{y} = (4, 5, 7)$ d. $\mathbf{x} = (1, 0, 0)$ and $\mathbf{y} = (4, 5, 7)$

6. Find the inner products $\langle \mathbf{x}, \mathbf{y} \rangle = 5x_1 y_1 + 7x_2 y_2 + 4x_3 y_3$ of the pairs of vectors $\mathbf{x} = (x_1, x_2, x_3)$ and $\mathbf{y} = (y_1, y_2, y_3)$ in Exercise 3.

7. Use *MATHEMATICA* to find the inner product trace $B^T A$ of the following pairs of matrices.

a. $A = \begin{pmatrix} 5 & 0 & -3 \\ 0 & 0 & -3 \end{pmatrix}$ and $B = \begin{pmatrix} -1 & 4 & 5 \\ 5 & -2 & -2 \end{pmatrix}$

b. $A = \begin{pmatrix} 5 & 0 & 1 \\ 6 & 2 & 6 \\ 1 & 4 & -4 \\ 4 & 2 & 6 \end{pmatrix}$ and $B = \begin{pmatrix} 3 & 1 & 4 \\ -6 & 3 & -2 \\ 1 & 3 & 0 \\ 1 & -8 & 3 \end{pmatrix}$

8. Use the pairs of vectors in Exercise 1 and verify the Cauchy-Schwarz inequality for the dot product.

9. Use the pairs of vectors in Exercise 3 and verify the Cauchy-Schwarz inequality for the inner product defined in Exercise 4.

10. Let $A = (a_{ij})$ be an $m \times n$ real matrix, and let $\mathcal{E} = \{e_1, \ldots, e_n\}$ be the standard basis of $\mathbb{R}^n$ and $\mathcal{E}' = \{e'_1, \ldots, e'_m\}$ the standard basis of $\mathbb{R}^m$. Show that $a_{ij} = \langle Ae_j, e'_i \rangle$ for all i, j in the standard inner product on $\mathbb{R}^m$.

11. Find a condition on r and s for which the *MATHEMATICA* function

$$\texttt{Dot[r\{x[1],x[2]\},s\{y[1],y[2]\}]}$$

is an inner product on $\mathbb{R}^2$.

Positive Definite Matrices

We saw in the last section that both the function

$$f\left[\begin{pmatrix} x_1 \\ x_2 \end{pmatrix}, \begin{pmatrix} y_1 \\ y_2 \end{pmatrix}\right] = x_1 y_1 + x_2 y_2$$

and the function

$$f\left[\begin{pmatrix} x_1 \\ x_2 \end{pmatrix}, \begin{pmatrix} y_1 \\ y_2 \end{pmatrix}\right] = 2x_1 y_1 + 3x_2 y_2$$

define an inner product on $\mathbb{R}^2$. In matrix notation, we can write these functions as

$$\begin{pmatrix} x_1 \\ x_2 \end{pmatrix}^T \begin{pmatrix} 1 & 0 \\ 0 & 1 \end{pmatrix} \begin{pmatrix} y_1 \\ y_2 \end{pmatrix} \quad \text{and} \quad \begin{pmatrix} x_1 \\ x_2 \end{pmatrix}^T \begin{pmatrix} 2 & 0 \\ 0 & 3 \end{pmatrix} \begin{pmatrix} y_1 \\ y_2 \end{pmatrix}$$

In other words, both inner products can be written in the form $x^T A y$, for some 2×2 matrix A. This leads us to search for conditions under which an inner product can be defined by a matrix. Since $\langle x, x \rangle$ must be positive for nonzero vectors x, we certainly require that $x^T A x > 0$ for all nonzero $x \in \mathbb{R}^n$. In the next theorem, we prove that the symmetry of $\langle x, y \rangle$ also requires that the matrix A is symmetric. We combine these two properties and define a real $n \times n$ symmetric matrix A as *positive definite* if $x^T A x > 0$ for all nonzero $x \in \mathbb{R}^n$.

THEOREM 7.5 (Inner product theorem)   *If A is a positive definite symmetric matrix, then the function $\langle \mathbf{x}, \mathbf{y} \rangle = \mathbf{x}^T A \mathbf{y}$ is an inner product on $\mathbb{R}^n$.*

Proof. The function $\mathbf{x}^T A \mathbf{y}$ is linear in the first variable since

$$\langle a\mathbf{u} + b\mathbf{v}, \mathbf{y} \rangle = (a\mathbf{u} + b\mathbf{v})^T A \mathbf{y}$$
$$= \left(a\mathbf{u}^T + b\mathbf{v}^T \right) A \mathbf{y}$$
$$= (a\mathbf{u})^T A\mathbf{z} + (b\mathbf{v})^T A \mathbf{y}$$
$$= a\langle \mathbf{u}, \mathbf{y} \rangle + b \langle \mathbf{v}, \mathbf{y} \rangle$$

Since A is symmetric, and since any real number z can be considered as a 1×1 symmetric matrix (z), we have $\mathbf{x}^T A \mathbf{y} = \left(\mathbf{x}^T A \mathbf{y} \right)^T$. It follows that $\langle \mathbf{x}, \mathbf{y} \rangle$ is symmetric:

$$\langle \mathbf{x}, \mathbf{y} \rangle = \mathbf{x}^T A \mathbf{y} = \left(\mathbf{x}^T A \mathbf{y} \right)^T = \mathbf{y}^T A^T \mathbf{x}^T = \mathbf{y}^T A \mathbf{x} = \langle \mathbf{y}, \mathbf{x} \rangle$$

The fact that A is positive definite means that $\langle \mathbf{x}, \mathbf{x} \rangle = \mathbf{x}^T A \mathbf{x} > 0$ for all nonzero $\mathbf{x} \in \mathbb{R}^n$. Moreover, $\langle \mathbf{0}, \mathbf{0} \rangle = \mathbf{0}^T A \mathbf{0} = 0$. Therefore $\langle \mathbf{x}, \mathbf{y} \rangle$ is an inner product on $\mathbb{R}^n$. ■

In Chapter 8, we will show that all real symmetric matrices are diagonalizable. It is an easy consequence of this fact that a symmetric matrix is positive definite if and only if its eigenvalues are positive. We call this theorem the *eigenvalue test* for positive definiteness.

EXAMPLE 7.17   ■ **A Positive Definite Matrix**

Use the eigenvalue test to show that the symmetric matrix

$$A = \begin{pmatrix} 9 & 0 & -2 \\ 0 & 7 & 0 \\ -2 & 0 & 1 \end{pmatrix}$$

is positive definite.

Solution. The following calculation shows that the eigenvalues of A are positive.

```
In[1]:= Eigenvalues[( 9   0  -2 )]
                     ( 0   7   0 )
                     (-2   0   1 )

Out[1]= {7, 5-2√5, 5+2√5}
```

The first and third eigenvalues are clearly positive, and since $20 < 25$, it follows that $2\sqrt{5} < 5$. Therefore, $0 < 5 - 2\sqrt{5}$. This shows that A is positive definite. ◄►

EXERCISES 7.5

1. Use *MATHEMATICA* to generate three random 2×1, 2×2, and 2×3 matrices A, B, and C. Form the 2×2 symmetric matrices AA^T, BB^T, and CC^T. Use the eigenvalue test to determine which of these matrices, if any, is positive definite.

2. Show that no 2×2 matrix $A^T A$ is positive definite if A is a 1×2 matrix.

3. Show that no 3×3 matrix $A^T A$ is positive definite if A is a 1×3 matrix.

4. Find conditions on a, b, c, d such that $A^T A$ is positive definite, where

$$A = \begin{pmatrix} a & b \\ c & d \end{pmatrix}$$

5. Use the eigenvalue test to determine which of the following symmetric matrices are positive definite.

$$\text{a.} \begin{pmatrix} 4 & 1 \\ 1 & -1 \end{pmatrix} \quad \text{b.} \begin{pmatrix} 1 & 1 & 0 \\ 1 & 2 & 0 \\ 0 & 0 & 4 \end{pmatrix} \quad \text{c.} \begin{pmatrix} 1 & -2 \\ -2 & 7 \end{pmatrix}$$

6. Show in detail that the products $\mathbf{x}^T A \mathbf{y}$ determined by the positive definite matrices in Exercise 1 are inner products.

7. Show that if A is the matrix

$$\begin{pmatrix} 50 & 79 \\ 56 & 49 \end{pmatrix}$$

then the symmetric matrix $A^T A$ is positive definite and $\mathbf{x} A^T A \mathbf{y}$ is an inner product on $\mathbb{R}^2$. Find the norm of the vectors $\mathbf{x}_1 = (1, -1)$, $\mathbf{x}_2 = (45, 46)$, and $\mathbf{x}_3 = (10, 0)$ in this inner product. Find the distance between $\mathbf{x}_1$ and $\mathbf{x}_2$, between $\mathbf{x}_1$ and $\mathbf{x}_3$, and between $\mathbf{x}_2$ and $\mathbf{x}_3$.

8. Show that if A is the matrix

$$\begin{pmatrix} 1 & 0 & -1 \\ 0 & 2 & 0 \\ -1 & 0 & 3 \end{pmatrix}$$

then $\mathbf{x} A^T A \mathbf{y}$ is an inner product on $\mathbb{R}^3$. Find the norm of the vectors $\mathbf{x}_1 = (1, -1, 3)$, $\mathbf{x}_2 = (45, 46, 47)$, and $\mathbf{x}_3 = (10, 0, -9)$ in this inner product. Find the distance between $\mathbf{x}_1$ and $\mathbf{x}_2$, between $\mathbf{x}_1$ and $\mathbf{x}_3$, and between $\mathbf{x}_2$ and $\mathbf{x}_3$.

9. Use the **Array** and **Random** functions of *MATHEMATICA* to generate three random 3×3 matrices A, B, and C. Determine whether the functions $\mathbf{x}A^T A\mathbf{y}$, $\mathbf{x}B^T B\mathbf{y}$, and $\mathbf{x}C^T C\mathbf{y}$ are inner products on $\mathbb{R}^3$.

Inner Product Norms

We are now in a position to unify the study of length, distance, and angles by using vector norms determined by inner products. The following theorem explains the construction of a norm from an inner product.

THEOREM 7.6 (Inner product norm theorem)   *If V is a real vector space with an inner product $\langle \mathbf{x}, \mathbf{y} \rangle$, then the function $\|\mathbf{x}\| = \sqrt{\langle \mathbf{x}, \mathbf{x} \rangle}$ is a norm on V.*

Proof. Since $\langle \mathbf{x}, \mathbf{x} \rangle = 0$ if $\mathbf{x}$ is $\mathbf{0}$ and is greater than 0 otherwise, $\|\mathbf{x}\| = \sqrt{\langle \mathbf{x}, \mathbf{x} \rangle} > 0$ for all $\mathbf{x} \neq \mathbf{0}$. In addition, $\sqrt{\langle \mathbf{0}, \mathbf{0} \rangle} = 0$. Hence the first axiom of norms is satisfied. Furthermore,

$$\|a\mathbf{x}\| = \sqrt{\langle a\mathbf{x}, a\mathbf{x} \rangle} = \sqrt{a^2 \langle \mathbf{x}, \mathbf{x} \rangle} = |a| \sqrt{\langle \mathbf{x}, \mathbf{x} \rangle} = |a|\, \|\mathbf{x}\|$$

Therefore the second axiom is also satisfied. It remains to verify the third axiom. For this we use Theorem 7.4, together with the bilinearity and symmetry of the inner product:

$$\begin{aligned} \|\mathbf{x} + \mathbf{y}\|^2 &= \langle \mathbf{x} + \mathbf{y}, \mathbf{x} + \mathbf{y} \rangle \\ &= \langle \mathbf{x}, \mathbf{x} \rangle + 2\langle \mathbf{x}, \mathbf{y} \rangle + \langle \mathbf{y}, \mathbf{y} \rangle \\ &\leq \|\mathbf{x}\|^2 + 2\|\mathbf{x}\|\, \|\mathbf{y}\| + \|\mathbf{y}\|^2 \\ &= (\|\mathbf{x}\| + \|\mathbf{y}\|)^2 \end{aligned}$$

If we take the square root of each side of this inequality, we get

$$\|\mathbf{x} + \mathbf{y}\| \leq \|\mathbf{x}\| + \|\mathbf{y}\|$$

Hence the function $\mathbf{x} \to \sqrt{\langle \mathbf{x}, \mathbf{x} \rangle}$ is a norm on V. ∎

DEFINITION 7.5   *If $\langle \mathbf{x}, \mathbf{y} \rangle$ is an inner product on a real vector space V, then the function $\|\mathbf{x}\| = \sqrt{\langle \mathbf{x}, \mathbf{x} \rangle}$ is an **inner product norm** on V.*

Although every inner product determines a norm in this way, there are norms that are not determined by an inner product. The infinity norm on $\mathbb{R}^2$, for example, is not an inner product norm. We will show this by showing that it fails to satisfy the **parallelogram law**, a law satisfied by all inner product norms. The law takes its name from ordinary geometry, where it says that the sum of the squares on the diagonals of a parallelogram is equal to twice the sum of the squares on its sides.

THEOREM 7.7 (Parallelogram law)   *If $\|\mathbf{z}\|$ is an inner product norm on a real vector space V, then $\|\mathbf{x} + \mathbf{y}\|^2 + \|\mathbf{x} - \mathbf{y}\|^2 = 2(\|\mathbf{x}\|^2 + \|\mathbf{y}\|^2)$ for all $\mathbf{x}, \mathbf{y} \in V$.*

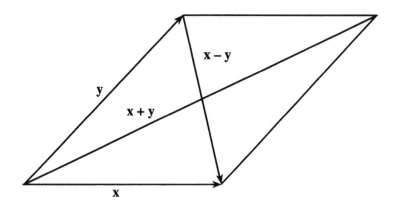

FIGURE 1   Diagonals of a Parallelogram.

Proof. Suppose that $\|\mathbf{z}\| = \sqrt{\langle \mathbf{z}, \mathbf{z} \rangle}$. Then $\|\mathbf{z}\|^2 = \langle \mathbf{z}, \mathbf{z} \rangle$. Therefore,

$$\|\mathbf{x} + \mathbf{y}\|^2 = \langle \mathbf{x} + \mathbf{y}, \mathbf{x} + \mathbf{y} \rangle = \langle \mathbf{x}, \mathbf{x} \rangle + 2 \langle \mathbf{x}, \mathbf{y} \rangle + \langle \mathbf{y}, \mathbf{y} \rangle$$

and

$$\|\mathbf{x} - \mathbf{y}\|^2 = \langle \mathbf{x} - \mathbf{y}, \mathbf{x} - \mathbf{y} \rangle = \langle \mathbf{x}, \mathbf{x} \rangle - 2 \langle \mathbf{x}, \mathbf{y} \rangle + \langle \mathbf{y}, \mathbf{y} \rangle$$

If we add the two equations, we get

$$\|\mathbf{x} + \mathbf{y}\|^2 + \|\mathbf{x} - \mathbf{y}\|^2 = 2 \left(\langle \mathbf{x}, \mathbf{x} \rangle + \langle \mathbf{y}, \mathbf{y} \rangle \right) = 2 \left(\|\mathbf{x}\|^2 + \|\mathbf{y}\|^2 \right)$$

for all $\mathbf{x}, \mathbf{y} \in V$. ∎

Figure 1 illustrates the geometric connection between the vectors connected by the parallelogram law.

EXAMPLE 7.18 ■ **The Infinity Norm on $\mathbb{R}^2$**

Show that the infinity norm

$$\left\| \begin{pmatrix} x_1 \\ x_2 \end{pmatrix} \right\|_{\infty} = \max(|x_1|, |x_2|)$$

is not an inner product norm.

Solution. It suffices to find two vectors in $\mathbb{R}^2$ for which the norm fails to satisfy the parallelogram law. Consider the vectors $\mathbf{x} = (1, 0)$ and $\mathbf{y} = (0, 1)$.

Since

$$\mathbf{x} + \mathbf{y} = \begin{pmatrix} 1 \\ 1 \end{pmatrix} \quad \text{and} \quad \mathbf{x} - \mathbf{y} = \begin{pmatrix} 1 \\ -1 \end{pmatrix}$$

we get

$$\|\mathbf{x}+\mathbf{y}\|^2 + \|\mathbf{x}-\mathbf{y}\|^2 = \max(|1|,|1|)^2 + \max(|1|,|-1|)^2 = 1+1 = 2$$

and

$$2\left(\|\mathbf{x}\|^2 + \|\mathbf{y}\|^2\right) = 2\left(\max(|1|,|0|)^2 + \max(|0|,|1|)^2\right) = 2(1+1) = 4$$

Hence the parallelogram law fails. ◀▶

EXERCISES 7.6

1. Use the inner product $\mathbf{x}^T\mathbf{y}$ on $\mathbb{R}^2$ to find the norm $\|\mathbf{x}\|$ of the following vectors.

     a. $\mathbf{x} = (4,5)$ b. $\mathbf{x} = (-1,2)$ c. $\mathbf{x} = (0,0)$

     d. $\mathbf{x} = (-4,5)$ e. $\mathbf{x} = (1,0)$ f. $\mathbf{x} = (-4,-5)$

2. Repeat Exercise 1 using the inner product $\mathbf{x}^T A\mathbf{y}$ determined by the matrix

$$A = \begin{pmatrix} 3 & 0 \\ 0 & 5 \end{pmatrix}$$

3. Use the standard inner product $\mathbf{x}^T\mathbf{y}$ on $\mathbb{R}^3$ to find the norm $\|\mathbf{x}\|$ of the following vectors.

     a. $\mathbf{x} = (4,5,7)$ b. $\mathbf{x} = (4,-5,-7)$ c. $\mathbf{x} = (1,0,0)$

     d. $\mathbf{x} = (-1,2,0)$ e. $\mathbf{x} = (0,1,1)$ f. $\mathbf{x} = (-1,-2,-3)$

4. Repeat Exercise 3 using the inner product $\mathbf{x}^T A\mathbf{y}$ determined by the matrix

$$A = \begin{pmatrix} 3 & 0 & 0 \\ 0 & 5 & 0 \\ 0 & 0 & 6 \end{pmatrix}$$

5. Show that the symmetric matrix

$$A = \begin{pmatrix} 3 & 0 & 1 \\ 0 & 5 & 0 \\ 1 & 0 & 6 \end{pmatrix}$$

     is positive definite, and use it to compute the norms of the vectors in Exercise 3, determined by the inner product $\mathbf{x}^T A\mathbf{y}$.

6. Use *MATHEMATICA* to verify the parallelogram law for the following pairs of vectors.

     a. $\mathbf{x} = (1,2,-2)$ and $\mathbf{y} = (4,5,7)$ b. $\mathbf{x} = (-1,2,0)$ and $\mathbf{y} = (4,5,7)$

     c. $\mathbf{x} = (1,0,0)$ and $\mathbf{y} = (0,5,7)$ d. $\mathbf{x} = (0,-2,-5)$ and $\mathbf{y} = (4,5,-2)$

     e. $\mathbf{x} = (1,2,3)$ and $\mathbf{y} = (0,0,0)$ f. $\mathbf{x} = (4,5,7)$ and $\mathbf{y} = (-4,-5,-7)$

7. For which pairs of vectors **x** and **y** in Exercise 6, does $\|\mathbf{x} + \mathbf{y}\|^2 = \|\mathbf{x}\|^2 + \|\mathbf{y}\|^2$? Use *MATHEMATICA* to determine your answers.

8. Prove that in any inner product space, the ***polar identity***

$$\langle \mathbf{x}, \mathbf{y} \rangle = \frac{1}{4} \left(\|\mathbf{x} + \mathbf{y}\|^2 - \|\mathbf{x} - \mathbf{y}\|^2 \right)$$

expresses the inner product in terms of the norm.

ANGLES

In ordinary geometry, the law of cosines says that the angle between two vectors **x** and **y** in the plane is related to the length of **x** and **y** by the formula

$$\|\mathbf{x} - \mathbf{y}\|^2 = \|\mathbf{x}\|^2 + \|\mathbf{y}\|^2 - 2 \|\mathbf{x}\| \|\mathbf{y}\| \cos \theta$$

Figure 2 illustrates this relationship.

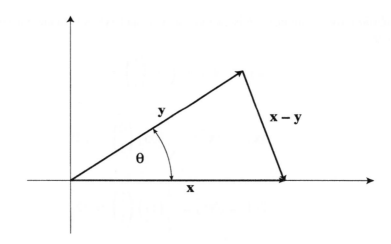

FIGURE 2   An Angle Between Two Vectors.

If we express $\|\mathbf{x} - \mathbf{y}\|^2$, $\|\mathbf{x}\|^2$, and $\|\mathbf{y}\|^2$ in terms of inner products, we get

$$\langle \mathbf{x} - \mathbf{y}, \mathbf{x} - \mathbf{y} \rangle = \langle \mathbf{x}, \mathbf{x} \rangle + \langle \mathbf{y}, \mathbf{y} \rangle - 2 \|\mathbf{x}\| \|\mathbf{y}\| \cos \theta$$

Therefore,

$$\langle \mathbf{x} - \mathbf{y}, \mathbf{x} - \mathbf{y} \rangle - \langle \mathbf{x}, \mathbf{x} \rangle - \langle \mathbf{y}, \mathbf{y} \rangle = -2 \|\mathbf{x}\| \|\mathbf{y}\| \cos \theta$$

Since inner products are linear in both variables, this equation converts to

$$\langle \mathbf{x}, \mathbf{x} \rangle - 2 \langle \mathbf{x}, \mathbf{y} \rangle + \langle \mathbf{y}, \mathbf{y} \rangle - \langle \mathbf{x}, \mathbf{x} \rangle - \langle \mathbf{y}, \mathbf{y} \rangle = -2 \, \|\mathbf{x}\| \, \|\mathbf{y}\| \cos \theta$$

If we simplify and solve for $\cos \theta$, we get $\cos \theta = \langle \mathbf{x}, \mathbf{y} \rangle / (\|\mathbf{x}\| \|\mathbf{y}\|)$. The right-hand side of this equation involves only the inner product $\langle \mathbf{x}, \mathbf{y} \rangle$ and the norms $\|\mathbf{x}\|$ and $\|\mathbf{y}\|$. Moreover, the norms $\|\mathbf{x}\| = \sqrt{\langle \mathbf{x}, \mathbf{x} \rangle}$ and $\|\mathbf{y}\| = \sqrt{\langle \mathbf{y}, \mathbf{y} \rangle}$ are both defined by the inner product $\langle \mathbf{x}, \mathbf{y} \rangle$.

This tells us how angles should be defined in terms of inner products. The idea is to define the cosine of an angle first and then define the angle θ as the inverse of $\cos \theta$.

DEFINITION 7.6   *The **cosine of the angle** θ $(0 \leq \theta < \pi)$ between two nonzero vectors* $\mathbf{x}$ *and* $\mathbf{y}$ *in a real inner product space V is*

$$\cos \theta = \frac{\langle \mathbf{x}, \mathbf{y} \rangle}{\|\mathbf{x}\| \, \|\mathbf{y}\|}$$

EXAMPLE 7.19   ■ The Cosine of an Angle

Use the standard inner product to find the cosine of the angle determined by the vectors $\mathbf{x} = (0, 5)$ and $\mathbf{y} = (1, 1)$.

Solution. By definition, $\cos \theta$ is the ratio of $\langle \mathbf{x}, \mathbf{y} \rangle$ and $\|\mathbf{x}\| \, \|\mathbf{y}\|$. In the standard inner product on $\mathbb{R}^2$,

$$\langle \mathbf{x}, \mathbf{y} \rangle = \mathbf{x}^T \mathbf{y} = \begin{pmatrix} 0 & 5 \end{pmatrix} \begin{pmatrix} 1 \\ 1 \end{pmatrix} = 5$$

$$\|\mathbf{x}\| = \sqrt{\mathbf{x}^T \mathbf{x}} = \sqrt{\begin{pmatrix} 0 & 5 \end{pmatrix} \begin{pmatrix} 0 \\ 5 \end{pmatrix}} = \sqrt{25} = 5$$

$$\|\mathbf{y}\| = \sqrt{\mathbf{y}^T \mathbf{y}} = \sqrt{\begin{pmatrix} 1 & 1 \end{pmatrix} \begin{pmatrix} 1 \\ 1 \end{pmatrix}} = \sqrt{2}$$

Therefore, $\cos \theta = \langle \mathbf{x}, \mathbf{y} \rangle / (\|\mathbf{x}\| \|\mathbf{y}\|) = 5/5\sqrt{2} = 1/\sqrt{2}$. ◀▶

DEFINITION 7.7   *The **angle** between two nonzero vectors* $\mathbf{x}$ *and* $\mathbf{y}$ *in an inner product space V is the unique real number θ for which*

$$\cos \theta = \frac{\langle \mathbf{x}, \mathbf{y} \rangle}{\|\mathbf{x}\| \, \|\mathbf{y}\|} \qquad (0 \leq \theta < \pi)$$

EXERCISES 7.7

1. Find the cosine of the angles between the following pairs of vectors in $\mathbb{R}^2$.

   a. $\mathbf{x} = (1, 2)$ and $\mathbf{y} = (4, 5)$ b. $\mathbf{x} = (4, 5)$ and $\mathbf{y} = (-1, 2)$
   c. $\mathbf{x} = (1, 0)$ and $\mathbf{y} = (4, 5)$ d. $\mathbf{x} = (4, 5)$ and $\mathbf{y} = (0, -2)$
   e. $\mathbf{x} = (1, 2)$ and $\mathbf{y} = (0, 0)$ f. $\mathbf{x} = (1, 0)$ and $\mathbf{y} = (0, 1)$

2. Find the angles between the pairs of vectors in Exercise 1.

3. Find the cosine of the angles between the following pairs of vectors in $\mathbb{R}^3$.

   a. $\mathbf{x} = (1, 2, 3)$ and $\mathbf{y} = (4, 5, 7)$ b. $\mathbf{x} = (4, 5, 7)$ and $\mathbf{y} = (-1, 2, 0)$
   c. $\mathbf{x} = (1, 0, 0)$ and $\mathbf{y} = (4, 5, 7)$ d. $\mathbf{x} = (4, 5, 7)$ and $\mathbf{y} = (0, -2, -3)$
   e. $\mathbf{x} = (1, 2, 3)$ and $\mathbf{y} = (0, 0, 0)$ f. $\mathbf{x} = (1, 0, 1)$ and $\mathbf{y} = (0, 1, 0)$

4. Find the angles between the pairs of vectors in Exercise 3.

5. Use the Frobenius norm to find the angle between the matrices

$$\mathbf{x} = \begin{pmatrix} 4 & 1 \\ -4 & -1 \end{pmatrix} \quad \text{and} \quad \mathbf{y} = \begin{pmatrix} 1 & 2 \\ -2 & 7 \end{pmatrix}$$

Angles in Statistics

In this section, we discuss an interesting conceptual connection between inner products and statistics. We show that *covariances* and *standard deviations* can be computed using dot products. We then use this information to interpret the *Pearson correlation coefficient* of two data sets as the cosine of the angle between these sets. In this interpretation, statistically independent data correspond to orthogonal vectors.

In statistics, the covariance is used to measure the extent to which two ranges of data $\mathbf{x}$ and $\mathbf{y}$ vary together: whether large entries of $\mathbf{x}$ correspond to large entries of $\mathbf{y}$ (positive covariance), or whether small entries of $\mathbf{x}$ correspond to large entries of $\mathbf{y}$ (negative covariance), or whether the entries of $\mathbf{x}$ and $\mathbf{y}$ are unrelated (covariance near zero). The standard deviations of $\mathbf{x}$ and $\mathbf{y}$, on the other hand, measure to what extent the data in $\mathbf{x}$ and $\mathbf{y}$ deviate from the arithmetic mean of their components. We begin with an example.

EXAMPLE 7.20   ■ Variances and Standard Deviations

Suppose that $\mathbf{x} = (x_1, x_2, x_3)$ and $\mathbf{y} = (y_1, y_2, y_3)$ are two numerical data vectors and that $\hat{x} = \frac{1}{3}(x_1 + x_2 + x_3)$ and $\hat{y} = \frac{1}{3}(y_1 + y_2 + y_3)$ are their arithmetic means. Then the values

$(x_i - \hat{x})$ are the deviations of the data x_i from the mean $\hat{x}$. Similarly, the quantities $(y_i - \hat{y})$ are the deviations of the data y_i from the mean $\hat{y}$. The arithmetic means

$$\frac{1}{3}\left[(x_1 - \hat{x})^2 + (x_2 - \hat{x})^2 + (x_3 - \hat{x})^2\right]$$

and

$$\frac{1}{3}\left[(y_1 - \hat{y})^2 + (y_2 - \hat{y})^2 + (y_3 - \hat{y})^2\right]$$

of these quantities are called the *variances* of **x** and **y**, and the square roots

$$\sigma_x = \sqrt{\frac{1}{3}\left[(x_1 - \hat{x})^2 + (x_2 - \hat{x})^2 + (x_3 - \hat{x})^2\right]}$$

and

$$\sigma_y = \sqrt{\frac{1}{3}\left[(y_1 - \hat{y})^2 + (y_2 - \hat{y})^2 + (y_3 - \hat{y})^2\right]}$$

of the variances are called the *standard deviations* of **x** and **y**. ◀▶

In some applications, a distinction is required between the variance and standard deviation over an entire set of n data and over a proper subset. In the latter case, the variance and standard deviation are averaged over $(n-1)$ instead of n.

EXAMPLE 7.21 ■ Covariances

The *covariance* of two data sets is calculated by looking at the products of the deviations, point by point, and taking their average. The covariance

$$\text{cov}(\mathbf{x}, \mathbf{y}) = \frac{1}{3}\left[(x_1 - \hat{x})(y_1 - \hat{y}) + (x_2 - \hat{x})(y_2 - \hat{y}) + (x_3 - \hat{x})(y_3 - \hat{y})\right]$$

of **x** and **y** is also denoted by σ_{xy}. ◀▶

The Pearson correlation coefficient r combines covariances and standard deviations into a single formula that measures the degree to which two sets of data are correlated.

DEFINITION 7.8 *Let* **x** *and* **y** *be two finite sets of data. Then the **Pearson correlation coefficient** of* **x** *and* **y** *is*

$$r = \frac{\text{cov}(\mathbf{x}, \mathbf{y})}{\sigma_x \sigma_y}$$

The coefficient r ranges from -1 to 1. If we write out the formula for r in full, we get

$$r = \frac{\frac{1}{n}\sum_{i \in \mathbf{n}}(x_i - \hat{x})(y_i - \hat{y})}{\sqrt{\frac{\sum_{i \in \mathbf{n}}(x_i - \hat{x})^2}{n}}\sqrt{\frac{\sum_{i \in \mathbf{n}}(y_i - \hat{y})^2}{n}}}$$

Since the factor $1/n$ in the denominator and numerator of r cancels out, the coefficient r becomes

$$\frac{\sum_{i\in n}\left(x_i - \hat{x}\right)\left(y_i - \hat{y}\right)}{\sqrt{\sum_{i\in n}\left(x_i - \hat{x}\right)^2}\sqrt{\sum_{i\in n}\left(y_i - \hat{y}\right)^2}}$$

This allows us to rewrite r is the cosine of an angle. Here are the calculations.

Let $\mathbf{u} = (x_1 - \hat{x}, \ldots, x_n - \hat{x})$ and $\mathbf{v} = (y_1 - \hat{y}, \ldots, y_n - \hat{y})$ be the vectors listing the deviations from the means of $\mathbf{x} = (x_1, \ldots, x_n)$ and $\mathbf{y} = (y_1, \ldots, y_n)$. Then the dot product of $\mathbf{u}$ and $\mathbf{v}$ and their two-norms are the following.

1. The dot product of $\mathbf{u}$ and $\mathbf{v}$ is

$$\mathbf{u} \cdot \mathbf{v} = \sum_{i\in n}\left(x_i - \hat{x}\right)\left(y_i - \hat{y}\right)$$

2. The two-norms of $\mathbf{u}$ and $\mathbf{v}$ are

$$\|\mathbf{u}\| = \sqrt{\mathbf{u} \cdot \mathbf{u}} = \sqrt{\sum_{i\in n}\left(x_i - \hat{x}\right)\left(x_i - \hat{x}\right)} = \sqrt{\sum_{i\in n}\left(x_i - \hat{x}\right)^2}$$

and

$$\|\mathbf{v}\| = \sqrt{\mathbf{v} \cdot \mathbf{v}} = \sqrt{\sum_{i\in n}\left(y_i - \hat{y}\right)\left(y_i - \hat{y}\right)} = \sqrt{\sum_{i\in n}\left(y_i - \hat{y}\right)^2}$$

Therefore,

$$r = \frac{\mathbf{u} \cdot \mathbf{v}}{\|\mathbf{u}\|\,\|\mathbf{v}\|}$$

is the cosine of the angle θ between the vectors $\mathbf{u}$ and $\mathbf{v}$ in $\mathbb{R}^n$.

EXAMPLE 7.22   ■ The Pearson Correlation Coefficient of a Set of Points

Use *MATHEMATICA* to find the Pearson correlation coefficient for the points $(1, 2)$, $(2, 2)$, $(3, 5)$, $(4, 6)$ in $\mathbb{R}^2$.

Solution. Using the given points we let $x_1 = 1$, $x_2 = 2$, $x_3 = 3$, $x_4 = 4$, $y_1 = 2$, $y_2 = 2$, $y_3 = 5$, and $y_4 = 6$. Therefore,

$$\hat{x} = \frac{1+2+3+4}{4} = \frac{5}{2} \quad \text{and} \quad \hat{y} = \frac{2+2+5+6}{4} = \frac{15}{4}$$

We now put

$$\mathbf{u} = (1 - \tfrac{5}{2}, 2 - \tfrac{5}{2}, 3 - \tfrac{5}{2}, 4 - \tfrac{5}{2}) \quad \text{and} \quad \mathbf{x} = (2 - \tfrac{15}{4}, 2 - \tfrac{15}{4}, 5 - \tfrac{15}{4}, 6 - \tfrac{15}{4})$$

Next we use *MATHEMATICA* to compute the cosine of the angle between $\mathbf{u}$ and $\mathbf{x}$. First we input $\mathbf{u}$ and $\mathbf{x}$.

```
In[1]:= u={{1-5/2},{2-5/2},{3-5/2},{4-5/2}};
```

```
In[2]:= x={{2-15/4},{2-15/4},{5-15/4},{6-15/4}};
```

We then calculate a numerical approximation of the cosine of the angle between **u** and **x**.

```
In[3]:= N[r=(Transpose[u].x)/

(Sqrt[Transpose[u].u] Sqrt[Transpose[x].x])]

Out[3]= {{0.939336}}
```

This shows that the coefficient r of the given set of points is approximately .93936. ◀▶

Let us now consider the problem of computing the Pearson correlation coefficient for two sets of data arranged in the form of two matrices.

Suppose $A = (a_{ij})$ and $B = (b_{ij})$ are two real $m \times n$ matrices. We show that r can also be expressed as the cosine of an angle, this time an angle determined by the trace inner product and the Frobenius norm. By definition, the Pearson correlation coefficient $r(A, B)$ between A and B is

$$\frac{\sum_{i \in \mathbf{m}} \sum_{j \in \mathbf{n}} (a_{ij} - \hat{a})(b_{ij} - \hat{b})}{\sqrt{\sum_{i \in \mathbf{m}} \sum_{j \in \mathbf{n}} (a_{ij} - \hat{a})^2} \sqrt{\sum_{i \in \mathbf{m}} \sum_{j \in \mathbf{n}} (b_{ij} - \hat{b})^2}}$$

where $\hat{a}$ is the arithmetic mean taken over all entries of the matrix A, and $\hat{b}$ is the arithmetic mean taken over all entries of the matrix B.

Let $\hat{A} = (\hat{a})$ be the constant $m \times n$ matrix with entries $\hat{a}$, and let $\hat{B} = (\hat{b})$ be the constant $m \times n$ matrix with entries $\hat{b}$. If we let $C = A - \hat{A}$ and $D = B - \hat{B}$, then the trace inner product of C and D is

$$\langle C, D \rangle = \text{trace } C^T D$$
$$= \text{trace } \left(A - \hat{A}\right)^T \left(B - \hat{B}\right)$$
$$= \sum_{i \in \mathbf{m}} \sum_{j \in \mathbf{n}} (a_{ij} - \hat{a})(b_{ij} - \hat{b})$$

and the squares of the Frobenius norms of C and D are

$$\|C\|_F^2 = \left\|A - \hat{A}\right\|_F^2 = \sum_{i \in \mathbf{m}} \sum_{j \in \mathbf{n}} (a_{ij} - \hat{a})^2$$

and

$$\|D\|_F^2 = \left\|B - \hat{B}\right\|_F^2 = \sum_{i \in \mathbf{m}} \sum_{j \in \mathbf{n}} (b_{ij} - \hat{b})^2$$

Therefore,

$$r(A, B) = \frac{\langle C, D \rangle}{\|C\|_F \|D\|_F}$$

is the cosine of the angle determined by the matrices C and D in $\mathbb{R}^{m \times n}$.

MATHEMATICA has built-in functions for computing Pearson correlation coefficients, covariances, and standard deviations. The measures are organized in matrix form.

If the given data sets consist of n vectors in $\mathbb{R}^m$, the ***correlation matrix*** $A = (a_{ij})$ is a symmetric matrix in which a_{ij} is the Pearson correlation coefficient of the data sets corresponding to the ith and jth columns of A. The ***covariance matrix*** $A = (a_{ij})$ is a symmetric matrix in which a_{ij} is the covariance of the data sets represented by the ith and jth columns of A.

EXAMPLE 7.23   ■ A Correlation Matrix for Three Data Sets

Use *MATHEMATICA* to calculate the Pearson correlation coefficients for the vectors $\mathbf{x} = (1, 2, 1, 1.5, 7)$, $\mathbf{y} = (3, -2, 1, 0, 7)$, and $\mathbf{z} = (2, 2, -1, 1, 9)$.

Solution. We first construct a 5×3 matrix whose columns are the vectors $\mathbf{x}$, $\mathbf{y}$, and $\mathbf{z}$. Then we apply the **CorrelationMatrix** function in the **MultiDescriptiveStatistics** package. The result is a 3×3 matrix $A = (a_{ij})$ in which a_{ij} is the Pearson correlation coefficient of ith and jth columns of A.

```
In[1]:= <<Statistics`MultiDescriptiveStatistics`
```

```
In[2]:= A={{1,3.0,2.0},{2,-2,2},{1,1,-1},{1.5,0,1},{7,7,-9}};
```

```
In[3]:= MatrixForm[CorrelationMatrix[A]]
Out[3]//MatrixForm=
```

$$\begin{pmatrix} 1. & .75969 & -.930501 \\ .75969 & 1. & -.835418 \\ -.930501 & -.835418 & 1. \end{pmatrix}$$

The value $a_{12} = .75969$ indicates that the vectors $\mathbf{x}$ and $\mathbf{y}$ are positively correlated; the value $a_{13} = -.930501$ indicates that $\mathbf{x}$ and $\mathbf{z}$ are negatively correlated; and the value $a_{23} = -.835418$ indicates that $\mathbf{y}$ and $\mathbf{z}$ are also negatively correlated. ◀▶

EXAMPLE 7.24 ■ **A Covariance Matrix for Three Data Sets**

Use *MATHEMATICA* to find the covariances cov($\mathbf{x}$, $\mathbf{y}$), cov($\mathbf{x}$, $\mathbf{z}$), and cov($\mathbf{y}$, $\mathbf{z}$) of the data sets represented by the columns of the matrix

$$A = \begin{pmatrix} 1 & 3.0 & 2.0 \\ 2 & -2 & 2 \\ 1 & 1 & -1 \\ 1.5 & 0 & 1 \\ 7 & 7 & -9 \end{pmatrix}$$

where $\mathbf{x}$, $\mathbf{y}$, and $\mathbf{z}$ are the first, second, and third columns of A, respectively.

Solution. We use the **CovarianceMatrix** function in the **MultiDescriptiveStatistics** package.

```
In[1]:= <<Statistics`MultiDescriptiveStatistics`
```

```
In[2]:= A={{1,3.0,2.0},{2,-2,2},{1,1,-1},{1.5,0,1},{7,7,-9}};
```

```
In[3]:= MatrixForm[CovarianceMatrix[A]]

Out[3]//MatrixForm=
```
$$\begin{pmatrix} 6.5 & 6.625 & -11. \\ 6.625 & 11.7 & -13.25 \\ -11. & -13.25 & 21.5 \end{pmatrix}$$

The value $a_{12} = 6.625$ is cov($\mathbf{x}$, $\mathbf{y}$); the value $a_{13} = -11$ is cov($\mathbf{x}$, $\mathbf{z}$); and the value $a_{23} = -13.25$ is cov($\mathbf{y}$, $\mathbf{z}$). ◀▶

EXAMPLE 7.25 ■ **The Standard Deviations of Three Data Sets**

Use *MATHEMATICA* to find the standard deviations of the data sets represented by the columns of the matrix A in Example 7.24.

Solution. We use the **StandardDeviation** function in the **MultiDescriptiveStatistics** package.

```
In[1]:= <<Statistics`MultiDescriptiveStatistics`
```

In[2]:= A={{1,3.0,2.0},{2,-2,2},{1,1,-1},{1.5,0,1},{7,7,-9}}

In[3]:= StandardDeviation[A]

Out[3]= {2.54951,3.42053,4.63681}

The value 2.54951 is the standard deviation of the first column; the value 3.42053 is the standard deviation of the second column; and the value 4.63681 is the standard deviation of the third column. ◄►

EXERCISES 7.8

1. Use the definition to find the Pearson correlation coefficient for each of the following data sets.

   a. $(x_1, y_1) = (1, -1)$, $(x_2, y_2) = (2, 1)$, $(x_3, y_3) = (3, 7)$, $(x_4, y_4) = (4, 5)$, $(x_5, y_5) = (5, 6)$

   b. $(x_1, y_1) = (1.5, 2.7)$, $(x_2, y_2) = (-3, 4)$, $(x_3, y_3) = (3, 4)$, $(x_4, y_4) = (5, 0)$, $(x_5, y_5) = (6.75, 3)$

   c. $(x_1, y_1) = (1, 0)$, $(x_2, y_2) = (2, 3)$, $(x_3, y_3) = (3, 0)$, $(x_4, y_4) = (4, 3)$, $(x_5, y_5) = (5, 0)$

2. Use the **CorrelationMatrix** function in *MATHEMATICA* to find the Pearson correlation coefficient of the data pairs in Exercise 1.

3. Use the **CovarianceMatrix** function in *MATHEMATICA* to find the covariance of the data sets in Exercise 1.

4. Use the definition to find the Pearson correlation coefficient for the following pairs of matrices.

   a. $\begin{pmatrix} 8 & 4 & -5 \\ -5 & 3 & -5 \end{pmatrix}, \begin{pmatrix} 8 & 5 & -1 \\ 0 & 3 & -4 \end{pmatrix}$ b. $\begin{pmatrix} 1 & 2 & 3 \\ 4 & 5 & 6 \end{pmatrix}, \begin{pmatrix} 2 & 4 & 6 \\ 8 & 10 & 12 \end{pmatrix}$

   c. $\begin{pmatrix} 1 & 2 & 3 \\ 4 & 5 & 6 \end{pmatrix}, \begin{pmatrix} -1 & -2 & -3 \\ -4 & -5 & -6 \end{pmatrix}$ d. $\begin{pmatrix} 1 & 2 & 3 \\ 4 & 5 & 6 \end{pmatrix}, \begin{pmatrix} 6 & 5 & 4 \\ 3 & 2 & 1 \end{pmatrix}$

5. Consider the following data.

	x	y	z
Observation 1	1.0	1.3	1.3
Observation 2	1.1	1.2	1.2
Observation 3	2.0	1.9	1.9
Observation 4	1.0	8.0	8.0
Observation 4	3.0	4.0	4.0
Observation 6	5.0	−6.0	6.0

a. Use the standard inner product on $\mathbb{R}^6$ to calculate the covariances $\mathrm{cov}(\mathbf{x}, \mathbf{y})$ and $\mathrm{cov}(\mathbf{x}, \mathbf{z})$ and the standard deviations $\sigma_{\mathbf{x}}, \sigma_{\mathbf{y}}$, and $\sigma_{\mathbf{z}}$ for the data sets $\mathbf{x}$, $\mathbf{y}$, and $\mathbf{z}$.

b. Find a nonzero vector $\mathbf{y} \in \mathbb{R}^6$ for which $.5 \leq \mathrm{cov}(\mathbf{x}, \mathbf{y}) \leq .75$.

c. Find a nonzero vector $\mathbf{y} \in \mathbb{R}^6$ for which $-.75 \leq \mathrm{cov}(\mathbf{x}, \mathbf{y}) \leq -.5$.

d. Find a nonzero vector $\mathbf{y} \in \mathbb{R}^6$ for which $\sigma_{\mathbf{y}} > 4$.

e. Find a nonzero vector $\mathbf{y} \in \mathbb{R}^6$ for which $\sigma_{\mathbf{y}} < .4$.

6. Use the **StandardDeviation** function in *MATHEMATICA* to find the standard deviations of the data sets in Exercise 4.

QUADRATIC FORMS

Inner products are special cases of functions $f : V \times V \to \mathbb{R}$ satisfying the equations

$$f(a_1\mathbf{x}_1 + a_2\mathbf{x}_2, \mathbf{y}) = a_1 f(\mathbf{x}_1, \mathbf{y}) + a_2 f(\mathbf{x}_2, \mathbf{y})$$
$$f(\mathbf{x}, b_1\mathbf{y}_1 + b_2\mathbf{y}_2) = b_1 f(\mathbf{x}, \mathbf{y}_1) + b_2 f(\mathbf{x}, \mathbf{y}_2)$$

for all vectors $\mathbf{x}, \mathbf{y} \in V$. Such functions are known as **bilinear forms**. A bilinear form is **symmetric** if $f(\mathbf{x}, \mathbf{y}) = f(\mathbf{y}, \mathbf{x})$ for all $\mathbf{x} \in V$. Thus inner products are symmetric bilinear forms. We now show that every symmetric bilinear form is determined by a symmetric matrix. We apply this result to study the quadratic equations $q(\mathbf{x}) = f(\mathbf{x}, \mathbf{x})$ of standard geometric objects determined by symmetric bilinear forms.

THEOREM 7.8 (Bilinear form theorem)   *For every bilinear form $f : V \times V \to \mathbb{R}$, there exists a matrix A for which $f(\mathbf{x}, \mathbf{y}) = \mathbf{x}^T A \mathbf{y}$.*

Proof. Suppose that $\mathcal{B} = \{\mathbf{v}_1, \mathbf{v}_2\}$ is a basis for V. Then bilinearity of f implies that for any two vectors $\mathbf{x} = a_1\mathbf{v}_1 + a_2\mathbf{v}_2$ and $\mathbf{y} = b_1\mathbf{v}_1 + b_2\mathbf{v}_2$, the scalar $f(\mathbf{x}, \mathbf{y})$ can be written as matrix products:

$$f(\mathbf{x}, \mathbf{y}) = \begin{pmatrix} a_1 & a_2 \end{pmatrix} \begin{pmatrix} f(\mathbf{v}_1, \mathbf{v}_1) & f(\mathbf{v}_1, \mathbf{v}_2) \\ f(\mathbf{v}_2, \mathbf{v}_1) & f(\mathbf{v}_2, \mathbf{v}_2) \end{pmatrix} \begin{pmatrix} b_1 \\ b_2 \end{pmatrix}$$

$$= \begin{pmatrix} a_1 \\ a_2 \end{pmatrix}^T \begin{pmatrix} f(\mathbf{v}_1, \mathbf{v}_1) & f(\mathbf{v}_1, \mathbf{v}_2) \\ f(\mathbf{v}_2, \mathbf{v}_1) & f(\mathbf{v}_2, \mathbf{v}_2) \end{pmatrix} \begin{pmatrix} b_1 \\ b_2 \end{pmatrix}$$

This shows that $f(\mathbf{x}, \mathbf{y}) = \mathbf{x}^T A \mathbf{y}$, where $A = (a_{ij})$ and $a_{ij} = f(\mathbf{v}_i, \mathbf{v}_j)$. The general case is analogous. ∎

EXAMPLE 7.26 ■ **A Bilinear Form on $\mathbb{R}^3$**

Represent the function $f : \mathbb{R}^3 \times \mathbb{R}^3 \to \mathbb{R}$, defined by

$$f\left[\begin{pmatrix} x_1 \\ x_2 \\ x_3 \end{pmatrix}, \begin{pmatrix} y_1 \\ y_2 \\ y_3 \end{pmatrix} \right] = 6x_1 y_1 - 2x_1 y_2 + 4x_2 y_1 - 8x_2 y_3 + 4x_3 y_2 - x_3 y_3$$

in the form $\mathbf{x}^T A \mathbf{y}$.

Solution. By multiplying out the right-hand side of the equation

$$f(\mathbf{x}, \mathbf{y}) = \begin{pmatrix} x_1 & x_2 & x_3 \end{pmatrix} \begin{pmatrix} 6 & -2 & 0 \\ 4 & 0 & -8 \\ 0 & 4 & -1 \end{pmatrix} \begin{pmatrix} y_1 \\ y_2 \\ y_3 \end{pmatrix}$$

we see that the matrix

$$A = \begin{pmatrix} 6 & -2 & 0 \\ 4 & 0 & -8 \\ 0 & 4 & -1 \end{pmatrix}$$

and $f(\mathbf{x}, \mathbf{y}) = \mathbf{x}^T A \mathbf{y}$. ◀▶

It follows easily from Theorem 7.8 that symmetric bilinear forms are represented by symmetric matrices.

THEOREM 7.9   *Every symmetric bilinear form can be represented by a symmetric matrix.*

Proof. Suppose that $f : V \times V \to \mathbb{R}$ is a symmetric bilinear form. Then $f(\mathbf{x}, \mathbf{y}) = f(\mathbf{y}, \mathbf{x})$ for all $\mathbf{x}, \mathbf{y} \in V$. By Theorem 7.8, there exists a matrix A for which $f(\mathbf{x}, \mathbf{y}) = \mathbf{x}^T A \mathbf{y}$ for all $\mathbf{x}, \mathbf{y} \in V$. Since $\mathbf{x}^T A \mathbf{y}$ is a scalar, $\mathbf{x}^T A \mathbf{y} = (\mathbf{x}^T A \mathbf{y})^T$. Therefore

$$f(\mathbf{x}, \mathbf{y}) = \mathbf{x}^T A \mathbf{y} = (\mathbf{x}^T A \mathbf{y})^T = \mathbf{y}^T A^T \mathbf{x} = f(\mathbf{y}, \mathbf{x}) = \mathbf{y}^T A \mathbf{x}$$

for all $\mathbf{x}, \mathbf{y} \in V$. This means that $\mathbf{y}^T A^T \mathbf{x} = \mathbf{y}^T A \mathbf{x}$ for all $\mathbf{x}, \mathbf{y} \in V$. Hence $A^T = A$. ∎

We now use symmetric bilinear forms to define **quadratic forms**.

DEFINITION 7.9   *A **quadratic form** $q : V \to \mathbb{R}$ on a real inner product space V is a function $q(\mathbf{x}) = f(\mathbf{x}, \mathbf{x})$ for some symmetric bilinear form $f : V \times V \to \mathbb{R}$.*

Since every symmetric bilinear form can be represented by a symmetric matrix, it is clear that every quadratic form can also be represented by a symmetric matrix. The study of quadratic forms is thus intimately connected with the study of symmetric matrices.

EXAMPLE 7.27 ■ **The Matrix of a Quadratic Form on** $\mathbb{R}^2$

Use *MATHEMATICA* to find the quadratic form $q(\mathbf{x}) : \mathbb{R}^2 \to \mathbb{R}$ determined by

$$A = \begin{pmatrix} 1 & 2 \\ 2 & 3 \end{pmatrix} \quad \text{and} \quad \mathbf{x} = \begin{pmatrix} x \\ y \end{pmatrix}$$

Solution. Since *MATHEMATICA* detects from the context whether a vector is a row or column vector, we can write $\mathbf{x}A\mathbf{x}$ in place of $\mathbf{x}^T A\mathbf{x}$.

```
In[1]:= Expand[q={x,y}.{{1,2},{2,3}}.{x,y}]
Out[1]= x² +4xy+3y²
```

This shows that $q(\mathbf{x}) = \mathbf{x}^T A\mathbf{x} = x^2 + 4xy + 3y^2 = q(x, y)$. ◄►

EXAMPLE 7.28 ■ **A Quadratic Form on** $\mathbb{R}^2$ **as a Matrix Product**

Use *MATHEMATICA* to convert the quadratic form $q(x, y) = 7x^2 - 9y^2$ into a matrix product.

Solution. We use *MATHEMATICA* to find

$$q(x, y) = q(\mathbf{x}) = \mathbf{x}^T A\mathbf{x} = \begin{pmatrix} x & y \end{pmatrix} \begin{pmatrix} a & b \\ b & c \end{pmatrix} \begin{pmatrix} x \\ y \end{pmatrix}$$

```
In[1]:= Expand[q={x,y}.{{a,b},{b,c}}.{x,y}]
Out[1]= ax² +2bxy+cy²
```

Therefore, $ax^2 + 2bxy + cy^2 = 7x^2 - 9y^2$. This means that $a = 7$, $b = 0$, and $c = -9$. Hence

$$q(\mathbf{x}) = \begin{pmatrix} x & y \end{pmatrix} \begin{pmatrix} 7 & 0 \\ 0 & -9 \end{pmatrix} \begin{pmatrix} x \\ y \end{pmatrix}$$

is the required matrix form. ◄►

EXAMPLE 7.29 ■ **A Quadratic Form on $\mathbb{R}^3$ as a Matrix Product**

Use *MATHEMATICA to* convert the quadratic form

$$q(\mathbf{x}) = q(x, y, z) = 3x^2 - 2xy + 3xz + 6y^2 + 5yz - 8z^2$$

to a matrix product.

Solution. We know from the definition of quadratic forms that the required matrix is symmetric. Suppose, therefore, that

$$A = \begin{pmatrix} d & a & b \\ a & e & c \\ b & c & f \end{pmatrix}$$

is the required matrix. We use *MATHEMATICA* to find $q(\mathbf{x})$.

```
In[1]:= Expand[q={x,y,z}.{{d,a,b},{a,e,c},{b,c,f}}.{x,y,z}]
Out[1]= d x²+2axy+ey²+2bxz+2cyz+fz²
```

Therefore,

$$dx^2 + 2axy + ey^2 + 2bxz + 2cyz + fz^2 = 3x^2 - 2xy + 3xz + 6y^2 + 5yz - 8z^2$$

By comparing coefficients, we can see that $d = 3$, $a = -1$, $e = 6$, $2b = 3$, $2c = 5$, and $f = -8$. This means that

$$\begin{pmatrix} x & y & z \end{pmatrix} \begin{pmatrix} 3 & -1 & \frac{3}{2} \\ -1 & 6 & \frac{5}{2} \\ \frac{3}{2} & \frac{5}{2} & -8 \end{pmatrix} \begin{pmatrix} x \\ y \\ z \end{pmatrix}$$

is the matrix form of $q(\mathbf{x})$. ◄►

EXERCISES 7.9

1. Rewrite the following bilinear forms as matrix products $\mathbf{x}^T A \mathbf{y}$.

   a. $f(u, v, x, y) = 6xu + 5xv + 5yu + 8vy$

   b. $f(u, v, x, y) = 6xu + 5xv + 8yu + 8vy$

   c. $f(u, v, x, y) = -6xu + 5xv + 8yu - 8vy$

2. Rewrite the following symmetric bilinear forms in equational form.

a. $f\left[\begin{pmatrix} u \\ v \end{pmatrix}, \begin{pmatrix} x \\ y \end{pmatrix}\right] = \begin{pmatrix} u & v \end{pmatrix}\begin{pmatrix} 6 & 0 \\ 0 & 8 \end{pmatrix}\begin{pmatrix} x \\ y \end{pmatrix}$

b. $f\left[\begin{pmatrix} u \\ v \end{pmatrix}, \begin{pmatrix} x \\ y \end{pmatrix}\right] = \begin{pmatrix} u & v \end{pmatrix}\begin{pmatrix} 6 & -1 \\ -1 & 8 \end{pmatrix}\begin{pmatrix} x \\ y \end{pmatrix}$

c. $f\left[\begin{pmatrix} u \\ v \end{pmatrix}, \begin{pmatrix} x \\ y \end{pmatrix}\right] = \begin{pmatrix} u & v \end{pmatrix}\begin{pmatrix} -6 & 2 \\ 2 & 8 \end{pmatrix}\begin{pmatrix} x \\ y \end{pmatrix}$

d. $f\left[\begin{pmatrix} u \\ v \end{pmatrix}, \begin{pmatrix} x \\ y \end{pmatrix}\right] = \begin{pmatrix} u & v \end{pmatrix}\begin{pmatrix} -6 & 2 \\ 2 & -8 \end{pmatrix}\begin{pmatrix} x \\ y \end{pmatrix}$

3. Show that the function $f : \mathbb{R}^2 \times \mathbb{R}^2 \to \mathbb{R}$ defined by

$$f\left[\begin{pmatrix} a \\ b \end{pmatrix}, \begin{pmatrix} c \\ d \end{pmatrix}\right] = \det\begin{pmatrix} a & c \\ b & d \end{pmatrix}$$

is a bilinear form.

4. Show that if $f : \mathbb{R}^2 \times \mathbb{R}^2 \to \mathbb{R}$ is a bilinear form for which $f(\mathbf{x}, \mathbf{y}) = -f(\mathbf{y}, \mathbf{x})$ for all $\mathbf{x}, \mathbf{y} \in \mathbb{R}^2$, then

$$f\left[\begin{pmatrix} a \\ b \end{pmatrix}, \begin{pmatrix} c \\ d \end{pmatrix}\right] = s \det\begin{pmatrix} a & c \\ b & d \end{pmatrix}$$

for some scalar s. (*Hint*: Use the standard basis of $\mathbb{R}^2$.)

5. Decide which of the following functions $\mathbb{R}^2 \to \mathbb{R}$ are quadratic forms. Explain your answer.

 a. $q(x, y) = x + y$ b. $q(x, y) = xy$

 c. $q(x, y) = x^2 + x + y^2$ d. $q(x, y) = 3x^2 + 2xy - y^2$

 e. $q(x, y) = 5y^2$ f. $q(x, y) = 5y^2 + 7$

 g. $q(x, y) = 0$ h. $q(x, y) = x^2 + xy$

6. Find the bilinear forms associated with the following quadratic forms.

 a. $q(x, y) = x^2 + y^2$ b. $q(x, y) = 3x^2 + 2y^2$

 c. $q(x, y) = x^2 - 2xy + y^2$ d. $q(x, y) = x^2$

7. Show that if q is a quadratic form determined by a symmetric bilinear form f, then

$$f(\mathbf{x}, \mathbf{y}) = \frac{1}{2}(q(\mathbf{x} + \mathbf{y}) - q(\mathbf{x}) - q(\mathbf{y}))$$

8. Use *MATHEMATICA* to find the quadratic form $q(\mathbf{x}) : \mathbb{R}^3 \to \mathbb{R}$ determined by

$$A = \begin{pmatrix} 1 & 2 & 3 \\ 2 & 4 & 5 \\ 3 & 5 & 6 \end{pmatrix} \quad \text{and} \quad \mathbf{x} = \begin{pmatrix} x \\ y \\ z \end{pmatrix}$$

The Principal Axis Theorem

In Chapter 8, we will prove that for every real symmetric matrix A there exists an orthogonal matrix Q and a diagonal matrix D for which $A = QDQ^T$. We now show how this result can be used to simplify the equations of standard geometric objects.

THEOREM 7.10 (Principal axis theorem)   *For every real quadratic form $q(\mathbf{x}) = \mathbf{x}^T A \mathbf{x}$ there exists an orthogonal matrix Q and a diagonal matrix D for which $q(\mathbf{y}) = \mathbf{y}^T D \mathbf{y}$.*

Proof. We know that the quadratic form $q(\mathbf{x})$ can be represented in the form $\mathbf{x}^T A \mathbf{x}$ for some symmetric matrix A. By Corollary 8.35, A is orthogonally diagonalizable. Therefore, there exists a diagonal matrix D and an orthogonal matrix Q such that $A = QDQ^T$. Let $\mathbf{y} = Q^T \mathbf{x}$. Then $\mathbf{x} = Q\mathbf{y}$ and

$$q(\mathbf{x}) = \mathbf{x}^T A \mathbf{x} = (Q\mathbf{y})^T A (Q\mathbf{y}) = \mathbf{y}^T Q^T A Q \mathbf{y} = \mathbf{y}^T D \mathbf{y}$$

Thus $q(\mathbf{y}) = \mathbf{y}^T D \mathbf{y}$ is the required quadratic form. ∎

Theorem 7.10 tells us that we can always change the coordinate vector $\mathbf{x}$ to a vector

$$\mathbf{y} = \begin{pmatrix} y_1 \\ \vdots \\ y_n \end{pmatrix} = Q\mathbf{x}$$

such that $D = \text{diag}(\lambda_1, \ldots, \lambda_n)$ and $q(\mathbf{y}) = \mathbf{y}^T D \mathbf{y} = \lambda_1 y_1^2 + \cdots + \lambda_n y_n^2$.

EXAMPLE 7.30   ■ **A Diagonal Representation of a Quadratic Form in $\mathbb{R}^3$**

Use *MATHEMATICA* to represent the quadratic form $q(x_1, x_2) = 3x_1^2 - 2x_1 x_2 + 3x_2^2$ by a diagonal matrix.

Solution. We begin by writing $q(\mathbf{x})$ in matrix form:

$$q(\mathbf{x}) = \mathbf{x}^T A \mathbf{x} = \begin{pmatrix} x_1 & x_2 \end{pmatrix} \begin{pmatrix} 3 & -1 \\ -1 & 3 \end{pmatrix} \begin{pmatrix} x_1 \\ x_2 \end{pmatrix}$$

Since A is orthogonally diagonalizable, $A = QDQ^T$ for some diagonal matrix D and some orthogonal matrix Q. We know that the diagonal entries of D must be the eigenvalues of A and the columns of Q must be the normalized eigenvectors associated with the eigenvalues of D. Moreover, $\mathbf{y} = Q\mathbf{x}$ and $q(\mathbf{y}) = \mathbf{y}^T Q^T A Q \mathbf{y}$. To represent $q(\mathbf{x})$ in the diagonal form $q(\mathbf{y})$, we must therefore find the matrices D and Q.

```
In[1]:= Eigensystem[{{3,-1},{-1,3}}]
```

```
Out[1]= {{2,4},{{1,1},{-1,1}}}
```

Thus 2 and 4 are the eigenvalues of A, and $(1, 1)$ and $(-1, 1)$ are corresponding eigenvectors. We note that $\begin{pmatrix} 1 & 1 \end{pmatrix} \begin{pmatrix} -1 \\ 1 \end{pmatrix} = 0$. This shows that the returned eigenvectors are orthogonal. To form the orthogonal matrix Q, we normalize these vectors in the two-norm. Since

$$\left\| \begin{pmatrix} 1 \\ 1 \end{pmatrix} \right\| = \sqrt{1^2 + 1^2} = \sqrt{2} = \sqrt{1^2 + (-1)^2} = \left\| \begin{pmatrix} -1 \\ 1 \end{pmatrix} \right\|$$

we let $y_1 = \frac{1}{\sqrt{2}}(1, 1)$ and $y_2 = \frac{1}{\sqrt{2}}(-1, 1)$. Therefore,

$$Q = \begin{pmatrix} \frac{1}{\sqrt{2}} & -\frac{1}{\sqrt{2}} \\ \frac{1}{\sqrt{2}} & \frac{1}{\sqrt{2}} \end{pmatrix} \quad \text{and} \quad Q^T = \begin{pmatrix} \frac{1}{\sqrt{2}} & \frac{1}{\sqrt{2}} \\ -\frac{1}{\sqrt{2}} & \frac{1}{\sqrt{2}} \end{pmatrix}$$

We now verify that $A = QDQ^T$.

```
In[2]:= Q= ( 1/√2  -1/√2
             1/√2   1/√2 );
```

```
In[3]:= Dg=DiagonalMatrix[{2,4}];
```

```
In[4]:= Qt=Transpose[Q];
```

```
In[5]:= Q.Dg.Qt=={{3,-1},{-1,3}}
Out[5]= True
```

This shows that $q(y_1, y_2) = 2y_1^2 + 4y_2^2$ is the required diagonal representation. ◄►

Graphically, the orthogonal diagonalization of a quadratic form corresponds to a rotation of the form. Since the matrix Q in the transformation $A \rightarrow Q^T A Q = D$ in the previous example is orthogonal, the transformation preserves the basic geometric properties of length and angles; it merely changes the position of graph of the form. This property explains in part the following definition: Two symmetric matrices A and B are said to be ***congruent*** if there exists an orthogonal matrix Q for which $A = Q^T B Q$. The matrices A and D in Example 7.30 are congruent.

EXAMPLE 7.31 ■ Congruence and Quadratic Forms

We know from Example 7.30 that the matrices

$$\begin{pmatrix} 3 & -1 \\ -1 & 3 \end{pmatrix} \quad \text{and} \quad \begin{pmatrix} 2 & 0 \\ 0 & 4 \end{pmatrix}$$

are congruent. Use *MATHEMATICA* to graph the quadratic equations

$$3x^2 - 2xy + y^2 = 10 \quad \text{and} \quad 2x^2 + 4y^2 = 10$$

determined by these matrices.

Solution. We begin by loading the **ImplicitPlot** package.

```
In[1]:= <<Graphics`ImplicitPlot`
```

Next we graph the two equations in the same coordinate system.

```
In[2]:= ImplicitPlot[{3x²-2x y+3y²==10,2x²+4y²==10},
{x,-4,4},{y,-4,4}]
```

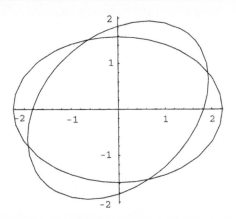

```
Out[2]= - Graphics -
```

As we can see, the shape and position of the two graphs suggest that we can obtain the graph of the equation $2x^2 + 4y^2 = 10$ by rotating the graph of the equation $3x^2 - 2xy + y^2 = 10$ about the y-axis. ◄►

Conic Sections

One of the benefits of the diagonal representation of quadratic forms is that the geometric interpretation of the associated quadratic equations is immediate. We know from geometry, for instance, that in $\mathbb{R}^2$, the equation

$$2y_1^2 + 4y_2^2 = 1$$

represents an ellipse since we know that in all cases, a quadratic equation $ax^2 + by^2 = 1$ represents an ellipse if $a, b > 0$ and $a \neq b$. It represents a circle if $a, b > 0$ and $a = b$. An interpretation of an equation of the form

$$ax^2 + bxy + cy^2 = 1$$

is less obvious. We therefore aim at finding an equivalent equation without an xy term. Similar comments apply to other geometric objects. Here are some examples.

EXAMPLE 7.32   ■ An Ellipse in $\mathbb{R}^2$

Use *MATHEMATICA* to show that the quadratic equation $3x^2 + 5y^2 = 1$ represents an ellipse.

Solution. We use an implicit plot to graph the equation.

```
In[1]:= <<Graphics`ImplicitPlot`
```

```
In[2]:= ImplicitPlot[3x²+5y²==1,{x,-√3,√3}]
```

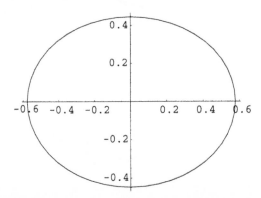

```
Out[2]= - Graphics -
```

This shows graphically that the equation $3x^2 + 5y^2 = 1$ represents an ellipse. ◄►

EXAMPLE 7.33 ■ **A Hyperbola in** $\mathbb{R}^2$ **Centered Around the** *y***-Axis**

Use *MATHEMATICA* to show that the quadratic equation $3x^2 - 5y^2 = 1$ represents a hyperbola centered around the *y*-axis.

Solution. We use an implicit plot to graph the equation.

In[1]:= <<**Graphics'ImplicitPlot'**

In[2]:=**ImplicitPlot[3x^2-5y^2==1,{x,-$\sqrt{3}$,$\sqrt{3}$}]**

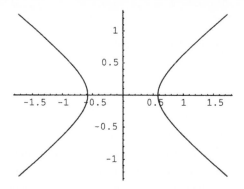

Out[2]= - Graphics -

This shows graphically that the equation $3x^2 - 5y^2 = 1$ represents a hyperbola centered around the *y*-axis. ◀▶

EXAMPLE 7.34 ■ **A Hyperbola in** $\mathbb{R}^2$ **Centered Around the** *x***-Axis**

Use *MATHEMATICA* to show that the quadratic equation $5y^2 - 3x^2 = 1$ represents a hyperbola centered around the *x*-axis.

Solution. We use an implicit plot to graph the equation.

In[1]:= <<**Graphics'ImplicitPlot'**

In[2]:= **ImplicitPlot[5y^2-3x^2==1,**

{x,-5,5},{y,-$\sqrt{5}$,$\sqrt{5}$},AxesOrigin $\rightarrow$ {0,0}]

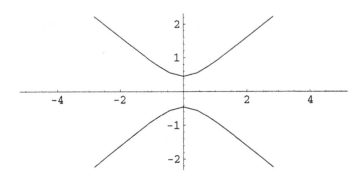

Out[2]= - ContourGraphics -

This shows graphically that the equation $5y^2 - 3x^2 = 1$ represents a hyperbola centered around the x-axis. ◄►

Quadric Surfaces

The following theorem, known as Sylvester's theorem, shows that all real quadratic forms can be written in a canonical way that reveals the properties of the geometric objects represented by the form. The theorem uses the fact that any real symmetric matrix is congruent to exactly one diagonal matrix with diagonal entries 1, -1, and 0, in which all positive diagonal entries precede all negative diagonal entries, and all negative diagonal entries precede all zero entries. The set of these diagonal matrices is a set of canonical forms for real symmetric matrices under the relation of **congruence**. Up to congruence, the next theorem therefore provides a complete finite classification of real quadratic forms.

THEOREM 7.11 (Sylvester's theorem)   *Every real quadratic form q has a unique representation of the form $q(x_1, \ldots, x_n) = x_1^2 + \cdots + x_s^2 - x_{s+1}^2 - \cdots - x_r^2$.*

Proof. By Corollary 8.35, we can represent q in the diagonal form

$$
q(y_1, \ldots, y_n) = \begin{pmatrix} y_1 & \cdots & y_n \end{pmatrix} D \begin{pmatrix} y_1 \\ \vdots \\ y_n \end{pmatrix}
$$

for some diagonal matrix $D = \operatorname{diag}(\lambda_1, \ldots, \lambda_n) \in \mathbb{R}^{n \times n}$. Since real numbers can be ordered, we can arrange the diagonal entries of D so that all positive $\lambda_1 \geq \cdots \geq \lambda_s$ precede all negative $\lambda_{s+1} \leq \cdots \leq \lambda_r$, and that these precede all $\lambda_{r+1} = \cdots = \lambda_n = 0$. Moreover, since every positive real number λ has a square root μ, we can put $\lambda_i y_i^2 = \mu_i^2 y_i^2 = x_i^2$ for $\lambda_i > 0$ and $\lambda_i y_i^2 = -\mu_i^2 y_i^2 = -x_i^2$ for $\lambda_i < 0$. This yields

$$
q(x_1, \ldots, x_n) = x_1^2 + \cdots + x_s^2 - x_{s+1}^2 - \cdots - x_r^2
$$

It remains to prove that this representation is unique. Suppose therefore that

$$q(z_1, \ldots, z_n) = z_1^2 + \cdots + z_t^2 - z_{t+1}^2 - \cdots - z_r^2$$

is another representation of q with t positive terms. Let $s \neq t$ and assume that $t < s$. Now consider the two subspaces

$$U = \{(x_1, \ldots, x_n) \in \mathbb{R}^n : x_1 = \cdots = x_t = 0\}$$
$$W = \{(z_1, \ldots, z_n) \in \mathbb{R}^n : z_{s+1} = \cdots = z_n = 0\}$$

of $\mathbb{R}^n$. The space U is of dimension $n - t$ and W is of dimension s. Moreover, the dimension of $U + W$ is at most $\dim \mathbb{R}^n = n$. By Theorem 4.35, we therefore have

$$\dim(U \cap W) = \dim U + \dim W - \dim(U + W)$$
$$\geq (n - t) + s - n$$
$$= s - t$$

Since $t < s$, we can find a nonzero vector $\mathbf{x} \in U \cap W$. The definition of U and the fact that

$$q(z_1, \ldots, z_n) = z_1^2 + \cdots + z_t^2 - z_{t+1}^2 - \cdots - z_r^2$$

imply that $q(\mathbf{x}) \leq 0$. On the other hand, the definition of W and the fact that

$$q(x_1, \ldots, x_n) = x_1^2 + \cdots + x_s^2 - x_{s+1}^2 - \cdots - x_r^2$$

imply that $q(\mathbf{x}) > 0$. This is a contradiction. Hence we must have $s = t$. ∎

Sylvester's theorem tells us that by choosing a suitable basis, we can make an ellipse look like a circle, but we can never choose a basis, for example, that makes an ellipse look like a hyperbola.

EXAMPLE 7.35   ■ Circles and Ellipses

Use *MATHEMATICA* to superimpose the graphs of two sections of the quadratic forms $q_1(x, y) = 2x^2 + 5y^2$ and $q_2(x, y) = x^2 + y^2$.

Solution. We begin by loading the **ImplicitPlot** package.

```
In[1]:= <<Graphics`ImplicitPlot`
```

Next we graph the two equations in the same coordinate system.

```
In[2]:= ImplicitPlot[{2x²+5y²==10,x²+y²==5},{x,-4,4},{y,-4,4}]
```

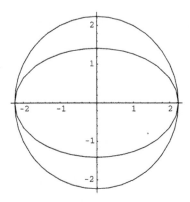

```
Out[2]= - Graphics -
```

As we can see, the graph consists of an ellipse inside of a circle. ◄►

The number of plus and minus signs in the canonical representation of a quadratic form provided by Sylvester's theorem is often called the **signature** of the form. Theorem 7.11 proves that the signature of a quadratic form is unique and can be used to classify these forms. We use *MATHEMATICA* to illustrate Sylvester's theorem by graphing the five possible canonical quadric surfaces determined by the signs of the eigenvalues λ_1 and λ_2. Let $q(x_1, x_n) = \lambda_1 x_1^2 + \lambda_2 x_2^2$, and consider the following cases:

1. The case where $\lambda_1 > 0$ and $\lambda_2 > 0$
2. The case where $\lambda_1 > 0$ and $\lambda_2 < 0$
3. The case where $\lambda_1 < 0$ and $\lambda_2 < 0$
4. The case where $\lambda_1 > 0$ and $\lambda_2 = 0$
5. The case where $\lambda_1 < 0$ and $\lambda_2 = 0$

We can think of these forms as representations of three-dimensional geometric objects and use the **Plot3D** function to graph them. To simplify the labeling, we will write x in place of x_1 and y in place of x_2.

EXAMPLE 7.36   ■ A Concave-Up Elliptic Paraboloid in $\mathbb{R}^3$

```
In[1]:= Plot3D[x²+y²,{x,-10,10},{y,-10,10}]
```

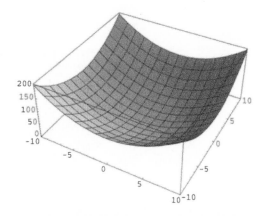

Out[1]= - SurfaceGraphics -

As we can see, the graph is concave up and has the shape of an elliptic paraboloid. ◄►

EXAMPLE 7.37 ■ **A Hyperbolic Paraboloid in $\mathbb{R}^3$**

```
In[1]:= Plot3D[x²-y², {x,-10,10}, {y,-10,10}]
```

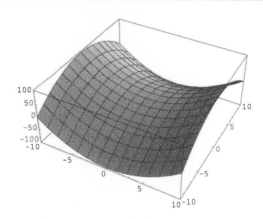

Out[1]= - SurfaceGraphics -

As we can see, the graph has the shape of a hyperbolic paraboloid. ◄►

EXAMPLE 7.38 ■ **A Concave-Down Elliptic Paraboloid in $\mathbb{R}^3$**

```
In[1]:= Plot3D[-x²-y², {x,-10,10}, {y,-10,10}]
```

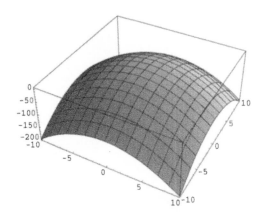

Out[1]= - SurfaceGraphics -

As we can see, the graph is concave down and has the shape of an elliptic paraboloid. ◀▶

EXAMPLE 7.39 ■ **A Concave-Up Parabolic Cylinder in** $\mathbb{R}^3$

In[1]:= **Plot3D[x², {x,-10,10}, {y,-10,10}]**

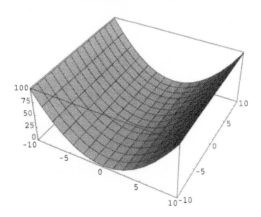

Out[1]= - SurfaceGraphics -

As we can see, the graph is concave up and has the shape of a parabolic cylinder. ◀▶

EXAMPLE 7.40 ■ **A Concave-Down Parabolic Cylinder in** $\mathbb{R}^3$

In[1]:= **Plot3D[-x², {x,-10,10}, {y,-10,10}]**

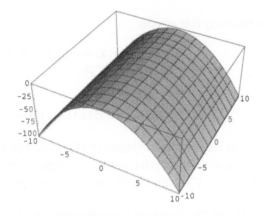

Out[1]= - SurfaceGraphics -

As we can see, the graph is concave up and has the shape of a parabolic cylinder. ◄►

EXERCISES 7.10

1. Decompose the following matrices into a product of the form QDQ^T, where Q is an orthogonal and D a diagonal matrix.

$$\text{a. } A = \begin{pmatrix} -5 & -2 \\ -2 & -1 \end{pmatrix} \quad \text{b. } A = \begin{pmatrix} 1 & 0 & 0 \\ 0 & 2 & 4 \\ 0 & 4 & 2 \end{pmatrix} \quad \text{c. } A = \begin{pmatrix} 9 & 5 \\ 5 & 3 \end{pmatrix}$$

2. Use *MATHEMATICA* to graph the following quadratic equations.

$$\text{a. } 4x^2 + 7y^2 = 1 \qquad \text{b. } 4x^2 - 7y^2 = 1$$
$$\text{c. } -x^2 + 10xy + 2y^2 = 1 \quad \text{d. } x^2 + 10xy + y^2 = 1$$

3. Use quadratic forms of your choice to illustrate Sylvester's theorem for quadratic forms in two, three, and four variables.

COMPLEX INNER PRODUCTS

Certain applications of linear algebra require complex-valued inner products. A *complex inner product* on a complex vector space V is a complex-valued function $(\mathbf{x}, \mathbf{y}) \to \langle \mathbf{x}, \mathbf{y} \rangle$

satisfying the following four axioms.

1. $\langle a\mathbf{x} + b\mathbf{y}, \mathbf{z} \rangle = a \langle \mathbf{x}, \mathbf{z} \rangle + b \langle \mathbf{y}, \mathbf{z} \rangle$
2. $\langle \mathbf{x}, \mathbf{y} \rangle = \overline{\langle \mathbf{y}, \mathbf{x} \rangle}$
3. $\langle \mathbf{x}, \mathbf{x} \rangle \geq 0$
4. $\langle \mathbf{x}, \mathbf{x} \rangle = 0$ if and only if $\mathbf{x} = \mathbf{0}$

The scalar $\overline{\langle \mathbf{y}, \mathbf{x} \rangle}$ is the complex conjugate of $\langle \mathbf{x}, \mathbf{y} \rangle$. Complex inner products are no longer symmetric since $\langle \mathbf{x}, \mathbf{y} \rangle$ must equal its complex conjugate. A **complex inner product space** is a complex vector space V equipped with a complex inner product.

EXAMPLE 7.41 ■ The Standard Inner Product on $\mathbb{C}^n$

Let $\mathbf{x} = (u_1, \ldots, u_n)$ and $\mathbf{y} = (v_1, \ldots, v_n)$ be two vectors in $\mathbb{C}^n$. Then the function defined by

$$\mathbf{x}^T \overline{\mathbf{y}} = \left(\begin{array}{ccc} u_1 & \cdots & u_n \end{array} \right) \left(\begin{array}{c} \overline{v}_1 \\ \vdots \\ \overline{v}_n \end{array} \right) = u_1 \overline{v}_1 + \cdots + u_n \overline{v}_n$$

is a complex inner product. ◄►

The inner product in Example 7.41 is called the **standard inner product** on $\mathbb{C}^n$. As in the real case, we can use this inner product to define a norm on $\mathbb{C}^n$.

EXAMPLE 7.42 ■ The Two-Norm on $\mathbb{C}^2$

Suppose that $\mathbf{z} = (z_1, z_2) = (x_1 + y_1 i, x_2 + y_2 i)$ is a vector in the coordinate space $\mathbb{C}^2$, determined by the real numbers $x_1, y_1, x_2,$ and y_2, and that $|z_k| = \sqrt{x_k^2 + y_k^2}$ is the modulus of z_k. Then the function

$$\|\mathbf{z}\|_2 = \sqrt{|z_1|^2 + |z_2|^2} = \sqrt{x_1^2 + y_1^2 + x_2^2 + y_2^2}$$

is a vector norm on $\mathbb{C}^2$. The space $\mathbb{C}^2$, equipped with this norm, known as the **complex Euclidean 2-space**. It is also know as a **unitary space**. ◄►

We will need complex inner product spaces in Chapter 8 to be able to prove that real symmetric matrices have real eigenvalues.

EXERCISES 7.11

1. Calculate the standard inner product of the following pairs of vectors in $\mathbb{C}^2$.

a. $\left(\begin{array}{c} 3 - 5i \\ -i \end{array} \right)$ and $\left(\begin{array}{c} 4 + i \\ 7 + 9i \end{array} \right)$ b. $\left(\begin{array}{c} -i \\ -4 + 4i \end{array} \right)$ and $\left(\begin{array}{c} 7 + 9i \\ 5 - 2i \end{array} \right)$

2. Find the standard inner product of the following pairs of complex vectors.

a. $\begin{pmatrix} 3 \\ 1 \\ -4 \end{pmatrix}$ and $\begin{pmatrix} 4 \\ 7 \\ 5 \end{pmatrix}$ b. $\begin{pmatrix} 3 - 5i \\ -i \\ -4 + 4i \end{pmatrix}$ and $\begin{pmatrix} 4 + i \\ 7 + 9i \\ 5 - 2i \end{pmatrix}$

3. Compute the two-norm of the vectors in Exercise 1.

4. Compute the two-norm of the vectors in Exercise 2.

5. Use the two-norm to compute the distance between the pairs of vectors in Exercise 1.

6. Use the two-norm to compute the distance between the pairs of vectors in Exercise 2.

REVIEW

KEY CONCEPTS ▶ Define and discuss each of the following.

Euclidean Spaces

Euclidean distance, Euclidean inner product, Euclidean n-space.

Geometry

Angle between two vectors, cosine of the angle between two vectors, distance between two vectors, length of a vector, parallelogram law, triangle inequality.

Inner Products

Complex inner product, dot product, Euclidean inner product, standard inner product.

Matrices

Positive definite matrix, symmetric matrix.

Matrix Norms

Frobenius norm of a matrix, infinity-norm of a matrix, one-norm of a matrix, two-norm of a matrix.

Vector Norms

Inner product norm of a vector, one-norm of a vector, two-norm of a vector, infinity-norm of a vector.

Vector Spaces

Normed vector space, real inner product space, unitary space.

Vectors

Unit vector.

KEY FACTS ▶ Explain and illustrate each of the following.

1. Let A be a real $n \times n$ matrix, and let $\mathbf{x} \in \mathbb{R}^n$. Then

$$\max \left\{ \frac{\|A\mathbf{x}\|}{\|\mathbf{x}\|} : \|\mathbf{x}\| \neq \mathbf{0} \right\} = \max \left\{ \|A\mathbf{x}\| : \|\mathbf{x}\| = 1 \right\}$$

2. The function $\|A\| = \max \left\{ \frac{\|A\mathbf{x}\|}{\|\mathbf{x}\|} : \|\mathbf{x}\| \neq \mathbf{0} \right\} = \max \left\{ \|A\mathbf{x}\| : \|\mathbf{x}\| = 1 \right\}$ is a matrix norm.

3. For every nonzero vector $\mathbf{x}$ in an inner product space V, the vector $\mathbf{y} = \mathbf{x}/\|\mathbf{x}\|$ is a unit vector.

4. If $(\mathbf{x}, \mathbf{y}) \to \langle \mathbf{x}, \mathbf{y} \rangle$ is an inner product on a real vector space V, then $|\langle \mathbf{x}, \mathbf{y} \rangle| \leq \|\mathbf{x}\| \|\mathbf{y}\|$ for all $\mathbf{x}, \mathbf{y} \in V$.

5. If A is a positive definite symmetric matrix, then the function $\langle \mathbf{x}, \mathbf{y} \rangle = \mathbf{x}^T A \mathbf{y}$ is an inner product on $\mathbb{R}^n$.

6. If V is a real vector space with an inner product $\langle \mathbf{x}, \mathbf{y} \rangle$, then the function $\|\mathbf{x}\| = \sqrt{\langle \mathbf{x}, \mathbf{x} \rangle}$ is a norm on V.

7. If $\|\mathbf{z}\|$ is an inner product norm on a real vector space V, then $\|\mathbf{x} + \mathbf{y}\|^2 + \|\mathbf{x} - \mathbf{y}\|^2 = 2\left(\|\mathbf{x}\|^2 + \|\mathbf{y}\|^2\right)$ for all $\mathbf{x}, \mathbf{y} \in V$.

8. For every bilinear form $f : V \times V \to \mathbb{R}$, there exists a matrix A for which $f(\mathbf{x}, \mathbf{y}) = \mathbf{x}^T A \mathbf{y}$.

9. Every symmetric bilinear form can be represented by a symmetric matrix.

10. For every real quadratic form $q(\mathbf{x}) = \mathbf{x}^T A \mathbf{x}$, there exist an orthogonal matrix Q and a diagonal matrix D for which $q(\mathbf{y}) = \mathbf{y}^T D \mathbf{y}$.

11. Every real quadratic form q has a unique representation of the form $q(x_1, \ldots, x_n) = x_1^2 + \cdots + x_s^2 - x_{s+1}^2 - \cdots - x_r^2$.

5. If A is a positive-definite symmetric matrix, then the function $\langle x, y \rangle = x^T A y$ defines an inner product on $\mathbb{R}^n$.

6. If T has real eigenvalues, v is an inner product on $\mathbb{R}^n$, then the function $\langle x, y \rangle = \frac{1}{2} \langle x, x \rangle$ is a norm on $\mathbb{R}^n$.

7. If A is an inner product matrix on a real vector space V, then $\langle x, y \rangle = x^T A y$ $\leq \sqrt{\langle x, x \rangle} \cdot \sqrt{\langle y, y \rangle}$ for all $x, y \in V$.

8. For every bilinear form $f : V \times V \to \mathbb{R}$, there exists a matrix A for which $f(x, y) = x^T A y$.

9. Every symmetric bilinear form can be represented by a symmetric matrix.

10. For every real quadratic form $q(x) = x^T A x$, there exist an orthogonal matrix Q and a diagonal matrix D for which $Q^T A Q = D$.

11. Every real quadratic form q has a unique representation of the form $q(x_1, \ldots, x_n) = \ldots$

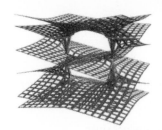

8

ORTHOGONALITY

Many applications of linear algebra are based on the geometric concept of orthogonality. Best possible approximate solutions of overdetermined linear systems, real inner products as dot products, fitting of curves to observational data, the representation of adjoint transformations using transposed matrices, and many other applications involve orthogonality in one form or another.

ORTHOGONAL VECTORS

Inner products provide us with the appropriate tool for defining and analyzing orthogonality.

DEFINITION 8.1   *Two vectors* $\mathbf{x}$ *and* $\mathbf{y}$ *in a real inner product space* V *are* **orthogonal** *with respect to an inner product* $\langle \mathbf{x}, \mathbf{y} \rangle$ *on* V *if* $\langle \mathbf{x}, \mathbf{y} \rangle = 0$.

The most famous theorem involving orthogonality is probably the Pythagorean theorem which tells us that the square on the longest side of a right triangle equals the sum of the squares on the other two sides of the triangle. The inner product version of this theorem relates the norms of two orthogonal vectors to the norm of their sum.

THEOREM 8.1 (Pythagorean theorem)   *If* $\mathbf{x}$ *and* $\mathbf{y}$ *are two orthogonal vectors in an inner product space* V, *then* $\|\mathbf{x} + \mathbf{y}\|^2 = \|\mathbf{x}\|^2 + \|\mathbf{y}\|^2$.

Proof. Since $\mathbf{x}$ and $\mathbf{y}$ are orthogonal vectors, $\langle \mathbf{x}, \mathbf{y} \rangle = \langle \mathbf{y}, \mathbf{x} \rangle = 0$. Therefore,

$$\begin{aligned}
\|\mathbf{x} + \mathbf{y}\|^2 &= \langle \mathbf{x} + \mathbf{y}, \mathbf{x} + \mathbf{y} \rangle \\
&= \langle \mathbf{x}, \mathbf{x} \rangle + 2\langle \mathbf{x}, \mathbf{y} \rangle + \langle \mathbf{y}, \mathbf{y} \rangle \\
&= \langle \mathbf{x}, \mathbf{x} \rangle + \langle \mathbf{y}, \mathbf{y} \rangle \\
&= \|\mathbf{x}\|^2 + \|\mathbf{y}\|^2
\end{aligned}$$

This proves the theorem. ∎

EXAMPLE 8.1   ■ **Two Orthogonal Vectors in** $\mathbb{R}^2$

Show that the vectors $\mathbf{x} = (1, 0)$ and $\mathbf{y} = (0, 1)$ are orthogonal in the standard inner product $\langle \mathbf{x}, \mathbf{y} \rangle = \mathbf{x}^T \mathbf{y} = x_1 y_1 + x_2 y_2$.

Solution. Since $\langle \mathbf{x}, \mathbf{y} \rangle = \mathbf{x}^T \mathbf{y} = 1 \times 0 + 0 \times 1 = 0$, the vectors are orthogonal. ◄►

EXAMPLE 8.2 ■ **Two Nonorthogonal Vectors in $\mathbb{R}^2$**

Show that the vectors $\mathbf{x} = (2, 1)$ and $\mathbf{y} = (-1, 2)$ are orthogonal in the standard inner product $\mathbf{x}^T \mathbf{y}$, but are not orthogonal in the inner product $\langle \mathbf{x}, \mathbf{y} \rangle = 2x_1 y_1 + 3x_2 y_2$.

Solution. The vectors $\mathbf{x}$ and $\mathbf{y}$ are orthogonal in the standard inner product since

$$\mathbf{x}^T \mathbf{y} = \begin{pmatrix} 2 & 1 \end{pmatrix} \begin{pmatrix} -1 \\ 2 \end{pmatrix} = 0$$

However, $\langle \mathbf{x}, \mathbf{y} \rangle = 2 \times 2 \times -1 + 3 \times 1 \times 2 = 2$. Since $\langle \mathbf{x}, \mathbf{y} \rangle \neq 0$, the vectors $\mathbf{x}$ and $\mathbf{y}$ are not orthogonal in the inner product $\langle \mathbf{x}, \mathbf{y} \rangle$. ◄►

The next theorem provides an important link between orthogonality and linear independence. We call it the **orthogonality test** for linear independence.

THEOREM 8.2 (Orthogonality test for linear independence)   *Orthogonal vectors in an inner product space are linearly independent.*

Proof. Let $\mathcal{S} = \{\mathbf{x}_1, \ldots, \mathbf{x}_n\}$ be a set of nonzero orthogonal vectors in an inner product space V. Then $\langle \mathbf{x}_i, \mathbf{x}_j \rangle = 0$ for all $1 \leq i, j \leq n$ for which $i \neq j$. Now suppose that $a_1 \mathbf{x}_1 + \cdots + a_n \mathbf{x}_n = \mathbf{0}$. It follows from the linearity of $\langle \mathbf{x}, \mathbf{y} \rangle$ in $\mathbf{y}$ and the orthogonality of the vectors $\mathbf{x}_i$ that $\langle \mathbf{x}_i, a_1 \mathbf{x}_1 + \cdots + a_n \mathbf{x}_n \rangle = a_i \langle \mathbf{x}_i, \mathbf{x}_i \rangle = 0$ for all $i \in \mathbf{n}$. Since $\mathbf{x}_i \neq \mathbf{0}$, we have $\langle \mathbf{x}_i, \mathbf{x}_i \rangle \neq 0$. Hence $a_i = 0$ for all $i \in \mathbf{n}$. Therefore $\mathcal{S}$ is linearly independent. ∎

COROLLARY 8.3   *If a basis $\mathcal{B} = \{\mathbf{x}_1, \ldots, \mathbf{x}_n\}$ for an inner product space V consists of orthogonal unit vectors, then $\mathbf{x} = \langle \mathbf{x}_1, \mathbf{x} \rangle \mathbf{x}_1 + \cdots + \langle \mathbf{x}_n, \mathbf{x} \rangle \mathbf{x}_n$ for any vector $\mathbf{x} \in V$.*

Proof. Since $\mathcal{B}$ is a basis, the vector $\mathbf{x}$ can be written as a unique linear combination

$$\mathbf{x} = a_1 \mathbf{x}_1 + \cdots + a_n \mathbf{x}_n$$

of the vectors in $\mathcal{B}$. Therefore,

$$\langle \mathbf{x}_i, \mathbf{x} \rangle = \langle \mathbf{x}_i, a_1 \mathbf{x}_1 + \cdots + a_n \mathbf{x}_n \rangle = a_i \langle \mathbf{x}_i, \mathbf{x}_i \rangle = a_i$$

This proves the corollary. ∎

Another useful consequence of Theorem 8.2 is the fact that only the zero vector can be orthogonal to all vectors of an orthogonal basis.

THEOREM 8.4 (Zero vector theorem)   *If $\mathcal{B} = \{\mathbf{x}_1, \ldots, \mathbf{x}_p\}$ is an orthogonal set of vectors in an inner product space V, and if $\mathbf{x}$ is a vector in the span of $\mathcal{B}$ and is orthogonal to all vectors in $\mathcal{B}$, then $\mathbf{x} = \mathbf{0}$.*

Proof. Since $\mathbf{x}$ is orthogonal to each vector $\mathbf{x}_i$, all $\langle \mathbf{x}, \mathbf{x}_i \rangle$ are 0. Hence

$$a_1 \langle \mathbf{x}, \mathbf{x}_1 \rangle + \cdots + a_p \langle \mathbf{x}, \mathbf{x}_p \rangle = 0$$

for all choices of $a_i \in \mathbb{R}$. The bilinearity of inner products therefore guarantees that

$$\langle \mathbf{x}, a_1 \mathbf{x}_1 + \cdots + a_p \mathbf{x}_p \rangle = 0$$

for all scalars a_i. By Theorem 8.2, the vectors $\mathbf{x}_i$ are linearly independent. Therefore, $\mathcal{B}$ is a basis for its span. This means that there exist scalars b_i for which $\mathbf{x} = b_1 \mathbf{x}_1 + \cdots + b_p \mathbf{x}_p$. Since

$$\langle \mathbf{x}, a_1 \mathbf{x}_1 + \cdots + a_p \mathbf{x}_p \rangle = 0$$

for all scalars a_i, we let $a_i = b_i$ and get

$$\langle b_1 \mathbf{x}_1 + \cdots + b_p \mathbf{x}_p, b_1 \mathbf{x}_1 + \cdots + b_p \mathbf{x}_p \rangle = 0$$

The definition of inner products implies that $b_1 \mathbf{x}_1 + \cdots + b_p \mathbf{x}_p = \mathbf{x} = \mathbf{0}$. ∎

EXAMPLE 8.3 ■ **The Linear Independence of Orthogonal Vectors**

Use *MATHEMATICA* and the orthogonality test to show that the set of vectors

$$\left\{ \mathbf{x} = \begin{pmatrix} -\frac{6}{19} \\ \frac{15}{19} \\ \frac{10}{19} \end{pmatrix}, \mathbf{y} = \begin{pmatrix} \frac{15}{19} \\ \frac{10}{19} \\ -\frac{6}{19} \end{pmatrix}, \mathbf{z} = \begin{pmatrix} \frac{10}{19} \\ -\frac{6}{19} \\ \frac{15}{19} \end{pmatrix} \right\}$$

is linearly independent.

Solution. By Theorem 8.2, it suffices to show that $\mathbf{x}$, $\mathbf{y}$, and $\mathbf{z}$ are orthogonal in the standard inner product. To do so, we use the **Dot** function.

```
In[1]:= x={-6/19,15/19,10/19};y={15/19,10/19,-6/19};
z={10/19,-6/19,15/19};
```

```
In[2]:= {Dot[x,y],Dot[x,z],Dot[y,z]}
Out[2]= {0,0,0}
```

Since the three dot products are zero, the vectors $\mathbf{x}$, $\mathbf{y}$, and $\mathbf{z}$ are orthogonal. They are therefore also linearly independent. ◂▸

EXAMPLE 8.4 ■ **A Linear Combination of Orthogonal Vectors**

Express the vector $\mathbf{y} = (3, 7, 1)$ as a linear combination $\mathbf{x} = a\mathbf{x}_1 + b\mathbf{x}_2 + c\mathbf{x}_3$ of the pairwise orthogonal vectors in the basis

$$\mathcal{B} = \left\{ \mathbf{x}_1 = \begin{pmatrix} -\frac{6}{19} \\ \frac{15}{19} \\ \frac{10}{19} \end{pmatrix}, \mathbf{x}_3 = \begin{pmatrix} \frac{15}{19} \\ \frac{10}{19} \\ -\frac{6}{19} \end{pmatrix}, \mathbf{x}_3 = \begin{pmatrix} \frac{10}{19} \\ -\frac{6}{19} \\ \frac{15}{19} \end{pmatrix} \right\}$$

and show that in the standard inner product on $\mathbb{R}^3$,

$$a = \langle \mathbf{x}_1, \mathbf{x} \rangle, \quad b = \langle \mathbf{x}_2, \mathbf{x} \rangle, \quad \text{and} \quad c = \langle \mathbf{x}_3, \mathbf{x} \rangle$$

Solution. Suppose that

$$\begin{pmatrix} 3 \\ 7 \\ 1 \end{pmatrix} = a \begin{pmatrix} -\frac{6}{19} \\ \frac{15}{19} \\ \frac{10}{19} \end{pmatrix} + b \begin{pmatrix} \frac{15}{19} \\ \frac{10}{19} \\ -\frac{6}{19} \end{pmatrix} + c \begin{pmatrix} \frac{10}{19} \\ -\frac{6}{19} \\ \frac{15}{19} \end{pmatrix}$$

Then

$$\begin{pmatrix} 3 \\ 7 \\ 1 \end{pmatrix} = \begin{pmatrix} -\frac{6}{19}a + \frac{15}{19}b + \frac{10}{19}c \\ \frac{15}{19}a + \frac{10}{19}b - \frac{6}{19}c \\ \frac{10}{19}a - \frac{6}{19}b + \frac{15}{19}c \end{pmatrix}$$

We compute the coefficients a, b, and c by solving the associated linear system.

```
In[1]:= Solve[{3,7,1}=={-6/19a+15/19b+10/19c,
15/19a+10/19b-6/19c,10/19a-6/19b+15/19c}]
Out[1]= {a→ 97/19 ,b→ 109/19 ,c→ 3/19 }
```

It remains to show that $\langle \mathbf{x}_1, \mathbf{x} \rangle = \frac{97}{19}$, $\langle \mathbf{x}_2, \mathbf{x} \rangle = \frac{109}{19}$, and $\langle \mathbf{x}_3, \mathbf{x} \rangle = \frac{3}{19}$.

```
In[2]:= Dot[{-6/19,15/19,10/19},{3,7,1}]
Out[2]= 97/19
```

```
In[3]:= Dot[{15/19,10/19,-6/19},{3,7,1}]
Out[3]= 109/19
```

```
In[4]:= Dot[{10/19,-6/19,+15/19},{3,7,1}]
Out[4]= 3/19
```

Therefore, $a = \langle \mathbf{x}_1, \mathbf{x} \rangle$, $b = \langle \mathbf{x}_2, \mathbf{x} \rangle$, and $c = \langle \mathbf{x}_3, \mathbf{x} \rangle$. ◀▶

EXERCISES 8.1

1. Use the standard inner product to verify the Pythagorean theorem for the following pairs of vectors in $\mathbb{R}^2$.

   a. $\mathbf{x} = (1, 2)$ and $\mathbf{y} = (-7, 8)$ b. $\mathbf{x} = (-91, 24)$ and $\mathbf{y} = (-7, 8)$
   c. $\mathbf{x} = (1, 0)$ and $\mathbf{y} = (-7, 8)$ d. $\mathbf{x} = (0, 0)$ and $\mathbf{y} = (-7, 8)$

2. Repeat Exercise 1 for the inner product $\mathbf{x}^T A \mathbf{y}$ determined by the diagonal matrix

$$A = \begin{pmatrix} 3 & 0 \\ 0 & 5 \end{pmatrix}$$

3. Use the standard inner product to verify the Pythagorean theorem for the following pairs of vectors in $\mathbb{R}^3$.

   a. $\mathbf{x} = (1, 2, 3)$ and $\mathbf{y} = (-7, 8, 5)$ b. $\mathbf{x} = (1, 0, 0)$ and $\mathbf{y} = (-7, 8, 5)$
   c. $\mathbf{x} = (0, 0, 0)$ and $\mathbf{y} = (-7, 8, 5)$ d. $\mathbf{x} = (-91, 24, 107)$ and $\mathbf{y} = (-7, 8, 5)$

4. Repeat Exercise 3 for the inner product $\mathbf{x}^T A \mathbf{y}$ determined by the positive definite matrix

$$A = \begin{pmatrix} 9 & 0 & -2 \\ 0 & 7 & 0 \\ -2 & 0 & 1 \end{pmatrix}$$

5. Use Theorem 8.2 to prove that the following pairs of vectors are linearly independent.

   a. $\mathbf{x} = (0, 24, 107, 0)$ and $\mathbf{y} = (1, 0, 0, 1)$
   b. $\mathbf{x} = (1, 24, 0, 1)$ and $\mathbf{y} = (-1, 0, 7, 1)$
   c. $\mathbf{x} = (0, 24, 0, 0)$ and $\mathbf{y} = (1, 0, 7, 1)$
   d. $\mathbf{x} = (1, -1, 1, -1)$ and $\mathbf{y} = (-1, 1, -1, 1)$

Orthogonal Projections

The basic operation for constructing orthogonal vectors is that of an *orthogonal projection*. Suppose that $\mathbf{x}$ and $\mathbf{y}$ are two vectors in an inner product space V. We would like to split the vector $\mathbf{x}$ into two orthogonal vectors $\text{proj}(\mathbf{x} \rightarrow \mathbf{y})$ and $\text{perp}(\mathbf{x} \perp \mathbf{y})$ so that the following hold.

   1. $\text{proj}(\mathbf{x} \rightarrow \mathbf{y}) + \text{perp}(\mathbf{x} \perp \mathbf{y}) = \mathbf{x}$
   2. $\text{proj}(\mathbf{x} \rightarrow \mathbf{y}) = a\mathbf{y}$
   3. $\langle \text{perp}(\mathbf{x} \perp \mathbf{y}), \mathbf{y} \rangle = 0$

This turns out to be quite straightforward. If the vectors $\text{proj}(\mathbf{x} \rightarrow \mathbf{y})$ and $\text{perp}(\mathbf{x} \perp \mathbf{y})$ are orthogonal, then

$$\langle \mathbf{x}, \mathbf{y} \rangle = \langle \text{proj}(\mathbf{x} \to \mathbf{y}) + \text{perp}(\mathbf{x} \perp \mathbf{y}), \mathbf{y} \rangle$$
$$= \langle \text{proj}(\mathbf{x} \to \mathbf{y}), \mathbf{y} \rangle + \langle \text{perp}(\mathbf{x} \perp \mathbf{y}), \mathbf{y} \rangle$$
$$= \langle a\mathbf{y}, \mathbf{y} \rangle + 0$$
$$= a\langle \mathbf{y}, \mathbf{y} \rangle$$

Therefore, a must equal $\langle \mathbf{x}, \mathbf{y} \rangle / \langle \mathbf{y}, \mathbf{y} \rangle$. Hence

$$\text{proj}(\mathbf{x} \to \mathbf{y}) = \frac{\langle \mathbf{x}, \mathbf{y} \rangle}{\langle \mathbf{y}, \mathbf{y} \rangle}\mathbf{y} \quad \text{and} \quad \text{perp}(\mathbf{x} \perp \mathbf{y}) = \mathbf{x} - \frac{\langle \mathbf{x}, \mathbf{y} \rangle}{\langle \mathbf{y}, \mathbf{y} \rangle}\mathbf{y}$$

Figure 1 illustrates this construction relative to the standard inner product.

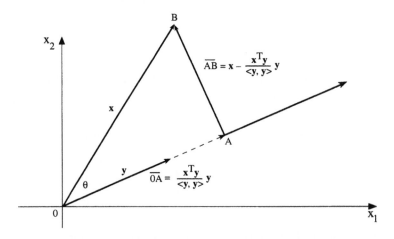

FIGURE 1   Projection of a Vector onto a Subspace.

The vector $\overline{OA} = \text{proj}(\mathbf{x} \to \mathbf{y})$ is the **orthogonal projection** of the vector $\mathbf{x}$ onto the vector $\mathbf{y}$, and the vector $\overline{AB} = \text{perp}(\mathbf{x} \perp \mathbf{y})$ is the **vector component** of $\mathbf{x}$ orthogonal to $\mathbf{y}$.

We can generalize this construction and define the orthogonal projection of a vector $\mathbf{x} \in V$ onto a subspace W of V. The construction is particularly simple relative to a basis $\mathcal{B} = \{\mathbf{x}_1, \ldots, \mathbf{x}_n\}$ for V consisting of orthogonal unit vectors. In that case, we know from Corollary 8.3 that every vector $\mathbf{x} \in V$ can be written as a unique linear combination of the form

$$\mathbf{x} = \langle \mathbf{x}_1, \mathbf{x} \rangle \mathbf{x}_1 + \cdots + \langle \mathbf{x}_n, \mathbf{x} \rangle \mathbf{x}_n$$

If $\mathcal{C} = \{\mathbf{y}_1, \ldots, \mathbf{y}_p\}$ is a subset of the basis S and $W = \text{span}(\mathcal{C})$, then we can form the projection

$$\text{proj}(\mathbf{x} \to W) = \langle \mathbf{y}_1, \mathbf{x} \rangle \mathbf{y}_1 + \cdots + \langle \mathbf{y}_p, \mathbf{x} \rangle \mathbf{y}_p$$

of $\mathbf{x}$ onto the subspace W, obtained from $\langle \mathbf{x}_1, \mathbf{x} \rangle \mathbf{x}_1 + \cdots + \langle \mathbf{x}_n, \mathbf{x} \rangle \mathbf{x}_n$ by deleting all terms $\langle \mathbf{x}_i, \mathbf{x} \rangle \mathbf{x}_i$ for which $\mathbf{x}_i \notin \mathcal{C}$. By construction, the vector $\text{proj}(\mathbf{x} \to W)$ belongs to W, and the vector $\text{perp}(\mathbf{x} \perp W) = \mathbf{x} - \text{proj}(\mathbf{x} \to W)$ is the vector component of $\mathbf{x}$ orthogonal to every vector in W. Therefore,

1. $\text{proj}(\mathbf{x} \to W) + \text{perp}(\mathbf{x} \perp \mathbf{W}) = \mathbf{x}$,

2. $\text{proj}(\mathbf{x} \to W) \in W$, and

3. $\langle \text{perp}(\mathbf{x} \perp W), \mathbf{y}_i \rangle = 0$ for each i.

The vector $\text{proj}(\mathbf{x} \to W)$ is the ***orthogonal projection*** of $\mathbf{x}$ onto the subspace W.

EXAMPLE 8.5   ■ An Orthogonal Projection onto a Subspace of $\mathbb{R}^2$

Let $\mathcal{E} = \{\mathbf{e}_1, \mathbf{e}_2\}$ be the standard basis of $\mathbb{R}^2$, let W_1 be the subspace of $\mathbb{R}^2$ spanned by $\{\mathbf{e}_1\}$, and let $\mathbf{x} = x_1\mathbf{e}_1 + x_2\mathbf{e}_2$ be any vector in $\mathbb{R}^2$. Find the vector $\text{proj}(\mathbf{x} \to W_1)$, and show that it is orthogonal to any vector in $W_2 = \text{span}\{\mathbf{e}_2\}$.

Solution. By definition, $\text{proj}(\mathbf{x} \to W_1) = x_1\mathbf{e}_1$. Let $\mathbf{y} = a\mathbf{e}_2$ be any vector in W_2. Then the calculation

$$\langle \text{proj}(\mathbf{x} \to W_1), \mathbf{y} \rangle = \langle x_1\mathbf{e}_1, a\mathbf{e}_2 \rangle = ax_1\langle \mathbf{e}_2, \mathbf{e}_1 \rangle = 0$$

shows that the vector $\text{proj}(\mathbf{x} \to W)$ is orthogonal to $\mathbf{y}$. ◄►

EXAMPLE 8.6   ■ An Orthogonal Projection onto a Subspace of $\mathbb{R}^3$

Let W be the subspace of $\mathbb{R}^3$ spanned by the subset $\{\mathbf{x}_2, \mathbf{x}_3\}$ of the basis

$$\mathcal{B} = \left\{ \mathbf{x}_1 = \begin{pmatrix} -\frac{1}{3} \\ \frac{2}{3} \\ \frac{2}{3} \end{pmatrix}, \mathbf{x}_2 = \begin{pmatrix} \frac{2}{3} \\ -\frac{1}{3} \\ \frac{2}{3} \end{pmatrix}, \mathbf{x}_3 = \begin{pmatrix} \frac{2}{3} \\ \frac{2}{3} \\ -\frac{1}{3} \end{pmatrix} \right\}$$

and let

$$\mathbf{x}_\mathcal{B} = \begin{pmatrix} x_1 \\ x_2 \\ x_3 \end{pmatrix}_\mathcal{B}$$

be any coordinate vector in the basis $\mathcal{B}$. Find the projection of $\mathbf{x}_\mathcal{B}$ onto the subspace W in the standard inner product.

Solution. We begin by showing that the basis $\mathcal{B}$ consists of orthogonal unit vectors.

```
In[1]:= x1={-1/3,2/3,2/3};x2={2/3,-1/3,2/3};x3={2/3,2/3,-1/3};
```

```
In[2]:= {Dot[x1,x2],Dot[x1,x3],Dot[x2,x3]}

Out[2]= {0,0,0}
```

This shows that the vectors x_1, x_2, and x_3 are orthogonal. Next we show that their length is 1. Since the coordinates of these vectors are permutations of each other, it is sufficient to show that any one of them has length 1. Let us choose x_1.

```
In[3]:= √((-1/3)² + (2/3)² + (2/3)²)
Out[3]= 1
```

Now consider any vector $x = x_1 x_1 + x_2 x_2 + x_3 x_3$ in $\mathbb{R}^3$, written in the basis $\mathcal{B}$. By Corollary 8.3, $\text{proj}(x \to W) = x_2 x_2 + x_3 x_3$. The calculation

$$\langle \text{proj}(x \to W), x_1 x_1 \rangle = \langle x_2 x_2 + x_3 x_3, x_1 x_1 \rangle$$
$$= \langle x_2 x_2, x_1 x_1 \rangle + \langle x_3 x_3, x_1 x_1 \rangle$$
$$= x_2 x_3 \langle x_2, x_3 \rangle + x_3 x_1 \langle x_3, x_1 \rangle$$
$$= 0$$

shows that $\text{proj}(x \to W)$ is orthogonal to any vector in $U = \text{span}\{x_1\}$. ◀▶

The **Orthogonalization** package of *MATHEMATICA* has a built-in **Projection** function for projecting vectors onto vectors. The following example illustrates its use.

EXAMPLE 8.7 ■ An Orthogonal Projection onto a Vector

Use the **Projection** function to find the orthogonal projection of the vector $x = (1, 2, 3)$ onto the vector $y = (5, 0, 3)$ in the standard inner product.

Solution. We first load the **Orthogonalization** package and then input the vectors x and y.

```
In[1]:= <<LinearAlgebra`Orthogonalization`
```

```
In[2]:= x={1,2,3}; y={5,0,3};
```

Next we calculate $\text{proj}(x \to y)$ and $\text{perp}(x \perp y)$.

```
In[3]:= v1=Projection[x,y]
Out[3]= {35/17, 0, 21/17}
```

```
In[4]:= v2=x-Projection[x,y]
Out[4]= {-18/17, 2, 30/17}
```

We conclude by verifying that proj$(\mathbf{x} \to \mathbf{y})$ + perp$(\mathbf{x} \perp \mathbf{y})$ = $\mathbf{x}$ and that the dot product of proj$(\mathbf{x} \to \mathbf{y})$ and perp$(\mathbf{x} \perp \mathbf{y})$ is zero.

```
In[5]:= v1+v2

Out[5]= {1,2,3}
```

```
In[6]:= Dot[v1,v2]

Out[6]= 0
```

This shows that proj$(\mathbf{x} \to \mathbf{y})$ + perp$(\mathbf{x} \perp \mathbf{y})$ = $\mathbf{x}$ is an orthogonal decomposition of the vector $\mathbf{x}$. ◂▸

EXERCISES 8.2

1. Use the **Projection** function of *MATHEMATICA* to find the vectors proj$(\mathbf{x} \to \mathbf{y})$ and perp$(\mathbf{x} \perp \mathbf{y})$ for the following pairs of vectors in the standard inner product.

   a. $\mathbf{x} = \begin{pmatrix} 1 \\ 1 \end{pmatrix}$ and $\mathbf{y} = \begin{pmatrix} 4 \\ 5 \end{pmatrix}$ b. $\mathbf{x} = \begin{pmatrix} -2 \\ 7 \end{pmatrix}$ and $\mathbf{y} = \begin{pmatrix} 4 \\ 5 \end{pmatrix}$

   c. $\mathbf{x} = \begin{pmatrix} 1 \\ 3 \\ 1 \end{pmatrix}$ and $\mathbf{y} = \begin{pmatrix} 4 \\ 5 \\ 0 \end{pmatrix}$ d. $\mathbf{x} = \begin{pmatrix} 1 \\ 3 \\ 1 \end{pmatrix}$ and $\mathbf{y} = \begin{pmatrix} -1 \\ -9 \\ 12 \end{pmatrix}$

2. Use the standard basis $\mathcal{E} = \{\mathbf{e}_1, \mathbf{e}_2\}$ for $\mathbb{R}^2$ and the standard inner product for parts (a)–(c).

   a. Find the orthogonal projection of the vector $(1, 2)$ onto the subspace $W = \text{span}(\mathbf{e}_1)$.

   b. Find the orthogonal projection of the vector $(1, 2)$ onto the subspace $W = \text{span}(\mathbf{e}_2)$.

   c. Explain what happens when you try to find the orthogonal projection of a vector (x, y) onto $\mathbb{R}^2$.

3. Repeat Exercise 2 for the basis

$$\mathcal{B} = \left\{ \mathbf{x}_1 = \frac{1}{\sqrt{2}} \begin{pmatrix} 1 \\ 1 \end{pmatrix}, \mathbf{x}_2 = \frac{1}{\sqrt{2}} \begin{pmatrix} 1 \\ -1 \end{pmatrix} \right\}$$

for $\mathbb{R}^2$. Use the standard inner product. The vector $(1, 2)$ is to be taken as a coordinate vector in the basis $\mathcal{B}$.

4. Use the standard basis $\mathcal{E} = \{\mathbf{e}_1, \mathbf{e}_2, \mathbf{e}_3\}$ for $\mathbb{R}^3$ and the standard inner product for parts (a)–(c).

   a. Find the orthogonal projection of the vector $(1, 2, 3)$ onto the subspace W spanned by the unit vector $\mathbf{e}_1$.

   b. Find the orthogonal projection of the vector $(1, 2, 3)$ onto the subspace W spanned by the unit vectors $\mathbf{e}_1$ and $\mathbf{e}_2$.

   c. Find the orthogonal projection of the vector $(1, 2, 3)$ onto the subspace W spanned by the unit vectors $\mathbf{e}_1$ and $\mathbf{e}_3$.

5. Repeat Exercise 4 for the basis

$$\mathcal{B} = \left\{ \mathbf{x}_1 = \frac{1}{\sqrt{2}} \begin{pmatrix} 1 \\ 0 \\ 1 \end{pmatrix}, \mathbf{x}_2 = \frac{1}{\sqrt{2}} \begin{pmatrix} -1 \\ 0 \\ 1 \end{pmatrix}, \mathbf{x}_3 = \begin{pmatrix} 0 \\ 1 \\ 0 \end{pmatrix} \right\}$$

for $\mathbb{R}^3$. Use the standard inner product. The vector $(1, 2, 3)$ is to be taken as a coordinate vector in the basis $\mathcal{B}$.

ORTHOGONAL BASES

In this section, we show that every basis of a finite-dimensional inner product space can be converted to a basis whose vectors are mutually orthogonal.

Gram-Schmidt Orthogonalization

In many applications, it is essential to work with bases whose vectors are orthogonal. We now show that this can be done. Every basis in an inner product space can be converted to one in which all pairs of vectors are orthogonal. The algorithm involved is known as the *Gram-Schmidt process*. The process iterates the orthogonal projection construction.

DEFINITION 8.2 *A basis $\mathcal{B} = \{\mathbf{v}_1, \ldots, \mathbf{v}_n\}$ for a finite-dimensional inner product space V is **orthogonal** if $\langle \mathbf{v}_i, \mathbf{v}_j \rangle = 0$ for each $1 \leq i, j \leq n$.*

The process of converting a basis to an orthogonal one is known as **orthogonalization**. We now use induction to describe this process.

THEOREM 8.5 (Gram-Schmidt orthogonalization process) *If $\mathcal{B} = \{\mathbf{x}_1, \ldots, \mathbf{x}_n\}$ is a basis for a real inner product space V with inner product $(\mathbf{x}, \mathbf{y}) \rightarrow \langle \mathbf{x}, \mathbf{y} \rangle$, then the set $\mathcal{C} = \{\mathbf{y}_1, \ldots, \mathbf{y}_n\}$ is an orthogonal basis for V, where y_i are the following vectors.*

1. $\mathbf{y}_1 \;=\; \mathbf{x}_1$

2. $\mathbf{y}_2 \;=\; \mathbf{x}_2 - \dfrac{\langle \mathbf{x}_2, \mathbf{y}_1 \rangle}{\langle \mathbf{y}_1, \mathbf{y}_1 \rangle} \mathbf{y}_1$

3. $\mathbf{y}_3 \;=\; \mathbf{x}_3 - \dfrac{\langle \mathbf{x}_3, \mathbf{y}_1 \rangle}{\langle \mathbf{y}_1, \mathbf{y}_1 \rangle} \mathbf{y}_1 - \dfrac{\langle \mathbf{x}_3, \mathbf{y}_2 \rangle}{\langle \mathbf{y}_2, \mathbf{y}_2 \rangle} \mathbf{y}_2$

$\vdots$

$n.\quad \mathbf{y}_n \;=\; \mathbf{x}_n - \dfrac{\langle \mathbf{x}_n, \mathbf{y}_1 \rangle}{\langle \mathbf{y}_1, \mathbf{y}_1 \rangle} \mathbf{y}_1 - \dfrac{\langle \mathbf{x}_n, \mathbf{y}_2 \rangle}{\langle \mathbf{y}_2, \mathbf{y}_2 \rangle} \mathbf{y}_2 - \cdots - \dfrac{\langle \mathbf{x}_n, \mathbf{y}_n \rangle}{\langle \mathbf{y}_n, \mathbf{y}_n \rangle} \mathbf{y}_n$

Proof. We prove the theorem by induction on the number n of vectors in $\mathcal{B}$. If $n = 1$, the result is obvious. Now suppose that we have found the orthogonal vectors $\{\mathbf{y}_1, \ldots, \mathbf{y}_k\}$ and that x_{k+1} is linearly independent of $\{\mathbf{y}_1, \ldots, \mathbf{y}_k\}$. We form the subspace

$$\operatorname{span}\{\mathbf{y}_1, \ldots, \mathbf{y}_k, \mathbf{x}_{k+1}\} \subseteq V$$

and look for a vector

$$\mathbf{y}_{k+1} \in \operatorname{span}\{\mathbf{y}_1, \ldots, \mathbf{y}_k, \mathbf{x}_{k+1}\}$$

that is orthogonal to the vectors $\{\mathbf{y}_1, \ldots, \mathbf{y}_k\}$. Since $\{\mathbf{y}_1, \ldots, \mathbf{y}_k, \mathbf{y}_{k+1}\}$ is a basis for

$$\operatorname{span}\{\mathbf{y}_1, \ldots, \mathbf{y}_k, \mathbf{x}_{k+1}\}$$

we can solve our problem by finding scalars for which

$$\mathbf{y}_{k+1} = a_1 \mathbf{y}_1 + \cdots + a_k \mathbf{y}_k + \mathbf{x}_{k+1}$$

and $\langle \mathbf{y}_i, \mathbf{y}_{k+1} \rangle = 0$ for $1 \le i \le k$. Since the vector $\mathbf{y}_{k+1}$ is orthogonal to the vectors $\{\mathbf{y}_1, \ldots, \mathbf{y}_k\}$ precisely when

$$
\begin{aligned}
\langle \mathbf{y}_i, \mathbf{y}_{k+1} \rangle &= \langle \mathbf{y}_i, a_1 \mathbf{y}_1 + \cdots + a_k \mathbf{y}_k + \mathbf{x}_{k+1} \rangle \\
&= a_i \langle \mathbf{y}_i, \mathbf{y}_i \rangle + \langle \mathbf{y}_i, \mathbf{x}_{k+1} \rangle \\
&= 0
\end{aligned}
$$

it is clear that $a_i = -\langle \mathbf{y}_i, \mathbf{x}_{k+1} \rangle / \langle \mathbf{y}_i, \mathbf{y}_i \rangle$ for all $1 \le i \le k$. ∎

EXAMPLE 8.8   ■ An Orthogonal Basis for $\mathbb{R}^2$

Use the induction steps of the Gram-Schmidt process to convert the basis

$$\mathcal{B} = \left\{ \mathbf{x}_1 = \begin{pmatrix} 1 \\ -1 \end{pmatrix}, \mathbf{x}_2 = \begin{pmatrix} 3 \\ 2 \end{pmatrix} \right\}$$

for $\mathbb{R}^2$ to an orthogonal basis $\mathcal{C} = \{\mathbf{y}_1, \mathbf{y}_2\}$ in the standard inner product on $\mathbb{R}^2$.

Solution. The standard inner product on $\mathbb{R}^2$ is defined by $\langle \mathbf{u}, \mathbf{x} \rangle = \mathbf{u}^T \mathbf{x}$. By the Gram-Schmidt process, the vectors

$$1. \quad \mathbf{y}_1 \quad = \quad \mathbf{x}_1 = \begin{pmatrix} 1 \\ -1 \end{pmatrix}$$

$$2. \quad \mathbf{y}_2 \quad = \quad \mathbf{x}_2 - \frac{\mathbf{x}_2^T \mathbf{y}_1}{\mathbf{y}_1^T \mathbf{y}_1} \mathbf{y}_1 = \begin{pmatrix} 3 \\ 2 \end{pmatrix} - \frac{\mathbf{x}_2^T \mathbf{y}_1}{\mathbf{y}_1^T \mathbf{y}_1} \begin{pmatrix} 1 \\ -1 \end{pmatrix}$$

are orthogonal and form the required basis $\mathcal{C}$.

We use *MATHEMATICA* to compute the coefficient $\mathbf{x}_2^T \mathbf{y}_1 / \mathbf{y}_1^T \mathbf{y}_1$.

```
In[1]:= Dot[{3,2}.{1,-1}]/Dot[{1,-1}.{1,-1}]

Out[1]= 1/2
```

Therefore,

$$\mathbf{y}_2 = \begin{pmatrix} 3 \\ 2 \end{pmatrix} - \frac{1}{2} \begin{pmatrix} 1 \\ -1 \end{pmatrix} = \begin{pmatrix} \frac{5}{2} \\ \frac{5}{2} \end{pmatrix}$$

By construction, the set $\mathcal{C} = \{\mathbf{y}_1, \mathbf{y}_2\}$ is an orthogonal basis for $\mathbb{R}^2$. ◄►

EXAMPLE 8.9   ■ An Orthogonal Basis for $\mathbb{R}^3$

Use the Gram-Schmidt process to convert the basis

$$\mathcal{B} = \{\mathbf{x}_1 = (1, -1, 0), \mathbf{x}_2 = (3, 0, 1), \mathbf{x}_3 = (3, 2, 1)\}$$

for $\mathbb{R}^3$ to an orthogonal basis $\mathcal{C} = \{\mathbf{y}_1, \mathbf{y}_2, \mathbf{y}_3\}$ in the standard inner product on $\mathbb{R}^3$.

Solution. By the Gram-Schmidt process, the required vectors are

$$1. \quad \mathbf{y}_1 \quad = \quad \mathbf{x}_1$$

$$2. \quad \mathbf{y}_2 \quad = \quad \mathbf{x}_2 - \frac{\mathbf{x}_2^T \mathbf{y}_1}{\mathbf{y}_1^T \mathbf{y}_1} \mathbf{y}_1$$

$$3. \quad \mathbf{y}_3 \quad = \quad \mathbf{x}_3 - \frac{\mathbf{x}_3^T \mathbf{y}_1}{\mathbf{y}_1^T \mathbf{y}_1} \mathbf{y}_1 - \frac{\mathbf{x}_3^T \mathbf{y}_2}{\mathbf{y}_2^T \mathbf{y}_2} \mathbf{y}_2$$

We begin by computing $\mathbf{y}_2$.

```
In[1]:= Dot[{3,0,1}.{1,-1,0}]/Dot[{1,-1,0}.{1,-1,0}]

Out[1]= 3/2
```

Therefore,

$$\mathbf{y}_2 = \begin{pmatrix} 3 \\ 0 \\ 1 \end{pmatrix} - \frac{3}{2}\begin{pmatrix} 1 \\ -1 \\ 0 \end{pmatrix} = \begin{pmatrix} \frac{3}{2} \\ \frac{3}{2} \\ 1 \end{pmatrix}$$

Next we compute $\mathbf{y}_3$.

```
In[2]:= Dot[{3,2,1},{1,-1,0}]/Dot[{1,-1,0},{1,-1,0}]
Out[2]= 1/2
```

```
In[3]:= Dot[{3,2,1},{3/2,3/2,1}]/Dot[{3/2,3/2,1},{3/2,3/2,1}]
Out[3]= 17/11
```

This means that

$$\mathbf{y}_3 = \mathbf{x}_3 - \frac{\mathbf{x}_3^T \mathbf{y}_1}{\mathbf{y}_1^T \mathbf{y}_1}\mathbf{y}_1 - \frac{\mathbf{x}_3^T \mathbf{y}_2}{\mathbf{y}_2^T \mathbf{y}_2}\mathbf{y}_2$$

$$= \begin{pmatrix} 3 \\ 2 \\ 1 \end{pmatrix} - \frac{1}{2}\begin{pmatrix} 1 \\ -1 \\ 0 \end{pmatrix} - \frac{17}{11}\begin{pmatrix} \frac{3}{2} \\ \frac{3}{2} \\ 1 \end{pmatrix} = \begin{pmatrix} \frac{2}{11} \\ \frac{2}{11} \\ -\frac{6}{11} \end{pmatrix}$$

Thus the set $\mathcal{C} = \{\mathbf{y}_1, \mathbf{y}_2, \mathbf{y}_3\}$ is an orthogonal basis for $\mathbb{R}^3$. ◄►

The Gram-Schmidt algorithm is built into the **Orthogonalization** package of *MATHE-MATICA*.

EXAMPLE 8.10 ■ Orthogonalizing a Basis for $\mathbb{R}^3$

Use the **GramSchmidt** function in *MATHEMATICA* to convert the basis

$$\mathcal{B} = \left\{ \mathbf{x}_1 = \begin{pmatrix} 1 \\ 1 \\ 0 \end{pmatrix}, \mathbf{x}_2 = \begin{pmatrix} 0 \\ 2 \\ 1 \end{pmatrix}, \mathbf{x}_3 = \begin{pmatrix} 3 \\ 0 \\ 1 \end{pmatrix} \right\}$$

to a basis $\mathcal{C} = \{\mathbf{y}_1, \mathbf{y}_2, \mathbf{y}_3\}$ whose vectors are mutually orthogonal unit vectors in the standard inner product.

Solution. First we load the **Orthogonalization** package.

```
In[1]:= <<LinearAlgebra`Orthogonalization`
```

Next we carry out the one-step conversion.

In[2]:= **GramSchmidt[{{1,1,0},{0,2,1},{3,0,1}}]**

Out[2]= $\{\{\frac{1}{\sqrt{2}},\frac{1}{\sqrt{2}},0\},\{-\frac{1}{\sqrt{3}},\frac{1}{\sqrt{3}},\frac{1}{\sqrt{3}}\},\{\frac{1}{\sqrt{6}},-\frac{1}{\sqrt{6}},\sqrt{\frac{2}{3}}\}\}$

The calculation

In[3]:= **{y1.y1,y2.y2,y3.y3,y1.y2,y1.y3,y2.y3}**

Out[3]= $\{1,1,1,0,0,0\}$

shows that the vectors in the set

$$\mathcal{C} = \left\{ \begin{pmatrix} \frac{1}{\sqrt{2}} \\ \frac{1}{\sqrt{2}} \\ 0 \end{pmatrix}, \begin{pmatrix} -\frac{1}{\sqrt{3}} \\ \frac{1}{\sqrt{3}} \\ \frac{1}{\sqrt{3}} \end{pmatrix}, \begin{pmatrix} \frac{1}{\sqrt{6}} \\ -\frac{1}{\sqrt{6}} \\ \sqrt{\frac{2}{3}} \end{pmatrix} \right\}$$

are mutually orthogonal. ◄►

EXERCISES 8.3

1. Use the inductive steps of the Gram-Schmidt process to convert the basis for $\mathbb{R}^2$ determined by the columns of the matrix

$$A = \begin{pmatrix} 1 & 1 \\ 1 & -1 \end{pmatrix}$$

to an orthogonal basis.

2. Use the **GramSchmidt** function of *MATHEMATICA* to verify your calculations in Exercise 1.

3. Use a nonstandard inner product, in conjunction with the **InnerProduct** option of the **GramSchmidt** function, to convert the columns of the matrix in Exercise 1 to an orthogonal basis.

4. Use the inductive steps of the Gram-Schmidt process to convert the basis for $\mathbb{R}^3$ determined by the columns of the matrix

$$A = \begin{pmatrix} 1 & 1 & 2 \\ 1 & -1 & 0 \\ 0 & 2 & 0 \end{pmatrix}$$

to an orthogonal basis.

5. Use the **GramSchmidt** function to verify your calculations in Exercise 4.

6. Use a nonstandard inner product, in conjunction with the **InnerProduct** option of the **GramSchmidt** function, to convert the columns of the matrix in Exercise 4 to an orthogonal basis.

7. Explain what happens when you use the Gram-Schmidt process to convert the standard basis $\mathcal{E} = \{\mathbf{e}_1, \mathbf{e}_2, \mathbf{e}_3\}$ for $\mathbb{R}^3$ to itself.

8. Use the vector $\mathbf{x} = (1, 1, 1)$ and the Gram-Schmidt process to extend the set

$$S = \left\{ \begin{pmatrix} 1 \\ 1 \\ 0 \end{pmatrix}, \begin{pmatrix} 1 \\ -1 \\ 2 \end{pmatrix} \right\}$$

to an orthogonal basis for $\mathbb{R}^3$.

ORTHONORMAL BASES

In addition to being orthogonal, the bases in many applications must also be unit vectors.

DEFINITION 8.3   *An orthogonal basis* $\mathcal{B} = \{\mathbf{v}_1, \ldots, \mathbf{v}_n\}$ *for an inner product space V is* ***orthonormal*** *if* $\|\mathbf{v}_i\| = 1$ *for all* $i \in \mathbf{n}$.

By Theorem 7.3, every nonzero vector can be normalized. We can therefore convert every orthogonal basis to an orthonormal basis by normalizing its vectors. The **GramSchmidt** function can be used to produce both a nonnormalized and a normalized orthogonal basis. The default is a normalized basis. The option **Normalized → False,** on the other hand, produces a basis whose vectors may not be normal. If no option is specified, *MATHEMATICA* sets the **Normalized** option to **True.**

EXAMPLE 8.11   ■ **An Orthonormal Basis for** $\mathbb{R}^3$

Verify that the vectors in the orthogonal basis $\mathcal{C} = \{\mathbf{x}_1, \mathbf{x}_2, \mathbf{x}_3\}$ produced by the **Gram-Schmidt** function from the basis

$$\mathcal{B} = \left\{ \begin{pmatrix} 1 \\ -1 \\ 1 \end{pmatrix}, \begin{pmatrix} 2 \\ 2 \\ 2 \end{pmatrix}, \begin{pmatrix} -3 \\ 1 \\ 3 \end{pmatrix} \right\}$$

are unit vectors.

Solution. We first load the **Orthogonalization** package.

```
In[1]:= <<LinearAlgebra`Orthogonalization`
```

Then we carry out the one-step conversion.

```
In[2]:= {x1,x2,x3}=GramSchmidt[{{1,-1,1},{2,2,2},{-3,1,3}}]
Out[2]= {{1/√3,-1/√3,1/√3},{1/√6,√(2/6),1/√6},{-1/√2,0,1/√2}}
```

Next we use the **Normalize** function to test whether the vectors x_1, x_2, and x_3 are normal.

```
In[3]:= Normalize[x1]=x1
Out[3]= True
```

```
In[4]:= Normalize[x2]=x2
Out[4]= True
```

```
In[5]:= Normalize[x3]=x3
Out[5]= True
```

This shows that in each case, the input vector equals the output vector. Hence the vectors are normal. ◀▶

In Example 8.11, the command

GramSchmidt[{{1,-1,1},{2,2,2},{-3,1,3}},Normalized → False]

would have produced the orthogonal basis

$$\{y_1, y_2, y_3\} = \left\{ \{1, -1, 1\}, \left\{ \frac{4}{3}, \frac{8}{3}, \frac{4}{3} \right\}, \{-3, 0, 3\} \right\}$$

However, since $\|y_1\| = \sqrt{1^2 + (-1)^2 + 1^2} = \sqrt{3} \neq 1$, for example, this basis is not normal.

EXERCISES 8.4

1. Use the induction steps of the Gram-Schmidt process to convert the basis

$$\mathcal{B} = \left\{ \begin{pmatrix} 1 \\ 2 \end{pmatrix}, \begin{pmatrix} 1 \\ -1 \end{pmatrix} \right\}$$

to a basis

$$C = \left\{ \begin{pmatrix} a \\ b \end{pmatrix}, \begin{pmatrix} c \\ d \end{pmatrix} \right\}$$

that is orthogonal in the standard inner product on $\mathbb{R}^2$. Convert C to an orthonormal basis.

2. Use the induction steps of the Gram-Schmidt process to convert the basis

$$B = \left\{ \begin{pmatrix} 1 \\ 2 \end{pmatrix}, \begin{pmatrix} 1 \\ -1 \end{pmatrix} \right\}$$

to a basis

$$C = \left\{ \begin{pmatrix} a \\ b \end{pmatrix}, \begin{pmatrix} c \\ d \end{pmatrix} \right\}$$

that is orthogonal in the inner product $\mathbf{x} A^T A \mathbf{y}$ on $\mathbb{R}^2$ determined by the matrix

$$A = \begin{pmatrix} 50 & 79 \\ 56 & 49 \end{pmatrix}$$

Convert C to an orthonormal basis.

3. Use the Gram-Schmidt process to convert the basis

$$B = \left\{ \begin{pmatrix} 1 \\ 2 \\ 1 \end{pmatrix}, \begin{pmatrix} 1 \\ -1 \\ 0 \end{pmatrix}, \begin{pmatrix} 0 \\ 1 \\ 0 \end{pmatrix} \right\}$$

to a basis

$$C = \left\{ \begin{pmatrix} a \\ b \\ c \end{pmatrix}, \begin{pmatrix} d \\ e \\ f \end{pmatrix}, \begin{pmatrix} g \\ h \\ i \end{pmatrix} \right\}$$

that is orthogonal in the standard inner product on $\mathbb{R}^2$. Convert C to an orthonormal basis.

4. Use the Gram-Schmidt process to convert the basis

$$B = \left\{ \begin{pmatrix} 1 \\ 2 \\ 1 \end{pmatrix}, \begin{pmatrix} 1 \\ -1 \\ 0 \end{pmatrix}, \begin{pmatrix} 0 \\ 1 \\ 0 \end{pmatrix} \right\}$$

to a basis

$$C = \left\{ \begin{pmatrix} a \\ b \\ c \end{pmatrix}, \begin{pmatrix} d \\ e \\ f \end{pmatrix}, \begin{pmatrix} g \\ h \\ i \end{pmatrix} \right\}$$

that is orthogonal in the inner product $\mathbf{x}A^TA\mathbf{y}$ on $\mathbb{R}^3$ determined by the matrix

$$A = \begin{pmatrix} 1 & 0 & 0 \\ 0 & 2 & 0 \\ 0 & 0 & 3 \end{pmatrix}$$

Convert $\mathcal{C}$ to an orthonormal basis.

5. Let $\langle \mathbf{x}, \mathbf{y} \rangle$ be an arbitrary inner product on $\mathbb{R}^2$, and let $\mathbf{x} = (3, 4)$ and $\mathbf{y} = (2, 5)$ be two vectors in $\mathbb{R}^2$. Construct an orthonormal basis $\mathcal{B}$ for $\mathbb{R}^2$, and find the coordinates

$$[\mathbf{x}]_\mathcal{B} = \begin{pmatrix} a \\ b \end{pmatrix} \quad \text{and} \quad [\mathbf{y}]_\mathcal{B} = \begin{pmatrix} c \\ d \end{pmatrix}$$

   in the basis $\mathcal{B}$. Show that $\langle \mathbf{x}, \mathbf{y} \rangle = ac + bd$.

6. Let $\langle \mathbf{x}, \mathbf{y} \rangle$ be an arbitrary inner product on $\mathbb{R}^2$, and let $\mathbf{x} = (3, 4, -2)$ and $\mathbf{y} = (2, 5, 7)$ be two vectors in $\mathbb{R}^2$. Construct an orthonormal basis $\mathcal{B}$ for $\mathbb{R}^2$, and find the coordinates

$$[\mathbf{x}]_\mathcal{B} = \begin{pmatrix} a \\ b \\ c \end{pmatrix} \quad \text{and} \quad [\mathbf{y}]_\mathcal{B} = \begin{pmatrix} d \\ e \\ f \end{pmatrix}$$

   in the basis $\mathcal{B}$. Show that $\langle \mathbf{x}, \mathbf{y} \rangle = ad + be + cf$.

7. Use the **GramSchmidt** function to convert the basis

$$\mathcal{B} = \left\{ \begin{pmatrix} 3 \\ 4 \end{pmatrix}, \begin{pmatrix} 4 \\ 3 \end{pmatrix} \right\},$$

   for $\mathbb{R}^2$ to an orthogonal basis whose vectors may not be normal.

8. Use the **Normalize** function to normalize the basis produced in Exercise 7.

9. Use the **GramSchmidt** function to convert the basis

$$\mathcal{B} = \left\{ \begin{pmatrix} 8 \\ -5 \\ 8 \end{pmatrix}, \begin{pmatrix} 4 \\ 3 \\ 5 \end{pmatrix}, \begin{pmatrix} -5 \\ -5 \\ -1 \end{pmatrix} \right\}$$

   for $\mathbb{R}^3$ to an orthogonal basis whose vectors may not be normal.

10. Use the **Normalize** function to normalize the basis produced in Exercise 9.

11. Use the **GramSchmidt** function to convert the basis in Exercise 9 to an orthonormal basis in one step.

Real Inner Products and Orthonormal Bases

It is quite remarkable that relative to orthonormal bases, all inner products on real inner product spaces can be computed as dot products. The difficult arithmetic involved in computing inner products is now hidden in the arithmetic required to convert a given basis to an orthonormal one.

Suppose $\mathcal{B} = \{\mathbf{x}_1, \mathbf{x}_2\}$ is a basis for an inner product space V, and let

$$\mathbf{x} = a_1\mathbf{x}_1 + a_2\mathbf{x}_2 \quad \text{and} \quad \mathbf{y} = b_1\mathbf{x}_1 + b_2\mathbf{x}_2$$

be two vectors in V. Then it follows from the linearity of $\langle \mathbf{x}, \mathbf{y} \rangle$ in both $\mathbf{x}$ and $\mathbf{y}$ that

$$\begin{aligned}\langle \mathbf{x}, \mathbf{y} \rangle &= \langle a_1\mathbf{x}_1 + a_2\mathbf{x}_2, b_1\mathbf{x}_1 + b_2\mathbf{x}_2 \rangle \\ &= a_1b_1\langle \mathbf{x}_1, \mathbf{x}_1 \rangle + a_1b_2\langle \mathbf{x}_1, \mathbf{x}_2 \rangle + a_2b_1\langle \mathbf{x}_2, \mathbf{x}_1 \rangle + a_2b_2\langle \mathbf{x}_2, \mathbf{x}_2 \rangle\end{aligned}$$

If $\mathcal{B}$ is orthonormal, then $\langle \mathbf{x}_1, \mathbf{x}_2 \rangle = \langle \mathbf{x}_2, \mathbf{x}_1 \rangle = 0$ and $\langle \mathbf{x}_1, \mathbf{x}_1 \rangle = \langle \mathbf{x}_2, \mathbf{x}_2 \rangle = 1$. Hence

$$\langle \mathbf{x}, \mathbf{y} \rangle = a_1b_1 \times 1 + a_1b_2 \times 0 + a_2b_1 \times 0 + a_2b_2 \times 1 = a_1b_1 + a_2b_2$$

takes the form of the dot product. The next theorem expresses this property for inner product spaces of arbitrary finite dimension.

THEOREM 8.6 (Orthonormal basis theorem) *If the function* $(\mathbf{x}, \mathbf{y}) \rightarrow \langle \mathbf{x}, \mathbf{y} \rangle$ *is an inner product on a finite-dimensional real vector space V and if $\mathcal{E} = \{\mathbf{e}_1, \ldots, \mathbf{e}_n\}$ is an orthonormal basis for V then* $\langle \mathbf{x}, \mathbf{y} \rangle = a_1b_1 + \cdots + a_nb_n$ *for all vectors* $\mathbf{x} = a_1\mathbf{e}_1 + \cdots + a_n\mathbf{e}_n$ *and* $\mathbf{y} = b_1\mathbf{e}_1 + \cdots + b_n\mathbf{e}_n$ *in V.*

Proof. Consider the inner product $\langle \mathbf{x}, \mathbf{y} \rangle = \langle a_1\mathbf{e}_1 + \cdots + a_n\mathbf{e}_n, b_1\mathbf{e}_1 + \cdots + b_n\mathbf{e}_n \rangle$. Since $\mathcal{E}$ is orthonormal and $\langle \mathbf{x}, \mathbf{y} \rangle$ is bilinear, we get

$$\langle \mathbf{x}, \mathbf{y} \rangle = \sum_{i,j \in \mathbf{n}} a_ib_j \langle \mathbf{e}_i, \mathbf{e}_j \rangle = a_1b_1 + \cdots + a_nb_n$$

Therefore, $\langle \mathbf{e}_i, \mathbf{e}_j \rangle = 0$ if $i \neq j$ and $\langle \mathbf{e}_i, \mathbf{e}_j \rangle = 1$ otherwise. ∎

COROLLARY 8.7 *If $\mathcal{B} = \{\mathbf{e}_1, \ldots, \mathbf{e}_n\}$ is an orthonormal basis for an inner product space V, then* $\mathbf{x} = \langle \mathbf{x}, \mathbf{e}_1 \rangle \mathbf{e}_1 + \cdots + \langle \mathbf{x}, \mathbf{e}_n \rangle \mathbf{e}_n$ *for any $\mathbf{x} \in V$.*

Proof. Let $\mathbf{x} = a_1\mathbf{e}_1 + \cdots + a_n\mathbf{e}_n$ and $\mathbf{y} = \mathbf{e}_i$. Then Theorem 8.6 shows that

$$\langle \mathbf{x}, \mathbf{e}_i \rangle = \langle a_1\mathbf{e}_1 + \cdots + a_n\mathbf{e}_n, \mathbf{e}_i \rangle = a_i$$

for all $i \in \mathbf{n}$. This proves the corollary. ∎

EXERCISES 8.5

1. Use *MATHEMATICA* to find an orthonormal basis $\mathcal{B}$ for $\mathbb{R}^2$ in which the real inner product $\mathbf{x}^T A \mathbf{y}$ determined by the positive definite matrix

$$A = \begin{pmatrix} 8 & 0 \\ 0 & 3 \end{pmatrix}$$

can be computed as a dot product. Illustrate your result with a numerical example.

2. Use *MATHEMATICA* to find an orthonormal basis $\mathcal{B}$ for $\mathbb{R}^2$ in which the real inner product $\mathbf{x}^T A \mathbf{y}$ determined by the positive definite matrix

$$A = \begin{pmatrix} 8 & 4 \\ 4 & 3 \end{pmatrix}$$

can be computed as a dot product. Illustrate your result with a numerical example.

3. Use *MATHEMATICA* to find an orthonormal basis $\mathcal{B}$ for $\mathbb{R}^3$ in which the real inner product $\mathbf{x}^T A \mathbf{y}$ determined by the positive definite matrix

$$A = \begin{pmatrix} 8 & 0 & 0 \\ 0 & 3 & 0 \\ 0 & 0 & 5 \end{pmatrix}$$

can be computed as a dot product. Illustrate your result with a numerical example.

4. Use *MATHEMATICA* to find an orthonormal basis $\mathcal{B}$ for $\mathbb{R}^3$ in which the real inner product $\mathbf{x}^T A \mathbf{y}$ determined by the positive definite matrix

$$A = \begin{pmatrix} 8 & 0 & 0 \\ 0 & 3 & -1 \\ 0 & -1 & 5 \end{pmatrix}$$

can be computed as a dot product. Illustrate your result with a numerical example.

QR DECOMPOSITION

In this section, we show how the Gram-Schmidt orthogonalization process can be used to decompose a matrix into a specific matrix product. The decomposition is called a *QR decomposition*. It has several powerful applications in numerical linear algebra. Among them are a procedure for solving linear systems $A\mathbf{x} = \mathbf{b}$ whose coefficient matrices A are invertible, a method for solving the *normal equations* $A^T A \mathbf{x} = A^T \mathbf{b}$ of overdetermined linear systems $A\mathbf{x} = \mathbf{b}$, and an algorithm for calculating approximations of eigenvalues and eigenvectors in situations where exact solutions are unavailable.

THEOREM 8.8 (*QR* decomposition theorem)   *If A is an m × n matrix whose n columns are linearly independent vectors in $\mathbb{R}^m$, then there exist an m × n matrix Q whose columns form an orthonormal set in $\mathbb{R}^m$ and an n × n upper-triangular invertible matrix R such that A = QR.*

Proof. Suppose that $\mathcal{B} = \{\mathbf{x}_1, \ldots, \mathbf{x}_n\}$ is the set of columns of A. Then $\mathcal{B}$ is a basis for a subspace W of $\mathbb{R}^m$. Let $\mathcal{C} = \{\mathbf{y}_1, \ldots, \mathbf{y}_n\}$ be the orthogonal basis for W obtained from $\mathcal{B}$ by the Gram-Schmidt process, and form the matrix $Q = (\mathbf{y}_1 \ \mathbf{y}_2 \ \cdots \ \mathbf{y}_n)$. Then the columns of Q are an orthonormal set in $\mathbb{R}^m$. Since $\mathcal{C}$ is a basis for W, every vector $\mathbf{x}_i \in \mathcal{B}$ is a unique linear combination $a_1\mathbf{y}_1 + \cdots + a_n\mathbf{y}_n$ of the vectors $\mathbf{y}_1, \ldots, \mathbf{y}_n \in \mathcal{C}$. In particular, it follows from the recursive definition of the Gram-Schmidt process that

$$
\begin{cases}
\mathbf{x}_1 &= r_{11}\mathbf{y}_1 \\
\mathbf{x}_2 &= r_{12}\mathbf{y}_1 + r_{22}\mathbf{y}_2 \\
\mathbf{x}_3 &= r_{13}\mathbf{y}_1 + r_{23}\mathbf{y}_2 + r_{33}\mathbf{y}_3 \\
&\vdots \\
\mathbf{x}_n &= r_{1n}\mathbf{y}_1 + \cdots + r_{nn}\mathbf{y}_n
\end{cases}
$$

with $r_{ii} \neq 0$ for all $i \in \mathbf{n}$. We use these equations to construct the coordinate vectors

$$\mathbf{r}_1 = (r_{11}, 0, 0, \ldots, 0), \mathbf{r}_2 = (r_{12}, r_{22}, 0, \ldots, 0), \ldots, \mathbf{r}_n = (r_{1n}, r_{2n}, \ldots, r_{nn})$$

and let R be the matrix $(\mathbf{r}_1 \ \mathbf{r}_2 \ \cdots \ \mathbf{r}_n)$. Then

$$
\begin{aligned}
QR &= \begin{pmatrix} \mathbf{y}_1 & \mathbf{y}_2 & \cdots & \mathbf{y}_n \end{pmatrix} \begin{pmatrix} \mathbf{r}_1 & \mathbf{r}_2 & \cdots & \mathbf{r}_n \end{pmatrix} \\
&= \begin{pmatrix} (r_{11}\mathbf{y}_1) & \cdots & (r_{12}\mathbf{y}_1 + r_{22}\mathbf{y}_2) & \cdots & (r_{1n}\mathbf{y}_1 + \cdots + r_{nn}\mathbf{y}_n) \end{pmatrix} \\
&= \begin{pmatrix} \mathbf{x}_1 & \mathbf{x}_2 & \cdots & \mathbf{x}_n \end{pmatrix} \\
&= A
\end{aligned}
$$

This proves the theorem. ■

MATHEMATICA has a built-in algorithm for computing QR decompositions. For any $m \times n$ matrix A of rank n, the command **{P,R}=QRDecomposition[A]** produces a decomposition $QR = P^TR$ of A.

EXAMPLE 8.12 ■ **A *QR* Decomposition of Square Matrix**

Use *MATHEMATICA* to find the QR decomposition of the matrix

$$
A = \begin{pmatrix} 1 & 3 & 0 \\ 0 & 5 & 7 \\ 2 & -8 & 4 \end{pmatrix}
$$

and verify that the matrix Q in the product QR is orthogonal.

Solution. We use the **QRDecomposition** function to decompose the matrix A into an orthogonal matrix Q and an upper-triangular matrix R.

$$\text{In[1]:= } \{P,R\}=\text{QRDecomposition}\left[\begin{pmatrix} 1 & 3 & 0 \\ 0 & 5 & 7 \\ 2 & -8 & 4 \end{pmatrix}\right];$$

In[2]:= **MatrixForm[P]**

Out[2]//MatrixForm=

$$\begin{pmatrix} \frac{1}{\sqrt{5}} & 0 & \frac{2}{\sqrt{5}} \\ \frac{28}{\sqrt{1605}} & 5\sqrt{\frac{5}{321}} & -\frac{14}{\sqrt{1605}} \\ -\frac{10}{\sqrt{321}} & \frac{14}{\sqrt{321}} & \frac{5}{\sqrt{321}} \end{pmatrix}$$

In[3]:= **MatrixForm[R]**

Out[3]//MatrixForm=

$$\begin{pmatrix} \sqrt{5} & -\frac{13}{\sqrt{5}} & \frac{8}{\sqrt{5}} \\ 0 & \sqrt{\frac{321}{5}} & \frac{119}{\sqrt{1605}} \\ 0 & 0 & \frac{118}{\sqrt{321}} \end{pmatrix}$$

We now show that P is orthogonal. This guarantees that its columns are orthonormal.

In[4]:= **Transpose[P].P==IdentityMatrix[3]**

Out[4]= True

Next we show that P and R yield the expected decomposition.

$$\text{In[5]:= } \text{Transpose[P].R}==\begin{pmatrix} 1 & 3 & 0 \\ 0 & 5 & 7 \\ 2 & -8 & 4 \end{pmatrix}$$

Out[5]= True

It is clear by inspection that R is an upper-triangular matrix. It therefore remains to show that R is invertible. We do so by verifying that the determinant of R is not zero.

```
In[6]:= Det[R]==0

Out[6]= False
```

Hence $P^T R$ is a QR decomposition of A. ◄►

EXAMPLE 8.13 ■ A QR Decomposition of a Rectangular Matrix

Use *MATHEMATICA* to find a QR decomposition of the matrix

$$A = \begin{pmatrix} 1 & 3 \\ 0 & 5 \\ 2 & -8 \end{pmatrix}$$

Solution. We use the **QRDecomposition** function to decompose the matrix A into a matrix Q whose set of columns is orthonormal and an invertible upper-triangular matrix R.

```
In[1]:= {P,R}=QRDecomposition[( 1   3
                                 0   5
                                 2  -8 )];
```

```
In[2]:= MatrixForm[Transpose[P]]

Out[2]//MatrixForm=
```
$$\begin{pmatrix} \frac{1}{\sqrt{5}} & \frac{28}{\sqrt{1605}} & 0 \\ 0 & 5\sqrt{\frac{5}{321}} & 0 \\ \frac{2}{\sqrt{5}} & -\frac{14}{\sqrt{1605}} & 0 \end{pmatrix}$$

```
In[3]:= MatrixForm[R]

Out[3]//MatrixForm=
```
$$\begin{pmatrix} \sqrt{5} & -\frac{13}{\sqrt{5}} \\ 0 & \sqrt{\frac{321}{5}} \\ 0 & 0 \end{pmatrix}$$

Moreover, the following calculation shows that $P^T R = A$.

```
In[4]:= Transpose[P].R
Out[4]= {{1,3},{0,5},{2,-8}}
```

If we delete the zero column from P^T and the zero row from R, we obtain two matrices $(P^T)'$ and R' that satisfy the conclusion of the theorem. ◄►

Linear Systems with Invertible Coefficient Matrices

The QR decomposition provides an efficient method for solving linear equations determined by invertible matrices. Suppose that A is invertible and we would like to solve the system $Ax = b$. Since A is invertible, it has a QR decomposition. Therefore, $Ax = QRx = b$. If we multiply both sides by Q^T, then the orthogonality of Q implies that $Q^T QRx = Rx = Q^T b$. It therefore remains to solve the equation $Rx = Q^T b$. Since R is upper triangular, this equation can be solved by back substitution.

EXAMPLE 8.14 ■ Solving a Linear System Using QR Decomposition

Use QR decomposition and back substitution to solve the linear system

$$Ax = \begin{pmatrix} 1 & 2 & 1 & 0 \\ 1 & 1 & 3 & 1 \\ 2 & 3 & 1 & 5 \\ 0 & 0 & 4 & 1 \end{pmatrix} \begin{pmatrix} w \\ x \\ y \\ z \end{pmatrix} = \begin{pmatrix} 2 \\ 1 \\ 0 \\ 5 \end{pmatrix} = b$$

Solution. We begin by finding a QR decomposition of the matrix A.

```
In[1]:= A:= (1 2 1 0 / 1 1 3 1 / 2 3 1 5 / 0 0 4 1); X:= (w / x / y / z); B:= (2 / 1 / 0 / 5);
```

```
In[2]:= {P,R}=QRDecomposition[A]; R.X; P.B;
```

```
In[3]:= MatrixForm[R.X]
```

Out[3]//MatrixForm=

$$
\begin{pmatrix}
\sqrt{6}w + 3\sqrt{\frac{3}{2}}x + \sqrt{6}y + \dfrac{11z}{\sqrt{6}} \\
\dfrac{x}{\sqrt{2}} - \sqrt{2}y - \dfrac{z}{\sqrt{2}} \\
\sqrt{19}\,y \\
\sqrt{\frac{19}{3}}\,z
\end{pmatrix}
$$

```
In[4]:= MatrixForm[P.B]
```

Out[4]//MatrixForm=

$$
\begin{pmatrix}
\sqrt{\frac{2}{3}} + \frac{1}{\sqrt{6}} \\
-\frac{1}{\sqrt{2}} + \sqrt{2} \\
\frac{23}{\sqrt{19}} \\
\sqrt{\frac{3}{19}}
\end{pmatrix}
$$

Since $R\mathbf{x} = P\mathbf{b}$, we can tell by back substitution that w, x, y, and z have the values $-121/19$, $68/19$, $23/19$, and $3/19$, respectively. ◄►

Normal Equations of Overdetermined Linear Systems

The QR decomposition can be used to find the least-squares solution to certain linear systems.

THEOREM 8.9   *If A is an $m \times n$ matrix of rank n, then the solution of the linear system $A^T A\mathbf{x} = A^T \mathbf{b}$ is $\widehat{\mathbf{x}} = R^{-1}Q^T\mathbf{b}$, where R and Q are the matrices obtained from A by QR decomposition.*

Proof. Consider the linear system $A\mathbf{x} = \mathbf{b}$. Since A is an $m \times n$ matrix of rank n, its columns must be linearly independent. By Theorem 8.8, A can therefore be decomposed into a product QR. Therefore,

$$
A^T A\mathbf{x} = (QR)^T (QR)\mathbf{x} = (QR)^T\mathbf{b}
$$

This means that

$$
R^T(Q^T Q)R\mathbf{x} = R^T Q^T\mathbf{b}
$$

Moreover, $Q^TQ = I$ since the columns of Q are orthonormal. It follows that

$$R^TR\mathbf{x} = R^TQ^T\mathbf{b}$$

We also know from Theorem 8.8 that R is invertible. Hence R^T is invertible. It follows that

$$(R^T)^{-1}R^TR\mathbf{x} = R\mathbf{x} = (R^T)^{-1}R^TQ^T\mathbf{b} = Q^T\mathbf{b}$$

By multiplying both sides by R^{-1}, we therefore get $\widehat{\mathbf{x}} = R^{-1}Q^T\mathbf{b}$. ∎

EXAMPLE 8.15 ■ **Solving a Normal Equation by QR Decomposition**

Use QR decomposition to solve the normal equation of the linear system

$$A\mathbf{x} = \begin{pmatrix} 1 & 2 & 3 \\ 2 & 2 & 2 \\ 1 & 1 & 3 \\ 1 & 2 & 4 \end{pmatrix} \begin{pmatrix} x \\ y \\ z \end{pmatrix} = \begin{pmatrix} 2 \\ -1 \\ 4 \\ 3 \end{pmatrix} = \mathbf{b}$$

Solution. If the matrix A has rank 3, then the required solution is $\widehat{\mathbf{x}} = R^{-1}Q^T\mathbf{b}$, where Q and R are the matrices obtained from A by QR decomposition. We use *MATHEMATICA* to verify that A has rank 3 and to calculate $\widehat{\mathbf{x}}$.

```
In[1]:= A := ( 1 2 3        X := ( x )   ; B := ( 2 )   ;
               2 2 2              y              -1
               1 1 3    ;         z              4
               1 2 4 )                           3
```

To calculate the rank of A, we use the **RowRank** function defined in Appendix E.

```
In[2]:= RowRank[A_]:=Length[A]-Length[NullSpace[Transpose[A]]]
```

```
In[3]:= RowRank[A]

Out[3]= 3
```

Hence the QR decomposition theorem applies.

```
In[4]:= {P,R}=QRDecomposition[A];
```

```
In[5]:= MatrixForm[Inverse[R].P.B]
```

Out[5]//MatrixForm=

$$\begin{pmatrix} -\frac{9}{37} \\ -\frac{84}{37} \\ \frac{77}{37} \end{pmatrix}$$

It is easy to verify that

$$\begin{pmatrix} 1 & 2 & 3 \\ 2 & 2 & 2 \\ 1 & 1 & 3 \\ 1 & 2 & 4 \end{pmatrix}^T \begin{pmatrix} 1 & 2 & 3 \\ 2 & 2 & 2 \\ 1 & 1 & 3 \\ 1 & 2 & 4 \end{pmatrix} \begin{pmatrix} -\frac{9}{37} \\ -\frac{84}{37} \\ \frac{77}{37} \end{pmatrix} = \begin{pmatrix} 7 \\ 12 \\ 28 \end{pmatrix} = \begin{pmatrix} 1 & 2 & 3 \\ 2 & 2 & 2 \\ 1 & 1 & 3 \\ 1 & 2 & 4 \end{pmatrix}^T \begin{pmatrix} 2 \\ -1 \\ 4 \\ 3 \end{pmatrix}$$

Hence the vector $\mathbf{x} = (7, 12, 28)$ solves the normal equation $A^T A \mathbf{x} = A^T \mathbf{b}$. ◄►

The *QR* Algorithm for Approximating Eigenvalues

The QR algorithm provides an efficient method for approximating the eigenvalues of certain invertible matrices. The algorithm is based on the construction of a sequence of similar matrices $A_1, A_2, \ldots, A_{k+1}$ for which the diagonal entries of A_{k+1}, for a large enough k, are good approximations of the eigenvalues of A. The sequence of matrices is defined recursively.

$$\begin{aligned} A = A_1 \quad &= Q_1 R_1 \\ A_2 \quad &= R_1 Q_1 = Q_2 R_2 \\ A_3 \quad &= R_2 . Q_2 = Q_3 R_3 \\ &\vdots \\ A_{k+1} &= R_k Q_k \end{aligned}$$

EXAMPLE 8.16   ■ Eigenvalues and the *QR* Algorithm

Use *MATHEMATICA* and the QR algorithm to find the matrices $A_1 = Q_1 R_1$, $A_2 = Q_2 R_2$, $A_3 = Q_3 R_3$, $A_4 = Q_4 R_4$, $A_5 = Q_5 R_5$, and $A_6 = R_5 Q_5$ for

$$A = \begin{pmatrix} 0 & 1 \\ 1 & 1 \end{pmatrix}$$

and estimate the difference between the eigenvalues of A and the diagonal entries of A_6.

Solution. We proceed recursively.

In[1]:= A = $\begin{pmatrix} 0 & 1 \\ 1 & 1 \end{pmatrix}$;

In[2]= **MatrixForm[A1=A]**

Out[2]//MatrixForm=

$$\begin{pmatrix} 0 & 1 \\ 1 & 1 \end{pmatrix}$$

Next we create A_2.

In[3]:= **{Q1,R1}=QRDecomposition[A1];**

In[4]:= **MatrixForm[A2=R1.Q1]**

Out[4]//MatrixForm=

$$\begin{pmatrix} 1 & 1 \\ 1 & 0 \end{pmatrix}$$

From A_2, we create A_3.

In[5]:= **{Q2,R2}=QRDecomposition[A2];**

In[6]:= **MatrixForm[A3=R2.Q2]**

Out[6]//MatrixForm=

$$\begin{pmatrix} \frac{3}{2} & \frac{1}{2} \\ \frac{1}{2} & -\frac{1}{2} \end{pmatrix}$$

From A_3, we create A_4, and so on.

```
In[7]:= {Q3,R3}=QRDecomposition[A3];
```

```
In[8]:= MatrixForm[A4=R3.Q3]
```
Out[8]//MatrixForm=
$$\begin{pmatrix} \frac{8}{5} & \frac{1}{5} \\ \frac{1}{5} & -\frac{3}{5} \end{pmatrix}$$

```
In[9]:= {Q4,R4}=QRDecomposition[A4];
```

```
In[10]:= MatrixForm[A5=R4.Q4]
```
Out[10]//MatrixForm=
$$\begin{pmatrix} \frac{21}{13} & \frac{1}{13} \\ \frac{1}{13} & -\frac{8}{13} \end{pmatrix}$$

```
In[11]:= {Q5,R5}=QRDecomposition[A5];
```

```
In[12]:= MatrixForm[A6=R5.Q5]
```
Out[12]//MatrixForm=
$$\begin{pmatrix} \frac{55}{34} & \frac{1}{34} \\ \frac{1}{34} & -\frac{21}{34} \end{pmatrix}$$

Next we find numerical approximations of the eigenvalues of A.

```
In[13]:= {v1,v2}=Eigenvalues[A]
```
Out[13]= $\{ \frac{1}{2}(1-\sqrt{5}), \frac{1}{2}(1+\sqrt{5}) \}$

```
In[14]:= N[v1]
```
Out[14]= -0.618034

In[15]:= N[v2]

Out[15]= 1.61803

The diagonal entries of the sequence $A_1, A_2, A_3, A_4, A_5, A_6$ determine the sequences of real numbers

$$s_1 = \left(1, 0, -\tfrac{1}{2}, -\tfrac{3}{5}, -\tfrac{8}{13}, -\tfrac{21}{34}\right) \quad \text{and} \quad s_2 = \left(0, 1, \tfrac{3}{2}, \tfrac{8}{5}, \tfrac{21}{13}, \tfrac{55}{34}\right)$$

Let us evaluate $-21/34$ and $55/34$ numerically and compare the results with the eigenvalues of A.

In[16]:= N[-21/34]-v1

Out[16]= 0.00038693

In[17]:= N[55/34]-v2

Out[17]= -0.00038693

These calculations show that if $n = 6$, the differences are less than $\pm.00039$. ◀▶

EXERCISES 8.6

1. Find QR decompositions of the following matrices. In each case, show that the columns of Q form an orthonormal set and that R is an invertible upper-triangular matrix.

a. $\begin{pmatrix} 1 & 1 \\ -1 & 0 \\ 0 & 2 \end{pmatrix}$ b. $\begin{pmatrix} 2 & 5 & 0 \\ 2 & 0 & 3 \\ 3 & 2 & 1 \end{pmatrix}$ c. $\begin{pmatrix} 2 & 5 & 0 \\ 2 & 0 & 3 \\ 3 & 2 & 1 \end{pmatrix}^{T}$

d. $\begin{pmatrix} 6 & 1 & 7 & 9 \\ 6 & 8 & 6 & 9 \\ 3 & 1 & 4 & 6 \\ 3 & 2 & 1 & 3 \end{pmatrix}$ e. $\begin{pmatrix} 6 & 1 & 7 \\ 6 & 8 & 6 \\ 3 & 1 & 4 \\ 3 & 2 & 1 \end{pmatrix}$ f. $\begin{pmatrix} 6 & 1 \\ 6 & 8 \\ 3 & 1 \\ 3 & 2 \end{pmatrix}$

2. Explain why Theorem 8.8 does or does not apply to the following matrices.

a. $\begin{pmatrix} 6 & 1 & 7 & 9 \\ 6 & 8 & 14 & 9 \\ 3 & 1 & 4 & 6 \\ 3 & 2 & 5 & 3 \end{pmatrix}$ b. $\begin{pmatrix} 6 & 1 & 7 \\ 6 & 8 & 6 \\ 3 & 1 & 4 \\ 3 & 2 & 1 \end{pmatrix}^T$ c. $\begin{pmatrix} 2 & 1 & 0 \\ 2 & 2 & 3 \\ 3 & 8 & 1 \end{pmatrix}$

d. $\begin{pmatrix} 2 & 5 & 0 & 1 \\ 2 & 0 & 3 & 2 \end{pmatrix}^T$ e. $\begin{pmatrix} 2 & 5 & 0 \\ 2 & 0 & 3 \\ 4 & 5 & 3 \end{pmatrix}$ f. $\begin{pmatrix} 2 & 3 & 0 & 1 \\ 2 & 0 & 3 & 1 \\ 3 & 8 & 1 & 1 \end{pmatrix}$

3. Use QR decomposition and back substitution to solve the following linear systems.

a. $\begin{pmatrix} 2 & 1 & 0 \\ 0 & 2 & 3 \\ 3 & 0 & 2 \end{pmatrix} \begin{pmatrix} x \\ y \\ z \end{pmatrix} = \begin{pmatrix} 5 \\ 4 \\ 9 \end{pmatrix}$ b. $\begin{pmatrix} 2 & 5 & 0 \\ 2 & 0 & 3 \\ 4 & 1 & 3 \end{pmatrix} \begin{pmatrix} x \\ y \\ z \end{pmatrix} = \begin{pmatrix} 1 \\ 1 \\ 1 \end{pmatrix}$

d. $\begin{pmatrix} 2 & 1 & 0 & -1 \\ 0 & 2 & 3 & 5 \\ 3 & 0 & 2 & 0 \\ -4 & 1 & 1 & 2 \end{pmatrix} \begin{pmatrix} w \\ x \\ y \\ z \end{pmatrix} = \begin{pmatrix} 5 \\ 4 \\ 9 \\ 1 \end{pmatrix}$ e. $\begin{pmatrix} 2 & 5 & 0 & 1 \\ 2 & 0 & 3 & 0 \\ 4 & 1 & 3 & 5 \\ 4 & 3 & 0 & 2 \end{pmatrix} \begin{pmatrix} w \\ x \\ y \\ z \end{pmatrix} = \begin{pmatrix} 1 \\ 1 \\ 1 \\ 1 \end{pmatrix}$

4. Use QR decomposition and back substitution to solve the normal equations of the following overdetermined linear systems.

a. $\begin{pmatrix} 2 & 1 \\ 0 & 2 \\ 3 & 0 \end{pmatrix} \begin{pmatrix} x \\ y \end{pmatrix} = \begin{pmatrix} 5 \\ 4 \\ 9 \end{pmatrix}$ b. $\begin{pmatrix} 2 & 0 \\ 2 & 3 \\ 4 & 3 \end{pmatrix} \begin{pmatrix} x \\ y \end{pmatrix} = \begin{pmatrix} 1 \\ 1 \\ 1 \end{pmatrix}$

d. $\begin{pmatrix} 2 & 1 & -1 \\ 0 & 2 & 5 \\ 3 & 0 & 0 \\ -4 & 1 & 2 \end{pmatrix} \begin{pmatrix} x \\ y \\ z \end{pmatrix} = \begin{pmatrix} 5 \\ 4 \\ 9 \\ 1 \end{pmatrix}$ e. $\begin{pmatrix} 2 & 5 & 0 \\ 2 & 0 & 3 \\ 4 & 1 & 3 \\ 4 & 3 & 0 \end{pmatrix} \begin{pmatrix} x \\ y \\ z \end{pmatrix} = \begin{pmatrix} 1 \\ 1 \\ 1 \\ 1 \end{pmatrix}$

5. Use the QR algorithm to construct the sequence of matrices $A = A_1, A_2, A_3$ for the matrix

$$A = \begin{pmatrix} 1 & 2 \\ 0 & -1 \end{pmatrix}$$

and calculate the differences between the eigenvalues of A and the diagonal entries of A_3.

6. Use the QR algorithm to construct the sequence of matrices $A = A_1, A_2, A_3$ for the matrix

$$A = \begin{pmatrix} 1 & 2 & 0 \\ 0 & -1 & 0 \\ 2 & -1 & 1 \end{pmatrix}$$

and calculate the differences between the eigenvalues of A and the diagonal entries of A_3.

7. The matrix

$$A = \begin{pmatrix} 0 & 1 \\ -1 & 0 \end{pmatrix}$$

has no real eigenvalues. Use the QR algorithm to construct the sequence of matrices $A = A_1, A_2, A_3$ for A and explain the result.

8. Explain why the following statement is false. "A matrix A has a QR decomposition if and only if its transpose A^T has a QR decomposition."

ORTHOGONAL MATRICES

The results in this section are based on the fact that all vectors in real inner product spaces can be arranged to be linear combinations of orthogonal unit vectors. Sets of vectors with this property are called **orthonormal sets**. It will be fundamental for our work that such subsets of $\mathbb{R}^n$ are bases if and only if their vectors determine orthogonal matrices. We know from Chapter 2 that a real matrix A is **orthogonal** if and only if $AA^T = A^TA = I$.

THEOREM 8.10 (Orthogonal matrix theorem) *A real square matrix is orthogonal if and only if its rows and columns are orthonormal sets in the standard inner product.*

Proof. We prove the theorem for column sets. By interchanging rows and columns, we obtain a proof for row sets. Let $A = (\mathbf{c}_1 \ \cdots \ \mathbf{c}_n)$ be a square matrix with columns $\mathbf{c}_i$, and let A^T be its transpose with rows $\mathbf{c}_i^T$. Then

$$A^T = \begin{pmatrix} \mathbf{c}_1^T \\ \vdots \\ \mathbf{c}_n^T \end{pmatrix}$$

and $A^TA = (a_{ij})$ is the identity matrix I if and only if

$$a_{ij} = \mathbf{c}_i^T \mathbf{c}_j = \begin{cases} 1 & \text{if } i = j \\ 0 & \text{if } i \neq j \end{cases}$$

This is so if and only if the columns of A are orthogonal and if $\mathbf{c}_i^T \mathbf{c}_j = \|\mathbf{c}_i\| = 1$ for all $i \in \mathbf{n}$. The case for AA^T is analogous. ∎

COROLLARY 8.11   *Every invertible real matrix A is the product QR of an orthogonal matrix Q and an upper-triangular invertible matrix R.*

Proof. Let BC be a QR decomposition of A. By Theorem 8.8, C is an upper-triangular invertible matrix. It remains to prove that B is orthogonal. Since A is invertible, Theorem 4.13 guarantees that the columns of A are linearly independent. By Theorems 8.8 and 8.10 B is therefore a matrix whose columns form an orthonormal set in the standard inner product. It therefore follows from Theorem 8.10 that B is orthogonal. ■

EXAMPLE 8.17   ■ Orthogonal Columns of a Matrix

Find the relationship between the entries of the real matrix

$$A = \begin{pmatrix} a_1 & a_2 & a_3 \\ b_1 & b_2 & b_3 \\ c_1 & c_2 & c_3 \end{pmatrix}$$

that makes A orthogonal in the standard inner product on $\mathbb{R}^3$.

Solution. If A is orthogonal, then the columns of A must be orthogonal. This means that

$$a_1 a_2 + b_1 b_2 + c_1 c_2 = a_1 a_3 + b_1 b_3 + c_1 c_3 = a_2 a_3 + b_2 b_3 + c_2 c_3 = 0$$

In addition, the columns must be normal. In other words,

$$\sqrt{a_1^2 + b_1^2 + c_1^2} = \sqrt{a_2^2 + b_2^2 + c_2^2} = \sqrt{a_3^2 + b_3^2 + c_3^2} = 1$$

If these conditions hold, then Theorem 8.10 guarantees that the matrix A is orthogonal. ◀▶

EXAMPLE 8.18   ■ An Orthonormal Sets of Vectors

Use Theorem 8.10 to convert the orthogonal matrix

$$A = \begin{pmatrix} \frac{1}{\sqrt{2}} & \frac{1}{\sqrt{3}} & \frac{1}{\sqrt{6}} \\ 0 & \frac{1}{\sqrt{3}} & -\sqrt{\frac{2}{3}} \\ -\frac{1}{\sqrt{2}} & \frac{1}{\sqrt{3}} & \frac{1}{\sqrt{6}} \end{pmatrix}$$

to an orthonormal basis for $\mathbb{R}^3$, and to convert the orthonormal basis

$$\mathcal{B} = \left\{ \begin{pmatrix} \frac{1}{\sqrt{2}} \\ 0 \\ -\frac{1}{\sqrt{2}} \end{pmatrix}, \begin{pmatrix} 0 \\ 1 \\ 0 \end{pmatrix}, \begin{pmatrix} \frac{1}{\sqrt{2}} \\ 0 \\ \frac{1}{\sqrt{2}} \end{pmatrix} \right\}$$

for $\mathbb{R}^3$ to an orthogonal matrix B.

Solution. By Theorem 8.10, the set

$$C = \left\{ \begin{pmatrix} \frac{1}{\sqrt{2}} \\ 0 \\ -\frac{1}{\sqrt{2}} \end{pmatrix}, \begin{pmatrix} \frac{1}{\sqrt{3}} \\ \frac{1}{\sqrt{3}} \\ \frac{1}{\sqrt{3}} \end{pmatrix}, \begin{pmatrix} \frac{1}{\sqrt{6}} \\ -\sqrt{\frac{2}{3}} \\ \frac{1}{\sqrt{6}} \end{pmatrix} \right\}$$

of columns of A is an orthonormal basis for $\mathbb{R}^3$, and the matrix

$$B = \begin{pmatrix} \frac{1}{\sqrt{2}} & \frac{1}{\sqrt{3}} & \frac{1}{\sqrt{6}} \\ 0 & \frac{1}{\sqrt{3}} & -\sqrt{\frac{2}{3}} \\ -\frac{1}{\sqrt{2}} & \frac{1}{\sqrt{3}} & \frac{1}{\sqrt{6}} \end{pmatrix}$$

is an orthogonal matrix. ◄►

The next theorem explains how the orthonormal bases B and C in this example are themselves connected by orthogonal matrices.

THEOREM 8.12 (Coordinate conversion and orthonormal bases theorem) *Every coordinate conversion matrix between orthonormal bases of a real inner product space is orthogonal.*

Proof. Let $B = \{x_1, \ldots, x_n\}$ and $C = \{y_1, \ldots, y_n\}$ be two orthonormal bases of an inner product space V. Then we can write the vectors in C as unique linear combinations of vectors in B. This means that

$$\begin{cases} x_1 & = & a_{11}y_1 + \cdots + a_{1n}y_n \\ & \vdots & \\ x_n & = & a_{n1}y_1 + \cdots + a_{nn}y_n \end{cases}$$

Let $P = (a_{ij})$ be the coefficient matrix of this system of equations. Then P^T is the coordinate conversion matrix from B to C. Since C is orthonormal,

$$\delta_{ij} = \langle y_i, y_j \rangle = a_{i1}a_{j1} + \cdots + a_{in}a_{jn}$$

where δ_{ij} is the Kronecker delta function defined in Chapter 4. Therefore, the rows of P form an orthonormal set. By Theorem 8.10, the matrix P is orthogonal. ∎

EXAMPLE 8.19   ■ An Orthogonal Coordinate Conversion Matrix

Let $B = \{x_1, x_2, x_2\}$ and $C = \{y_1, y_2, y_2\}$ be the orthonormal bases in Example 8.18. Then the coordinate conversion matrix P from B to C is the transpose of the coefficient matrix of the linear system

$$\begin{cases} x_1 = a_{11}y_1 + a_{12}y_2 + a_{13}y_3 \\ x_2 = a_{21}y_1 + a_{22}y_2 + a_{23}y_3 \\ x_3 = a_{31}y_1 + a_{32}y_2 + a_{33}y_3 \end{cases}$$

First we note that if

$$\mathbf{x}_1 = \begin{pmatrix} \frac{1}{\sqrt{2}} \\ 0 \\ -\frac{1}{\sqrt{2}} \end{pmatrix} = a_{11} \begin{pmatrix} \frac{1}{\sqrt{2}} \\ 0 \\ -\frac{1}{\sqrt{2}} \end{pmatrix} + a_{12} \begin{pmatrix} \frac{1}{\sqrt{3}} \\ \frac{1}{\sqrt{3}} \\ \frac{1}{\sqrt{3}} \end{pmatrix} + a_{13} \begin{pmatrix} \frac{1}{\sqrt{6}} \\ -\sqrt{\frac{2}{3}} \\ \frac{1}{\sqrt{6}} \end{pmatrix}$$

then $a_{11} = 1$, and $a_{12} = a_{13} = 0$. Secondly, if

$$\mathbf{x}_2 = \begin{pmatrix} 0 \\ 1 \\ 0 \end{pmatrix} = a_{21} \begin{pmatrix} \frac{1}{\sqrt{2}} \\ 0 \\ -\frac{1}{\sqrt{2}} \end{pmatrix} + a_{22} \begin{pmatrix} \frac{1}{\sqrt{3}} \\ \frac{1}{\sqrt{3}} \\ \frac{1}{\sqrt{3}} \end{pmatrix} + a_{23} \begin{pmatrix} \frac{1}{\sqrt{6}} \\ -\sqrt{\frac{2}{3}} \\ \frac{1}{\sqrt{6}} \end{pmatrix}$$

then $a_{21} = 0$, $a_{22} = \frac{1}{3}\sqrt{3}$, and $a_{23} = -\frac{1}{3}\sqrt{2}\sqrt{3}$. Finally, if

$$\mathbf{x}_3 = \begin{pmatrix} \frac{1}{\sqrt{2}} \\ 0 \\ \frac{1}{\sqrt{2}} \end{pmatrix} = a_{31} \begin{pmatrix} \frac{1}{\sqrt{2}} \\ 0 \\ -\frac{1}{\sqrt{2}} \end{pmatrix} + a_{32} \begin{pmatrix} \frac{1}{\sqrt{3}} \\ \frac{1}{\sqrt{3}} \\ \frac{1}{\sqrt{3}} \end{pmatrix} + a_{33} \begin{pmatrix} \frac{1}{\sqrt{6}} \\ -\sqrt{\frac{2}{3}} \\ \frac{1}{\sqrt{6}} \end{pmatrix}$$

then $a_{31} = 0$, $a_{32} = \frac{1}{3}\sqrt{2}\sqrt{3}$, and $a_{33} = \frac{1}{3}\sqrt{3}$.

Using this information, we let P be the matrix

$$P = \begin{pmatrix} 1 & 0 & 0 \\ 0 & \frac{1}{3}\sqrt{3} & \frac{1}{3}\sqrt{2}\sqrt{3} \\ 0 & -\frac{1}{3}\sqrt{2}\sqrt{3} & \frac{1}{3}\sqrt{3} \end{pmatrix}$$

It is easy to verify that

$$PP^T = \begin{pmatrix} 1 & 0 & 0 \\ 0 & \frac{1}{3}\sqrt{3} & \frac{1}{3}\sqrt{2}\sqrt{3} \\ 0 & -\frac{1}{3}\sqrt{2}\sqrt{3} & \frac{1}{3}\sqrt{3} \end{pmatrix} \begin{pmatrix} 1 & 0 & 0 \\ 0 & \frac{1}{3}\sqrt{3} & -\frac{1}{3}\sqrt{2}\sqrt{3} \\ 0 & \frac{1}{3}\sqrt{2}\sqrt{3} & \frac{1}{3}\sqrt{3} \end{pmatrix} = \begin{pmatrix} 1 & 0 & 0 \\ 0 & 1 & 0 \\ 0 & 0 & 1 \end{pmatrix}$$

Thus P is the orthogonal coordinate conversion matrix from the basis $\mathcal{B}$ to the basis $\mathcal{C}$. ◀▶

EXERCISES 8.7

1. Prove that the determinant of an orthogonal matrix is ± 1.

2. Verify that the matrix

$$A = \begin{pmatrix} \frac{1}{\sqrt{2}} & -\frac{1}{\sqrt{2}} \\ \frac{1}{\sqrt{2}} & \frac{1}{\sqrt{2}} \end{pmatrix}$$

is orthogonal by showing that the rows and columns of A are orthonormal sets in the standard inner product.

3. Repeat Exercise 1 for the matrix

$$A = \begin{pmatrix} \frac{1}{\sqrt{6}} & \frac{1}{\sqrt{2}} & -\frac{1}{\sqrt{3}} \\ -\frac{1}{\sqrt{6}} & \frac{1}{\sqrt{2}} & \frac{1}{\sqrt{3}} \\ \frac{2}{\sqrt{6}} & 0 & \frac{1}{\sqrt{3}} \end{pmatrix}$$

4. Show that the coordinate conversion matrix P from the standard basis

$$\mathcal{E} = \left\{ \begin{pmatrix} 1 \\ 0 \\ 0 \end{pmatrix}, \begin{pmatrix} 0 \\ 1 \\ 0 \end{pmatrix}, \begin{pmatrix} 0 \\ 0 \\ 1 \end{pmatrix} \right\}$$

to the orthonormal basis

$$\mathcal{B} = \left\{ \begin{pmatrix} \frac{2}{3} \\ \frac{2}{3} \\ -\frac{1}{3} \end{pmatrix}, \begin{pmatrix} -\frac{1}{3} \\ \frac{2}{3} \\ \frac{2}{3} \end{pmatrix}, \begin{pmatrix} \frac{2}{3} \\ -\frac{1}{3} \\ \frac{2}{3} \end{pmatrix} \right\}$$

is orthogonal.

5. Construct the Householder matrix A determined by the vector $\mathbf{u} = (3, 7)$, and show that A is orthogonal.

6. Repeat Exercise 4 for the vector $\mathbf{u} = (5, 0, 1)$.

7. Use induction to prove that if A is a real $n \times n$ matrix whose eigenvalues are all real, then there exist an orthogonal matrix Q and an upper-triangular matrix U for which $A = QUQ^T$. (*Hint*: Let $\lambda_1, \ldots, \lambda_r$ be the eigenvalue of A and let

$$\mathbf{x}_1 = \begin{pmatrix} \lambda_1 \\ 0 \\ \vdots \\ 0 \end{pmatrix}$$

Then $A\mathbf{x}_1 = \lambda_1 \mathbf{x}_1$. Use the Gram-Schmidt algorithm to find column vectors $\mathbf{x}_2, \ldots, \mathbf{x}_r$ for which $Q_1 = (\mathbf{x}_1 \ \mathbf{x}_2 \ \cdots \ \mathbf{x}_r)$ is an orthogonal matrix. Form the matrix $Q_1^T A Q_1$. Since $\mathbf{x}_1$ is an eigenvector of A belonging to λ_1, you get

$$Q_1^T A Q_1 = \begin{pmatrix} \mathbf{x}_1^T \\ \vdots \\ \mathbf{x}_n^T \end{pmatrix} (A\mathbf{x}_1 \ A\mathbf{x}_2 \ \cdots \ A\mathbf{x}_r)$$

$$= \begin{pmatrix} \mathbf{x}_1^T \\ \vdots \\ \mathbf{x}_n^T \end{pmatrix} (\lambda_1 \mathbf{x}_1 \ A\mathbf{x}_2 \ \cdots \ A\mathbf{x}_n)$$

$$= \begin{pmatrix} \lambda_1 & * \\ \mathbf{0} & B \end{pmatrix}$$

Show that there therefore exists an orthogonal matrix Q_2 and an upper-triangular matrix U_1 for which $Q_2^T B Q_2 = U_1$. The diagonal entries of U_1 are $\lambda_2, \ldots, \lambda_r$ and are eigenvalues of A. Form the orthogonal matrix

$$Q_3 = \begin{pmatrix} 1 & \mathbf{0} \\ \mathbf{0} & Q_2 \end{pmatrix}$$

and use Q_1 and Q_3 to form

$$Q_3^T Q_1^T A Q_1 Q_3 = \begin{pmatrix} 1 & \mathbf{0} \\ \mathbf{0} & Q_2^T \end{pmatrix} \begin{pmatrix} \lambda_1 & * \\ \mathbf{0} & B_1 \end{pmatrix} \begin{pmatrix} 1 & \mathbf{0} \\ \mathbf{0} & Q_2 \end{pmatrix}$$

$$= \begin{pmatrix} \lambda_1 & \mathbf{0} \\ \mathbf{0} & Q_2^T B_1 Q_2 \end{pmatrix}$$

$$= \begin{pmatrix} \lambda_1 & \mathbf{0} \\ \mathbf{0} & U_1 \end{pmatrix}$$

The matrix $Q = Q_1 Q_3$ has the property that

$$Q^T A Q = \begin{pmatrix} \lambda_1 & t_{12} & t_{13} & \cdots & t_{1r} \\ 0 & \lambda_2 & t_{23} & \cdots & t_{2r} \\ 0 & 0 & \lambda_3 & \cdots & t_{3r} \\ \vdots & \vdots & \vdots & \ddots & \vdots \\ 0 & 0 & 0 & \cdots & \lambda_r \end{pmatrix} = U$$

is an upper-triangular matrix. This result is known as **Schur's lemma**.)

8. Compute the conjugate transpose of the following matrices.

a. $\begin{pmatrix} 1 & -i \\ 0 & 3 \end{pmatrix}$ b. $\begin{pmatrix} 1 & 5 \\ 0 & 6 - 14i \\ 1 + 2i & 4i \end{pmatrix}$ c. $\begin{pmatrix} 1 & -i & 5 \\ 0 & 3 & 6 - 14i \\ 1 + 2i & 23 & 4i \end{pmatrix}$

9. Recall from Chapter 2 that a **unitary matrix** is a complex matrix A with the property that $A^{-1} = A^H$, where A^H is the conjugate transpose of A. Use *MATHEMATICA* to verify that the following matrices are unitary.

a. $A = \begin{pmatrix} \frac{1}{2} i \sqrt{2} & \frac{1}{2} \sqrt{2} \\ \frac{1}{2} \sqrt{2} & \frac{1}{2} i \sqrt{2} \end{pmatrix}$ b. $A = \begin{pmatrix} \frac{\sqrt{2}}{2} & 0 & \frac{\sqrt{2}}{2} \\ 0 & 1 & 0 \\ \frac{i}{2}\sqrt{2} & 0 & -\frac{i}{2}\sqrt{2} \end{pmatrix}$

10. Determine which of the following matrices are unitary.

a. $\begin{pmatrix} 1 & -i \\ i & 3 \end{pmatrix}$
b. $\begin{pmatrix} 5 & 0 & 0 \\ 0 & \frac{1}{2}\sqrt{2} & \frac{1}{2}\sqrt{2} \\ 0 & \frac{1}{2}+\frac{1}{2}i & -\frac{1}{2}-\frac{1}{2}i \end{pmatrix}$
c. $\begin{pmatrix} \frac{1}{2}\sqrt{2} & \frac{1}{2}\sqrt{2} \\ \frac{1}{2}+\frac{1}{2}i & -\frac{1}{2}-\frac{1}{2}i \end{pmatrix}$

11. Use the fact that a matrix A is unitary if and only if its columns determine an orthonormal set in the standard inner product on $\mathbb{C}^4$ to create a 4×4 unitary matrix.

ORTHOGONAL SUBSPACES

We now extend the idea of orthogonality from vectors to subspaces. Sometimes this will enable us to think of a space as a sum of orthogonal subspaces and to describe the properties of a linear transformation on the whole space in terms of the actions of the transformation on the subspaces.

DEFINITION 8.4 *Two subspaces U and W of an inner product space V are **orthogonal** with respect to an inner product $\langle \mathbf{x}, \mathbf{y} \rangle$ on V if $\langle \mathbf{x}, \mathbf{y} \rangle = 0$ for all $\mathbf{x} \in U$ and all $\mathbf{y} \in W$.*

EXAMPLE 8.20 ■ **The Orthogonal Zero Subspace**

The zero subspace $\{\mathbf{0}\}$ of an inner product space V is orthogonal to every other subspace U of V since $\langle \mathbf{0}, \mathbf{y} \rangle = \langle \mathbf{00} + \mathbf{00}, \mathbf{y} \rangle = 0 \langle \mathbf{0} + \mathbf{0}, \mathbf{y} \rangle = 0$ for all $\mathbf{y} \in U$. ◀▶

EXAMPLE 8.21 ■ **Two Proper Orthogonal Subspaces**

Let $U = \{(x, 0, 0) : x \in \mathbb{R}\}$ and $W = \{(0, y, z) : y, z \in \mathbb{R}\}$ be two subspaces of $\mathbb{R}^3$. Then the fact that

$$\begin{pmatrix} x & 0 & 0 \end{pmatrix} \begin{pmatrix} 0 \\ y \\ z \end{pmatrix} = 0$$

for all vectors in U and W shows that the subspaces U and W are orthogonal in the standard inner product. ◀▶

The fundamental subspaces of real matrices are among our prime examples of orthogonal subspaces. We now prove that these spaces are pairwise orthogonal.

THEOREM 8.13 (Orthogonal subspace theorem) *If A is an $m \times n$ real matrix, then the null space Nul A is orthogonal to the row space Row A in the standard inner product on $\mathbb{R}^n$.*

Proof. Suppose that $\mathbf{x} \in$ Nul A and $\mathbf{y} \in$ Row $A =$ Col A^T. Then $A\mathbf{x} = \mathbf{0}$ and $\mathbf{y} = A^T\mathbf{z}$ for some $\mathbf{z} \in \mathbb{R}^n$. Therefore, $\mathbf{y}^T = \mathbf{z}^T A$. This means that $\langle \mathbf{y}^T, \mathbf{x} \rangle = \langle \mathbf{z}^T A, \mathbf{x} \rangle = \mathbf{z}^T A\mathbf{x} = \mathbf{z}^T \mathbf{0} = 0$ in the standard inner product on $\mathbb{R}^n$. ■

By exchanging the matrices A and A^T in Theorem 8.13, we immediately obtain the result that the left null space Nul A^T is orthogonal to the column space Col A in the standard inner product on $\mathbb{R}^m$.

EXERCISES 8.8

1. Show that the spaces $U = \text{span}\{(1, 2, 0), (0, 4, 3)\}$ and $V = \text{span}\{(6, -3, 4)\}$ are orthogonal subspaces of $\mathbb{R}^3$.

2. Find a subspace of $\mathbb{R}^3$ that is orthogonal to the space U spanned by the vectors $(1, 2, 1)$ and $(1, -1, 3)$.

3. Find a subspace of $\mathbb{R}^4$ that is orthogonal to the space U spanned by the vectors $(1, 2, 1, 2)$ and $(1, -1, 1, 1)$.

4. Describe all subspaces of $\mathbb{R}^2$ that are orthogonal to the space $U = \text{span}\{(1, 2)\}$.

5. Find the fundamental subspaces of the matrix

$$A = \begin{pmatrix} 9 & -9 & 4 & 0 \\ -8 & -5 & 4 & 7 \end{pmatrix}$$

and discuss their orthogonality.

6. Find the fundamental subspaces of the matrix

$$A = \begin{pmatrix} -7 & 1 & 9 & 0 & 0 \\ 1 & -1 & 8 & -9 & -9 \\ 0 & 0 & 0 & 0 & 0 \\ 3 & -9 & -6 & 3 & 6 \\ 9 & -9 & -5 & 6 & 0 \end{pmatrix}$$

and discuss their orthogonality.

7. Find the two nonzero subspaces of $\mathbb{R}^5$ of different dimensions that are orthogonal to the space $U = \text{span}\{(-7, 1, 0, 3, 9), (1, -1, 0, -9, -9)\}$ and discuss their orthogonality.

8. Prove that if A is an $m \times n$ real matrix, then Nul A^T is orthogonal to Col A in the standard inner product on $\mathbb{R}^m$.

Orthogonal Complements

Many of the pairs of orthogonal subspaces discussed in the previous section have the property that every vector in the ambient space is a sum of vectors in the orthogonal subspaces. In a sense, each subspace is a set-theoretical complement of the other. This property leads to the following definition, not just for subspaces, but for subsets in general. We will see later that orthogonal complements play an important role in the direct sum decomposition of inner product spaces.

DEFINITION 8.5 *The **orthogonal complement** $S^\perp$ of a subset S of an inner product space V is the set of all vectors $\mathbf{x} \in V$ with the property that $\langle \mathbf{x}, \mathbf{y} \rangle = 0$ for all $\mathbf{y} \in S$.*

It is remarkable that even if the set S is not a subspace of V, its orthogonal complement always is.

THEOREM 8.14 (Orthogonal subspace theorem) *The orthogonal complement $S^\perp$ of a subset S of an inner product space V is a subspace of V.*

Proof. If $S = \emptyset$, then $\langle \mathbf{x}, \mathbf{y} \rangle = 0$ for all $\mathbf{y} \in S$ holds for all $\mathbf{x} \in V$, so that $S^\perp = V$. Suppose, therefore, that $S \neq \emptyset$. Since $\langle \mathbf{0}, \mathbf{y} \rangle = 0$ for all $\mathbf{y} \in S$, it follows that $\mathbf{0} \in S^\perp$. Moreover, we know that if $x_1, x_2 \in S^\perp$ and $a, b \in \mathbb{R}$, then

$$\langle a\mathbf{x}_1 + b\mathbf{x}_2, \mathbf{y} \rangle = \langle a\mathbf{x}_1, \mathbf{y} \rangle + \langle b\mathbf{x}_2, \mathbf{y} \rangle$$
$$= a\langle \mathbf{x}_1, \mathbf{y} \rangle + b\langle \mathbf{x}_2, \mathbf{y} \rangle$$
$$= a0 + b0 = 0$$

for all $\mathbf{y} \in S$. Hence $ax_1 + bx_2 \in S^\perp$. ∎

It is customary to read $S^\perp$ as "S perp." The **perp** operation can be considered as a function from the set of all subsets of V to the set of all subspaces of V. It has some interesting inclusion properties.

THEOREM 8.15 *The perp function from the subsets to subspaces of an inner product space V has the following properties: $\emptyset^\perp = V$; $V^\perp = \{\mathbf{0}\}$; $S \subseteq S^{\perp\perp}$; and if $S_1 \subseteq S_2$, then $S_2^\perp \subseteq S_1^\perp$.*

Proof. We must verify four conditions:

1. The fact that $\emptyset^\perp = V$ was observed in the proof of Theorem 8.14.

2. If $\mathbf{x} \in V^\perp$, then $\langle \mathbf{x}, \mathbf{y} \rangle = 0$ for all $\mathbf{y} \in V$. This can be so only if $\mathbf{x} = \mathbf{0}$.

3. If $\mathbf{x} \in S$, then $\langle \mathbf{x}, \mathbf{y} \rangle = 0$ for all $\mathbf{y} \in S^\perp$. Hence $\mathbf{x} \in S^{\perp\perp}$, so that $S \subseteq S^{\perp\perp}$.

4. If $\langle \mathbf{x}, \mathbf{y} \rangle = 0$ for all $\mathbf{y} \in S_2$ and $S_2 \supseteq S_1$, it holds that $\langle \mathbf{x}, \mathbf{y} \rangle = 0$ for all $\mathbf{y} \in S_1$. Hence $x \in S_2^\perp$ implies that $x \in S_1^\perp$, so that $S_2^\perp \subseteq S_1^\perp$. ∎

The next theorem describes a fundamental decomposition property of orthogonal complements.

THEOREM 8.16 (Orthogonal complement theorem)   *If V is a finite-dimensional inner product space and W is a subspace of V, then $V = W \oplus W^\perp$.*

Proof. First we observe that $W \cap W^\perp = \{0\}$ since $\mathbf{x} \in W \cap W^\perp$ implies that $\langle \mathbf{x}, \mathbf{x} \rangle = 0$. This implies that $\mathbf{x} = 0$. Next we use the Gram-Schmidt process to construct an orthogonal basis $\mathcal{B} = \{\mathbf{x}_1, \ldots, \mathbf{x}_p\}$ for W. By Theorem 4.12 and a second application of the Gram-Schmidt process, we can extend $\mathcal{B}$ to an orthogonal basis $\mathcal{C} = \{\mathbf{x}_1, \ldots, \mathbf{x}_p, \mathbf{y}_1, \ldots, \mathbf{y}_q\}$ for V. Since $\mathcal{C}$ is an orthogonal basis, we know that $\langle \mathbf{x}_i, \mathbf{y}_j \rangle = 0$ for all i and j. Therefore, $\mathbf{y}_j \in W^\perp$ for all j. A dimension argument shows that $\mathcal{D} = \{\mathbf{y}_1, \ldots, \mathbf{y}_q\}$ is a basis for $W^\perp$. This means that the basis $\mathcal{C}$ for V is the disjoint union of the bases $\mathcal{B}$ and $\mathcal{D}$. We can therefore write every vector in V as $\mathbf{x} + \mathbf{y}$, with $\mathbf{x} \in W$ and $\mathbf{y} \in W^\perp$. Therefore, $V = W \oplus W^\perp$. ∎

COROLLARY 8.17   *If W is a subspace of a finite-dimensional real inner product space V and $\mathcal{B} = \{\mathbf{x}_1, \ldots, \mathbf{x}_p\}$ is an orthogonal basis for W, then every vector $\mathbf{x} \in V$ can be written uniquely in the form $\widehat{\mathbf{x}} + \mathbf{y}$, where $\mathbf{y} = \mathbf{x} - \widehat{\mathbf{x}}$ and*

$$\widehat{\mathbf{x}} = \frac{\langle \mathbf{x}, \mathbf{x}_1 \rangle}{\langle \mathbf{x}_1, \mathbf{x}_1 \rangle}\mathbf{x}_1 + \cdots + \frac{\langle \mathbf{x}, \mathbf{x}_p \rangle}{\langle \mathbf{x}_p, \mathbf{x}_p \rangle}\mathbf{x}_p$$

Proof. It is clear from Theorem 8.16 that it suffices to show that $\widehat{\mathbf{x}} \in W$, that $\mathbf{x} - \widehat{\mathbf{x}} \in W^\perp$, and that $\mathbf{x} = \widehat{\mathbf{x}} + \mathbf{y}$. Since $\widehat{\mathbf{x}}$ is a linear combination of the basis vectors for W, it is clear that $\widehat{\mathbf{x}} \in W$. Moreover, since the basis $\mathcal{B}$ is orthogonal, the calculation

$$\langle \mathbf{x} - \widehat{\mathbf{x}}, \mathbf{x}_1 \rangle = \langle \mathbf{x}, \mathbf{x}_1 \rangle - \langle \widehat{\mathbf{x}}, \mathbf{x}_1 \rangle = \langle \mathbf{x}, \mathbf{x}_1 \rangle - \langle \mathbf{x}, \mathbf{x}_1 \rangle = 0$$

shows that $\mathbf{y} = \mathbf{x} - \widehat{\mathbf{x}} \in W^\perp$. Therefore, $\widehat{\mathbf{x}} + \mathbf{y} = \widehat{\mathbf{x}} + \mathbf{x} - \widehat{\mathbf{x}} = \mathbf{x}$. To show that the decomposition is unique, suppose that $\widehat{\mathbf{x}} + \mathbf{y} = \widehat{\mathbf{u}} + \mathbf{v}$, with $\widehat{\mathbf{u}} \in W$ and $\mathbf{v} \in W^\perp$. Then $\widehat{\mathbf{x}} - \widehat{\mathbf{u}} = \mathbf{y} - \mathbf{v} \in (W \cap W^\perp)$. Since W and $W^\perp$ are disjoint, $\widehat{\mathbf{x}} - \widehat{\mathbf{u}} = \mathbf{y} - \mathbf{v} = \mathbf{0}$. Hence $\widehat{\mathbf{x}} = \widehat{\mathbf{u}}$ and $\mathbf{y} = \mathbf{v}$. ∎

COROLLARY 8.18   *If W is a subspace of a finite-dimensional real inner product space V and $\mathcal{B} = \{\mathbf{x}_1, \ldots, \mathbf{x}_p\}$ is an orthonormal basis for W, then every vector $\mathbf{x} \in V$ can be written in the form $\widehat{\mathbf{x}} + \mathbf{y}$, where $\widehat{\mathbf{x}} = \langle \mathbf{x}, \mathbf{x}_1 \rangle x_1 + \cdots + \langle \mathbf{x}, \mathbf{x}_p \rangle x_p$ and $\mathbf{y} = \mathbf{x} - \widehat{\mathbf{x}}$.*

Proof. Since $\langle \mathbf{x}_1, \mathbf{x}_1 \rangle = \cdots = \langle \mathbf{x}_p, \mathbf{x}_p \rangle = 1$, the result follows from Corollary 8.17. ∎

The vector $\widehat{\mathbf{x}}$ in this corollary is the orthogonal projection $\text{proj}(\mathbf{x} \to W)$ discussed earlier. This vector is useful because it enables us to decompose a vector into a part in W and a complementary part in $W^\perp$.

THEOREM 8.19 (Orthogonal projection theorem)   *If $\mathcal{B} = \{\mathbf{x}_1, \ldots, \mathbf{x}_p\}$ is an orthonormal basis for a subspace W of $\mathbb{R}^n$ in the standard inner product and if $P = (\mathbf{x}_1 \ \cdots \ \mathbf{x}_p)$ is the matrix whose columns are the vectors in $\mathcal{B}$, then $\text{proj}(\mathbf{x} \to W) = PP^T\mathbf{x}$ for any $\mathbf{x} \in \mathbb{R}^n$.*

Proof. We prove the theorem for $V = \mathbb{R}^3$ and $p = 2$. The general case is analogous. We let

$$
\mathcal{B} = \left\{ \mathbf{x}_1 = \begin{pmatrix} a_1 \\ a_2 \\ a_3 \end{pmatrix}, \mathbf{x}_2 = \begin{pmatrix} b_1 \\ b_2 \\ b_3 \end{pmatrix} \right\}
$$

be an orthonormal basis for a two-dimensional subspace for $\mathbb{R}^3$, and we let $\mathbf{x} = (x, y, z)$ be an arbitrary vector in $\mathbb{R}^3$. Then

$$
P = \begin{pmatrix} a_1 & b_1 \\ a_2 & b_2 \\ a_3 & b_3 \end{pmatrix} \quad \text{and} \quad P^T = \begin{pmatrix} a_1 & a_2 & a_3 \\ b_1 & b_2 & b_3 \end{pmatrix}
$$

Therefore,

$$
P^T \mathbf{x} = \begin{pmatrix} a_1 & a_2 & a_3 \\ b_1 & b_2 & b_3 \end{pmatrix} \begin{pmatrix} x \\ y \\ z \end{pmatrix} = \begin{pmatrix} a_1 x + a_2 y + a_3 z \\ b_1 x + b_2 y + b_3 z \end{pmatrix} = \begin{pmatrix} \mathbf{x}_1^T \mathbf{x} \\ \mathbf{x}_2^T \mathbf{x} \end{pmatrix}
$$

and

$$
PP^T \mathbf{x} = \begin{pmatrix} a_1 & b_1 \\ a_2 & b_2 \\ a_3 & b_3 \end{pmatrix} \begin{pmatrix} \mathbf{x}_1^T \mathbf{x} \\ \mathbf{x}_2^T \mathbf{x} \end{pmatrix} = \begin{pmatrix} a_1 \mathbf{x}_1^T \mathbf{x} + b_1 \mathbf{x}_2^T \mathbf{x} \\ a_2 \mathbf{x}_1^T \mathbf{x} + b_2 \mathbf{x}_2^T \mathbf{x} \\ a_3 \mathbf{x}_1^T \mathbf{x} + b_3 \mathbf{x}_2^T \mathbf{x} \end{pmatrix}
$$

$$
= \begin{pmatrix} a_1 \mathbf{x}_1^T \mathbf{x} \\ a_2 \mathbf{x}_1^T \mathbf{x} \\ a_3 \mathbf{x}_1^T \mathbf{x} \end{pmatrix} + \begin{pmatrix} b_1 \mathbf{x}_2^T \mathbf{x} \\ b_2 \mathbf{x}_2^T \mathbf{x} \\ b_3 \mathbf{x}_2^T \mathbf{x} \end{pmatrix} = \mathbf{x}_1^T \mathbf{x} \begin{pmatrix} a_1 \\ a_2 \\ a_3 \end{pmatrix} + \mathbf{x}_2^T \mathbf{x} \begin{pmatrix} b_1 \\ b_2 \\ b_3 \end{pmatrix}
$$

$$
= \left(\mathbf{x}_1^T \mathbf{x} \right) \mathbf{x}_1 + \left(\mathbf{x}_2^T \mathbf{x} \right) \mathbf{x}_2 = \langle \mathbf{x}_1, \mathbf{x} \rangle \mathbf{x}_1 + \langle \mathbf{x}_2, \mathbf{x} \rangle \mathbf{x}_2
$$

By Corollary 8.18, $PP^T \mathbf{x} = \text{proj}(\mathbf{x} \to W)$. ∎

One of the remarkable properties of the fundamental subspaces determined by a real matrix is the fact (proved in Theorem 8.13) that they are pairwise complementary subspaces.

THEOREM 8.20 (Fundamental subspace theorem) *If A is an $m \times n$ real matrix, then $\text{Nul } A \oplus \text{Col } A^T = \mathbb{R}^n$ and $\text{Nul } A^T \oplus \text{Col } A = \mathbb{R}^m$.*

Proof. In Theorem 8.13, we showed that $\text{Nul } A$ is orthogonal to $\text{Col } A^T$ and that $\text{Nul } A^T$ is orthogonal to $\text{Col } A$. Moreover, we know from Theorems 4.28 and 4.29 that if A has rank r, then the dimension of $\text{Col } A^T$ is r and the dimension of $\text{Nul } A$ is $n - r$. Since these spaces are disjoint, we can form their direct sum $\text{Nul } A \oplus \text{Col } A^T$ and obtain a subspace of $\mathbb{R}^n$ of dimension $(n - r) + r = n$. This shows that $\text{Nul } A \oplus \text{Col } A^T$ must equal $\mathbb{R}^n$. An analogous

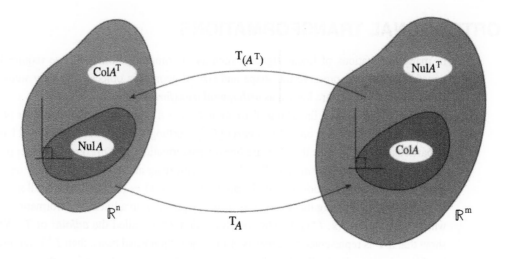

FIGURE 2   Orthogonal Complements.

argument shows that Nul $A^T \oplus$ Col $A = \mathbb{R}^m$ since the disjoint subspaces Col A and Nul A^T of $\mathbb{R}^m$ have dimensions r and $m - r$. ∎

Figure 2 illustrates the theorem.

EXERCISES 8.9

1. Find the orthogonal complement of the set $S = \{(1, 3)\}$ in $\mathbb{R}^2$.

2. Find the orthogonal complement $S^\perp$ of the set $S = \{(1, 3, 2), (4, 12, 8)\}$ in $\mathbb{R}^3$. What is the dimension of $S^\perp$? Explain your answer.

3. Find the orthogonal complement of the set $S = \{(1, 3, 2), (2, 1, 0)\}$ in $\mathbb{R}^3$.

4. Use first principles to find a basis for the complement of the column space of the matrix

$$A = \begin{pmatrix} 5 & 0 & 5 & 0 \\ 1 & 2 & 3 & 0 \\ 0 & 2 & 2 & 0 \end{pmatrix}$$

Compare your basis with that obtained by using the **NullSpace** function.

5. Use first principles to find a basis for the complement of the null space of the matrix

$$A = \begin{pmatrix} 5 & 0 & -2 & 0 \\ 1 & 2 & 0 & 0 \end{pmatrix}$$

Compare your basis with that obtained by using the method illustrated in Example 4.44.

ORTHOGONAL TRANSFORMATIONS

In many applications of linear algebra, such as in computer animation, we require linear transformations that preserve the shape and size of geometric objects. We now study such transformations. They are known as *orthogonal transformations*.

We will show that a linear transformation T on a finite-dimensional real inner product space is orthogonal if and only if its matrix $[T]_\mathcal{B}^\mathcal{B}$ is orthogonal in some orthonormal basis $\mathcal{B}$. In particular, we will show that they are linear transformations that *preserve* the inner product on V. By a linear transformation $T : V \to V$ *preserving* an inner product $\langle \mathbf{x}, \mathbf{y} \rangle$ on V, we mean a linear transformation T with the property that $\langle \mathbf{x}, \mathbf{y} \rangle = \langle T(\mathbf{x}), T(\mathbf{y}) \rangle$ for all $\mathbf{x}, \mathbf{y} \in V$.

We begin by showing that every T is naturally paired with a linear transformation T^* for which $\langle T(\mathbf{x}), \mathbf{y} \rangle = \langle \mathbf{x}, T^*(\mathbf{y}) \rangle$. The transformation T^* is called the *adjoint* of T. We will show that if T is represented by a matrix A in some orthonormal basis, then T^* is represented by A^T in the same basis. We then discuss the case of linear transformations that are equal to their adjoints, the *self-adjoint transformations*, and use their properties to show that a linear transformation T preserves an inner product $\langle \mathbf{x}, \mathbf{y} \rangle$ if and only if $TT^* = T^*T = I$.

Adjoint Transformations

In this section, we show that adjoints are represented by transposes. To prove this fact, we need to establish a fundamental property of linear functionals. We recall from earlier work that a linear transformation f from a real vector space V to the space $\mathbb{R}$ of real numbers is called a *linear functional*.

LEMMA 8.21 *If $f : V \to \mathbb{R}$ is a linear functional on a finite-dimensional inner product space V, then there exists a unique vector $\mathbf{y} \in V$ such that $f(\mathbf{x}) = \langle \mathbf{x}, \mathbf{y} \rangle$ for all $\mathbf{x} \in V$.*

Proof. Using the Gram-Schmidt process, we construct an orthonormal basis $\mathcal{S} = \{\mathbf{v}_1, \ldots, \mathbf{v}_n\}$ for V. We then use the functional f to define the vector $\mathbf{y}$ as a linear combination of the given basis vectors:

$$\mathbf{y} = f(\mathbf{v}_1)\mathbf{v}_1 + \cdots + f(\mathbf{v}_n)\mathbf{x}_n$$

Next we construct the linear functional $\vec{\mathbf{y}}$ that agrees with f on the orthonormal basis $\mathcal{S}$:

$$\vec{\mathbf{y}}(\mathbf{x}_i) = \langle \mathbf{v}_i, \mathbf{y} \rangle = \langle \mathbf{v}_i, f(\mathbf{v}_1)\mathbf{x}_1 + \cdots + f(\mathbf{v}_n)\mathbf{v}_n \rangle = f(\mathbf{x}_i) \langle \mathbf{v}_i, \mathbf{v}_i \rangle = f(\mathbf{x}_i).$$

It remains to establish the uniqueness of $\mathbf{y}$. Suppose that $\mathbf{z} \in V$ also has the property that $f(\mathbf{x}) = \langle \mathbf{x}, \mathbf{z} \rangle$ for all $\mathbf{x} \in V$. Then $\langle \mathbf{x}, \mathbf{y} \rangle = \langle \mathbf{x}, \mathbf{z} \rangle$, so that

$$\langle \mathbf{x}, \mathbf{y} \rangle - \langle \mathbf{x}, \mathbf{z} \rangle = \langle \mathbf{x}, \mathbf{y} - \mathbf{z} \rangle = 0$$

Let $\mathbf{x} = \mathbf{y} - \mathbf{z}$. Then $\langle \mathbf{y} - \mathbf{z}, \mathbf{y} - \mathbf{z} \rangle = 0$. Since the inner product of a vector with itself is 0 if and only if the vector is $\mathbf{0}$, it follows that $\mathbf{y} - \mathbf{z} = \mathbf{0}$, so that $\mathbf{y} = \mathbf{z}$. ∎

We use this lemma to prove the existence of adjoints.

THEOREM 8.22 (Adjoint existence theorem) *If $T : V \rightarrow V$ is a linear transformation on a finite-dimensional real inner product space V, then there exists a linear transformation $T^* : V \rightarrow V$ for which $\langle T(\mathbf{x}), \mathbf{y} \rangle = \langle \mathbf{x}, T^*(\mathbf{y}) \rangle$.*

Proof. Let $\mathbf{y}$ be a fixed vector in V. Then the function $f(\mathbf{x}) = \langle T(\mathbf{x}), \mathbf{y} \rangle$ is a linear functional $f : V \rightarrow \mathbb{R}$. Lemma 8.21 guarantees that there exists a unique vector $\mathbf{z} \in V$ such that $f(\mathbf{x}) = \langle \mathbf{x}, \mathbf{z} \rangle$. We can therefore put $T^*(\mathbf{y}) = \mathbf{z}$ and obtain for all $\mathbf{x}, \mathbf{y} \in V$, $\langle T(\mathbf{x}), \mathbf{y} \rangle = \langle \mathbf{x}, T^*(\mathbf{y}) \rangle$. The uniqueness of $\mathbf{z}$ ensures that T^* is a function on V.

It remains to verify that T^* is linear. Since

$$
\begin{aligned}
\langle \mathbf{x}, T^*(a\mathbf{u} + b\mathbf{v}) \rangle &= \langle T(\mathbf{x}), a\mathbf{u} + b\mathbf{v} \rangle \\
&= \langle T(\mathbf{x}), a\mathbf{u} \rangle + \langle T(\mathbf{x}), b\mathbf{v} \rangle \\
&= a \langle T(\mathbf{x}), \mathbf{u} \rangle + b \langle T(\mathbf{x}), \mathbf{v} \rangle \\
&= a \langle \mathbf{x}, T^*(\mathbf{u}) \rangle + b \langle \mathbf{x}, T^*(\mathbf{v}) \rangle \\
&= \langle \mathbf{x}, aT^*(\mathbf{u}) \rangle + \langle \mathbf{x}, bT^*(\mathbf{v}) \rangle \\
&= \langle \mathbf{x}, aT^*(\mathbf{u}) + bT^*(\mathbf{v}) \rangle
\end{aligned}
$$

for all $x, y \in V$ and all $a, b \in \mathbb{R}$, the linearity of T^* follows. ■

The next theorem tells us that the two matrices A and B representing T and T^* in an orthonormal basis have a very simple connection.

THEOREM 8.23 (Adjoint representation theorem) *If A represents a linear transformation $T : V \rightarrow V$ on a real inner product space V in an orthonormal basis $\mathcal{B} = \{\mathbf{x}_1, \ldots, \mathbf{x}_n\}$, then the matrix B representing the adjoint $T^* : V \rightarrow V$ of T in the basis $\mathcal{B}$ is A^T.*

Proof. Suppose that the matrix A represents T and that B represents T^* in the orthonormal basis $\mathcal{B}$. It follows from Corollary 8.3 that for each $1 \leq j \leq n$,

$$
T(\mathbf{x}_j) = \langle T(\mathbf{x}_j), \mathbf{x}_1 \rangle \mathbf{x}_1 + \cdots + \langle T(\mathbf{x}_j), \mathbf{x}_n \rangle \mathbf{x}_n
$$

By Theorem 5.11, the jth column of A is therefore the coordinate vector

$$
T(\mathbf{x}_j) = \begin{pmatrix} \langle T(\mathbf{x}_j), \mathbf{x}_1 \rangle \\ \vdots \\ \langle T(\mathbf{x}_j), \mathbf{x}_n \rangle \end{pmatrix}
$$

Similarly,

$$
\begin{aligned}
T^*(\mathbf{x}_j) &= \langle T^*(\mathbf{x}_j), \mathbf{x}_1 \rangle \mathbf{x}_1 + \cdots + \langle T^*(\mathbf{x}_j), \mathbf{x}_n \rangle \mathbf{x}_n \\
&= \langle \mathbf{x}_j, T(\mathbf{x}_1) \rangle \mathbf{x}_1 + \cdots + \langle \mathbf{x}_j, T(\mathbf{x}_n) \rangle \mathbf{x}_n \\
&= \langle T(\mathbf{x}_1), \mathbf{x}_j \rangle \mathbf{x}_1 + \cdots + \langle T(\mathbf{x}_n), \mathbf{x}_j \rangle \mathbf{x}_n
\end{aligned}
$$

Hence the jth column of B is the coordinate vector

$$T^*(\mathbf{x}_j) = \begin{pmatrix} \langle T(\mathbf{x}_1), \mathbf{x}_j \rangle \\ \vdots \\ \langle T(\mathbf{x}_n), \mathbf{x}_j \rangle \end{pmatrix}$$

This shows that the columns of A are the rows of B. Hence $B = A^T$. ∎

EXAMPLE 8.22   ■ The Adjoint of a Linear Transformation

Use *MATHEMATICA* to illustrate Theorem 8.23.

Solution. We create the matrices

$$A = \begin{pmatrix} 1 & 2 & 3 \\ 4 & 5 & 6 \\ 7 & 8 & 9 \end{pmatrix} \quad \text{and} \quad A^T = \begin{pmatrix} 1 & 4 & 7 \\ 2 & 5 & 8 \\ 3 & 6 & 9 \end{pmatrix}$$

and compare the inner products

$$\left\langle A \begin{pmatrix} x \\ y \\ z \end{pmatrix}, \begin{pmatrix} u \\ v \\ w \end{pmatrix} \right\rangle \quad \text{and} \quad \left\langle \begin{pmatrix} x \\ y \\ z \end{pmatrix}, A^T \begin{pmatrix} u \\ v \\ w \end{pmatrix} \right\rangle$$

```
In[1]:= A={{1,2,3},{4,5,6},{7,8,9}};
```

```
In[2]:= X={{x},{y},{z}};
```

```
In[3]:= Y={{u},{v},{w}};
```

```
In[4]:= Transpose[A.X].Y
Out[4]= {{u (x+2 y+3 z)+v (4 x+5 y+6 z)+w (7 x+8 y+9 z)}}
```

```
In[5]:= Transpose[X].Transpose[A].Y
Out[5]= {{u (x+2 y+3 z)+v (4 x+5 y+6 z)+w (7 x+8 y+9 z)}}
```

By combining these results, we see that

$$\left\langle A \begin{pmatrix} x \\ y \\ z \end{pmatrix}, \begin{pmatrix} u \\ v \\ w \end{pmatrix} \right\rangle = \left\langle \begin{pmatrix} x \\ y \\ z \end{pmatrix}, A^T \begin{pmatrix} u \\ v \\ w \end{pmatrix} \right\rangle$$

holds for all x, y, z, u, v, w. Hence the linear transformations represented by the matrices A and A^T are adjoints. ◄►

EXERCISES 8.10

1. Use Theorem 8.23 to find the adjoint of the linear transformation $T : \mathbb{R}^2 \to \mathbb{R}^2$ defined by $T(x, y) = (2x - 3y, x + y)$.

2. Use Theorem 8.23 to find the adjoint of the linear transformation $T : \mathbb{R}^3 \to \mathbb{R}^3$ defined by $T(x, y, z) = (2x - 3y + z, x + y, z)$.

3. Suppose that T is the linear transformation represented by the matrix

$$A = \begin{pmatrix} 1 & 0 & 2 \\ 4 & 3 & 0 \\ 0 & 0 & 3 \end{pmatrix}$$

and T^* is the linear transformation represented by A^T in the standard basis of $\mathbb{R}^3$. Verify that $\langle T(\mathbf{x}), \mathbf{y} \rangle = \langle \mathbf{x}, T^*(\mathbf{y}) \rangle$ for all $\mathbf{x}, \mathbf{y} \in \mathbb{R}^3$.

4. Suppose that T is the linear transformation represented by the matrix

$$A = \begin{pmatrix} -3 & 3 & 5 \\ 3 & 5 & -2 \\ 5 & -2 & -3 \end{pmatrix}$$

and T^* is the linear transformation represented by A^T in the standard basis of $\mathbb{R}^3$. Verify that $\langle T(\mathbf{x}), \mathbf{y} \rangle = \langle \mathbf{x}, T^*(\mathbf{y}) \rangle$ for all $\mathbf{x}, \mathbf{y} \in \mathbb{R}^3$. What is the connection between T and T^*?

Self-Adjoint Linear Transformations

If a matrix $A = [T]_\mathcal{B}^\mathcal{B}$ represents a linear transformation T on a real inner product space V in an orthonormal basis $\mathcal{B} = \{\mathbf{x}_1, \ldots, \mathbf{x}_n\}$, then we know from Theorem 8.23 that the matrix A^T represents the adjoint T^* of T in the basis $\mathcal{B}$. Therefore, $T = T^*$ if and only if $A = A^T$. For this reason, we say that T is *self-adjoint* if it can be represented by a real symmetric matrix in some orthonormal basis. In this section, we show that a linear transformation T is orthogonal if and only if the associated transformation $TT^* - I$ is self-adjoint. We begin with two examples.

EXAMPLE 8.23 ■ Rotations and Self-Adjointness

Show that the rotation T through the angle $\theta = \pi$ is self-adjoint.

Solution. We know that in the standard basis $\mathcal{E}$ of $\mathbb{R}^2$, the transformation T is represented by the matrix

$$A = \begin{pmatrix} \cos \pi & -\sin \pi \\ \sin \pi & \cos \pi \end{pmatrix} = \begin{pmatrix} -1 & 0 \\ 0 & -1 \end{pmatrix}$$

Moreover, by Theorem 8.23, the matrix A^T represents the adjoint T^* of T. But $A = A^T$. Hence $T = T^*$. ◄►

Another important family of self-adjoint transformations are the reflections.

EXAMPLE 8.24 ■ Reflections and Self-Adjointness

Show that the reflection T about the lines $y = x$ is a self-adjoint linear transformation.

Solution. We know that in the standard basis $\mathcal{E}$ of $\mathbb{R}^2$, the transformation T is represented by the matrix

$$A = \begin{pmatrix} 0 & 1 \\ 1 & 0 \end{pmatrix}$$

and Theorem 8.23 tells us that the adjoint T^* of T is represented by A^T. But $A = A^T$. Therefore, $T = T^*$. This means that T is self-adjoint. ◄►

We now use self-adjointness to characterize orthogonal transformations.

LEMMA 8.24   *For any linear transformation T on a real inner product space V, the transformation $T^*T - I$ is self-adjoint.*

Proof. For all $\mathbf{x} \in V$ it holds that

$$\begin{aligned}
\langle (T^*T - I)(\mathbf{x}), \mathbf{x} \rangle &= \langle (T^*T)(\mathbf{x}), \mathbf{x} \rangle - \langle I(\mathbf{x}), \mathbf{x} \rangle \\
&= \langle T(\mathbf{x}), T(\mathbf{x}) \rangle - \langle \mathbf{x}, I(\mathbf{x}) \rangle \\
&= \langle \mathbf{x}, T^*T(\mathbf{x}) \rangle - \langle \mathbf{x}, I(\mathbf{x}) \rangle \\
&= \langle \mathbf{x}, (T^*T - I)(\mathbf{x}) \rangle
\end{aligned}$$

Hence $T^*T - I$ is self-adjoint. ■

LEMMA 8.25   *If T is a self-adjoint linear transformation on a finite-dimensional real inner product space V and $\langle T(\mathbf{x}), \mathbf{x} \rangle = 0$ for all $\mathbf{x} \in V$, then $T = 0$.*

Proof. Since T is self-adjoint and $\langle T(\mathbf{x}), \mathbf{x} \rangle = \langle T(\mathbf{y}), \mathbf{y} \rangle = 0$, we have

$$\begin{aligned}
\langle T(\mathbf{x} + \mathbf{y}), \mathbf{x} + \mathbf{y} \rangle &= \langle T(\mathbf{x}) + T(\mathbf{y}), \mathbf{x} + \mathbf{y} \rangle \\
&= \langle T(\mathbf{x}), \mathbf{x} + \mathbf{y} \rangle + \langle T(\mathbf{y}), \mathbf{x} + \mathbf{y} \rangle \\
&= \langle T(\mathbf{x}), \mathbf{x} \rangle + \langle T(\mathbf{x}), \mathbf{y} \rangle + \langle T(\mathbf{y}), \mathbf{x} \rangle + \langle T(\mathbf{y}), \mathbf{y} \rangle \\
&= \langle T(\mathbf{x}), \mathbf{y} \rangle + \langle T(\mathbf{y}), \mathbf{x} \rangle \\
&= 0
\end{aligned}$$

for all $\mathbf{x}, \mathbf{y} \in V$. It follows that if $\mathbf{y} = T(\mathbf{x})$, then

$$\langle T(\mathbf{x}), T(\mathbf{x}) \rangle + \langle T(T(\mathbf{x})), \mathbf{x} \rangle = \langle T(\mathbf{x}), T(\mathbf{x}) \rangle + \langle T(T(\mathbf{x})), \mathbf{x} \rangle$$
$$= \langle T(\mathbf{x}), T(\mathbf{x}) \rangle + \langle T(\mathbf{x}), T(\mathbf{x}) \rangle$$
$$= 2 \langle T(\mathbf{x}), T(\mathbf{x}) \rangle$$
$$= 0$$

Therefore, $\langle T(\mathbf{x}), T(\mathbf{x}) \rangle = 0$ for all $\mathbf{x} \in V$. This means that $T(\mathbf{x}) = 0$ for all $\mathbf{x} \in V$. It follows that the transformation T itself is 0. ∎

We now apply Lemmas 8.24 and 8.25 to characterize orthogonal transformations.

THEOREM 8.26 (Orthogonal transformation theorem)   *If T is a linear transformation on a finite-dimensional real inner product space V, then the following conditions are equivalent:*

1. $\langle T(\mathbf{x}), T(\mathbf{y}) \rangle = \langle \mathbf{x}, \mathbf{y} \rangle$ for all $\mathbf{x}, \mathbf{y} \in V$
2. $\|T(\mathbf{x})\| = \|\mathbf{x}\|$ for all $\mathbf{x} \in V$
3. $TT^* = T^*T = I$

Proof. Suppose that $\langle T(\mathbf{x}), T(\mathbf{y}) \rangle = \langle \mathbf{x}, \mathbf{y} \rangle$ for all $\mathbf{x}, \mathbf{y} \in V$. Then

$$\|T(\mathbf{x})\| = \sqrt{\langle T(\mathbf{x}), T(\mathbf{x}) \rangle} = \sqrt{\langle \mathbf{x}, \mathbf{x} \rangle} = \|\mathbf{x}\|$$

Moreover, if $\|T(\mathbf{x})\| = \|\mathbf{x}\|$ for all $\mathbf{x} \in V$, then

$$\langle T^*T(\mathbf{x}), \mathbf{x} \rangle = \langle T(\mathbf{x}), T(\mathbf{x}) \rangle = \langle \mathbf{x}, \mathbf{x} \rangle = \langle I(\mathbf{x}), \mathbf{x} \rangle$$

This means that

$$\langle (T^*T - I)(\mathbf{x}), \mathbf{x} \rangle = 0$$

for all $\mathbf{x} \in V$. By Lemma 8.24, $T^*T - I$ is self-adjoint. It therefore follows from Lemma 8.25 that $T^*T - I = 0$. An analogous calculation shows that $TT^* - I = 0$. Therefore,

$$\langle T(\mathbf{x}), T(\mathbf{y}) \rangle - \langle \mathbf{x}, \mathbf{y} \rangle = \langle \mathbf{x}, T^*T(\mathbf{y}) \rangle - \langle \mathbf{x}, I(\mathbf{y}) \rangle$$
$$= \langle \mathbf{x}, (T^*T - I)(\mathbf{y}) \rangle$$
$$= \langle \mathbf{x}, 0(\mathbf{y}) \rangle$$
$$= 0$$

We have shown that $\langle T(\mathbf{x}), T(\mathbf{y}) \rangle = \langle \mathbf{x}, \mathbf{y} \rangle$ for all $\mathbf{x}, \mathbf{y} \in V$. ∎

COROLLARY 8.27 (Orthogonal transformations and orthogonal matrices)   *A matrix A represents an orthogonal transformation on a finite-dimensional real inner product space in an orthonormal basis if and only if A is orthogonal.*

Proof. Suppose A represents an orthogonal transformation T with respect to an orthonormal basis $\mathcal{B}$. By Theorem 8.23, T has an adjoint T^* represented by the transpose A^T of A. By Theorem 8.26, $TT^* = T^*T = I$, so that A^T must be A^{-1}. ∎

EXERCISES 8.11

1. Prove or disprove the following statements.

   a. All rotations $T : \mathbb{R}^2 \to \mathbb{R}^2$ are self-adjoint in the standard inner product on $\mathbb{R}^2$.

   b. Some rotations $T : \mathbb{R}^2 \to \mathbb{R}^2$ are self-adjoint in the standard inner product on $\mathbb{R}^2$.

   c. All reflections $T : \mathbb{R}^2 \to \mathbb{R}^2$ are self-adjoint in the standard inner product on $\mathbb{R}^2$.

   d. Some reflections $T : \mathbb{R}^2 \to \mathbb{R}^2$ are self-adjoint in the standard inner product on $\mathbb{R}^2$.

   e. The zero transformation $T : \mathbb{R}^2 \to \mathbb{R}^2$ is self-adjoint in all inner products on $\mathbb{R}^2$.

   f. If $T_1, T_2 : \mathbb{R}^2 \to \mathbb{R}^2$ are self-adjoint linear transformations, then $T_1 + T_2 : \mathbb{R}^2 \to \mathbb{R}^2$ is a self-adjoint linear transformation.

   g. If $T : \mathbb{R}^2 \to \mathbb{R}^2$ is a self-adjoint linear transformation and $a \in \mathbb{R}$, then $(aT) : \mathbb{R}^2 \to \mathbb{R}^2$ is a self-adjoint linear transformation.

   h. The self-adjoint linear transformations (in the standard inner product) form a subspace of $\mathrm{Hom}(\mathbb{R}^2, \mathbb{R}^2)$.

   i. All orthogonal transformations $T : \mathbb{R}^2 \to \mathbb{R}^2$ are either rotations or reflections.

   j. All orthogonal transformations $T : \mathbb{R}^3 \to \mathbb{R}^3$ are either rotations or reflections.

   k. If $T_1, T_2 : \mathbb{R}^2 \to \mathbb{R}^2$ are two rotations, then their composition $T_1 \circ T_2 : \mathbb{R}^2 \to \mathbb{R}^2$ is a rotation.

   l. If $T_1, T_2 : \mathbb{R}^2 \to \mathbb{R}^2$ are two reflections, then their composition $T_1 \circ T_2 : \mathbb{R}^2 \to \mathbb{R}^2$ is a reflection.

   m. All reflections $T : \mathbb{R}^2 \to \mathbb{R}^2$ have a real eigenvalue.

   n. Some reflections $T : \mathbb{R}^2 \to \mathbb{R}^2$ have a real eigenvalue.

   o. All rotations $T : \mathbb{R}^2 \to \mathbb{R}^2$ have a real eigenvalue.

   p. Some rotations $T : \mathbb{R}^2 \to \mathbb{R}^2$ have a real eigenvalue.

   q. Some orthogonal transformations $T : \mathbb{R}^2 \to \mathbb{R}^2$ are both a rotation and a reflection.

   r. If $\mathbf{x}$ and $\mathbf{y}$ are two unit vectors in a two-dimensional real inner product space, then there exists a rotation $T : \mathbb{R}^2 \to \mathbb{R}^2$ for which $T(\mathbf{x}) = \mathbf{y}$.

2. Find an angle θ for which the matrix

$$A = \begin{pmatrix} -\frac{1}{2} & -\frac{1}{2}\sqrt{3} \\ \frac{1}{2}\sqrt{3} & -\frac{1}{2} \end{pmatrix}$$

represents a rotation $T : \mathbb{R}^2 \to \mathbb{R}^2$ in the standard inner product.

3. Prove that the eigenvalues of an orthogonal matrix are ± 1. Show by means of a counterexample that the converse is not true: Not every matrix whose eigenvalues are ± 1 is orthogonal.

4. Let $A = (a_{ij})$ be an $m \times n$ real matrix, and let $\mathcal{E} = \{\mathbf{e}_1, \ldots, \mathbf{e}_n\}$ be the standard basis of $\mathbb{R}^n$ and $\mathcal{E}' = \{\mathbf{e}'_1, \ldots, \mathbf{e}'_m\}$ the standard basis of $\mathbb{R}^m$. Use the fact that $a_{ij} = \langle A\mathbf{e}_j, \mathbf{e}'_i \rangle$ for all i, j in the standard inner product on $\mathbb{R}^m$ to show that if T_A is self-adjoint, then A is a real symmetric matrix.

The Spectral Theorem

A real square matrix A is ***orthogonally diagonalizable*** if there exists a diagonal matrix D and an orthogonal matrix Q for which $A = QDQ^T$. In this section, we prove that every real symmetric matrix is orthogonally diagonalizable. This fact is the basis for the singular value decomposition of arbitrary rectangular real matrices studied in Chapter 9. We begin by establishing some preliminary results needed to prove the existence of D and Q.

THEOREM 8.28 (Orthogonal eigenvector theorem)   *If T is a self-adjoint linear transformation and $\mathbf{x}$ and $\mathbf{y}$ are eigenvectors belonging to distinct eigenvalues λ_1 and λ_2 of T, then $\mathbf{x}$ and $\mathbf{y}$ are orthogonal.*

Proof. Suppose that $T(\mathbf{x}) = \lambda_1\mathbf{x}$ and $T(\mathbf{y}) = \lambda_2\mathbf{y}$, and that $\lambda_1 \neq \lambda_2$. Then

$$\lambda_1\langle \mathbf{x}, \mathbf{y} \rangle = \langle \lambda_1\mathbf{x}, \mathbf{y} \rangle = \langle T(\mathbf{x}), \mathbf{y} \rangle = \langle \mathbf{x}, T(\mathbf{y}) \rangle = \langle \mathbf{x}, \lambda_2\mathbf{y} \rangle = \lambda_2\langle \mathbf{x}, \mathbf{y} \rangle$$

Therefore, $\lambda_1\langle \mathbf{x}, \mathbf{y} \rangle - \lambda_2\langle \mathbf{x}, \mathbf{y} \rangle = (\lambda_1 - \lambda_2)\langle \mathbf{x}, \mathbf{y} \rangle = 0$. Since $\lambda_1 - \lambda_2 \neq 0$, we must have $\langle \mathbf{x}, \mathbf{y} \rangle = 0$. Hence $\mathbf{x}$ and $\mathbf{y}$ are orthogonal. ∎

COROLLARY 8.29   *The eigenvectors of a real symmetric matrix A belonging to distinct eigenvalues are orthogonal.*

Proof. Let $T_A : \mathbb{R}^n \to \mathbb{R}^n$ be the matrix transformation represented by A in the standard orthonormal basis of $\mathbb{R}^n$. Then Theorem 8.23 tells us that the adjoint T_A^* is represented by A^T. Since A is symmetric, $T_A = T_A^*$. This means that T_A is self-adjoint. The corollary therefore follows from Theorem 8.28. ∎

COROLLARY 8.30   *Eigenvectors belonging to distinct eigenvalues of a self-adjoint linear transformation are linearly independent.*

Proof. Suppose that $\mathbf{x}$ and $\mathbf{y}$ are two eigenvectors belonging to distinct eigenvalues of a self-adjoint transformation T. Then Theorem 8.28 ensures that $\mathbf{x}$ and $\mathbf{y}$ are orthogonal. Theorem 8.2 therefore tells us that $\mathbf{x}$ and $\mathbf{y}$ are linearly independent. ∎

The next result requires a brief excursion into complex inner product spaces and uses the fundamental theorem of algebra. (This theorem is discussed in Appendix A.)

THEOREM 8.31 (Symmetric matrix theorem)   *The eigenvalues of a real symmetric matrix A are real.*

Proof. Let $c_A(t)$ be the characteristic polynomial of A, considered as a complex polynomial in $\mathbb{C}[t]$. Then it follows from the fundamental theorem of algebra that $c_A(t) = (t - \lambda)q(t)$ for some complex number λ. Since λ is an eigenvalue of A, there exists a nonzero vector $\mathbf{x}$ for which $A\mathbf{x} = \lambda\mathbf{x}$. We now show that λ must be real.

Consider A as a matrix transformation A on the complex vector space $\mathbb{C}^n$ and let $\langle \mathbf{x}, \mathbf{y} \rangle = \mathbf{x}^T \bar{\mathbf{y}}$. We know from Chapter 7 that $\langle \mathbf{x}, \mathbf{y} \rangle$ is an inner product on $\mathbb{C}^n$. Since A is a symmetric matrix, $A = A^T$, and since A is real, $A = A^H$, where A^H is the conjugate transpose of A. Therefore,

$$\langle A\mathbf{x}, \mathbf{x} \rangle = (A\mathbf{x})^T \bar{\mathbf{x}} = \mathbf{x}^T A^T \bar{\mathbf{x}} = \mathbf{x}^T A^H \bar{\mathbf{x}} = \mathbf{x}^T \overline{A\mathbf{x}} = \langle \mathbf{x}, A\mathbf{x} \rangle$$

It follows that if $\mathbf{x}$ is an eigenvector associated with an eigenvalue λ, then

$$\lambda \langle \mathbf{x}, \mathbf{x} \rangle = \langle \lambda\mathbf{x}, \mathbf{x} \rangle = \langle A\mathbf{x}, \mathbf{x} \rangle = \langle \mathbf{x}, A\mathbf{x} \rangle = \langle \mathbf{x}, \lambda\mathbf{x} \rangle = \bar{\lambda} \langle \mathbf{x}, \mathbf{x} \rangle$$

Since $\mathbf{x}$ is nonzero, we have $\langle \mathbf{x}, \mathbf{x} \rangle \neq 0$. Consequently, $\lambda = \bar{\lambda}$. The eigenvalue λ is therefore real. ∎

COROLLARY 8.32   *The eigenvalues of a self-adjoint linear transformation T on a finite-dimensional real inner product space V are real.*

Proof. It is clear from our introduction to self-adjoint linear transformations that there exists an orthonormal basis for V in which the matrix of T is symmetric. The result therefore follows from Theorem 8.31. ∎

Next we show that the eigenvectors of a self-adjoint linear transformation on a finite-dimensional real inner product space span the space.

THEOREM 8.33 (Orthonormal basis theorem)   *Let T be a self-adjoint linear transformation on a finite-dimensional real inner product space V. Then there exists an orthonormal basis for V consisting of eigenvectors of T.*

Proof. We prove the theorem by an induction on the dimension of V.

If $\dim V = 1$, Theorem 8.31 guarantees the existence of a nonzero eigenvector $\mathbf{x}$ of T. Since every nonzero vector can be normalized, we may assume that $\mathbf{x}$ is a normal eigenvector. Therefore, $\mathcal{B} = \{\mathbf{x}\}$ is the required orthonormal basis.

Now suppose that $\dim V = n > 1$, and that the theorem holds for spaces of dimension $n - 1$. We know from Theorem 8.31 that T has at least one real eigenvalue λ. Let $\mathbf{x}$ be a normal eigenvector belonging to λ, and let $W = \text{span}(\mathbf{x})$ be the one-dimensional subspace of V spanned by $\mathbf{x}$. Then it follows from the properties of orthogonal complements that $V = W \oplus W^\perp$. Moreover, since T is self-adjoint,

$$\langle T(\mathbf{y}), \mathbf{x} \rangle = \langle \mathbf{y}, T(\mathbf{x}) \rangle = \langle \mathbf{y}, \lambda\mathbf{x} \rangle = \lambda \langle \mathbf{y}, \mathbf{x} \rangle = \lambda 0 = 0$$

for all $\mathbf{y} \in W^\perp$. Therefore, $T(\mathbf{y}) \in W^\perp$, so that T is a self-adjoint linear transformation on $W^\perp$. By the induction hypothesis, there therefore exists an orthonormal basis $\mathcal{B} = \{\mathbf{x}_2, \ldots, \mathbf{x}_n\}$ for $W^\perp$ consisting of eigenvectors of T. Let $\mathcal{C}$ be the set of eigenvectors $\{\mathbf{x}, \mathbf{x}_2, \ldots, \mathbf{x}_n\}$ of T. Since $W \oplus W^\perp$ is a direct sum, the vector $\mathbf{x}$ is linearly independent of the vectors $\mathbf{x}_2, \ldots, \mathbf{x}_n$, and since $\mathcal{B}$ is a basis for $W^\perp$, the set $\mathcal{C}$ is therefore a basis for $W \oplus W^\perp = V$. Since the vectors in $\mathcal{B}$ are normal and orthogonal eigenvectors of T, the set $\mathcal{C}$ is an orthonormal basis for the whole space consisting entirely of eigenvectors of T. ∎

COROLLARY 8.34   *Every self-adjoint transformation T on a finite-dimensional real inner product space can be represented by a diagonal matrix whose diagonal entries are the eigenvalues of T.*

Proof. By Theorem 8.33, the inner product space V has an orthonormal basis $\mathcal{B} = \{\mathbf{x}_1, \ldots, \mathbf{x}_n\}$ consisting of eigenvectors of T. Let $[T(\mathbf{x}_i)]_\mathcal{C}$ be the coordinate vector of $T(\mathbf{x}_i)$ in the basis $\mathcal{B}$. Then $[T(\mathbf{x}_i)]_\mathcal{C}$ is the column vector whose ith entry is λ_i and whose other coordinates are 0. Hence the matrix $([T(\mathbf{x}_1)]_\mathcal{C} \cdots [T(\mathbf{x}_n)]_\mathcal{C})$ is diagonal and represents T in $\mathcal{B}$. ∎

Corollary 8.34 yields the *spectral theorem* for real symmetric matrices.

COROLLARY 8.35 (Spectral theorem)   *For every real symmetric matrix A, there exists an orthogonal matrix Q such that $D = Q^T A Q$ is a diagonal matrix whose entries are the eigenvalues of A.*

Proof. Consider $T : \mathbb{R}^n \to \mathbb{R}^n$ as a linear transformation defined by $T(\mathbf{x}) = A\mathbf{x}$ in some orthonormal basis $\mathcal{B}$. We know from Theorem 8.33 that there exists an orthonormal basis $\mathcal{C}$ for $\mathbb{R}^n$ consisting of eigenvectors of A. Therefore, the matrix $[T]_\mathcal{C} = D$ is orthogonal. Let Q be the coordinate conversion matrix from $\mathcal{C}$ to $\mathcal{B}$. Then $D = Q^{-1}AQ$. By Theorem 8.12, the matrix Q is orthogonal. Hence $D = Q^T A Q$. ∎

We note that if $A = QDQ^T$, then

$$
A = \begin{pmatrix} \mathbf{u}_1 & \cdots & \mathbf{u}_n \end{pmatrix} \begin{pmatrix} \lambda_1 & & 0 \\ & \ddots & \\ 0 & & \lambda_n \end{pmatrix} \begin{pmatrix} \mathbf{u}_1^T \\ \vdots \\ \mathbf{u}_n^T \end{pmatrix}
$$

$$
= \begin{pmatrix} \lambda_1 \mathbf{u}_1 & \cdots & \lambda_n \mathbf{u}_n \end{pmatrix} \begin{pmatrix} \mathbf{u}_1^T \\ \vdots \\ \mathbf{u}_n^T \end{pmatrix}
$$

$$
= \lambda_1 \mathbf{u}_1 \mathbf{u}_1^T + \cdots + \lambda_n \mathbf{u}_n \mathbf{u}_n^T
$$

Since the set of eigenvalues of a matrix A is called the **spectrum** of A, the equation

$$
A = \lambda_1 \mathbf{u}_1 \mathbf{u}_1^T + \cdots + \lambda_n \mathbf{u}_n \mathbf{u}_n^T
$$

is called the ***spectral decomposition*** of A. It follows from our earlier analysis of the matrices $\mathbf{u}_i \mathbf{u}_i^T$ that for all $\mathbf{x} \in \mathbb{R}^n$, the vectors $\left(\mathbf{u}_i \mathbf{u}_i^T\right)\mathbf{x}$ are the orthogonal projections $\text{proj}(\mathbf{x} \to \text{span}\{\mathbf{u}_i\})$ of $\mathbf{x}$ onto the subspace of $\mathbb{R}^n$ spanned by $\mathbf{u}_i$.

The spectral theorem implies a characterization of positive definite matrices.

THEOREM 8.36 (Positive definiteness test)   *A real symmetric matrix is positive definite if and only if its eigenvalues are positive.*

Proof. Suppose that A is a positive definite real $n \times n$ matrix and that λ is an eigenvalue of A. For all eigenvectors $\mathbf{x} \in \mathbb{R}^n$ associated with λ,

$$0 < \mathbf{x}^T A \mathbf{x} = \mathbf{x}^T \lambda \mathbf{x} = \lambda \mathbf{x}^T \mathbf{x} = \lambda \|\mathbf{x}\|^2$$

where $\|\mathbf{x}\| = \|\mathbf{x}\|_2$ is the Euclidean norm on $\mathbb{R}^n$. Since $0 < \lambda\|\mathbf{x}\|^2$ and $0 < \|\mathbf{x}\|$, the eigenvalue λ must be positive.

Suppose now that all eigenvalues of A are positive. By Corollary 8.35, $A = QDQ^T$, where D is a diagonal matrix whose diagonal entries are the eigenvalues of A, and Q is an orthogonal matrix whose columns are associated eigenvectors. Let $\mathbf{x} = (x_1, \ldots, x_n)$ and $\mathbf{y} = (y_1, \ldots, y_n)$ be two nonzero vectors $\mathbb{R}^n$, related by $\mathbf{x} = Q\mathbf{y}$. Then

$$\mathbf{x}^T A \mathbf{x} = (Q\mathbf{y})^T A (Q\mathbf{y}) = \mathbf{y}^T Q^T A Q \mathbf{y} = \mathbf{y}^T D \mathbf{y} = \lambda_1 y_1^2 + \cdots + \lambda_n y_n^2$$

Since all $\lambda_1, \ldots, \lambda_n$ are positive and since $\mathbf{y}$ is nonzero, the real number $\lambda_1 y_1^2 + \cdots + \lambda_n y_n^2$ must be positive. Therefore, $\mathbf{x}^T A \mathbf{x}$ is positive. ∎

EXAMPLE 8.25   ■ Orthogonal Diagonalization of a Real Symmetric Matrix

Use the **Orthogonalization** package of *MATHEMATICA* to decompose the matrix

$$A = \begin{pmatrix} 3 & -4 \\ -4 & -3 \end{pmatrix}$$

into a product QDQ^T, where Q is an orthogonal and D is a diagonal matrix.

Solution. We begin by loading the **Orthogonalization** package.

```
In[1]:= <<LinearAlgebra`Orthogonalization`
```

Next we create the matrix A.

```
In[2]:= A={{3,-4},{-4,-3}};
```

Since A is a real symmetric matrix, Corollary 8.35 guarantees that there exist eigenvalues and associated eigenvectors for which $A = QDQ^T$. We use *MATHEMATICA* to find them.

```
In[3]:= S=Eigensystem[A]

Out[3]= {{-5,5},{{1,2},{-2,1}}}
```

It follows from the definition of the **Eigensystem** function that -5 and 5 are eigenvalues of A. We can therefore construct a diagonal matrix D.

```
In[4]:= Dm=DiagonalMatrix[S[[1]]]

Out[4]= {{-5,0},{0,5}}
```

We also know from the definition of the **Eigensystem** function that the vectors

$$\mathbf{u} = \begin{pmatrix} 1 \\ 2 \end{pmatrix} \quad \text{and} \quad \mathbf{v} = \begin{pmatrix} -2 \\ 1 \end{pmatrix}$$

are eigenvectors associated with the eigenvalues -5 and 5, respectively. They are orthogonal in the standard inner product on $\mathbb{R}^2$ since

$$\mathbf{u}^T\mathbf{v} = \begin{pmatrix} 1 \\ 2 \end{pmatrix}^T \begin{pmatrix} -2 \\ 1 \end{pmatrix} = 0$$

We use *MATHEMATICA* to normalize the vectors **u** and **v**.

```
In[5]:= x = Normalize[S[[2]][[1]]]

Out[5]= {1/√5, 2/√5}
```

```
In[6]:= y = Normalize[S[[2]][[2]]]

Out[6]= {-2/√5, 1/√5}
```

Next we construct the matrix Q. Since the normalized eigenvectors must form the *columns* of Q, we need to use the transpose in the construction.

```
In[7]:= Q=Transpose[{x,y}]

Out[7]= {{1/√5, -2/√5}, {2/√5, 1/√5}}
```

We conclude by showing that $A = QDQ^T$.

```
In[8]:= A= =Q.Dm.Transpose[Q]
Out[8]= True
```

We have therefore produced the required orthogonal diagonalization. ◀▶

EXERCISES 8.12

1. Show that there exists a basis $\mathcal{B} = \{x_1, x_2\}$ for $\mathbb{R}^2$ with the property that x_1 and x_2 are eigenvectors of each of the matrices

$$A = \begin{pmatrix} 1 & 2 \\ 2 & 1 \end{pmatrix} \quad B = \begin{pmatrix} 1 & 1 \\ 1 & 1 \end{pmatrix} \quad C = \begin{pmatrix} 0 & 3 \\ 3 & 0 \end{pmatrix}$$

2. Calculate the rank of the symmetric matrices $A^T A$, $B^T B$, and $C^T C$, where

$$A = \begin{pmatrix} 2 & 3 \\ 5 & 2 \\ 4 & 2 \end{pmatrix} \quad B = \begin{pmatrix} 2 & 5 & 4 \\ 3 & 2 & 2 \end{pmatrix} \quad C = \begin{pmatrix} 2 & 5 & 4 \end{pmatrix}$$

and use this information to conclude that the columns of $A^T A$ are linearly independent, but those of $B^T B$, and $C^T C$ are not. Show that it is still possible, however, to construct two bases for $\mathbb{R}^3$ consisting of eigenvectors of $B^T B$ and $C^T C$.

3. Prove that the dimension of an eigenspace of a real symmetric matrix equals the algebraic multiplicity of the corresponding eigenvalue of the matrix.

4. Use *MATHEMATICA* to find the eigenvalues of the matrix

$$A = \begin{pmatrix} 2 & 1 & 4 & 6 & 6. \\ 1 & 8 & 7 & 9 & 4 \\ 4 & 7 & 9 & 2 & 8 \\ 6 & 9 & 2 & 10 & 1 \\ 6 & 4 & 8 & 1 & 5 \end{pmatrix}$$

by computing the roots of the characteristic polynomial of A. Find an eigenvector for each eigenvalue and show that these vectors are *approximately* orthogonal. Explain why your result is not exact.

5. Use the method applied in Example 8.25 to diagonalize the following symmetric matrices.

a. $\begin{pmatrix} 4 & 3 \\ 3 & 2 \end{pmatrix}$ b. $\begin{pmatrix} -2 & 0 & 0 \\ 0 & 4 & 3 \\ 0 & 3 & 2 \end{pmatrix}$ c. $\begin{pmatrix} -2 & 0 & 0 & 1 \\ 0 & 4 & 3 & 0 \\ 0 & 3 & 2 & 0 \\ 1 & 0 & 0 & 0 \end{pmatrix}$

6. Apply the techniques used in the proofs of Theorems 8.35 and 8.36 to write the quadratic form

$$\begin{pmatrix} x & y \end{pmatrix} \begin{pmatrix} 1 & 3 \\ 3 & 2 \end{pmatrix} \begin{pmatrix} x \\ y \end{pmatrix}$$

as a sum of squares.

7. Apply the techniques used in the proof of Theorems 8.35 and 8.36 to write the quadratic form

$$\begin{pmatrix} x & y & z \end{pmatrix} \begin{pmatrix} 1 & 0 & 0 \\ 0 & 2 & 4 \\ 0 & 4 & 2 \end{pmatrix} \begin{pmatrix} x \\ y \\ z \end{pmatrix}$$

as a sum of squares.

THE METHOD OF LEAST SQUARES

In some applications of linear algebra we are faced with the problem of having to solve overdetermined linear systems. Since exact solutions are usually impossible, we look for best approximations. In this section, we explain what is meant by "best" and develop the ***method of least squares*** for finding such approximations.

Suppose A is an $m \times n$ real matrix. We know from Chapter 5 that A corresponds to a matrix transformation $T_A : \mathbb{R}^n \to \mathbb{R}^m$ defined by the rule $T_A(\mathbf{x}) = A\mathbf{x}$. Moreover, Theorem 5.15 tells us that $A\mathbf{x} \in \text{Col } A$. Hence the statement that $A\mathbf{x} = \mathbf{b}$ asserts that the vector $\mathbf{b}$ belongs to Col A. If $\mathbf{b} \notin \text{Col } A$ for any $\mathbf{x} \in \mathbb{R}^n$, then the equation $A\mathbf{x} = \mathbf{b}$ has no solution. However, we can still ask whether there is a vector $\widehat{\mathbf{x}} \in \mathbb{R}^n$ for which the vector $A\widehat{\mathbf{x}} - \mathbf{b}$ is as short as possible. Figure 3 illustrates this situation.

A best possible approximation of a solution of the equation $A\mathbf{x} = \mathbf{b}$ consists of a vector $\widehat{\mathbf{x}} \in \mathbb{R}^n$ for which the distance between $A\widehat{\mathbf{x}}$ and $\mathbf{b}$ is a minimum. We use the Euclidean norm to measure this distance. The vector $\widehat{\mathbf{x}}$ is called the ***least-squares solution*** of the equation $A\mathbf{x} = \mathbf{b}$. Thus we are trying to find a vector $\widehat{\mathbf{x}} \in \mathbb{R}^n$ for which $\| A\widehat{\mathbf{x}} - \mathbf{b} \|_2$ is a minimum.

Suppose that

$$A\widehat{\mathbf{x}} - \mathbf{b} = \begin{pmatrix} y_1 \\ \vdots \\ y_m \end{pmatrix}$$

Then $\| A\widehat{\mathbf{x}} - \mathbf{b} \| = \sqrt{y_1^2 + \cdots + y_m^2}$. Therefore, $\| A\widehat{\mathbf{x}} - \mathbf{b} \|$ is a minimum if and only if $y_1^2 + \cdots + y_m^2$ is a minimum. This explains the term "method of least squares."

THEOREM 8.37 (Best approximation theorem) *Let W be a subspace of $\mathbb{R}^m$ and let $\mathbf{b}$ be a vector in $\mathbb{R}^m$. Then the vector $\text{proj}(\mathbf{b} \to W)$ is the best approximation of $\mathbf{b}$ by vectors in W.*

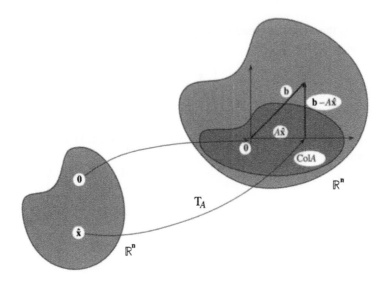

FIGURE 3   A Least Squares Approximation.

Proof. We show that

$$\|\mathbf{b} - \text{proj}(\mathbf{b} \to W)\| < \|\mathbf{b} - \mathbf{w}\|$$

for all $\mathbf{w} \in W$, provided that $\mathbf{w} \neq \text{proj}(\mathbf{b} \to W)$. By the Pythagorean theorem,

$$\|\mathbf{b} - \text{proj}(\mathbf{b} \to W)\|^2 + \|\text{proj}(\mathbf{b} \to W) - \mathbf{w}\|^2 = \|\mathbf{b} - \mathbf{w}\|^2$$

Hence $\|\mathbf{b} - \text{proj}(\mathbf{b} \to W)\|^2 < \|\mathbf{b} - \mathbf{w}\|^2$ if $\mathbf{w} \neq \text{proj}(\mathbf{b} \to W)$. ∎

We can reword Theorem 8.37 in terms of matrices. If A is any $m \times n$ real matrix and $W = \text{Col } A$, then for any vector $\mathbf{b} \in \mathbb{R}^m$, a vector $\mathbf{x} \in \mathbb{R}^n$ minimizes the distance $\|A\mathbf{x} - \mathbf{b}\|$ if and only if $A\mathbf{x} = \text{proj}(\mathbf{b} \to \text{Col } A)$.

THEOREM 8.38 *For any $m \times n$ real matrix A, a vector $\widehat{\mathbf{x}} \in \mathbb{R}^n$ is a least-squares solution of the equation $A\mathbf{x} = \mathbf{b}$ if and only if $\widehat{\mathbf{x}}$ is a solution of the normal equation $A^T A\mathbf{x} = A^T \mathbf{b}$.*

Proof. Suppose that $\widehat{\mathbf{x}}$ is a least-squares solution of the equation $A\mathbf{x} = \mathbf{b}$ and let $W = \text{Col } A$. Then $A\widehat{\mathbf{x}} = \text{proj}(\mathbf{b} \to W)$. We know from Section 8, that the vector $\mathbf{b} - A\widehat{\mathbf{x}}$ is orthogonal to any vector $A\mathbf{x} \in W$. Hence $(A\mathbf{x})^T (\mathbf{b} - A\widehat{\mathbf{x}}) = 0$. Theorem 8.23 tells us that

$$(A\mathbf{x})^T (\mathbf{b} - A\widehat{\mathbf{x}}) = \mathbf{x} A^T (\mathbf{b} - A\widehat{\mathbf{x}})$$

for all $\mathbf{x} \in \mathbb{R}^n$. Therefore, $\mathbf{x} A^T (\mathbf{b} - A\widehat{\mathbf{x}}) = 0$ for all $\mathbf{x} \in \mathbb{R}^n$. It follows that $A^T(\mathbf{b} - A\widehat{\mathbf{x}}) = \mathbf{0}$. In other words, $A^T \mathbf{b} - A^T A\widehat{\mathbf{x}} = \mathbf{0}$. Hence $\widehat{\mathbf{x}}$ is a solution of the normal equation $A^T A\widehat{\mathbf{x}} = A^T \mathbf{b}$. Conversely, suppose that $A^T A\widehat{\mathbf{x}} = A^T \mathbf{b}$ for some $\widehat{\mathbf{x}} \in \mathbb{R}^n$. Then $A^T(\mathbf{b} - A\widehat{\mathbf{x}}) = \mathbf{0}$. Therefore, $\mathbf{b} - A\widehat{\mathbf{x}}$ is orthogonal to each row of A^T. Since the rows of A^T are the columns of

A, the vector $\mathbf{b} - A\widehat{\mathbf{x}}$ is orthogonal to all vectors in W. Hence it is clear from Section 8 that $\mathbf{b} = A\widehat{\mathbf{x}} + (\mathbf{b} - A\widehat{\mathbf{x}})$ is a decomposition of $\mathbf{b}$, with $A\widehat{\mathbf{x}} \in W$ and $\mathbf{b} - A\widehat{\mathbf{x}} \in W^{\perp}$. By Corollary 8.17 the decomposition is unique. Hence $A\widehat{\mathbf{x}} = \text{proj}(\mathbf{b} \to W)$. ∎

Theorem 8.38 implies that if the matrix $A^T A$ is invertible, the vector $\widehat{\mathbf{x}} = (A^T A)^{-1} A^T \mathbf{b}$ is the least-squares solution of $A\mathbf{x} = \mathbf{b}$.

EXAMPLE 8.26 ■ Fitting a Straight Line to a Set of Data

Use the method of least squares to fit a straight line to the points $(x_1, y_1) = (2, 3)$, $(x_2, y_2) = (3, 5)$, $(x_3, y_3) = (4, 3)$, and $(x_4, y_4) = (5, 6)$.

Solution. Suppose the required line is defined by the equation $y = b + mx$. Then the given points lie on this line if and only if

$$\begin{cases} y_1 &= b + mx_1 \\ y_2 &= b + mx_2 \\ y_3 &= b + mx_3 \\ y_4 &= b + mx_4 \end{cases}$$

In other words, the points satisfy the matrix vector equation $\mathbf{y} = A\mathbf{x}$, given by

$$\begin{pmatrix} y_1 \\ y_2 \\ y_3 \\ y_4 \end{pmatrix} = \begin{pmatrix} 1 & x_1 \\ 1 & x_2 \\ 1 & x_3 \\ 1 & x_4 \end{pmatrix} \begin{pmatrix} b \\ m \end{pmatrix}$$

If we think of the matrix

$$A = \begin{pmatrix} 1 & x_1 \\ 1 & x_2 \\ 1 & x_3 \\ 1 & x_4 \end{pmatrix}$$

as a matrix transformation from $A : \mathbb{R}^2 \to \mathbb{R}^4$, then the equation $\mathbf{y} = A\mathbf{x}$ asserts that $\mathbf{y}$ lies in the image of A. If $\mathbf{y} \neq A\mathbf{x}$, we would like to find an approximate solution by locating a vector $\mathbf{y} \in \mathbb{R}^4$ whose distance from the subspace $\text{im } A = \{A\mathbf{x} : \mathbf{x} \in \mathbb{R}^2\}$ of $\mathbb{R}^2$ is a minimum. In that case, $\mathbf{y}$ must be orthogonal to the image of A. By the definition of the Euclidean inner product, this is so if and only if

$$(A\mathbf{x})^T (\mathbf{y} - A\mathbf{x}) = \left(\mathbf{x}^T A^T\right)(\mathbf{y} - A\mathbf{x}) = \mathbf{x}^T \left(A^T \mathbf{y} - A^T A\mathbf{x}\right) = 0$$

Since this equation holds for all $\mathbf{x}$ it follows that $A^T \mathbf{y} - A^T A\mathbf{x} = \mathbf{0}$. Therefore, $A^T \mathbf{y} = A^T A\mathbf{x}$. If $A^T A$ is invertible, we can solve this equation for $\mathbf{x}$ and obtain the required minimizing vector

$$\mathbf{x} = \left(A^T A\right)^{-1} A^T \mathbf{y}$$

Since the given points are not collinear, the matrix A^TA is invertible. Therefore,

$$\mathbf{x} = \left[\begin{pmatrix} 1 & 1 & 1 & 1 \\ x_1 & x_2 & x_3 & x_4 \end{pmatrix} \begin{pmatrix} 1 & x_1 \\ 1 & x_2 \\ 1 & x_3 \\ 1 & x_4 \end{pmatrix} \right]^{-1} \begin{pmatrix} 1 & 1 & 1 & 1 \\ x_1 & x_2 & x_3 & x_4 \end{pmatrix} \begin{pmatrix} y_1 \\ y_2 \\ y_3 \\ y_4 \end{pmatrix}$$

$$= \left[\begin{pmatrix} 1 & 1 & 1 & 1 \\ 2 & 3 & 4 & 5 \end{pmatrix} \begin{pmatrix} 1 & 2 \\ 1 & 3 \\ 1 & 4 \\ 1 & 5 \end{pmatrix} \right]^{-1} \begin{pmatrix} 1 & 1 & 1 & 1 \\ 2 & 3 & 4 & 5 \end{pmatrix} \begin{pmatrix} 3 \\ 5 \\ 3 \\ 6 \end{pmatrix}$$

$$= \begin{pmatrix} \frac{9}{5} \\ \frac{7}{10} \end{pmatrix}$$

$$= \begin{pmatrix} b \\ m \end{pmatrix}$$

We conclude that the required straight line is $y = \frac{9}{5} + \frac{7}{10}x$. We can use the **Fit** function of *MATHEMATICA* to confirm this result.

```
In[1]:= data={{2,3},{3,5},{4,3},{5,6}};
```

```
In[2]:= Fit[data,{1,x},x]
Out[2]= 1.8+0.7x
```

It is easy to verify that $1.8 + 0.7x$ is approximately equal to $\frac{9}{5} + \frac{7}{10}x$. ◄►

EXAMPLE 8.27 ■ Fitting a Parabola to a Set of Data

Use the method of least squares to fit a parabola to the points $(x_1, y_1) = (2, 3)$, $(x_2, y_2) = (3, 5)$, $(x_3, y_3) = (4, 3)$, and $(x_4, y_4) = (5, 6)$.

Solution. Suppose that the required parabola is given by $y = a + bx + cx^2$. Then the given points lie on this parabola if and only if

$$\begin{cases} y_1 = a + bx_1 + cx_1^2 \\ y_2 = a + bx_2 + cx_2^2 \\ y_3 = a + bx_3 + cx_3^2 \\ y_4 = a + bx_4 + cx_4^2 \end{cases}$$

This is so if and only if

$$\begin{pmatrix} y_1 \\ y_2 \\ y_3 \\ y_4 \end{pmatrix} = \begin{pmatrix} 1 & x_1 & x_1^2 \\ 1 & x_2 & x_4^2 \\ 1 & x_3 & x_4^2 \\ 1 & x_4 & x_4^2 \end{pmatrix} \begin{pmatrix} a \\ b \\ c \end{pmatrix}$$

If we think of this equation as $\mathbf{y} = A\mathbf{x}$ and reason as in Example 8.26, we see that if $\mathbf{y} \neq A\mathbf{x}$ then $\|\mathbf{y} - A\mathbf{x}\|$ is a minimum in $\mathbb{R}^4$ if and only if $\mathbf{x} = (A^T A)^{-1} A^T \mathbf{y}$. This means that

$$\mathbf{x} = \left[\begin{pmatrix} 1 & 1 & 1 & 1 \\ x_1 & x_2 & x_3 & x_4 \\ x_1^2 & x_2^2 & x_2^2 & x_2^2 \end{pmatrix} \begin{pmatrix} 1 & x_1 & x_1^2 \\ 1 & x_2 & x_2^2 \\ 1 & x_3 & x_3^2 \\ 1 & x_4 & x_4^2 \end{pmatrix} \right]^{-1} \begin{pmatrix} 1 & 1 & 1 & 1 \\ x_1 & x_2 & x_3 & x_4 \\ x_1^2 & x_2^2 & x_2^2 & x_2^2 \end{pmatrix} \begin{pmatrix} y_1 \\ y_2 \\ y_3 \\ y_4 \end{pmatrix}$$

$$= \left[\begin{pmatrix} 1 & 1 & 1 & 1 \\ 2 & 3 & 4 & 5 \\ 2^2 & 3^2 & 4^2 & 5^2 \end{pmatrix} \begin{pmatrix} 1 & 2 & 2^2 \\ 1 & 3 & 3^2 \\ 1 & 4 & 4^2 \\ 1 & 5 & 5^2 \end{pmatrix} \right]^{-1} \begin{pmatrix} 1 & 1 & 1 & 1 \\ 2 & 3 & 4 & 5 \\ 2^2 & 3^2 & 4^2 & 5^2 \end{pmatrix} \begin{pmatrix} 3 \\ 5 \\ 3 \\ 6 \end{pmatrix}$$

$$= \begin{pmatrix} \frac{91}{20} \\ -\frac{21}{20} \\ \frac{1}{4} \end{pmatrix} = \begin{pmatrix} a \\ b \\ c \end{pmatrix}$$

The required parabola is therefore $y = \frac{91}{20} - \frac{21}{20}x + \frac{1}{4}x^2$.

We use the **Fit** function of *MATHEMATICA* to confirm this result.

```
In[1]:= data={{2,3},{3,5},{4,3},{5,6}};
```

```
In[2]:= Fit[data,{1,x,x²},x]
Out[2]= 4.55-1.05x+0.25x²
```

It is easy to verify that this polynomial is approximately equal to $\frac{91}{20} - \frac{21}{20}x + \frac{1}{4}x^2$. ◂▸

EXERCISES 8.13

1. Convert the overdetermined linear system

$$\begin{cases} x = 1 \\ x = 2 \end{cases}$$

to the matrix form $A\mathbf{x} = \mathbf{b}$ and use the normal equation $A^T A\mathbf{x} = A^T\mathbf{b}$ to find the least-squares solution $\widehat{\mathbf{x}}$ of the given system. Explain your result.

2. Repeat Exercise 1 using the equations $x = a$ and $x = b$.

3. Express the arithmetic mean of the numbers $1, 2, 5, 6, 7$ as the least-squares solution of an overdetermined system.

4. Convert the overdetermined linear system

$$\begin{cases} x + y = 1 \\ x - y = 2 \\ 2x + y = 7 \end{cases}$$

to the matrix form $A\mathbf{x} = \mathbf{b}$ and use the normal equation $A^T A\mathbf{x} = A^T\mathbf{b}$ to find the least-squares solution $\widehat{\mathbf{x}}$ of the given system.

5. Show that if A is an $m \times n$ real matrix whose columns form an orthonormal set in $\mathbb{R}^m$, then $\mathbf{x} = A^T\mathbf{b}$ is a least-squares solution of the linear system $A\mathbf{x} = \mathbf{b}$.

6. Let y_1, y_2, and y_3 be three scalars. Find an overdetermined system whose least-squares solution is

$$\frac{y_1 a + y_2 b + y_3 c}{3}$$

Explain your result.

7. Let $A\mathbf{x} = \mathbf{b}$ be the linear system

$$\begin{pmatrix} 1 & -2 \\ 2 & 1 \end{pmatrix} \begin{pmatrix} x \\ y \end{pmatrix} = \begin{pmatrix} -7 \\ 5 \end{pmatrix}$$

Find the least-squares solution of the system and show that it coincides with the usual solution of the system.

8. Let $A\mathbf{x} = \mathbf{b}$ be the linear system

$$\begin{pmatrix} 1 & 3 & 0 \\ 0 & -1 & 8 \\ 1 & 2 & 4 \end{pmatrix} \begin{pmatrix} x \\ y \\ z \end{pmatrix} = \begin{pmatrix} 1 \\ 2 \\ 3 \end{pmatrix}$$

Find the least-squares solution $\mathbf{x} = \left(A^T A \right)^{-1} A^T\mathbf{b}$ of the system and show that it coincides with the usual solution.

9. Explain why the method used in Exercise 8 does not work for the system

$$\begin{pmatrix} 1 & 3 & 2 \\ 0 & -1 & 0 \\ 1 & 2 & 2 \end{pmatrix} \begin{pmatrix} x \\ y \\ z \end{pmatrix} = \begin{pmatrix} 1 \\ 2 \\ 3 \end{pmatrix}$$

10. Find the solutions of the normal equation of the linear system

$$\begin{pmatrix} 1 & 3 & 0 \\ 0 & -1 & 8 \end{pmatrix} \begin{pmatrix} x \\ y \\ z \end{pmatrix} = \begin{pmatrix} 1 \\ 2 \end{pmatrix}$$

11. Use the method of least squares to fit a curve $y = a + bx + cx^2 + dx^3 + ex^4$ to each of the following sets of points.

a. $(x_1, y_1) = (1, 6)$, $(x_2, y_2) = (2, 9)$, $(x_3, y_3) = (3, 10)$, $(x_4, y_4) = (4, 21)$, $(x_5, y_5) = (5, 30)$

b. $(x_1, y_1) = (-2, 7)$, $(x_2, y_2) = (-1, -2)$, $(x_3, y_3) = (0, -5)$, $(x_4, y_4) = (1, -1)$, $(x_5, y_5) = (2, 7)$

c. $(x_1, y_1) = (-2, 1)$, $(x_2, y_2) = (-1, 2)$, $(x_3, y_3) = (0, 1)$, $(x_4, y_4) = (1, 2)$, $(x_5, y_5) = (2, 1)$

REVIEW

KEY CONCEPTS ▶ Define and discuss each of the following.

Matrices
Orthogonal matrix, orthogonally diagonalizable matrix.

Orthogonal
Basis, complement, projection, subspaces, vectors.

Orthogonal Methods
Gram-Schmidt process, method of least squares, QR decomposition, perp function.

Orthogonal Transformations
Adjoint transformation, self-adjoint transformation.

Orthonormal
Basis, set.

KEY FACTS ▶ Explain and illustrate each of the following.

1. If $\mathbf{x}$ and $\mathbf{y}$ are two orthogonal vectors in an inner product space V, then $\|\mathbf{x} + \mathbf{y}\|^2 = \|\mathbf{x}\|^2 + \|\mathbf{y}\|^2$.

2. Orthogonal vectors in an inner product space are linearly independent.

3. If a basis $\mathcal{B} = \{\mathbf{x}_1, \ldots, \mathbf{x}_n\}$ for an inner product space V consists of orthogonal unit vectors, then $\mathbf{x} = \langle \mathbf{x}_1, \mathbf{x}\rangle \mathbf{x}_1 + \cdots + \langle \mathbf{x}_n, \mathbf{x}\rangle \mathbf{x}_n$ for any vector $\mathbf{x} \in V$.

4. If $\mathcal{B} = \{\mathbf{x}_1, \ldots, \mathbf{x}_p\}$ is an orthogonal set of vectors in an inner product space V, and if $\mathbf{x}$ is a vector in the span of $\mathcal{B}$ and is orthogonal to all vectors in $\mathcal{B}$, then $\mathbf{x} = \mathbf{0}$.

5. If $\mathcal{B} = \{\mathbf{x}_1, \ldots, \mathbf{x}_n\}$ is a basis for a real inner product space V with inner product $(\mathbf{x}, \mathbf{y}) \to \langle \mathbf{x}, \mathbf{y}\rangle$ and if

$$1. \quad \mathbf{y}_1 \quad = \quad \mathbf{x}_1$$

$$2. \quad \mathbf{y}_2 \quad = \quad \mathbf{x}_2 - \frac{\langle \mathbf{x}_2, \mathbf{y}_1\rangle}{\langle \mathbf{y}_1, \mathbf{y}_1\rangle}\mathbf{y}_1$$

$$3. \quad \mathbf{y}_3 \quad = \quad \mathbf{x}_3 - \frac{\langle \mathbf{x}_3, \mathbf{y}_1\rangle}{\langle \mathbf{y}_1, \mathbf{y}_1\rangle}\mathbf{y}_1 - \frac{\langle \mathbf{x}_3, \mathbf{y}_2\rangle}{\langle \mathbf{y}_2, \mathbf{y}_2\rangle}\mathbf{y}_2$$

$$\vdots$$

$$n. \quad \mathbf{y}_n \quad = \quad \mathbf{x}_n - \frac{\langle \mathbf{x}_n, \mathbf{y}_1\rangle}{\langle \mathbf{y}_1, \mathbf{y}_1\rangle}\mathbf{y}_1 - \frac{\langle \mathbf{x}_n, \mathbf{y}_2\rangle}{\langle \mathbf{y}_2, \mathbf{y}_2\rangle}\mathbf{y}_2 - \cdots - \frac{\langle \mathbf{x}_n, \mathbf{y}_n\rangle}{\langle \mathbf{y}_n, \mathbf{y}_n\rangle}\mathbf{y}_n$$

then the set $\mathcal{C} = \{\mathbf{y}_1, \ldots, \mathbf{y}_n\}$ is an orthogonal basis for V.

6. If the function $(\mathbf{x}, \mathbf{y}) \rightarrow \langle \mathbf{x}, \mathbf{y} \rangle$ is an inner product on a finite-dimensional real vector space V and if $\mathcal{E} = \{\mathbf{e}_1, \ldots, \mathbf{e}_n\}$ is an orthonormal basis for V, then $\langle \mathbf{x}, \mathbf{y} \rangle = a_1 b_1 + \cdots + a_n b_n$ for all vectors $\mathbf{x} = a_1 \mathbf{e}_1 + \cdots + a_n \mathbf{e}_n$ and $\mathbf{y} = b_1 \mathbf{e}_1 + \cdots + b_n \mathbf{e}_n$ in V.

7. If $\mathcal{B} = \{\mathbf{e}_1, \ldots, \mathbf{e}_n\}$ is an orthonormal basis for an inner product space V, then $\mathbf{x} = \langle \mathbf{x}, \mathbf{e}_1 \rangle \mathbf{e}_1 + \cdots + \langle \mathbf{x}, \mathbf{e}_n \rangle \mathbf{e}_n$ for any $\mathbf{x} \in V$.

8. If A is an $m \times n$ matrix whose n columns are linearly independent vectors in $\mathbb{R}^m$, then there exist an $m \times n$ matrix Q whose columns form an orthonormal set in $\mathbb{R}^m$ and an $n \times n$ upper-triangular invertible matrix R such that $A = QR$.

9. If A is an $m \times n$ matrix of rank n, then the solution of the linear system $A^T A \mathbf{x} = A^T \mathbf{b}$ is $\widehat{\mathbf{x}} = R^{-1} Q^T \mathbf{b}$, where R and Q are the matrices obtained from A by QR decomposition.

10. A real square matrix is orthogonal if and only if its rows and columns are orthonormal sets in the standard inner product.

11. Every invertible real matrix A is the product QR of an orthogonal matrix Q and an upper-triangular invertible matrix R.

12. Every coordinate conversion matrix between orthonormal bases of a real inner product space is orthogonal.

13. If A is an $m \times n$ real matrix, then the null space Nul A is orthogonal to the row space Row A in the standard inner product on $\mathbb{R}^n$.

14. By exchanging the matrices A and A^T in Theorem 8.13, we immediately obtain the result that the left null space Nul A^T is orthogonal to the column space Col A in the standard inner product on $\mathbb{R}^m$.

15. The orthogonal complement $S^\perp$ of a subset S of an inner product space V is a subspace of V.

16. The perp function from the subsets to subspaces of an inner product space V has the following properties: $\emptyset^\perp = V$; $V^\perp = \{\mathbf{0}\}$; $S \subseteq S^{\perp\perp}$; and if $S_1 \subseteq S_2$, then $S_2^\perp \subseteq S_1^\perp$.

17. If V is a finite-dimensional inner product space and W is a subspace of V, then $V = W \oplus W^\perp$.

18. If W is a subspace of a finite-dimensional real inner product space V and $\mathcal{B} = \{\mathbf{x}_1, \ldots, \mathbf{x}_p\}$ is an orthogonal basis for W, then every vector $\mathbf{x} \in V$ can be written in the form $\widehat{\mathbf{x}} + \mathbf{y}$, where $\mathbf{y} = \mathbf{x} - \widehat{\mathbf{x}}$ and

$$\widehat{\mathbf{x}} = \frac{\langle \mathbf{x}, \mathbf{x}_1 \rangle}{\langle \mathbf{x}_1, \mathbf{x}_1 \rangle} \mathbf{x}_1 + \cdots + \frac{\langle \mathbf{x}, \mathbf{x}_p \rangle}{\langle \mathbf{x}_p, \mathbf{x}_p \rangle} \mathbf{x}_p$$

19. If W is a subspace of a finite-dimensional real inner product space V and $\mathcal{B} = \{\mathbf{x}_1, \ldots, \mathbf{x}_p\}$ is an orthonormal basis for W, then every vector $\mathbf{x} \in V$ can be written in the form $\widehat{\mathbf{x}} + \mathbf{y}$, where $\widehat{\mathbf{x}} = \langle \mathbf{x}, \mathbf{x}_1 \rangle \mathbf{x}_1 + \cdots + \langle \mathbf{x}, \mathbf{x}_p \rangle \mathbf{x}_p$ and $\mathbf{y} = \mathbf{x} - \widehat{\mathbf{x}}$.

20. If $\mathcal{B} = \{\mathbf{x}_1, \ldots, \mathbf{x}_p\}$ is an orthonormal basis for a subspace W of $\mathbb{R}^n$ in the standard inner product and if $P = (\mathbf{x}_1 \cdots \mathbf{x}_p)$ is the matrix whose columns are the vectors in $\mathcal{B}$, then $\text{proj}(\mathbf{x} \rightarrow W) = PP^T\mathbf{x}$ for any $\mathbf{x} \in \mathbb{R}^n$.

21. If A is an $m \times n$ real matrix, then $\text{Nul } A \oplus \text{Col } A^T = \mathbb{R}^n$ and $\text{Nul } A^T \oplus \text{Col } A = \mathbb{R}^m$.

22. If $f : V \rightarrow \mathbb{R}$ is a linear functional on a finite-dimensional inner product space V, then there exists a unique vector $\mathbf{y} \in V$ such that $f(\mathbf{x}) = \langle \mathbf{x}, \mathbf{y} \rangle$ for all $\mathbf{x} \in V$.

23. If $T : V \rightarrow V$ is a linear transformation on a finite-dimensional real inner product space V, then there exists a linear transformation $T^* : V \rightarrow V$ for which $\langle T(\mathbf{x}), \mathbf{y} \rangle = \langle \mathbf{x}, T^*(\mathbf{y}) \rangle$.

24. If A represents a linear transformation $T : V \rightarrow V$ on a real inner product space V in an orthonormal basis $\mathcal{B} = \{\mathbf{x}_1, \ldots, \mathbf{x}_n\}$, then the matrix B representing the adjoint $T^* : V \rightarrow V$ of T in the basis $\mathcal{B}$ is A^T.

25. For any linear transformation $T : V \rightarrow V$ on a real inner product space V, the transformation $T^*T - I$ is self-adjoint.

26. If T is a self-adjoint linear transformation on a finite-dimensional real inner product space V and $\langle T(\mathbf{x}), \mathbf{x} \rangle = 0$ for all $\mathbf{x} \in V$, then $T = 0$.

27. If T is a linear transformation on a finite-dimensional real inner product space V, then the following conditions are equivalent.

   a. $\langle T(\mathbf{x}), T(\mathbf{y}) \rangle = \langle \mathbf{x}, \mathbf{y} \rangle$ for all $\mathbf{x}, \mathbf{y} \in V$

   b. $\|T(\mathbf{x})\| = \|\mathbf{x}\|$ for all $\mathbf{x} \in V$

   c. $TT^* = T^*T = I$

28. A matrix A represents an orthogonal transformation on a finite-dimensional real inner product space in an orthonormal basis if and only if A is orthogonal.

29. If T is a self-adjoint linear transformation and $\mathbf{x}$ and $\mathbf{y}$ are eigenvectors belonging to distinct eigenvalues λ_1 and λ_2 of T, then $\mathbf{x}$ and $\mathbf{y}$ are orthogonal.

30. The eigenvectors of a real symmetric matrix A belonging to distinct eigenvalues are orthogonal.

31. Eigenvectors belonging to distinct eigenvalues of a self-adjoint linear transformation are linearly independent.

32. The eigenvalues of a real symmetric matrix A are real.

33. The eigenvalues of a self-adjoint linear transformation T on a finite-dimensional real inner product space V are real.

34. Let T be a self-adjoint linear transformation on a finite-dimensional real inner product space V. Then there exists an orthonormal basis for V consisting of eigenvectors of T.

35. Every self-adjoint transformation T on a finite-dimensional real inner product space can be represented by a diagonal matrix whose diagonal entries are the eigenvalues of T.

36. For every real symmetric matrix A, there exists an orthogonal matrix Q such that $D = Q^T A Q$ is a diagonal matrix whose entries are the eigenvalues of A.

37. A real symmetric matrix is positive definite if and only if its eigenvalues are positive.

38. Let W be a subspace of $\mathbb{R}^m$, and let $\mathbf{b}$ be a vector in $\mathbb{R}^m$. Then the vector $\text{proj}(\mathbf{b} \to W)$ is the best approximation of $\mathbf{b}$ by vectors in W.

39. Suppose that A is an $m \times n$ real matrix of rank n and $\mathbf{x} \in \mathbb{R}^m$. Then the vector $\widehat{\mathbf{x}} = (A^T A)^{-1} A^T \mathbf{b}$ is a solution of the normal equation $A^T A \mathbf{x} = A^T \mathbf{b}$.

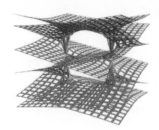

9

SINGULAR VALUES AND SINGULAR VECTORS

In this chapter, we show that every rectangular real matrix A can be decomposed into a product UDV^T consisting of two orthogonal matrices U and V and a generalized diagonal matrix D. The product UDV^T is called the *singular value decomposition* of A.

The construction of UDV^T is based on the fact that for all real matrices A, the matrix $A^T A$ is symmetric, and that by Corollary 8.35 there exists an orthogonal matrix Q and a diagonal matrix D for which $A^T A = QDQ^T$. We know from our earlier work that the diagonal entries of D are the eigenvalues of $A^T A$. We now show that they are nonnegative in all cases and that their square roots, called the *singular values* of A, can be used to construct UDV^T.

Due to the nature of the singular value decomposition algorithm, all numerical results in this chapter are approximations.

SINGULAR VALUES

We begin by showing that the eigenvalues of all symmetric matrices of the form $A^T A$ are nonnegative.

LEMMA 9.1   *For all real $m \times n$ matrices A, the eigenvalues of $A^T A$ are nonnegative.*

Proof. The $n \times n$ matrix $A^T A$ is symmetric since

$$(A^T A)^T = A^T A^{TT} = A^T A$$

By Corollary 8.35, it can therefore be diagonalized relative to an orthonormal basis $\mathcal{S} = \{\mathbf{v}_1, \ldots, \mathbf{v}_n\}$ for $\mathbb{R}^n$. Let $\lambda_1, \ldots, \lambda_n$ be the associated eigenvalues. Then

$$\begin{aligned}
\|A\mathbf{v}_i\|_2^2 &= (A\mathbf{v}_i)^T (A\mathbf{v}_i) \\
&= \mathbf{v}_i^T A^T A \mathbf{v}_i \\
&= \mathbf{v}_i^T \lambda_i \mathbf{v}_i \\
&= \lambda_i \mathbf{v}_i^T \mathbf{v}_i \\
&= \lambda_i
\end{aligned}$$

Therefore, $\lambda_i \geq 0$ for all $i \in \mathbf{n}$. ∎

Since the eigenvalues of the matrix $A^T A$ are nonnegative real numbers, we can use their square roots to study the decomposition properties of the matrix A.

DEFINITION 9.1 *The **singular values** of a real matrix A are the square roots of the eigenvalues of the matrix $A^T A$.*

Here are two examples.

EXAMPLE 9.1 ■ **The Singular Values of a Square Matrix**

Use *MATHEMATICA* to find the singular values of the matrix

$$A = \begin{pmatrix} 1 & 3 \\ 0 & 2 \end{pmatrix}$$

Solution. The singular values of A are the square roots of the eigenvalues of $A^T A$. We therefore compute these values.

```
In[1]:= A= ( 1  3
            0  2 ) ;
```

```
In[2]:= Eigenvalues[Transpose[A].A]
Out[2]= {7-3√5,7+5√3}
```

Therefore, the singular values of A are $\sqrt{7 - 3\sqrt{5}}$ and $\sqrt{7 + 3\sqrt{5}}$. ◄►

EXAMPLE 9.2 ■ **The Singular Values of a Rectangular Matrix**

Use *MATHEMATICA* to find the singular values of the matrix

$$A = \begin{pmatrix} 0 & 1 & 0 & 0 & 0 \\ -1 & 0 & 0 & 0 & 0 \\ 0 & 0 & 0 & 0 & 1 \end{pmatrix}$$

and of its transpose A^T.

Solution. We first find the eigenvalues of the matrices $A^T A$ and $\left(A^T\right)^T A^T = A A^T$.

```
In[1]:= A= ( 0  1  0  0  0
            -1  0  0  0  0
             0  0  0  0  1 ) ;
```

```
In[2]:= Eigenvalues[Transpose[A].A]

Out[2]= {0,0,1,1,1}
```

```
In[3]:= Eigenvalues[A.Transpose[A]]

Out[3]= {1,1,1}
```

Thus A has the singular values 0 and 1, whereas A^T has only the singular value 1. However, they both have the same positive singular values. ◄►

We can use Lemma 9.1 and Corollary 8.35 to prove that the two-norm of a real matrix is its largest singular value.

THEOREM 9.2 (Two-norm theorem) *The two-norm of any real square matrix A is the largest singular value of A.*

Proof. Let A be a real $n \times n$ matrix. By Lemma 9.1, we can assume that the eigenvalues of $A^T A$ are $\lambda_1 \geq \cdots \geq \lambda_n \geq 0$. Let $\mathbf{v}_1, \ldots, \mathbf{v}_n$ be eigenvectors associated with $\lambda_1, \ldots, \lambda_n$. Then

$$\|A\mathbf{v}_i\|_2^2 = (A\mathbf{v}_i)^T (A\mathbf{v}_i) = \mathbf{v}_i^T A^T A \mathbf{v}_i = \mathbf{v}_i^T \lambda_i \mathbf{v}_i = \lambda_i \mathbf{v}^T \mathbf{v}_i = \lambda_i \|\mathbf{v}_i\|_2^2$$

Therefore,

$$\sigma_i = \sqrt{\lambda_i} = \frac{\|A\mathbf{v}_i\|_2}{\|\mathbf{v}_i\|_2}$$

This implies that

$$\|A\|_2 = \max \left\{ \frac{\|A\mathbf{v}\|_2}{\|\mathbf{v}\|_2} : \|\mathbf{v}\|_2 \neq 0 \right\} \geq \frac{\|A\mathbf{v}_1\|_2}{\|\mathbf{v}_1\|_2} = \sigma_1$$

Let us now show that $\|A\|_2 \leq \sigma_1$. By Corollary 8.35, $A^T A$ can be diagonalized relative to an orthonormal basis $\mathcal{S} = \{\mathbf{v}_1, \ldots, \mathbf{v}_n\}$ for $\mathbb{R}^n$. Therefore, any vector $\mathbf{v} \in \mathbb{R}^n$ is a unique linear combination $a_1 \mathbf{v}_1 + \cdots + a_n \mathbf{v}_n$ of orthogonal unit vectors. Moreover, the linearity of the inner product $\langle \mathbf{x}, \mathbf{y} \rangle$ implies that

$$\|\mathbf{v}\|_2^2 = \langle \mathbf{v}, \mathbf{v} \rangle = \langle a_1 \mathbf{v}_1 + \cdots + a_n \mathbf{v}_n, a_1 \mathbf{v}_1 + \cdots + a_n \mathbf{v}_n \rangle$$
$$= \sum_{i \in \mathbf{n}} \sum_{j \in \mathbf{n}} a_i a_j \langle \mathbf{v}_i, \mathbf{v}_j \rangle = \sum_{i \in \mathbf{n}} a_i^2$$

and that

$$\|A\mathbf{v}\|_2^2 = \langle A\mathbf{v}, A\mathbf{v} \rangle = (A\mathbf{v})^T A\mathbf{v} = \mathbf{v} A^T A \mathbf{v} = \langle \mathbf{v}, A^T A \mathbf{v} \rangle$$
$$= \sum_{i \in \mathbf{n}} a_i^2 \lambda_i \leq \sum_{i \in \mathbf{n}} a_i^2 \lambda_1 = \lambda_1 \sum_{i \in \mathbf{n}} a_i^2 = \lambda_1 \|\mathbf{v}\|_2^2$$

Therefore, $\|A\mathbf{v}\|_2 \leq \sqrt{\lambda_1} \|\mathbf{v}\|_2$. It follows that $\|A\|_2 \leq \sigma_1$. ∎

EXERCISES 9.1

1. Use the **Eigenvalues** function of *MATHEMATICA* to show that each of the following matrices has the same positive singular values as its transpose.

$$
\text{a. } \begin{pmatrix} 0 & 1 \\ -1 & 0 \end{pmatrix}
\qquad
\text{b. } \begin{pmatrix} 0 & 1 & 0 \\ -1 & 0 & 0 \end{pmatrix}
\qquad
\text{c. } \begin{pmatrix} 0 \\ -1 \\ 0 \end{pmatrix}
$$

2. Use *MATHEMATICA* to find the singular values of the matrix

$$
A = \begin{pmatrix} 1 & 1 & 4. & 0. \\ 1 & 0 & 0 & 0. \\ 4 & 0 & 0 & -1 \end{pmatrix}
$$

and of its transpose A^T. Explain the difference.

3. Use *MATHEMATICA* to calculate the singular values of the following matrices and explain their accuracy.

$$
\text{a. } \begin{pmatrix} 2 & 4 \\ 0 & 1.0 \\ 0 & 1 \\ 1 & 1 \end{pmatrix}
\qquad
\text{b. } \begin{pmatrix} 2 & 0 & 1 & 4 \\ 5.0 & 0 & 1 & 5 \end{pmatrix}
\qquad
\text{c. } \begin{pmatrix} 2.0 & 3 & 0 & 0 & 0 \\ 5 & 3 & 5 & 2 & 0 \\ -1 & 1 & 1 & 0 & 2 \end{pmatrix}
$$

4. Show that the singular values of the following matrices are the same as the eigenvalues of the matrices.

$$
\text{a. } \begin{pmatrix} 4 & 2 & 1 \\ 2 & 8 & 0 \\ 1 & 0 & 8 \end{pmatrix}
\qquad
\text{b. } \begin{pmatrix} 4 & 0 & 1 \\ 0 & 8 & 0 \\ 1 & 0 & 8 \end{pmatrix}
\qquad
\text{c. } \begin{pmatrix} 4 & 0 & 0 \\ 0 & 8 & 0 \\ 0 & 0 & 9 \end{pmatrix}
$$

5. Prove that all singular values of an orthogonal matrix are 1.

6. Find conditions on the entries of the matrix

$$
A = \begin{pmatrix} a & b \\ c & d \end{pmatrix}
$$

for which $AA^T = A^TA$.

7. Use Theorem 9.2 to calculate the two-norm of the following matrices.

$$
\text{a. } A = \begin{pmatrix} -1 & -2 & 0 \\ 1 & -1 & 1 \\ -2 & 1 & -1 \end{pmatrix}
\qquad
\text{b. } \begin{pmatrix} -2 & 2 & -2 & 1 \\ 0 & -1 & 0 & -2 \\ 0 & 0 & -1 & 2 \\ 0 & 0 & 0 & 0 \end{pmatrix}
\qquad
\text{c. } \begin{pmatrix} 0 & 2 & 9 \\ -2 & 0 & 0 \\ -9 & 0 & 0 \end{pmatrix}
$$

SINGULAR VALUE DECOMPOSITION

Following are some of the properties that make singular value decompositions useful.

1. All real matrices have singular value decompositions.

2. A real square matrix is invertible if and only if all of its singular values are nonzero.

3. If A is an $n \times n$ real invertible matrix, the ratio $\text{cond } A = \sigma_1/\sigma_n$ of the largest and smallest singular values tells us how close A is computationally to being singular relative to the arithmetic precision of a computing device. A large ratio indicates that a computer may not be able to detect that the matrix is invertible.

4. For any $m \times n$ real rectangular matrix A, the number of nonzero singular values of A is equal to the rank of A.

5. If $A = UDV^T$ is a singular value decomposition of an invertible matrix A, then $A^{-1} = VD^{-1}U^T$.

6. The Frobenius norm $\|A\|_F$ of a real matrix is the square root of the sum of the squares of the singular values of A.

7. For positive definite symmetric matrices, the orthogonal decomposition QDQ^T and the singular value decomposition UDV^T coincide.

THEOREM 9.3 (Singular value decomposition theorem)   *For every $m \times n$ real matrix A, there exists an orthogonal $m \times m$ matrix U, an orthogonal $n \times n$ matrix V, and an $m \times n$ generalized diagonal matrix D such that $A = UDV^T$.*

Proof. Let A be a real $m \times n$ matrix of rank r and form the real symmetric matrix A^TA. By Corollary 8.35, the matrix A^TA is an orthogonally diagonalizable matrix. It therefore has n real eigenvalues $\lambda_1, \ldots, \lambda_n$. We know from Lemma 9.1 that these eigenvalues are nonnegative. We can therefore assume that $\lambda_1 \geq \cdots \geq \lambda_n \geq 0$. We use this information to construct an $n \times n$ orthogonal matrix V.

For each λ_1 we choose an eigenvector $\mathbf{v}_i$ in such a way that the set $\{\mathbf{v}_1, \ldots, \mathbf{v}_n\}$ is an orthonormal basis for $\mathbb{R}^n$. We then combine the column vectors $\mathbf{v}_i$ into the matrix

$$V = (\mathbf{v}_1 \ \cdots \ \mathbf{v}_n)$$

We know from Theorem 8.10 that V is orthogonal. The columns of V are called the ***right singular vectors*** of A.

Next we construct the diagonal matrix D. Since the rank of A is r, we know that the eigenvalues $\lambda_1, \ldots, \lambda_r$ are positive. We can therefore form the diagonal matrix

$$D_r = \text{diag}(\sigma_1, \ldots, \sigma_r)$$

whose diagonal entries σ_r are the square roots of the positive eigenvalues λ_i. We then form the $m \times n$ generalized diagonal matrix D by adding zero rows and zero columns to D_r if necessary.

We conclude by constructing the orthogonal matrix U. We recall from the proof of Lemma 9.1 that $\|A\mathbf{v}_i\|^2 = \lambda_i$ for all $1 \leq i \leq n$, so that $\|A\mathbf{v}_i\| = \sqrt{\lambda_i} = \sigma_i$ for all $1 \leq i \leq n$. For $\sigma_1 \geq \ldots \geq \sigma_r > 0$, we can therefore form the unit vectors

$$\mathbf{u}_i = \frac{A\mathbf{v}_i}{\|A\mathbf{v}_i\|} = \frac{A\mathbf{v}_i}{\sigma_i} \qquad (1 \leq i \leq r)$$

The set $W_r = \{\mathbf{u}_1, \ldots, \mathbf{u}_r\}$ is orthonormal since

$$\langle A\mathbf{v}_i, A\mathbf{v}_j \rangle = \langle A^T A\mathbf{v}_i, \mathbf{v}_j \rangle = \langle \lambda_i \mathbf{v}_i, \mathbf{v}_j \rangle = \lambda_i \langle \mathbf{v}_i, \mathbf{v}_j \rangle = 0$$

for all $i \neq j$ and since $\lambda_i \neq 0$. By Theorem 8.2, W_r is linearly independent and forms an orthonormal basis for Col A. Using Theorems 4.12 and 8.5, we extend W_r to an orthonormal basis $W_m = \{\mathbf{u}_1, \ldots, \mathbf{u}_r, \ldots, \mathbf{u}_m\}$ for $\mathbb{R}^m$. The vectors $\mathbf{u}_i$ are column vectors in $\mathbb{R}^m$, and we use them to form the $m \times m$ matrix

$$U = (\mathbf{u}_1 \ \cdots \ \mathbf{u}_m)$$

By Theorem 8.10, the matrix U is orthogonal. The columns of U are called the ***left singular vectors*** of A.

We conclude by showing that $A = UDV^T$. By Theorem 2.1, we have

$$\begin{aligned} UD &= (\mathbf{u}_1 \ \cdots \ \mathbf{u}_m)D \\ &= (\sigma_i \mathbf{u}_1 \ \cdots \ \sigma_m \mathbf{u}_m) \\ &= (\sigma_i \mathbf{u}_1 \ \cdots \ \sigma_r \mathbf{u}_r \ \mathbf{0} \ \cdots \ \mathbf{0}) \\ &= (A\mathbf{v}_1 \ \cdots \ A\mathbf{v}_r \ \mathbf{0} \ \cdots \ \mathbf{0}) \end{aligned}$$

and

$$\begin{aligned} AV &= A(\mathbf{v}_1 \ \cdots \ \mathbf{v}_n) \\ &= (A\mathbf{v}_1 \ \cdots \ A\mathbf{v}_n) \\ &= (A\mathbf{v}_1 \ \cdots \ A\mathbf{v}_r \mathbf{0} \ \cdots \ \mathbf{0}) \end{aligned}$$

Hence $AV = UD$. Since V is orthogonal, the matrix VV^T is the identity and we get $A = AI = AVV^T = UDV^T$. This proves the theorem. ∎

We can use the built-in **SingularValues** function to compute singular value decompositions. For any $m \times n$ matrix A of rank r, the output of **SingularValues[A]** consists of three parts: An $r \times m$ matrix P, the list $\lambda_1 \geq \ldots \geq \lambda_r > 0$ of positive singular values of A, and an $r \times n$ matrix Q. The product $P^T D_r Q$ is the ***reduced singular value decomposition*** of A, with $D_r = \mathrm{diag}(\lambda_1, \ldots, \lambda_r)$.

As indicated at the beginning of this chapter, the usual notation for the full singular value decomposition of a matrix A is UDV^T. However, *MATHEMATICA* calculates this decomposition in reduced form as $A = P^T D_r Q$. In the theoretical part of this chapter, we use the more familiar UDV^T format to make it easier to compare the material with that in other sources. The difference between the two notations is a result of the specific way in which the singular decomposition algorithm is implemented in *MATHEMATICA* and has no theoretical significance.

We can build the complete decomposition from $P^T D_r Q$ by adding appropriate zero rows and zero columns to the diagonal matrix D_r, and augmenting the rows of P with the vectors in an orthonormal basis for Nul A^T and the columns of Q with the vectors in an orthonormal basis for Nul A. However, in most applications of singular value decomposition, the reduced form of the decomposition is actually all that is needed.

It is important to note that the arguments of the **SingularValues** function must be specified to be real numbers. Moreover, the function removes all singular values smaller than a number specified by the **Tolerance** option. The tolerance is the largest singular value multiplied by a positive real number.

EXAMPLE 9.3   ■ A Singular Value Decomposition of a 2 × 2 Matrix

Use *MATHEMATICA* to find the singular value decomposition of the matrix

$$A = \begin{pmatrix} 0 & 5.0 \\ 2 & 2 \end{pmatrix}$$

Solution. It is easy to see that the matrix A is a square matrix of full rank. Therefore we can expect the **SingularValues** function to produce a complete singular value decomposition.

```
In[1]:= A= ⎛ 0  5.0 ⎞ ;
           ⎝ 2   2  ⎠
```

```
In[2]:= {P,W,Q}=SingularValues[A]

Out[2]= {{{-0.907648,-0.419733},{0.419733,-0.907648}},
{5.44283,1.83728},{{-0.154233,-0.988034},
{-0.988034,0.154233}}}
```

We can use *MATHEMATICA* to reconstruct the matrix A from P, W, and Q.

```
In[3]:= Transpose[Q].DiagonalMatrix[W].Q

Out[3]= {{2.22045 × 10⁻¹⁶, 5.}, {2., 2.}}
```

Since 2.22045×10^{-16} is approximately 0, the matrix product P^T diag(5.44283, 1.83728)Q
is approximately equal to A. ◄►

We have already seen that the two-norm of a square matrix A can be described in terms
of singular values of A. We now show that the singular value decomposition can also be used
to describe the Frobenius norm of A.

THEOREM 9.4   *If $\sigma_1, \ldots, \sigma_n$ are the singular values of a real $n \times n$ matrix A, then the
square of the Frobenius norm $\|A\|_F^2$ is the sum $\sigma_1^2 + \cdots + \sigma_n^2$ of the singular values of A.*

Proof. We know from Chapter 7 that $\|A\|_F^2 = \text{trace } A^T A$. It therefore suffices to prove that
trace $A^T A = \sigma_1^2 + \cdots + \sigma_n^2$.
Let $A = UDV^T$ be a singular value decomposition of A. Then

$$\text{trace } A^T A = \text{trace } \left(UDV^T\right)^T \left(UDV^T\right)$$

$$= \text{trace } VDU^T UDV^T$$

$$= \text{trace } VD^2 V^T$$

$$= \text{trace } VV^T D^2$$

$$= \text{trace } D^2$$

We know from Theorem 9.3 that $D = \text{diag}(\sigma_1, \ldots, \sigma_n)$. Therefore,

$$D^2 = \text{diag}(\sigma_1^2, \ldots, \sigma_n^2)$$

It follows that trace $D^2 = \sigma_1^2 + \cdots + \sigma_n^2$. ∎

EXAMPLE 9.4   ■ **A Frobenius Norm**

Use *MATHEMATICA* and Theorem 9.3 to approximate the Frobenius norm of the matrix

$$A = \begin{pmatrix} 9 & 2 & -2 & 4.0 \\ 5 & 1 & 1 & 2 \\ 6 & 1 & 2 & -5 \\ -2 & 3 & 2 & 7 \end{pmatrix}$$

and compare the result with the value trace $A^T A$.

Solution. We begin by computing the singular values of A.

$$\text{In[1]:= } A = \begin{pmatrix} 9 & 2 & -2 & 4.0 \\ 5 & 1 & 1 & 2 \\ 6 & 1 & 2 & -5 \\ -2 & 3 & 2 & 7 \end{pmatrix};$$

Next we compute the trace of $A^T A$.

```
In[2]:= Tr[Transpose[A].A]
Out[2]= 268
```

```
In[3]:= {P,W,Q}=SingularValues[A];
```

```
In[4]:= Sum[W[[i]]^2,{i,1,4}]
Out[4]= 268
```

As we can see, the results are the same. Both results yield $\|A\|_F = \sqrt{268} = 16.3707.$ ◄►

Next we illustrate the calculation of a reduced singular value decomposition.

EXAMPLE 9.5 ■ **A Singular Value Decomposition of a 4 × 5 Matrix**

Construct a 4 × 5 matrix of rank 3 and use *MATHEMATICA* to find the singular value decomposition of the matrix.

Solution. The first three columns of the matrix

$$A = \begin{pmatrix} 1 & 0 & 0 & 0 & 1. \\ 0 & 1 & 0 & 1 & 0 \\ 0 & 0 & 1 & 1 & 1 \\ 0 & 0 & 0 & 0 & 0 \end{pmatrix}$$

are linearly independent and the last two columns are linear combinations of the first three. Therefore, A has rank 3.

We use the **SingularValues** function to decompose A.

$$\text{In[1]:= } A = \begin{pmatrix} 1 & 0 & 0 & 0 & 1.0 \\ 0 & 1 & 0 & 1 & 0 \\ 0 & 0 & 1 & 1 & 1 \\ 0 & 0 & 0 & 0 & 0 \end{pmatrix};$$

```
In[2]:= {P,W,Q}=SingularValues[A]
Out[2]= {{{-0.408248,-0.816497,-0.408248,0.},
{-0.707107,1.28975×10⁻¹⁶,0.707107,0.},
{-0.57735,0.57735,-0.57735,0.}},
{2.,1.41421,1.},{{-0.204124,-0.408248,
-0.204124,-0.612372,-0.612372},
{-0.5,1.53805×10⁻¹⁶,0.5,-0.5,0.5},
{-0.57735,0.57735,-0.57735,8.11984×10⁻¹⁷,1.18095×10⁻¹⁷}}}
```

Let us verify that A is approximately equal to $P^T W Q$.

```
In[3]:= Transpose[P].DiagonalMatrix[W].Q
Out[3]= {{1.,-3.33067×10⁻¹⁶,4.44089×10⁻¹⁶,1.,2.15226×10⁻¹⁶},
{-2.77556×10⁻¹⁶,1.,-3.33067×10⁻¹⁶,1.,1.},
{-1.11022×10⁻¹⁶,-2.77556×10⁻¹⁶,1.,6.41424×10⁻¹⁷,1.},
{0.,0.,0.,0.,0.}}
```

As we can see, this matrix is approximately equal to the matrix A. ◄►

EXAMPLE 9.6 ■ **A Singular Value Decomposition of a Matrix Without Eigenvalues**

Use *MATHEMATICA* to find the singular values decomposition of the rotation matrix

$$A = \begin{pmatrix} 0 & -1 \\ 1 & 0 \end{pmatrix}$$

Solution. Here are the calculations.

$$\text{In[1]:= \{P,W,Q\}=SingularValues}\left[\begin{pmatrix} 0 & -1 \\ 1 & 0 \end{pmatrix}\right]$$

$$\text{Out[1]= \{\{\{0.,-1.\},\{-1.,0.\}\},\{1.,1.\},\{\{-1.,0.\},\{0.,1.\}\}\}}$$

Thus

$$\begin{pmatrix} 0 & -1 \\ 1 & 0 \end{pmatrix} = \begin{pmatrix} 0 & -1 \\ -1 & 0 \end{pmatrix}\begin{pmatrix} 1 & 0 \\ 0 & 1 \end{pmatrix}\begin{pmatrix} -1 & 0 \\ 0 & 1 \end{pmatrix}$$

is the desired singular value decomposition. ◀▶

EXERCISES 9.2

1. Show that if A is a positive definite matrix, then A has a singular value decomposition of the form QDQ^T.

2. Use Exercise 1 to conclude that if A is a positive definite matrix, then the eigenvalues and the singular values of A coincide.

3. Show that if $\mathbf{u}_i$ is a left singular vector and $\mathbf{v}_i$ is a right singular vector belonging to the singular value σ_i of a rectangular matrix A of rank r, then $A^T\mathbf{u}_i = \sigma_i\mathbf{v}_i$ for all $1 \le i \le r$.

4. Use *MATHEMATICA* to show that the following matrices are positive definite and use the singular value decomposition to find their approximate eigenvalues.

   a. $\begin{pmatrix} 4 & 0 & 0 & 0 & 0 \\ 0 & 9 & 0 & -2 & 0 \\ 0 & 0 & 7 & 0 & 0 \\ 0 & -2 & 0 & 1 & 0 \\ 0 & 0 & 0 & 0 & 3 \end{pmatrix}$ b. $\begin{pmatrix} 2.7 & 1 & 0 & 0 & 0 \\ 1 & 2.5 & 1 & 0 & 0 \\ 0 & 1 & 3.8 & 1 & 0 \\ 0 & 0 & 1 & 1 & -.3 \\ 0 & 0 & 0 & -.3 & 2 \end{pmatrix}$

5. Use Exercise 1 to explain why the following matrices are not positive definite.

   a. $\begin{pmatrix} 0 & 1 \\ 1 & 0 \end{pmatrix}$ b. $\begin{pmatrix} 2 & .5 & -1 \\ .5 & 3 & .5 \\ -1 & .5 & 1 \end{pmatrix}$ c. $\begin{pmatrix} 2 & 5 \\ 5 & 3 \end{pmatrix}$

6. Show that if A is an $m \times n$ real matrix of rank r and $A = UDV^T$ is a singular value decomposition of A with $U = (\mathbf{u}_1 \cdots \mathbf{u}_m)$ and $V = (\mathbf{v}_1 \cdots \mathbf{v}_n)$ and positive singular values $\sigma_1, \ldots, \sigma_r$, then $A = \sigma_1\mathbf{u}_1\mathbf{v}_1^T + \cdots + \sigma_r\mathbf{u}_r\mathbf{v}_r^T$.

7. Use *MATHEMATICA* to generate three random 4×4 invertible matrices and show that their singular values are nonzero.

8. Compute the ratio σ_1/σ_n of the largest and smallest singular value for each of the matrices in Exercise 7.

9. Use *MATHEMATICA* to generate three random 4×6 real matrices and show that the number of nonzero singular values of each matrix equals the rank of the matrix.

10. Compute the singular value decompositions of the following matrices and find their inverses in terms of V, D^{-1}, and U^T.

$$\text{a. } A = \begin{pmatrix} 2 & -3 \\ -3 & -3 \end{pmatrix} \quad \text{b. } \begin{pmatrix} -1 & -1 & 2 \\ -1 & -3 & 3 \\ 2 & 3 & -1 \end{pmatrix} \quad \text{c. } A = \begin{pmatrix} 5 & 3 & 3 \\ -1 & 5 & -1 \\ -2 & -1 & 3 \end{pmatrix}$$

11. Use singular values to compute the Frobenius norm of the following matrices:

$$\text{a. } A = \begin{pmatrix} 2 & -3 \\ -3 & -3 \end{pmatrix} \quad \text{b. } \begin{pmatrix} 9 & 0 & -2 \\ -2 & 0 & 1 \end{pmatrix} \quad \text{c. } A = \begin{pmatrix} 5 & 3 & 3 \\ -1 & 5 & -1 \\ -2 & -1 & 3 \end{pmatrix}$$

APPLICATIONS

Fundamental Subspaces

An elegant application of the singular value decomposition is the calculation of orthonormal bases for the fundamental subspaces of real matrices.

THEOREM 9.5 (Singular-vector bases theorem)   *Suppose that $\mathbf{u}_1, \ldots, \mathbf{u}_m$ are the left singular vectors and $\mathbf{v}_1, \ldots, \mathbf{v}_n$ are the right singular vectors of a real $m \times n$ matrix A whose singular values are $\sigma_1, \ldots, \sigma_n$ and whose rank is r. Then*

    1. $U_r = \{\mathbf{u}_1, \ldots, \mathbf{u}_r\}$ is an orthonormal basis for Col A

    2. $U_{m-r} = \{\mathbf{u}_{r+1}, \ldots, \mathbf{u}_m\}$ is an orthonormal basis for Nul A^T

    3. $V_r = \{\mathbf{v}_1, \ldots, \mathbf{v}_r\}$ is an orthonormal basis for Col A^T

    4. $V_{n-r} = \{\mathbf{v}_{r+1}, \ldots, \mathbf{v}_n\}$ is an orthonormal basis for Nul A

Proof. We begin by recalling from the proof of Theorem 9.3 that the set of left singular vectors

$$U = \{\mathbf{u}_1, \ldots, \mathbf{u}_r, \mathbf{u}_{r+1}, \ldots, \mathbf{u}_m\}$$

is an extension of an orthonormal basis W_r for Col A to a basis W for $\mathbb{R}^m$. Theorem 4.38 implies that the complement,

$$U_{m-r} = \{\mathbf{u}_{r+1}, \ldots, \mathbf{u}_m\}$$

of U_r in U is an orthonormal basis for Nul A^T since $\mathbb{R}^m = \text{Col } A \oplus \text{Nul } A^T$.

Next we consider the right singular case. Let

$$V = \{\mathbf{v}_1, \ldots, \mathbf{v}_r, \mathbf{v}_{r+1}, \ldots, \mathbf{v}_n\}$$

be the basis for $\mathbb{R}^n$ consisting of the right singular vectors of A. We show that the restriction

$$V_{n-r} = \{\mathbf{v}_{r+1}, \ldots, \mathbf{v}_n\}$$

is an orthonormal basis for Nul A. Theorem 4.38 will then imply that

$$V_r = \{\mathbf{v}_1, \ldots, \mathbf{v}_r\}$$

is an orthonormal basis for Col A^T since $\mathbb{R}^n = \text{Nul } A \oplus \text{Col } A^T$.

Since V is orthonormal, it suffices to show that

$$V_{n-r} = \{\mathbf{v}_{r+1}, \ldots, \mathbf{v}_n\}$$

is a basis for Nul A. Since $\lambda_i = 0$ for $i > r$ and $A\mathbf{v}_i = \lambda \mathbf{u}_i$, it follows that

$$A\mathbf{v}_{r+1} = \cdots = A\mathbf{v}_n = 0$$

Hence $\mathbf{v}_{r+1}, \ldots, \mathbf{v}_n \in \text{Nul } A$. Moreover, by Corollary 8.30, the vectors $\mathbf{v}_{r+1}, \ldots, \mathbf{v}_n$ are linearly independent. It remains to show that they span Nul A. Since V spans $\mathbb{R}^n$, every $\mathbf{v} \in \text{Nul } A$ can be written as a unique linear combination

$$\mathbf{v} = a_1\mathbf{v}_1 + \cdots + a_r\mathbf{v}_r + a_{r+1}\mathbf{v}_{r+1} + \cdots + a_n\mathbf{v}_n$$

Therefore,

$$\mathbf{u} = \mathbf{v} - a_1\mathbf{v}_1 - \cdots - a_r\mathbf{v}_r = a_{r+1}\mathbf{v}_{r+1} + \cdots + a_n\mathbf{v}_n = 0$$

This means that

$$\begin{aligned}
A(\mathbf{u}) &= A(\mathbf{v} - a_1\mathbf{v}_1 - \cdots - a_r\mathbf{v}_r) \\
&= A\mathbf{v} - a_1 A\mathbf{v}_1 - \cdots - a_r A\mathbf{v}_r \\
&= 0 - a_1\lambda_1\mathbf{v}_1 - \cdots - a_r\lambda_r\mathbf{v}_r \\
&= 0
\end{aligned}$$

Since $\lambda_1 \geq \cdots \geq \lambda_r > 0$, this can only hold if $a_1 = \cdots = a_r = 0$. Hence

$$\mathbf{v} = a_{r+1}\mathbf{v}_{r+1} + \cdots + a_n\mathbf{v}_n$$

It follows that V_{n-r} is a basis for Nul A. ∎

EXERCISES 9.3

1. Use singular value decompositions to find orthonormal bases for the fundamental sub-spaces of the following matrices.

$$
\text{a. } A = \begin{pmatrix} 0 & 0 & 0 & 0 \\ 1 & 0 & 2 & 3 \\ 0 & 0 & 0 & 0 \end{pmatrix}
\quad
\text{b. } A = \begin{pmatrix} 1 & 2 & 4 \\ 0 & 0 & 1 \\ 3 & 6 & 12 \end{pmatrix}
\quad
\text{c. } A = \begin{pmatrix} 1 & 2 & 4 & 0 & 1 \\ 0 & 0 & 1 & 9 & 0 \\ 0 & 2 & 1 & 0 & 0 \\ 0 & 0 & 0 & 3 & 0 \\ 0 & 0 & 0 & 0 & 3 \end{pmatrix}
$$

2. Let $U = (\mathbf{u}_1 \ \mathbf{u}_2 \ \mathbf{u}_3)$ and $V = (\mathbf{v}_1 \ \mathbf{v}_2 \ \mathbf{v}_3)$ be the orthogonal matrices in a singular value decomposition of the matrix

$$
A = \begin{pmatrix} 1 & 2 & 4 \\ 0 & 0 & 1 \\ 3 & 6 & 12 \end{pmatrix}
$$

Show that for all positive singular values σ of A, $A\mathbf{v}_i = \sigma_i \mathbf{u}_i$.

3. Show that if $A = UDV^T$ is a singular value decomposition of A, then VDU^T is a singular value decomposition of A^T. Use this fact to show that A and A^T have the same nonzero singular values.

4. Verify that the left singular vectors of the matrix

$$
A = \begin{pmatrix} 0 & 1 & -1 & 1 \\ -1 & 2 & 0 & 1 \end{pmatrix}
$$

form an orthonormal basis for Col A. What conclusion can you draw about Nul A^T?

5. Verify that the right singular vectors of the matrix

$$
A = \begin{pmatrix} 0 & 1 & -1 & 1 \\ -1 & 2 & 0 & 1 \end{pmatrix}
$$

form an orthonormal basis for Col A^T. What conclusion can you draw about Nul A?

Pseudoinverses

The reduced singular value decompositions produced by *MATHEMATICA* have several pow-erful applications. Recall that if A is an $m \times n$ real matrix of rank r and UDV^T is a singular value decomposition of A, then we can think of UDV^T as the product

$$
A = \begin{pmatrix} U_r & U_{m-r} \end{pmatrix} \begin{pmatrix} D_r & 0 \\ 0 & 0 \end{pmatrix} \begin{pmatrix} V_r^T \\ V_{n-r}^T \end{pmatrix} = U_r D_r V_r^T
$$

where U_r is the matrix $(\mathbf{u}_1 \ \cdots \ \mathbf{u}_r)$ and V_r is the matrix $(\mathbf{v}_1 \ \cdots \ \mathbf{v}_r)$. Suppose that $r > 0$. Then D_r is invertible since it is a diagonal matrix with nonzero diagonal entries. We can therefore define the matrix

$$A^+ = V_r D_r^{-1} U_r^T$$

by transposing the product $U_r D_r V_r^T$ and replacing D_r by its inverse, D_r^{-1}. The matrix A^+ is called the *pseudoinverse* of A. If A is invertible, then $A^+ = A^{-1}$ since

$$AA^+ = AV_r D_r^{-1} U_r^T = U_r D_r V_r^T V_r D_r^{-1} U_r^T = I$$

We can use the result of the *MATHEMATICA* command

$$\textbf{\{P,Q,R\}=SingularValues[A]}$$

to calculate the pseudoinverse A^+ of A using the formula

$$\textbf{P.Inverse[DiagonalMatrix[Q]].Transpose[R]}$$

As pointed out earlier, the *MATHEMATICA* notation corresponds to the transpose of the usual notation. However, *MATHEMATICA* has a built-in **PseudoInverse** function for computing A^+. The command **PseudoInverse[A]** produces an approximation of A^+ directly.

EXAMPLE 9.7 ■ **A Pseudoinverse**

Use *MATHEMATICA* in two ways to find a pseudoinverse for the matrix

$$A = \begin{pmatrix} 1 & 0 & 0 \\ 0 & 2.0 & 0 \\ 0 & 0 & 3 \end{pmatrix}$$

Solution. First we use the **SingularValues** function. To do so, we must explicitly designate one of the entries of A as a real number.

```
In[1]:= A= ( 1    0  0
             0  2.0  0  ;
             0    0  3 )
```

```
In[2]:= {P,Q,R}=SingularValues[A];
```

```
In[3]:= Aplus=P.Inverse[DiagonalMatrix[Q]].Transpose[R]
Out[3]= {{1.,0.,0.},{0.,0.5,0.},{0.,0.,0.333333}}
```

Next we use the **PseudoInverse** function.

```
In[4]:= PseudoInverse[A]

Out[4]= {{1.,0.,0.},{0.,0.5,0.},{0.,0.,0.333333}}
```

As we can see, *MATHEMATICA* produces two identical results. ◄►

Pseudoinverses provide a powerful method for finding least-squares solutions for linear systems.

THEOREM 9.6 (Pseudoinverse theorem) *For any linear system $A\mathbf{x} = \mathbf{b}$, the vector $\widehat{\mathbf{x}} = A^{+}\mathbf{b}$ is a least-squares solution of the system.*

Proof. By definition, $A^{+} = V_r D_r^{-1} U_r^T$. Therefore $A^{+}\mathbf{b} = V_r D_r^{-1} U_r^T \mathbf{b}$. If we multiply both sides of this equation by A, we get

$$AA^{+}\mathbf{b} = AV_r D_r^{-1} U_r^T \mathbf{b}$$
$$= \left(U_r D_r V_r^T\right)\left(V_r D_r^{-1} U_r^T\right)\mathbf{b}$$
$$= U_r U_r^T \mathbf{b}$$

Moreover,

$$U_r U_r^T \mathbf{b} = (\mathbf{u}_1 \ \cdots \ \mathbf{u}_r) \begin{pmatrix} \mathbf{u}_1^T \\ \vdots \\ \mathbf{u}_r^T \end{pmatrix} \mathbf{b} = (\mathbf{u}_1 \ \cdots \ \mathbf{u}_r) \begin{pmatrix} \mathbf{u}_1^T \mathbf{b} \\ \vdots \\ \mathbf{u}_r^T \mathbf{b} \end{pmatrix}$$

$$= (\mathbf{u}_1 \ \cdots \ \mathbf{u}_r) \begin{pmatrix} \mathbf{b} \cdot \mathbf{u}_1 \\ \vdots \\ \mathbf{b} \cdot \mathbf{u}_r \end{pmatrix} = (\mathbf{b} \cdot \mathbf{u}_1)\mathbf{u}_1 + \cdots + (\mathbf{b} \cdot \mathbf{u_r})\mathbf{u_r}$$

Since $(\mathbf{b} \cdot \mathbf{u}_1)\mathbf{u}_1 + \cdots + (\mathbf{b} \cdot \mathbf{u_r})\mathbf{u_r} = \text{proj}(\mathbf{b} \to \text{span}\{\mathbf{u}_1, \ldots, \mathbf{u}_r\})$, it follows from Theorem 8.37 that $A\widehat{\mathbf{x}}$ is the least-squares solution of the system $A\mathbf{x} = \mathbf{b}$. ∎

EXAMPLE 9.8   ■ A Least-Squares Solution

Use the **PseudoInverse** function to find a least-squares solution of the linear system

$$\begin{cases} 2x + 3y + z = 4 \\ x - y - z = 1 \\ x + y + 3z = 3 \\ 2x - 5y + z = 6 \end{cases}$$

Solution. The matrix equation $A\mathbf{x} = \mathbf{b}$ of the given linear system is

$$\begin{pmatrix} 2 & 3 & 1 \\ 1 & -1 & -1 \\ 1 & 1 & 3 \\ 2 & -5 & 1 \end{pmatrix} \begin{pmatrix} x \\ y \\ z \end{pmatrix} = \begin{pmatrix} 4 \\ 1 \\ 3 \\ 6 \end{pmatrix}$$

By Theorem 9.6, the least-squares solution of $A\mathbf{x} = \mathbf{b}$ is $A^+\mathbf{b}$.

```
In[1]:= A= ( 2  3  1       4
             1 -1 -1  ; b= 1  ;
             1  1  3       3
             2 -5  1 )     6
```

```
In[2]:=x=MatrixForm[PseudoInverse[A].b]
```
Out[2]//MatrixForm=

$$\begin{pmatrix} \frac{660}{337} \\ -\frac{87}{337} \\ \frac{190}{337} \end{pmatrix}$$

Therefore, the vector $A\widehat{\mathbf{x}}$ is

$$\begin{pmatrix} 2 & 3 & 1 \\ 1 & -1 & -1 \\ 1 & 1 & 3 \\ 2 & -5 & 1 \end{pmatrix} \begin{pmatrix} \frac{660}{337} \\ -\frac{87}{337} \\ \frac{190}{337} \end{pmatrix} = \begin{pmatrix} \frac{1249}{337} \\ \frac{557}{337} \\ \frac{1143}{337} \\ \frac{1945}{337} \end{pmatrix}$$

The distance between $A\widehat{\mathbf{x}}$ and $\mathbf{b}$ is

$$\|A\widehat{\mathbf{x}} - \mathbf{b}\|^2 = \left(\frac{1249}{337} - 4\right)^2 + \left(\frac{557}{337} - 1\right)^2 + \left(\frac{1143}{337} - 3\right)^2 + \left(\frac{1945}{337} - 6\right)^2 = \frac{242}{337}$$

Hence the distance between the vectors $A\widehat{\mathbf{x}}$ and $\mathbf{b}$ is $\sqrt{242/337}$ units. ◂▸

EXAMPLE 9.9 ■ **A Reduced Singular Value Decomposition of a 3 × 4 Matrix**

Use *MATHEMATICA* to find the reduced singular value decomposition of the matrix

$$A = \begin{pmatrix} 1 & 2 & 0 & 1 & 4 \\ 0 & 3 & 0 & 2 & 0 \\ 1 & 1 & 0 & 0 & 1 \end{pmatrix}$$

Solution. By definition, the **SingularValues** function produces reduced singular value decompositions.

In[1]:= {P,W,Q}=SingularValues$\left[\begin{pmatrix} 1 & 2 & 0 & 1 & 4 \\ 0 & 3 & 0 & 2 & 0 \\ 1 & 1 & 0 & 0 & 1 \end{pmatrix}\right]$;

We now examine the matrices P, W, and Q.

In[2]:= **MatrixForm[P]**

Out[2]//MatrixForm=

$$\begin{pmatrix} -0.834954 & 0.462109 & 0.298843 \\ -0.473395 & -0.880023 & 0.0381578 \\ -0.280622 & 0.109611 & -0.953539 \end{pmatrix}$$

In[3]:= **MatrixForm[DiagonalMatrix[W]]**

Out[3]//MatrixForm=

$$\begin{pmatrix} 5.37479 & 0 & 0 \\ 0 & 2.90266 & 0 \\ 0 & 0 & 0.828323 \end{pmatrix}$$

In[4]:= **MatrixForm[Q]**

Out[4]//MatrixForm=

$$\begin{pmatrix} -0.207557 & -0.627134 & 0. & -0.3315 & -0.673596 \\ 0.196964 & -0.553368 & 0. & -0.447154 & 0.674568 \\ -0.790387 & -0.291407 & 0. & 0.452914 & 0.291957 \end{pmatrix}$$

Thus P is a 3×3 matrix, W is a 3×3 diagonal matrix, and Q is a 3×5 matrix. We show that P, W, and Q determine a reduced singular value decomposition of A.

```
In[5]= Transpose[P].DiagonalMatrix[W].Q//MatrixForm
```

Out[5]//MatrixForm=

$$
\begin{pmatrix}
1. & 2. & 0. & 1. & 4. \\
-9.02056 \times 10^{-17} & 3. & 0. & 2. & -1.99493 \times 10^{-16} \\
1. & 1. & 0. & 3.33067 \times 10^{-16} & 0.
\end{pmatrix}
$$

This shows that the matrix **Transpose[P]** is approximately equal to the matrix U_r, that the matrix **DiagonalMatrix[W]** is approximately equal to D_r, and that the matrix Q is approximately equal to V_r^T. ◄►

EXAMPLE 9.10 ■ Least-Squares Solution Equals Exact Solution

Use *MATHEMATICA* to show that the least-squares and exact solutions of the linear equation

$$
\begin{pmatrix}
1 & 2 & 1 \\
2 & 4 & 1 \\
3 & -3 & 1
\end{pmatrix}
\begin{pmatrix}
x \\
y \\
z
\end{pmatrix}
=
\begin{pmatrix}
4 \\
5 \\
6
\end{pmatrix}
$$

coincide.

Solution. We will show that the exact solution produced by the **Solve** function and the least-squares solution produced by the **PseudoInverse** function are identical. We begin by creating the matrix A.

```
In[1]:= A= ( 1   2   1
            2   4   1   ;
            3  -3   1 )
```

The calculation

```
In[2]:= Solve[A.{x,y,z}=={4,5,6}]

Out[2]= {{x → 1,y → 0,z → 3}}
```

shows that $z = 1$, $y = 0$, $z = 3$ is an exact solution of the equation. Moreover, the calculation

```
In[3]:= PseudoInverse[A].{4,5,6}

Out[3]= {1,0,3}
```

produces the vector $(x, y, z) = (1, 0, 3)$. These results are obviously equivalent. ◄►

The real difference between the two methods shows up in examples that have no exact solution. If we ask *MATHEMATICA* to solve the linear system

$$\begin{pmatrix} 1 & 2 & 1 \\ 2 & 4 & 2 \\ 3 & -3 & 1 \end{pmatrix} \begin{pmatrix} x \\ y \\ z \end{pmatrix} = \begin{pmatrix} 4 \\ 5 \\ 6 \end{pmatrix}$$

for example, then the

$$\textbf{Solve[A.\{x,y,z\}==\{4,5,6\}]}$$

command produces the empty output {}. We can conclude that the vector $(4, 5, 6)$ does not belong to the column space of A. However, the

$$\textbf{PseudoInverse[A].\{4,5,6\}}$$

command produces the output $(x, y, z) = (\frac{19}{11}, \frac{14}{275}, \frac{267}{275})$. This means that of all the vectors in Col A, the vector

$$\begin{pmatrix} 1 & 2 & 1 \\ 2 & 4 & 2 \\ 3 & -3 & 1 \end{pmatrix} \begin{pmatrix} \frac{19}{11} \\ \frac{14}{275} \\ \frac{267}{275} \end{pmatrix}$$

is closest to the vector $(4, 5, 6)$ in the Euclidean distance.

EXERCISES 9.4

1. Suppose that $Ax_1 = b_1$ and $Ax_2 = b_2$ are the two linear systems determined by

a. $A = \begin{pmatrix} 1 & 2 \\ 3 & 4 \\ 5 & 6 \end{pmatrix} \quad x_1 = \begin{pmatrix} -6 \\ 6 \end{pmatrix} \in \mathbb{R}^2 \quad b_1 = \begin{pmatrix} 6 \\ 6 \\ 6 \end{pmatrix} \in \text{Col } A$

b. $A = \begin{pmatrix} 1 & 2 \\ 3 & 4 \\ 5 & 6 \end{pmatrix} \quad x_2 = \begin{pmatrix} -\frac{16}{3} \\ \frac{67}{12} \end{pmatrix} \in \mathbb{R}^2 \quad b_2 = \begin{pmatrix} 6 \\ 6 \\ 7 \end{pmatrix} \notin \text{Col } A$

Use the **Solve** function to show that $Ax_1 = b_1$ holds, but that $Ax_2 = b_2$ does not hold. Then use the **PseudoInverse** function to calculate $x_1 = A^+b_1$ and $x_2 = A^+b_2$.

2. Use the definition of A^+ to find a least-squares solution of the linear system $Ax = b$ determined by

$$A = \begin{pmatrix} 1 & 2 & 0 & 1 & 4. \\ 0 & 3 & 0 & 2 & 0 \\ 1 & 1 & 0 & 0 & 1 \end{pmatrix} \quad \text{and} \quad b = \begin{pmatrix} 1 \\ 2 \\ 3 \end{pmatrix}$$

3. Compute the pseudoinverse A^+ for the following matrices and test whether A^+ is an actual inverse.

a. $\begin{pmatrix} -1 & 1 & 0 \\ 1 & -1 & 2 \\ -2 & 1 & 0 \end{pmatrix}$ b. $\begin{pmatrix} -1 & 1 & -2 \\ 1 & -1 & 2 \\ -2 & 1 & 0 \end{pmatrix}$ c. $\begin{pmatrix} 3 & 2 \\ 2 & 1 \\ 7 & 0 \end{pmatrix}$ d. $\begin{pmatrix} -2 & 2 & 0 & 0 \\ -2 & -1 & 1 & 1 \\ -1 & 0 & 1 & 1 \end{pmatrix}$

4. Use *MATHEMATICA* to find a singular value decomposition $\{P, Q, R\}$ for the inverse of an invertible matrix A and show that

Inverse[A]==Transpose[R].Inverse[DiagonalMatrix[Q]].P.

5. Use *MATHEMATICA* to generate a random 5×6 matrix and find a singular value decomposition for the matrix.

6. It can be proved that for any given matrix A, the pseudoinverse A^+ is the unique matrix B for which $ABA = A$, $BAB = B$, $(AB)^T = AB$, and $(BA)^T = BA$. Use these conditions to prove that $(A^T)^+ = (A^+)^T$ and that $A^{++} = A$.

Effective Rank

In many applications of linear algebra it is important to determine the rank of a matrix. One way of finding the rank is to use Gaussian elimination and reduce a matrix to its row echelon form. The number of nonzero rows of the echelon matrix is the rank of the matrix. If the matrix is large, the computer version of Gaussian elimination may fail to produce the rank of the matrix.

Accumulated round-off errors may introduce nonzero entries in what should be zero rows. A more reliable method is to find the singular value decomposition of the matrix and then discard very small singular values. The **Tolerance** option of the **SingularValues** function makes it possible to decide which singular values in a given problem to discard. The rank obtained in this way is known as the *effective rank* of a matrix with respect to a given tolerance.

EXAMPLE 9.11   ■ The Effective Rank of a Matrix

Use *MATHEMATICA* to compute the effective rank of the matrix

$$A = \begin{pmatrix} 2.0 & 0 & 0 & 0 \\ 0 & 1.8 & 0 & 0 \\ 0 & 0 & 1.7 & 0 \end{pmatrix}$$

for the following tolerances: 1, .9, and .8.

Solution. Since A is in row echelon form and has three nonzero rows, its exact rank is 3. We can also tell that its singular values are its nonzero entries. A tolerance of 1 means that all singular values smaller than $2.0 \times 1 = 2$ are discarded. A tolerance of .9 means that all singular values smaller than $2.0 \times .9 = 1.8$ are discarded. And a tolerance of .8 means that all singular values smaller than $2.0 \times .8 = 1.6$ are discarded.

Let us verify that this is indeed the case.

$$\text{In}[1]:= A = \begin{pmatrix} 2.0 & 0 & 0 & 0 \\ 0 & 1.8 & 0 & 0 \\ 0 & 0 & 1.7 & 0 \end{pmatrix};$$

In[2]:= SingularValues[A, Tolerance → 1][[2]]

Out[2]= {2.}

In[3]:= SingularValues[A, Tolerance → .9][[2]]

Out[3]= {2., 1.8}

In[4]:= SingularValues[A, Tolerance → .8][[2]]

Out[4]= {2., 1.8, 1.7}

As expected, *MATHEMATICA* keeps only the singular values determined by the specified tolerances. ◄►

EXERCISES 9.5

1. Construct a matrix whose two smallest singular values are $.8 \times 10^{-6}$ and $.9 \times 10^{-6}$ and whose largest singular value is 5. Find a tolerance that will eliminate the value $.9 \times 10^{-6}$ but retain the value $.8 \times 10^{-6}$. Justify your claim.

2. Construct a 4×4 matrix with four nonzero singular values and specify a tolerance that will retain all singular values. Explain your reasoning.

3. Construct a 4×4 matrix with four nonzero singular values and specify a tolerance that eliminates all four singular values. Explain your reasoning.

Orthogonal Matrices

Orthogonal matrices play an important role in computer graphics. They are the matrices that preserve length and angles relative to orthonormal bases. They are also the matrices whose inverses are easy to compute. Matrix transposition involves considerably fewer arithmetic operations than matrix inversion.

As we saw in Chapter 2, animation of graphic objects can be achieved through the multiplication of matrices. Usually the operations involved must be invertible. Often they also need to preserve length and angles. Since orthogonal matrices have both of these properties, they are at the center of computer-based graphic applications. Unfortunately, round-off errors often destroy the orthogonality of the matrices required in the process. For example, the numerical approximation

$$A = \begin{pmatrix} .40825 & .7071 & -.57735 \\ -.40825 & .7071 & .57735 \\ .8165 & 2.122 \times 10^{-6} & .57735 \end{pmatrix}$$

of the orthogonal matrix

$$Q = \begin{pmatrix} \frac{1}{\sqrt{6}} & \frac{1}{\sqrt{2}} & -\frac{1}{\sqrt{3}} \\ -\frac{1}{\sqrt{6}} & \frac{1}{\sqrt{2}} & \frac{1}{\sqrt{3}} \\ \frac{2}{\sqrt{6}} & 0 & \frac{1}{\sqrt{3}} \end{pmatrix}$$

is not orthogonal since the product

$$A^T A = \begin{pmatrix} .99999 & -1.067\,5 \times 10^{-5} & 4.603 \times 10^{-6} \\ -1.067\,5 \times 10^{-5} & .99999 & -1.602 \times 10^{-6} \\ 4.603 \times 10^{-6} & -1.602 \times 10^{-6} & 1.0 \end{pmatrix}$$

is not the identity matrix I_3. Therefore, it is often necessary to replace a matrix that should be orthogonal by the closest orthogonal matrix to it. The Frobenius norm is used to measure the distance between these matrices.

The calculation of orthogonal approximations is based on two facts: (1) the singular value decomposition UDV^T involves two orthogonal matrices U and V, and (2) that the singular values of orthogonal matrices are 1. If A is an orthogonal matrix, then

$$A = UDV^T = UIV^T = UV^T$$

On the other hand, if A is a square matrix that is not orthogonal, then the matrix $B = UV^T$ is the closest orthogonal matrix to it. Thus if $\|A - B\|_F$ is sufficiently small, it may be possible to use B in place of A.

EXAMPLE 9.12 ■ A Best Approximation

Find the best orthogonal approximation of the matrix

$$A = \begin{pmatrix} 0.001 & -0.999 \\ 1.0001 & 0.0003 \end{pmatrix}$$

Solution. First we find the singular value decomposition of A.

```
In[1]:= {P,W,Q}=SingularValues[A=( 0.001  -0.999 )]
                                    ( 1.0001  0.0003 )
Out[1]= {{{-0.2799,-0.960029},{-0.960029,0.2799}},{1.0002,
0.998898},{{-0.960211,0.279275},{0.279275,0.960211}}}
```

We then let B be the matrix

$$B = P^T Q = \begin{pmatrix} -0.2799 & -0.960029 \\ -0.960029 & 0.2799 \end{pmatrix} \begin{pmatrix} -0.960211 & 0.279275 \\ 0.279275 & 0.960211 \end{pmatrix}$$

$$= \begin{pmatrix} 6.5096 \times 10^{-4} & 1.0 \\ -1.0 & 6.5096 \times 10^{-4} \end{pmatrix}$$

and note that $BB^T = I$. Hence B is orthogonal. An easy calculation shows that the distance $\|A - B\|_F$ between A and B thus is approximately 8 units. ◄►

EXERCISES 9.6

1. Find orthogonal matrices that are closest to the following matrices in the Frobenius norm.

a. $\begin{pmatrix} .7075 & -.7071 \\ .7072 & .7079 \end{pmatrix}$ b. $\begin{pmatrix} .7 & -.8 \\ .9 & .7 \end{pmatrix}$ c. $\begin{pmatrix} .42 & .707 & -.58 \\ -.41 & .707 & .58 \\ .81 & 0 & .58 \end{pmatrix}$

2. Use the matrix

$$A = \begin{pmatrix} .7075 & -.7071 \\ .7072 & .7079 \end{pmatrix}$$

to translate the points $\mathbf{x}$ on the unit circle $x^2 + y^2 = 1$ to the points $A\mathbf{x}$. Find an orthogonal matrix B that is closest to A in the Frobenius norm, and use B to translate the points $\mathbf{x}$ on the unit circle to the points $B\mathbf{x}$. Describe the difference in the locations of the corresponding points

$$\text{a. } A\begin{pmatrix} 0 \\ 1 \end{pmatrix} \quad \text{b. } A\begin{pmatrix} 1 \\ 0 \end{pmatrix} \quad \text{c. } A\begin{pmatrix} 0 \\ -1 \end{pmatrix} \quad \text{d. } A\begin{pmatrix} -1 \\ 0 \end{pmatrix}$$

and

$$\text{e. } B\begin{pmatrix} 0 \\ 1 \end{pmatrix} \quad \text{f. } B\begin{pmatrix} 1 \\ 0 \end{pmatrix} \quad \text{g. } B\begin{pmatrix} 0 \\ -1 \end{pmatrix} \quad \text{h. } B\begin{pmatrix} -1 \\ 0 \end{pmatrix}$$

3. Use *MATHEMATICA*, homogeneous coordinates, and matrix multiplication to move the triangle determined by the vertices

$$\begin{pmatrix} 0 \\ 0 \end{pmatrix} \quad \begin{pmatrix} 0 \\ 1 \end{pmatrix} \quad \begin{pmatrix} 2 \\ 0 \end{pmatrix}$$

through the following locations.

a. Rotate the triangle counterclockwise through .3658 radians.
b. Reflect the resulting triangle about the line $y = x$.
c. Translate the new triangle by $(h, k) = (12, -3)$.
d. Rotate the triangle counterclockwise through 1.7523.
e. Finally reverse all of these steps. Use the **Inverse** function to compute the inverses of the matrices in steps (a), (b), and (d).
f. Use singular value decomposition to compute the orthogonal matrices closest to the inverse matrices required in steps (a), (b), and (d).
g. Repeat the exercise by using the orthogonal substitutes of the inverse matrices involved.
h. Discuss the difference between the triangles resulting by the two processes.

Condition Numbers

If A is an invertible real $n \times n$ matrix, then there exists a unique $n \times n$ matrix A^{-1} such that $AA^{-1} = A^{-1}A = I_n$. If A is large, computer calculations of A^{-1} may produce a matrix that is actually singular due to round-off errors. The fact that this can happen leads to the following classification of invertible matrices using condition numbers. The *condition number* of a matrix is used to measure the extent to which an invertible matrix is sensitive to round-off errors.

DEFINITION 9.2   *Relative to a matrix norm* $\|A\|$, *the **condition number** cond A of an invertible matrix A is* $\|A\| \, \|A^{-1}\|$.

We use this definition to divide invertible matrices into three kinds. An invertible matrix is *perfectly conditioned* if cond $A = 1$, *well-conditioned* if cond A is slightly bigger than 1, and *ill-conditioned* if cond A is much larger than 1. The following property of condition number is as expected.

LEMMA 9.7   *The condition number of an identity matrix is 1.*

Proof. By definition, cond $I = \|I\| \|I^{-1}\| = \|I\| \|I\| = 1 \times 1 = 1.$ ∎

The next theorem provides a lower bound on condition numbers.

THEOREM 9.8   *The condition number* cond A *of any invertible matrix is greater than or equal to* 1.

Proof. If follows from the definition of condition numbers and the properties of matrix norms that

$$\text{cond } A = \|A\| \left\|A^{-1}\right\| \geq \left\|AA^{-1}\right\| \geq \|I\| = 1$$

Hence 1 is a lower bound for condition numbers. ∎

EXAMPLE 9.13   ■ **A Condition Number in the Infinity-Norm**

Use the infinity-norm to compute the condition number of the matrix

$$A = \begin{pmatrix} 2 & -1 & 0 \\ 2 & -4 & -1 \\ -1 & 0 & 2 \end{pmatrix}$$

Solution. First we compute A^{-1}.

```
In[1]:=MatrixForm[Inverse[A={{2,-1,0},{2,-4,-1},{-1,0,2}}]]

Out[1]//MatrixForm=
```
$$\begin{pmatrix} \frac{8}{13} & -\frac{2}{13} & -\frac{1}{13} \\ \frac{3}{13} & -\frac{4}{13} & -\frac{2}{13} \\ \frac{4}{13} & -\frac{1}{13} & \frac{6}{13} \end{pmatrix}$$

Since the infinity-norm of a matrix is the maximum absolute row sum,

$$\|A\|_\infty = \max\{|2| + |-1| + |0|, |2| + |-4| + |-1|, |-1| + |0| + |2|\} = 7$$

and

$$\left\|A^{-1}\right\|_\infty = \max\left\{ \left|\frac{8}{13}\right| + \left|\frac{-2}{13}\right| + \left|\frac{-1}{13}\right|, \left|\frac{3}{13}\right| + \left|\frac{-4}{13}\right| + \left|\frac{-2}{13}\right|, \left|\frac{4}{13}\right| + \left|\frac{-1}{13}\right| + \left|\frac{6}{13}\right| \right\} = \frac{11}{13}$$

Therefore,

$$\text{cond } A = \|A\|_\infty \left\|A^{-1}\right\|_\infty = 7 \times \frac{11}{13} \approx 5.9231$$

Depending on the application, we might consider this number to be reasonably small and conclude that A is reasonably well-conditioned. ◄►

Some matrices are notoriously ill-conditioned. For example, consider the 3×3 Hilbert matrix

$$A = \begin{pmatrix} 1 & \frac{1}{2} & \frac{1}{3} \\ \frac{1}{2} & \frac{1}{3} & \frac{1}{4} \\ \frac{1}{3} & \frac{1}{4} & \frac{1}{5} \end{pmatrix}$$

whose entries are defined by $a_{ij} = 1/(i + j - 1)$. We show that the condition number of this matrix is quite large.

EXAMPLE 9.14 ■ The Condition Number of a Hilbert Matrix

Find the condition number of the 3×3 Hilbert matrix relative to the one-norm.

Solution. We begin by finding the inverse of A.

```
In[1]:= Inverse[A=( 1    1/2  1/3
                    1/2  1/3  1/4
                    1/3  1/4  1/5 )]; MatrixForm[Inverse[A]]
```

```
Out[1]//MatrixForm=
        (    9   -36    30  )
        (  -36   192  -180  )
        (   30  -180   180  )
```

The one-norm of a matrix is the maximum absolute column sum of the matrix. Therefore $\|A\|_1$ is

$$\max\left\{|1| + \left|\frac{1}{2}\right| + \left|\frac{1}{3}\right|, \left|\frac{1}{2}\right| + \left|\frac{1}{3}\right| + \left|\frac{1}{4}\right|, \left|\frac{1}{3}\right| + \left|\frac{1}{4}\right| + \left|\frac{1}{5}\right|\right\} = \frac{11}{6}$$

and $\left\|A^{-1}\right\|_1$ is

$$\max\left\{|9| + |-36| + |30|, |-36| + |192| + |-180|, |30| + |-180| + |180|\right\} = 408$$

Hence the condition number cond A of A is $\|A\|_1 \left\|A^{-1}\right\|_1 = \frac{11}{6} \times 408 = 748$. As we can see, this number is quite large. ◄►

By adapting Example 9.14, we can easily confirm that the condition numbers of Hilbert matrices increase rapidly as the size of the matrices increases. Large Hilbert matrices are therefore considered to be extremely ill-conditioned.

We might think that if the determinant of a matrix is close to zero, then the matrix is ill-conditioned. However, this is false. Consider the matrix

$$A = \begin{pmatrix} 10^{-6} & 0 \\ 0 & 10^{-6} \end{pmatrix}$$

for which $\det A = 10^{-12} \approx 0$. In the one-norm,

$$\text{cond } A = \|A\|_1 \left\| A^{-1} \right\|_1$$

$$= \max \left\{ \left|10^{-6}\right|, \left|10^{-6}\right| \right\} \times \max \left\{ \left|10^6\right|, \left|10^6\right| \right\} = \left|10^{-6}\right| \left|10^6\right| = 1$$

The matrix A is therefore perfectly conditioned. Thus a small determinant is necessary but not sufficient for a matrix to be ill-conditioned.

The next theorem establishes a connection between singular values and condition numbers. It is often taken as the definition of condition numbers.

THEOREM 9.9 (Condition number theorem)   *Let A be any invertible matrix. Then the condition number of A is $\max \sigma_i / \min \sigma_i$.*

Proof. Suppose that A is an $n \times n$ invertible matrix. Then its eigenvalues are nonzero and so are the eigenvalues of $A^T A$. Let

$$\lambda_1 \geq \lambda_2 \geq \cdots \geq \lambda_n > 0$$

be the eigenvalue of $A^T A$. Then it follows from Theorem 6.9 that the eigenvalues of $\left(A^T A\right)^{-1}$ are

$$0 < \frac{1}{\lambda_1} \leq \frac{1}{\lambda_2} \leq \cdots \leq \frac{1}{\lambda_n}$$

Therefore, $\sigma_1 = \sqrt{\lambda_1} \geq \cdots \geq \sigma_n = \sqrt{\lambda_n} > 0$ are the singular values of A and $0 < \frac{1}{\sigma_1} \leq \cdots \leq \frac{1}{\sigma_n}$ are the singular values of A^{-1}. We know from Chapter 7 that the two-norm $\|A\|_2$ of A is the largest singular value of A and that the two-norm of A^{-1} is the largest singular value of A^{-1}.

By the definition of condition numbers, we therefore have

$$\text{cond } A = \|A\|_2 \left\| A^{-1} \right\|_2 = \frac{\sigma_1}{\sigma_n}$$

This proves the theorem. ∎

Note that if a square matrix A has a zero singular value, then the smallest singular value σ_n is 0. In that case, we can define its condition number to be infinity. This means that A is singular. Hence we can tell whether A is invertible by simply looking at the singular value decomposition UDV^T of A. If a diagonal entry of D is 0, then A is not invertible.

EXAMPLE 9.15 ■ **The Singular Value Decomposition of a Singular Matrix**

Use the singular value decomposition to show that the matrix

$$A = \begin{pmatrix} 2 & -1 & 0 \\ 2 & -1 & 0 \\ -1 & 0 & 2. \end{pmatrix}$$

is singular.

Solution. We use *MATHEMATICA* to find the singular value decomposition of A.

In[1]:= {P,W,Q}=SingularValues[A= $\begin{pmatrix} 2 & -1 & 0 \\ 2 & -1 & 0 \\ -1 & 0 & 2. \end{pmatrix}$];

In[2]:= DiagonalMatrix[W]

Out[2]= {{3.35781,0},{0,1.93005}}

Since the diagonal matrix produced by the singular value decomposition is 2×2 and A is 3×3, *MATHEMATICA* is telling us that 0 is a singular value of A. Hence A is a singular matrix. The condition number of A is therefore infinite. ◀▶

EXAMPLE 9.16 ■ **The Condition Number of the 4 × 4 Identity Matrix**

Use *MATHEMATICA* to find the condition number of the 4×4 identity matrix.

Solution. We use singular values to find the condition number of I_4.

In[1]:= {P,W,Q}=SingularValues[N[IdentityMatrix[4]]];

In[2]:= W

Out[2]= {1.,1.,1.,1.}

Therefore cond $I_4 = 1.0/1.0 = 1$. As the condition number of 1 indicates, the identity matrix I_4 is perfectly conditioned. ◀▶

Next we take the singular matrix

$$A = \begin{pmatrix} 1 & -1 \\ -1 & 1 \end{pmatrix}$$

and convert it to an invertible matrix by adding a small quantity to a_{22}. We will add .00001.

EXAMPLE 9.17   ■ An Ill-Conditioned Matrix

Use *MATHEMATICA* to find the condition number of the matrix

$$A = \begin{pmatrix} 1 & -1 \\ -1 & 1.00001 \end{pmatrix}$$

Solution. We compute the condition number of A as the ratio of the largest and smallest singular values of A.

In[1]:= A = $\begin{pmatrix} 1 & -1 \\ -1 & 1.00001 \end{pmatrix}$

Out[1]= {{1,-1},{-1,1.00000000000001}}

In[2]:= {P,W,Q}=SingularValues[A];

In[3]:= DiagonalMatrix[W]

Out[3]= {{2.00001,0},{0,4.99999×10⁻¹⁶}}

Therefore,

$$\operatorname{cond} A = \frac{2.00001}{4.99999 \times 10^{-16}} = 4.0 \times 10^{15}$$

This is a large number. If we were to ask *MATHEMATICA* to compute the inverse of A, we would get the message

```
''Result for Inverse of badly conditioned matrix
{{1,-1},{-1,1.00000000000001}} may contain significant
numerical errors.''
```

Thus *MATHEMATICA* recognizes A as an ill-conditioned matrix. ◄►

EXERCISES 9.7

1. Compute the condition numbers of the following matrices.

a. $\begin{pmatrix} 4 & 0 & 0 \\ 0 & 5 & 0 \\ 0 & 0 & 4 \end{pmatrix}$ b. $\begin{pmatrix} 4 & 0 & 0 \\ 2 & 5 & 0 \\ 3 & 2 & 4 \end{pmatrix}$ c. $\begin{pmatrix} \frac{1}{\sqrt{2}} & \frac{1}{\sqrt{2}} & 0 \\ -\frac{1}{\sqrt{2}} & \frac{1}{\sqrt{2}} & 0 \\ 0 & 0 & 1 \end{pmatrix}$

2. Use the **HilbertMatrix** function in the **MatrixManipulation** package of *MATHEMATICA* to generate the 2×2, 3×3, 4×4, and 5×5 Hilbert matrices and compute their condition numbers. What can you say about the rate at which these numbers increase?

3. Use the **SingularValues** function of *MATHEMATICA* to calculate an approximation B of the matrix

$$A = \begin{pmatrix} 9 & 7 & 7 \\ 4 & 8 & 5 \\ 7 & 1 & 2 \end{pmatrix}$$

Compute the condition numbers of A and B. Compare the matrix B with the matrix M obtained from A by using the **PseudoInverse** function.

4. Show that the matrix

$$A = \begin{pmatrix} 1.0000000000000001 & 1 \\ 1 & 1 \end{pmatrix}$$

is ill-conditioned.

REVIEW

KEY CONCEPTS ▶ Define and discuss each of the following.

Invertible Matrices

Condition number of an invertible matrix, ill-conditioned invertible matrix, perfectly conditioned invertible matrix, well-conditioned invertible matrix.

Rectangular Matrices

Pseudoinverse of a rectangular matrix, reduced singular value decomposition of a matrix, singular value decomposition of a matrix, singular values of a rectangular matrix.

Singular Vectors

Left singular vector, right singular vector.

KEY FACTS ▶ Explain and illustrate each of the following.

1. For all real $m \times n$ matrices A, the eigenvalues of $A^T A$ are nonnegative.

2. The two-norm of any real square matrix A is the largest singular value of A.

3. For every $m \times n$ real matrix A, there exists an orthogonal $m \times m$ matrix U, an orthogonal $n \times n$ matrix V, and an $m \times n$ generalized diagonal matrix D such that $A = UDV^T$.

4. If $\sigma_1, \ldots, \sigma_n$ are the singular values of a real $n \times n$ matrix A, then the square of the Frobenius norm $\|A\|_F^2$ is the sum $\sigma_1^2 + \cdots + \sigma_n^2$ of the singular values of A.

5. Suppose that $\mathbf{u}_1, \ldots, \mathbf{u}_m$ are the left singular vectors and $\mathbf{v}_1, \ldots, \mathbf{v}_n$ are the right singular vectors of a real $m \times n$ matrix A whose singular values are $\sigma_1, \ldots, \sigma_n$ and whose rank is r. Then the following hold.

    a. $U_r = \{\mathbf{u}_1, \ldots, \mathbf{u}_r\}$ is an orthonormal basis for Col A.

    b. $U_{m-r} = \{\mathbf{u}_{r+1}, \ldots, \mathbf{u}_m\}$ is an orthonormal basis for Nul A^T.

    c. $V_r = \{\mathbf{v}_1, \ldots, \mathbf{v}_r\}$ is an orthonormal basis for Col A^T.

    d. $V_{n-r} = \{\mathbf{v}_{r+1}, \ldots, \mathbf{v}_n\}$ is an orthonormal basis for Nul A.

6. For any linear system $A\mathbf{x} = \mathbf{b}$, the vector $\widehat{\mathbf{x}} = A^+\mathbf{b}$ is a least-squares solution of the system.

7. The condition number of an identity matrix is 1.

8. The condition number cond A of any invertible matrix is ≥ 1.

9. Let A be any invertible matrix. Then the condition number cond A is $\max \sigma_i / \min \sigma_i$.

APPENDICES

A

THE FUNDAMENTAL THEOREM OF ALGEBRA

In Chapter 8, we used the fundamental theorem of algebra to prove that real symmetric matrices have real eigenvalues. To do so, we had to work in complex vector spaces. We now provide a summary of the properties of complex numbers required for the definition of complex vector and complex inner product spaces.

The evolution of the usual number systems $\mathbb{N} \subset \mathbb{Z} \subset \mathbb{Q} \subset \mathbb{R} \subset \mathbb{C}$ can in many ways be explained as a step-by-step enrichment of mathematical systems in response to new needs. The names chosen for the systems in this chain also reveal some of the attitudes of mathematicians at these stages of evolution to the augmented systems.

The **natural** numbers $\mathbb{N} = \{1, 2, 3, \ldots\}$ were needed for counting and were considered to be altogether *natural*. The **integers** $\mathbb{Z} = \{0, 1, -1, 2, -2, 3, -3, \ldots\}$ are the whole numbers. The introduction of the number zero and of negative numbers was a major conceptual step forward in the history of mathematics. The **rational** numbers $\mathbb{Q}$ made up of ratios of integers p/q was probably an obvious extension of $\mathbb{N}$ required for measurements. The discovery that not all distances could be measured with rational numbers was a profound realization. The ruler-and-compass construction of $\sqrt{2}$ and the subsequent proof that $\sqrt{2} \notin \mathbb{Q}$ led to the introduction of **irrational** numbers such as $\sqrt{2}$. Moreover, it was realized that other quantities occurring in nature, such as π—the ratio of the circumference of a circle to its diameter—are not rational numbers. Hence there was a need to extend $\mathbb{Q}$ to a more complete system. The result is the set $\mathbb{R}$ of **real** numbers. We have assumed throughout the text that the points on a straight line are in one-one correspondence with the real numbers. An actual construction of $\mathbb{R}$ starting from $\mathbb{Q}$ is usually discussed in courses in calculus. In linear algebra, the existence and the algebraic properties of $\mathbb{R}$ are taken for granted.

Unfortunately, the eigenvalue problem forces us to find roots of real polynomials and some very simple polynomials have no real roots. We therefore sometimes need a more complete number system in which all polynomials have roots. The system $\mathbb{C}$ of **complex** numbers is that system. It is obtained from $\mathbb{R}$ by adding a single new object i, which has the property that $i^2 = -1$. We refer to i as the square root of -1 and write $i = \sqrt{-1}$.

Consider the polynomial $p(t) = t^2 + 1$ and suppose that $p(t) = (t - \lambda_1)(t - \lambda_2)$. Then

$$(t - \lambda_1)(t - \lambda_2) = t^2 - t\lambda_2 - \lambda_1 t + \lambda_1\lambda_2 = t^2 + 1$$

Therefore, $\lambda_1 + \lambda_2 = 0$ and $\lambda_1 \lambda_2 = 1$. It follows that $\lambda_1(-\lambda_1) = \lambda_1^2 = -1$. Hence λ_1 must be i.

DEFINITION A.1   *A **complex number** is any expression of the form $x + iy$, where x and y are real numbers and i is a symbol with the property that $i^2 = -1$.*

The sum and product of two complex numbers are easily defined.

DEFINITION A.2   *If $z_1 = x_1 + iy_1$ and $z_2 = x_2 + iy_2$ are two complex numbers, then $z_1 + z_2 = (x_1 + x_2) + i(y_1 + y_2)$ and $z_1 z_2 = (x_1 x_2 - y_1 y_2) + i(x_1 y_2 + y_1 x_2)$.*

It is easy to verify that these operations on the set $\mathbb{C}$ of complex numbers satisfy the usual axioms of arithmetic.

The main theorem about complex numbers required for linear algebra guarantees that $\mathbb{C}$ is rich enough to allow us to solve all polynomial equations.

THEOREM 1 (Fundamental theorem of algebra)   *For every polynomial $p(t) \in \mathbb{C}[t]$ there exists a $\lambda \in \mathbb{C}$ for which $p(\lambda) = 0$.* ∎

EXAMPLE A.1   ■ **The Roots of a Complex Polynomial**

Use *MATHEMATICA* to find the roots of the polynomial $p(t) = t^4 + t^2 + 1$.

Solution. We use the **Solve** function to find the roots of $p(t)$.

```
In[1]:= Solve[t⁴+4t²+1==0]

Out[1]={{t → -√-2+√5}, {t → √-2+√5}, {t → -i√2+√5},
{t → i√2+√5}}
```

Thus $p(t)$ factors into the product

$$\left(t + \sqrt{-2 + \sqrt{5}}\right)\left(t - \sqrt{-2 + \sqrt{5}}\right)\left(t + i\sqrt{2 + \sqrt{5}}\right)\left(t - i\sqrt{2 + \sqrt{5}}\right)$$

over $\mathbb{C}$. Two of the roots of $p(t)$ are real and two are complex. ◄►

If $p(t)$ is the characteristic polynomial of a real symmetric matrix, then Theorem 8.31 tells us that all roots of $p(t)$ are real.

EXAMPLE A.2   ■ The Eigenvalues of a Symmetric Matrix

Use *MATHEMATICA* to find the eigenvalues of the matrix

$$A = \begin{pmatrix} 8. & 4 & -5 \\ 4 & -5 & 3 \\ -5 & 3 & -5 \end{pmatrix}$$

Solution. We use *MATHEMATICA* to find approximate eigenvalues of A.

```
In[1]:= A = ( 8.   4  -5
              4  -5   3  );
             -5   3  -5
```

```
In[2]:= Eigenvalues[A]
Out[2]= {10.2589, -10.2291, -2.02974}
```

Let us check that these real numbers are approximate roots of the characteristic polynomial of A.

```
In[3]:= CharacteristicPolynomial[A, t]
Out[3]= 213. + 105. t - 2. t² - 1. t³
```

```
In[4]:= p[t_] := 213. + 105. t - 2. t² - 1. t³;
```

```
In[5]:= p[10.2589]
Out[5]= -0.00378858
```

```
In[6]:= p[-10.2291]
Out[6]= -0.00784459
```

```
In[7]:= p[-2.02974]
Out[7]= -0.000175826
```

As we can see, the approximate eigenvalues of A are approximate roots of $p(t)$. ◄►

At various stages in this text, we showed that several concepts for real vector spaces had complex analogues. The relevant constructions required the idea of the **conjugate** of a complex number and of its **modulus**.

DEFINITION A.3 *For every complex number $z = x + iy$, the complex number $\bar{z} = x - iy$ is the **conjugate** of z.*

It is easy to show that conjugates preserve addition and multiplication:

$$\overline{z_1 + z_2} = \overline{z_1} + \overline{z_2} \quad \text{and} \quad \overline{z_1 z_2} = \overline{z_1} \overline{z_2}$$

EXAMPLE A.3 ■ **The Conjugates of Complex Numbers**

Calculate the complex conjugates of the complex numbers $z_1 = 3 - 4i$, $z_2 = 7$, and $z_3 = 9i$.

Solution. By definition,

$$\overline{z_1} = \overline{3 - 4i} = 3 + 4i$$
$$\overline{z_2} = \overline{7 + 0i} = 7$$
$$\overline{z_3} = \overline{9i} = -9i$$

The fact that $\overline{7} = 7$ is an example of the fact that a complex number z is real if and only if $z = \bar{z}$. ◄►

EXAMPLE A.4 ■ **The Conjugate Transpose of a Complex Matrix**

Use *MATHEMATICA* to calculate the conjugate transpose A^H of the matrix

$$A = \begin{pmatrix} 4 + i & 7 - 2i \\ i & 5 \end{pmatrix}$$

Solution. We use the nested **Transpose** and **Conjugate** functions.

```
In[1]:= A= ( 4+i  7-2i )
           ( i    5    ) ;
```

```
In[2]:= Transpose[Conjugate[A]]//MatrixForm

Out[2]//MatrixForm=
           ( 4-i   -i )
           ( 7+2i   5 )
```

As we can see, we can calculate the conjugate transpose of A in two stages. First we form the conjugate of each entry of A and then we transpose the resulting matrix. ◀▶

DEFINITION A.4   *The modulus $|z|$ of a complex number $z = x + iy$ is the real number $\sqrt{x^2 + y^2}$.*

It is easy to show that the modulus is a distance function in the sense of Chapter 7. It is the length of the arrow determined by the point (x, y) in the xy-plane.

EXAMPLE A.5   ■ **Moduli of Complex Numbers**

Calculate the modulus of the complex numbers $z_1 = 3 - 4i$, $z_2 = 7$, and $z_3 = 9i$.

Solution. By definition,

$$|z_1| = |3 - 4i| = \sqrt{3^2 + 4^2} = 5$$
$$|z_2| = |7 + 0i| = 7$$
$$|z_3| = |9i| = -9i = \sqrt{(-9)^2} = 9$$

are the required moduli. ◀▶

The modulus of a complex number can also be computed using the standard inner product on $\mathbb{C}$.

EXAMPLE A.6   ■ **The Modulus and the Standard Inner Product**

Use the standard inner product on $\mathbb{C}$ to calculate the modulus of $z = 3 - 4i$.

Solution. Consider $\mathbb{C}$ as a complex vector space. Then the dimension of $\mathbb{C}$ is one. Hence every element of $\mathbb{C}$ is a 1×1 complex matrix with the property that $z^T = z$. According to the definition in Chapter 7, the square $\|z\|^2$ of the norm of z is therefore

$$z^T \bar{z} = (3 - 4i)(3 + 4i) = 25$$

Hence $\|z\| = \sqrt{25} = 5$, as expected. As we can see, the Euclidean norm of the vector z and the modulus of the complex number z coincide. ◀▶

Every complex number $z = x + iy$ determines an angle θ, with $-\pi < \theta \leq \pi$ called the *principal argument* arg (z) of z. The angle θ is the angle between the positive x-axis and the line segment terminating at the point (x, y). We know from geometry that if $r = |z|$, then $x = r\cos\theta$ and $y = r\sin\theta$. Therefore, $z = r(\cos\theta + i\sin\theta)$. If we require θ to satisfy $-\pi < \theta \leq \pi$, then θ is uniquely determined by z.

EXAMPLE A.7   ■ **The Principal Argument of a Complex Number**

Calculate the principal argument of $z = \sqrt{3} + i$.

Solution. The modulus r of z is $\sqrt{3 + 1} = 2$. Since $x = r\cos\theta$, we have $\sqrt{3} = 2\cos\theta$. Therefore, $\theta = \cos^{-1}\frac{\sqrt{3}}{2} = \frac{1}{6}\pi$. ◀▶

In Example A.6, we used the fact that the set $\mathbb{C}$ of complex numbers is a complex vector space of dimension 1. The set $\mathbb{C}$ can also be made into a real vector space of dimension 2. Let $S = \{\mathbf{x}_1 = 1, \mathbf{x}_2 = i\}$. Then S is linearly independent since no real multiple of 1—in other words, no real number $a = a \times 1$—is equal to the imaginary number i and no real multiple ai of i yields a nonzero real number. However, every complex number $z = a_1 + ia_2$ is a linear combination of the form $a_1\mathbf{x}_1 + a_2\mathbf{x}_2$. This example shows that the same set of vectors can give rise to very different vector spaces if the set of scalars is changed.

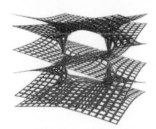

B

NUMERICAL CALCULATIONS

The scalars of the vector spaces studied in this text were assumed to be real or, sometimes, complex numbers. When calculating with such numbers, *MATHEMATICA* provides us with two choices: calculating symbolically and calculating numerically. Thus $\pi + \sqrt{2}$ is a symbolic name for a specific real number, and in many situations we can use this symbol for that number. For example, we know that $\sin \pi = 0$. So any occurrence of $\sin \pi$ in a calculation can be replaced by 0. No approximation of π is necessary. We also know, for example, that $e^{\pi i} = -1$. So there is no need to approximate e and π in any calculation involving $e^{\pi i}$.

In real-world applications, however, we usually need numerical approximations for both rational and irrational numbers such as $1/3$ and $\pi + \sqrt{2}$. Moreover, no computer can calculate with more than a finite number of the digits of any decimal expansion of $\pi + \sqrt{2}$. Thus every numerical calculation with $\pi + \sqrt{2}$ involves round-off errors. In this section, we briefly discuss two examples of the impact of such errors on numerical calculations in linear algebra.

All computers store real numbers as ***floating point numbers*** using finite representations such as

$$a = \pm .d_1 d_2 \ldots d_m \times 10^n$$

where $d_i \in \{0, 1, 2, 3, 4, 5, 6, 7, 8, 9\}$ and $n = \pm 0, 1, 2, 3, \ldots$, with $d_1 \neq 0$ if $a \neq 0$. Table 1 gives some examples.

TABLE 1   Floating-Point Numbers.

Symbolic Names	Numerical Names
2.4^{45}	$.128678 \times 10^{18}$
$2.3 + 5.63$	$.793 \times 10^1$
π	$.314159 \times 10^1$
$\sin[45]$	$.850904 \times 10^0$
$2/333$	$.600601 \times 10^{-2}$

The actual floating-point numbers used as approximations of a given real number depend on the application and on the specific computer and computer software involved. Depending on the accuracy required, for example, we often use $.314159 \times 10^1$, $.31416 \times 10^1$, $.314 \times 10^1$, or other approximations of π in calculations involving this real number.

The following two examples illustrate how round-off errors can produce what appear to be erroneous results.

EXAMPLE B.1   ■ $a + b = a$ and $b \neq 0$

Suppose we need to add the floating-point numbers $a = .451235 \times 10^4$ and $b = .123456 \times 10^{-2}$ and expect the answer to be accurate to within six decimal places. Then

$$a + b = .451235 \times 10^4 + .000000123456 \times 10^4$$
$$= .451235123456 \times 10^4$$
$$= .451235 \times 10^4$$
$$= a$$

Thus, $b \neq 0$ and yet $a + b = a$. In *MATHEMATICA*, this calculation takes the following form.

```
In[1]:= a=.451235*10⁴

Out[1]= 4512.35
```

```
In[2]:= b=.123456*10⁻²

Out[2]= 0.00123456
```

```
In[3]:= N[a+b,6]

Out[3]= 4512.35
```

Thus, $a + b = a$ and $b \neq 0$. ◀▶

In this example, the round-off error is perfectly transparent. When multiple calculations are nested, however, the source and the nature of these errors may be more difficult to detect and their consequences may be more elusive.

The next example shows that there are times when a round-off error causes *MATHEMAT-ICA* to produce a singular approximation of a nonsingular matrix.

EXAMPLE B.2   ■ **An Approximately Singular Matrix**

Use *MATHEMATICA* to show that for the real numbers $r = .123456 \times 10^5$ and $s = .123456 \times 10^{-30}$, round-off errors cause the diagonal matrix

$$A = \begin{pmatrix} (r + s) - r & 0 \\ 0 & (r + s) - r \end{pmatrix}$$

to be approximately singular relative to the default accuracy settings.

Solution. We first construct the floating-point real number $(r + s) - r$ and then use it to create the matrix A.

```
In[1]:= r=.123456×10⁵;
```

```
In[2]:= s=.123456×10⁻³⁰;
```

```
In[3]:= r+s;
```

```
In[4]:= MatrixForm[A={{(r+s)-r,0},{0,(r+s)-r}}]
Out[4]//MatrixForm=
```
$$\begin{pmatrix} 0. & 0 \\ 0 & 0. \end{pmatrix}$$

```
In[5]:= Det[A]
Out[5]= 0
```

Since the determinant of A is 0, the matrix A is singular. *MATHEMATICA* confirms this.

```
In[6]:= Inverse[A]
Inverse::sing:  Matrix {{0.,0},{0,0.}} is singular.
Out[6]= Inverse[{{0.,0},{0,0.}}]
```

But in the absence of round-off errors, $(r + s) - r$ equals s, and for the matrix

$$B = \begin{pmatrix} s & 0 \\ 0 & s \end{pmatrix}$$

MATHEMATICA certainly calculates a nonzero determinant, as shown next.

```
In[7]:= MatrixForm[B={{s,0},{0,s}}]
Out[7]//MatrixForm=
```
$$\begin{pmatrix} 1.23456 \times 10^{-31} & 0 \\ 0 & 1.23456 \times 10^{-31} \end{pmatrix}$$

In[8]:= **Inverse[B]**

Out[8]//MatrixForm=

$$
\begin{pmatrix}
8.10005 \times 10^{30} & 0 \\
0 & 8.10005 \times 10^{30}
\end{pmatrix}
$$

Since the determinant of B is not zero, the matrix B is nonsingular. ◀▶

By adjusting the powers of 10, we can always modify these examples to create round-off errors similar to those in the given examples, no matter what accuracy settings for *MATHE-MATICA* calculations are chosen.

Most linear algebra problems in science, engineering, the social sciences, and business are solved by computers using floating-point arithmetic. The impact of this arithmetic on the efficiency and reliability of algorithms used in these calculations is studied in *numerical linear algebra.*

Except for our work in Chapter 9, we have avoided any discussion of these difficulties inherent in using computers to solve linear algebra problems. In practice, however, this issue is particularly acute for eigenvalues since it has been proved that there are no algorithms for finding the roots of arbitrary polynomials of degree 5 or higher, even if their coefficients are integers.

The nonexistence of a suitable algorithm forces computer programs such as *MATHE-MATICA* to use numerical methods to calculate roots of polynomials. What is important, therefore, is that the computation processes used for this purpose are *stable*. This means that the approximations generated by the process must *converge*. If s is the exact solution of a problem, and s_n is an approximation of s, then s_n must approach s as n increases. In this text, we have relied on the assumption that the numerical methods implemented in *MATHEMATICA* are stable.

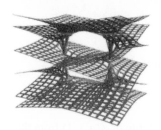

C

MATHEMATICAL INDUCTION

Mathematical induction is a powerful procedure for proving that certain statements involving natural numbers are true no matter which natural numbers are involved. For example, we might have a formula $\varphi(n)$ for computing the sum $1+2+\cdots+n$, and we want to prove that the formula works for all n. If $n = 15$, then the formula $\varphi(15)$ computes the sum $1+2+\cdots+15$; if $n = 1001$, then $\varphi(1001)$ computes the sum $1 + 2 + \cdots + 1001$; and so on. We would like to *prove* the fact that we can use the formula $\varphi(n)$ to find the sum $1 + \cdots + n$ for any natural number n. Mathematical induction allows us to do precisely that.

Behind mathematical induction is the reasoning that the set $\mathbb{N}$ of natural numbers can be characterized as the set whose elements are either 0 or of the form $n+1 = (1+1+\cdots+1)+1$. The principle of mathematical induction uses this property to assert that if we can prove that the formula $\varphi(0)$ holds, and if we can deduce the validity of $\varphi(n + 1)$ from the validity of $\varphi(n)$, then the formula $\varphi(m)$ holds for all $m \in \mathbb{N}$.

A proof by mathematical induction consists of two parts. The first part, called the ***induction basis***, consists of the proof that $\varphi(0)$ holds. Depending on the specific claim expressed by the formula $\varphi(n)$, the statement may actually only make sense for some $p > 0$. In that case, the induction basis is the proof that $\varphi(p)$ holds. We may also have to prove the validity of the first case or two by different means, depending on the nature of the formula $\varphi(n)$. If the induction basis involves more than one initial proof, the induction involved is known as *complete induction*.

The second part of the proof, called the ***induction step***, consists of a calculation that shows that if $\varphi(q)$ holds for an arbitrary $q \in N$, then $\varphi(q+1)$ holds. The assumption that $\varphi(q)$ holds is called the ***induction hypothesis***.

The following example illustrates the two parts involved in a proof by mathematical induction. It is well known that

$$1 + 2 + \cdots + n = \frac{n(n + 1)}{2} = \varphi(n)$$

for all $n \in \mathbb{N}$. The statement $\varphi(2)$ holds, for example, because

$$1 + 2 = 3 = \frac{2(2 + 1)}{2}$$

Suppose now that $\varphi(q)$ holds. This means that we are assuming that

$$1 + 2 + \cdots + q = \frac{q(q+1)}{2}$$

for an arbitrary natural number q. We would like to deduce the validity of $\varphi(q+1)$ from the assumed validity of $\varphi(q)$. This is done as follows.

$$
\begin{aligned}
1 + 2 + \cdots + q + (q+1) &= \frac{q(q+1)}{2} + (q+1) \\
&= \frac{q(q+1)}{2} + \frac{2(q+1)}{2} \\
&= \frac{(q+2)(q+1)}{2} \\
&= \frac{(q+1)(q+2)}{2}
\end{aligned}
$$

The principle of mathematical induction now allows us to conclude that $\varphi(n)$ holds for all n. (In this case the statement $\varphi(n)$ holds for all $n \geq 2$, since the expression makes no sense for $n = 0$ or $n = 1$.)

In finite-dimensional linear algebra, statements of the form $\varphi(n)$ are often statements asserting that a certain fact φ holds for spaces of all dimensions. Other types of statements proved by mathematical induction are statements about linear combinations of arbitrary length, or about polynomials of arbitrary degree. Good illustrations of proofs involving these kinds of statements are the proofs of the uniqueness of the reduced row echelon form, of the fact that determinants can be computed using elementary products, of the fact that eigenvectors belonging to distinct eigenvalues are linearly independent, of the fact that finite-dimensional real inner product spaces have orthonormal bases consisting of eigenvectors of self-adjoint linear transformations, and of the Gram-Schmidt process.

Using induction to define certain formulas is called a **definition by induction**. Sometimes it is also called a **recursive definition**. The factorial function $n!$, for example, is defined by induction. First we define $0!$ to be 1. Then we define $(n+1)!$ to be $n! \times (n+1)$. The proof that this process defines the factorial function is by mathematical induction. The definition of the Laplace expansion of the determinant of a matrix is another example of a definition by induction.

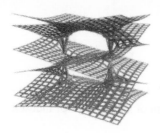

D

SIGMA NOTATION

Another important application of mathematical induction is the definition of the *sigma notation*. Arithmetic statements in linear algebra are often about sums made up of a fixed but arbitrary number of terms. There are two basic ways of dealing with this situation. We can either use an *ellipsis* (three dots) to represent missing elements, as in $\mathbf{x}_1 + \cdots + \mathbf{x}_n$, or we can use variable subscripts and the capital sigma and write $\sum_{i \in \mathbf{n}} \mathbf{x}_i$, where $\mathbf{n}$ denotes the standard set $\{1, \ldots, n\}$. The sigma notation is defined by mathematical induction.

DEFINITION D.1   *If* $\mathbf{n} = 1$, *then* $\sum_{i \in \mathbf{1}} \mathbf{x}_i = \mathbf{x}_1$; *and if* $\mathbf{n} = \mathbf{k}+1$, *then* $\sum_{i \in \mathbf{k}+1} \mathbf{x}_i = \left(\sum_{i \in \mathbf{k}} \mathbf{x}_i \right) + \mathbf{x}_{k+1}$.

However, we usually write the sum $\sum_{i \in \mathbf{n}} \mathbf{x}_i$ less formally as "$\mathbf{x}_1 + \cdots + \mathbf{x}_n$," using an ellipsis.

For any $m \times n$ matrix $A = (a_{ij})$, the expression $\sum_{i \in \mathbf{m}} a_{ij}$ denotes the sum of entries in the jth column of A, and the expression $\sum_{j \in \mathbf{n}} a_{ij}$ denotes the sum of the entries in the ith row of A.

EXAMPLE D.1   ■ A Sum of Vectors

The expressions $\mathbf{x}_1 + \mathbf{x}_2 + \mathbf{x}_3$ and $\sum_{i \in \mathbf{3}} \mathbf{x}_i$ denote the same sum of vectors. ◄►

EXAMPLE D.2   ■ A Linear Combination of Vectors

The expressions $a_1\mathbf{x}_1 + a_2\mathbf{x}_2 + a_3\mathbf{x}_3$ and $\sum_{i \in \mathbf{3}} a_i\mathbf{x}_i$ denote the same linear combination of the vectors $\mathbf{x}_1, \mathbf{x}_2$, and $\mathbf{x}_3$. ◄►

EXAMPLE D.3   ■ A Column Sum

If $A = (a_{ij})$ is a $3 \times n$ matrix, then the two expressions $a_{1j} + a_{2j} + a_{3j}$ and $\sum_{i \in \mathbf{3}} a_{ij}$ both denote the sum of the entries in the jth column of A. ◄►

EXAMPLE D.4   ■ A Row Sum

If $A = (a_{ij})$ is an $m \times 3$ matrix, then the two expressions $a_{i1} + a_{i2} + a_{i3}$ and $\sum_{j \in \mathbf{3}} a_{ij}$ both denote the sum of the entries in the ith row of A. ◄►

The set of all permutations on a standard set $\mathbf{n}$ is denoted by S_n, and $\sum_{\pi \in S_n} \varphi(\pi)$ denotes the sum of all terms $\varphi(\pi)$ determined by the permutations π.

EXAMPLE D.5   ■ A Sum Indexed by a Set of Permutations

Let $\mathbf{n} = 2$ and $S_n = \{\pi_1 = \{(1, 1), (2, 2)\}, \pi_2 = \{(1, 2), (2, 1)\}\}$. Then the two expressions $\varphi(\pi_1) + \varphi(\pi_2)$ and $\sum_{\pi \in S_2} \varphi(\pi)$ both denote the same sum of instances of φ, indexed by the set of permutations S_n. ◄►

Although the use of ellipses is often avoidable, there are situations where they are an essential part of the notation. For example, we frequently need to refer to *finite sequences* of vectors of unspecified length n, such as $\mathbf{v}_1, \ldots, \mathbf{v}_n$, or to *finite arrays* of elements of unspecified dimension, such as

$$\begin{pmatrix} a_{11} & \cdots & a_{1n} \\ \vdots & \ddots & \vdots \\ a_{m1} & \cdots & a_{mn} \end{pmatrix}$$

Here the horizontal, vertical, and diagonal ellipses stand for possibly missing terms.

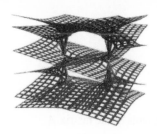

E

MATHEMATICA PACKAGES AND DEFINED FUNCTIONS

Many of the *MATHEMATICA* functions used in this text are defined in special packages that must be loaded once *MATHEMATICA* is running. We list the required packages and the relevant functions defined in them.

PACKAGES

The AffineMaps Package

`<<ProgrammingInMathematica'AffineMaps'`

- **AffineMap[**$\varphi, \psi, r, s, e, f$**]** generates an affine map with rotation angles φ, ψ, scale factors r, s, and translation components e, f.

- **AffineMap[***matrix***]** uses a 2×3 matrix to define the affine map.

The Arrow Package

`<<Graphics'Arrow'`

- **Arrow[***start,finish,opts***]** draws a two-dimensional arrow from the coordinates *start* to *finish*.

The Combinatorica Package

`<<DiscreteMath'Combinatorica'`

- **Inversions[**$\{i_1, \ldots, i_n\}$**]** counts the number of inversions of the permutation $\{i_1, \ldots, i_n\}$ of the standard list **Range[**n**]**.

- **Cofactor[**$A, \{i, j\}$**]** calculates the cofactor of the submatrix $A[i, j]$ obtained from a matrix A by deleting the ith row and jth column.

The Gaussian Elimination Package

`<<LinearAlgebra`GaussianElimination``

- **LUFactor** [A] gives the LU decomposition and the list of pivots of the matrix A.

- **LUSolve** [lu, b] solves the linear system represented by lu and right-hand side b.

The ImplicitPlot Package

`<<Graphics`ImplicitPlot``

- **ImplicitPlot** [eqn, $\{x,xmin,xmax\}$] plots the solution of the equation eqn using the **Solve** function, with x ranging from $xmin$ to $xmax$.

The Legend Package

`<<Graphics`Legend``

- **ShadowBox** [$pos,size,opts$] gives the graphics primitives for a box with a drop shadow.

The Matrix Manipulation Package

`<<LinearAlgebra`MatrixManipulation``

- **AppendColumns** [$A_1, A_2, \ldots, A_n$] joins the columns in matrices $A_1, A_2, \ldots, A_n$.

- **AppendRows** [$A_1, A_2, \ldots, A_n$] joins the rows in matrices $A_1, A_2, \ldots, A_n$.

- **BlockMatrix** [$blocks$] joins the rows and columns of the submatrices in *blocks* to form a new matrix.

- **TakeRows** [A, n] takes the first n rows of A.

- **TakeRows** [$A, -n$] takes the last n rows of A.

- **TakeRows** [$A, \{m, n\}$] takes rows m through n of A.

- **TakeColumns** [A, n] takes the first n columns of A.

- **TakeColumns** [$A, -n$] takes the last n columns of A.

- **TakeColumns** [$A, \{m, n\}$] takes columns m through n of A.

- **TakeMatrix** [A,pos_1,pos_2] takes the submatrix of A between entries at positions pos_1 and pos_2.

- **SubMatrix** [A,pos,dim] takes the submatrix of A of dimension *dim* starting at position pos.

- **UpperDiagonalMatrix**[A, n] creates an $n \times n$ matrix with entries $A[i, j]$ above the diagonal.

- **LowerDiagonalMatrix**[A, n] creates an $n \times n$ matrix with entries $A[i, j]$ below the diagonal.

- **ZeroMatrix**[n] creates an $n \times n$ matrix of zeros.

- **ZeroMatrix**[m, n] creates an $m \times n$ matrix of zeros.

- **HilbertMatrix**[n] creates an $n \times n$ Hilbert matrix, with entries given by $1/(i+j+1)$.

- **HilbertMatrix**[m, n] creates an $m \times n$ Hilbert matrix.

- **LinearEquationsToMatrices** [*eqns,vars*] gives a list of the form {*mat,vec*}, where *mat* is the matrix of coefficients of the linear equations in the specified variables, and *vec* is the vector of the right-hand side.

- **SquareMatrixQ**[A] tests whether A is a square matrix.

The MultiDescriptiveStatistics Package

<<Statistics`MultiDescriptiveStatistics`

- **CorrelationMatrix**[A] computes the correlation matrix of the entries of the matrix A.

- **CovarianceMatrix**[A] computes the covariance matrix of the entries of the matrix A.

- **StandardDeviation**[A] computes the standard deviation of the entries of the matrix A.

The Orthogonalization Package

<<LinearAlgebra`Orthogonalization`

- **Gram-Schmidt** [{$x_1, x_2, \ldots, x_n$}] generates an orthonormal set from the given list of real vectors.

- **Normalize**[x] normalizes the vector x.

- **Projection**[x, y] gives the orthogonal projection of x onto y.

- **Gram-Schmidt** [{$x_1, x_2, \ldots, x_n$},**InnerProduct** $\rightarrow$ *func*] generates an orthonormal set using the inner product given by func.

- **Normalize** [*x*,**InnerProduct** → *func*] normalizes the vector *x* using the inner product given by func.

- **Projection** [*x*, *y*,**InnerProduct** → *func*] gives the orthogonal projection of *x* onto *y* using the inner product given by func.

- **Gram-Schmidt** [{$x_1, x_2, \ldots, x_n$},**Normalized** → **False**] generates an orthonormal set from the given list of real vectors, but does not normalize the vectors.

The Rotations Package

`<<Geometry'Rotation'`

- **RotationMatrix2D** [*theta*] gives the matrix for rotation by an angle θ in two dimensions.

- **Rotate2D** [*vec,theta*] rotates the vector vec clockwise by an angle θ.

- **Rotate2D** [*vec,theta*,{*x, y*}] rotates the vector vec about the point {*x, y*} by an angle θ.

- **RotationMatrix3D** [*psi,theta,phi*] gives the matrix for rotation by the specified Euler angles ψ, θ, and φ in three dimensions.

- **Rotate3D** [*vec,psi,theta,phi*] rotates the vector vec by the specified Euler angles.

- **Rotate3D** [*vec,psi,theta,phi*,{*x, y, z*}] rotates the vector vec about the point {*x, y, z*} by the specified Euler angles.

DEFINED FUNCTIONS

The following functions are used in the solutions of some of the exercises. Some of them require the **MatrixManipulation** package.

1. `mf[A_]:=MatrixForm[A]`

2. `id[n_]:=IdentityMatrix[n]`

3. `erop1[M_,i_,s_]:=If[s!=0,Block[{list=Part[M,i]},`
   `ReplacePart[M,s*list,i]],''Error: scalar must be nonzero.'']`

4. `erop2[M_,i_,j_]:=Block[{list1=Part[M,i],list2=Part[M,j]},`
   `ReplacePart[ReplacePart[M,list1,j],list2,i]]`

5. `erop3[M_,i_,j_,s_]:=Block[{list=Part[M,i]+s*Part[M,j]},`
   `ReplacePart[M,list,i]]`

6. `elm1[n_,i_,s_]:=erop1[id[n],i,s]`

7. `elm2[n_,i_,j_]:=erop2[id[n],i,j]`

8. `elm3[n_,i_,j_,s_]:=erop3[id[n],i,j,s]`

9. `mkaug[list_]:=Block[{A=Apply[LinearEquationsToMatrices,`
   `list]},AppendRows[First[A],Partition[Last[A],1]]]`

10. `lineqparam[a_,b_]:=Block[{x,y,t},{x,y}==a+t(b-a)]`

11. `lineqstd[a_,b_]:=Block[{x,y,n=Cross[b-a]},n.{x,y}==n.a]`

12. `mtrace[A_]:=If[SquareMatrixQ[A],Block[{n=Length[A]},`
    `Sum[A[[i,i]],{i,n}],''Error: Not a square matrix.'']`

13. `minor[A_,i_,j_]:=Block[{bnd=Length[A]-1,m=Delete[A,i]},`
    `Map[Delete[Part[m,#],j]&,Range[bnd]]`

14. `cofactor[A_,i_,j_]:=(-1)^(i+j)*Det[minor[A,i,j]]`

15. `invertiblem[A_]:=SquareMatrixQ[A]&&(Det[A]!=0)`

16. `varlist[x_,n_]:=Table[x[i],{i,n}]`

17. `vander[vars_]:=Module[{r=(Range[Length[vars]]-1)},`
    `Map[#^r&,vars]]`

18. `randpoly[t_,n_]:=Array[Random[Integer,{-15,15}]&,n].`
    `t^Range[0,n-1]`

19. `randinvert[n_]:=Block[{A=Array[Random[Integer,{-9,9}]&,`
    `{n,n}]},If[Det[A]!=0,A,randinvert[n]]]`

20. `randsim[A_]:=Block[{P=randinvert[Length[A]]},`
    `{P.(A.Inverse[P]),P}]`

21. `randnoninvert[n_]:=Block[{A=Array[Random[Integer,{-9,9}]&,`
    `{n,n}]},If[Det[A]==0,A,randnoninvert[n]]]`

22. `randmtx[n_]:=Array[Random[Real,{-5,5},3]&,{n,n}]`

23. `polyiso[dim_,poly_]  :=  Block[{list=Array[0&,dim],val,`
    `polylist},polylist[i,pol_]:=If[i>dim,list,val=pol/.t → 0;`

24. `list=ReplacePart[list,val,i]; polylist[i+1,`
    `Simplify[(pol-val)/t]]];polylist[1,poly]]`

25. `polyeval[poly_,A_]  := Block[{dim=Length[A],list,output},`
    `list=polyiso[1+dim,poly];output=ZeroMatrix[dim];`
    `Map[(output=A.(output+#*id[dim]))&,Reverse[Rest[list]]];`
    `mf[output+First[list]*id[dim]]]`

26. `unit[x_]:=Block[{l=x.x},If[l!=0,x/Sqrt[l],''`
    `Error: x must be nonzero.'']]`

27. `cosangle[x_,y_]:=If[Min[x.x,y.y]!=0,unit[x].unit[y],`
    `''Error'']`

28. `proj[x_,y_]:=(x.y)/(y.y)y`

29. `mtxip[A_,x_,y_]:=x.(Transpose[A].A).y`

30. `singvals[A_]:=Sqrt[Eigenvalues[Transpose[A].A]]`

31. `norm1[vect_]:=Apply[Plus,Map[Abs,vec]]`

32. `norm2[vect_]:=Sqrt[vect.vect]`

33. `norminfinity[vect_]:=Apply[Max,Map[Abs,vect]]`

34. `ColumnRank[A_]:=Length[Transpose[A]]-Length[NullSpace[A]]`

35. `RowRank[A_]:=Length[A]-Length[NullSpace[Transpose[A]]]`

36. `Rank[A_]:=RowRank[A]`

37. `RowSpace[A_]:=Take[RowReduce[A],Rank[A]]`

38. `ColumnSpace[A_]:=RowSpace[Transpose[A]]`

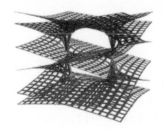

F

ANSWERS TO SELECTED EXERCISES

CHAPTER 1

EXERCISES 1.1

1. $0 = 5$; $y = 5$; $2y = 5$; $x = 5$; $x + y = 5$; $x + 2y = 5$; $2x = 5$; $2x + y = 5$; $2x + 2y = 5$.

3. `Solve[3x[1]+7x[2]-9x[3]==12,{x[1]}].`

5. `Solve[3x[1]+7x[2]-9x[3]==12,{x[1]}].`

6. `Solve[{3,7,-9}.{x[1],x[2],x[3]}==12,{x[2]}].`

7. We select the pair $(3, 2)$ at random. Since $(3, 2) \cdot (9, -7) = 13$, one such equation is $3x + 2y = 13$.

9. There are infinitely many solutions all of the form $12x + by = 12$, where b may be selected at random, say $b = \pi$, to get $12x + \pi y = 12$.

EXERCISES 1.2

1. `ListPlot[{{0,0},{1,2},{4,5},{-6,3},{1,2}+{4,5},{1,2}+{-6,3}},`
`Prolog→AbsolutePointSize[3]].`

2. `Show[Graphics[{Arrow[{0,0},{1,2}],Arrow[{0,0},{4,5}],`
`Arrow[{0,0},{-6,3}],Arrow[{0,0},{1,2}+{4,5}],`
`Arrow[{0,0},{1,2}+{-6,3}]}],Axes → True]].`

3. `{0,0},{1,2},{4,5},{-6,3},{1,2}+{4,5},{1,2}+{-6,3}.`

5. `{8,-2},{7,3},{-15,-1}.`

8. `ImplicitPlot[{x+y==0,x-y==0,x+y==1,3x-4y==2}.`
`{x,-5,5},{y,-5,5}].`

9. (a) `Plot3D[Part[Solve[x+y+z==0,{z}],1,1,2],{x,-5,5},{y,-10,10}].`

10. The equation of a line through a pair of points in $\mathbb{R}^2$ can be expressed in a variety of ways:
`Map[{Apply[lineqstd,#],Apply[lineqparam,#]}&,`
`{{{2,2},{1,-1}},{{2,2},{-1,1}},{{3,7},{2,2}},{{0,5},{8,1}}}].`

11. `Map[Block[{a,b,c},({a,b,c}=#);`
`ParametricPlot3D[t*b+u*c-(t+u-1)*a,{t,-5,5},{u,-5,5},`
`BoxRatios→{1,1,1}]]&,{{{2,2,3},{7,6,5},{1,-1,1}},`
`{{2,2,-2},{7,6,5},{1,-1,1}},{{2,2,6},{7,6,5},{-1,1,0}},`
`{{0,5,0},{8,1,3},{7,6,5}}}].`

13. `Map[Function[a,{-28,32,4}.a+128==0],`
`{{8,3,1},{0,5,-72},{6,5,-30}}].`

15. Three vectors $\mathbf{x}$, $\mathbf{y}$, and $\mathbf{z}$ are colinear if and only if the vector $\mathbf{y} - \mathbf{x}$ determines an angle of 0 or π with the vector $\mathbf{z} - \mathbf{x}$. (a) True; (b) False; (c) False; (d) False.

16. $(x, y) = (4, -2) + t(7, 12)$. `ParametricPlot[{4+7t,-2+12t},{t,-2,2}].`

19. The three planes intersect to enclose the tetrahedron with vertices $(7, 0, 0)$, $(0, 7, 0)$, $(0, 0, 7)$, $\frac{7}{3}(1, 1, 1)$.
`Show[Map[Plot3D[#,{x,0,7},{y,0,7},PlotPoints → 20,`
`BoxRatios→{1,1,1},PlotRange→{0,7}]&,{7-(x+y),`
`7-(3x+y),7-(x+3y),(7-(x+y))/3}]].`

EXERCISES 1.3

2. The system $\{x + y + z + w = 100, x + y + z = 50, y + z + w = 75\}$ has more unknowns than equations and is therefore underdetermined. For $x = 25$ and $w = 50$, we get the system $\{y + z = 25, y + z = 25, y + z = 25\}$. The system is consistent with solutions of the form $(25, 25 - z, z, 50)$.

5. (a) $2 \neq 6$; (b) $2 \neq 6$; (c) Adding the last two equations shows that $2x = 4$. Hence by the first equation $y = 0$. Returning to the last two equations we get $1 = 3$.

EXERCISES 1.4

4. $x - y + 2z = 9$. The system is inconsistent because the planes have no points in common.
5. `ImplicitPlot[{x+2y==2,x-y==4},{x,0,5},`
`PlotLabel→''Solution for (a) at` $\frac{1}{3}(10,-2)$`''];`
`ImplicitPlot[{x+y==2,-2x-2y==-4},{x,0,5},`
`PlotLabel→''All points on the line are solutions for (b)''];`
`ImplicitPlot[{x+2y==2,6x+12y==18},{x,0,5},PlotLabel →`
`''There are no solutions for (c)''].`

6. $(x, y, z) = (4, 8, 3) + t(3, 2, -7)$.

EXERCISES 1.5

2. (a) `Solve[{2x+y-z==2,x-3y+z==1,2x+y-2z==6},{x,y,z}].`
3. (a) $(4, 1)$; (b) $\frac{1}{13}(73, -7)$; (c) $(5, 2)$; (d) $(9, 9)$; (e) $(5, 21)$.

EXERCISES 1.6

1. `mf[Array[Power,{4,5}]]`.

2. `mf[Array[(-1)^(#1+#2)&,{4,5}]]`.

3. `Block[{fun3},fun3[1,1]:=3;fun3[1,2]:=1;fun3[1,3]:=4;`
`fun3[1,4]:=1;fun3[1,5]:=5;fun3[2,1]:=9;fun3[2,2]:=2;`
`fun3[2,3]:=6;fun3[2,4]:=5;fun3[2,5]:=3;fun3[3,1]:=5;`
`fun3[3,2]:=8;fun3[3,3]:=9;fun3[3,4]:=7;fun3[3,5]:=9;`
`fun3[4,1]:=3;fun3[4,2]:=2;fun3[4,3]:=3;fun3[4,4]:=8;`
`fun3[4,5]:=4;mf[Array[fun3,{4,5}]]]`.

EXERCISES 1.7

1. `Map[mf[Table[#,{i,4},{j,5}]]&,{i-j,(i+j)/i,i^j/i!}]`.

2. `Map[mf[Table[#,{i,4},{j,5}]]&,{(-1)^(i+j),`
`Sqrt[2]*Sin[Pi/4+Pi*j/2],i!/Fibonacci[j]}]`,
`Map[MatrixForm,{Table[Random[Integer,{-1-i,j+1}],{i,4},{j,5}],`
`Table[i*Random[Integer,{-10,10}],{i,4},{j,5}],`
`Table[Fibonacci[Random[Integer,{0,i*j}]],{i,4},{j,5}]}]`.

EXERCISES 1.8

2. `Map[mf,{{{1,0},{0,1}},{{-85},{97},{49}},{{-85,97,49}},`
`{{0,1,0},{1,0,0},{0,0,1}}}]`.

4. (a) We define individual values for $A : \mathbf{3} \times \mathbf{4} \to \mathbb{R}$.
`A[1,1]:=-85;A[1,2]:=-55;A[1,3]:=-37;A[1,4]:=-35;A[2,1]:=97;`
`A[2,2]:=50;A[2,3]:=79;A[2,4]:=56;A[3,1]:=49;A[3,2]:=63;`
`A[3,3]:=57;A[3,4]:=-59.`
(b) $A[i, j] = 1$ if $i = j$, $A[i, j] = 0$ otherwise. (c) $A[i, j] = 3 + j - i$.

6. `Map[Range,{1,2,5,6}]`.

EXERCISES 1.9

1. (a) `{{1,2,-1}}`; (b) `{{-8,2,99}}`; (c) `{{x[1],x[2],x[3]}}`.

4. `3{{1},{2},{-1}}-7{{-8},{2},{99}}+{{1},{0},{-100}}`.

EXERCISES 1.10

1. (a) $\begin{pmatrix} 2 & 1 & -1 & 2 \\ 1 & -3 & 1 & 1 \\ 2 & 1 & -2 & 6 \end{pmatrix}$ (e) $\begin{pmatrix} 1 & 0 & -1 & 1 & 2 \\ 0 & -1 & 1 & -7 & 4 \\ 2 & 1 & 0 & 0 & 5 \end{pmatrix}$

3. `Map[mf[mkaug[#]]&,{{{x+y==3,y==37/5},{x,y}},`
`{{x+y+z==3,y+z==37/5},{x,y,z}},`
`{{x+2y-z==2,x-y+z==4,2x+y-z==5},{x,y,z}}}].`

4. (a) `Block[{a=mkaug[{{x+2y-z==2,x-y+z==4,2x+y-z==5},{x,y,z}}]},`
`mf[a]→mf[a=erop3[erop3[a,2,1,-1],3,1,-2]]→mf[a=erop3[a,3,2,-1]]].`

The last matrix is in reduced form. The following operations correspond to the back substitution.

`Block[a={{1,2,-1,2},{0,-3,2,2},{0,0,-1,-1}},`
`mf[a=erop3[erop3[a,2,3,2],1,3,-1]] →`
`mf[a=erop1[erop1[a,2,-1/3],3,-1]]].`

EXERCISES 1.11

1. (a) $\mathbf{r}_1(A) = (1, 0, 0, 0, 3)$, $\mathbf{r}_2(A) = (1, 2, 0, 5, 3)$, $\mathbf{r}_3(A) = (0, 0, 0, 4, -1)$.

3. (a) `Block[a={{1,0,0,0,3},{1,2,0,5,3},{0,0,0,4,-1}},`
`{mf[erop1[a,3,4]],mf[erop3[a,2,1,7]], mf[erop2[a,2,3]]}].`

6. (a) The matrix is not in row echelon form. There are two leading nonzero entries in the first column. (b) The matrix is in row echelon form. Each leading nonzero entry appears to the right of the leading nonzero entry in the row above. (c) The matrix is not in row echelon form. The leading nonzero entry in the second row appears to the left of the leading nonzero entry in the first row.

EXERCISES 1.12

1. The following calculation shows that the matrices (a) and (c) are row equivalent:
`Block[a={{4,-13,5,0,5},{4,4,2,5,-5},{1,0,-3,-5,-8},{-5,4,7,-9,5}},`
`mf[a] → mf[a=erop3[a,1,2,1]] → mf[erop3[a,3,1,1]]].`

2. `A1={{2,-3,0},{0,4,3},{1,0,5},{2,5,2}};`
`A2={{0,0,0,0,3},{0,2,0,5,3},{1,0,0,4,-1}}`
`Map[Function[list,mf[RowReduce[list]]],{A1,A2}].`

EXERCISES 1.13

1. (a) Not in reduced row echelon form. The leading nonzero entry in the second row appears below that in the row above. (b) Not in reduced row echelon form. There is a 5 above the pivot in the third row. (c) Not in reduced row echelon form. The leading nonzero entry in the second row appears to the left of that in the row above. (d) Not in reduced row echelon form. There are nonzero entries above the pivots. (e) and (f) are in reduced row echelon form.

3. In each row echelon matrix produced, the pivots are 1s and are the only nonzero entries in their columns.

```
A1={{-1,1,4,1},{-3,3,5,5},{-2,-3,3,-3}}
A2={{2,-3,0},{0,4,3},{1,0,5},{2,5,2}}
A3={{5,-1,3},{-1,0,-4},{3,-4,-5}}
```

```
Print[''The reduced forms are:''];Map[mf[RowReduce[#]&,
{A1,A2,A3}].
```

4. In each case row echelon form can be achieved in at most two steps, and the pivot columns can be identified from that form. (a) The pivot columns are c_1, c_2, and c_4: Subtract the first row from the second. (b) The pivot columns are c_2, c_4, and c_5: Make two interchanges. (c) The pivot columns are c_1, c_2, and c_4: Interchange the first two rows. (d) The pivot column is c_1: Subtract the first row from the rows below. (e) The pivot columns are c_1 and c_2. (f) The pivot column is c_1.

EXERCISES 1.14

2. `Module[{p=(Inverse[vander[Range[6]]].{1,-1,2,3,-2,5}).` `Table[x^i,{i,0,5}]},Plot[p,{x,0.2,6.1},PlotPoints → 50];` `Print[''y='',p]].`

3. Two nonparallel planes will intersect in the given line provided that (1) $(4, 8, 3)$ lies in both planes, and (2) their normals are perpendicular to $(3, 2, 7)$. But this is easily arranged. From an infinity of choices, we select the equations $(2, -3, 0) \cdot (\mathbf{x} - (4, 8, 3)) = 0$ and $(0, 7, -2) \cdot (\mathbf{x} - (4, 8, 3)) = 0$ to get the matrix equation

$$\begin{pmatrix} 2 & -3 & 0 \\ 0 & 7 & -2 \end{pmatrix} \begin{pmatrix} x \\ y \\ z \end{pmatrix} = \begin{pmatrix} -16 \\ 50 \end{pmatrix}$$

4. $$\begin{pmatrix} 1 & 0 & 0 & -3 \\ 0 & 1 & 0 & -2 \\ 0 & 0 & 1 & -7 \end{pmatrix} \begin{pmatrix} x \\ y \\ z \\ t \end{pmatrix} = \begin{pmatrix} 4 \\ 8 \\ 3 \end{pmatrix}$$

EXERCISES 1.15

1. (a) All columns are pivot columns. (b) Only the last column is not a pivot column because the reduced form does not contain a leading nonzero entry. (c) The last two columns are not pivot columns for the same reason as in (b).

2. (a) There are no free variables. (b) There are no free variables. (c) Here w is given as a free variable because its column is not a pivot column. Any one of x, y, z, or w may be taken as a free variable. This is most easily seen by permuting the variable columns and then using *MATHEMATICA* to row reduce the matrix. In every case, the first three columns are pivot columns and so the variable corresponding to the fourth column may be taken as free.

5. `{Solve[{5x+4y+2z==0,5x+2y+z==0},{x,y}],`
`Solve[{5x+4y+2z+s(5x+2y+z)==0,5x+2y+z==0},{x,y}]}.`

EXERCISES 1.16

1. (a) $N_2 + 3H_2 \rightarrow 2NH_3$. (c) $2CH_4 + 2NH_3 + 3O_2 \rightarrow 2HCN + 6H_2O$.

7. The number of females in the different age groups are approximately 60,075, 18,214, 10,463, 1,269, and 891.

CHAPTER 2

EXERCISES 2.1

1. The following table lists the ingredients required for the three meals:

	Meat	Vegetables	Spices	Other
Filet Dinner	80	100	2	3
Irish Stew	100	300	4	5
Hungarian Goulash	150	350	5	7

Therefore, the vector

$$\begin{pmatrix} 80 & 100 & 2 & 3 \\ 100 & 300 & 4 & 5 \\ 150 & 350 & 5 & 7 \end{pmatrix} \begin{pmatrix} 8 \\ 3 \\ 9 \\ 4 \end{pmatrix} = \begin{pmatrix} 970 \\ 1756 \\ 2323 \end{pmatrix}$$

lists the total cost of each meal: The filet mignon meal costs the hotel $970, the Irish stew meal costs $1,756, and the Hungarian goulash meal costs $2,323. The total cost of the three meals is therefore $5,049.

EXERCISES 2.2

1. The equality $A + B = B + A$ follows from the fact that

$$a_{ij} + b_{if} = b_{ij} + a_{ij}$$

for all i, j.

3. $\begin{pmatrix} r_1c_1 & r_1c_2 & r_1c_3 \\ r_2c_1 & r_2c_2 & r_2c_3 \\ r_3c_1 & r_3c_2 & r_3c_3 \end{pmatrix} = \begin{pmatrix} 39 & 89 & 72 \\ 39 & 106 & 53 \\ 35 & 67 & 98 \end{pmatrix}$

5. The definition of matrix multiplication places a size restriction on the operation: To calculate the components of AB, the length of the rows of A must be identical to the length of the columns of B, but here the lengths are 5 and 4, respectively.

7. For example, if

$$A = \begin{pmatrix} 1 & 5 & 7 \\ 6 & 2 & -2 \\ -1 & -4 & 3 \end{pmatrix} \quad \text{and} \quad B = \begin{pmatrix} 2 & 0 & 0 \\ 0 & 2 & 0 \\ 0 & 0 & 2 \end{pmatrix}$$

then $AB = BA$. To find a nontrivial example requires solving nine equations in nine unknowns. If we change B to $\begin{pmatrix} 2 & 0 & 0 \\ 0 & 2 & 0 \\ 0 & 0 & 3 \end{pmatrix}$ the equality fails.

8. `Block[{a=Array[Random[Integer,{-9,9}]&,{3,5}],`
`b=Array[Random[Integer,{-9,9}]&,{5,3}]},`
`Print[''For A = '', MatrixForm[a],'', and B = '',`
`MatrixForm[b], '', we have AB = '',MatrixForm[a.b]];`
`Print['' whereas BA = '', MatrixForm[b.a],''.'']].`

9. The equality of the second and fourth vectors is the subject of Theorem 2.5.

10. The equality of the first and last matrices is the subject of Theorem 2.9, Part 4.

11. `Block[{A={{1,2,3},{2,1,0}},x={3,0,3},y={5,6,2},a,b},`
`mf[Expand[A.(a*x+b*y)]]==mf[a*(A.x)]+mf[b*(A.y)]].`

13. This equality follows from the fact that $a_{ij} + 0 = a_{ij}$ for all i, j.

15. This equality follows from the fact that $s(a_{ij} + b_{ij}) = sa_{ij} + sb_{ij}$ for all i, j.

19. Let the ith row of A be $\mathbf{x}$ and the jth columns of B and C be $\mathbf{y}$ and $\mathbf{z}$, respectively. Then the ijth entry of $A(B+C)$ is $\mathbf{x}(\mathbf{y} + \mathbf{z}) = \mathbf{xy} + \mathbf{xz}$, which is exactly the sum of the ijth entries in AB and AC.

24. The command `Tr[Array[Random[Real,{-9,9},3]&,{12,12}]]` computes the required trace.

27. Exercise 26 provides a counterexample: $428 \neq -55 \times 187$.

EXERCISES 2.3

1. (a) Upper triangular; (b) All of the properties; (c) None of the properties; (d) All of the properties; (e) All of the properties; (f) Symmetric.

2. `Block[{a=Table[Random[Integer,{-15,15}],{i,1,4},{j,1,4}],`
`b=Table[Random[Real,{-5,5},3],{i,1,4},{j,1,4}],`
`c=Table[Random[Integer,{-15,15}]*t^(j-1),{i,1,4},{j,1,4}]},`
`Map[mf,{a,b,c}]].`

3. `Block[{a=Array[Function[{i,j},If[j>i,0,`
`Random[Integer,{-15,15}]]],{5,5}],`
`b=Array[Function[{i,j},If[j>i,0,`
`Random[Real,{-5,5},3]]],{5,5}],`
`c=Array[Function[{i,j},If[j>i,0,`
`Random[Integer,{-15,15}]*t^(j-1)]],{5,5}]},`
`Map[mf,{a,b,c}]].`

4. `Block[{a=Array[Function[{i,j},If[j<i,0,`
`Random[Integer,{-15,15}]]],{5,5}],`
`b=Array[Function[{i,j},If[j<i,0,`
`Random[Real,{-5,5},3]]],{5,5}],`
`func:=Array[Random[Integer,{-15,15}]&,3],`
`c=Array[Function[{i,j},If[j<i,0,{1,t,t^2}.func]],{5,5}]},`
`Map[mf,mlist={a,b,c}]].`

5. `Block[{a=Array[Function[{i,j},If[i!=j,0,`
`Random[Integer,{-15,15}]]],{6,6}],`
`b=Array[Function[{i,j},If[i!=j,0,`
`Random[Real,{-5,5},3]]],{6,6}],`
`func:=Array[Random[Integer,{-9,9}]&,3],c=Array[ Function[{i,j},`
`If[i!=j,0,{1,t,t^2}.func]],{6,6}]},Map[mf,{a,b,c}]].`

7. We use the list `mlist` in Exercise 4.
`Map[Function[m,mf[m+Transpose[m]]],mlist],Clear[mlist].`

8. A small change to the solution of Exercise 5 does the trick.
`Block[{a=Array[Function[{i,j},If[i!=j,0,`
`Random[Integer,{-15,15}]]],{7,4}],b=Array[Function[{i,j},`
`If[i!=j,0,Random[Integer,{-15,15}]]],{5,6}]},Map[mf,{a,b}]].`

9. `Block[a={{3,0,1,7,1},{-1,2,0,-1,1},{0,-2,21,1,1}},`
`Map[mf,{TakeMatrix[a,{1,2},{2,2}],SubMatrix[a,{1,2},{2,2}],`
`TakeMatrix[a,{1,2},{3,4}],SubMatrix[a,{1,2},{3,4}]}]].`

10. `Block[{a=Array[Random[Integer,{-15,15}]&,{2,2}],`
`b=Array[Random[Integer,{-15,15}]&,{2,2}]},`
`{mf[a],mf[b],a.b==b.a}].`

14. Making use of the reversibility of the elementary row operations, we see that a matrix is an elementary matrix if and only if it can be reduced to the identity matrix by a single elementary row operation. Accordingly, (a) is elementary (multiply the first row by 1); (b) is elementary (exchange the first and third rows); (c) is not elementary (all four rows must be moved to get the identity matrix and this requires at least two steps).

16. `Block[{a=Array[Random[Integer,{-15,15}]&,{7,7}]},`
`Map[MatrixForm,{a,Transpose[a]}]].`

17. `Block[{a=Array[Random[Integer,{-15,15}]&,{6,6}],`
`t=Transpose[a]},{mf[a],a.t==Transpose[a.t],`
`t.a==Transpose[t.a]}].`

19. Let $A = \begin{pmatrix} 2 & 2 & 0 \\ 0 & 0 & 0 \\ 0 & 0 & 0 \end{pmatrix}$ and $B = \begin{pmatrix} 1 & 0 & 0 \\ -1 & 0 & 0 \\ 0 & 0 & 0 \end{pmatrix}$.

21. We use *MATHEMATICA*'s **Block** construction.
`Block[{a={{0,-2,9,-3},{2,0,-7,-8},{-9,7,0,5},{3,8,-5,0}},`
`a==-Transpose[a]].`

23. In the following, the symbol "b" represents an undefined function, thus providing variables for the entries of the matrix.
`Block[{b,a=Array[b,{3,4}],t=Transpose[a]},{MatrixForm[a],`
`a.t==Transpose[a.t]}].`

26. Every 3×3 submatrix is generated by deleting a row and a column uniquely determined by the entry at which they intersect:
`mf[Array[Function[{i,j},minor[{{19,12,17,13},{25,12,20,12},`
`{17,20,12,18},{24,24,22,13}},i,j]],{4,4}]].`

EXERCISES 2.4

1. (b) $\begin{pmatrix} 6 & 0 \\ 4 & 6 \end{pmatrix} \rightarrow \begin{pmatrix} 1 & 0 \\ 4 & 6 \end{pmatrix} \rightarrow \begin{pmatrix} 1 & 0 \\ 0 & 6 \end{pmatrix} \rightarrow \begin{pmatrix} 1 & 0 \\ 0 & 1 \end{pmatrix}$.

4. Each of the two columns of the reduced echelon form of the matrices contains a pivot.

9. (b) AB, where

$$A = \begin{pmatrix} 4 & 0 & 0 \\ 0 & 1 & 0 \\ 0 & 0 & 1 \end{pmatrix} \begin{pmatrix} 1 & 0 & 0 \\ 2 & 1 & 0 \\ 0 & 0 & 1 \end{pmatrix} \begin{pmatrix} 1 & 0 & 0 \\ 0 & 1 & 0 \\ 3 & 0 & 1 \end{pmatrix}$$

$$B = \begin{pmatrix} 1 & 0 & 0 \\ 0 & 5 & 0 \\ 0 & 0 & 1 \end{pmatrix} \begin{pmatrix} 1 & 0 & 0 \\ 0 & 1 & 0 \\ 0 & 2 & 1 \end{pmatrix} \begin{pmatrix} 1 & 0 & 0 \\ 0 & 1 & 0 \\ 0 & 0 & 4 \end{pmatrix}$$

10. The row reductions in Exercise 8 show that the matrices of Exercise 7 have pivots in each column.

12. `Block[{a=Array[Random[Integer,{-15,15}]&,{4,4}],b},`
`b=AppendRows[a,IdentityMatrix[4]];Map[mf,{a,b,RowReduce[b]}]].`

EXERCISES 2.5

3. An easy counterexample is

$$A = \begin{pmatrix} 1 & 1 \\ 0 & 0 \end{pmatrix} \quad B = \begin{pmatrix} 1 & 2 \\ -1 & -2 \end{pmatrix} \quad C = \begin{pmatrix} 0 & 3 \\ 0 & 4 \end{pmatrix}$$

Even if all of the matrices are invertible, the desired equation is unlikely to hold because matrix multiplication is not commutative:

```
Block[{a=Table[Random[Integer,{-9,9}],{i,1,3},{j,1,3}],
b=Table[Random[Integer,{-9,9}],{i,1,3},{j,1,3}],
c=(a.b).Inverse[a],a1,b1,c1,d1,e1},{a1,b1,c1,d1,e1} =
Map[mf,{a,b,c,a.b,c.a}];
Print[''For A = '',a1,'', B = '',b1,'' and C = '',c1,'' we have
AB = '',d1, '' and CA = '',e1]].
```

7. Block[{A={{1,-3,-3,1,0,0},{-3,1,3,0,1,0},{-3,3,-2,0,0,1}}},
mf[A] → mf[A=erop3[erop3[A,2,1,3],3,1,3]] →
mf[A=erop1[erop1[A,1,6],2,6]] →
mf[A=erop3[erop3[A,1,3,-3],2,3,-8]] →
mf[A=erop1[erop1[erop1[A,1,26/3],2,1/2],3,26]]→
mf[A=erop3[erop3[A,1,2,-5],3,2,11]] →
mf[A=erop1[erop1[A,2,2],3,-1/3]] → mf[erop2[A,2,3]]].

8. mf[Factor[Inverse[{{1,a,a^2},{1,b,b^2},{1,c,c^2}}]]].

EXERCISES 2.6

1. Suppose that A and B are orthogonal. Then $(AB)^{-1} = B^{-1}A^{-1} = B^T A^T = (AB)^T$.

2. Orthogonality is clear by inspection. To show that $A\mathbf{x}$ represents counterclockwise rotation through an angle θ, we must show two things: (1) that $A\mathbf{x}$ is equal to $\mathbf{x}$ in length and (2) that the angle made between $A\mathbf{x}$ and the positive x-axis exceeds that made by $\mathbf{x}$ and the positive x-axis by exactly θ. For convenience, we set $\mathbf{x} = (x, y)$. The length of $A\mathbf{x}$ is given by

$$\sqrt{(x\cos\theta - y\sin\theta)^2 + (x\sin\theta + y\cos\theta)^2} = \sqrt{x^2 + y^2}$$

which is the length of $\mathbf{x}$. Moreover, if $\mathbf{y} = \mathbf{x}/\sqrt{x^2 + y^2}$, then $\mathbf{y} = (\cos\varphi, \sin\varphi)$, where φ is the angle between $\mathbf{x}$ and the positive x-axis. This means that

$$A\mathbf{y} = (\cos\varphi\cos\theta - \sin\varphi\sin\theta, \cos\varphi\sin\theta + \sin\varphi\cos\theta)$$
$$= (\cos(\varphi + \theta), \sin(\varphi + \theta))$$

3. Consider reflections in the plane across the coordinate axes and the lines $y = x$ and $y = -x$. They are given by the matrices

$$\begin{pmatrix} 1 & 0 \\ 0 & -1 \end{pmatrix} \begin{pmatrix} -1 & 0 \\ 0 & 1 \end{pmatrix} \begin{pmatrix} 0 & -1 \\ -1 & 0 \end{pmatrix} \begin{pmatrix} 0 & 1 \\ 1 & 0 \end{pmatrix}$$

Each is symmetric and its own inverse.

5. We note that all of these matrices are symmetric, so the question reduces to testing the equation $A^2 = I$.

```
Block[{a=1/5{{4,-3},{-3,-4}},b={{-1,0},{0,1}},
c=1/19{{15,-10,6},{-10,-6,15},{6,15,10}}},Map[mf,{a.a,b.b,c.c}]].
```

EXERCISES 2.7

1. (a) Yes; (b) No (not square); (c) Yes.

3. Take A=Array[Random[Integer,{-8,8}]&,{3,3}] and
B=Array[Random[Integer,{1,5}]&,{3,3}].

10. (b) $\begin{pmatrix} 1 & 0 & 0 \\ \frac{3}{2} & 1 & 0 \\ 0 & 0 & 1 \end{pmatrix} \begin{pmatrix} 1 & 0 & 0 \\ 0 & 1 & 0 \\ \frac{1}{2} & 0 & 1 \end{pmatrix} \begin{pmatrix} 1 & 0 & 0 \\ 0 & 1 & 0 \\ 0 & 1 & 1 \end{pmatrix} \begin{pmatrix} 2 & 2 & 1 & 0 & 3 \\ 0 & 2 & -\frac{3}{2} & 0 & -\frac{9}{2} \\ 0 & 0 & 2 & 0 & 5 \end{pmatrix}$

11. (a) $\begin{pmatrix} 0 & 3 \\ 4 & 0 \\ 5 & 5 \end{pmatrix} = \begin{pmatrix} 0 & 1 & 0 \\ 1 & 0 & 0 \\ 0 & 0 & 1 \end{pmatrix} \begin{pmatrix} 1 & 0 & 0 \\ 0 & 1 & 0 \\ 0 & \frac{5}{3} & 1 \end{pmatrix} \begin{pmatrix} 1 & 0 & 0 \\ 0 & 1 & 0 \\ \frac{5}{4} & 0 & 1 \end{pmatrix} \begin{pmatrix} 4 & 0 \\ 0 & 3 \\ 0 & 0 \end{pmatrix}$

13. (b) M=LUDecomposition[{{1,1,4},{0,4,1},{4,5,4}}];
M1={{1,0,0},{4,1,0},{0,4,1}}.{{1,1,4},{0,1,-12},{0,0,49}},
M2={{1,0,0},{0,0,1},{0,1,0}}.M1.

14.

(a) $\left\{ x = -\frac{1}{2}, y = \frac{3}{8}, z = \frac{3}{4} \right\};$

(b) $\left\{ x = \frac{1}{2}, y = \frac{15}{8}, z = -\frac{1}{4} \right\};$

(c) $\left\{ x = 3, y = -\frac{31}{4}, z = \frac{3}{2} \right\}.$

EXERCISES 2.8

1. For

$$A = \begin{pmatrix} 2 & 1-i & 2+3i \\ 1+i & -3 & 7 \end{pmatrix} \quad \text{and} \quad B = \begin{pmatrix} -1 & -5+i & 3 \\ 3+2i & -5 & -2-i \end{pmatrix}$$

we have

$$A + B = \begin{pmatrix} 1 & -4 & 5+3i \\ 4+3i & -8 & 5-i \end{pmatrix}$$

2. If

$$A^H = \begin{pmatrix} 2 & 1-i \\ 1+i & -3 \\ 2-3i & 7 \end{pmatrix} \quad \text{and} \quad B^H = \begin{pmatrix} -1 & 3-2i \\ -5-i & -5 \\ 3 & -2+i \end{pmatrix}$$

then

$$(A + B)^H = \begin{pmatrix} 1 & 4-3i \\ -4 & -8 \\ 5-3i & 5+i \end{pmatrix}$$

5. The matrix $A = \begin{pmatrix} 1+i+(3-2i)t+t^2 & 2 \\ (1-i)t & it^2 \end{pmatrix}$ is one such matrix.

6. Here are two such matrices:

$$\begin{pmatrix} 2 & 2-i \\ 2+i & -3 \end{pmatrix} \quad \begin{pmatrix} 1 & 3+4i & i \\ 3-4i & 2 & 2 \\ -i & 2 & 1 \end{pmatrix}$$

The routine
```
Block[{a={{2,2-I},{2+I,-3}},b={{1,3+4I,I},{3-4I,2,2},{-I,2,1}},
Map[Conjugate[Transpose[#]&,{a,b}]=={a,b}]
```
shows that these matrices are Hermitian.

EXERCISES 2.9

1. The multiplications modify the horizontal component of **x** :
```
Show[Graphics[{Arrow[{0,0},{9,6}],Arrow[{0,0},{{3,0},{0,1}}.{9,6}],
Arrow[{0,0},{{1/3,0},{0,1}}.{9,6}]},Axes→True,
Epilog→{Text[''Ax'',{15,3}],Text[''Bx'',{3.5,5}]}]].
```
3. Multiplication by A adds the vector $(0, 2 \times 4)$ to **x**, and multiplication by B adds the vector $(2 \times -1, 0)$:
```
Show[Graphics[{Arrow[{0,0},{4,-1}],Arrow[{4,-1},{4,7}],
Arrow[{4,-1},{2,-1}],Arrow[{0,0},{{1,2},{0,1}}.{4,-1}],
Arrow[{0,0},{{1,0},{2,1}}.{4,-1}]},Axes→True,
AspectRatio→Automatic,PlotRange→{{-0.5,4.5},{-2,8}},
Epilog→{Text[''x'',{3.2,-0.5}],Text[''Bx'',{2.6,5}],
Text[''Ax'',{1,-1}]}]].
```

EXERCISES 2.10

1. $(3, 5, 1)$, $(-8, 0, 1)$, $(112, 115, 1)$.

2. $(4, 5)$, $(-8, -6)$, $(\pi, -\pi)$.

3. $\begin{pmatrix} 1 & 0 & 2 \\ 0 & 1 & -5 \\ 0 & 0 & 1 \end{pmatrix} \begin{pmatrix} \cos\frac{\pi}{4} & -\sin\frac{\pi}{4} & 0 \\ \sin\frac{\pi}{4} & \cos\frac{\pi}{4} & 0 \\ 0 & 0 & 1 \end{pmatrix} \begin{pmatrix} 1 & \frac{1}{3} & 0 \\ 0 & 1 & 0 \\ 0 & 0 & 1 \end{pmatrix} = \begin{pmatrix} \frac{1}{2}\sqrt{2} & -\frac{1}{3}\sqrt{2} & 2 \\ \frac{1}{2}\sqrt{2} & \frac{2}{3}\sqrt{2} & -5 \\ 0 & 0 & 1 \end{pmatrix}$

4. $(3, 5, 7, 1)$, $(-8, 0, 1, 1)$, $(112, 115, 0, 1)$.

5. $(4, 5, 2)$, $(-8, -6, -1)$, $(\pi, -\pi, 1)$.

6. $\begin{pmatrix} 1 & 0 & 0 & h \\ 0 & 1 & 0 & k \\ 0 & 0 & 1 & j \\ 0 & 0 & 0 & 1 \end{pmatrix}$

7. The equation

$$\begin{pmatrix} 1 & 0 & h \\ 0 & 1 & k \\ 0 & 0 & 1 \end{pmatrix}^T \begin{pmatrix} 1 & 0 & h \\ 0 & 1 & k \\ 0 & 0 & 1 \end{pmatrix} = \begin{pmatrix} 1 & 0 & 0 \\ 0 & 1 & 0 \\ 0 & 0 & 1 \end{pmatrix}$$

holds if and only if $h = k = 0$.

EXERCISES 2.11

1. The matrix

$$\begin{pmatrix} 19 & 15 & 13 & 5 & 27 & 16 & 5 & 4 & 16 & 12 & 5 & 27 & 12 \\ 9 & 11 & 5 & 27 & 2 & 1 & 19 & 5 & 2 & 1 & 12 & 12 & 27 \\ 1 & 14 & 4 & 27 & 19 & 15 & 13 & 5 & 27 & 12 & 9 & 11 & 5 \\ 27 & 6 & 15 & 15 & 20 & 2 & 1 & 12 & 12 & 28 & 0 & 0 & 0 \end{pmatrix}$$

represents the given message. The sequence built by concatenating the rows of the matrix

$$\begin{pmatrix} 44 & 50 & 24 & 204 & 7 & -9 & 108 & 47 & 40 & 49 & 63 & 1 & 109 \\ -10 & -28 & -4 & 62 & 15 & -15 & -26 & 33 & 70 & 97 & -25 & -127 & -121 \\ 250 & 76 & 130 & 154 & 110 & -10 & 78 & 90 & 8 & 150 & 46 & 82 & 166 \\ 249 & 162 & 155 & 257 & 254 & 108 & 127 & 106 & 186 & 216 & 86 & 178 & 166 \end{pmatrix}$$

is the encoded message.

EXERCISES 2.12

1. $\mathbf{c}_1 = \begin{pmatrix} .2 \\ .1 \\ .2 \\ 0 \end{pmatrix}, \mathbf{c}_2 = \begin{pmatrix} .1 \\ .2 \\ .3 \\ .2 \end{pmatrix}, \mathbf{c}_3 = \begin{pmatrix} 0 \\ .2 \\ .2 \\ .2 \end{pmatrix}, \mathbf{c}_4 = \begin{pmatrix} .1 \\ .3 \\ .2 \\ .1 \end{pmatrix}.$

2. $\begin{pmatrix} .8 & -.1 & 0 & -.1 \\ -.1 & .8 & -.2 & -.3 \\ -.2 & -.3 & .8 & -.2 \\ 0 & -.2 & -.2 & .9 \end{pmatrix} \begin{pmatrix} x_1 \\ x_2 \\ x_3 \\ x_4 \end{pmatrix} = \begin{pmatrix} 150 \\ 200 \\ 100 \\ 300 \end{pmatrix}.$

4. $\mathbf{d} = \mathbf{x} - C\mathbf{x}.$

EXERCISES 2.13

1. Let the ijth entry refer to an edge between vertices v_i and v_j. Then

$$M(G) = \begin{pmatrix} 0 & 1 & 0 & 1 \\ 1 & 0 & 2 & 0 \\ 0 & 2 & 0 & 0 \\ 1 & 0 & 0 & 0 \end{pmatrix} \quad \text{and} \quad A(G) = \begin{pmatrix} 0 & 1 & 0 & 1 \\ 1 & 0 & 1 & 0 \\ 0 & 1 & 0 & 0 \\ 1 & 0 & 0 & 0 \end{pmatrix}$$

2. There are six paths of length 2: $e_1 \sim e_2 \sim e_1$, $e_1 \sim e_2 \sim e_3$, $e_2 \sim e_1 \sim e_2$, $e_2 \sim e_3 \sim e_2$, $e_3 \sim e_2 \sim e_1$, and $e_3 \sim e_2 \sim e_3$.

10. The graph constructed in Exercise 1, is an example of a strongly connected graph. Every vertex can be reached from every other vertex by a path of length 3 or less. Therefore the

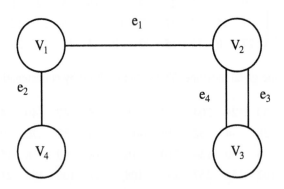

FIGURE 1

matrix

$$A(G) + A(G)^2 + A(G)^3 = \begin{pmatrix} 2 & 4 & 1 & 3 \\ 4 & 2 & 3 & 1 \\ 1 & 3 & 1 & 1 \\ 3 & 1 & 1 & 1 \end{pmatrix}$$

has no zero entry.

CHAPTER 3

EXERCISES 3.1

1. (a) 940; (b) 146,321; (c) 0; (d) 1944; (e) 7182; (f) 12.

2. `Block[{a=Array[Random[Real,{-9,9},3]&,{3,3}]},`
`Print[''For the matrix A = '',mf[a], '','','' Det A = '',`
`Det[a], ''but Sum[a[[i,i]]*C[[i,i]],{i,1,3}] = '',`
`Simplify[Sum[(a[[,i,]]*cofactor[a,i,i]),{i,1,3}]]]].`

4. `Block[{a=Array[Random[Integer,{-9,9}]&,{3,3}],`
`b=Array[Random[Integer,{-9,9}]&,{3,3}]},`
`Print[''For A='', mf[a],'',B ='',mf[b], '',Det[A]='',`
`Det[a],'',Det[B] = '', Det[b],''.'']`
`Print[''Det[A+B] = '',Det[a+b], '', so Det[A+B]≠Det[A]+Det[B]`
`and '',''Det[5A]= '',Det[5a], '' so Det[5A]≠ 5Det[A].'']].`

EXERCISES 3.2

1. (a) 6; (b) 0; (c) −32; (e) 0; (f) −91, 708; (g) −160.

4. (a) Not invertible; (b) Determinant is -2, so the matrix has an inverse; (c) Invertible.

5. `Block[{mtx},For[i=1,i<4,i++,mtx=Array[Random[Integer,{-9,9}]&,`
`{5,5}]; Print[''For A = '',mf[mtx],'', we have det A= '',`
`Det[mtx], ''and detA`^T` = '', Det[Transpose[mtx]]]]].`

6. `Block[{mtx,a},mtx=Table[a[i,j],{i,4},{j,4}];`
`Print[''For A = '', mf[mtx],'', we have det A - detA`^T` = '',`
`Det[mtx]- Det[Transpose[mtx]]]].`

7. `Block[{mtxaa=Array[Random[Real,{-9,9},4]&,{3,3}],`
`mtxbb=Array[Random[Real,{-9,9},4]&,{3,3}]},`
`Print[''For A = '',mf[mtxaa],'' and B = '',mf[mtxbb],'',`
`we have det A det B = '',Det[mtxaa], '' × '',Det[mtxbb],'' = '',`
`Det[mtxaa.mtxbb],'' = det AB.'']].`

9. This follows from the definition of a permutation matrix and Corollary 3.7.

EXERCISES 3.3

1. (a) $\begin{pmatrix} 2 & 1 \\ 1 & -3 \end{pmatrix} \begin{pmatrix} x \\ y \end{pmatrix} = \begin{pmatrix} 2 \\ 1 \end{pmatrix}$, $\det \begin{pmatrix} 2 & 1 \\ 1 & -3 \end{pmatrix} = -7$, and

$\begin{pmatrix} x \\ y \end{pmatrix} = -\frac{1}{7} \begin{pmatrix} 1 \\ 0 \end{pmatrix}$; (b) $\begin{pmatrix} 2 & 1 \\ -2 & -1 \end{pmatrix} \begin{pmatrix} x \\ y \end{pmatrix} = \begin{pmatrix} 2 \\ 1 \end{pmatrix}$,

$\det \begin{pmatrix} 2 & 1 \\ -2 & -1 \end{pmatrix} = 0$, no solution.

EXERCISES 3.4

1. $\begin{pmatrix} 6 & 3 \\ 4 & 4 \end{pmatrix} \begin{pmatrix} \frac{1}{3} & -\frac{1}{4} \\ -\frac{1}{3} & \frac{1}{2} \end{pmatrix} = \frac{1}{12} \begin{pmatrix} 6 & 3 \\ 4 & 4 \end{pmatrix} \begin{pmatrix} 4 & -3 \\ -4 & 6 \end{pmatrix} = \frac{1}{12} \begin{pmatrix} 12 & 0 \\ 0 & 12 \end{pmatrix} = I.$

EXERCISES 3.5

2. (a) 0; (b) 160.

5. The matrix $\begin{pmatrix} 4 & 0 & 0 \\ 0 & 3 & 1 \\ 0 & 1 & 1 \end{pmatrix}$ is an example.

EXERCISES 3.6

1. The command

$$\text{Map[Det,} \left\{ \begin{pmatrix} x & y & 1 \\ 3 & 4 & 1 \\ 2 & 7 & 1 \end{pmatrix}, \begin{pmatrix} x & y & 1 \\ 3 & 4 & 1 \\ 2 & -7 & 1 \end{pmatrix}, \begin{pmatrix} x & y & 1 \\ 3 & 4 & 1 \\ -2 & 7 & 1 \end{pmatrix}, \begin{pmatrix} x & y & 1 \\ 3 & 4 & 1 \\ -2 & -7 & 1 \end{pmatrix} \right\} \text{]}$$

yields the expressions

$$\{13 - 3x - y, -29 + 11x - y, 29 - 3x - 5y, -13 + 11x - 5y\}$$

Therefore, the required equations are: (a) $13 - 3x - y = 0$; (b) $-29 + 11x - y = 0$; (c) $29 - 3x - 5y = 0$; (d) $-13 + 11x - 5y = 0$.

2. (a) The determinant

$$\det \begin{pmatrix} x^2 & x & y & 1 \\ 81 & 9 & 4 & 1 \\ 4 & 2 & 7 & 1 \\ 9 & 3 & 4 & 1 \end{pmatrix} = 0$$

yields $3x^2 - 36x + 107 = 7y$; (b) $5x^2 - 88x + 303 = 21y$; (c) $3x^2 + 18x - 125 = -11y$; (d) $x^2 + 4x - 33 = -3y$.

3. (a) $3x^2 + 3y^2 - 36x - 40y + 193 = 0$; (d) $x^2 + y^2 + 10x - 6y - 31 = 0$.

4. (a) 18; (b) 28; (c) 17; (d) 64.

5. The absolute value of the determinant

$$\det \begin{pmatrix} a & c & 0 \\ b & d & 0 \\ 1 & 1 & 1 \end{pmatrix}$$

gives the volume of the parallelepiped with slant height 1. Its cross section (in the plane $z = 1$) is a parallelogram that is the image under H (see Section 2) of the parallelogram determined by the vertices $(0, 0)$, (a, b), and (c, d). The volume of a cylinder is given by *slant height $\times$ cross-sectional area*. Therefore, the given equation implies that H preserves areas.

EXERCISES 3.7

4. (b) f:=x⁴+y²+8x+9y;H={{D[f,x,x],D[f,x,y]},{D[f,y,x],D[f,y,y]}};Det[H]
The point $(-\sqrt[3]{2}, -9/2)$ is a relative minimum since Det[H] is positive at $(-\sqrt[3]{2}, -9/2)$.

7. (a) The Wronskian determinant is

$$W(e^x, e^{2x}, e^{3x}) = \begin{pmatrix} e^x & e^{2x} & e^{3x} \\ D_x e^x & D_x e^{2x} & D_x e^{3x} \\ D_x^2 e^x & D_x^2 e^{2x} & D_x^2 e^{3x} \end{pmatrix} = \begin{pmatrix} e^x & e^{2x} & e^{3x} \\ e^x & 2e^{2x} & 3e^{3x} \\ e^x & 4e^{2x} & 9e^{3x} \end{pmatrix}$$

If $x = 0$, then $\det W(e^x, e^{2x}, e^{3x})(0) = 2 \neq 0$. Hence the functions are linearly independent.

CHAPTER 4

EXERCISES 4.1

1. $a\mathbf{u} + (-a)\mathbf{u} = (a - a)\mathbf{u} = \mathbf{0}$ and $a\mathbf{u} + a(-\mathbf{u}) = a(\mathbf{u} + (-\mathbf{u})) = \mathbf{0}$, both by Theorem 4.1.

3. The eight axioms hold for addition and multiplication of real numbers. It is easy to verify that the given operations preserve the truth of each axiom. For example, the calculation

$$\mathbf{x} + \mathbf{y} = (x_1, x_2) + (y_1, y_2) = (x_1 + y_1, x_2 + y_2)$$
$$= (y_1 + x_1, y_2 + x_2) = (y_1, y_2) + (x_1, x_2) = \mathbf{y} + \mathbf{x}$$

shows that vector addition is commutative.

5. (a) The line is nonempty, consisting of all points $(x, y) = (x, 3x) = x(1, 3)$. This collection of points is closed under addition and scalar multiplication and so is a vector space in the plane. (b) The plane is nonempty, containing all points $\mathbf{x} = (x, y, z)$ for which $(1, 2, 3) \cdot \mathbf{x} = 0$. This set is again closed under addition and scalar multiplication and therefore is a vector space in $\mathbb{R}^3$.

6. This set of points does not include the zero vector and so cannot be a vector space with respect to the usual operations on vectors in the plane.

7. If two such polynomials are added, the coefficient of t^3 in the sum is 2. This set is therefore not closed under addition.

8. If we let $A = \begin{pmatrix} 1 & 2 \\ 3 & 4 \end{pmatrix}$ and $a = \sqrt{2}$, then A is in the set but aA is not. Therefore, this set is not closed under scalar multiplication.

EXERCISES 4.2

2. Adapt Example 4.3.

EXERCISES 4.3

2. (a) The set $\mathbb{R}$ is a set of numbers, whereas the set $\mathbb{R}^{1 \times 1}$ is the set of functions from $\{(1, 1)\}$ into $\mathbb{R}$. (b) $\mathbb{R}^3$ and $\mathbb{R}^{3 \times 1}$ are identical. By definition, $\mathbb{R}^3$ is the set of 3×1 matrices, also called column vectors. The set $\mathbb{R}^{1 \times 3}$ is the set of all 1×3 matrices, also called row vectors. (d) $\mathbb{R}^{2 \times 3}$ is the collection of functions on the set

$$\{(1, 1), (1, 2), (1, 3), (2, 1), (2, 2), (2, 3)\}$$

whereas $\mathbb{R}^{3 \times 2}$ is the collection of functions on the set

$$\{(1, 1), (1, 2), (2, 1), (2, 2), (3, 1), (3, 2))\}$$

3. The "vectors" do not have inverses with respect to the usual addition.

EXERCISES 4.4

1. The vector space operations of $\mathbb{R}_3[t]$ are the following:

If $p(t) = (a_0 + a_1 t + a_2 t^2 + a_3 t^3)$ and $q(t) = (b_0 + b_1 t + b_2 t^2 + b_3 t^3)$ and s is any scalar, then

$$p(t) + q(t) = (a_0 + b_0) + (a_1 + b_1)t + (a_2 + b_2)t^2 + (a_3 + b_3)t^3$$
$$s\, p(t) = (sa_0) + (sa_1)t + (sa_2)t^2 + (sa_3)t^3$$

EXERCISES 4.5

2. We abbreviate $(a_1, a_2, \ldots, a_n, \ldots)$ as (a_i). Then the first axiom holds because

$$(a_i) + ((b_i) + (c_i)) = (a_i) + (b_i + c_i)$$
$$= (a_i + (b_i + c_i))$$
$$= ((a_i + b_i) + c_i)$$

$$= (a_i + b_i) + (c_i)$$
$$= ((a_i) + (b_i)) + (c_i)$$

9. The zero vectors for these spaces are (a) $(0, 0)$; (b) $0t + 0t^2 + 0t^3 + 0t^4$;

(c) $\begin{pmatrix} 0 & 0 & 0 \\ 0 & 0 & 0 \end{pmatrix}$; (e) $0 : \mathbb{R} \to \mathbb{R}$ defined by $0(t) = 0$; (f) $0 : \mathbb{N} \to \mathbb{R}$ defined by $0(n) = 0$.

EXERCISES 4.6

1. Amounts of food ingredients are commonly given as amounts per unit weight. Other alternatives are to give the amounts as percentages of the weight, or as absolute weights for the particular package of food. All of these alternatives are easily interconvertible, so we may fix on percentages as our standard. Thus if the ingredients are $F_1, \ldots, F_n$, then a food product can be expressed as a sum $\sum_{i \in \mathbf{n}} c_i F_i$, where $\sum_{i \in \mathbf{n}} c_i = 1$. Since the various food ingredients are not precisely defined and since the ingredient lists are usually incomplete, the linear combination $\sum_{i \in \mathbf{n}} c_i F_i$ must in practice be treated as approximative information.

3. (a) All of $\mathbb{R} \times \mathbb{R}$; (b) All of $\mathbb{R}^2$; (c) The "line" in $\mathbb{R}^{2 \times 2}$ given by

$$A = t \begin{pmatrix} 1 & 2 \\ 3 & 4 \end{pmatrix}$$

with $t \in \mathbb{R}$.

5. (a) Independent; (b) Dependent; (c) Independent; (d) Dependent.

6. (a) Invertible; (b) Not invertible.

7. (a) Independent; (b) Independent; (c) Dependent.

EXERCISES 4.7

1. Any set $\{a\}$ with $a \in \mathbb{R}$ and $a \neq 0$ is a basis.

3. The set spans if and only if

$$x_1 \begin{pmatrix} 1 & 2 \\ 3 & 4 \end{pmatrix} + x_2 \begin{pmatrix} 2 & 0 \\ 0 & 5 \end{pmatrix} + x_3 \begin{pmatrix} 0 & 2 \\ 5 & 0 \end{pmatrix} + x_4 \begin{pmatrix} 1 & 1 \\ 1 & 1 \end{pmatrix} = \begin{pmatrix} a & b \\ c & d \end{pmatrix}$$

has a solution for all a, b, c, d. This is so if and only if the linear system

$$\begin{cases} x_1 + 2x_2 + x_4 = a \\ 2x_1 + 2x_3 + x_4 = b \\ 3x_1 + 5x_2 + x_4 = c \\ 4x_1 + 5x_2 + x_4 = d \end{cases}$$

is consistent. Moreover, the coefficient matrix

$$\begin{pmatrix} 1 & 2 & 0 & 1 \\ 2 & 0 & 2 & 1 \\ 3 & 0 & 5 & 1 \\ 4 & 5 & 0 & 1 \end{pmatrix}$$

is invertible. Therefore, the system has a unique solution. The uniqueness of the solution shows that the set is linearly independent.

6. $\{(108, -28, -57, 20)\}$.

7. A basis is the set $\{f\}$ consisting of the function $f : \mathbb{R} \to \mathbb{R}$ defined by $f(t) = c^{-t}$. See Example 4.10.

EXERCISES 4.8

1. `Map[mf[Table[{KroneckerDelta[i-#]},{i,4}]]&,Range[4]]`.

4. `Map[mf[Table[KroneckerDelta[#+3-(3i+j)],{i,2},{j,3}]]&,Range[6]]`.

5. The virtue of the **KroneckerDelta** function as given in *MATHEMATICA* is that it allows us to construct the characteristic function of various sets of entries in a matrix. Here are two examples.
(a) The diagonal entries in a matrix can be selected by defining
`kdelta[i_,j_]:=KroneckerDelta[i-j]`
(b) The entries above or on the main diagonal can be selected by
`uptriangle[i_,j_]:=KroneckerDelta[Max[i,j]-j]`.

EXERCISES 4.9

2. The absence of free variables guarantees four pivots in the coefficient matrix, which, by Theorem 4.3, implies the linear independence of the columns. The invertibility of the matrix is also implied, and so the columns span $\mathbb{R}^4$.

EXERCISES 4.10

1. `invertgen[n_]:= If[invertiblem[A=`
`Array[Random[Integer,{-9,9}]&,{n,n}]],mf[A],invertgen[n]]`.

3. Both spanning and linear independence are guaranteed by the fact that A has a pivot in each column.

EXERCISES 4.11

2. The set $\{(1, 2, 3), (1, 0, 0), (0, 1, 0)\}$ is a possible extension.

4. $\{(1, 2, 3)\}$ and $\{(1, 2, 3), (1, 0, 0), (0, 1, 0)\}$.

8. $\{1, t, t^2\}$.

14. The vector $(1, 0, 0)$ is not included in the span of $\mathcal{S}$.

15. The set $\{1\}$ spans $\mathbb{R}$.

EXERCISES 4.12

1. $(3, 7, -1, 0)$, $(4, 0, 0, 1)$, $(0, 0, 21, 0)$, $(1, 1, 1, 0)$, and $(0, 2, 0, 3)$ with respect to the standard basis $\{1, t, t^2, t^3\}$.

2. If we define $\mathbf{b}_k$ by `Array[Function[{i,j},KroneckerDelta[k+3-(3i+j)]],` `{2,3}]]` and let $\mathcal{B} = \{\mathbf{b}_1, \ldots, \mathbf{b}_6\}$, then the coordinate transformation is determined by $([\mathbf{b}_k])_{\mathcal{B}} = \mathbf{e}_k$.

4. $\mathbb{R}[t]$ has an infinite standard basis $\{1, t, t^2, \ldots, t^n, \ldots\}$. Therefore, the notion of a coordinate vector is not covered by Definition 4.8. Although the degree of each polynomial is finite, no n is large enough to allow us to represent all polynomials using the set $\{1, t, t^2, \ldots, t^n\}$.

EXERCISES 4.13

3. (a) `Solve[2x-y+3z==0,{x,z}]` yields $\{\{x \rightarrow y/2 - 3z/2\}\}$. If $y = z = 2$, then $(-2, 2, 2)$ is a solution. If $y = 0$ and $z = 2$, then $(-3, 0, 2)$ is a solution. The set $\mathcal{B} = \{(-2, 2, 2), (-3, 0, 2)\}$ is a basis.

5. The set of polynomials of even degree is not a vector space. Let $p(t) = t + t^2$ and $q(t) = t - t^2$. Then $p(t) + q(t) = 2t$ is a polynomial of odd degree.

10. (a) Let $U = \{(x, y) : x = y\}$ and $V = \{(x, y) : x = 2y\}$. Then $(1, 1) \in U$ and $(2, 1) \in V$. But $(1, 1) + (2, 1) = (3, 1)$ belongs to neither U nor V.

EXERCISES 4.14

8. The four subspaces are generated by the following bases. The null space is the span of $\{(-2, 1, 0\}$, the column space is the span of $\{(1, 0), (0, 1)\}$, the left null space is the span of $\emptyset$, and the row space is the span of $\{(1, 2, 0), (0, 0, 1)\}$. Thus the left null space is the zero subspace.

10. The two spaces have the following bases. The column space of A is the span of the set $\{(1, 0, 3), (0, 1, 0)\}$ and the null space of B is the span of the set $\{(-1, 0, 1), (0, 1, 0)\}$. Since the vectors $(1, 0, 3)$ and $(-1, 0, 1)$ are linearly independent, the intersection of the two spaces is the span of $(0, 1, 0)$. Geometrically, it corresponds to the y-axis in $\mathbb{R}^3$.

11. Let A be a 3×4 matrix in row echelon form, and consider the column rank of A. The rank of A corresponds to the number of pivot columns of A. If A has no pivot columns, A must be the zero matrix. If A has rank 1, it has one pivot column; if A has rank 2, it has two pivot columns; and if it has rank 3, it has three pivot columns.

EXERCISES 4.15

1. The vectors $(1, 3)$ and $(1, -7)$ are linearly independent and span $\mathbb{R}^2$. The first one satisfies $y = 3x$ and the second one $y = -7x$.

3. The following are disjoint subspaces of $\mathbb{R}^3$. The row space of A is the span of the set $\{(1, 0, 1)\}$ and the null space of A is the span of the set $\{(-1, 0, 1), (0, 1, 0)\}$.

EXERCISES 4.16

4. The set consists of all pairs

$$(a + ib) \begin{pmatrix} 1 \\ 0 \end{pmatrix} + (c + id) \begin{pmatrix} 0 \\ 1 \end{pmatrix} = \begin{pmatrix} a + ib \\ c + id \end{pmatrix}$$

of complex numbers. Hence $\mathrm{span}(S) = \mathbb{C}^2$.

5. The set consists of pairs

$$(a) \begin{pmatrix} 1 \\ 0 \end{pmatrix} + (c) \begin{pmatrix} 0 \\ 1 \end{pmatrix} = \begin{pmatrix} a \\ c \end{pmatrix}$$

of real numbers. Hence $\mathrm{span}(S) = \mathbb{R}^2$, considered as a subspace of $\mathbb{C}^2$.

6. The set consists of all pairs

$$a \begin{pmatrix} 1 \\ 0 \end{pmatrix} + b \begin{pmatrix} i \\ 0 \end{pmatrix} + c \begin{pmatrix} 0 \\ 1 \end{pmatrix} + d \begin{pmatrix} 0 \\ i \end{pmatrix} = \begin{pmatrix} a + ib \\ c + id \end{pmatrix}$$

of complex numbers. Hence span S coincides with the space $\mathbb{C}^2$ described in Exercise 4.

CHAPTER 5

EXERCISES 5.1

2. By Theorem 5.2, T is linear if and only if $h = k = 0$.

3. For all functions except (c) we have $T(\mathbf{0}) \neq \mathbf{0}$. For (c) $T(2\mathbf{x}) \neq 2T(\mathbf{x})$, hence none is linear.

5. All functions except (d) and (e) can be written as matrix multiplications and thus are linear. For (d), $T(2\mathbf{x}) \neq 2T(\mathbf{x})$, and for (e), $T(\mathbf{0}) \neq \mathbf{0}$, so neither of these is linear.

7. $\det(2A) \neq 2 \det A$, so the determinant in (a) is not linear. The rest are all linear.

12. The points are the vertices of a vertical isosceles triangle with base 6 and height 5, and the center of the base translated from the origin to the point $(4, 5)$.

17. This function can be written as matrix multiplication

$$f(\mathbf{x}) = \left(\begin{array}{ccc} 5 & 2 & -7 \end{array}\right)\mathbf{x}$$

and is therefore linear. Here is the *MATHEMATICA* check.
```
Block[{f,A=(5 2 -7),X,x,Y,y,a,b},f[vect_]:=Part[A.vect,1];
X=Array[x,3]; Y=Array[y,3];Simplify[f[a*X+b*Y]==a*f[X]+b*f[Y]]].
```
19. `Block[{f,X0,x0,X,x,Y,y,a,b},X0=Array[x0,5];X=Array[x,5];`
`Y=Array[y,5]; f[vect_] :=X0.vect;`
`Simplify[f[a*X+b*Y]==a*f[X]+b*f[Y]]].`

EXERCISES 5.2

2. Define $T_1, T_2 : \mathbb{R}^2 \to \mathbb{R}^2$ by

$$T_1(\mathbf{x}) = \left(\begin{array}{cc} 1 & 2 \\ 3 & 4 \end{array}\right)\mathbf{x} \quad T_2(\mathbf{x}) = \left(\begin{array}{cc} -3 & 2 \\ 3 & 0 \end{array}\right)\mathbf{x}$$

5. The claim is false. Consider the transformation $T : \mathbb{R} \to \mathbb{R}^2$ defined by

$$T(x) = \left(\begin{array}{c} x \\ 0 \end{array}\right)$$

6. The claim is false. Consider the transformation $T : \mathbb{R}^2 \to \mathbb{R}$ defined by

$$T\left(\begin{array}{c} x \\ y \end{array}\right) = x$$

8. T is given by the $m \times n$ matrix

$$A = \left(\begin{array}{ccc} T(\mathbf{e}_1) & \cdots & T(\mathbf{e}_n) \end{array}\right)$$

as $T(x) = Ax$. The image of T is spanned by the columns of A so that $n \geq m$. Applying the same argument to T^{-1}, we have $m \geq n$. Hence $m = n$.

9. The image of T is the subspace V of $\mathbb{R}^m$ spanned by the columns of

$$A = \left(\begin{array}{ccc} T(\mathbf{e}_1) & \cdots & T(\mathbf{e}_n) \end{array}\right)$$

Since T is one-one, it preserves linear independence and the set

$$\mathcal{B} = \{\mathbf{b}_1 = T(\mathbf{e}_1), \ldots, \mathbf{b}_n = T(\mathbf{e}_n)\}$$

is therefore a basis for V. The result then follows from Exercise 7.

12. $\text{Hom}(V, W)$ is a real vector space provided the vector space operations are induced by the operations on $W : (f + g)(\mathbf{x}) = f(\mathbf{x}) + g(\mathbf{x})$ and $(af)(\mathbf{x}) = af(\mathbf{x})$. With these definitions, $\text{Hom}(V, W)$ is closed under addition and scalar multiplication. The vector space axioms hold because they hold in W.

EXERCISES 5.3

1. Since T is linear, we have $T(\mathbf{x}) = A\mathbf{x}$, with

$$A = \left(\begin{array}{ccc} T(\mathbf{e}_1) & \cdots & T(\mathbf{e}_n) \end{array} \right)$$

This means that T is singular if and only if A is singular. Since T is not one-one, $\det A = 0$. Therefore, A is not invertible.

2. $\det A = \det B = 0$. The transformations are therefore singular. $\det(A + B) = 5$. Hence T_{A+B} is nonsingular.

3. Let A and B be the $n \times n$ matrices

$$A = \left(\begin{array}{cccc} \mathbf{e}_1 & \mathbf{0} & \cdots & \mathbf{0} \end{array} \right) \quad \text{and} \quad B = \left(\begin{array}{cccc} \mathbf{0} & \mathbf{e}_2 & \cdots & \mathbf{e}_n \end{array} \right)$$

Then $I = T_A + T_B$.

6. This follows from the fact that if A is a square matrix and $\det A \neq 0$, then $\det A^n = (\det A)^n \neq 0$.

EXERCISES 5.4

1. The standard matrices are the following:

(a) $\begin{pmatrix} 4 & 8 \\ 2 & 3 \end{pmatrix}$ (b) $\begin{pmatrix} 4 & 1 & 8 \\ 2 & 1 & 3 \\ 1 & 1 & 4 \end{pmatrix}$ (c) $\begin{pmatrix} 4 & 0 \\ \frac{5}{3} & \frac{1}{3} \end{pmatrix}$ (d) $\begin{pmatrix} \frac{17}{3} & \frac{61}{9} & \frac{1}{3} \\ \frac{11}{3} & \frac{46}{9} & \frac{1}{3} \\ \frac{8}{3} & \frac{28}{9} & \frac{1}{3} \end{pmatrix}$

The following *MATHEMATICA* routine confirms these results.

```
basis2std[basis_,vals_]:=If[SquareMatrixQ[basis]&&(Det[basis]!=0),
vals.Inverse[basis],''Error.''];stdrelbases[std_,basis1_,basis2_]:=
Inverse[basis2].(std.basis1).
```

3. $\begin{pmatrix} -\frac{1}{2} & 0 \\ \frac{1}{2} & \frac{1}{3} \end{pmatrix}$.

5. $\begin{pmatrix} -1 & -1 & 4 \\ 0 & 0 & 4 \\ 0 & 1 & 2 \end{pmatrix}$.

8. (a) $\begin{pmatrix} 0 & 1 & \frac{3}{5} \\ \frac{5}{4} & \frac{1}{4} & \frac{1}{4} \\ 0 & 0 & -\frac{1}{2} \end{pmatrix}$; (b) $\begin{pmatrix} -\frac{1}{5} & \frac{4}{5} & \frac{4}{25} \\ 1 & 0 & \frac{6}{5} \\ 0 & 0 & -2 \end{pmatrix}$.

EXERCISES 5.5

4. The product

$$\begin{pmatrix} 1 & 0 \\ 7 & 1 \end{pmatrix} \begin{pmatrix} 1 & 0 \\ 0 & -6 \end{pmatrix} \begin{pmatrix} 1 & 2 \\ 0 & 1 \end{pmatrix} = E_1 E_2 E_3$$

is the required decomposition. T_{E_1} is a horizontal shear in which the first coordinate is augmented by twice the second coordinate. The effect of T_{E_2} is to multiply the second coordinate by 6 and then reflect across the horizontal axis. T_{E_3} is a vertical shear in which the second coordinate is augmented by 7 times the first coordinate.

5. From one point of view the geometric effect of these transformations is nil since no shapes are distorted. Only the orientation of all figures is affected. This may also be expressed by saying that the effect is to relabel the coordinate axes without shifting any figures. Thus E_B reverses the roles of the first and third axes, while E_A reverses the roles of the second and third axes. The composition rotates the axes: third to first, first to second, and second to third.

EXERCISES 5.6

2. The inverse of T_A is T_B, where $B = A^{-1} = \frac{1}{4} \begin{pmatrix} -3 & 8 \\ 2 & -4 \end{pmatrix}$.

4. A is orthogonal if and only if $A^T = A^{-1}$.

EXERCISES 5.7

1. (a) The image is $\mathbb{R}^2$. The kernel is the two-dimensional subspace of $\mathbb{R}^4$ determined by the basis $\mathcal{B} = \{(30, 1, 0, -5), (10, -7, 5, 0)\}$. (b) The image is the two-dimensional subspace of $\mathbb{R}^4$ determined by the basis $\mathcal{B} = \{(1, 0, -2, 6), (0, 5, 7, 1)\}$. The kernel is $\{0\}$. (c) The matrix is invertible. Therefore, the image is $\mathbb{R}^3$. The kernel is $\{0\}$. (d) The set $\mathcal{B} = \{e_1, e_2, e_3\}$ is a basis for the image in $\mathbb{R}^4$. The kernel is $\{0\}$. (e) The image is the xy-plane in $\mathbb{R}^3$. The kernel is the subspace of $\mathbb{R}^4$ generated by the basis $\mathcal{B} = \{e_3, e_4\}$. (f) The image is the line $\mathbf{x} = t\mathbf{e}_1$, with $t \in \mathbb{R}$. The kernel is the subspace of $\mathbb{R}^4$ generated by the basis $\mathcal{B} = \{(9, 0, 0, -1), e_2, e_3\}$.

3. We take $T(\mathbf{x}) = \begin{pmatrix} -8 & 3 & 11 \\ 2 & 77 & 75 \\ 3 & 5 & 8 \end{pmatrix} \mathbf{x}$. There are infinitely many other possibilities.

6. The image is $\mathbb{R}_3[t]$. The kernel is the set of constant polynomials in $\mathbb{R}_4[t]$.

7. The image is the set of all polynomials $p(t)$ with $p(0) = 0$ of degree less than or equal to 4. The kernel is $\{0\}$.

8. If f is nonconstant, we can use the mean value theorem to show that $D(f) \neq 0$. This shows that ker D is the set of constant functions.

9. If $f \in \ker D^2$, then Exercise 8 shows that $f'(t) = a_1$, so that $f(t) = a_0 + a_1 t$. Therefore, the kernel of D^2 is the set of all linear functions $\mathbb{R} \to \mathbb{R}$.

11. The image is the subspace of $\mathbb{R}^6$ generated by the basis $\mathcal{B} = \{\mathbf{e}_1, \mathbf{e}_2, \mathbf{e}_3\}$. The kernel of T is the subspace of $\mathbb{R}^{2\times 3}$ generated by the basis

$$\mathcal{B} = \left\{ \begin{pmatrix} 0 & 0 & 0 \\ 1 & 0 & 0 \end{pmatrix}, \begin{pmatrix} 0 & 0 & 0 \\ 0 & 1 & 0 \end{pmatrix}, \begin{pmatrix} 0 & 0 & 0 \\ 0 & 0 & 1 \end{pmatrix} \right\}$$

EXERCISES 5.8

1. (a) rank $= 2$, nullity $= 2$. The dimension of the domain of T is 4. (b) rank $= 2$, nullity $= 1$. The dimension of the domain of T is 3. (c) rank $= 2$, nullity $= 1$. The dimension of the domain of T is 3. (d) rank $= 3$, nullity $= 0$. The dimension of the domain of T is 3. (e) rank $= 2$, nullity $= 2$. The dimension of the domain of T is 4. (f) rank $= 1$, nullity $= 3$. The dimension of the domain of T is 4. If the rank of T_A is n, then the columns of A form a basis for $\mathbb{R}^n$. Therefore, T_A is onto and hence an isomorphism by Theorem 5.27. Conversely, any isomorphism is onto. Therefore, the span of the columns of A must have dimension n.

4. If the nullity of T_A is 0, then T_A is one-one by Theorem 5.22 and an isomorphism by Theorem 5.27. Conversely, any isomorphism is one-one. Therefore, the nullity of T_A must be 0.

EXERCISES 5.9

1. The absolute value of $\frac{1}{2} \det \begin{pmatrix} 2 & 5 \\ 3 & -1 \end{pmatrix} = \frac{17}{2}$.

2. The absolute value of $2 \det \begin{pmatrix} 2 & 1 \\ 2 & -1 \end{pmatrix} = 8$.

3. `ListPlot[{{1,2},{5,-3},{12,2},{7,8},{1,2},{12,2}},`
`Prolog→AbsolutePointSize[3],PlotJoined→True,AxesOrigin → {0,0}].`

5. `Block[{A={{1,0,2},{0,1,0},{-6,0,1},{0,0,0}}},`
`colspace[mtx_]:=Block[{B=RowReduce[ Transpose[mtx]]},`
`Map[ mf[Partition[#,1]]&,`
`DeleteCases[ RowReduce[B], Array[0&,Length[First[B]]]]]];`
`Print[ ''The image of T`$_A$` is the subspace of ''`
`StringReplacePart[''R`^{n}` '', ToString[ Length[A]], {-3,-3}],`
`''with basis '',colspace[A]]].`

EXERCISES 5.10

3. If A is similar to the zero matrix $\mathbf{0}$, then for some invertible matrix P, we have $A = P\mathbf{0}P^{-1} = \mathbf{0}$.

4. If A is similar to the identity matrix I, then for some invertible matrix P, we have $A = PIP^{-1} = I$.

6. Consider the following matrices.

$$A = \begin{pmatrix} 3 & 2 \\ -1 & 1 \end{pmatrix}, \quad B = \tfrac{1}{42}\begin{pmatrix} -23 & -73 \\ 181 & 191 \end{pmatrix}, \quad C = \begin{pmatrix} 1 & 1 \\ 0 & 0 \end{pmatrix},$$

$$D = \begin{pmatrix} 1 & 0 \\ -\tfrac{8}{9} & 0 \end{pmatrix}, \quad P_1 = \begin{pmatrix} -3 & 5 \\ -\tfrac{8}{9} & -1 \end{pmatrix}, \quad P_2 = \begin{pmatrix} -9 & -9 \\ 8 & 0 \end{pmatrix}$$

It is easy to check that $B = P_1 A P_1^{-1}$ and $D = P_2 C P_2^{-1}$. Moreover, $A + C$ and $B + D$ are dissimilar by the determinant test, while AC and BD are dissimilar by the trace test. There are infinitely many other possibilities.

9. If $BA = PABP^{-1}$, then $BAP = PAB$. Since matrix multiplication is not commutative, we can construct easy examples for which this equality fails.

10.
```
getnonsim[]:=Block[{p={{j,k,l},{m,n,q},{r,s,t}},
a=randmtrx[3],b=randmatx[3],eqs=Map[#==0&,Flatten[p.(a.b)-(b.a).p]],
coeff=First[LinearEquationsToMatrices[eqs,Flatten[p]]]},
If[Det[coeff]!=0,{mf[a],mf[b]},Print[''0''];
getnonsim[]]]eqs=Map[#==0&,Flatten[p.(a.b)-(b.a).p]],
coeff=First[LinearEquationsToMatrices[eqs,Flatten[p]]]},
If[Det[coeff]!=0,{mf[a],mf[b]},Print[''0''];getnonsim[]]].
```

CHAPTER 6

EXERCISES 6.1

1.
```
Block[{rect},rect[x_,y_]:=ListPlot[{{0,y},{x,y},{x,0}},
PlotJoined → True];Show[{rect[1,1],rect[3,4],rect[-3,4],
rect[3,-4],rect[-3,-4]},
{AspectRatio→Automatic,PlotRange→{{-3.5,3.5},{-4.5,4.5}},
Ticks→{Range[-3,3],Range[-4,4]},Epilog→{Text[''C'',{0.5,0.5}],
Text[''T(C) for a)'',{1.5,3}],Text[''T(C) for b)'',{-1.5,3}],
Text[''T(C) for c)'',{1.5,-3}],Text[''T(C) for d)'',{-1.5,-3}]}}]].
```

2.
```
Block[{box},box[pnt_]:=WireFrame[Graphics3D[Cuboid[{0,0,0},
pnt]]];Show[{box[{1,1,1}],box[{3,4,5}],box[{-3,4,5}],
box[{3,-4,5}],box[{3,4,-5}]},
{Ticks→{Range[-3,3],Range[-4,4],Range[-5,5]},
Axes → True,AxesLabel→{''x'',''y'',''z''},Boxed → False }]].
```

3. That 5 and -2 are eigenvalues is clear from the definition. Sample eigenvectors for 5 are e_1 and $2e_1$. Sample eigenvectors for -2 are e_2 and $2e_2$.

5. This question is equivalent to asking whether the equations $Ax = -5x$ and so on have nontrivial solutions. This in turn is equivalent to asking whether the determinants $\det(A + 5I)$

and so on are equal to zero. Using this test, we conclude that -5, -4, and 3 are eigenvalues. The following *MATHEMATICA* routine performs the required check and calculation.

```
Block[{checkeigen,A={{0,0,60},{1,0,7},{0,1,-6}}},
checkeigen[x_]:=If[Det[A-x*IdentityMatrix[3]]==0,
Print[x,} is an eigenvalue.''],Print[x,} is not an eigenvalue.'']];
checkeigen[-5];checkeigen[-4];checkeigen[3]].
```

7. ```Block[{p=randinvert[3]},
Eigenvalues[p.{{0,0,60},{1,0,7},{0,1,-6}}.Inverse[p]]].```

8. *MATHEMATICA* produces the eigenvalues 1, 2, 3, 4, and 5. We then solve each linear system $A - \lambda I = 0$, setting the free variable equal to 1. We conclude by testing whether the five solutions form a basis.

```
Block[{A={{0,0,0,0,120},{1,0,0,0,-274},{0,1,0,0,225},{0,0,1,0,-
85},
{0,0,0,1,15}},evals=Eigenvalues[A],syssolve},
syssolve[mtx_,lambda_]:={0,0,0,0,1}-
Flatten[TakeColumns[RowReduce[mtx-lambda*id[5]],-1]];
Det[Map[syssolve[A,#]&,evals]]].
```

9. The eigenvalues are 1, 2, and 3. Hence we can tell immediately that the first two systems have only the trivial solution, whereas the last system has as solutions all eigenvectors of 3.

12. Let **x** be an eigenvector for λ. Then $T^2(\mathbf{x}) = T(T(\mathbf{x})) = \lambda^2\mathbf{x}$.

13. The possibility of negative eigenvalues settles the question. See Exercise 6(b).

16. $\det(A^T - \lambda I) = \det(A - \lambda I)^T = \det(A - \lambda I)$.

18. For each λ, the matrix $A - \lambda K$ is upper triangular, with zeros on the main diagonal. Hence $\det(A - \lambda I) = 0$.

EXERCISES 6.2

1. (a) $\begin{pmatrix} 1-t & 2 \\ 0 & -t \end{pmatrix}$; (b) $\begin{pmatrix} -t & -1 \\ 1 & -t \end{pmatrix}$; (d) $\begin{pmatrix} 1-t & 0 & -2 \\ 0 & 5-t & 7 \\ 0 & 0 & -t \end{pmatrix}$.

2. (a) `CharacteristicPolynomial[{{1,2},{0,0}},t].`
(b) `CharacteristicPolynomial[{{0,-1},{1,0}},t].`
(d) `CharacteristicPolynomial[{{1,0,-2},{0,5,7},{0,0,0}},t].`

3. The following command verifies that $c_A(A) = \mathbf{0}$, where A are the matrices in (a), (b), and (d).

```
Block[{matrices={{{1,2},{0,0}},{{0,-1},{1,0}},{{1,0,-2},
{0,5,7},{0,0,0}}},polys},polys=
Map[CharacteristicPolynomial[#,t]&, matrices];
MapThread[polyeval[#1,#2]&,{polys,matrices}]].
```

6. (a) $1 \pm 2\sqrt{2}$; (b) 1, 3.

8. The matrices

$$\begin{pmatrix} 1 & 0 & 0 \\ 0 & 2 & 0 \\ 0 & 0 & 3 \end{pmatrix} \quad \text{and} \quad \begin{pmatrix} 1 & -1 & 3 \\ 0 & 2 & 4 \\ 0 & 0 & 3 \end{pmatrix}$$

for example. The first has $\{(1,0,0), (0,1,0), (0,0,1)\}$ among its eigenvectors, whereas the second has $\{(1,0,0), (-1,1,0), (-1,8,2)\}$.

10. Suppose that λ is a nonzero eigenvalue of AB. Then $(AB)\mathbf{x} = \lambda\mathbf{x}$ for some nonzero vector $\mathbf{x}$. Since λ and $\mathbf{x}$ are both nonzero, the vector $\lambda\mathbf{x}$ is nonzero. Since $(AB)\mathbf{x} = A(B\mathbf{x}) \neq \mathbf{0}$, it follows that $B\mathbf{x} \neq \mathbf{0}$. Therefore,

$$(BA)(B\mathbf{x}) = (BAB)\mathbf{x} = B(AB\mathbf{x}) = B\lambda\mathbf{x} = \lambda(B\mathbf{x})$$

Hence λ is a nonzero eigenvalue of BA. Similarly the other way around. Now suppose that 0 is an eigenvalue of AB. Then AB is singular. Over the real numbers, this means that $\det AB = 0$. But $\det BA = \det AB$. Hence BA is singular. Hence 0 is an eigenvalue of BA. However, 0 is an eigenvalue of a matrix M if and only if M is singular.

EXERCISES 6.3

1. `Block[{matrices={{{1,2},{3,4}},{{97,50},{79,56}},{{0,2,1}, {1,0,0},{-1,0,3}}},polys},`
`polys=Map[CharacteristicPolynomial[#,t]&, matrices];`
`MapThread[polyeval[#1,#2]&,{polys,matrices}]].`

3. By the Cayley-Hamilton theorem, we have

$$-4{,}587{,}563{,}299i + 122{,}673{,}722A + 539{,}812A^2 + 17{,}584A^3 - 104A^4 - A^5 = 0$$

From this we conclude that

$$A^{-1} = \frac{1}{4{,}587{,}563{,}299}\left(122{,}673{,}722i + 539{,}812A + 17{,}584A^2 - 104A^3 - A^4\right)$$

4. Take $A = \begin{pmatrix} 1 & 0 & 0 \\ 0 & -2 & 0 \\ 0 & 0 & 4 \end{pmatrix}$, for example. Then $c_A(A^2) \neq 0$. Moreover, $c_{A^2}(A^2) \neq 0$.

This contradicts (b).

EXERCISES 6.4

2. `Block[{randhaus,lst},randhaus[n_]:=`
`Block[{u=Array[Random[Integer,{-5,5}]&,n]},`

```
id[n]-(2/(u.u))Transpose[{u}].{u}];
1st=Map[randhaus,{3,4,5}];Map[{mf[#],#==
Transpose[#],#==Inverse[#]}&,1st]].
Block[{randhess},randhess[n_]:=Array[If[#1<2+#2,
Random[Integer,{-5,5}],0]&,{n,n}];Map[mf[randhess[#]]&,{3,4,5}]].
```

EXERCISES 6.5

2. $n - 5$.

3. Subspaces are closed under scalar multiplication and $A^n\mathbf{x} = \lambda^n\mathbf{x}$.

4. If dim $E_\lambda = 3$, then $E_\lambda = \mathbb{R}^3$. Therefore, there exists a basis $\mathcal{B} = \{\mathbf{e}_1, \mathbf{e}_2, \mathbf{e}_3\}$ for $\mathbb{R}^3$ consisting of eigenvectors belonging to λ. Thus $A\mathbf{e}_i = \lambda\mathbf{e}_i = \lambda I\mathbf{e}_i$. Hence $A = \lambda I$.

5. The matrix $\begin{pmatrix} 2 & -1 & 0 \\ 3 & 2 & 0 \\ 0 & 0 & 3 \end{pmatrix}$, for example, has 3 as its only real eigenvalue. The eigenvectors all have the form $k\mathbf{e}_3$.

EXERCISES 6.6

1. `Map[MemberQ[Eigenvalues[#],0]&,{{{1,2},{3,0}},` `{{1,2,1},{3,0,2},{1,0,1}}}]`.

2. `Map[Eigenvalues[#]^(-1)&,{{{3,0},{4,-4}},` `{{8,4,-8},{0,7,2},{0,0,8}}}]`.

EXERCISES 6.7

1. (a) Distinct eigenvalues 0 and 1. Corresponding eigenvectors are $(0, 1)$ and $(1, 0)$. (b) Distinct eigenvalues -1 and 1. Corresponding eigenvectors are $(-1, 1)$ and $(1, 1)$. The routine `Map[Eigensystem,{{{1,0},{0,0}},{{0,1},{1,0}}}]` confirms this fact.

7. By Theorem 6.10, we have n linearly independent vectors in $\mathbb{R}^n$.

EXERCISES 6.8

1. Each system has a pair of distinct eigenvalues: (a) 0, 2; (b) 1, 3; (c) $2\pm\sqrt{26}$; (d) $-1\pm\sqrt{41}$.

2. The first three matrices have just one eigenvalue and the resulting eigenspaces are one-dimensional. The last matrix has no real eigenvalues.

3. (a) The matrix has eigenvalues 1 and 2. Moreover, the row reduced form of the matrix $(A - I)$ yields two free variables and therefore two linearly independent eigenvectors for a total of three. (b) Has the three distinct eigenvalues $2, \frac{1}{2}(1 \pm \sqrt{11})$. (c) Has eigenvalues 1 and 2. Moreover, the row reduced form of the matrix $(A - I)$ yields two free variables and therefore two linearly independent eigenvectors for a total of three.

4. (a) Has eigenvalues 1 and 2. Moreover, the row reduced form of the matrices $(A - I)$ and $(A - 2I)$ yield one free variable each for a total of two linearly independent eigenvectors. (b) Has eigenvalues 1 and 2. Moreover, the row reduced form of the matrices $(A - I)$ and $(A - 2I)$ yield one free variable each for a total of two linearly independent eigenvectors. (c) Has only the eigenvalue 2 and the row reduced form of the matrix $(A - 2I)$ yields one free variable and so only one linearly independent eigenvector.

7. (a) For all positive values of a there are two distinct eigenvalues, so the matrix is diagonalizable. If a is negative, there are none. If $a = 0$, then the only eigenvalue is 1 and the row reduced form of $(A - I)$ has only one free variable.

9. By Theorem 6.11, A is diagonalizable. Hence

$$A = PDP^{-1} = P(\lambda I)P^{-1} = \lambda PP^{-1} = \lambda I$$

11. Take $P = \begin{pmatrix} 12 & 8 & 6 \\ -7 & -6 & -5 \\ 1 & 1 & 1 \end{pmatrix}$, for example. Then $P^{-1}AP = \text{diag}(2, 3, 4)$.

12. The following *MATHEMATICA* routine will produce companion matrices.
```
compmtx[rt1_,rt2_,rt3_,rt4_]:=
Block[{p=Expand[(t-rt1)(t-rt2)(t-rt3)(t-rt4)]},
AppendRows[Array[If[(#1-1)!=#2,0,1]&,{4,3}],
Partition[polyiso[4,t^4-p],1]]].
```

EXERCISES 6.9

1. Any $\lambda \in \mathbb{R}$ is an eigenvalue with eigenvectors $ke^{\lambda t}$. There are no others. The dimension of the eigenspace is 1.

2. First solution: We apply Exercise 1 and observe that the only eigenvalue is 0, with the corresponding one-dimensional eigenspace of constant polynomials. By Theorem 6.11, D is not diagonalizable. Second solution: We apply the standard isomorphism to the basis $\{1, t, t^2\}$ and obtain the matrix transformation $T_A : \mathbb{R}^3 \to \mathbb{R}^3$ defined by

$$A = \begin{pmatrix} 0 & 1 & 0 \\ 0 & 0 & 2 \\ 0 & 0 & 0 \end{pmatrix}$$

The characteristic polynomial is λ^3. The only eigenvalue is 0 with eigenvectors $p(t) = k$.

EXERCISES 6.10

1. `Expand[(t-2)(t-3)(t-4)];A={{0,0,24},{1,0,-26},{0,1,9}};`
`X[k_]:=MatrixPower[A,k].{{a},{b},{c}};X[9].`

5. Each number in the sequence

$$1, 1, 2, 3, 5, 8, 13, 21, 34, 55, \ldots a, b, a + b, a + 2b, \ldots$$

is the sum of the two previous numbers. The following multiplication turns a, b into b, $a + b$.

$$\begin{pmatrix} 1 & 1 \\ 1 & 0 \end{pmatrix} \begin{pmatrix} b \\ a \end{pmatrix} = \begin{pmatrix} a + b \\ b \end{pmatrix}$$

EXERCISES 6.11

3. (a) Yes; (b) Yes; (c) No; (d) No.

4. (a) `A={{.5,.3},{.5,.7}};A.{x,y}-{{x},{y}};`
`Solve[{-0.5x+0.3y==0,0.5x-0.3y==0},{x,y}]; A.{0.6y,y}.`

7. (a) Yes; (b) No; (c) Yes; (d) Yes.

CHAPTER 7

EXERCISES 7.1

1. The one-norms are 7, 0, 7, 1, 20, and 20; the two-norms are 5, 0, 5, 1, $\sqrt{202}$, and $\sqrt{202}$; the infinity-norms are 4, 0, 4, 1, 11, and 11.
`Block[{vlst={{3,4},{0,0},{-3,4},{1,0},{-9,11},{-9,-11}}},`
`{Map[norm1,vlst],Map[norm2,vlst],Map[norminfinity,vlst]}].`

3. $\frac{1}{12}(3, 4, 5)$, $\frac{1}{12}(-3, 4, 5)$, $\frac{1}{2}(1, 0, 1)$, $(0, 0, 1)$, and $\frac{1}{25}(-9, -11, -5)$.
`Block[{vlst={{3,4,5},{-3,4,5},{1,0,1},{0,0,1},{-9,-11,-5}}},`
`Map[(1/norm1[#])*#&,vlst]].`

4. $\frac{1}{5\sqrt{2}}(3, 4, 5)$, $\frac{1}{5\sqrt{2}}(-3, 4, 5)$, $\frac{1}{\sqrt{2}}(1, 0, 1)$, $(0, 0, 1)$, and $\frac{1}{\sqrt{227}}(-9, -11, -5)$.
`Block[{vlst={{3,4,5},{-3,4,5},{1,0,1},{0,0,1},{-9,-11,-5}}},`
`Map[(1/norm2[#])*#&,vlst]].`

5. $\frac{1}{5}(3, 4, 5)$, $\frac{1}{5}(-3, 4, 5)$, $(1, 0, 1)$, $(0, 0, 1)$, and $\frac{1}{11}(-9, -11, -5)$.
`Block[{vlst={{3,4,5},{-3,4,5},{1,0,1},{0,0,1},{-9,-11,-5}}},`
`Map[(1/norminfinity[#])*#&,vlst]].`

EXERCISES 7.2

1. 6, 8, 11, 6, 3, and 2.

2. $3\sqrt{2}$, $\sqrt{34}$, $\sqrt{65}$, $\sqrt{26}$, $\sqrt{5}$, and $\sqrt{2}$.

3. 3, 5, 7, 5, 2, and 1.

7. The third axiom fails.

EXERCISES 7.3

1. The set AS of images of S under A consists of the vectors $A\mathbf{x}_1 = (18, 4)$, $A\mathbf{x}_2 = (-9, -17)$, $A\mathbf{x}_3 = (7, 1)$, $A\mathbf{x}_4 = (0, 0)$, and $A\mathbf{x}_5 = (19, -13)$. The one-norms of these vectors are $\|A\mathbf{x}_1\|_1 = 22$, $\|A\mathbf{x}_2\|_1 = 26$, $\|A\mathbf{x}_3\|_1 = 8$, $\|A\mathbf{x}_4\|_1 = 0$, $\|A\mathbf{x}_5\|_1 = 32$. Hence $\max\{22, 26, 8, 0, 32\} = 32$.

2. Using the set AS from Exercise 1, we get norms $\|A\mathbf{x}_1\|_2 = 2\sqrt{85}$, $\|A\mathbf{x}_2\|_2 = \sqrt{370}$, $\|A\mathbf{x}_3\|_2 = 5\sqrt{2}$, $\|A\mathbf{x}_4\|_2 = 0$, $\|A\mathbf{x}_5\|_2 = \sqrt{530}$. Hence $\max\left\{2\sqrt{85}, \sqrt{370}, 5\sqrt{2}, 0, \sqrt{530}\right\} = \sqrt{530}$.

3. Using the set AS from Exercise 1, we get $\|A\mathbf{x}_1\|_3 = 18$, $\|A\mathbf{x}_2\|_3 = 17$, $\|A\mathbf{x}_3\|_3 = 7$, $\|A\mathbf{x}_4\|_3 = 0$, $\|A\mathbf{x}_5\|_3 = 19$. Hence $\max\{18, 17, 7, 0, 19\} = 19$.

4. (a) The one-norm is $\|A\|_1 = \max\{3 + 1, 4 + 2\} = 6$. We use *MATHEMATICA* to compute the two-norm. The command
`Eigenvalues[Transpose[{{3,4},{-1,2}}].{{3,4},{-1,2}}]` yields

$$\left\{5\left(3 - \sqrt{5}\right), 5\left(3 + \sqrt{5}\right)\right\}$$

Thus $\|A\|_2 = \max\left\{\sqrt{15 + 5\sqrt{5}}\right\}$. Moreover, $\|A\|_\infty = \max\{3, 4, 1, 2\} = 4$ and $\|A\|_F = \sqrt{3^2 + 4^2 + 1^2 + 2^2} = \sqrt{30}$.

5. $\|A\|_F = \sqrt{a^2 + 2^2 + 3^2 + 4^2}$ and $\|B\|_F = \sqrt{b^2 + 2^2 + 3^2 + 4^2}$. Furthermore, $|a| = \sqrt{a^2}$ and $|b| = \sqrt{b^2}$. Therefore, $\sqrt{a^2} < \sqrt{b^2}$ if and only if $a^2 < b^2$.

EXERCISES 7.4

3. (a) 14; (b) 6; (c) -10; (d) 0.

6. (a) 174; (b) 50; (c) -154; (d) 20.

7. (a) -14; (b) -4.

9. (a) $174 \le \sqrt{31,119}$; (b) $50 \le 11\sqrt{123}$; (c) $154 \le 8\sqrt{451}$; (d) $20 \le \sqrt{2255}$.

10. For any vector $\mathbf{x}$ and any matrix $A = (a_{ij})$, the scalar $\langle \mathbf{x}, \mathbf{e}_k \rangle$ is the kth entry of $\mathbf{x}$, and $A\mathbf{e}_k$ is the kth column of A, whenever the products are defined. Therefore, $\langle A\mathbf{e}_j, \mathbf{e}'_i \rangle$ is the ith entry in the jth column.

EXERCISES 7.5

2. Let $A = \begin{pmatrix} a & b \end{pmatrix}$.
If $a = 0$, then

$$\begin{pmatrix} 1 & 0 \end{pmatrix} \begin{pmatrix} 0 & b \end{pmatrix}^T \begin{pmatrix} 0 & b \end{pmatrix} \begin{pmatrix} 1 \\ 0 \end{pmatrix} = 0$$

If $a \neq 0$, then

$$\begin{pmatrix} -y\frac{b}{a} & y \end{pmatrix} \begin{pmatrix} a & b \end{pmatrix}^T \begin{pmatrix} a & b \end{pmatrix} \begin{pmatrix} -y\frac{b}{a} \\ y \end{pmatrix} = 0$$

In both cases, we can therefore find a nonzero vector $\mathbf{x}$ for which $\mathbf{x}^T A \mathbf{x} = 0$. Hence A is not positive definite.

EXERCISES 7.6

1. (a) $\sqrt{41}$; (b) $\sqrt{5}$; (c) 0; (d) $\sqrt{41}$; (e) 1; (f) $\sqrt{41}$.

4. (a) $\sqrt{467}$; (b) $\sqrt{467}$; (c) $\sqrt{3}$; (d) $\sqrt{23}$; (e) $\sqrt{11}$; (f) $\sqrt{77}$.

5. Suppose that $\mathbf{x} = (x, y, z) \neq (0, 0, 0)$. Then

$$\begin{pmatrix} x \\ y \\ z \end{pmatrix}^T \begin{pmatrix} 3 & 0 & 1 \\ 0 & 5 & 0 \\ 1 & 0 & 6 \end{pmatrix} \begin{pmatrix} x \\ y \\ z \end{pmatrix}$$

equals $(3x + z)x + 5y^2 + (x + 6z)z = 3x^2 + 2xz + 5y^2 + 6z^2$. If either $x = 0$ or $z = 0$ or $xz > 0$, then $\mathbf{x}^T A \mathbf{x} > 0$. It remains to consider the case where $xz < 0$. At worst, $y = 0$. In that case,

$$\mathbf{x}^T A \mathbf{x} = 3x^2 + 2xz + 6z^2 = \left(\sqrt{3}x + \sqrt{6}z\right)^2 - \left(6\sqrt{2} - 2\right)xy > 0$$

Hence A is positive definite. The function $\mathbf{x}^T A \mathbf{x}$ is therefore a norm on $\mathbb{R}^3$. Applied to the vectors in Exercise 3, it yields (a) $\sqrt{523}$; (b) $\sqrt{411}$; (c) $\sqrt{3}$; (d) $\sqrt{23}$; (e) $\sqrt{11}$; (f) $\sqrt{83}$.

6. We define a *MATHEMATICA* function using the standard inner product on $\mathbb{R}^3$ and check Theorem 7.7.

```
Block[{prlst={{{1,2,3},{4,5,7}},{{-1,2,0},{4,5,7}},{{1,0,0},{4,5,7}},
{{0,-2,-3},{4,5,7}},{{1,2,3},{0,0,0}},{{4,5,7},{-4,-5,-7}}},chkthm7},
chkthm7[x_,y_]:=(x+y).(x+y)+(x-y).(x-y)==2(x.x+y.y);
Map[Apply[chkthm7,#]&,prlst]].
```

The returned values are {True, True, True, True, True, True}.

7. As a variation to the solution in Exercise 6, we define a *MATHEMATICA* returning the relevant pairs of values. In each case, the first value of a pair is less than or equal to the second.

```
Block[{prlst={{{1,2,3},{4,5,7}},{{-1,2,0},{4,5,7}},{{1,0,0},{4,5,7}},
{{0,-2,-3},{4,5,7}},{{1,2,3},{0,0,0}},
{{4,5,7},{-4,-5,-7}}},chktri},chktri[x_,y_]:=
{norm2[x+y],norm2[x]+norm2[y]};N[Map[Apply[chktri,#]&,prlst]]].
```

EXERCISES 7.7

1. (a) `x={1,2};y={4,5};Cosine=Dot[x,y]/(Sqrt[Dot[x,x]]Sqrt[Dot[y,y]]).`

2. The command `N[ArcCos[Cosine]]` computes the angle of the cosine in Exercise 1(a).

5. `A={{4,1},{-4,-1}};B={{1,2},{-2,7}};`
`a=Tr[Transpose[B].A];b=Tr[Transpose[A].A];`
`c=Tr[Transpose[B].B];Cosine=a/(b c).`

EXERCISES 7.8

2. `<<Statistics`MultiDescriptiveStatistics`;`
`A ={{1,-1},{2,1},{3,7},{4,5}};MatrixForm[CorrelationMatrix[A]].`

4. (a) `P={{8,4,-5},{-5,3,-5}};Q ={{8,5,-1},{0,3,-4}};`
`p=1/6(8+4-5-5+3-5); q=1/6((8+5-1+3-4));`
`m[i_,j_]:=q;Av(Q)=Array[m,{2,3}];R=Q-Av(Q);`
`a=Tr[Transpose[P].R];b=Sqrt[Tr[Transpose[P].P]];`
`c=Sqrt[Tr[Transpose[R].R]];r=a/(b c).`

EXERCISES 7.9

1. (a) $\begin{pmatrix} u & v \end{pmatrix} \begin{pmatrix} 6 & 5 \\ 5 & 8 \end{pmatrix} \begin{pmatrix} x \\ y \end{pmatrix}$; (b) $\begin{pmatrix} u & v \end{pmatrix} \begin{pmatrix} 6 & 8 \\ 5 & 8 \end{pmatrix} \begin{pmatrix} x \\ y \end{pmatrix}$;

(c) $\begin{pmatrix} u & v \end{pmatrix} \begin{pmatrix} -6 & 8 \\ 5 & -8 \end{pmatrix} \begin{pmatrix} x \\ y \end{pmatrix}$.

2. (a) $f(u, v, x, y) = 6xu + 8yv$; (b) $f(u, v, x, y) = 6xu - xv - yu + 8yv$.

5. A quadratic form on $\mathbb{R}^2$ cannot have nonzero linear terms, so (a), (c), and (f) are immediately ruled out. The remaining functions are quadratic forms determined by the following matrices.

(b) $\begin{pmatrix} 0 & \frac{1}{2} \\ \frac{1}{2} & 0 \end{pmatrix}$; (d) $\begin{pmatrix} 3 & 1 \\ 1 & -1 \end{pmatrix}$; (e) $\begin{pmatrix} 0 & 0 \\ 0 & 5 \end{pmatrix}$; (g) $\begin{pmatrix} 0 & 0 \\ 0 & 0 \end{pmatrix}$; (h) $\begin{pmatrix} 1 & \frac{1}{2} \\ \frac{1}{2} & 0 \end{pmatrix}$.

6. (a) $f(x, y) = (x_1, x_2) \begin{pmatrix} 1 & 0 \\ 0 & 1 \end{pmatrix} \begin{pmatrix} y_1 \\ y_2 \end{pmatrix} = x_1y_1 + x_2y_2$;

(b) $f(x, y) = (x_1, x_2) \begin{pmatrix} 3 & 0 \\ 0 & 2 \end{pmatrix} \begin{pmatrix} y_1 \\ y_2 \end{pmatrix} = 3x_1y_1 + 2x_2y_2$;

(d) $f(x, y) = (x_1, x_2) \begin{pmatrix} 1 & 0 \\ 0 & 0 \end{pmatrix} \begin{pmatrix} y_1 \\ y_2 \end{pmatrix} = x_1y_1$.

EXERCISES 7.10

1. (a) $D = \begin{pmatrix} -3 - 2\sqrt{2} & 0 \\ 0 & -3 + 2\sqrt{2} \end{pmatrix}$ and $Q = \begin{pmatrix} \frac{1+\sqrt{2}}{\sqrt{2(2+\sqrt{2})}} & \frac{1-\sqrt{2}}{\sqrt{4-2\sqrt{2}}} \\ \frac{1}{\sqrt{2(2+\sqrt{2})}} & \frac{1}{\sqrt{4-2\sqrt{2}}} \end{pmatrix}$.

EXERCISES 7.11

1. (a) $-2 - 30i$; (b) $-37 + 5i$.

2. (a) -1; (b) $-30 - 18i$.

4. (a) $\sqrt{26}$ and $3\sqrt{10}$; (b) $\sqrt{67}$ and $4\sqrt{11}$.

6. We define two auxiliary *MATHEMATICA* functions that compute a complex inner product and the associate complex norm:
```
cip[x_,y_]:=x.Conjugate[y];norm2c[x_]:=Sqrt[cip[x,x]].
```
Using these functions, we get the following distance: (a) $\sqrt{118}$; (b) $\sqrt{303}$.

CHAPTER 8

EXERCISES 8.1

5. In each case, the vectors are orthogonal with respect to the usual inner product. By Theorem 8.2, they are therefore linearly independent.

EXERCISES 8.2

1. (a) $\frac{9}{41}(4, 5)$, $\frac{1}{41}(5, -4)$; (b) $\frac{27}{41}(4, 5)$, $\frac{38}{41}(-5, 4)$; (c) $\frac{19}{41}(4, 5, 0)$, $\frac{1}{41}(-35, 28, 41)$; (d) $\frac{8}{113}(1, 9, -12)$, $\frac{1}{113}(105, 267, 209)$.

2. (a) $(1, 0)$; (b) $(0, 2)$; (c) Assuming an orthogonal basis, we obtain the vector expressed in that basis.

4. (a) $(1, 0, 0)$; (b) $(1, 2, 0)$; (c) $(1, 0, 3)$.

5. (a) $(1, 0, 0)_B$; (b) $(1, 2, 0)_B$; (c) $(1, 0, 3)_B$.

EXERCISES 8.3

2. `GramSchmidt[{{1,1},{1,-1}},Normalized→False].`

3. `Block[{m={{3,1},{2,-1}},f},m=Transpose[m].m;`
`Print[mf[m],Eigenvalues[m]];f[x_,y_]:=x.m.y;`
`GramSchmidt[{{1,1},{1,-1}},InnerProduct→f]].`

4. $\left\{ (1, 1, 0), (1, -1, 2), \frac{2}{3}(1, -1, -1) \right\}.$

5. `GramSchmidt[{{1,1,0},{1,-1,2},{2,0,0}},Normalized → False].`

6. `Block[{m={{3,1,0},{2,-1,0},{0,0,1}},f},m=Transpose[m].m;`
`Print[mf[m],Eigenvalues[m]];f[x_,y_]:=x.m.y;`
`GramSchmidt[{{1,1,0},{1,-1,2},{2,0,0}}, InnerProduct → f]].`

EXERCISES 8.4

1. $(1, -1) + \frac{1}{5}(1, 2) = \frac{1}{5}(6, -3)$. So the desired basis is $\{(1, 2), (2, -1)\}$. The associated orthonormal basis is $\left\{\frac{1}{\sqrt{5}}(1, 2), \frac{1}{\sqrt{5}}(2, -1)\right\}$.

3. $(1, -1, 0) + \frac{1}{6}(1, 2, 1) = \frac{1}{6}(7, -4, 1)$. We therefore begin with $\{(1, 2, 1), (7, -4, 1)\}$. Furthermore, $(0, 1, 0) - \frac{1}{3}(1, 2, 1) + \frac{2}{33}(7, -4, 1) = \frac{1}{11}(3, 3, -9) = \frac{1}{11}(1, 1, -3)$. Thus $\{(1, 2, 1), (7, -4, 1), (1, 1, -3)\}$ is the desired orthogonal basis. The associated orthonormal basis is $\left\{\frac{1}{\sqrt{6}}(1, 2, 1), \frac{1}{\sqrt{66}}(7, -4, 1), \frac{1}{\sqrt{11}}(1, 1, -3)\right\}$.

EXERCISES 8.5

1. Let $\mathcal{B} = \{\mathbf{x} = (a, b), \mathbf{y} = (c, d)\}$ be the required orthonormal basis. Then

$$\langle \mathbf{x}, \mathbf{y} \rangle = \begin{pmatrix} a \\ b \end{pmatrix}^T \begin{pmatrix} 8 & 0 \\ 0 & 3 \end{pmatrix} \begin{pmatrix} c \\ d \end{pmatrix} = 8ac + 3bd = 0$$

$$\langle \mathbf{x}, \mathbf{x} \rangle = \begin{pmatrix} a \\ b \end{pmatrix}^T \begin{pmatrix} 8 & 0 \\ 0 & 3 \end{pmatrix} \begin{pmatrix} a \\ b \end{pmatrix} = 8a^2 + 3b^2 = 1$$

$$\langle \mathbf{y}, \mathbf{y} \rangle = \begin{pmatrix} c \\ d \end{pmatrix}^T \begin{pmatrix} 8 & 0 \\ 0 & 3 \end{pmatrix} \begin{pmatrix} c \\ d \end{pmatrix} = 8c^2 + 3d^2 = 1$$

If we solve these equations for a, b, c, and d, we get

$$\mathbf{x} = \begin{pmatrix} \frac{1}{\sqrt{8}} \\ 0 \end{pmatrix}, \quad \mathbf{y} = \begin{pmatrix} 0 \\ \frac{1}{\sqrt{3}} \end{pmatrix}, \quad \mathbf{u} = \begin{pmatrix} 1 \\ 1 \end{pmatrix}, \quad \mathbf{v} = \begin{pmatrix} 2 \\ 5 \end{pmatrix}$$

and

$$\langle \mathbf{u}, \mathbf{v} \rangle = \begin{pmatrix} 1 \\ 1 \end{pmatrix}^T \begin{pmatrix} 8 & 0 \\ 0 & 3 \end{pmatrix} \begin{pmatrix} 2 \\ 5 \end{pmatrix} = 31$$

We verify that $\langle \mathbf{u}, \mathbf{v} \rangle$ is a dot product if $\mathbf{u}$ and $\mathbf{v}$ are written in the basis $\mathcal{B}$.

1. $\mathbf{u} = r \begin{pmatrix} \frac{1}{\sqrt{8}} \\ 0 \end{pmatrix} + s \begin{pmatrix} 0 \\ \frac{1}{\sqrt{3}} \end{pmatrix} = \begin{pmatrix} \frac{1}{4}r\sqrt{2} \\ \frac{1}{3}s\sqrt{3} \end{pmatrix} = \begin{pmatrix} 1 \\ 1 \end{pmatrix}$. Therefore, $r = 2\sqrt{2}$ and $s = \sqrt{3}$.

2. $\mathbf{v} = p \begin{pmatrix} \frac{1}{\sqrt{8}} \\ 0 \end{pmatrix} + q \begin{pmatrix} 0 \\ \frac{1}{\sqrt{3}} \end{pmatrix} = \begin{pmatrix} \frac{1}{4}p\sqrt{2} \\ \frac{1}{3}q\sqrt{3} \end{pmatrix} = \begin{pmatrix} 2 \\ 5 \end{pmatrix}$. Therefore, $p = 4\sqrt{2}$ and $q = 5\sqrt{3}$.

3. $[\mathbf{u}]_B = \begin{pmatrix} r \\ s \end{pmatrix} = \begin{pmatrix} 2\sqrt{2} \\ \sqrt{3} \end{pmatrix}$ and $[\mathbf{v}]_B = \begin{pmatrix} p \\ q \end{pmatrix} = \begin{pmatrix} 4\sqrt{2} \\ 5\sqrt{3} \end{pmatrix}$.

4. $[\mathbf{u}]_B^T [\mathbf{v}]_B = \begin{pmatrix} 2\sqrt{2} \\ \sqrt{3} \end{pmatrix}^T \begin{pmatrix} 4\sqrt{2} \\ 5\sqrt{3} \end{pmatrix} = 31.$

EXERCISES 8.6

1. (a) $\begin{pmatrix} 1 & 1 \\ -1 & 0 \\ 0 & 2 \end{pmatrix} = \begin{pmatrix} \frac{1}{2}\sqrt{2} & \frac{1}{6}\sqrt{2} & \frac{2}{3} \\ -\frac{1}{2}\sqrt{2} & \frac{1}{6}\sqrt{2} & \frac{2}{3} \\ 0 & \frac{2}{3}\sqrt{2} & -\frac{1}{3} \end{pmatrix} \begin{pmatrix} \sqrt{2} & \frac{1}{2}\sqrt{2} \\ 0 & \frac{3}{2}\sqrt{2} \\ 0 & 0 \end{pmatrix}$.

A ={{1,1},{-1,0},{0,2}}.{Q,R}=QRDecomposition[A].Dot[Q[[1]],Q[[2]]].

2. (a) The determinant is zero. The matrix does not have four linearly independent columns.
(b) Since $n > m$, the matrix cannot have n linearly independent columns.

5. $A_1 = \begin{pmatrix} 1 & 2 \\ 0 & -1 \end{pmatrix}$, $A_2 = \begin{pmatrix} 1 & -2 \\ 0 & -1 \end{pmatrix}$, $A_3 = \begin{pmatrix} 1 & 2 \\ 0 & -1 \end{pmatrix}$.

EXERCISES 8.7

1. (a) $1 = \det I = \det A \det A^T = \det A \det A = d^2$. Hence $d = \pm 1$.

2. (a) $\begin{pmatrix} \frac{1}{\sqrt{2}} \\ \frac{1}{\sqrt{2}} \end{pmatrix}^T \begin{pmatrix} -\frac{1}{\sqrt{2}} \\ \frac{1}{\sqrt{2}} \end{pmatrix} = 0$; $\sqrt{\left(\frac{1}{\sqrt{2}}\right)^2 + \left(\frac{1}{\sqrt{2}}\right)^2} = 1$; $\sqrt{\left(-\frac{1}{\sqrt{2}}\right)^2 + \left(\frac{1}{\sqrt{2}}\right)^2} = 1$.

4. A=IdentityMatrix[2];s=2/(Dot[{3,7},{3,7}]);v={{3},{7}};
W =v.Transpose[v];A-sW.

8. (a) $\begin{pmatrix} 1 & 0 \\ i & 3 \end{pmatrix}$; (b) $\begin{pmatrix} 1 & 0 & 1-2i \\ 5 & 6+14i & -4i \end{pmatrix}$; (c) $\begin{pmatrix} 1 & 0 & 1-2i \\ i & 3 & 23 \\ 5 & 6+14i & -4i \end{pmatrix}$.

9. A={{1/2 Sqrt[2]I,1/2 Sqrt[2]},{1/2 Sqrt[2],1/2 Sqrt[2]I}};
Inverse[A]==Conjugate[Transpose[A]].

10. (a) No; (b) No; (c) Yes.

EXERCISES 8.8

1. Dot[{1,2,0},{6,-3,4}]; Dot[{0,4,3},{6,-3,4}].
2. u=Dot[{1,2,1},{a,b,c}];v=Dot[{1,-1,3},{a,b,c}];
Solve[{u==0, v==0},{a,b,c}].

5. A={{9,-9,4,0},{-8,-5,4,7}}; NullSpace[A];
ColumnSpace=Take[RowReduce[Transpose[A]],2];
NullSpace[Transpose[A]];RowSpace=RowReduce[A].

The command `NullSpace[Transpose[A]]` yields the empty set. Hence the left null space is the zero subspace. Nul $A \oplus \text{Col } A^T = \mathbb{R}^4$. Nul $A^T \oplus \text{Col } A = \text{Col } A = \mathbb{R}^2$.

EXERCISES 8.9

1. `Solve[Dot[{1,3},{a,b}]==0,{a,b}];Dot[{1,3},{-3b,b}].`

4. `A={{5,0,5,0},{1,2,3,0},{0,2,2,0}};`
`R=RowReduce[Transpose[A]];B=Take[R,2];`
`Solve[{B[[1]].{a,b,c}==0,B[[2]].{a,b,c}==0},{a,b,c}];`
`Comp={1,-5,5};Dot[B[[1]],Comp];Dot[B[[2]],Comp].`

EXERCISES 8.10

1. Since $T(1,0) = (2,1)$ and $T(0,1) = (-3,1)$, the matrix $[A]_{\mathcal{E}}$ is $\begin{pmatrix} 2 & -3 \\ 1 & 1 \end{pmatrix}$.

`A={{2,-3},{1,1}};B=Transpose[A];`
`Expand[Dot[A.{x,y},{u,v}]]==Expand[Dot[{x,y},B.{u,v}]].`

EXERCISES 8.11

2. $\theta = 2\pi/3$.

3. Let A be an $n \times n$ orthogonal matrix and assume that $\mathbf{x}$ is an eigenvector of A belonging to the eigenvalue λ. Then

$$0 \neq \langle \mathbf{x}, \mathbf{x} \rangle = \langle A\mathbf{x}, A\mathbf{x} \rangle = \langle \lambda\mathbf{x}, \lambda\mathbf{x} \rangle = \lambda^2 \langle \mathbf{x}, \mathbf{x} \rangle$$

Hence $\lambda^2 = 1$. Therefore, $\lambda = \pm 1$. Let $p(t) = (t-1)^2$. Then $p(t)$ is the characteristic polynomial of its companion matrix

$$A = \begin{pmatrix} 0 & -1 \\ 1 & 2 \end{pmatrix}$$

The matrix A is not orthogonal since $A^{-1} \neq A^T$.

EXERCISES 8.12

1. (a) The eigenvalues of A are -1 and 3, those of B are 0 and 2, and that of C is 3. (b) The basis $\mathcal{S} = \{(1,1), (-1,1)\}$ is a basis consisting of eigenvectors of A, of B, and of C.

2. The rank of $A^T A$ and $B^T B$ is 2, and that of $C^T C$ is 1. The columns of $A^T A$ form a basis for $\mathbb{R}^2$. `Eigensystem[Transpose[B].B]`. Since `C` is a protected symbol, we use the lower case `c` and define $C^T C$ as `Eigensystem[Transpose[c].c]`. We then extract the eigenvectors required to form the two bases.

4. `A={{2,1,4,6,6.},{1,8,7,9,4},{4,7,9,2,8},{6,9,2,10,1},{6,4,8,1,5}};`

Eigensystem[A];p=CharacteristicPolynomial[A,t];Roots[p==0,t].

5. A={{1,0,0},{0,2,4},{0,4,2}};Eigensystem[A];
Q={{0,-1/Sqrt[2],1/Sqrt[2]},{1,0,0},{0,1/Sqrt[2],1/Sqrt[2]}};
Inverse[Q]==Transpose[Q];{u,v,w}=Q{x,y,z}.

EXERCISES 8.13

1. (a) $\begin{pmatrix} 1 \\ 1 \end{pmatrix} \begin{pmatrix} x \end{pmatrix} = \begin{pmatrix} 1 \\ 2 \end{pmatrix}$;

(b) $\begin{pmatrix} x \end{pmatrix} = \left[\begin{pmatrix} 1 \\ 1 \end{pmatrix}^T \begin{pmatrix} 1 \\ 1 \end{pmatrix} \right]^{-1} \begin{pmatrix} 1 \\ 1 \end{pmatrix}^T \begin{pmatrix} 1 \\ 2 \end{pmatrix} = \frac{3}{2}$.

The least-squares solution of the given system is the arithmetic mean of the components of the given vector of constants.

9. The matrix $A^T A$ is not invertible.

10. The linear system determined by the normal equation

$$\begin{pmatrix} 1 & 3 & 0 \\ 0 & -1 & 8 \end{pmatrix}^T \begin{pmatrix} 1 & 3 & 0 \\ 0 & -1 & 8 \end{pmatrix} \begin{pmatrix} x \\ y \\ z \end{pmatrix} = \begin{pmatrix} 1 & 3 & 0 \\ 0 & -1 & 8 \end{pmatrix}^T \begin{pmatrix} 1 \\ 2 \end{pmatrix}$$

is $\{x + 3y = 1, 3x + 10y - 8z = 1, -8y + 64z = 16\}$. It solutions are $(x, y, z) = (-24z + 7, 8z - 2, z)$.

CHAPTER 9

EXERCISES 9.1

1. The sets of singular values of the matrices are $\{1, 1\}$, $\{0, 1, 1\}$, and $\{1\}$. The singular values of their transposes are $\{1, 1\}$, $\{1, 1\}$, and $\{0, 0, 1\}$.

2. The commands

```
singvals[{{1,1,4.,0},{1,0,0,0.},{4,0,0,-1}}]
singvals[Transpose[{{1,1,4.,0},{1,0,0,0.},{4,0,0,-1}}]]
```

produce the outputs

$$\{4.70054, 3.72148, 0.235701, 2.26059 \times 10^{-8}\}$$
$$\{4.70054, 3.72148, 0.235701\}$$

If 2.26059×10^{-8} is considered to be essentially zero, this result says that the two matrices have the same nonzero approximate singular values.

3. For the first two matrices exact values can be obtained by converting to integers. For the third we need to solve a quintic. In general this can only be done using approximations.

```
Block[{lst1={{{2,4},{0,1.},{0,1},{1,1}},{{2,0,1,4},{5.,0,1,5}},
{{2.,3,0,0,0},{5,3,5,2,0},{-1,1,1,0,2}}},
lst2={{{2,4},{0,1},{0,1},{1,1}},{{2,0,1,4},{5,0,1,5}},
{{2,3,0,0,0},{5,3,5,2,0},{-1,1,1,0,2}}}},
{Map[singvals,lst1],Map[singvals,lst2]}].
```

5. Suppose that A is an orthogonal matrix. Then $A^T A = I$.

6. Either $b = c$ or $b = -c \neq 0$ and $a = d$.

7. (a) Approximately 2.97621; (b) $\sqrt{17}$; (c) $\sqrt{85}$.

EXERCISES 9.2

4. (a) The routine
```
Block[{A={{4.,0,0,0,0},{0,9,0,-2,0},{0,0,7,0,0},{0,-2,0,1,0},
{0,0,0,0,3}},x,y},y=Array[x,5];
{Expand[y.A.y],Part[SingularValues[A],2]}]
```
shows that

$$x^T A x = 4x[1]^2 + 9x[2]^2 + 7x[3]^2 - 4x[2]x[4] + x[4]^2 + 3x[5]^2$$

A simple calculation establishes that $x^T A x > 0$ for all nonzero x. Furthermore the calculation also returns the eigenvalues {9.47214, 7., 4., 3., 0.527864}. Since all eigenvalues are positive, Theorem 8.36 confirms that A is positive definite.

EXERCISES 9.3

4. ```A={{0,1,-1.,1},{-1,2,0,1}};
ColumnRank[A]:=Length[Transpose[A]]-Length[NullSpace[A]]=2;
{P,Q,R}=SingularValues[A];S={Transpose[U][[1]],Transpose[U][[2]]};
Dot[Transpose[U][[1]],Transpose[U][[2]]]=0```
shows S is linearly independent. Hence is a basis for Col A.

EXERCISES 9.4

1. (a) ```A={{1,2.},{3,4},{5,6}};x1={{-6},{6.}};
b1={{6},{6},{6}};B=PseudoInverse[A];x1==B.b1;
x2={{-16/3},{N[67/12]}};b2 ={{6},{6},{7}};B.b2; x2==B.b2.```

4. ```A={{2,0,0},{0,3.,0},{0,0,3}};{P,Q,R}=SingularValues[A];
Inverse[A]==Transpose[R].Inverse[DiagonalMatrix[Q]].P.```

EXERCISES 9.5

3. The singular values of

$$A = \begin{pmatrix} 1 & 0 & 0 & 0 \\ 0 & 2. & 0 & 0 \\ 0 & 0 & 3 & 0 \\ 0 & 0 & 0 & 4 \end{pmatrix}$$

are $1, 2, 3$, and 4. To suppress all four values, we must choose an x for which $4x > 4$. `Tolerance` $\rightarrow$ `1.1` will do since $4 \times 1.1 = 4.4 > 4$.

EXERCISES 9.6

1. (b) `A={{.7,-.8},{.9,.7}};{P,Q,R}=SingularValues[A];`
`B=Transpose[P].R;Inverse[B].`

EXERCISES 9.7

1. (b) `M=SingularValues[{{4.,0,0},{2,5,0},{3,2,4}}];`
`Condition number = M[[2]][[1]]/M[[2]][[3]].`

3. `A={{4.,0,0},{0,5,0},{0,0,4}};{P,Q,R}=SingularValues[A];`
`Aplus=P.Inverse[DiagonalMatrix[Q]].Transpose[R].PseudoInverse[A].`

INDEX

Printed and bound by CPI Group (UK) Ltd, Croydon, CR0 4YY

03/10/2024

01040315-0008